Texts in Theoretical Computer Science
An EATCS Series

Editors: W. Brauer G. Rozenberg A. Salomaa
On behalf of the European Association
for Theoretical Computer Science (EATCS)

Advisory Board: G. Ausiello M. Broy C. Calude
S. Even J. Hartmanis N. Jones T. Leighton M. Nivat
C. Papadimitriou D. Scott

Springer
Berlin
Heidelberg
New York
Hong Kong
London
Milan
Paris
Tokyo

Peter Clote • Evangelos Kranakis

Boolean Functions and Computation Models

With 19 Figures

Springer

Authors

Prof. Dr. Peter Clote
Boston College
Department of Computer Science
and Department of Biology
Fulton Hall 410 B
140 Commonwealth Avenue
Chestnut Hill, MA 02467, USA
clote@cs.bc.edu

Prof. Dr. Evangelos Kranakis
Carleton University
School of Computer Science
1125 Colonel By Drive
Ottawa, Ontario, K1S 5B6, Canada
kranakis@scs.carleton.ca

Series Editors

Prof. Dr. Wilfried Brauer
Institut für Informatik
Technische Universität München
Arcisstrasse 21, 80333 München, Germany
Brauer@informatik.tu-muenchen.de

Prof. Dr. Grzegorz Rozenberg
Leiden Institute of Advanced Computer Science
University of Leiden
Niels-Bohrweg 1, 2333 CA Leiden, The Netherlands
rozenber@liacs.nl

Prof. Dr. Arto Salomaa
Turku Centre for Computer Science
Lemminkäisenkatu 14 A, 20 520 Turku, Finland
asalomaa@utu.fi

Library of Congress Cataloging-in-Publication Data

Clote, Peter.
Boolean functions and computation models/Peter Clote, Evangelos Kranakis.
p. cm. – (Texts in theoretical computer science)
Includes bibliographical references and index.

1. Computational complexity. 2. Algebra, Boolean. I. Kranakis, Evangelos. II. Title.
III. Series.

QA267.7 .C58 2001
511.3–dc21 2201031128

ACM Computing Classification (1998): F.1.1, F.4.1, F.1.3

ISBN 978-3-642-08217-7

This work is subject to copyright. All rights are reserved, whether the whole or part of the material is concerned, specifically the rights of translation, reprinting, reuse of illustrations, recitation, broadcasting, reproduction on microfilm or in any other way, and storage in data banks. Duplication of this publication or parts thereof is permitted only under the provisions of the German Copyright Law of September 9, 1965, in its current version, and permission for use must always be obtained from Springer-Verlag. Violations are liable for prosecution under the German Copyright Law.

Springer-Verlag Berlin Heidelberg New York,
a member of BertelsmannSpringer Science+Business Media GmbH

© Springer-Verlag Berlin Heidelberg 2010
Printed in Germany

The use of general descriptive names, trademarks, etc. in this publication does not imply, even in the absence of a specific statement, that such names are exempt from the relevant protective laws and regulations and therefore free for general use.

Cover Design: KünkelLopka, Heidelberg

Dedicated to our parents:

Mary Ann and Paul J. Clote
Stamatia and Kostantinos Kranakis

Preface

The foundations of computational complexity theory go back to Alan Turing in the 1930s who was concerned with the existence of automatic procedures deciding the validity of mathematical statements. The first example of such a problem was the undecidability of the Halting Problem which is essentially the question of debugging a computer program: Will a given program eventually halt? Computational complexity today addresses the quantitative aspects of the solutions obtained: Is the problem to be solved tractable? But how does one measure the intractability of computation? Several ideas were proposed: A. Cobham [Cob65] raised the question of what is the right model in order to measure a "computation step", M. Rabin [Rab60] proposed the introduction of axioms that a complexity measure should satisfy, and C. Shannon [Sha49] suggested the boolean circuit that computes a boolean function.

However, an important question remains: What is the nature of computation? In 1957, John von Neumann [vN58] wrote in his notes for the Silliman Lectures concerning the nature of computation and the human brain that

> ... logics and statistics should be primarily, although not exclusively, viewed as the basic tools of 'information theory'. Also, that body of experience which has grown up around the planning, evaluating, and coding of complicated logical and mathematical automata will be the focus of much of this information theory. The most typical, but not the only, such automata are, of course, the large electronic computing machines.
>
> Let me note, in passing, that it would be very satisfactory if one could talk about a 'theory' of such automata. Regrettably, what at this moment exists — and to what I must appeal — can as yet be described only as an imperfectly articulated and hardly formalized 'body of experience'.

With almost a half century after von Neumann's death, the theory of computation and automata is now a well-developed and sophisticated branch of mathematics and computer science. As he forecasted, the principal tools have proven to come from the fields of mathematical logic, combinatorics, and probability theory.

Using these tools, we have attempted to give a survey of the present state of research in the study of boolean functions, formulas, circuits, and propositional proof systems. All of these subjects are related to the overriding concern of how computation can be modeled, and what limitations and interrelations there are between different computation models.

This text is structured as follows. We begin with methods for the construction of boolean circuits which compute certain arithmetic and combinatorial functions, and investigate upper and lower bounds for circuit families. The techniques used are from combinatorics, probability and finite group theory. We then survey steps taken in a program initiated by S.A. Cook of investigating non-deterministic polynomial time, from a proof-theoretic viewpoint. Specifically, lower bounds are presented for lengths of proofs for families of propositional tautologies, when proven in certain proof systems. Techniques here involve both logic and finite combinatorics and are related to constant depth boolean circuits and to monotone arithmetic circuits.

Outline of the book

A more detailed breakdown of the book is as follows. In Chapter 1, circuits are constructed for data processing (string searching, parsing) and arithmetic (multiplication, division, fast Fourier transform). This material is intended to provide the reader with concrete examples, before initiating a more abstract study of circuit depth and size.

Chapter 2 presents a sampling of techniques to prove size lower bounds for certain restricted classes of circuits – constant depth or monotonic. These include Razborov's elegant constructive proof of the Håstad switching lemma, the Haken–Cook monotonic real circuit lower bound for the *broken moskito screen* problem, Razborov's algebraic approximation method for *majority*, and Smolensky's subsequent generalization to finite fields.

Chapter 3 studies symmetric boolean functions and related invariance groups. A characterization is given of those symmetric functions computable by constant depth polysize circuits. Invariance groups of boolean functions are characterized by a condition concerning orbit structure, and tight upper bounds are given for almost symmetric functions. Applications are given to anonymous networks such as rings and hypercubes. Most of these results are due to P. Clote and E. Kranakis.

Chapter 4 concerns the empirically observed *threshold* phenomenon concerning clause density $r = \frac{m}{n}$, where with high probability random formulas in k-CNF form having m clauses over n variables are satisfiable (unsatisfiable) if r is less (greater) than a threshold limit. The results of this chapter include a proof of an analytic upper bound, a result due to M. Kirousis, E. Kranakis and D. Krizanc.

Chapter 5 studies propositional proof systems, which have relevance to complexity theory, since $\mathrm{NP} = co{-}\mathrm{NP}$ if and only if there exists a polynomially bounded propositional proof system. In obtaining exponential lower bounds for increasingly stronger proof systems, new techniques have been developed, such as random restriction, algebraic and bottleneck counting methods – these techniques may ultimately play a role in separating complexity classes, and in any case are of interest in themselves. The proof systems include resolution, cutting planes, threshold logic, Nullstellensatz system, polynomial calculus, constant depth Frege, Frege, extended Frege, and substitution Frege systems.

In Chapter 6 we define various computation models including uniform circuit families, Turing machines and parallel random access machines, and illustrate some features of parallel computation by giving example programs. We then give characterizations of different parallel and sequential complexity classes in terms of function algebras – i.e., as the smallest class of functions containing certain initial functions and closed under certain operations. In the early 1960's, A. Cobham first defined polynomial time P and argued its robustness on the grounds of his machine independent characterization of P via function algebras.

With the development that certain programming languages now admit polymorphism and higher type functionals, using function algebras, complexity theory can now be lifted in a natural manner to higher types, a development which is the focus of Chapter 7. In that chapter, a new yet unpublished characterization of type 2 NC^1 functionals (due to the first author) is given in terms of a natural function algebra and related lambda calculus.

How to use the book

This text is to be of use to students as well as researchers interested in the emerging field of *logical complexity theory* (also called *implicit complexity theory*). The chapters of the book can be read as independent units. However one semester courses can be given as follows:

Semester Course	**Chapters**
Boolean Functions & Complexity	1, 2, 3
Proof Systems & Satisfiability	5, 4
Machine Models, Function Algebras & Higher Types	6, 7

At the end of every chapter, there are several exercises: some are simple extensions of results in the book while others constitute the core result of a research article. The various levels of difficulty are indicated with an asterisk placed before more difficult problems, and two asterisks for quite challenging and sometimes open research problems. The reader is invited to attempt them all.

Acknowledgments

Writing this book would have been impossible without the financial support of various research foundations, and without the exchange of ideas from many colleagues and friends.

Peter Clote is indebted to the NSF (National Science Foundation), CNRS (Centre National pour la Recherche Scientifique), Czech Academy of Science and Volkswagen Foundation for financial support of work on this text. In particular, thanks to J.-P. Ressayre for arranging a visit to Université Paris VII, and to D. Thérien for arranging a visit to the Barbados Complexity Theory Workshop, where some of the material from this text was presented. Evangelos Kranakis is indebted to NSERC (Natural Sciences and Engineering Research Council of Canada), and NWO (Netherlands Organization for the Advancement of Research) for partial support during the preparation of the book.

While holding the Gerhard Gentzen Chair of Theoretical Computer Science at the University of Munich, the first author (P. Clote) gave several courses using parts of the current text and would like to thank his students for the feedback. We would like to thank A. Abel, D. Achlioptas, T. Altenkirch, P. Beame, S. Bellantoni, E. Ben-Sasson, S. Buss, N. Danner, M. Hofmann, R. Impagliazzo, J. Johannsen, J. Krajíček, L. M. Kirousis, D. Krizanc, K.-H. Niggl, P. Pudlák, H. Schwichtenberg, Y. Stamatiou, T. Strahm, H. Straubing, G. Takeuti and J. Woelki for comments and suggestions, although of course the authors are solely responsible for any remaining errors. In particular, any omitted or erroneous references are purely unintentional. We are deeply grateful to Sam Buss, Jan Krajíček, Pavel Pudlák, and Gaisi Takeuti, with whom the first author collaborated over a period of years, and who have established many of the deepest results in propositional proof systems, as well as L. M. Kirousis and D. Krizanc with whom the second author has spent many enjoyable discussions.

Finally, we would like to express our deepest appreciation to Dr. Hans Wössner, Executive Editor for Computer Science of Springer-Verlag, who never lost faith in our project.

This book was type set using LaTeX with additional macros developed by S.R. Buss for typesetting proof figures.

Boston
Ottawa
July 2002

Peter Clote
Evangelos Kranakis

Contents

1. Boolean Functions and Circuits

Mathematical facts worthy of being studied are those which, by their analogy with other facts, are capable of conducting us to the knowledge of a mathematical law, ... *H. Poincaré [Poi52]*

1.1 Introduction

Many fundamental questions in complexity theory can be formulated as a language recognition problem, where by language L we understand a set of strings over a finite alphabet, say $\Sigma = \{0, 1\}$. The basic question is then to determine the computational resources (such as time, space, number of processors, energy consumption of a chip, etc.) needed in determining whether or not $x \in L$, for any string $x \in \Sigma^*$. In turn, if for any n we let f_n denote the characteristic function of $L \cap \{0,1\}^n$, then the previous question can be rephrased as a problem of determining the computational resources needed to compute the boolean function $f_n(x)$, for arbitrary x, n.

There are many natural models for computing a boolean function: boolean circuits, threshold and modular counting circuits, boolean formula, switching networks, branching programs, VLSI circuits, etc. In this chapter, we introduce some elementary concepts concerning boolean functions, formulas and circuits, then proceed to construct efficient boolean circuits for parsing and integer arithmetic, and then consider general construction methods for arbitrary boolean functions.

Apart from the obvious fact that boolean circuits underlie computer hardware, two important justifications for their study are:

1. Boolean circuit families with unbounded fan-in AND/OR gates of certain size and depth correspond exactly to parallel random access machines (PRAM) with certain processor and time bounds [SV84]. If one neglects the complexity induced by message passing, then the virtual machine for certain massively parallel machines [HS86] is essentially the PRAM. Thus boolean circuit theory has applications in parallel algorithm design.
2. Boolean circuit lower bounds are intimately related to the P = NP question, for if an NP-complete problem such as satisfiability SAT of boolean

formulas can be shown to have superpolynomial lower bound, then deterministic polynomial time P is different from nondeterministic polynomial time NP.

1.2 Boolean Functions and Formulas

In this section we introduce the notions of DNF, CNF, term, clause, and characterize complete sets of boolean functions.

A *boolean function* in the variables $x_1, \ldots, x_n$ is a map $f : \{0,1\}^n \to \{0,1\}$. The collection of all n-ary boolean functions is denoted $\mathcal{B}_n$; clearly $|\mathcal{B}_n| = 2^{2^n}$.

Some examples of boolean functions include

- the 0-ary constant functions 0 and 1,
- the unary function $\neg$ (negation),
- the binary functions $\vee$ (OR), $\wedge$ (AND), $\oplus$ (EXCLUSIVE OR), $\to$ (implication, where $x \to y$ is defined by $\neg x \vee y$), and $\equiv$ (equivalence, also called bi-implication).

Frequently, we will use the alternative notation $+$ and $\cdot$ for $\oplus$ and $\wedge$, respectively. We also use the symbols $\bigvee, \bigwedge, \bigoplus$ for the multivariable OR, AND, EXCLUSIVE OR, respectively. For the EXCLUSIVE OR function we may occasionally use the summation symbol.

A *literal* is a boolean variable x_i or its negation $\neg x_i$, where the latter is often denoted $\overline{x}_i$. The negation of literal $\overline{x}_i$ is x_i. A *propositional formula* (or *boolean formula* over the *De Morgan basis* $\{\wedge, \vee, \neg, 0, 1\}$) is defined recursively as follows:

- The constants (i.e., 0-place connectives) 0 (FALSE) and 1 (TRUE) are propositional formulas.
- A boolean variable is a propositional formula.
- If F and G are propositional formulas, then $\neg F$ (*negation*), $(F \wedge G)$ (*conjunction*) and $(F \vee G)$ (*disjunction*) are propositional formulas.

The connectives $\wedge$ and $\vee$ associate to the right, so that $F \vee G \vee H$ means $F \vee (G \vee H)$. Let $x^1 = x$ and $x^0 = \neg x$. A conjunction of the form $x_1^{a_1} \wedge \cdots \wedge x_n^{a_n}$, where $a_i \in \{0,1\}$, is called a *term* . Dually, a disjunction of the form $x_1^{a_1} \vee \cdots \vee x_n^{a_n}$, where $a_i \in \{0,1\}$, is called a *clause* . An *s-term* (*s-clause*) is a term (clause) of at most s literals. A variable x_i is said to *occur* in, or be *mentioned* in a term or clause if either x_i or $\overline{x}_i$ appears in the term or clause.

A *truth assignment* for a boolean formula F with variables $x_1, \ldots, x_n$ is a mapping $\sigma : \{x_1, \ldots, x_n\} \to \{0,1\}$. Recall that 0 represents FALSE and 1 represents TRUE. The truth assignment σ yields a truth value for F as defined recursively: $\sigma(\neg G) = 1 - \sigma(G)$, $\sigma(G \wedge H) = \min(\sigma(G), \sigma(H))$, $\sigma(G \vee H) = \max(\sigma(G), \sigma(H))$. If $\sigma(F) = 1$ (or 0), then σ is said to be a *satisfying* (*falsifying*) truth assignment for F, and this is sometimes written $\sigma \models F$

($\sigma \not\models F$). A boolean formula F is a *tautology* if all truth assignments on the variables occurring in F satisfy F.

Two boolean formulas F, G are equivalent, written $F \equiv G$, if F and G have the same truth values under every truth assignment. Similarly, a boolean function f is *equivalent* to a boolean formula F, written $f \equiv F$, if for all truth assignments σ, $f(\sigma(x_1), \ldots, \sigma(x_n)) = \sigma(F)$. As in the case for boolean formulas, if f is a boolean function, then $\sigma \models f$ ($\sigma \not\models f$) means that $f(\sigma(x_1), \ldots, \sigma(x_n)) = 1$ (0).

A formula F is in *disjunctive normal form* (DNF) if F is a disjunction of terms (i.e., of conjunctions of literals). Similarly, F is in *conjunctive normal form* (CNF) if F is a conjunction of clauses (i.e., of disjunctions of literals). A formula is in $s-$DNF(s-CNF) if it is a disjunction (conjunction) of conjunctions (disjunctions) of at most s literals. We will try to reserve s-disjunction (s-conjunction) for a disjunction (conjunction) of at most s variables, as contrasted with s literals in the definition of s-DNF and s-CNF . A DNF formula F equivalent to a given boolean function f can be obtained from the truth table of f by forming a disjunction

$$\bigvee_{\sigma \models f} x_1^{\sigma(x_1)} \wedge \cdots \wedge x_n^{\sigma(x_n)}$$

of signed literals $x_i^1 = x_i$, $x_i^0 = \overline{x}_i$ corresponding to truth assignments σ which set f to be true.

Example 1.2.1. Suppose that $f \in \mathcal{B}_3$ is equivalent to the formula $(x_1 \vee x_2) \wedge \overline{x}_3$. The truth table of f is as follows.

x_1	x_2	x_3	f
0	0	0	0
0	0	1	0
0	1	0	1
0	1	1	0
1	0	0	1
1	0	1	0
1	1	0	1
1	1	1	0

A DNF formula equivalent to f is then

$$(\overline{x}_1 \wedge x_2 \wedge \overline{x}_3) \vee (x_1 \wedge \overline{x}_2 \wedge \overline{x}_3) \vee (x_1 \wedge x_2 \wedge \overline{x}_3).$$

Given arbitrary f, the CNF formula G

$$\bigwedge_{\sigma \not\models f} x_1^{1-\sigma(x_1)} \vee \cdots \vee x_n^{1-\sigma(x_n)}$$

equivalent to f can be obtained by finding the DNF of $1-f$, and then applying *De Morgan's rules*:

$$\neg(F \wedge G) \equiv \neg F \vee \neg G$$
$$\neg(F \vee G) \equiv \neg F \wedge \neg G$$

along with the rule for double negation $\neg\neg F \equiv F$. Using the previous example for f, this leads to $(x_1 \vee x_2 \vee x_3) \wedge (x_1 \vee x_2 \vee \overline{x}_3) \wedge (x_1 \vee \overline{x}_2 \vee \overline{x}_3) \wedge (\overline{x}_1 \vee x_2 \vee \overline{x}_3) \wedge (\overline{x}_1 \vee \overline{x}_2 \vee \overline{x}_3)$.

An alternative notation for boolean formulas is that of polynomials over the two element field $GF(2)$ or $\mathbf{Z}_2$. The EXCLUSIVE-OR $x \oplus y$ of x, y is defined as $(x \wedge \overline{y}) \vee (\overline{x} \wedge y)$. Then $\oplus, \wedge$ are respectively addition and multiplication in $GF(2)$. The negation $\neg x$ is equivalent to $1 \oplus x$, and $(x \vee y) \equiv \neg(\neg x \wedge \neg y)$, so that every boolean formula is equivalent to a multivariate polynomial over $GF(2)$. Translation of the previous canonical DNF form leads to

$$\sum_{\sigma \models f} \prod_{i=1}^{n} (\sigma(x_i) \oplus 1 \oplus x_i)$$

and using distribution of multiplication over addition to a *sum-of-products* normal form

$$\sum_{i \in I} x_1^{a_{i,1}} x_2^{a_{i,2}} \cdots x_n^{a_{i,n}}$$

for a set $I \subseteq \{0,1\}^n$. Here x^i denotes exponentiation, so that $x^1 = x$ and $x^0 = 1$. Context will distinguish between this notation and the earlier convention $x^1 = x$, $x^0 = \overline{x}$. Moreover, since $x_i^2 = x_i$ in $GF(2)$, we may assume the $a_{i,j} \in \{0,1\}$. It follows that a sum-of-products normal form is either 0 or a sum of multivariate monomials with coefficient 1; i.e., for some $\emptyset \neq I \subset \{1,\ldots,n\}$ of the form

$$\sum_{A \in I} \prod_{i \in A} x_i. \tag{1.1}$$

Using the previous example of f, this yields

$$(1 \oplus x_1)(x_2)(1 \oplus x_3) + (x_1)(1 \oplus x_2)(1 \oplus x_3) + (x_1)(x_2)(1 \oplus x_3)$$

which is equal to

$$x_2(1 + x_1 + x_3 + x_1x_3) + x_1(1 + x_2 + x_3 + x_2x_3) + x_1x_2 + x_1x_2x_3$$

hence to

$$x_1 + x_2 + x_1x_2 + x_1x_3 + x_2x_3 + x_1x_2x_3.$$

A set $\{f_1, \ldots, f_s\}$ of boolean functions is *complete* for the class of boolean functions if every boolean function f can be obtained as the composition of these functions. A complete set is called a *basis* if in addition it is minimal, i.e., no proper subset is complete.[1] The following are complete sets of boolean functions (see Exercise 1.13.2):

- $\{\vee, \wedge, \neg, 0, 1\}$,
- $\{\vee, \neg, 0, 1\}$,
- $\{\wedge, \neg, 0, 1\}$,
- $\{\oplus, \wedge, 0, 1\}$,
- $\{|, 0, 1\}$, where $|$ denotes NOR; i.e., $x|y = \neg x$ if $x = y$, and $= 0$ otherwise.

In this section, we characterize the complete sets of boolean functions. To this end, we distinguish the following characteristic classes of boolean functions:

- T_n^0: The class of functions $f \in \mathcal{B}_n$ satisfying $f(0^n) = 0$.
- T_n^1: The class of functions $f \in \mathcal{B}_n$ satisfying $f(1^n) = 1$.
- D_n (*Self-dual*): The class of functions $f \in \mathcal{B}_n$ satisfying

$$f(x_1, \ldots, x_n) = \neg f(\neg x_1, \ldots, \neg x_n).$$

- M_n (*Monotone*): The class of functions $f \in \mathcal{B}_n$ satisfying

$$x \preceq y \Rightarrow f(x) \leq f(y),$$

where for n tuples $x = (x_1, \ldots, x_n), y = (y_1, \ldots, y_n)$ we define $x \preceq y$ if and only if $(\forall i \leq n)(x_i \leq y_i)$.[2]
- L_n (*Linear*): The class of functions $f \in B_n$ of the form

$$f(x) = \sum_{i=1}^{k} b_i x_i \bmod 2,$$

where $b_i \in \{0, 1\}$, for $i = 1, \ldots, k$.

The following result characterizes those sets of functions which are complete.

Theorem 1.2.1. *A class $\mathcal{C} \subseteq \mathcal{B}_n$ of boolean functions is complete if and only if $\mathcal{C}$ is not a subset of any of the classes $T_n^0, T_n^1, D_n, M_n, L_n$.*

Proof. The condition $\mathcal{C} \not\subseteq T_n^0, T_n^1, D_n, M_n, L_n$ is clearly necessary. To prove sufficiency choose boolean functions

$$f_T^i \in \mathcal{C} \setminus T_n^i, f_D \in \mathcal{C} \setminus D_n, f_M \in \mathcal{C} \setminus M_n, f_L \in \mathcal{C} \setminus L_n.$$

Since

$$\{f_T^0, f_T^1, f_D, f_M, f_L\} \subseteq \mathcal{C}$$

[1] By an abuse of terminology, a non-minimal complete set of connectives is often called a basis as well.

[2] Throughout we write $(\forall i \leq n)$ rather than the more cumbersome, but correct form $(\forall 1 \leq i \leq n)$ or $(\forall i \in \{1, \ldots, n\})$.

it is enough to show that the set $\{f_T^0, f_T^1, f_D, f_M, f_L\}$ is complete. In turn this will follow from the fact that these functions can generate the constants $0, 1$ and the functions $\neg x, x \wedge y$.

By definition of f_D, there exist $a_1, \ldots, a_n \in \{0,1\}$ such that

$$f_D(\neg a_1, \ldots, \neg a_n) = f_D(a_1, \ldots, a_n).$$

Define the unary function g by

$$g(x) = f_D(x \oplus a_1, \ldots, x \oplus a_n)$$

so that we have $g(0) = g(1)$. Hence g is one of the constant functions 0 or 1.

Now $f_{T_n^0}(0^n) = 1$ and $f_{T_n^1}(1^n) = 0$. If $g \equiv 0$, then

$$h(x) = f_{T_n^0}(g(x), \ldots, g(x)) = 1$$

so $h \equiv 1$. If $g \equiv 1$, then

$$h(x) = f_{T_n^1}(g(x), \ldots, g(x)) = 0$$

so $h \equiv 0$. Hence we have the constant functions $0, 1$.

From the constants $0, 1$ and the function f_M we can generate $\neg$ as follows. We claim that there must exist $a, b \in \{0,1\}^n$ such that $a \preceq b$ and $1 = f_M(a) > f_M(b) = 0$, having Hamming distance $\rho(a, b) = 1$, where *Hamming distance* $\rho(a_1 \cdots a_n, b_1 \cdots b_n)$ is defined to be $|\{i : 1 \leq i \leq n, a_i \neq b_i\}|$.[3] Indeed, if not, then whenever $c = c_1 \cdots c_n \preceq d_1 \cdots d_n = d$ and $\rho(c, d) = 1$, then either $f_M(c) = 0$ or $f_M(c) = 1 = f_M(d)$. Then $f_M(a) = 1$, and making bit changes stepwise to intermediate words yields $f_M(b) = 1$, a contradiction. Thus there exists such a pair a, b with $a \preceq b$, $\rho(a, b) = 1$, $f_M(a) = 1$ and $f_M(b) = 0$. Now, suppose that $a_{i_0} \neq b_{i_0}$, but $a_i = b_i$ for all other values of i. It clearly follows that

$$f_M(a_1, \ldots, a_{i_0-1}, x, a_{i_0+1}, \ldots, a_n) = \neg x.$$

It remains to generate the function $\wedge$. By sum-of-products normal form, it follows that

$$f_L(x_1, \ldots, x_n) = \sum_{1 \leq i_1, \ldots, i_n \leq n} a_{i_1, \ldots, i_n} x_{i_1} \cdots x_{i_n} \tag{1.2}$$

for some coefficients $a_{i_1, \ldots, i_n} \in \{0,1\}$. Since $f_L \notin L_n$, there exist two variables, say x_1, x_2, such that $f_L(x_1, \ldots, x_n)$ can be rewritten in the form

$$x_1 x_2 g_1(x_3, \ldots, x_n) + x_1 g_2(x_3, \ldots, x_n) + x_2 g_3(x_3, \ldots, x_n) + g_4(x_3, \ldots, x_n)$$

with $g_1 \not\equiv 0$. Thus there exist $a_3, \ldots, a_n \in \{0,1\}$ such that $g_1(a_3, \ldots, a_n) = 1$. Substituting this in equation (1.2), we obtain

[3] The cardinality of a set A is denoted by $|A|$, whereas later the length of the binary representation of an integer n is also denoted by $|n|$. The intended meaning will be clear from the context.

$$f_L(x_1, x_2, a_3, \ldots, a_n) = x_1 x_2 + c x_1 + d x_2 + e$$

for some boolean constants c, d, e. It is now easy to check that

$$x_1 x_2 = f_L(x_1 + d, x_2 + c, a_3, \ldots, a_n) + cd + e.$$

1.3 Circuit Model

The number of operations required to compute a boolean function is of extreme interest in complexity theory. This can be formalized by the notion of circuit size of a certain depth. A *circuit* is a directed acyclic graph. The sources are called *input nodes* and are labeled with $x_1, \ldots, x_n, 0, 1$. Non-input nodes are called *gates* and are labeled by a boolean function, whose arity is the in-degree of the node. The in-degree (out-degree) of a gate is called the *fan-in* (*fan-out*). *Sink* nodes have fan-out 0 and are called *output nodes.* With few exceptions, we usually consider circuits having a single output node. Boolean formulas, earlier defined, are simply fan-out 1 boolean circuits over the De Morgan basis $\{\wedge, \vee, \neg, 0, 1\}$. A circuit is *leveled* if the gates can be arranged in levels so that all inputs are at level 0, while gates at level $s+1$ have inputs only from level s gates. A leveled boolean circuit over the De Morgan basis is *alternating* if the input nodes are labeled with $x_1, \overline{x}_1, \ldots, x_n, \overline{x}_n, 0, 1$ and gates at the same level are either all ORs or all ANDs, where OR (AND) gates at level s are followed by AND (OR) gates at level $s+1$. The *size* of a circuit is the number of gates, while the *depth* is the length of the longest path from an input to output node. Circuit size of C is sometimes defined in the literature to be the number of subcircuits of C; however, as defined above, we take size to be the number of gates (non-input subcircuits).

Usually boolean circuits are depicted with the leaves (input nodes) at the bottom. For instance, the leveled circuit in Figure 1.1 has depth 2, size 5 and corresponds to the DNF form for the function $f(x_1, x_2, x_3) = x_1 \oplus x_2 \oplus x_3$.

Circuits with n input nodes compute a boolean function on n variables in the obvious manner. Formally, an input node v labeled by x_i computes the boolean function $f_v(x_1, \ldots, x_n) = x_i$. A node v having in-edges from $v_1, \ldots, v_m$, and labeled by the m-place function g from the basis set, computes the boolean function $f_v(x_1, \ldots, x_n) = g(f_{v_1}(x_1, \ldots, x_n), \ldots, f_{v_m}(x_1, \ldots, x_n))$.

For any finite set Ω of connectives, and any boolean function $f \in \mathcal{B}_n$, we define $C_\Omega(f)$ to be the minimum size of a *circuit* with inputs $x_1, \ldots, x_n, 0, 1$ and connectives from Ω which computes f (unless otherwise indicated, fan-in is usually assumed arbitrary, but we will try to keep to the convention that multivariable connectives $\bigwedge$, $\bigvee$, $\bigoplus$ are explicitly written, to distinguish from the fan-in 2 connectives $\wedge$, $\vee$, $\oplus$. By $L_\Omega(f)$ we mean the minimum size of a boolean *formula* (circuit with fan-out 1) over connectives from Ω which computes f. No super-linear lower bounds for unrestricted circuits over any basis have been proved. Despite this, as presented in the next chapter, much

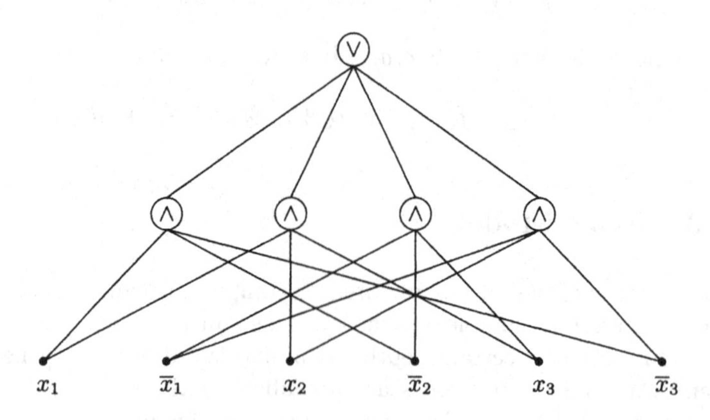

Fig. 1.1. A leveled circuit for $x_1 \oplus x_2 \oplus x_3$

progress has been made for restricted classes of circuits (monotonic circuits, and constant depth circuits). To formalize the notion of size for constant depth circuits, we define $L_k^{\wedge,\vee,\neg}(f)$ to be the minimum $size(C)$, such that C is a depth k, alternating, leveled boolean circuit over the De Morgan basis computing f, where inputs are $x_1, \overline{x}_1, \ldots, x_n, \overline{x}_n, 0, 1$, and AND/OR gates have arbitrary fan-in. A similar definition is possible for $\{1, \oplus, \wedge\}$, or indeed any set Ω of gates thus giving the minimum size $L_k^{\Omega}(f)$ for a depth k unbounded fan-in, alternating, leveled circuit to compute f over Ω.

1.4 Basic Functions and Reductions

If $f : \{0,1\}^* \to \{0,1\}$ then we denote by $\{f_n\}$ the sequence of functions $f_n = f \restriction \{0,1\}^n$. The following basic functions often arise in the study of circuits.

- $\bigvee$ outputs 1 if and only if at least one of the inputs is 1.
- $\bigwedge$ outputs 1 if and only if all the inputs are 1.
- $\neg$ negates the input.
- Majority: outputs 1 if and only if the majority of the inputs are 1,

$$\mathrm{MAJ}_n(x_1, \ldots, x_n) = \begin{cases} 1 \text{ if } \sum_{1 \le i \le n} x_i \ge n/2 \\ 0 \text{ otherwise.} \end{cases}$$

- Threshold: for fixed k, outputs 1 if at least k inputs are 1,

$$\mathrm{TH}_k^n(x_1, \ldots, x_n) = \begin{cases} 1 \text{ if } \sum_{1 \le i \le n} x_i \ge k \\ 0 \text{ otherwise.} \end{cases}$$

- Exact: for fixed k outputs 1 if exactly k inputs are 1,

$$\text{EXACT}_k^n(x_1,\ldots,x_n) = \begin{cases} 1 \text{ if } |\{i : x_i = 1\}| = k \\ 0 \text{ otherwise.} \end{cases}$$

- For $s < p$, $\text{MOD}_{s,p}^n$ outputs 1 if the sum of the inputs is s modulo p,

$$\text{MOD}_{s,p}^n(x_1,\ldots,x_n) = \begin{cases} 1 \text{ if } \sum_{i=1}^n x_i \equiv s \bmod p \\ 0 \text{ otherwise.} \end{cases}$$

- MOD_p^n outputs 0 if the sum of the inputs is 0 modulo p,

$$\text{MOD}_p^n(x_1,\ldots,x_n) = \begin{cases} 0 \text{ if } \sum_{i=1}^n x_i \equiv 0 \bmod p \\ 1 \text{ otherwise.} \end{cases}$$

  Thus it is clear that $\text{MOD}_p^n(x_1,\ldots,x_n) = \neg(\text{MOD}_{0,p}^n(x_1,\ldots,x_n))$. In addition, MOD_2^n is usually denoted by PAR_n and called the *parity* function.

In the superscript n is omitted, then we understand the corresponding family of boolean functions; i.e., $\text{MOD}_p : \{0,1\}^* \to \{0,1\}$ is defined by $\text{MOD}_p(x_1,\ldots,x_n) = \text{MOD}_p^n(x_1,\ldots,x_n)$.

A boolean function $f \in \mathcal{B}_n$ is *symmetric* if

$$f(x_1,\ldots,x_n) = f(x_{\sigma(1)},\ldots,x_{\sigma(n)})$$

for every permutation σ in the group S_n of permutations on n letters. Examples of symmetric functions are $\neg$, $\bigwedge$, $\bigvee$, $\bigoplus$, MOD_p, TH_k^n, EXACT_k^n, etc. Symmetry, however, is not preserved under composition of boolean functions. Indeed, as seen in the next chapter, composition (related to depth of composed circuits) is very difficult to analyze. In a later chapter, we'll consider the symmetry or invariance group of a boolean function.

A recurring theme in computational complexity is whether a given boolean function is "harder" than another. To make this precise we define the notion of AC^0-reduction.

Following [CSV84], a boolean function f is AC^0*-reducible* to the set $\mathcal{C}$ of boolean functions, denoted by $f \in \text{AC}^0(\mathcal{C})$, if there is a constant depth, unbounded fan-in polynomial-size circuit with basic operations from $\mathcal{C}$ and output f.[4] It is clear that AC^0-reducibility is a transitive relation; moreover the following elementary theorem holds.[5]

---

[4] To be precise, a family $\{f_n : n \in \mathbf{N}\}$ of boolean functions is AC^0*-reducible* to $\mathcal{C}$, if there is a family $\{C_n : n \in \mathbf{N}\}$ of constant depth, unbounded fan-in polynomial size circuits, where C_n has basic operations from $\mathcal{C}$ and outputs $f_n = f \upharpoonright \{0,1\}^n$.

[5] By an abuse of notation, we will often omit the superscripts when we refer to these functions.

Theorem 1.4.1 (AC^0**-reductions)**. *The following statements hold*

1. *Every symmetric function is in* $\mathrm{AC}^0(\{0,1,\vee,\wedge\}\cup\{\mathrm{EXACT}_k : 0 \le k \le n\})$.
2. $\mathrm{EXACT}_k \in \mathrm{AC}^0(\{\wedge,\vee,\neg,\mathrm{TH}_k,\mathrm{TH}_{k+1}\})$.
3. $\mathrm{EXACT}_k \in \mathrm{AC}^0(\{\wedge,\vee,\neg,\mathrm{MAJ}\})$.
4. *Every symmetric function is in* $\mathrm{AC}^0(\{1,\oplus,\wedge,\mathrm{MAJ}\})$.
5. $\mathrm{MOD}_{s,p} \in \mathrm{AC}^0(\{\wedge,\vee,\neg,\mathrm{MOD}_p\})$, *for all* $s < p$.
6. $\mathrm{MOD}_p \in \mathrm{AC}^0(\{\wedge,\vee,\neg\}\cup\{\mathrm{EXACT}_k : 1 \le k \le n\})$.
7. $\mathrm{MOD}_p \in \mathrm{AC}^0(\{\wedge,\vee,\neg,\mathrm{MAJ}\})$.
8. $a|b \Rightarrow \mathrm{MOD}_a \in \mathrm{AC}^0(\{\wedge,\vee,\neg,\mathrm{MOD}_b\})$.
9. $\mathrm{MOD}_{p^m} \in \mathrm{AC}^0(\{\wedge,\vee,\neg,\mathrm{MOD}_p\})$.

Proof. (1) Let f be a symmetric function. For each input x the value $f(x)$ depends only on the number $|x|_1$ of 1s occurring in the components x_i of x, also called the *weight* of x. If the weight of x is j then $\mathrm{EXACT}_k(x)$ holds, so f is equivalent to

$$\bigvee_{k=0}^{n} (a_k \wedge \mathrm{EXACT}_k(x))$$

for appropriate boolean constants a_k.
(2), (3) Notice that $x \wedge \neg x$ $(x \vee \neg x)$ is equivalent to 0 (1) and that

$$\mathrm{EXACT}_k(x_1,\ldots,x_n) = \mathrm{TH}_k(x_1,\ldots,x_n) \wedge \neg\mathrm{TH}_{k+1}(x_1,\ldots,x_n)$$

and

$$\mathrm{TH}_k(x_1,\ldots,x_n) = \mathrm{MAJ}_{2n}(x_1,\ldots,x_n,0^k,1^{n-k}).$$

(4) Follows from (1), (2) and (3), since the disjunction in (1) is exclusive.
(5) Notice that for $s < p$,

$$\mathrm{MOD}_{s,p}(x_1,\ldots,x_n) = \neg(\mathrm{MOD}_p(x_1,\ldots,x_n,1^{p-s})).$$

(6), (7) Notice that

$$\mathrm{MOD}_p(x_1,\ldots,x_n) = \neg\left(\bigvee_{k\,\equiv\,0 \bmod p,\; k\le n} \mathrm{EXACT}_k\right),$$

while (7) follows from (6) and (3).
(8) Assume that $a|b$, i.e., a divides b. Then we have the identity

$$\mathrm{MOD}_a(x_1,\ldots,x_n) = \mathrm{MOD}_b(\overbrace{\overbrace{x_1,\ldots,x_n},\ldots,\overbrace{x_1,\ldots,x_n}}^{b/a}).$$

(9) Given $x_1,\ldots,x_n$ let us define $y_1,\ldots,y_n$ as follows:

$$y_i = \begin{cases} 1 \text{ if } x_i = 1 \text{ and } \mathrm{MOD}_{p^{m-1}}(0,\ldots,0,x_i,\ldots,x_n) = 0 \\ 0 \text{ otherwise.} \end{cases}$$

Clearly, the sequence $y_1, \ldots, y_n$ is obtained from the sequence $x_1, \ldots, x_n$ by retaining only the kp^m-th 1 (for $kp^m \leq n$) in a sequence of p^m 1s. It is then clear that

$$\text{MOD}_{p^m}(x_1, \ldots, x_n) = \text{MOD}_{p^{m-1}}(x_1, \ldots, x_n) \vee \text{MOD}_p(y_1, \ldots, y_n)$$

and the desired assertion follows by induction on m.

1.5 Nomenclature

In the following sections, to build up intuition we present boolean circuits for parsing and integer arithmetic. As a warm-up, consider the question of transitive closure of a binary relation R, such as the edge relation in a graph. Let $G = (V, E)$ be a directed graph, where the vertex set $V = \{1, \ldots, n\}$ and edge set $E \subseteq V \times V$. Define the *adjacency matrix* for G to be an $n \times n$ boolean matrix $M = (m_{i,j})$, where $m_{i,j} = 1$ iff (i, j) is an edge in G. Define the *reachability matrix* $N = (n_{i,j})$ by $n_{i,j} = 1$ iff there is a directed path from vertex i to j. We will sketch the design of a logarithmic depth, polynomial size unbounded fan-in boolean circuit D which computes N from M, where D has n^2 input nodes $x_{i,j}$ and n^2 output nodes $n_{i,j}$.

Define the boolean matrix product $C = (c_{i,j}) = A \cdot B$, for $n \times n$ boolean matrices A, B by

$$c_{i,j} = \bigvee_{k=1}^{n} a_{i,k} \wedge b_{k,j}$$

for $1 \leq i, j \leq n$. It is clear that there is a constant depth, $O(n^3)$ size, unbounded fan-in boolean circuit computing the boolean matrix product. Let $(I \vee M)$ be the modified adjacency matrix for G, obtained from M by setting diagonal elements of the matrix to 1. Since every path between two vertices has length at most $n - 1$, and $2^{\lceil \log n \rceil} \geq n$, the reachability matrix N can be obtained by $\lceil \log n \rceil$ repeated squarings as follows.[6]

```
A = (I ∨ M)
for i = 1 to ⌈log n⌉ do
    A = A · A
N = A
```

Putting together the $\lceil \log n \rceil$ many constant depth circuits results in a logarithmic depth, polynomial size circuit for computing the transitive closure.

To simplify description of size, depth and fan-in for circuits we introduce the following definitions. NC^k denotes the family of all functions $f : \{0,1\}^* \to \{0,1\}$ such that f is computed by a family $\langle C_n : n \in \mathbf{N} \rangle$ of fan-in 2 boolean circuits, where *depth*(C_n) is $O(\log^k(n))$ and *size*(C_n) is $n^{O(1)}$. Similarly, AC^k is defined as in NC^k, but with unbounded fan-in AND/OR gates in place of fan-in 2 gates. In terms of the earlier notation for AC^0-reductions, the function f :

[6] Throughout this book we follow the convention in C, where a variable assignment is indicated by = rather than :=.

$\{0,1\}^* \to \{0,1\}$ is in AC^0 if each $f_n \in \text{AC}^0(\{\wedge,\vee,\neg\})$, where $f_n = f \restriction \{0,1\}^n$. The class NC designates $\cup_{k=1}^{\infty}\text{NC}^k$. Clearly $\text{NC}^k \subseteq \text{AC}^k$, and by replacing an unbounded fan-in AND/OR gate by a binary tree of fan-in 2 gates of depth logarithmic in the fan-in of the simulated gate, it follows that $\text{AC}^k \subseteq \text{NC}^{k+1}$, and so $\text{NC} = \cup_{k=1}^{\infty}\text{AC}^k$. Subclasses of NC^1 of much current interest are the class TC^0 of constant depth, polynomial size circuits having unbounded fan-in threshold gates, and the classes $\text{ACC}(p)$ of constant depth, polynomial size circuits, which in addition to negation and unbounded fan-in AND, OR gates have gates outputting 1 if the sum of the inputs is 0 modulo p. Here $p > 1$ is not necessarily prime, and $\text{ACC} = \cup_{p>1}\text{ACC}(p)$.

Later we consider *uniform* circuit families $\langle C_n : n \in \mathbf{N}\rangle$, where some uniformity criterion is given for the construction of the circuits C_n, though in this chapter we concentrate only on non-uniform circuits. It turns out that uniform AC^k corresponds exactly to $O(\log^k n)$ time on a parallel random access machine with polynomially many active processors, so that unbounded fan-in circuit design has applications for parallel algorithms.

1.6 Parsing Regular and Context-Free Languages

Parsing is an important concept with applications in compilers, formula evaluation, natural language understanding, etc. Here we begin by considering whether an expression belongs to a given regular language. Following that, we consider context-free language recognition, an important topic with applications ranging from programming language debuggers to RNA secondary structure prediction.

Recall that a finite-state automaton M is defined to be a quintuple $\langle Q, \Sigma, q_0, \delta, F\rangle$, where $Q = \{q_0, \ldots, q_{m-1}\}$ is a finite set of states, $\Sigma = \{a_1, \ldots, a_k\}$ is a finite alphabet, q_0 is the initial state, $\delta : Q \times \Sigma \to Q$ is the transition function and $F \subseteq Q$ is a set of accepting states. The *extended transition* function $\hat{\delta} : Q \times \Sigma^* \to Q$ is defined by: $\hat{\delta}(q, \lambda) = q$, where λ is the empty word, and, for $a \in \Sigma$, $w \in \Sigma^*$, $\hat{\delta}(q, aw) = \hat{\delta}(\delta(q,a), w)$. The language accepted by M, denoted by $L(M)$, is $\{w \in \Sigma^* : \hat{\delta}(q_0, w) \in F\}$.

Let $L \subseteq \Sigma^*$ be a regular language. Then L is recognized by a finite state automaton $M = \langle Q, \Sigma, q_0, \delta, F\rangle$. For each word w of Σ^*, we associate the mapping $f_w : Q \to Q$ obtained by repeatedly applying the transition function δ on the letters from w; i.e., if $w = w_1 \cdots w_n$ then

$$f_w(q) = \delta(\cdots \delta(\delta(q, w_1), w_2), \ldots, w_n) = \hat{\delta}(q, w).$$

When M is the minimal finite state automaton recognizing the regular language L, then the (finite) collection $\{f_w : w \in \Sigma^*\}$, constructed as above from M, is called the *syntactic monoid* of L. There is a rich algebraic structure theory of such syntactic monoids. In particular, the Krohn-Rhodes theorem [Arb68] states that every finite monoid is a homomorphic image of a wreath product of finite simple groups and possibly of three particular non-groups called *units*.[7] The following proposition yields NC^1 circuits for recognizing a regular language $L(M)$ accepted by finite state automaton M. We assume that the description of M's transition function δ is in the form of an $k \times m$ matrix, where Σ has k elements and Q has m states. If we denote the transition function matrix by A, then $A[i,j] = \delta(q_i, a_j)$.

Proposition 1.6.1. *Let $M = \langle Q, \Sigma, q_0, \delta, F \rangle$ be a finite state automaton, and let $L(M) \subseteq \Sigma^*$ be the language accepted by M. Then $L(M) \in$ NC^1.*

Proof. We describe the construction of the n-th circuit C_n, whose inputs are words $w = w_1 \cdots w_n$ in Σ^*, and whose output is 0 (1), according to whether $w \notin L(M)$ ($w \in L(M)$). Since Σ may not be $\{0,1\}$, the circuit C_n is not necessarily boolean. However, using a suitable encoding of the finite alphabet Σ, a corresponding boolean circuit B_n may be given, whose depth and size are within a constant factor of the depth and size of C_n.

The circuit C_n implements a binary function composition tree T as follows. First, we represent a finite function $g : Q \to Q$ by the $|Q| \times |Q|$ boolean matrix $M_g = (m_{i,j})$, where $m_{i,j} = 1$ exactly when $g(i) = j$. Note that the matrix $M_{g \circ h}$ associated with the composition $g \circ h$ is just the matrix product $M_g \times M_h$. Now, for each $1 \leq i \leq n$, associate boolean matrix $M_{f_{w_i}}$ with the i-th leaf of T, and let each non-leaf node of T compute the composition $g \circ h$ of its two children g, h. The root of T then contains $f_w = f_{w_n} \circ \cdots \circ f_{w_1}$. It follows that $w \in L(M)$ if and only if $f_w(q_0) \in F$. Since Q is fixed throughout, computing the matrix product for the composition requires constant size circuitry, so the depth of the boolean circuit implementing this binary, function composition tree is logarithmic in n, its size polynomial in n, and its fan-in is 2.

Thus every regular language belongs to NC^1. A related implementation of this algorithm for recognizing (and parsing) a regular language has been implemented in *LISP on the parallel Connection Machine by [SH86].

We now show that *context-free* languages are recognizable in AC^1, a result due to W.L. Ruzzo [Ruz80].[8]

Definition 1.6.1. *A context-free grammar is given by $G = (V, \Sigma, R, S)$, where V is a finite set of nonterminal symbols (also called variables), Σ is a disjoint finite set of terminal symbols, $S \in V$ is the* start *nonterminal, and*

$$R \subset V \times (V \cup \Sigma)^*$$

---

[7] See Chapter 5 of [Arb68]. The prime decomposition theorem of the algebraic theory of machines is by K. B. Krohn, J. Rhodes and B. Tilson.

[8] In [Ruz80] W.L. Ruzzo proved that *LOGCFL*, the logspace closure of context-free languages, is in logspace uniform NC^2. When general interest in the AC^k classes arose, it was immediately noted that his construction actually shows containment in AC^1. We present Ruzzo's construction as embodied in Lemma 2 of [Ruz80], but do not discuss uniformity issues here.

is a finite set of production rules. Elements of R are usually denoted by $A \to w$, rather than (A, w).

If $x, y \in (V \cup \Sigma)^*$ and $A \to w$ is a rule, then by replacing the occurrence of A in xAy we obtain xwy. Such a derivation in one step is denoted by $xAy \Rightarrow_G xwy$, while the reflexive, transitive closure of $\Rightarrow_G$ is denoted $\Rightarrow_G^*$. The language generated by context-free grammar G is denoted by $L(G)$, and defined by

$$L(G) = \{w \in \Sigma^* : S \Rightarrow_G^* w\}.$$

The grammar G is in Chomsky normal form when all rules in R are of the form $A \to BC$, or $A \to a$, where $A, B, C \in V$ and $a \in \Sigma$.

It is a classical result that every context-free language, which does not contain the empty word λ, is generated by a context-free grammar in Chomsky normal form (see Exercise 1.13.7). Before giving the AC^1 algorithm, we present the following so-called 2-3 Lemma. If T is a binary tree and x is a node in T, then define T_x to be the subtree of T rooted at x

$$T_x = \{y \in T : x \text{ is an ancestor of } y\}$$

and let $||T_x||$ denote the number of leaves in T_x.

Lemma 1.6.1 ([LSH65]). *If T is a binary tree with $n > 1$ leaves, then there exists a node $x \in T$ such that*

$$\lceil \frac{n}{3} \rceil < ||T_x|| \leq \lfloor \frac{2n}{3} \rfloor.$$

Proof. Define the sequence $x_0, x_1, \ldots, x_d$ where x_0 is the root of T, x_d is a leaf of T, and having the property that for all $i < d$, x_{i+1} is that child of x_i such that $||T_{x_{i+1}}|| \geq ||T_{x_i}||$. Let $i_0 \in \{0, \ldots, d\}$ be the least index i such that $||T_{x_i}|| \leq \lceil \frac{n}{3} \rceil$ and let $i_1 = i_0 - 1$. Then by minimality of i_0, we have $\lceil \frac{n}{3} \rceil < ||T_{x_{i_1}}||$. Since $T_{x_{i_0}}$ contains at least half the number of leaves that $T_{x_{i_1}}$ contains, we have

$$||T_{x_{i_1}}|| \leq 2 \cdot \lceil \frac{n}{3} \rceil \leq \lfloor \frac{2n}{3} \rfloor.$$

This establishes the lemma.

Theorem 1.6.1 ([Ruz80]). *If L is a context-free language, then $L \in \mathrm{AC}^1$.*

Proof. We describe a family $\langle D_n : n \in \mathbf{N} \rangle$ of circuits, for which $L \cap \Sigma^n = \{w_1 \cdots w_n : D_n(w_1, \ldots, w_n) \text{ outputs } 1\}$.

Though our definition of boolean circuit allowed only $0, 1$ inputs, by a suitable encoding of the alphabet Σ, the D_n can be replaced by appropriate boolean circuits, whose depth and size are within a constant factor of the depth and size of D_n.

Let D_0 be the constant 1 or 0, depending on whether the empty word λ belongs to L. By Exercise 1.13.7, there is a *Chomsky normal form grammar*

$G = (V, \Sigma, R, S)$ which generates $L - \{\lambda\}$. All rules in G are of the form $A \to BC$ or $A \to a$, where A, B, C are nonterminals and a is a terminal symbol.

If $w = w_1 \cdots w_n$ is a word of length n in $L(G)$, where G is in Chomsky normal form, then a *parse tree* for w is a *binary* tree T, such that:

1. w is the word formed by reading from left to right the leaves of T.
2. The root of T is labeled by S, the "start" variable for the grammar G.
3. If a node of T is labeled by A, then
   a) either that node has only one child, which is labeled a and $A \to a$ is a rule of G,
   b) or that node has two children, which are labeled B, C and $A \to BC$ is a rule of G.

Now let $L = L(G)$, where $G = (V, \Sigma, R, S)$ is a context-free grammar in Chomsky normal form. For each n, we describe a circuit C_n with input nodes $x_1, \ldots, x_n$, which decides whether word $w_1 \cdots w_n \in \Sigma^n$ belongs to L, when w_i is placed at input node x_i. As before, using an encoding of Σ into $\{0,1\}^{\lceil \log \Sigma \rceil}$, a corresponding boolean circuit can then be described, which simulates C_n.

The circuit C_n attempts to guess a valid parse tree for a derivation of $x_1 \cdots x_n$ from S, using Chomsky normal form grammar G. Non-leaf nodes of C_n are associated with assertions of the form $A \Rightarrow_G^* u_1 \cdots u_m$, where $u_1 \cdots u_m \in (V \cup \{x_1, \ldots, x_n\})^*$.

There are two main ideas in Ruzzo's proof. The first idea is to apply the 2-3 Lemma to divide an assertion of the form $A \Rightarrow_G^* u_1 \cdots u_m$ into two smaller assertions $B \Rightarrow_G^* u_i \cdots u_j$ and $A \Rightarrow_G^* u_1 \cdots u_{i-1} B u_{j+1} \cdots u_m$, where $\lceil m/3 \rceil \leq j - i + 1 \leq \lfloor 2m/3 \rfloor$. Thus the underlying parse subtree with root labeled by $A \Rightarrow_G^* u_1 \cdots u_m$ is replaced by two smaller trees, of roughly half the size. Logarithmic circuit depth is guaranteed by iterating this idea. The subtree whose root is labeled by $A \Rightarrow_G^* u_1 \cdots u_{i-1} B u_{j+1} \cdots u_m$, is said to be *scarred.* The second idea is to ensure that every word $u_1 \cdots u_m \in (V \cup \{x_1, \ldots, x_n\})^*$, which appears in a label $A \Rightarrow_G^* u_1 \cdots u_m$ of a node of circuit C_n, is actually one of the forms

- $x_i \cdots x_j$
- $x_i \cdots x_{i_1 - 1} A x_{j_1 + 1} \cdots x_j$
- $x_i \cdots x_{i_1 - 1} A x_{j_1 + 1} \cdots x_{i_2 - 1} B x_{j_2 + 1} \cdots x_j$
- $x_i \cdots x_{i_1 - 1} A x_{j_1 + 1} \cdots x_{i_2 - 1} B x_{j_2 + 1} \cdots x_{i_3 - 1} C x_{j_3 + 1} \cdots x_j$

where $A, B, C \in V$ and $i \leq i_1 < j_1 < i_2 < j_2 < i_3 < j_3 \leq j$. This can be done as follows. Note that given any three scars of a binary tree, there is an ancestor node of exactly two of the three scars. Supposing that node D is the ancestor of A, B, but not C, replace

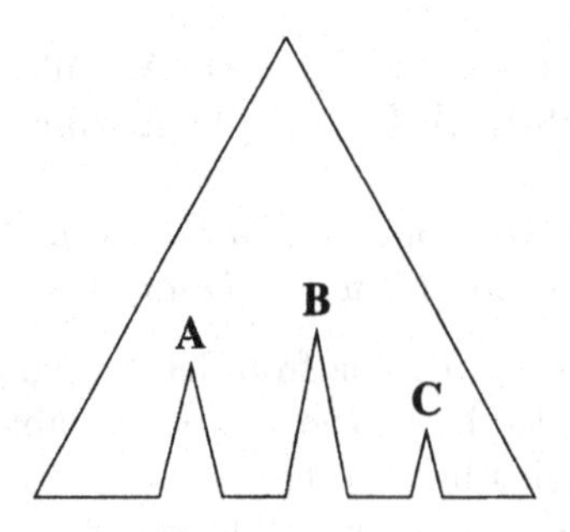

by the trees

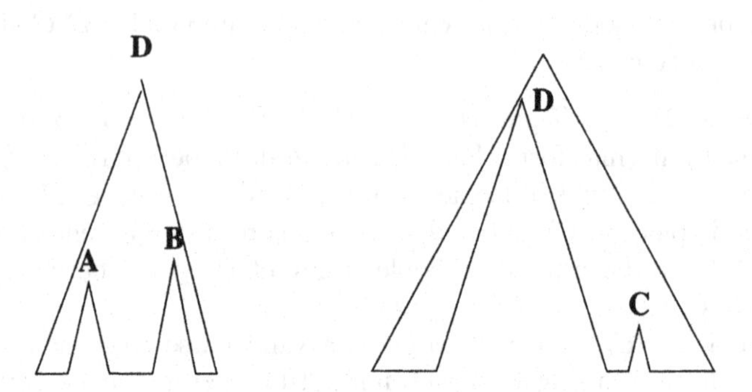

It follows that all labels in circuit C_n are of the form $A \Rightarrow_G^* u_1 \cdots u_m$, where $u_1 \cdots u_m$ is obtained from a convex subword $x_i \cdots x_j$ from input $x_1 \cdots x_n$ by possible replacement of at most three subwords of $x_i \cdots x_j$ by nonterminals. The placement of these nonterminals can be indicated by $i \leq i_1 < j_1 < i_2 < j_2 < i_3 < j_3 \leq j$, so there are at most $O(|V|^4 \cdot n^8)$ many possible labels. Now $|V|$ is constant, so the size of circuit C_n is polynomial in n.

We now describe the construction of C_n, the idea being to alternate between applications of the 2-3 Lemma and of scar reduction.

1. The top (output) node of C_n is an OR-gate labeled by $S \Rightarrow_G^* x_1 \cdots x_n$.
2. Suppose that the OR-gate g of the circuit constructed so far has label $A \Rightarrow_G^* u$, where $A \in V$ and $u \in (V \cup \{x_1, \ldots, x_n\})^*$, $u \neq \lambda$.
   a) If $|u| = 1$, then g has a single child, labeled by 1 if $u = w$ and there is a rule $A \to w$, otherwise labeled by 0.
   b) If $u = BC$, where B, C are nonterminals, then g has a single child, labeled by 1 if $A \to BC$ is a rule of G, otherwise labeled by 0.
   c) If $|u| > 1$, u has fewer than three nonterminals, and the previous case does not apply, then for each $B \in V$ and $1 \leq i \leq j \leq |u|$ satisfying $\lceil |u|/3 \rceil \leq j - i + 1 \leq \lfloor 2|u|/3 \rfloor$, there is an AND-gate h labeled by (B, u, i, j), whose parent is g. Note the fan-in of gate g is $O(|V| \cdot |u|^2)$. This case corresponds to an application of the 2-3 Lemma. Nodes h are not labeled by something of the form $A \Rightarrow_G^* u$, as we described before the formal construction of circuit C_n, but

rather labeled by something of that form along with $i, j \leq n$. Thus there are still only polynomially many possible labels, hence it will follow that C_n has size polynomial in n.

d) If $|u| > 1$ and u has three nonterminals, then u has the form

$$x_i \cdots x_{i_1-1} A x_{j_1+1} \cdots x_{i_2-1} B x_{j_2+1} \cdots x_{i_3-1} C x_{j_3+1} \cdots x_j.$$

For each $B \in V$ and $1 \leq i < j \leq m$ such that there are at most two nonterminals appearing in $u_i \cdots u_j$, there is an AND-gate h labeled by (B, u, i, j), whose parent is g. As in the previous case, the fan-in of g is $O(|V| \cdot |u|^2)$, and though the labels of the h are not of the form $A \Rightarrow_G^* w$, there are only polynomially many labels, so that C_n will have polynomial size.

3. Suppose that the AND-gate g of the circuit constructed so far has label (B, u, i, j), where $B \in V$, $u \in (V \cup \{x_1, \ldots, x_n\})^*$, $u \neq \lambda$, and $1 \leq i \leq j \leq |u|$. Suppose that the parent of g is labeled by $A \Rightarrow^* u$. Then the gate g has two children, both OR-gates, labeled respectively by $B \Rightarrow^* u_i \cdots u_j$ and $A \Rightarrow^* u_1 \cdots u_{i-1} B u_{j+1} \cdots u_{|u|}$.

It follows from the 2-3 Lemma that C_n has depth $O(\log n)$; moreover, C_n has size $n^{O(1)}$, since there are at most polynomially many labels. Clearly C_n outputs 1 on input $w_1 \cdots w_n \in \Sigma^n$ if and only if $w_1 \cdots w_n$ has a parse tree if and only if $w_1 \cdots w_n \in L(G)$.

1.7 Circuits for Integer Arithmetic

Current VLSI design supports only the implementation of bounded fan-in gates, hence it is of the greatest interest to find fan-in 2 boolean circuits of small depth and size for arithmetic operations. In this section, we present circuits for integer addition and multiplication.

For certain algorithms, such as the RSA cryptosystem used in electronic money transfer, specialized chips are employed to perform certain arithmetic operations, such as modular powering $a^b \bmod m$. In this regard, it is interesting to note that despite much work, it is still an open problem whether modular powering and greatest common denominator gcd computations can be performed by circuits of small (polylogarithmic) depth and feasible (polynomial) size.

1.7.1 Circuits for Addition and Multiplication

We begin by an example showing that the maximum of n integers, each of n bits, is in AC^0. The minimum can be similarly computed.

Example 1.7.1. The function $\max(a_0, \ldots, a_{n-1})$ of n integers, each of size at most m, can be computed by a boolean circuit as follows. Assume the integers

a_i are distinct (a small modification is required for non-distinct integers). Then the k-th bit of $\max(a_0, \ldots, a_{n-1})$ is 1 exactly when

$$(\exists i < n)(\forall j < n)(j \neq i \to a_j \leq a_i \wedge \text{BIT}(k, a_i) = 1).$$

This bounded quantifier formula is translated into a boolean circuit by

$$\bigvee_{i<n} \bigwedge_{j<n, j\neq i} \bigvee_{\ell<n} \bigwedge_{\ell<p<m} \text{BIT}(p, a_j) = \text{BIT}(p, a_i) \wedge \text{BIT}(\ell, a_j) = 0 \\ \wedge \text{BIT}(\ell, a_i) = 1.$$

The usual algorithm for addition can be implemented by fan-in 2 circuits of linear size and depth.

Proposition 1.7.1. *Addition of two n-bit numbers can be performed by a circuit of size $O(n)$ and depth $O(n)$.*

Proof. Given two n-bit numbers $x = \sum_{i=0}^{n-1} x_i \cdot 2^i$ and $y = \sum_{i=0}^{n-1} y_i \cdot 2^i$, we construct a circuit for computing their sum $z = x + y = \sum_{i=0}^{n} z_i \cdot 2^i$ as well as the corresponding sequence $c_0, \ldots, c_{n-1}, c_n$ of carries, where $c_0 = 0$ and c_{i+1} is the carry when adding x_i, y_i. This is based on the obvious inductive formula

$$z_i = x_i \oplus y_i \oplus c_i, \quad c_{i+1} = x_i y_i \vee x_i c_i \vee y_i c_i,$$

where $c_0 = 0$.

Still employing the usual algorithm, but with slightly more care, integer addition can be computed in constant depth, polynomial size circuits of unbounded fan-in.

Theorem 1.7.1. *Integer addition is in* AC^0.

Proof. Suppose that x, y are two n-bit integers. We design a family $\langle C_{2n,i} : n \in \mathbf{N}, 0 \leq i \leq n \rangle$ of circuits, such that $C_{2n,i}$ outputs the i-th bit of the sum $z = x + y$. The idea is that the i-th carry bit is 1 exactly if there is a previous bit where a carry is generated, and for all intermediate bits the carry is propagated. Let $x = \sum_{i=0}^{n-1} x_i \cdot 2^i$ and $y = \sum_{i=0}^{n-1} y_i \cdot 2^i$. Define the carry bit by letting $c_0 = 0$ and setting c_{i+1} to be

$$\bigvee_{0 \leq j \leq i} \left[(x_j \wedge y_j) \wedge \bigwedge_{j<k<i} (x_k \vee y_k) \right].$$

Then we have

$$z_i = x_i \oplus y_i \oplus c_i.$$

The depth of the circuit $C_{2n,i}$ is constant, and its size is $O(n^2)$. Thus altogether a circuit of constant depth and size $O(n^3)$ having multiple output nodes can compute $z = x + y$.

Multiplication is easily seen to be in AC^1 with a little more work, it can be shown to be in NC^1.

Theorem 1.7.2. *Integer multiplication is in* NC^1.

Proof. We give two proofs. From the elementary school algorithm for multiplication, it is clear that multiplication of two n-bit integers can be reduced to the problem of adding n many $2n$-bit numbers. We indicate how to add n many n-bit integers.

(1) The first proof uses a recursive application of the "3 for 2" trick due to Ofman and Wallace [Wal64]. This technique is called *carry-save addition.* Given three n-bit integers x, y, z, there are NC^0 circuits to compute two $n+1$-bit integers u, v for which $x+y+z = u+v$. Namely, u will be the bits formed by the exclusive-or of the bits of x, y, z and v will be the bits formed by the carry:

$$u_i = x_i \oplus y_i \oplus z_i$$

and

$$v_0 = 0, v_{i+1} = \begin{cases} 1 \text{ if } (x_i \wedge y_i) \vee (x_i \wedge z_i) \vee (y_i \wedge z_i) \\ 0 \text{ otherwise.} \end{cases}$$

An elementary calculation shows that the circuit obtained by recursively applying the 3 for 2 trick has fan-in 2 and depth $O(\log_{3/2} n) = O(\log n)$.

(2) In the second proof due to Avizienis [Avi61], a *redundant notation* is used in order to add two numbers in constant depth, polynomial size, fan-in 2 circuits. Then a binary logarithmic depth addition tree places multiplication in NC^1. An integer is in *balanced 4-ary notation* if its "digits" are among $0, \pm 1, \pm 2$. When adding two n-bit integers $x = \sum_{i<n} x_i \cdot 4^i$, $y = \sum_{i<n} y_i \cdot 4^i$ in balanced 4-ary notation, we will generate a carry c_{i+1} of 1 (-1) whenever $x_i + y_i > 2$ ($x_i + y_i < -2$). By this means, we ensure that carries do not propagate and that the sum of balanced 4-ary integers can be computed in NC^0. Note for example that when performing the following additions in balanced 4-ary, we have

$$(1)_4 + (2)_4 = (1 \quad -1)_4$$

and

$$(-2)_4 + (-1)_4 = (-1 \quad 1)_4.$$

The restriction of digits to $0, \pm 1, \pm 2$ is important. For example, the sum $(33)_4 + (33)_4$ is $(2 \quad -1 \quad 2)_4$, and the carry is then propagated. But this is just what we wanted to avoid.

Algorithm 1.7.1. NC^0 circuit for addition in balanced 4-ary notation.
Input. Integers $x = \sum_{i<n} x_i \cdot 4^i$, $y = \sum_{i<n} y_i \cdot 4^i$, in balanced 4-ary balanced notation of length at most n.
Output. The sum $z = \sum_{i \leq n} z_i \cdot 4^i$ in 4-ary balanced notation.

Let $c_0 = 0$ and for $1 \leq i < n$ let

$$c_{i+1} = \begin{cases} 1 & \text{if } x_i + y_i > 2 \\ -1 & \text{if } x_i + y_i < -2 \\ 0 & \text{otherwise} \end{cases}$$

and for $0 \leq i \leq n$ let

$$z_i = \begin{cases} x_i + y_i + c_i - 4 & \text{if } x_i + y_i > 2 \\ x_i + y_i + c_i + 4 & \text{if } x_i + y_i < -2 \\ x_i + y_i + c_i & \text{otherwise.} \end{cases}$$

Clearly appropriate circuits can be constructed for the above.

It is now clear why we insisted that a carry be generated whenever the absolute value of $x_i + y_i$ is greater than 2. By this means, we are sure that there is room for a possible carry from the previous bits of x and y, so that carries do not propagate.

Algorithm 1.7.2. AC^0 circuit for conversion from ordinary 4-ary to balanced 4-ary notation.
Input. 4-ary representation $\sum_{i<n} b_i \cdot 4^i$ of integer x (digits $0, 1, 2, 3$).
Output. Balanced 4-ary representation $\sum_{i<n} d_i \cdot 4^i$ of integer x (digits $0, \pm 1, \pm 2$).

$$d_i = \begin{cases} b_i - 3 & \text{if } b_i \geq 2 \text{ and } \bigvee_{0 \leq j < i} \bigwedge_{j<k<i} (b_j = 3 \wedge b_k \geq 2) \\ b_i + 1 & \text{if } b_i < 2 \text{ and } \bigvee_{0 \leq j < i} \bigwedge_{j<k<i} (b_j = 3 \wedge b_k \geq 2) \\ b_i & \text{if } b_i \leq 2 \text{ and } \neg \bigvee_{0 \leq j < i} \bigwedge_{j<k<i} (b_j = 3 \wedge b_k \geq 2) \\ -1 & \text{if } b_i = 3 \text{ and } \neg \bigvee_{0 \leq j < i} \bigwedge_{j<k<i} (b_j = 3 \wedge b_k \geq 2) \end{cases}$$

Algorithm 1.7.3. AC^0 circuit for conversion from balanced 4-ary to ordinary 4-ary notation.
Input. Balanced 4-ary representation

$$\sum_{i<n} d_i \cdot 4^i \text{ of integer } x \text{ (digits } 0, \pm 1, \pm 2).$$

Output. 4-ary representation $\sum_{i<n} b_i \cdot 4^i$ of integer x (digits $0, 1, 2, 3$).

$$b_i = \begin{cases} d_i - 1 & \text{if } d_i > 0 \text{ and } \bigvee_{0 \leq j < i} \bigwedge_{j<k<i} (d_j < 0 \wedge d_k = 0) \\ d_i + 3 & \text{if } d_i \leq 0 \text{ and } \bigvee_{0 \leq j < i} \bigwedge_{j<k<i} (d_j < 0 \wedge d_k = 0) \\ d_i & \text{if } d_i \geq 0 \text{ and } \neg \bigvee_{0 \leq j < i} \bigwedge_{j<k<i} (d_j < 0 \wedge d_k = 0) \\ d_i + 4 & \text{if } d_i < 0 \text{ and } \neg \bigvee_{0 \leq j < i} \bigwedge_{j<k<i} (d_j < 0 \wedge d_k = 0) \end{cases}$$

This leads to the following circuit design to compute $\sum_{i<n} a_i$, where $|a_i| \leq n$.

Algorithm 1.7.4. NC^1 circuits for iterated addition.
Input. $a_0, \ldots, a_{n-1}$, each a_i in binary notation and of length at most n.
Output. $b = \sum_{i<n} a_i$, in binary.

(1) for $i = 0$ to $n-1$ do in parallel
  convert a_i from binary into 4-ary notation b_i
(2) for $i = 0$ to $n-1$ do in parallel
  convert b_i into balanced 4-ary notation c_i
(3) compute $d = \sum_{i<n} c_i$ as a balanced 4-ary sum
(4) convert d into ordinary 4-ary notation e
(5) convert e into binary notation

Step (1) is in NC^0, as it can be done by taking pairs of bits of the binary representation and representing the pair by a digit $0, 1, 2, 3$. Step (2) is in AC^0. Step (3) can be done by using a binary addition tree of depth $\log(n)$. Since adding two balanced 4-ary integers is in NC^0, this step is in NC^1. Step (4) is in AC^0, and step (5) in NC^0.

Example 1.7.2. Suppose we must compute the sum of the decimal numbers $15, 3, 7, 15$ having ordinary 4-ary representations $(33)_4$, $(3)_4$, $(13)_4$, $(33)_4$. In step (2), we compute the balanced 4-ary representations

$$(1 \quad 0 \quad -1), (1 \quad -1), (2 \quad -1), (1 \quad 0 \quad -1).$$

Summing the first two and the last two, we obtain balanced 4-ary representations $(1 \quad 1 \quad -2)$, $(1 \quad 2 \quad -2)$. Summing these, we obtain balanced 4-ary representation $(3 \quad -2 \quad 0)$, which is $(2 \quad 2 \quad 0)$ in ordinary 4-ary notation. This yields the binary notation $(101000)_2$, which in decimal is 40.

1.7.2 Division Using Newton Iteration

We next consider integer division: given integers x, y determine $\lfloor x/y \rfloor$. We first present a simple NC^2 division algorithm due to Brent [Bre74] and later a sophisticated NC^1 division algorithm due to Beame, Cook and Hoover [BCH86]. The first algorithm depends on Newton iteration, while the second depends on computing the product $\prod_{i=1}^n a_i$ modulo many primes p by using discrete logarithms to replace iterated multiplication by iterated addition (by the previous section, iterated addition is in NC^1) and then computing the true product from the residues by the Chinese remainder theorem.

The principal step in Brent's algorithm concerns the computation of the quantity $\mathit{inverse}(u, n)$, an approximation to $1/u$ satisfying

$$\left| \frac{1}{u} - \mathit{inverse}(u, n) \right| < 1/2^n.$$

To that end, we will use Newton iteration to approximate the root of the function $f(x) = 1/x - u$.

Algorithm 1.7.5. Algorithm for Newton iteration.
Input. Integers u, n.
Output. $inverse(u, n)$, an approximation to $1/u$ which agrees on the first n bits.

$k = |u|$
$v = 1/2^k$
for $i = 0$ to $|n|$ do
  $v = 2v - uv^2$
$inverse(u, n) = v$

There are $\log(n)$-many iterations, in each of which two products and one sum are formed. Integer addition and multiplication belong to NC^1, so Newton iteration can be done in NC^2.

To justify the previous algorithm, we define a sequence $\langle v_0, v_1, v_2, \ldots \rangle$ converging to $1/u$, such that for every i, we have $v_i < v_{i+1} < 1/u$. Suppose that $k = |u|$, and let $v_0 = 1/2^k$. Since $|u| = k$, we have $2^{k-1} \leq u < 2^k$, and so $1/2^k < 1/u \leq 1/2^{k-1}$. Given v_i, let v_{i+1} be the x-axis intercept of the line passing through the point $(v_i, f(v_i))$ having slope $f'(v_i)$. Thus

$$f'(v_i) = \frac{0 - f(v_i)}{v_{i+1} - v_i}.$$

Since by induction, $f'(v_i) \neq 0$, we have

$$\begin{aligned} v_{i+1} &= v_i - \frac{f(v_i)}{f'(v_i)} \\ &= v_i + \frac{1 - uv_i}{v_i} / \frac{-1}{v_i^2} \\ &= 2v_i - uv_i^2. \end{aligned}$$

Note that for $0 < v_i < 1/u$, we have $f(v_i) > 0$ and $f'(v_i) < 0$, so that $v_i < v_{i+1}$.

We now give a bound on the error term $\epsilon_i = 1/u - v_i$. First, $v_0 = 1/2^n < 1/u \leq 1/2^{n-1}$, so

$$\epsilon_0 = 1/u - v_0 = 1/u - 1/2^n = \frac{2^k - u}{u \cdot 2^k} < \frac{2^{k-1}}{u \cdot 2^k} = 1/2u.$$

Next,

$$\begin{aligned} \epsilon_{i+1} &= 1/u - v_{i+1} \\ &= 1/u - (2v_i - uv_i^2) \\ &= u(1/u^2 - 2v_i/u + v_i^2) \\ &= u(1/u - v_i)^2 = u \cdot \epsilon_i^2. \end{aligned}$$

Thus we have $\epsilon_0 < 1/2u$, $\epsilon_1 < 1/4u$, ..., $\epsilon_i < 1/2^{2^i} \cdot u$. It follows that for all i, $v_i < 1/u$ and that $k = |u|$ many iterations suffice for an approximation within $1/2^k$ of $1/u$.

This yields the following integer division algorithm.

Algorithm 1.7.6. Brent's integer division.
Input. Integers x, y.
Output. $\lfloor x/y \rfloor$.

$n = |x|$
$z = inverse(y, n)$
  if $x - y \cdot \lfloor x \cdot z \rfloor > y$ then
    output $\lfloor x \cdot z \rfloor + 1$
  else
    output $\lfloor x \cdot z \rfloor$

To prove the correctness of this algorithm, we have the following claim.
CLAIM. $0 \leq x - y \cdot (\lfloor x \cdot inverse(y, n) \rfloor) < y + 1$.

Proof. Since $n = |x|$ we have $2^{n-1} \leq x < 2^n$, so $2/2^n < 1$. As well, recall that

$$inverse(y, n) \leq 1/y < inverse(y, n) + \frac{1}{y \cdot 2^n}$$

so

$$\frac{x}{y} - x \cdot inverse(y, n) < \frac{x}{y \cdot 2^n} < \frac{1}{y}.$$

Letting $z = inverse(x, k)$, by multiplying the previous inequality by y, it follows that

$$x - y \cdot (\lfloor x \cdot z \rfloor + x \cdot z - \lfloor x \cdot z \rfloor) < 1$$

so

$$\begin{aligned} x - y \cdot (\lfloor x \cdot z \rfloor) &< 1 + y \cdot (x \cdot z - \lfloor x \cdot z \rfloor) \\ &< 1 + y. \end{aligned}$$

The claim now follows.
If

$$x - y \cdot (\lfloor x \cdot inverse(y, k) \rfloor) < y,$$

then

$$x/y - \lfloor x \cdot inverse(y, k) \rfloor < 1,$$

otherwise

$$x/y - (\lfloor x \cdot inverse(y, k) \rfloor + 1) < 1.$$

This justifies the previous algorithm.

1.7.3 Division Using Iterated Product

The overall outline of the Beame, Cook and Hoover NC^1 division algorithm[9] is that to compute x/y, we compute a good approximation to $1/y$ by using the fact that for $0 < \epsilon < 1$, $\frac{1}{1-\epsilon} = \sum_{i=0}^{\infty} \epsilon^i$. Division thus reduces to computing a sum of powers of ϵ, and hence to iterated product. To compute the product $\prod_{i<n} a_i$ where $|a_i| = n$, we compute the residues $(\prod_{i<n} a_i) \bmod p$ for the first n^2 many primes p and then apply the Chinese remainder theorem. For the modular computations $(\prod_{i<n} a_i) \bmod p$, we build discrete logarithm tables into the n-th circuit and perform iterated addition modulo p. Since iterated addition is in NC^1, the entire algorithm is in NC^1.

Suppose that we have an NC^1 algorithm for computing the iterated powers $u^0, u^1, \ldots, u^n$ of an integer u. We then have an NC^1 algorithm to compute integer division $\lfloor x/y \rfloor$ as follows.

Algorithm 1.7.7. Division using iterated product.
Input. Non-negative integers x, y.
Output. $\lfloor x/y \rfloor$.

```
if y = 0 then
    output "undefined"
else if y = 1 then
    output x
else
    k = |y|
    ε = 1 − y/2^k
    for i = 0 to n − 1
        compute ε^i
    z = 2^(−k) · Σ_{i<n} ε^i
    compute x · z
    if x − y · ⌊x · z⌋ < y then
        output ⌊x · z⌋
    else
        output ⌊x · z⌋ + 1
    end if
end if
```

[9] A sharper analysis of this algorithm shows that it can be implemented in the class TC^0, i.e., constant depth, polynomial size threshold circuits. In this chapter, we have ignored the question of uniformity of circuit families. The Beame–Cook–Hoover construction is polynomial time uniform, but not known to be logspace uniform. Brent's construction yields logspace uniform $O(\log^2 n)$ size circuits. It is an open problem if integer division is computable in logspace; here, it should be noted that logspace uniform NC^1 is contained in logspace, by a careful use of depth first search. This remark illustrates the delicate issues raised by uniformity of circuits.

To get a feel for how the above algorithm works, suppose that $y = 5$, which has binary representation $(101)_2$. Move the "decimal point" to the left to obtain .101 and let $\epsilon = 1 - (.101)_2$. Then perform the powering of ϵ and trace the algorithm.

Let $k = |y|$ and suppose that $k > 1$. Then $2^{k-1} \leq y < 2^k$, so $1/2 \leq y/2^k < 1$ and $0 < 1 - y/2^k \leq 1/2$. Setting $\epsilon = 1 - y/2^k$, it follows that $0 < \epsilon \leq 1/2$ and $1/y = \frac{2^{-k}}{1-(1-y/2^k)} = \frac{2^{-k}}{1-\epsilon}$. We have

$$\frac{1}{1-\epsilon} = \sum_{i=0}^{\infty} \epsilon^i = \sum_{i<n} \epsilon^i + \epsilon^n \cdot S,$$

where $S = \sum_{i=0}^{\infty} \epsilon^i \leq 2$. Thus

$$\epsilon^n \cdot S \leq \epsilon^n \cdot 2 \leq \epsilon^n \cdot \frac{1}{\epsilon} \leq \epsilon^{n-1} \leq 1/2^{n-1}.$$

So

$$\begin{aligned} \text{abs}\left(1/y - 2^{-k} \cdot \sum_{i<n} \epsilon^i\right) &= \text{abs}\left(\frac{2^{-k}}{1-\epsilon} - 2^{-k} \cdot \sum_{i<n} \epsilon^i\right) \\ &= 2^{-k} \cdot \text{abs}\left(\frac{1}{1-\epsilon} - \sum_{i<n} \epsilon^i\right) \\ &\leq 2^{-k} \cdot 2^{-(n-1)} = 2^{-(n+k-1)}. \end{aligned}$$

As $k \geq 2$, we have that $\text{abs}(1/y - 2^{-k} \cdot \sum_{i<n} \epsilon^i) < 2^{-n}$, so $2^{-k} \cdot \sum_{i<n} \epsilon^i$ is an n-bit approximation to $1/y$. The justification of the selection statement is as in Brent's algorithm. Thus the above integer division algorithm is correct.

An implementation of the circuit C for inputs x, y of size n goes as follows.

1. Compute $\epsilon = 1 - y/2^n$. Subtraction can be reduced to addition by the 2's-complement trick, and addition is in AC^0.
2. Construct subcircuits $C_0, \ldots, C_{n-1}$ on the same level where the output of C_i is ϵ^i. For fixed $i < n$, an NC^1 circuit for computing the i-th power of its input will be constructed later. Assume for the present that C_i is given by an oracle gate.
3. Compute integer quotient $q = 2^n \cdot x \cdot \sum_{i<n} \epsilon^i$. Iterated sum and multiplication are in NC^1.
4. If remainder $r = x - q \cdot y < y$ then output q else $q + 1$. Multiplication is in NC^1, and subtraction and order comparison are in AC^0.

Assuming NC^1 circuits for computing powers, by connecting these subcircuits, it follows that C is a logarithmic depth, polynomial size circuit.

Recall the following notions from elementary number theory. For x, y, m integers, we write $x \equiv y \pmod{m}$ to mean that m divides $x - y$. As standard practice in programming languages, we use the operation mod m and write $x = y \bmod m$ to mean that $0 \leq x < m$ and $x \equiv y \pmod{m}$. The greatest common divisor of integers x, y is written (x, y), or sometimes $\gcd(x, y)$. The integers x, y are *relatively prime* if $(x, y) = 1$. If $(x, m) = 1$ then the Euler–Fermat theorem states that $x^{\phi(m)} \equiv 1 \pmod{m}$, where the Euler *totient* function $\phi(m)$ is defined to be the number of positive integers less than m which are relatively prime to m. In particular, for p prime, $\phi(p) = p-1$ and so $x^{p-1} \equiv 1 \pmod{p}$. If $(x, m) = 1$, then the *exponent* of x modulo m, written $\exp_m(x)$, is the smallest positive integer e satisfying $x^e \equiv 1 \pmod{m}$, i.e., the order of the cyclic group generated by the powers of x.. The positive integer $g < m$ is a *primitive root* modulo m if $\exp_m(g) = \phi(m)$. It is not difficult to see from these definitions that $g < m$ is a primitive root modulo m if and only if g is a generator for the multiplicative group $\mathbf{Z}_m^* = \{1 \leq i < m : (i, m) = 1\}$; that is, $\mathbf{Z}_m^*$ is formed by the residues modulo m of $\{1, g, g^2, \ldots, g^{\phi(m)-1}\}$. If x and m are relatively prime and g is a primitive root modulo m, then there exists a unique integer k, called the *index* of x to the base g modulo m, with the property that $0 \leq k < \phi(m)$ and $x \equiv g^k \pmod{m}$. In this case, we write $k = \text{ind}_g(x)$. In computer science, the index of x is usually called the *discrete logarithm* of x, and a primitive root modulo m is called a *discrete logarithm base* modulo m.

Let $\pi(x)$ be the function which counts the number of primes less than x. The celebrated prime number theorem states that the asymptotic density of the prime numbers is $1/\ln(x)$:

$$\lim_{x \to \infty} \frac{\pi(x)}{x/\ln(x)} = 1.$$

The prime number theorem implies a bound on the n-th prime number p_n; namely that $p_n = O(n \cdot \ln(n))$.

The *Chinese remainder theorem* states that if $m_1, \ldots, m_n$ are pairwise relatively prime, then for any integers $r_1, \ldots, r_n$ there is a unique integer $x < \prod_{i=1}^n m_i$ such that $x \equiv r_i \pmod{m_i}$. In other words, there is a unique solution to any system of n modular linear equations, provided that the moduli are pairwise relatively prime. See, e.g., [Apo76] for more background material on number theory.

Computing the i-th power of an integer of length n, where $i < n$, is clearly solved by an algorithm which computes the iterated product $\prod_{i=1}^n x_i$ of n many n-bit integers. The following algorithm uses discrete logarithms to reduce iterated product modulo a prime p to iterated sum, which we know to belong to NC^1. By finding the iterated product modulo sufficiently many distinct primes, an application of the Chinese remainder theorem yields the desired product.

Algorithm 1.7.8. Iterated multiplication.
Input. Integers $x_1, \ldots, x_n$, with $|x_i| \leq n$ for each i.
Output. The product $y = \prod_{1 \leq i \leq n} x_i$.

1. Compute the first n^2 many primes $p_1, \ldots, p_{n^2}$, and for each such prime, compute a generator $gen(p)$ and a table of discrete logarithms

$$z_0 = gen(p)^0, \ldots, z_{p-2} = gen(p)^{p-2}.$$

2. $P = \prod_{1 \leq i \leq n^2} p_i$
   for $i = 1$ to n^2
   $(P/p_i) = \prod_{1 \leq j \leq n^2, j \neq i} p_j$
   compute $(P/p_j)^{-1}$, the multiplicative inverse of $(P/p_j) \bmod p_j$
   i.e., $(P/p_j)^{-1} < p_j$ and $P/p_j \cdot (P/p_j)^{-1} \equiv 1 \bmod p_j$

3. for $i = 1$ to n
   for $j = 1$ to n^2
   $b_{i,j} = x_i \bmod p_j$

4. for $j = 1$ to n^2
   $b_j = (\prod_{1 \leq i \leq n} b_{i,j}) \bmod p_j$

5. Output

$$y = \sum_{1 \leq j \leq n^2} b_j \cdot (P/p_j) \cdot (P/p_j)^{-1} \bmod P$$

where $(P/p_j)^{-1}$ is the multiplicative inverse of P/p_j modulo p_j.

Steps (1) and (2) are externally computed for fixed n and hardwired into the circuit under construction (it is this part of the construction which raises uniformity problems). In Algorithm 1.7.9, NC^1 circuits will be given for computing the remainders modulo small primes in step (3). A total of n^3 such subcircuits must be constructed, one for each $1 \leq i \leq n, 1 \leq j \leq n^2$, but the subcircuits are placed at the same level allowing parallel computation. The computations $b_j = (\prod_{1 \leq i \leq n} b_{i,j}) \bmod p_j$ in step (4) are performed by n^2 subcircuits, one for each b_j, all placed at the same level. Each computation is done by table look-up to find the discrete logarithms $\text{ind}_{p_j}(b_{i,j})$ of the $b_{i,j}$, applying iterated addition and then doing table look-up to find the inverse logarithm to compute b_j. All of step (4) can be done in NC^1. Step (5), called *Chinese remaindering*, produces $y < P$ satisfying $y \equiv b_j \pmod{p_j}$, for each $1 \leq j \leq n^2$. Since $x = \prod_{1 \leq i \leq n} x_i < 2^{n^2} \leq \prod_{1 \leq j \leq n^2} p_j = P$ has the same residues as y modulo p_j, by uniqueness in the Chinese remainder theorem, $x = y$.

First we give a few details concerning the use of discrete logarithm tables and Chinese remaindering. To compute

$$b_j = (\prod_{1 \leq i \leq n} b_{i,j}) \bmod p_j$$

proceed as follows. Since the multiplicative group $\mathbf{Z}_p^*$ is cyclic, a primitive root or generator $1 \leq g < p$ of $\mathbf{Z}_{p_j}^*$ satisfies

$$\{1, 2, \ldots, p-1\} = \{1, g, g^2 \bmod m, \ldots, g^{p-2} \bmod m\}.$$

Since there are in fact $\phi(p-1)$ many such generators, where ϕ is the Euler totient function, for specificity we take g_p to be the smallest *generator* of $\mathbf{Z}_p^*$. By $\log_{g_p}(a) = b$, we mean that $0 < a, b < p$ and that $g_p^a \equiv b \bmod p$. We compute

$$d_j = \sum_{1 \le i \le n} \log_{g_p}(b_{i,j})$$

and then

$$e_j = d_j \bmod (p_j - 1)$$

and finally

$$b_j = g_{p_j}^{e_j} \bmod p_j.$$

By the Euler–Fermat theorem, $a^{p-1} \equiv 1 \pmod p$ holds for every prime p, hence $g_{p_j}^{d_j} \equiv g_{p_j}^{e_j} \pmod{p_j}$. It follows that $b_j \equiv \prod_{i=1}^n b_{i,j} \pmod{p_j}$.

For Chinese remaindering, since P/p_j and p_j are relatively prime, let $(P/p_j)^{-1}$ be the multiplicative inverse of P/p_j modulo p_j. Then

$$(b_j \cdot (P/p_j) \cdot (P/p_j)^{-1}) \bmod p_i = \begin{cases} b_j & \text{if } i = j \\ 0 & \text{otherwise} \end{cases}$$

so that

$$(\sum_{1 \le j \le n^2} b_j \cdot (P/p_j) \cdot (P/p_j)^{-1}) \bmod P$$

satisfies all the requisite congruences and must be the product of $x_1, \ldots, x_n$.

Define

$$y_i = b_j \cdot (P/p_j) \cdot (P/p_j)^{-1}$$

and

$$y = \sum_{1 \le j \le n^2} y_i.$$

Computing y by hardwiring the P/p_j and $(P/p_j)^{-1}$ (the multiplicative inverse of P/p_j modulo p_j) into the circuit and using multiplication and iterated addition, it follows that y is computable by NC^1 circuits. Let $\ell = 2^{\lceil \log_2 n^2 \rceil}$ be the least power of 2 greater than or equal to n^2. Then

$$\begin{aligned} \frac{y}{P} &= \sum_{j=1}^{n^2} \frac{y_i}{P} = \sum_{j=1}^{n^2} \frac{b_j}{p_j} \cdot (P/p_j)^{-1} \\ &= \frac{1}{\ell} \sum_{j=1}^{n^2} \frac{\ell b_j}{p_j} \cdot (P/p_j)^{-1} \\ &\ge \frac{1}{\ell} \sum_{j=1}^{n^2} \left\lfloor \frac{\ell b_j}{p_j} \cdot (P/p_j)^{-1} \right\rfloor . \end{aligned}$$

Letting s designate the last term, we have

$$0 \leq y - s \cdot P = \frac{1}{\ell} \sum_{j=1}^{n^2} \left(\frac{\ell b_j}{p_j} \cdot (P/p_j)^{-1} - \left\lfloor \frac{\ell b_j}{p_j} \cdot (P/p_j)^{-1} \right\rfloor \right) \cdot P.$$

Since $z - \lfloor z \rfloor < 1$ holds for all non-negative z, it follows that

$$y - s \cdot P < \frac{1}{\ell} \sum_{j=1}^{n^2} P = \frac{n^2 P}{\ell} \leq P$$

and so $0 \leq y - \lfloor s \rfloor \cdot P < 2P$. From this, one can test whether $y - \lfloor s \rfloor \cdot P < P$, in which case

$$\lfloor y/P \rfloor = \lfloor s \rfloor = \left\lfloor \left(\frac{1}{\ell} \sum_{j=1}^{n^2} \left\lfloor \frac{\ell b_j}{p_j} \cdot (P/p_j)^{-1} \right\rfloor \right) \right\rfloor .$$

If $y - \lfloor s \rfloor \cdot P \geq P$, then

$$\lfloor y/P \rfloor = \lfloor s \rfloor + 1 = \left\lfloor \left(\frac{1}{\ell} \sum_{j=1}^{n^2} \left\lfloor \frac{\ell b_j}{p_j} \cdot (P/p_j)^{-1} \right\rfloor \right) \right\rfloor + 1.$$

It now follows that step (5)

$$y - P \cdot \lfloor y/P \rfloor = \left(\sum_{j=1}^{n^2} b_j \cdot (P/p_j) \cdot (P/p_j)^{-1}) \bmod P \right)$$

can be computed by NC^1 circuits.

Returning to step (3), we need to compute $z \bmod p$ for the first n^2 prime numbers. By the prime number theorem, the n-th prime $p_n = O(n \cdot \ln(n))$. Since the input $x_1, \ldots, x_n$, with $|x_i| \leq n$ is of length at most n^2, we show that this operation is in NC^1 for $p \leq n^{O(1)}$.

Algorithm 1.7.9. Remainder modulo small integers.
Input. Integers x, k, m, n with $n = |x|$ and $m \leq n^k$.
Output. $x \bmod m$.
Suppose that $x = \sum_{i<n} x_i \cdot 2^i$.

(1) for $\bar{m} = 1$ to n^k
  for $i = 0$ to $n - 1$
    compute $a_{i,\bar{m}} = 2^i \bmod \bar{m}$
(2) compute $y = \sum_{i<n} a_{i,m} \cdot x_i$, noting that $y \leq n \cdot m$.
(3) compute $s = \min\{i < n : y - i \cdot m \geq 0\}$
(4) output $y - s \cdot m$.

For fixed n, step (1) is externally computed and the values $a_{i,\bar{m}} = 2^i \bmod \bar{m}$ are hardwired into the circuit. Step (2) is in NC^1 by iterated addition. Steps (3) and (4) require multiplication, subtraction, order comparison and computing the minimum of at most n many $O(n)$ bit integers. Multiplication is in NC^1, while the other operations are in AC^0, so these two steps are in NC^1 (these steps can actually be implemented in AC^0, since s, y, m are bounded by n^k). Thus we have proved the following.

Theorem 1.7.3 ([BCH86]). *Integer division is in* NC^1.

See the article of Shankar and Ramachandran [SR88] for improvements in the size of logarithmic depth fan-in 2 circuits for division.

1.8 Synthesis of Circuits

We now study the more general problem of designing boolean circuits for arbitrary boolean functions. In Section 1.8.1 we present several elementary methods which are useful in many practical situations, while in Section 1.8.2 we present Shannon's method leading to a $O(2^n/n)$ upper bound. In Section 1.8.3 we give Lupanov's method leading to a $2^n/n + o(2^n/n)$ upper bound, which, as we'll see in the next chapter, is optimal.

1.8.1 Elementary Methods

A standard technique used in the synthesis of circuits computing a given boolean function on n variables is to write f in disjunctive normal form

$$f(x_1, \ldots, x_n) = \bigvee_{\substack{a_1,\ldots,a_n \in \{0,1\} \\ f(a_1,\ldots,a_n)=1}} x^{a_1} \wedge \cdots \wedge x^{a_n}. \tag{1.3}$$

Then we use $n-1$ AND-gates to construct circuits computing the conjuncts $x^{a_1} \wedge \cdots \wedge x^{a_n}$ which we connect with OR-gates in order to construct the required circuit for f. Here $\neg$ gates are at level 0. This gives rise to the following three elementary but useful methods for the composition of circuits computing arbitrary boolean functions.

1. We consider an appropriate combination of the basic components x_i and $\neg x_i$, for $i = 1, \ldots, n$, depending on the value of $a_1, \ldots, a_n$ (such that $f(a_1, \ldots, a_n) = 1$) and join them together with $n-1$ AND-gates in order to form a circuit computing $x^{a_1} \wedge \cdots \wedge x^{a_n}$. If s is the number of disjuncts in (1.3) then we need $s-1$ $\vee$-gates to connect the previous circuits into a circuit for f. The total number of gates used is at most $n+(n-1)(s-1) \leq n2^n$.

2. A simple recursive method can be used in order to get a more compact representation of f. By induction on n we show how to construct circuits for the family $\{x^{a_1} \wedge \cdots \wedge x^{a_n} : a_1, \ldots, a_n \in \{0,1\}\}$. Assuming we have a circuit family for $n-1$ we construct a circuit family for n. Connect x_n and $\neg x_n$ and join them via an $\wedge$-gate with each circuit of the family $\{x^{a_1} \wedge \cdots \wedge x^{a_{n-1}}\}$. If ℓ_n is the total number of gates used at the nth stage of the construction then we have that $\ell_1 = 1$ and
$$\ell_n \leq \ell_{n-1} + 2 \cdot 2^{n-1} + 1,$$
which implies that $\ell_n \leq 2 \cdot 2^n + n - 4$. To form a circuit for f we connect the above circuits with $s-1$ $\vee$-gates to get a circuit with a total of $2 \cdot 2^n + s - 1 + n - 4 \leq 3 \cdot 2^n + n - 5$ gates.
3. A simple 2^n upper bound on the circuit complexity of boolean functions in $\mathcal{B}_n$ can be proved as follows. For a boolean function $f \in \mathcal{B}_n$ let $f_b \in \mathcal{B}_{n-1}$ be obtained from f by substituting $b \in \{0,1\}$ for the variable x_n, i.e., $f_b = f(\cdot, b)$. It is then obvious that we can compute a circuit for f inductively by using the circuits for f_0, f_1 as implied by the identity
$$f = (\neg x_n \wedge f_0) \vee (x_n \wedge f_1).$$
If $L(f)$ denotes the size of the circuit computing f as above then it follows easily that $L(f) \leq L(f_0) + L(f_1) + 3$. Let $s_n = \max_{f \in \mathcal{B}_n} L(f)$. Hence $s_n \leq 2s_{n-1} + 3$. Using the fact that $s_2 = 1$ we obtain $s_n \leq 2^n - 3$.

However, it is interesting to note that none of the above simple techniques comes close to the bound $2^n/n$, which by Shannon's Theorem 2.2.1 from the next chapter is the lower bound for almost all boolean functions. In what follows we will show that in fact $2^n/n$ is also an asymptotic upper bound.

1.8.2 Shannon's Method

In this section, we prove that every n-ary boolean function can be computed by a fan-in 2 boolean circuit of size $O(2^n/n)$. We begin with a lemma which proves the existence of universal circuits.

Lemma 1.8.1. *There is a circuit with multiple outputs and size $O(2^{2^n})$ which for any boolean function $f \in \mathcal{B}_n$ has an output computing f.*

Proof. We use induction on n. The result is trivial for $n = 1$. Assume the lemma is true for $n-1$. As in (3) of Subsection 1.8.1 above, let $f_b \in \mathcal{B}_{n-1}$ be obtained from $f \in \mathcal{B}_n$ by substitution of $b \in \{0,1\}$ for the variable x_n. Divide the boolean functions $f \in \mathcal{B}_n$ into three classes, depending on whether

1. both f_0 and f_1 are $\equiv 0$,
2. either $f_0 \not\equiv 0$ and $f_1 \equiv 0$ or $f_1 \not\equiv 0$ and $f_0 \equiv 0$,
3. both f_0 and f_1 are $\not\equiv 0$.

There is one function of type (1), $2(2^{2^{n-1}}-1)$ functions of type (2) and $2^{2^n}-2(2^{2^{n-1}}-1)-1$ functions of type (3). We compose a universal circuit for n-ary boolean functions by using the circuit implied by the identity

$$f = (\neg x_n \wedge f_0) \vee (x_n \wedge f_1)$$

in conjunction with the universal circuit for $(n-1)$-ary boolean functions. This requires $2(2^{2^{n-1}}-1)$ many $\wedge$-gates and $2^{2^n}-2(2^{2^{n-1}}-1)-1$ many $\vee$-gates. Using the induction hypothesis, we obtain the proof of the lemma.

Theorem 1.8.1. *Every n-ary boolean function can be computed by a fan-in 2 boolean circuit of size $O(2^n/n)$.*

Proof. The proof is based on the techniques of Section 1.8.1. Let $f \in \mathcal{B}_n$ be a given boolean function, and fix $k \leq n$. Clearly, f can be decomposed as follows:

$$f(x_1,\ldots,x_n) = \bigvee_{a_1,\ldots,a_k\in\{0,1\}} x_1^{a_1} \wedge \cdots \wedge x_k^{a_k} \wedge f(a_1,\ldots,a_k,x_{k+1},\ldots,x_n).$$

We construct a family of circuits computing the conjuncts

$$\{x_1^{a_1} \wedge \cdots \wedge x_k^{a_k} : a_1,\ldots,a_k \in \{0,1\}\}$$

by induction on k. For each fixed sequence $a_1,\ldots,a_k \in \{0,1\}$ we connect this via an $\wedge$-gate to a circuit for the function (of $n-k$ variables)

$$f(a_1,\ldots,a_k,\cdot,\ldots,\cdot).$$

A multiple circuit of size $O(2^{2^{n-k}})$ computing all these functions exists (see Lemma 1.8.1). Combining these with $\vee$-gates we get a circuit computing f of size $O(2^k)+O(2^{2^{n-k}})$. If we choose k in such a way that $n-k = \lfloor \log(n-2\log n) \rfloor$ then we get the desired result.

1.8.3 Lupanov's Method

Theorem 1.8.2 ([Lup58]). *Every n-ary boolean function can be computed by a fan-in 2 boolean circuit of size $2^n/n + o(2^n/n)$.*

Proof. We use the technique of (k,s)-Lupanov representations. Fix $k \leq n$, $s \leq 2^k$ and let $p = \lceil 2^k/s \rceil$. We write a given function $f \in \mathcal{B}_n$ as a disjunction

$$f(x_1,\ldots,x_n) = \bigvee_{i,v} f^1_{i,v}(x_1,\ldots,x_k) \wedge f^2_{i,v}(x_{k+1},\ldots,x_n), \tag{1.4}$$

where the functions $f^j_{i,v}$ are defined as follows (the range of the indices will be specified below). We construct a table representing the values of f for different inputs consisting of 2^k rows one row for each $(x_1,\ldots,x_k) \in \{0,1\}^k$, and 2^{n-k}

columns for the different values of $(x_{k+1}, \ldots, x_n) \in \{0,1\}^{n-k}$. Partition the rows into p sets $A_1, \ldots, A_p$, where A_i, $i \leq p-1$, consists of s consecutive rows and the last A_p consists of the last $s' = 2^k - (p-1)s$ consecutive rows such that $1 \leq s' \leq s$. Now we can define $f_i \in \mathcal{B}_n$ as follows for $x = (x_1, \ldots, x_n)$:

$$f_i(x) = \begin{cases} f(x) \text{ if } (x_1, \ldots, x_k) \in A_i \\ 0 \quad \text{otherwise.} \end{cases}$$

Hence $f = f_1 \vee \cdots \vee f_p$. Let $C_{i,v}$ be those columns of the table whose intersection with A_i is the vector $v \in \{0,1\}^s$ (if $i = p$ then $v \in \{0,1\}^{s'}$). Again define

$$f_{i,v}(x) = \begin{cases} f_i(x) \text{ if } (x_{k+1}, \ldots, x_n) \in C_{i,v} \\ 0 \quad \text{otherwise.} \end{cases}$$

It is again clear that $f_i = \bigvee_v f_{i,v}$. If we represent the values of $f_{i,v}$ as a $2^k \times 2^{n-k}$ table as above, then we see that rows outside A_i consist only of 0s while the rows of A_i contain two types of columns: those all of whose entries are 0, and those, which are 0 except where they intersect A_i and have value exactly v. This shows that each $f_{i,v}$ can be written as the conjunction of the boolean functions $f^1_{i,v} \in \mathcal{B}_k, f^2_{i,v} \in \mathcal{B}_{n-k}$ as follows:

$$f_{i,v}(x_1, \ldots, x_n) = f^1_{i,v}(x_1, \ldots, x_k) \wedge f^2_{i,v}(x_{k+1}, \ldots, x_n),$$

where

$$f^1_{i,v}(x_1, \ldots, x_k) = \begin{cases} 1 \text{ if } \exists j (v_j = 1 \text{ and} \\ \quad (x_1, \ldots, x_k) \text{ is the } j\text{th row of } A_i) \\ 0 \text{ otherwise} \end{cases}$$

$$f^2_{i,v}(x_{k+1}, \ldots, x_n) = \begin{cases} 1 \text{ if } (x_{k+1}, \ldots, x_n) \in C_{i,v} \\ 0 \text{ otherwise.} \end{cases}$$

This proves identity (1.4). Using this representation we can give a circuit computing a given function $f \in \mathcal{B}_n$. We construct circuits for the sets of formulas

$$\{x_1^{a_1} \wedge \cdots \wedge x_k^{a_k} : a_1, \ldots, a_k \in \{0,1\}\}$$

and

$$\{x_{k+1}^{a_{k+1}} \wedge \cdots \wedge x_n^{a_n} : a_{k+1}, \ldots, a_n \in \{0,1\}\}$$

which can be realized with $2^k(1+\epsilon)$ and $2^{n-k}(1+\epsilon)$ gates, respectively, where $\epsilon > 0$.

To realize $f^1_{i,v}$, in disjunctive normal form requires a disjunction of at most s terms, since $f^1_{i,v}$ takes the value 1 on at most s inputs. Thus to realize all the $f^1_{i,v}$ requires at most $(p-1)(s-1) + (s'-1) \cdot 2^s$ $\vee$-gates. Since $p \leq \frac{2^k}{s} + 1$ we need a total of at most 2^{k+s} $\vee$-gates.

To realize $f_{i,v}^2$, in disjunctive normal form, notice that since $C_{i,v} \cap C_{i,v'} = \emptyset$, for $v \neq v'$, they contain at most 2^{n-k} columns together. Hence to realize $f_{i,v}^2$ for all v requires no more that 2^{n-k} $\vee$-gates and for $i = 1, \ldots, p$, this number is at most $p2^{n-k}$.

Next, form the conjunction $f_{i,v}^1 \text{AND} f_{i,v}^2$ with $p2^s$ many AND-gates and realize f with an additional $p2^s$ many OR-gates. Using the inequality $p \leq \frac{2^k}{s} + 1$, the total number of gates used is at most

$$(2^k + 2^{n-k})(1 + \epsilon) + p(2^{n-k} + 2^s)$$

which in turn is at most

$$(2^k + 2^{n-k})(1 + \epsilon) + 2^{k+s} + \frac{2^n}{s} + \frac{2^{k+s}}{s}.$$

Choosing $k = \lceil 3 \log n \rceil$, and $s = \lceil n - 5 \log n \rceil$, we conclude that all terms are small compared to $2^n/s$ which is bounded by $2^n/n$.

1.8.4 Symmetric Functions

As we will see in the next chapter, a simple counting argument due to Shannon shows that almost all boolean functions on n variables have circuit complexity $2^n/n$. This contrasts sharply with the case of symmetric functions which can be computed with circuits of size linear in the number of inputs. However, this linear bound is not surprising since a symmetric function is essentially a boolean function on $\log n$ inputs, since the function depends solely on the *weight* of its inputs, where for input $x = (x_1, \ldots, x_n)$, the weight $|x|_1$ of x is the number of 1s in x, i.e., $|x|_1 = \sum_{i=1}^n x_i$. The following lemma shows how to construct to construct circuits to compute the weight.

Lemma 1.8.2. *For each n there is a fan-in 2 boolean circuit of size $O(n)$ computing the function $x \mapsto |x|_1$.*

Proof. Let C_n be the circuit computing this function on n inputs and let $size(C_n)$ denote its size. First of all we observe that we may assume without loss of generality that $n = 2^m$ is a power of 2. Indeed, let m be such that $n \leq 2^m < 2n$. Then it is enough to compute circuits for the following function:

$$x = (x_1, \ldots, x_n) \mapsto x0^{2^m - n} \mapsto |x0^{2^m - n}|_1 = |x|_1.$$

It follows that

$$size(C_n) \leq size(C_{2^m}) + 2^m - n = O(2^m) + n = O(n).$$

Now we prove Lemma 1.8.2 by induction on m. The assertion is trivial for $m = 0$. Assume it is true for m. We prove it for $m + 1$, i.e., $n = 2^{m+1}$. To compute the weight of an input $(x_1, \ldots, x_n)$ it is enough to compute the weights of $(x_1, \ldots, x_{n/2})$ and $(x_{n/2+1}, \ldots, x_n)$ and then add the resulting weights. The size of the resulting circuit is $2 \cdot O(n/2)$, for the circuits of the weight functions (this uses the induction hypothesis on m, since $n/2 = 2^m$) plus the size of the circuit for adding two inputs each of length $\log n$, which is $O(\log n)$ by Proposition 1.7.1.

From this, we have an upper bound on circuit size for symmetric boolean functions.

Theorem 1.8.3. *Every symmetric boolean function $f \in \mathcal{B}_n$ can be computed by a fan-in 2 boolean circuit of size $O(n)$.*

Proof. Symmetric functions depend only on the weight of the input, rather than on the input itself. Hence a circuit for f will consist of

- a circuit for the function $x \mapsto |x|_1$, which outputs the binary representation of $|x|_1$ for any input x, and
- a circuit with $\lceil \log n \rceil$ inputs for determining whether or not f is 1 on inputs with a given weight.

The circuit for the function $x \mapsto |x|_1$ was constructed in the proof of the previous lemma. Define the boolean function $g \in \mathcal{B}_{\lceil \log n \rceil}$ by $g(w) = f(x)$, where w is the binary representation for $|x|_1$. By the results of Section 1.8.1, g can be computed with a circuit of size $O(2^{\log n}) = O(n)$. Combining these circuits, we obtain a circuit for the function f. This completes the proof of the theorem.

By using the NC^1 algorithm for iterated addition from Theorem 1.7.2, we can compute the weight in logarithmic depth. Hence in a similar manner we have the following:

Theorem 1.8.4. *Let $L \subseteq \{0,1\}^*$ be a symmetric language. Then $L \in \mathrm{NC}^1$.*

Actually symmetric functions can be realized by circuits of size $O(n)$ and depth $O(\log n)$. We will not prove this result here. However, in Chapter 3, we will generalize the notion of symmetric boolean function, by introducing the concept of the invariance group of a boolean function.

1.9 Reducing the Fan-out

For boolean circuits of fan-in 2, with at most a linear increase in size and depth, one can find an equivalent boolean circuit with fan-out 2. This is given by the following.

Theorem 1.9.1 ([HKP84]). *Let the boolean circuit B of fan-in 2 have n input and m output gates, and let $s(B)$ denote the size of B (number of internal gates) and $d(B)$ the depth of B. Then there is an equivalent fan-in 2 boolean circuit C with fan-out 2 such that $s(C) + n \leq 3 \cdot [s(B) + n]$ and $d(C) \leq 2 \cdot d(B) + \log m$.*

Proof. For the scope of this proof, we let gate mean either input or internal gate, and assume that B has $s = s(B)+n$ gates and depth $d = d(B)$. Without loss of generality we may assume that B is a leveled circuit. Otherwise, add identity gates having one input and one output, so as to level the circuit. The argument given below transforms B into C by replacing gates of fan-out $r > 2$ by an appropriate binary tree, where AND-gates (OR-gates) are replaced by a tree of ANDs (ORs). At the end of the construction, the identity gates can be then removed, thus producing the desired circuit.

The circuit B is transformed into C, level by level, from the output downwards (recall that the circuit inputs are at the bottom, with output gates at the top). A gate v at level $s+1$ with fan-out $r > 2$ to the gates $v_1, \ldots, v_r$ at level s is replaced by an appropriate binary tree. The structure of the tree is determined by the depth of the v_i in the circuit C being constructed. To describe the tree, we need the following.

Lemma 1.9.1. *Let $a_1, \ldots, a_r$ be non-negative integers, for $r \geq 1$. There exists a binary tree T with r leaves and labeling ℓ, which map the nodes of T into $\mathbf{N}$ and satisfies the conditions:*

1. *The leaves $l_1, \ldots, l_r$ of T are labeled by $a_1, \ldots, a_r$, i.e.,*

$$\ell(l_i) = a_i, \text{ for } 1 \leq i \leq r.$$

2. *Each internal node v of T has label*

$$\ell(v) = 1 + \max(\ell(u), \ell(w))$$

where v is the parent of u, w.

3. *The label of the root v of T satisfies*

$$2^{\ell(v)} < \sum_{i=1}^{r} 2^{a_i+1}.$$

Proof. By induction on r. If $r = 1$, let T be the tree with one node labeled by a_1. Suppose now $r > 1$.

Case 1. $a_1 < \cdots < a_r$.

In this case, let T be the degenerate binary tree of the form given in Figure 1.2, where the smaller integers a_i are placed lower (further from the root) in the tree. By this placement, the parent of the leaves labeled by a_1, a_2 has label $a_2 + 1$, the parent of this node has label $a_3 + 1$, etc. and so the root has label $a_r + 1$. Clearly $2^{a_r+1} < \sum_{i=1}^{r} 2^{a_i+1}$, establishing the labeling requirement for T.

Case 2. $a_1 \leq a_2 \leq \cdots \leq a_{i-1} < a_i = a_{i+1} = \cdots = a_j < a_{j+1} \leq \cdots \leq a_r$.

In this case, let b denote the common value of $a_i, \ldots, a_j$, and set $k = \lceil \log j - i + 1 \rceil$. If $j - i + 1 = 2^k$, then form the complete binary tree T_0 of depth k with $j - i + 1$ leaves. Otherwise form an almost complete binary tree T_0 by placing 2^{k-1} leaves at depth k from the root, and $j - i + 1 - 2^{k-1}$

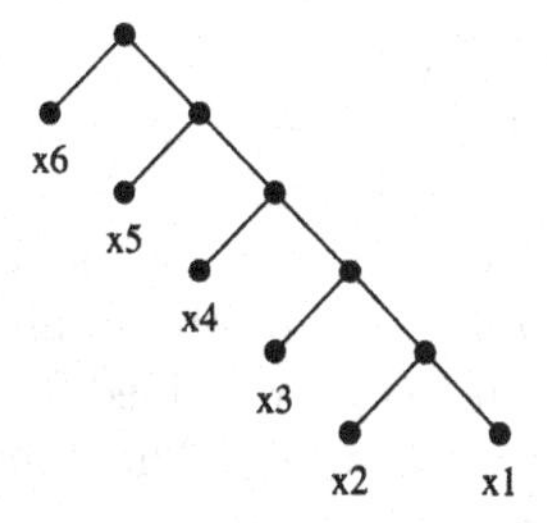

Fig. 1.2. The binary tree used in the circuit transformation

leaves at depth $k-1$ from the root. In either situation, label each leaf by the common value b. Let t denote the root of tree T_0, and note that the label of t satisfies $\ell(t) = 2^{k+b}$. We have

$$\begin{aligned} 2^{k+b} &= 2^{k-1} \cdot 2^{b+1} \\ &\leq 2^{\lfloor \log j-i+1 \rfloor} \cdot 2^{b+1} \\ &\leq (j-i+1) \cdot 2^{b+1} \\ &= \sum_{\ell=i}^{j} 2^{a_\ell+1}. \end{aligned}$$

Equality between the first and second line holds only if $j-i+1$ is not a power of 2; equality between the second and third line holds only if $j-i+1$ is a power of 2. Thus $2^{k+b} < \sum_{\ell=i}^{j} 2^{a_\ell+1}$ and so T_0 satisfies the labeling requirements for placement of $a_i, \ldots, a_j$.

By the induction hypothesis, obtain a tree T_1 for the set

$$a_1, \ldots, a_{i-1}, t, a_{j+1}, \ldots, a_r,$$

but note that t may fall in a different ordering position among the a's. Define T to be the result of replacing one leaf of T_1 having label t by the subtree T_0. Since T_0 satisfies the labeling requirements for $a_i, \ldots, a_j$ and T_1 for $a_1, \ldots, a_{i-1}, t, a_{j+1}, \ldots, a_r$, it follows that the composite tree T satisfies the labeling requirements. This completes the proof of the lemma.

We now describe the construction of circuit C from B in stages s. If v is a gate of B, then $B(v)$ ($C(v)$) will be defined to be the subcircuit of B (C) whose root is v, i.e., gates between v and the output gates, inclusive. For the remainder of the proof, let B_s be the set of gates in B at level s from the output nodes.

Stage 0. For each output node v, let $C(v) = v$ and $\ell(v) = 0$. The output nodes of C are thus exactly those of B.

Stage $s+1$. For each gate v of B at level $s+1$ from the output, do the following. If v has fan-out $r \leq 2$, then let $C(v)$ be obtained from $B(v)$ by retaining gate v and replacing $B(u), B(w)$ by $C(u), C(w)$, where u, w are the children of v. If v has fan-out $r > 2$ then suppose there is an edge from v to $v_1, \ldots, v_r$, each v_i at level s of B. For each $1 \leq i \leq r$ the subcircuit $C(v_i)$ of C has already been constructed, so let a_i be the depth of $C(v_i)$. Apply the preceding lemma to the set $a_1, \ldots, a_r$ to obtain binary tree T_v whose leaves are labeled by $a_1, \ldots, a_r$. Then $C(v)$ is obtained from $T(v)$ by replacing each leaf labeled by a_i by the subcircuit $C(v_i)$. This completes the construction.

Circuit C is obtained at the last stage $s = d(B)$. For each gate v of B, let $\ell(v)$ be the depth of v in circuit C.

CLAIM. For each $0 \leq s \leq d(B)$, $\sum_{v \in B_s} 2^{\ell(v)} \leq 4^s \cdot m$.

Proof of Claim. By induction on s. When $s = 0$, B_0 is the set of output gates. By hypothesis, there are m output gates, and $\sum_{v \in B_0} 2^{\ell(v)} = 4^0 \cdot m$. Now consider the case $s+1$. Suppose that $v \in B_{s+1}$ and v has an edge to $v_1, \ldots, v_r$, with each $v_i \in B_s$. If $r > 2$, then by the preceding lemma

$$2^{\ell(v)} < \sum_{i=1}^{r} 2^{\ell(v_i)+1}.$$

If $r \leq 2$, this inequality holds as well, as is easily checked. Since circuit B has fan-in 2, each $v_i \in B_s$ can have at most two parent gates in B_{s+1}, so can appear at most twice in the sum of the preceding expression over all $v \in B_{s+1}$. It follows that

$$\begin{aligned}\sum_{v \in B_{s+1}} 2^{\ell(v)} &< 2 \cdot \sum_{u \in B_s} 2^{\ell(u)+1} \\ &= 4 \cdot \sum_{u \in B_s} 2^{\ell(u)} \\ &\leq 4 \cdot 4^s \cdot m = 4^{s+1} \cdot m.\end{aligned}$$

This establishes the claim.

Recalling that d denotes the depth of B, we have

$$\sum_{v \in B_d} 2^{\ell(v)} < 4^d \cdot m.$$

Every input node v of B has depth d, and so

$$\ell(v) = \log 2^{\ell(v)} \leq \log 4^d m = 2d + \log m.$$

This proves the depth of C is at most $2d + \log m$. To establish a bound on the size of C, note first the following fact, easily established by induction.

Fact 1.9.1. If T is a binary tree with r leaves, then T has $r - 1$ internal nodes.

In constructing $C(v)$ at each stage, if v has edges in B to $v_1, \ldots, v_r$ with $r > 2$, then $v, v_1, \ldots, v_r$ are placed in C (these come from B), and by the preceding fact, $r-2$ new gates are added to C (the root v is one of the $r-1$ internal nodes, but has already been counted). Since B has fan-in 2, each gate at level s can have an edge to at most two gates at level $s+1$ (its parents), so the number of edges in the circuit B is at most twice the number s of gates of B. Thus the total number $s(C) - s(B)$ of new gates is bounded by the number of edges in the circuit B, hence by $2 \cdot s$. It follows that the total number of gates in C is at most $3 \cdot s$.

1.10 Relating Formula Size and Depth

A boolean formula over a finite basis of binary connectives can be viewed as a binary tree. If the tree is well-balanced, then its depth is clearly logarithmic in the number of leaves. Spira's theorem states that even in the non-balanced case we can find an equivalent balanced boolean formula with this property, whose size is polynomial in the given formula. The *leafsize* of a formula is the number of leaves in its formation tree, i.e., the number of occurrences of propositional variables in the formula.

Theorem 1.10.1 ([Spi71]). *Let F be a boolean formula of leafsize m over the complete basis $\Omega = \mathcal{B}_0 \cup \mathcal{B}_1 \cup \mathcal{B}_2$. Then there is an equivalent formula F' over the basis $\wedge, \vee, \neg$ such that $depth(F') \leq 1 + 2 \cdot \log_{3/2} m$.*

Proof. By induction on leafsize m. For $m = 1$, F has one leaf labeled by a variable x_i or constant $0, 1$, hence F must be equivalent to $0, 1, x_i$ or $\neg x_i$, all of depth at most 1. For the inductive case $m > 1$, by Lemma 1.6.1 let G be a subformula of F of leafsize s satisfying

$$\left\lceil \frac{m}{3} \right\rceil < s \leq \left\lfloor \frac{2m}{3} \right\rfloor .$$

Let F_0 (F_1) be F with the distinguished subformula G replaced by 0 (1). Clearly F is equivalent to

$$(F_0 \wedge \neg G) \vee (F_1 \wedge G).$$

By the induction hypothesis, F_0, F_1, G, $\neg G$ are equivalent respectively to F_0', F_1', G' and $(\neg G)'$ all of depth at most 1 plus the logarithm base 3/2 of their respective leafsizes. Let F' denote

$$(F_0' \wedge (\neg G)') \vee (F_1' \wedge G').$$

Then $F \equiv F'$ and

$$\begin{aligned} depth(F') &\leq 2 + \max\{depth(F'_0), depth(F'_1), depth(G'), depth((\neg G)')\} \\ &\leq 2 + 1 + 2 \cdot \log_{3/2}\left(\frac{2m}{3}\right) \\ &\leq 1 + 2 \cdot \log_{3/2} m. \end{aligned}$$

Recall that the size of a circuit is the number of internal gates, so the size of a boolean formula as a circuit with fan-out 1 is the number of subformulas (not counting variables). Without loss of generality assume that F' has at most one negation occurring on every path from root to leaf, which appears immediately above the variable occurring at the leaf. By Fact 1.9.1 which states that the number of leaves of a binary tree exceeds the number of internal nodes by 1, it follows that

$$\begin{aligned} size(F') &\leq 2^{depth(F')} \\ &\leq 2^{1+2 \cdot \log_{3/2}(m)} \\ &\leq 2^{1+2 \cdot \log m \cdot \log_{3/2}(2)} \\ &\leq 2m^{2 \cdot \log 3/2(2)} \end{aligned}$$

which is polynomial in m.

Spira [Spi71] in fact showed that the leafsize of F' is at most m^α, where α satisfies $\frac{1+2^\alpha}{3^\alpha} \leq \frac{1}{2}$ ($\alpha \geq 2.1964$ suffices). This result was sharpened by M.-L. Bonet and S.R. Buss [BB94]. By using an analogous lemma for k-ary trees in place of the Lemma 1.6.1, it is straightforward (see Exercise 1.13.8) to generalize Spira's theorem to show that for any finite basis Ω of k-ary connectives and formula F over basis Ω with leafsize m, there is an equivalent formula G over $\wedge, \vee, \neg$ whose depth is logarithmic in m. It follows that NC^1 equals the class of functions having polynomial size boolean formulas over any fixed finite basis Ω.

Definition 1.10.1. *If $f : \{0,1\}^n \to \{0,1\}$ is a boolean function, then $D(f)$ is the minimal depth of a fan-in 2 boolean circuit computing f, and $L(f)$ is the minimal size of a boolean formula computing f. If f is monotonic, then $D_m(f)$ ($L_m(f)$) denotes the minimal depth (size) of a fan-in 2 monotonic boolean circuit (monotonic formula) computing f.*

The following result, due to Spira, summarizes the relation between formula size and fan-in 2 circuit depth. The monotonic version of Spira's theorem is due to I. Wegener [Weg87].

Theorem 1.10.2 ([Spi71], [Weg87]).

1. $\log(L(f) + 1) \leq D(f) \leq k \cdot \log(L(f) + 1)$.
2. *If f is monotonic, then* $\log_m(L(f) + 1) \leq D_m(f) \leq k \cdot \log(L_m(f) + 1)$

where logarithms are to base 2, *and* $k = \frac{3}{\log_2(3)-1} \approx 5.13$.

Proof. To see that $\log(L(f)+1) \leq D(f)$, given a circuit of depth d which computes f, by unraveling and duplicating subcircuits, it is clear that f is computed by a formula of size at most $2^d - 1$, since the formula is a binary tree of depth d. Reason similarly for the monotonic case.

The previous theorem establishes the bound $D(f) \leq k \cdot \log(L(f)+1)$. The proof of the monotonic case is similar, with the following observation. In the proof of the previous theorem, let F_0 (F_1) be F with the distinguished subformula G replaced by 0 (1). Clearly F is equivalent to

$$(F_0 \wedge \neg G) \vee (F_1 \wedge G)$$

but in the monotonic case, $F_0 \leq F_1$; i.e., $F_0 \rightarrow F_1$ is tautologous.

CLAIM. For monotonic F,

$$F_0 \vee (F_1 \wedge G) \Leftrightarrow (F_0 \wedge \neg G) \vee (F_1 \wedge G).$$

Proof of Claim. Let ϕ denote $(F_0 \wedge \neg G) \vee (F_1 \wedge G)$. Suppose that truth assignment α satisfies F_0. If $\alpha \models \neg G$, then $\alpha \models \phi$; if $\alpha \models G$, then since $F_0 \leq F_1$, $\alpha \models F_1 \wedge G$, hence $\alpha \models \phi$. It follows that $F_0 \vee (F_1 \wedge G)$ implies ϕ. The converse is clearly true, thus establishing the claim. □

Now continue the argument as in the previous theorem. The consideration of constant k is left to the reader.

The following result is due to M. Bonet and S.R. Buss [BB94]; see also N.H. Bshouty, R. Cleve and W. Eberly [BCE91]].

Theorem 1.10.3 ([BB94]). *Consider a boolean formula F over basis $\wedge$, $\vee$, $\neg$ of leafsize m. For all $k \geq 2$, there exists an equivalent boolean formula F' over the basis $\wedge, \vee, \neg$ such that depth$(F') \leq 1 + 3k \ln m$ and the leafsize of F' is at most m^α, where $\alpha = 1 + \frac{1}{1+\log(k-1)}$.*

Proof. We present the elegant proof of [BB94]. Let C_0 (C_1) be the result of replacing a distinguished occurrence of D in C by 0 (1) and then collapsing the gate having constant 0 (1). Unlike the previous general case in Spira's theorem, the extra gate collapse occurring in C_0 and C_1 is possible because of the De Morgan basis. The key observation of [BB94] is the following.

Case 1. Suppose that D occurs *positively* in C (i.e., there are an even number of negations occurring between subformula D and the root of the tree).

By induction on the number of connectives lying on the path between distinguished subformula D and the root of the formula tree of C, one can prove that

$$C_0 \rightarrow C_1$$

is a tautology. Then $C \equiv (C_0 \wedge \neg D) \vee (C_1 \wedge D)$ and, moreover, the following implications hold:

$$\begin{aligned}(C_0 \wedge \neg D) \vee (C_1 \wedge D) &\rightarrow C_0 \vee (D \wedge C_1) \\ &\rightarrow (C_0 \vee D) \wedge C_1 \\ &\rightarrow (C_0 \wedge C_1) \vee (D \wedge C_1) \\ &\rightarrow (C_0 \wedge \neg D) \vee (C_1 \wedge D).\end{aligned}$$

The first implication holds, as $C_0 \wedge \neg D$ implies C_0. The second implication holds, since if σ is a truth assignment satisfying $C_0 \vee (D \wedge C_1)$ then either σ satisfies C_0 (and as $C_0 \rightarrow C_1$, also satisfies C_1), or σ satisfies $D \wedge C_1$. In both cases, σ satisfies $(C_0 \vee D) \wedge C_1$. The third implication holds because of distributivity, and the fourth implication holds by a similar consideration of truth assignment σ satisfying either $(C_0 \wedge C_1)$ or $(D \wedge C_1)$. In the former case, either σ satisfies D or its negation, resulting in the fourth implication. It now follows that C is equivalent to both the formulas

$$C_0 \vee (D \wedge C_1) \quad \text{and} \quad (C_0 \vee D) \wedge C_1.$$

Case 2. Suppose that D occurs *negatively* in C (i.e., there are an odd number of negations occurring between subformula D and the root of the tree).

In this case, a similar analysis yields that

$$C_1 \rightarrow C_0$$

is a tautology and, moreover,

$$\begin{aligned}(C_0 \wedge \neg D) \vee (C_1 \wedge D) &\rightarrow (C_0 \wedge \neg D) \vee C_1 \\ &\rightarrow (C_1 \vee \neg D) \wedge C_0 \\ &\rightarrow (C_0 \wedge C_1) \vee (C_0 \wedge \neg D) \\ &\rightarrow (C_0 \wedge \neg D) \vee (C_1 \wedge D)\end{aligned}$$

and so C is equivalent to both

$$C_1 \vee (\neg D \wedge C_0) \quad \text{and} \quad (C_1 \vee \neg D) \wedge C_0.$$

Using this observation, in the recursion step from Spira's theorem, we can remove one occurrence of the subformula D. As in the analysis of quicksort, it is now most efficient to ensure that the leafsize of D is larger than that of C_0 or C_1, so that maximum progress will be made in the recursion step. This is formalized as follows, where we follow the proof of [BB94] very closely.

For formula F, denote leafsize of F by $||F||$. Let D be a subformula of C of minimal leafsize at least $\frac{k-1}{k}m$, so that left and right subformulas D_L, D_R have leafsize less than $\frac{k-1}{k}m$. Let C_0 (C_1) be obtained from C by replacing the distinguished occurrence of D by 0 (1) and collapsing the gate. Then $||C_0||$ and $||C_1||$ are both at most $\frac{1}{k}m$. By the induction hypothesis, let C_0', C_1', D_L', D_R' be respectively equivalent to C_0, C_1, D_L, D_R and such that

$$\begin{aligned}
||C_0'||, ||C_1'|| &\leq \left(\frac{m}{k}\right)^\alpha \\
||D_L'||, ||D_R'|| &< \left(\frac{k-1}{k} m\right)^\alpha \\
depth(C_0'), depth(C_1') &\leq 3k \ln\left(\frac{m}{k}\right) \\
depth(D_L'), depth(D_R') &< 3k \ln\left(\frac{k-1}{k} m\right).
\end{aligned}$$

Denoting the gate type at D's root by $\circ$, if the distinguished occurrence of D in C is positive (case 1) then define C' to be

$$C_0' \vee ((D_L' \circ D_R') \wedge C_1')$$

while if negative (case 2) then define C' to be

$$C_1' \vee ((\neg D_L' \circ \neg D_R') \wedge C_0').$$

Then $depth(C')$ is the maximum of $depth(D_L')+3$, $depth(D_R')+3$, $depth(C_0')+2$, $depth(C_1')+2$, hence

$$\begin{aligned}
depth(C') &< 1 + 3k \ln\left(\frac{k-1}{k} m\right) + 3 \\
&= 1 + 3k \ln m + 3k \ln\left(\frac{k-1}{k}\right) + 3 \\
&< 1 + 3k \ln m + 3k\left(-\frac{1}{k}\right) + 3 \\
&= 1 + 3k \ln m
\end{aligned}$$

where the third inequality follows since $\ln\left(\frac{k-1}{k}\right) = \ln(1-\frac{1}{k}) < -\frac{1}{k}$. This establishes the depth requirement of C'.

We have

$$||C'|| \leq 2(m - ||D_L|| - ||D_R||)^\alpha + ||D_L||^\alpha + ||D_R||^\alpha.$$

Letting $b = ||D|| = ||D_L|| + ||D_R||$, consider the function

$$f(||D_L||) = 2(m-b)^\alpha + ||D_L||^\alpha + (b - ||D_L||)^\alpha.$$

Taking the second derivative establishes that f is concave up, hence takes its maximum at the left or right endpoints of the interval $D_L \in [0, \frac{k-1}{k} m]$, i.e., either $D_L = 0, D_R = \frac{k-1}{k} m$ or $D_L = \frac{k-1}{k} m, D_R = 0$. In both cases, $f(||D_L||)$ is bounded by

$$2(m - ||D||)^\alpha + \left(\frac{k-1}{k} m\right)^\alpha + \left(||D|| - \frac{k-1}{k} m\right)^\alpha.$$

This expression, as a function in $||D||$ is concave up, hence takes its maximum at one of the endpoints $||D|| = m$ or $||D|| = \frac{k-1}{k} m$. Clearly, the maximum is achieved when $||D|| = \frac{k-1}{k}m$, and so the worst case is $||C_0|| = ||C_1|| = \frac{m}{k}$, $||D_L|| = \frac{k-1}{k}m$ and $D_R = 0$. Hence we have that

$$||C'|| \le 2\left(\frac{m}{k}\right)^\alpha + \left(\frac{k-1}{k}m\right)^\alpha.$$

CLAIM. For $\alpha = 1 + \frac{1}{\log(k-1)+1}$, we have

$$2\left(\frac{m}{k}\right)^\alpha + \left(\frac{k-1}{k}m\right)^\alpha \le m^\alpha.$$

We establish that

$$2\left(\frac{1}{k}\right)^\alpha + \left(\frac{k-1}{k}\right)^\alpha \le 1.$$

for the above choice of α. The left-hand side of this inequality is decreasing in α, and the expression is larger than 1 when $\alpha = 0$. Let α_0 be the unique value for which

$$2\left(\frac{1}{k}\right)^{\alpha_0} + \left(\frac{k-1}{k}\right)^{\alpha_0} = 1$$

We show that $\alpha_0 < 1 + \frac{1}{\log(k-1)+1}$. Multiply the defining equation of α_0 by k^{α_0} to obtain

$$k^{\alpha_0} - (k-1)^{\alpha_0} = 2. \tag{1.5}$$

Note that $\alpha_0 < 2$, since $k \ge 2$ and $k^2 - (k-1)^2 = 2k - 1 > 2$. Define $g_0(k) = k^{\alpha_0}$. From equation (1.5) by the mean value theorem, there exists x in the interval $(k-1, k)$ with

$$g'_{\alpha_0}(x) = \alpha_0 x^{\alpha_0 - 1} = 2.$$

Since g'_{α_0} is increasing, we have $g'_{\alpha_0}(k-1) = \alpha_0(k-1)^{\alpha_0-1} < 2$. Take logarithms to obtain

$$\begin{aligned} \log(\alpha_0(k-1)^{\alpha_0-1}) &< \log 2 = 1 \\ \log\alpha_0 + (\alpha_0 - 1)\log(k-1) &< 1 \end{aligned}$$

Now it must be the case that

$$(\alpha_0 - 1)(\log(k-1) + 1) < 1$$

since $\alpha_0 - 1 < \log\alpha_0$ for α_0 in the interval $(1, 2)$. Hence

$$\begin{aligned} \alpha_0 - 1 &< \frac{1}{\log(k-1)+1} \\ \alpha_0 &< 1 + \frac{1}{\log(k-1)+1} \end{aligned}$$

This completes the proof of the theorem.

1.11 Other Models

In this section we briefly discuss several other non-uniform models of computation used in the study of the complexity of boolean functions. Uniform models of computation, like the Turing machine, parallel random access machine, and stack register machine, will be introduced in later chapters.

1.11.1 Switching Networks

There are two types of switching networks: *directed* (also known as *contact networks*, or contact gating schema) and *undirected* (also known as gating schema).

In the first case it is a directed acyclic graph with two distinguished nodes (input and output). Some edges (not necessarily all) are labeled by $x_1, \ldots, x_n$ and $\neg x_1, \ldots, \neg x_n$. The input $a_1, \ldots, a_n$ is accepted if the computation takes a path from the input node to the output node and consisting of edges labeled by x_i if $a_i = 0$ and $\neg x_i$ if $a_i = 1$. The second case is exactly as before except that now the network is an undirected acyclic graph. As a measure of complexity we use the number of edges of the network.

1.11.2 VLSI Circuits

A VLSI circuit is a planar circuit (i.e., a circuit embedded in the plane so that no two edges cross each other). Reflecting on the topological limitations of packing components on a wafer, there is a constant $\lambda > 0$ such that the minimum distance between wires is λ and each gate occupies area λ^2. VLSI circuits are laid on rectangular chips of length $\ell\lambda$ and width $w\lambda$, for some ℓ, w, and can be represented by a three-dimensional array of cells consisting of h levels with wires crossing each other only at different levels. Each cell may contain a gate, a wire or branching of a wire.

VLSI circuits are sequential synchronous machines with input ports (located at the border of the chip) and output ports. Fundamental parameters are the area $A = \ell w$ and T the number of clock cycles from the initial reading of an input to the production of the last output.

1.11.3 Energy Consumption

This model is due to G. Kissin and it emphasizes the energy consumption in executing computations. VLSI circuits demand the packing of thousands or even millions of transistors onto a chip of small area (usually 1 cm^2 or less). Circuits consume energy for their operation and this energy is in turn transformed into heat. However, since heat may be produced more rapidly than it can be dissipated, this can cause chip failures. Thus, as packing densities increase it becomes of vital importance to reduce the amount of heat dissipated by designing energy efficient VLSI circuits. There are basically two energy models [Kis87]:

1. the uniswitch model, which measures the differences between two states of a circuit, thus providing a lower bound on the total energy consumed, and
2. the multiswitch model, which accounts for the changes that occur when the circuit is transformed from one state to another, passing through intermediate states.

The multiswitch model is more sensitive to timing issues that can cause wires and gates to switch more than once.

The formal definitions for the uniswitch model are as follows. A legal state is a function s assigning the values $0, 1$ to the nodes and wires. Each input node has some value and non-input nodes have values consistent with the input and the labeling of the nodes, e.g. $s(0 \vee 1) = 1, s(\neg 0) = 1$, etc. Thus the state depends on the assignment of values to the inputs. If s_0 is an initial state, s_1 is a new state, and W is the set of wires then the wire energy $E(s_0, s_1)$ is defined by

$$E(s_0, s_1) = \sum_{\substack{w \in W \\ s_0(w) \neq s_1(w)}} |w|$$

where $|w|$ denotes the length of wire w. If C is a circuit computing a boolean function f then we define

$$E^L_{worst}(C) = \min_{s_0} \max_{s_1} E(C, s_0, s_1)$$
$$E^U_{worst}(C) = \max_{s_0, s_1} E(C, s_0, s_1).$$

For more details on the definition of the multiswitch model, as well as for the construction of circuits computing some basic boolean functions, the reader should consult [Kis87, Kis82, Kis90].

1.11.4 Boolean Cellular Automata

A *boolean cellular automaton* B is given by $(G, \mathbf{f}, \mathbf{x})$, where

- $G = (V, E)$ is a directed graph, $V = \{1, \ldots, n\}$, and $E \subseteq V \times V$ (so that loops but no multiple directed edges are allowed),
- $\mathbf{f} = (f_1, \ldots, f_n)$, where, for each $1 \leq i \leq n$, the fan-in of vertex i is $m(i)$ and $f_i : \{0,1\}^{m(i)} \to \{0,1\}$,
- $\mathbf{x} = (x_1, \ldots, x_n) \in \{0,1\}^n$.

Suppose that vertex i of fan-in $m(i)$ has in-edges from vertices $1 \leq v_{i,1} < \cdots < v_{i,m(i)} \leq n$. Then for state $\mathbf{y} = (y_1, \ldots, y_n) \in \{0,1\}^n$ define

$$B(\mathbf{y}) = (f_1(y_{v_{1,1}}, \ldots, y_{v_{1,m(1)}}), \ldots, f_n(y_{v_{n,1}}, \ldots, y_{v_{n,m(n)}})).$$

The state of $B = (G, \mathbf{f}, \mathbf{x})$ at time 0 is the *initial state* $\mathbf{x}$. At time 1, the state of B is $B(\mathbf{x})$, and generally at time t the state of B is $B^{(t)}(\mathbf{x}) = B(B(\cdots B(\mathbf{x}) \cdots))$ where there are t occurrences of B.

Boolean cellular automata were simulated on computer by S. Kauffman (see [Kau70]), who reported a surprising stability manifested by random automata. The first formal proofs of certain stable behavior were worked out by Cohen and Łuczak [LC91]. Sharp probabilistic analysis of this phenomenon was undertaken by J. Lynch [Lyn93, Lyn95] for the case where all directed graphs have fan-in 2. To state Lynch's results, some definitions are necessary.

A boolean function f of n variables *depends* on argument x_i if for some values $a_1, \ldots, a_{i-1}, a_{i+1}, \ldots, a_n$ in $\{0, 1\}$, we have

$$f(a_1, \ldots, a_{i-1}, 0, a_{i+1}, \ldots, a_n) \neq f(a_1, \ldots, a_{i-1}, 1, a_{i+1}, \ldots, a_n).$$

A boolean function f on n variables is *canalyzing* (a notion due to Kauffman [Kau70]), if there exists some $i \in \{1, \ldots, n\}$ and some values $u, v \in \{0, 1\}$ such that, for all $a_1, \ldots, a_{i-1}, a_{i+1}, \ldots, a_n$ in $\{0, 1\}$,

$$f(a_1, \ldots, a_{i-1}, u, a_{i+1}, \ldots, a_n) = v.$$

That is, if input i has the value u, then the output of f becomes v regardless of the values of any other inputs. For boolean functions of two variables, the only non-canalyzing functions are *exclusive-or* $\oplus$ and *equivalence* $\equiv$. Let a (c) be the probability that a gate is assigned a constant (non-canalyzing) 2-ary function. In [Lyn93], it was shown that if $a > c$, then with probability asymptotic to 1, a random boolean cellular automaton, all of whose vertices have fan-in 2, manifests stable behavior; specifically

1. almost all gates *stabilize*, where gate i stabilizes if
$$(\exists t_0)(\forall t \geq t_0)[f_i^{(t_0)}(\mathbf{x}) = f_i^{(t)}(\mathbf{x})],$$
2. almost all gates are *weak*, where gate i is weak if
$$(\exists t_0)(\exists d)(\forall t \geq t_0)[B^{(t)}(\mathbf{x}) = B^{(t+d)}(\text{toggle}(\mathbf{x}, i))],$$
where $\text{toggle}(\mathbf{x}, i) = (x_1, \ldots, x_{i-1}, 1 - x_i, x_{i+1}, \ldots, x_n)$, i.e., where the i-th bit has been flipped,
3. the state cycle is bounded in size, where the size of state cycle is the least d such that
$$(\exists t_0)(\forall t \geq t_0)[B^{(t)}(\mathbf{x}) = B^{(t+d)}(\mathbf{x})].$$

In [Lyn95], J. Lynch showed that chaotic behavior begins at the threshold $a = c$, where a (c) is the probability of assigning a constant (non-canalyzing) function to a gate; specifically with probability asymptotic to 1, almost all gates are stable and weak, but the average state cycle size is superpolynomial (strongly non-bounded) in the number of gates. It is worth pointing out that the superpolynomial average size of the state cycle contradicts Kauffman's claim that the average size is on the order of $\sqrt{n}$.

1.11.5 Branching Programs

Branching programs were first introduced by C.Y. Lee [Lee59], where they were called "binary decision programs", later considered by W. Masek [Mas76], where they were called "decision graphs", and have since been vigorously investigated in complexity theory (see A.A. Razborov's survey article [Raz91]). In the next chapter, shallow decision trees (or *read-once branching programs*, whose underlying graph is a tree) representing boolean functions play an important role in proofs of circuit size lower bounds for certain combinatorial problems. A restriction thereof, *boolean decision diagrams* (called BDDs or *ordered read-once branching programs*) are efficient data structures for computer implementation of boolean functions, and are actually used in circuit verification for chips.

A *branching program* is a directed acyclic graph with a single source node and distinguished sink nodes labeled as "accepting" or "rejecting". Internal nodes are labeled by variables among $x_1, \ldots, x_n$, and have two out-edges labeled by 0 and 1. A branching program computes a boolean function $f : \{0,1\}^n \to \{0,1\}$ in the obvious fashion; namely, for bits $a_1, \ldots, a_n$, $f(a_1, \ldots, a_n) = 1$ iff the computation, which starts at the source and follows the out-edges labeled a_i at nodes labeled x_i, terminates in an accepting sink node. A branching program is *read-once* if along every path from the source to a sink node, each x_i labels at most one node. As defined, a branching program is a deterministic object; see [AK96], where randomized read-once branching programs are shown to be more powerful than deterministic read-once branching programs, when size is restricted to polynomial in the length of the input.

A *decision tree* is a read-once branching program, whose underlying graph is a tree, directed from the root to the leaves, and will be used as a tool for circuit size lower bounds in the next chapter. Figure 1.3 illustrates a tree of depth 3 decision for the boolean function corresponding to the propositional formula $(x_1 \vee x_2) \wedge \overline{x}_3$.

Note that a DNF formula for boolean function f can immediately be read off from a decision tree for f, by writing the disjunction of the conjunctions of corresponding literals over all branches whose leaves are labeled by 1 (positive literal x_i is taken if the branch edge at node x_i is labeled by 1, otherwise negative literal $\overline{x}_i$). By taking the negation of the disjunction of conjunctions over branches whose leaves are labeled by 0, one has a CNF formula for f. From DNF and CNF formulas for a boolean function, one can also find a decision tree for f. We follow [Kra95][page 15] in the presentation of this lemma.[10]

[10] According to R. Impagliazzo, this result is known as the "Blum" trick, but was independently discovered by many people in the mid 1980s, including M. Blum, R. Impagliazzo, J. Hartmanis and L. Hemachandra, and G. Tardos. **Remark:** by the way, L. Hemachandra changed his name to L. Hemaspaandra.

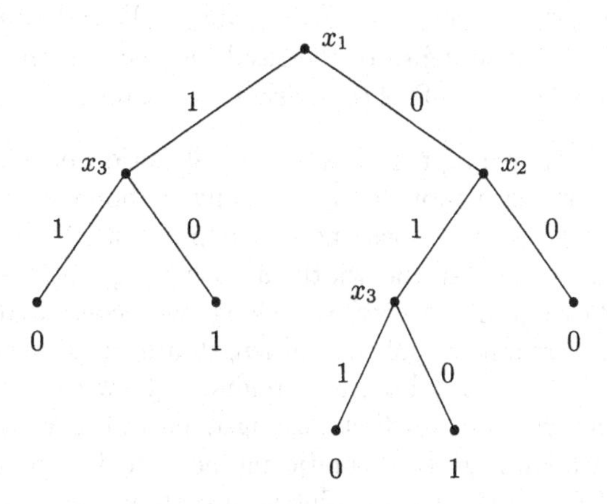

Fig. 1.3. A decision tree for $(x_1 \vee x_2) \wedge \overline{x}_3$

Lemma 1.11.1. *Suppose that f has a decision tree of depth s. Then f can be expressed in s-DNF and s-CNF . Suppose that the boolean function f can be expressed in s-CNF and s-DNF form. Then f has a decision tree of depth s^2.*

Proof. The first claim has already been proved above. If boolean function f can be expressed in s-DNF and s-CNF form, then $f \equiv \bigvee_{i \in I} A_i$ and $1 - f \equiv \bigvee_{j \in J} B_j$, where the A_i, B_j are s-terms (conjunctions of at most s literals).

CLAIM. For all $i \in I, j \in J$, there is a literal in A_i, whose negation appears in B_j.

The claim is obvious, for otherwise there is a truth assignment satisfying $\bigvee_{i \in I} A_i$ and $\bigvee_{j \in J} B_j$ at the same time. We now prove the lemma by induction on $s \geq 1$. When $s = 1$, by the claim it follows that $||I|| = 1 = ||J||$ and so $f \equiv \alpha$ for some literal α, and there is a decision tree for f of depth 1. Consider the inductive case $s > 1$. By the claim, for each B_j with $j \in J$, there is a literal of A_1 whose negation appears in B_j, and for each A_i with $i \in I$, there is a literal of B_1 whose negation appears in A_i. Altogether there are $2s - 1$ variables involved (s variables mentioned in A_1, and at most $s - 1$ variables mentioned in B_1, since one variable mentioned in A_1 is already mentioned in B_1). Without loss of generality assume that the variables involved are $x_1, \ldots, x_{2s-1}$. Form the complete binary tree T of depth $2s - 1$, where nodes at distance $k - 1$ from the root are labeled by x_k. For every branch b in T, let A_i^b, B_j^b be the $s - 1$ terms obtained respectively from A_i, B_j by instantiating the variables x_k, for $1 \leq k \leq 2s-1$, according to the label $b(x_k)$ occurring on b. Apply the induction hypothesis to obtain a decision tree T_b of depth at most $(s-1)^2$ for $f[x_1 = b(x_1), \ldots, x_{2s-1} = b(x_{2s-1})]$, and append T_b to the leaf of branch b in T. The depth of the resulting decision tree $\widetilde{T}$ for f is at most $2s - 1 + (s-1)^2 = s^2$. This establishes the lemma.

A leveled branching program has a natural notion of width. A family of boolean functions is computed by a family of bounded width branching programs, if there is a constant w independent of the number n of boolean variables, such that the width of each program is at most w. In [Bar89], D.A. Barrington[11] proved that languages computed by bounded width branching programs of polynomial length are exactly those computed by logarithmic depth fan-in 2 boolean circuits, i.e., $\text{BWBP} = \text{NC}^1$. To prove this result, the above model is first massaged into an equivalent algebraic form which then allows group theoretic methods to be applied. Formally, identify w with $\{0, \ldots, w-1\}$ and define an instruction to be a triple (i, f, g) where $1 \leq i \leq n$ and $f, g \in w^w$ are maps from w to w. The interpretation of such an instruction is that if $x_i = 0$ apply f, else apply g. Following [Bar89], a $w-\text{BP}$ of length L is a sequence of L such instructions. A boolean function is computed in the natural manner, i.e., $f(x_1, \ldots, x_n) = 1$ iff the composition of the maps obtained by applying each of the instructions yields the identity group permutation $id \in S_w$. In a similar fashion, branching programs may be defined over any finite group G, namely, by requiring $f, g \in G$, where G is a subgroup of the group S_w of permutations on w letters.

Recall a few definitions from elementary group theory. If G is a group, then the *commutator* of elements $a, b \in G$ is the element $aba^{-1}b^{-1}$. The *commutator subgroup* of G, denoted by $[G, G]$, is the subgroup of G generated by all the commutators of G. For any group G, define $G^{(0)} = G$, and $G^{(n+1)} = [G^n, G^n]$. By definition, a group G is *solvable* if there is a finite series $G = G^{(0)} \geq G^{(1)} \geq \cdots \geq G^{(n)} = \{e\}$. If G is finite, then G is non-solvable if and only if $G = G^{(0)} \geq G^{(1)} \geq \cdots \geq G^{(n)} = G^{(n+1)} \neq \{e\}$, i.e., $G^{(n)}$ is non-trivial and equal to its commutator subgroup. For example, the groups A_k, S_k for $k \geq 5$ are non-solvable.

Lemma 1.11.2 ([Bar89]). *Let G be any finite non-solvable permutation group, and let $L \in \text{NC}^1$. There exist polynomial length branching programs over G, which compute the characteristic functions of $L \cap \{0,1\}^n$, for $n \in \mathbf{N}$.*

Proof. Assume now that G is a non-solvable group with series

$$G = G^{(0)} \geq \cdots \geq G^{(n)} = H,$$

and that H is non-trivial and equal to its commutator subgroup $[H, H]$; i.e., there exists m such that every element of H can be expressed as a product

[11] Barrington's surname was later changed to Mix Barrington, so later articles appear under the latter name.

$\prod_{i=1}^{m} a_i b_i a_i^{-1} b_i^{-1}$ of m commutators of H. Using this observation, we show how to represent conjunctions and disjunctions as a word problem over H.

Given a non-identity element $g \in H$ and an alternating, leveled boolean circuit C with n inputs, describe a word w_n in the elements of H such that

$$w_C = \begin{cases} e \text{ if } C \text{ accepts } x_1, \ldots, x_n \\ g \text{ otherwise.} \end{cases}$$

This is done by induction on depth of node A in C. Recall that $H = [H, H]$ and every element of H can be written as the product of m commutators of H. Observe that if $A = (B \vee C)$ then

$$\begin{aligned} w_A(g) &= w_{B \vee C}(g) \\ &= \prod_{i<m} w_B(b_i) w_C(c_i) w_B(b_i^{-1}) w_C(c_i^{-1}). \end{aligned}$$

$B \wedge C$ and $\neg B$ can be expressed similarly. Inductively one forms the word w_C whose product equals e exactly when C accepts x.

The converse direction, BWBP $\subseteq$ NC^1, is straightforward by using a binary tree whose leaves are labeled by functions of G and whose internal nodes are labeled by the composition operation. For a fixed finite group G, it is clear that such a binary tree can be transformed into an equivalent logarithmic depth fan-in 2 boolean circuit. This yields the following.

Theorem 1.11.1 ([Bar89]). BWBP = NC^1.

We now briefly discuss Bryant's [Bry86] important application of branching programs to circuit verification. A *boolean decision graph* or *ordered read-once branching program* G is a rooted, directed acyclic graph with vertex set V, edge set $E \subseteq V \times V$, together with a labeling $\ell : V \cup E \to \{0, 1, \ldots, n\}$ of both vertices and edges, which satisfies the following conditions.

1. Nodes $w \in V$ with no fan-out (leaves) have label 0 or 1, so that $\ell(w) \in \{0, 1\}$. In this case, $\ell(v)$ is called the value of v.
2. Non-leaf nodes $w \in V$ have two children.
3. If u, v are non-leaf nodes connected by a directed path, then $\ell(u) \neq \ell(v)$.
4. Edges $e \in E$ have label 0 or 1. If $w \in V$ has children u, v, then edges (w, u) and (w, v) have unequal edge labels.

An *ordered boolean decision graph* or *ordered read-once branching program* G is a boolean decision graph which satisfies an additional requirement in (2):

- If w's children are u, v, and u (v) is not a leaf, then $\ell(w) < \ell(u)$ ($\ell(w) < \ell(v)$).

A boolean decision graph G *represents* a boolean function f as follows. If $w \in V$ is a leaf with value 0 (1) then $f_w = 0$ (1); if $w \in V$ has label i and children u, v, such that the label of edge (w, u) ((w, v)) is 0 (1), then

$$f_w = \overline{x}_i \cdot f_u + x_i \cdot f_v.$$

The function represented by G is then f_w, where w is the root of G.

Ordered boolean decision graphs were first defined and investigated by R.E. Bryant [Bry86], where they were called *function graphs.* Bryant's function graphs allowed for children u, v of w to be identical, but with this exception, his definition of function graph is equivalent to that given above.

Ordered boolean decision graphs $G = (V, E, \ell)$, $G' = (V', E', \ell')$ are *isomorphic* if there is a bijection $\phi : V \to V'$ satisfying the following conditions. For $w \in V$, let w' denote $\phi(w)$, and if w (w') is not a leaf, let u, v (u', v') denote the children of w (w'). Then for all $w \in V$,

1. $\ell(w) = \ell(w')$.
2. Either w, w' are both leaves, or both are non-leaves.
3. If w, w' are non-leaves, where edges $(w, u) \in E$ and $(w', u') \in E'$ have label 0 and edges $(w, v) \in E$ and $(w', v') \in E'$ have label 1, then $\phi(u) = u'$, $\phi(v) = v'$.

An ordered boolean decision graph G is *reduced* if for distinct $u, v \in V$, the subgraphs G_u, G_v rooted at u,v are not isomorphic. A *boolean decision diagram* (BDD) is a reduced ordered boolean decision graph. These structures uniquely correspond to boolean functions.

Theorem 1.11.2 ([Bry86]). *If f is a boolean function, then there exists a unique (up to isomorphism)* BDD *representing f. Moreover, every other ordered boolean decision graph representing f has more vertices.*

Using Shannon's contact networks, symmetric n-ary boolean functions can be shown to have BDDs with $O(n^2)$ vertices. Efficient algorithms for manipulating BDDs were given in [Bry86].

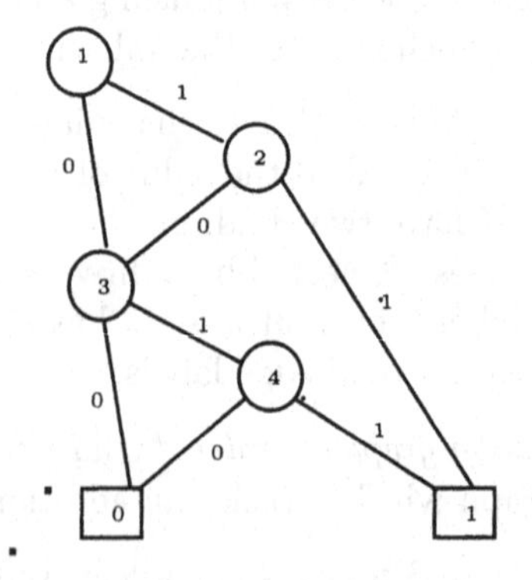

Fig. 1.4. A succinct boolean decision diagram

Examples of BDDs for $x_1x_2 + x_3x_4$ and $x_1x_3 + x_2x_4$ are given in Figures 1.4 and 1.5. Note that the first has six nodes, whereas the second has

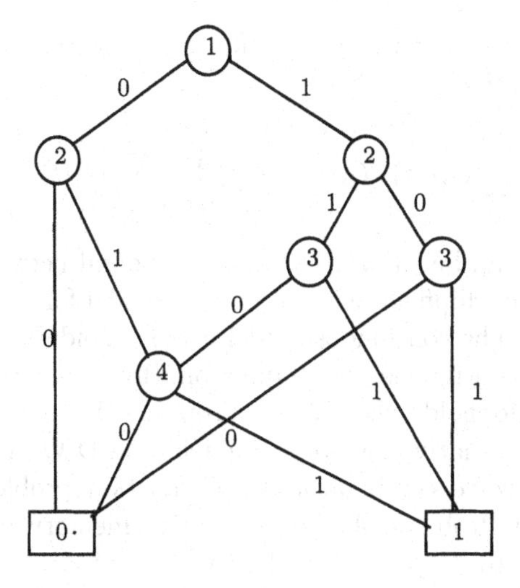

Fig. 1.5. A less succinct boolean decision diagram

eight nodes. This indicates the importance of ordering the boolean variables. Bryant generalizes this, indicating that

$$x_1x_2 + \cdots + x_{2n-1}x_{2n}$$

can be represented by a BDD of $2n+2$ vertices, while

$$x_1x_{n+1} + \cdots + x_nx_{2n}$$

requires 2^{n+1} vertices.

1.11.6 Hopfield Nets

A *Hopfield net* is a weighted directed graph $G = (V, E, w)$ with vertex set $V = \{1, \ldots, n\}$, edge set $E = \{(i,j) : i,j \in V, i \neq j\}$, and symmetric weight function $w : E \to \mathbf{R}$ where $w(i,j) = w(j,i)$ for all distinct vertices i, j. For each $1 \leq i \leq n$, there is an external input I_i and threshold value θ_i. For each $1 \leq j \leq n$ and time $t \geq 0$, the output $O_j(t)$ is defined as follows: $O_j(0)$ is a random initial value,

$$O_j(t+1) = \begin{cases} 1 & I_j + \sum_{i \neq j} w(i,j) \cdot O_j(t) > \theta_j \\ 0 & I_j + \sum_{i \neq j} w(i,j) \cdot O_j(t) < \theta_j \\ O_j(t) & \text{otherwise.} \end{cases}$$

The net G then computes the sequence $\mathbf{O}(0), \mathbf{O}(1), \mathbf{O}(2), \ldots$ where $\mathbf{O}(t) = (O_1(t), \ldots, O_n(t))$. An optimized solution to a combinatorial problem is achieved at time t when $\mathbf{O}(t) = \mathbf{O}(t+1)$.

Using the symmetry of the weight function together with a *Lyapunov network energy function*

$$E(t) = -\frac{1}{2}\left(\sum_{i,j} w(i,j) \cdot O_i(t) \cdot O_j(t)\right) - \sum_j O_j(t) \cdot (I_j - \theta_j)$$

stability of the net can be shown. For more on neural networks, see [Fu94].

As defined above, Hopfield net is a discrete model for combinatorial optimization problems. The continuous model uses sigmoidal real-valued threshold functions and a continuous energy function. Hopfield nets were introduced in [Hop82] by J.J. Hopfield, who initially proposed them as a possible biological model for an associative memory. Hopfield and D.W. Tank [HT86] later used such nets to solve combinatorial optimization problems such as the Traveling Salesman Problem. As an associative memory model, Hopfield's concluding remarks to [Hop82] are as follows:

> Much of the architecture of regions of the brains of higher animals must be made from a proliferation of simple local circuits with well-defined functions. The bridge between simple circuits and the complex computational properties of higher nervous systems may be the spontaneous emergence of new computational capabilities from the collective behavior of large numbers of simple processing elements.

1.11.7 Communication Complexity

This measure of complexity was first proposed by C. Papadimitriou and M. Sipser [PS82a] because it provides a lower bound for the area-delay product AT^2 of any chip recognizing a language L.

Suppose that a language L in the alphabet $\{0,1\}$ is to be recognized by two different processors. Each processor receives half the input bits and the computation proceeds using some communication protocol between the two processors. The minimum number of bits (minimized over all partitions of the input bits into two equal parts considered as a function of n) that has to be exchanged in order to recognize $L \cap \{0,1\}^{2n}$ is called the *communication complexity* of L.

Duris et al. [DGS84] prove several communication complexity results relating to number of rounds and partitions while Hajnal et al. [HMT88] consider communication complexity bounds for several graph-theoretic properties (connectivity, *st*-connectivity, bipartiteness). For more information on VLSI issues consult [Ull84], [MC80], [Len90]. A recent book on Communication Complexity is [Hro97].

1.11.8 Anonymous Networks

By an *anonymous* network we mean a network for which the following assumptions hold (see also [Ang80]):

1. The processors know the network topology and the size of the network (i.e., total number of processors).
2. The processors are anonymous, i.e., they do not know either the identities of themselves or of the other processors).
3. The processors are identical (i.e., they all run the same algorithm).
4. The processors are deterministic.
5. The network is asynchronous.
6. The network is labeled (by a labeling we mean a global, consistent labeling of the network links).
7. The network links are FIFO.

Let N be the number of nodes in the network. Each processor p is given a bit, say b_p, and a boolean function $f \in B_N$. The processors exchange bits according to the above rules by executing a certain protocol. The goal is for all the processors to compute the value $f(< b_p : 1 \leq p \leq N >)$. As a measure of complexity we use the total number of bits exchanged by all processors throughout the computation. Since the processors are "identical" it is clear that this model of computation is more suitable for distributed computation, where it is important to use decentralized resources and all processors must execute identical algorithms (given identical data). If the processors had identitities then every boolean function on N variables would be computable in this model. But under the assumptions previously outlined not every N-variable boolean function is computable. In fact the boolean functions which can be computed depend in a specific manner on the group of automorphisms of the network. For more details on this model see Section 3.9 as well as the survey article [Kra97b].

1.12 Historical and Bibliographical Remarks

Boolean functions are named after George Boole who introduced boolean algebras in his study of the logical operations governing the laws of thought [Boo58]. Boole's original ideas were later developed and refined by various mathematicians [Sik69, Hal74]. However, it is since the introduction of bistable devices in the construction of modern electronic computers that Boole's concepts have secured a permanent place in theoretical computer science. A strong relationship, on the one hand between boolean algebra and binary arithmetic, and on the other hand between boolean algebra and logic, has played a fundamental role in the development of electronic computers.

The circuit model presented in this chapter is a mathematical abstraction of the combinations of electronic devices and conductors, that when interconnected form a conducting path which implements a certain logic function. Elementary logic devices, like AND, OR, NAND, NOR gates and inverters, combine to form more complicated logic functions (see e.g. [Bar80]).

The theory of propositional logic was developed extensively by E. Post [Pos21] who also proved that every class of boolean functions which is closed under composition has a finite basis, and that there are a countable number of classes of boolean functions which are closed under composition [Pos41].

This chapter offers only an introduction to boolean functions and circuit complexity. An important topic not covered in this text is the design of efficient circuits for algebraic operations, for which we refer the reader to [Weg87]. Unpublished course notes of S.A. Cook, M. Sipser and E. Kaltofen were helpful in the preparation of this chapter. In particular, our presentation of Brent's division algorithm was influenced by Kaltofen's notes, and our proof of Theorem 1.9.1 by B. Poizat's stimulating monograph [Poi95]. Savitch's text [Sav70] is a classic introduction to circuit complexity, which is complemented by the excellent survey article [BS90]. See [Pud87] for a brief survey concerning some of the complexity measures introduced in Section 1.11.

In this chapter, we have been concerned only with *non-uniform* circuit families, so that in the definition of NC^1, etc. no effective procedure is required, which given n, constructs the n-th circuit of a circuit family. If uniformity is considered, then the construction of [BCH86] yields polynomial time uniform circuits, where polynomial time, rather than logarithmic space, is required to compute the product P in Step 2 of Algorithm 1.7.8. Recently, by working directly with the Chinese remainder representation, A. Chiu, G. Davida and B. Litow [CDL99] have proved the existence of logspace uniform circuits for integer division. Their result solves a long-standing open question, thus showing that integer division is in logspace.

A potentially rich and new direction for circuit complexity is sketched out in B. Poizat's monograph [Poi95], where notions of boolean circuit and Turing machine are generalized to arbitrary structures.[12] Standard complexity theory then corresponds to the structure $\{0,1\}^*$, while the Blum–Shub–Smale theory of real complexity [BSS89] corresponds to the structure $(\mathbf{R}, +, -, \times, =, <)$. This generalization permits one to consider complexity theory over the real and complex numbers with certain operations, i.e., structures such as $(\mathbf{R}, +, -, =)$ and $(\mathbf{C}, +, -, \times, =)$, etc. Such theories may be very natural when implementing particular kinds of algorithms (e.g. numerical integration over $\mathbf{R}$). See [Koi97, Koi94, FK98] for further references.

1.13 Exercises

Exercise 1.13.1. Show that any boolean function can be computed with an alternating leveled circuit over the De Morgan basis $\vee, \wedge, \neg, 0, 1$ where the inputs are either variables or their negation.

[12] This approach has little in common with finite model theory, a branch of complexity theory not covered in this text.

Exercise 1.13.2. Prove that the following sets of boolean functions are complete: $\{\vee, \wedge, \neg, 0, 1\}$, $\{\vee, \neg, 0, 1\}$, $\{\wedge, \neg, 0, 1\}$, $\{\oplus, \wedge, 0, 1\}$, $\{|, 0, 1\}$, where the NOR connective is defined so that $x|y$ is equivalent to $\neg(x \vee y)$. Using only the connective $|$, explicitly define the NAND connective, where x NAND y is equivalent to $\neg(x \wedge y)$. Can $\neg$ be defined from $\equiv, \wedge$?

Exercise 1.13.3. Show that every minimal complete system has at most four elements.
HINT: In the notation of the proof of Theorem 1.2.1, either f_T^0 is not self-dual, in which case $f_T^0(0^n) = f_T^0(1^n)$, or f_T^0 is not monotone.

Exercise 1.13.4 ([Mee87]). Show that the class T_n^1 is generated by the functions $\rightarrow, \wedge$, where $x \rightarrow y \equiv \neg x \vee y$. Prove a similar result for T_n^0.
HINT: Let $\mathcal{F}_n$ be the set of formulas generated by $\rightarrow, \wedge$. Clearly $\mathcal{F}_n \subseteq T_n^1$. It remains to prove $T_n^1 \subseteq \mathcal{F}_n$. Let $f \in T_n^1$. First notice that in view of the identity

$$p \vee q \equiv [(p \rightarrow (p \wedge q)) \rightarrow q]$$

the class $\mathcal{F}_n$ is closed under $\vee$. Without loss of generality we can assume that f is in conjunctive normal form $f_1 \wedge \cdots \wedge f_s$, in which case it is enough to prove that each conjunct f_i is in $\mathcal{F}_n$. This follows from the fact that each f_i has at least one positive literal.

Exercise 1.13.5. Show that the class M_n of monotone functions is generated by the constants $0, 1$ and the functions $\vee, \wedge$.
HINT: Let f be monotone and let $f_1, \ldots, f_s$ be the class of all maximal functions $g \leq f$ which are the conjunction of positive literals. Clearly, $f_1 \vee \cdots \vee f_s \leq f$. We claim that in fact $f_1 \vee \cdots \vee f_s = f$. Otherwise, there exists an input $a \in \{0,1\}^n$ such that $f(a) = 1$ but each $f_i(a) = 0$. Let $g(x)$ be the conjunction of those x_i for which $a_i = 1$. If $g(x) = 1$ then $a \leq x$ and hence $f(x) = 1$. Consequently, $g \leq f$, which is a contradiction.

Exercise 1.13.6. Show that

1. $|T_n^0| = |T_n^1| = \frac{1}{2} \cdot 2^{2^n}$.
2. $|D_n| = 2^{2^{n-1}}$.
3. $|M_n| \geq 2^{k(n)}$, where $k(n) = \binom{n}{n/2}$.

HINT: Let $\mathcal{E}$ be the set of subsets of $\{1, \ldots, n\}$ of size $n/2$. For each $T \subseteq \mathcal{E}$ consider the monotone function $f_T = \bigvee_{A \in T} \left(\bigwedge_{i \in A} x_i \right)$.

Exercise 1.13.7. For every context-free language L not containing the empty word λ, there is a context-free grammar G in Chomsky normal form for which $L = L(G)$.
HINT: See, for instance, [LP81] or [HU79] for a proof.

Exercise 1.13.8. Generalize the $2-3$ Lemma 1.6.1 to k-ary trees, each node of which has most k children; i.e., if T is a k-ary tree with leafsize $n \geq k$, then there exists a node x of T, such that the subtree

$$T_x = \{y \in T : x \text{ is an ancestor of } y\}$$

of T rooted at x contains m leaves, where $\lceil \frac{n}{k+1} \rceil < m \leq \lfloor \frac{kn}{k+1} \rfloor$. The tree obtained from T by removing the "scar" T_x is called the *scarred tree.* Modify Spira's Theorem 1.10.1 to show that for any finite basis Ω of k-place connectives, $L \subseteq \{0,1\}^*$ has polynomial size boolean formulas over Ω iff $L \in \text{NC}^1$.

Exercise 1.13.9. A pioneering theorem of S.A. Cook states that the collection SAT of all satisfiable propositional formulas is NP-complete.

1. Assuming Cook's theorem, show that the collection 3-SAT is NP- complete, where 3-SAT denotes the class of all satisfiable propositional formulas in conjunctive normal form, each conjunct of which is a disjunction of three literals.
2. Show that 2-SAT $\in \text{AC}^1$, where 2-SAT is the set of all satisfiable propositional formulas in conjunctive normal form, each conjunct of which is a disjunction of three literals.

HINT: Given the formula Θ in 2-CNF form, associate the directed graph $G_\Theta = (V, E)$, where the vertex set V and edge set E satisfy

$$V = \{p, \overline{p} : p \text{ or } \overline{p} \text{ appears in } \Theta\},$$
$$E = \{(x,y) : x, y \text{ literals and } \overline{x} \vee y \text{ or } x \vee \overline{y} \text{ appears in } \Theta\}$$

By transitive closure of the edge relation, in AC^1 obtain connected components of G_Θ. Show that

$$\Theta \text{ is unsatisfiable } \iff (\exists p, \overline{p})(p \text{ connected to } \overline{p} \wedge \overline{p} \text{ connected to } p).$$

(See [CL88] for details, where additionally a satisfying assignment is constructed for satisfiable $2 - \text{CNF}$ formulas.)

Exercise 1.13.10. By hand, work through the Beame–Cook–Hoover iterated multiplication algorithm in order to compute the product $4 \cdot 5 \cdot 7$. Note that the product of three 3–bit integers requires discrete logarithms for the first nine primes to be constructed.

Exercise 1.13.11. This exercise concerns some of the elementary number theory notions used in the Beame–Cook–Hoover division algorithm.

1. Prove that if a, b are integers with greatest common divisor $d = (a, b)$, then there exist integers x, y such that $ax + by = d$. (This is known as Bezout's lemma.)
2. Using Bezout's lemma, prove that if a, b are relatively prime, then for any integers u, v there exists a unique $x < ab$ such that

$$x \equiv u \pmod{a}$$
$$x \equiv v \pmod{b}.$$

Generalize this to the statement of the Chinese remainder theorem given in the text.

3. Let p_n denote the n-th prime number. From the prime number theorem
$$\lim_{x\to\infty} \frac{\pi(x)}{x/\ln(x)} = 1$$
show that there are constants c, d for which $p_n < c \cdot (n \ln(n)) + d$ holds for all n.
4. Prove that the modular powering algorithm given in the text (using repeated squaring) correctly computes $a^b \bmod m$.
HINT: Use induction on notation, which states that if $P(0)$ holds (*base case*), and if $P(\lfloor x/2 \rfloor)$ implies $P(x)$ (*inductive case*), then, for all x, $P(x)$ holds.
5. ($\star$) Prove that if a, b are integers of length n and $m \leq n$, then $a^b \bmod m$ is computable in logspace uniform NC^1 (see [BCH86]).

Exercise 1.13.12. Consider the boolean cellular automaton B consisting of the directed graph shown in Figure 1.6 on vertex set $\{1,2,3\}$ with $f_1(x,y) = x \vee y$, $f_2(x,y) = x \oplus y$, $f_3(x,y) = x \wedge y$, and $x_1 = 0$, $x_2 = 1$, $x_3 = 1$. Which

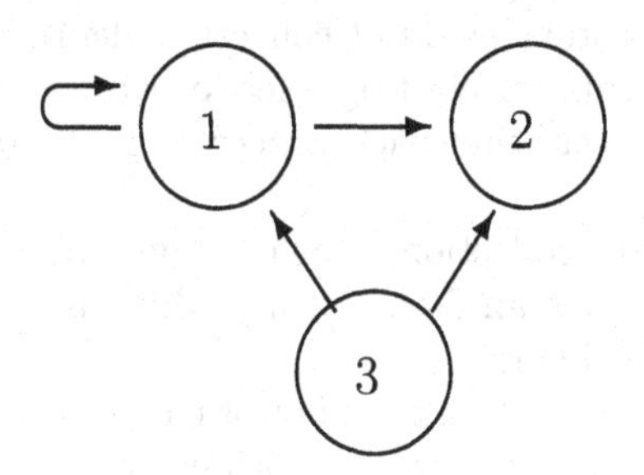

Fig. 1.6. A boolean cellular automaton with vertices 1, 2, 3

gates stabilize? Which gates are weak? What is the size d of the state space of B? Try different initial states.

Exercise 1.13.13. Design a boolean decision diagram (BDD) which outputs 1 if integer $a > b$, and otherwise outputs 0.

Exercise 1.13.14 (Kranakis–Krizanc, unpublished). ($\star$) We are given an anonymous network $\mathcal{N}$ (say, ring, torus, hypercube, etc.) consisting of N processors. A boolean function f on N variables is ϵ-computable if there is an algorithm AL such that for each input I to the network the ratio of processors with output equal to $f(I)$ is at least ϵ, i.e.,
$$\frac{|\{p \in \mathcal{N} : Out_p(I) = f(I)\}|}{N} \geq \epsilon,$$
where by $Out_p(I)$ we denote the output (bit) of processor p when the processors execute the algorithm AL with input I. Show that for any anonymous network $\mathcal{N}$, $(\frac{1}{2} + \frac{1}{N})$-computable boolean functions are computable.

Exercise 1.13.15 (Kranakis–Krizanc, unpublished). ($\star$) Let $\mathcal{F}(\mathcal{N})$ be the class of N-ary boolean functions computable in the anonymous network $\mathcal{N}$. Show that there is no network $\mathcal{N}$ such that $\mathcal{F}(\mathcal{N})$ is exactly the class of symmetric functions on N-variables.

Exercise 1.13.16. ($\star$) Consider the unidirectional ring of $N \geq 3$ processors, i.e., each processor has exactly two neighbors and the messages can travel in the clockwise direction. A leader is a distinguished processor among the N given processors.

1. Give an input collection algorithm for electing a leader (among the processors) assuming that the N processors have distinct identities.
2. Now suppose that the processors have no distinct identities but instead they have a random source for generating random bits indpendently of each other. Assume that the probability of generating the bit 0 is exactly $1/2$.
   a) Give an algorithm for electing a leader with probability at least $1 - 1/N$ such that each processor generates $O(\log N)$ bits.
   HINT: Use the previous input collection algorithm.
   b) Give an algorithm for electing a leader with probability at least $1 - 1/N$ in such a way that each processor generates only one random bit.
   HINT: Use input collection to collect the random bits. Observe that each processor has an N-bit string which is a circular shift of the string of its neighbors.
3. Give the trade-off of the total number of messages and the total number of random bits generated in the algorithms above.

2. Circuit Lower Bounds

> Abstraction is what makes mathematics work. If you concentrate too closely on too limited an application of a mathematical idea, you rob the mathematician of his most important tools: analogy, generality and simplicity. Mathematics is the ultimate in technology transfer. *I. Stewart [Ste89]*

2.1 Introduction

What is the smallest size circuit of a certain depth which computes an arbitrary boolean function? Later we will see that depth of an arbitrary fan-in boolean circuit corresponds to time on a parallel computer, whereas circuit size corresponds to the number of active processors, so the question of circuit size and depth has additional significance.

Work by Shannon and Lupanov completely answers the question of size of fan-in 2 circuits (with no constraint on depth) for "random" n-ary boolean functions. Though there is currently no super-linear lower bound known for arbitrary fan-in circuit size (with no constraint on depth) for any NP-complete problem, progress has been made for restricted classes of circuits: monotonic (boolean and real) circuits of bounded fan-in, constant depth circuits with unbounded fan-in boolean, modular counting and threshold gates. Here, by "constant depth" we mean a constant independent of input size n.

Symmetric functions are particularly simple, and by Theorem 1.8.3 have linear size circuits. Do all boolean functions have linear or polynomial size circuits? Essentially since there are 2^{n-1} mutually orthogonal satisfying truth assignments, every DNF formula representing $\mathrm{PAR}_n(x_1, \ldots, x_n) = x_1 \oplus \cdots \oplus x_n$ must have 2^{n-1} terms. Can this exponential lower bound for the size of depth 2 circuits be extended to depths $d > 2$? In this chapter, we study a variety on techniques for circuit size lower bounds, ranging from combinatorial to probabilistic and algebraic.

We begin by presenting Shannon's exponential lower bound for the circuit size of "most" boolean functions which matches Lupanov's optimal upper bound from Section 1.8.3. Section 2.3 then presents Nechiporuk's lower bound for formula depth, whose technique is interesting, though the lower bound result has since been superceded. We then prove the exponential lower bound for monotonic, bounded fan-in real circuits in Section 2.4. This lower bound is for the *Broken Mosquito Screen* problem, a symmetric version of the NP-complete problem CLIQUE, and is due to A. Haken and S.A. Cook (independently, at the same time P. Pudlák [Pud97] proved that A.A. Razborov's lower bound for monotonic boolean circuits for CLIQUE could be extended to monotonic real circuits). As we will see in chapter 5, this lower bound will directly yield a lower bound for the cutting planes proof system in propositional logic, by applying an interpolation theorem due to Pudlák [Pud97]. A.A. Razborov's superpolynomial lower bound for monotonic fan-in 2 boolean circuit size for CLIQUE (and *a fortiori* for the *Broken Mosquito Screen* problem) can be interpreted as a proof that monotonic P is different than monotonic NP. In [KW90], M. Karchmer and A. Wigderson proved an $n^{\Omega \log n}$ lower bound for fan-in 2 monotone circuit size for the problem of *st-connectivity*. Here we present a different proof of J. Johannsen's recent extension of this result to fan-in 2 monotonic real circuit size. The Karchmer–Wigderson lower bound can be interpreted as a proof that monotonic NC^1 is properly contained in monotonic AC^1, and Pudlák's interpolation theorem then leads from the circuit lower bound to a separation between *tree*-like and *dag*-like resolution and cutting planes proofs, presented in a later chapter.

The remainder of the chapter, beginning with Section 2.5, concerns constant depth, unbounded fan-in circuits. A major breakthrough in the study of such circuits was the independent work of Ajtai [Ajt83] and Furst, Saxe, and Sipser [FSS84], who proved a superpolynomial lower bound for PAR_n, namely $L_k^{\vee,\wedge,\neg}(\text{PAR}_n) = \Omega(n^{\log^{(3(k-2))} n})$. The technique has come to be known as the *random restriction* method. The idea is that by randomly setting a certain number (say $n^{1/d}$) of the input variables to 0 and 1 with a certain probability, then a circuit of depth d of a certain size will simplify into a circuit of depth $d-1$. Using similar techniques, Yao [Yao85] later improved the superpolynomial lower bound for parity [Ajt83, FSS84] to the exponential bound $L_k^{\vee,\wedge,\neg}(\text{PAR}_n) = 2^{\Omega(n^{1/4k})}$, though earlier Boppana [Bop84] had shown that $L_k^{\vee,\wedge}(\text{MAJ}_n) = 2^{\Omega(n^{1/(k-1)})}$ for monotone circuits.

In [Hås87], Håstad gave a dramatic simplification of Yao's argument, in distilling out the essence of the proof in his Switching Lemma, which yields the bound $L_k^{\vee,\wedge,\neg}(\text{PAR}_n) = 2^{(1/2c)^{k/(k-1)}}$. Later, in trying to understand which formal principles of reasoning were required to prove circuit lower bounds, A.A. Razborov gave a striking simplification of the Switching Lemma, whose proof used only simple combinatorics. This simplification allowed recent research in lower bounds to progress significantly further, in particular, in the direction of establishing Switching Lemmas in situations where the boolean variables are not necessarily independent.

First, we begin with M. Sipser's illustration of the random restriction technique applied to infinite boolean circuits [Sip85a], and continue in Section 2.6.1 with Håstad 's probabilistic lower bound method. In Section 2.6.2, we present M. Sipser's separation result of depth-k versus depth-$(k-1)$ circuits. In the following section, we present Razborov's elegant proof of the

Switching Lemma, using the formalization of decision trees following [Bea94]. Decision trees have since become a preferred approach to the random restriction method. We then return to the problem of *st*-connectivity, this time considering small depth boolean circuits, rather than monotone circuits. Here we present a hybrid form of switching lemma, with application to an improved lower bound for small distance connectivity, due to Beame, Impagliazzo and Pitassi.

In the next section we present Razborov's beautiful algebraic lower bound techniques for the majority function. Razborov's result has in a sense been superceded by Smolensky's algebraic lower bound for the modulo function. Nevertheless the elegance and originality of Razborov's mathematical ideas mandate their presentation together with their non-trivial generalizations as these evolved in Smolensky's work. We also present Smolensky's lower bound technique and give an exponential lower bound on the circuit complexity of computing the function MOD_r even when MOD_p gates are allowed (where p is prime and r is not a power of p).

Smolensky's main result gives no information when MOD_m gates are allowed, where m is composite. Section 2.8 refines Smolensky's idea even further by studying the polynomial method in circuit complexity. Here we study the power of MOD_m gates (where m is not necessarily prime) and present the recent work of Mix Barrington, Beigel, and Rudich as well its extension by Tsai. In particular we obtain lower bounds on the MOD_m complexity of the functions $\text{MOD}_p, \neg\text{MOD}_p$ (p prime) as well as the threshold functions TH_k.

Next we present the work of Fagin, Klawe, Pippenger and Stockmeyer who give necessary as well as sufficient conditions (in terms of the least number of variables that must be set to constants in order the resulting function be a constant) for a family of boolean functions to be computable with constant depth, polynomial size circuits. Finally, in the last section we prove the surprising result of Ajtai and Ben-Or that if a family of boolean functions is computable by constant depth probabilistic circuits with error $\frac{1}{(\log n)^r}$, for some $r \geq 1$, then in fact it is computable by deterministic constant depth circuits.

We also include in Section 2.10 a surprising result of R. Beigel which shows how to reduce the number of majority gates in a constant depth circuit to a single one.

2.2 Shannon's Lower Bound

In this section we consider a general lower bound on circuit size to compute boolean functions. We defined a boolean circuit to have inputs $x_1, \ldots, x_n$, and counted only internal (non-input) gates in the definition of size. By applying De Morgan's rules, negation may be pushed to the leaves, so that $\neg \bigwedge x_i \equiv \bigvee \neg x_i \equiv \bigvee \overline{x_i}$. It follows that every boolean circuit can be transformed to a circuit having gates $\bigwedge$, $\bigvee$ on inputs $x_1, \overline{x_1}, \ldots, x_n, \overline{x_n}$, without increasing size. Throughout this chapter, we may alternate between one and the other formulation of circuit, as appropriate for a particular argument.

Theorem 2.2.1 ([Sha49, Mul56]). *Let $F(s,n)$ be the number of boolean functions $f \in \mathcal{B}_n$ which can be computed by a fan-in 2 circuit over basis $\mathcal{B}_2$ of size s, with input nodes $x_1, \ldots, x_n, 0, 1$ and one output gate. Then, for $s \leq 2^n/n$,*

$$\lim_{n \to \infty} \frac{F(s,n)}{2^{2^n}} = 0.$$

In other words, for sufficiently large n, almost all boolean functions in $\mathcal{B}_n$ require fan-in 2 boolean circuit size at least $2^n/n$.

Proof. The following argument counts the number of fan-in 2 circuits in n variables and of size s over the basis $\mathcal{B}_2$. A similar proof will work for any constant number of gates. For fixed circuit size s, we have the following:

- There are s gates of 16 possible types.
- For each gate, we can choose either of its two predecessors in at most $s+n+2$ ways (namely, s gates, n variables, 2 constants).
- A circuit of size s has $s!$ different numberings.

From these properties, we obtain

$$F(s,n) \leq \frac{[16(s+n+2)^2]^s}{s!}.$$

Using Stirling's formula $s! = \Omega((s/e)^s \sqrt{s})$, the fact that $s \geq n+2$ and simplifying, we get

$$F(s,n) \leq c^s \cdot s^s$$

for some constant $c > 1$, and sufficiently large n. It follows that

$$\log F(s,n) \leq s \log c + s \log s.$$

From the last inequality, if $s \leq 2^n/n$, then

$$\log F(s,n) \leq 2^n \left(1 - \frac{\log n}{n} + \frac{\log c}{n}\right)$$

and hence

$$\lim_{n \to \infty} \frac{F(s,n)}{2^{2^n}} = 0.$$

Define the *k-th slice* S_k^n of $\{0,1\}^n$ to be the set of length n bit vectors of weight k; i.e., $S_k^n = \{x \in \{0,1\}^n : |x|_1 = k\}$. A boolean function $f : 2^n \to \{0,1\}$ is a *k-slice function* for $0 \leq k \leq n$ if

$$f(x) = \begin{cases} 0 & \text{if } |x|_1 < k \\ 1 & \text{if } |x|_1 > k. \end{cases}$$

Note that f may take on any value for elements of the k-th slice. If f is a k-slice function, for some $0 \leq k \leq n$, then we say that f is a *slice function.* Let $\mathcal{S}_k^n$ designate the collection of all k-slice functions. See [Weg87] for results on slice functions, where in particular a theorem of Berkowitz is presented, which states that monotonic circuit size and circuit size coincide on slice functions.

Corollary 2.2.1 ([Ros97]). *With $F(s,n)$ as in the previous theorem, for $s \leq \frac{2^{n/2}}{10n}$,*

$$\lim_{n\to\infty} \frac{F(s,n)}{|\mathcal{S}^n_{n/2}|} = 0$$

and hence, for sufficiently large n, almost all monotonic boolean functions in $\mathcal{B}_n$ require fan-in 2 monotonic boolean circuit size at least $\frac{2^{n/2}}{10n}$.

Proof. From the proof of the previous theorem, there exists $c > 1$ such that for sufficiently large n,

$$\log F(s,n) \leq s(\log s + \log c).$$

Letting $s \leq \frac{2^{n/2}}{10n}$, we have

$$\begin{aligned} \log F(s,n) &\leq s(\log s + \log c) \\ &\leq \frac{2^{n/2}}{10n}(n/2 - \log 10 - \log n + \log c) \\ &\leq 2^{n/2}\left(\frac{1}{20} - \frac{\log 10}{10n} - \frac{\log n}{10n} + \frac{\log c}{10n}\right) \end{aligned}$$

Since $|\mathcal{S}^n_{n/2}| = 2^{\binom{n}{n/2}} \geq 2^{2^{n/2}}$, it follows that for $s \leq \frac{2^{n/2}}{10n}$,

$$\lim_{n\to\infty} \frac{F(s,n)}{|\mathcal{S}^n_{n/2}|} = 0.$$

2.3 Nechiporuk's Bound

A standard way to compute a boolean function on n variables is via a propositional formula on n variables $x_1, \ldots, x_n$ using a predetermined basis. A possible complexity measure is then the least number $\mathcal{N}(f, V)$ of *occurrences* of variables from $V = \{x_1, \ldots, x_n\}$ in a propositional formula representing f. Since the number of leaves is one greater than the number of internal nodes in a binary tree, this measure is equivalent to *formula size*, or circuit size for fan-in 2, fan-out 1 circuits, so that $L(f) = \mathcal{N}(f, V) - 1$. Nechiporuk's technique yields a lower bound for the number of occurrences of variables in a propositional formula over the basis $\{0, 1, \oplus, \wedge\}$ (and hence for fan-out 1 circuit size), although the technique can be adapted to any fixed basis.

Let $V = \{x_1, \ldots, x_n\}$ be a set of propositional variables. For each $X \subseteq V$ let $\mathcal{B}_X$ be the set of boolean functions on the variables X. Inductively, define *propositional formulas* over $\{0, 1, \oplus, \wedge\}$ as follows: constants $0, 1$ are formulas; variables $x_1, x_2, \ldots$ are formulas; if ψ, θ are formulas, then so are $(\psi \oplus \theta)$ and $(\psi \wedge \theta)$.

For any propositional formula ϕ over $\{0, 1, \oplus, \wedge\}$ and any $X \subseteq V$ let $\mathcal{N}(\phi, X)$ be the number of occurrences of variables $x \in X$ in the formula ϕ. Formally, define

$$\mathcal{N}(\phi, X) = \begin{cases} 0 & \text{if } \phi \in \{0,1\} \text{ or } \phi \text{ is a variable not in } X \\ 1 & \text{if } \phi \text{ is a variable in } X \\ \mathcal{N}(\phi_1, X) + \mathcal{N}(\phi_2, X) & \text{if } \phi \text{ is } (\phi_1 \oplus \phi_2) \text{ or } (\phi_1 \wedge \phi_2) \end{cases}$$

A propositional formula ϕ built up from $\{0, 1, \oplus, \wedge\}$ and the variables $x_1, \ldots, x_n$ *represents* the boolean function $f \in \mathcal{B}_n$ if, for all $a_1, \ldots, a_n \in \{0, 1\}$, $f(a_1, \ldots, a_n) = 1$ if and only if $\sigma \models \phi$, where the truth assignment $\sigma(x_i) = a_i$ for $1 \leq i \leq n$. For $X \subseteq V$, the boolean function $g \in \mathcal{B}_X$ is a *subfunction* of f if, for some assignment of the variables in $V - X$, the boolean function resulting from assigning these values to the corresponding variables in f is equal to g. For $X \subseteq V$, let $E(f, X)$ be the set of *non-constant* subfunctions of f.

The following theorem is due to Nechiporuk [Nec66], as improved by M. Paterson (see [BS90]), and makes explicit the intuition that if a boolean function f has many subfunctions, then the formula size $L(f)$ of f must be large.

Theorem 2.3.1. *Let $V_1, \ldots, V_m$ be a partition of V and ϕ a propositional formula in the basis $\{0, 1, \oplus, \wedge\}$ representing a given boolean function $f \in \mathcal{B}_V$. If $|E(f, V_i)| \geq 1$, for all $i = 1, \ldots, m$, then*

$$\mathcal{N}(\phi, V) > \sum_{i=1}^{m} \log_5 |E(f, V_i)| \tag{2.1}$$

and so

$$L(f) \geq \sum_{i=1}^{m} \log_5 |E(f, V_i)|.$$

Proof. Since V_i, $i \leq m$, forms a partition of V we have that $\mathcal{N}(\phi, V) = \sum_{i=1}^{m} \mathcal{N}(\phi, V_i)$, from which the theorem would follow if we could prove that for all $i \leq m$, $\mathcal{N}(\phi, V_i) > \log_5 |E(f, V_i)|$. Hence it is enough to prove that for all $X \subseteq V$,

$$\mathcal{N}(\phi, X) > \log_5 |E(f, X)|.$$

For any $X \subseteq V$, define $S(f,X)$ to be

$$\{g \in \mathcal{B}_X : g \text{ is constant or for some } h \in E(f,X),\ h \oplus g \text{ is constant}\}$$

and define $S'(f,X)$ to consist of those *non-constant* elements of $S(f,x)$. Note that if $g \in E(f,X)$, then both g and $1-g$ belong to $S'(f,X)$, so that

$$\begin{aligned} |E(f,X)| &\le 2 \cdot |S'(f,X)| \\ &= 2 \cdot (|S(f,X)| - 2) \\ &= 2 \cdot |S(f,X)| - 4 \\ &< 2 \cdot |S(f,X)| - 3. \end{aligned}$$

It suffices thus to prove that for all $X \subseteq V$, $2 \cdot |S(f,X)| - 3 \le \mathcal{N}(\phi, X)$; i.e.,

$$|S(f,X)| \le \frac{3 + 5^{\mathcal{N}(\phi,X)}}{2}. \tag{2.2}$$

For any propositional formula ϕ let $\hat{\phi}$ denote the boolean function represented. Inequality (2.2) is proved by induction on the construction of ϕ.

If $\phi \in \{0,1\}$, or ϕ is a variable not in X, then $\mathcal{N}(\phi,X) = 0$, so $\frac{3+5^{\mathcal{N}(\phi,X)}}{2} = 2$, while $S(f,X) = \{0,1\}$, and so (2.2) holds. If ϕ is a variable in $x_i \in X$, then $S(f,X) \subseteq \{0,1,x_i,\neg x_i\}$, while $\mathcal{N}(\phi,X) = 1$, so that $|S(f,X)| \le 4 = \frac{3+5^{\mathcal{N}(\phi,X)}}{2}$, and so (2.2) holds. Suppose now that $\phi = \phi_1 \circ \phi_2$, where $\circ$ temporarily denotes either of the two operations $\oplus, \wedge$. In this case, $\hat{\phi} = \hat{\phi}_1 \circ \hat{\phi}_2$. Let $e_i = |S(\hat{\phi}_i, X)|$, for $i = 1,2$. If $g \in S(\hat{\phi}, X)$ then either g is the constant 0,1 or else $g = (g_1 \circ g_2) \oplus c$, for some $g_i \in S(\hat{\phi}_i, X)$ and constant c. We distinguish the following cases.

Case 1. If either g is constant or $g = (g_1 \Box g_2) \oplus c$ and both g_1, g_2 are constant, then g is always a constant. In this case there are only two constant mappings.

Case 2. If $g = (g_1 \Box g_2) \oplus c$ and g_1 is constant, while g_2 is not constant then there are $e_2 - 2$ nonconstant functions in $S(\hat{\phi}_2, X)$ and hence also in $S(\hat{\phi}, X)$. The case where g_2 is constant, while g_1 is not constant is treated exactly as before.

Case 3. Finally, if neither g_1 nor g_2 is constant then there are at most $2(e_1 - 2)(e_2 - 2)$ possibilities.

Combining the three cases above, we obtain

$$|S(\hat{\phi}, X)| \le 2 + (e_1 - 2) + (e_2 - 2) + 2(e_1 - 2)(e_2 - 2) = \frac{1}{2}[(2e_1 - 3)(2e_2 - 3) + 3].$$

Now use the induction hypothesis to conclude that

$$|S(\hat{\phi}, X)| \le \frac{1}{2}\left[5^{\mathcal{N}(\phi_1,X)} 5^{\mathcal{N}(\phi_2,X)} + 3\right] = \frac{5^{\mathcal{N}(\phi,X)} + 3}{2}.$$

This completes the proof of the theorem.

The *element distinctness* problem can be formalized by defining the boolean function $f \in \mathcal{B}_n$, for $n = 2m \log m$, which on an input $s_1, \ldots, s_m$ of m binary strings each of length $2 \log m$, outputs 1 exactly if all s_i are distinct. Applying Nechiporuk's technique yields a lower bound of $\Omega(n^2/\log^2 n)$ for formula size for element distinctness (see [BS90] for a related lower bound on branching program size for element distinctness).

2.4 Monotonic Real Circuits

In this section, we prove the Haken–Cook [HC99, Hak95] exponential lower bound for monotonic real circuits of bounded fan-in which compute the Broken Mosquito Screen problem BMS. Additionally, we present Johannsen's extension [Joh98] to monotonic real circuits of an older lower bound for *st*-connectivity on monotonic boolean circuits due to Karchmer and Wigderson [Kar93].

2.4.1 Broken Mosquito Screen

It is customary to represent directed graphs G on n vertices by encoding the adjacency matrix of G by a vector $\langle a_{i,j} : 1 \leq i, j \leq n \rangle$ of 0s and 1s, where $a_{i,j} = 1$ exactly if there is a directed edge from i to j. Undirected graphs are similarly encoded by $\binom{n}{2}$ bit vectors whose appropriate bit is 1 exactly if $\{i, j\}$ is an undirected edge of G. In this section, all graphs are undirected and have no loops or multiple edges.

Definition 2.4.1. *A binary monotonic real gate is a non-decreasing real function $f : \mathbf{R}^2 \to \mathbf{R}$; i.e., for all x, y, x', y'*

$$x \leq x', y \leq y' \Rightarrow f(x, y) \leq f(x', y').$$

A unary monotonic real gate is a non-decreasing unary function $f : \mathbf{R} \to \mathbf{R}$; i.e., for all x, x'

$$x \leq x' \Rightarrow f(x) \leq f(x').$$

Definition 2.4.2 (Broken Mosquito Screen BMS$_m$). *Let $m \geq 3$, and $n = m^2 - 2$. An undirected graph G on n vertices is called good (bad) if the n vertices can be partitioned into $m - 1$ rows of m elements together with one row of $m - 2$ elements, such that each row is a clique (anti-clique); i.e., if x, y are distinct vertices in the same row, then $\{x, y\}$ is an edge (is not an edge) of G. BMS$_m$ is the problem of separating good from bad graphs, for graphs whose vertex set has size $n = m^2 - 2$.*

Fix parameter $n = m^2 - 2$ throughout. Let G_0 (B_0) designate the collection of *minimal* good (*maximal* bad) graphs. Thus G_0 contains only those edges between distinct vertices in the same row, while B_0 contains all edges between vertices in different columns. Note that there are graphs which are neither good nor bad, so that BMS_s is not a language recognition problem, but rather a problem of separating two disjoint NP sets. The Karchmer–Wigderson theorem states that this separation cannot be done by simple circuits, where simple means monotonic NC^1. More generally, one can ask, by analogy with the situation in computability theory, whether there exist disjoint NP sets having no separating set in P.[1] The principal theorem of this section shows that when G_0, B_0 both NP disjoint sets, cannot be separated by any polynomial size monotonic circuit family. Intuitively, this can be expressed as: *monotonic* NP *is different from monotonic* P. This separation was first proved by A.A. Razborov [Raz87b] in the case of monotonic boolean circuits, for the related NP-complete problem CLIQUE. A. Haken later discovered the *bottleneck counting* argument described below, again for monotonic boolean circuits. The extension of Haken's argument to monotonic real circuits was given by S.A. Cook at about the same time as P. Pudlák discovered an extension of Razborov's original argument to the case of monotonic real circuits.

Lemma 2.4.1. *For $m \geq 3$, no graph is both good and bad.*

Proof. Suppose that G is good. Let $V_1, \ldots, V_m$ be cliques where V_i, for $1 \leq i < m$, has m elements, and V_m has $m-2$ elements. If G is bad then let $W_1, \ldots, W_m$ be anti-cliques such that W_i, for $1 \leq i < m$, has m elements and W_m has $m-2$ elements. Each W_i contains one element from each of $V_1, \ldots, V_m$ for $i = 1, \ldots, m-1$. But W_{m-1} cannot contain m elements, since all $m-2$ elements of V_m have already been placed in $W_1, \ldots, W_{m-2}$. So G cannot be a bad graph.

It is even simpler to see that no graph can simultaneously belong to G_0 and B_0; namely, since a minimal good graph has $(m-1) \cdot \binom{m}{2} + \binom{m-2}{2} = O(m^3)$ edges, while a maximal bad graph has $\binom{m^2-2}{2} - (m-1) \cdot \binom{m}{2} - \binom{m-2}{2}$ or $\Omega(m^4)$ edges.[2]

Minimal good (maximal bad) graphs can be constructed by placing elements from $1, \ldots, n = m^2 - 2$ into an $m \times m$ grid, where the last row (column) contains only $m-2$ elements, and then by adding all edges between all vertices in the same row (not in the same row). The number of such unordered partitions is given by the multinomial coefficient

$$\binom{m^2-2}{m \quad m \quad \cdots \quad m \quad m-2}$$

[1] It follows from Gödel's Incompleteness Theorem that in Peano arithmetic, the set A of provable formulas is disjoint from the set B of refutable formulas (F is refutable if $\neg F$ is provable), both sets are recursively enumerable, but there is no recursive separating set.

[2] Remark due to T. Altenkirch (personal communication).

which equals

$$\frac{(m^2-2)!}{(m!)^{m-1}(m-2)!(m-1)!}.$$

Another way to visualize the latter quantity is to pass from ordered partitions (in this case $n!$ permutations) to unordered partitions by dividing by the factor $(m!)^{m-1}(m-2)!(m-1)!$, where m elements in each of the full rows can be permuted among elements of that row, $m-2$ elements of the last row can be permuted among themselves, and the $m-1$ full rows can be permuted row-wise. Later counting arguments use a similar mental picture: one repeatedly places elements into an $m \times m$ grid, while ensuring a particular property, and divides out by an appropriate factor to pass from ordered to unordered partitions. These considerations prove the following lemma.

Lemma 2.4.2. *The number $|G_0|$ of minimal good graphs is given by*

$$\frac{\binom{m^2-2}{m\ \ m\ \ \cdots\ \ m\ \ m-2}}{(m-1)!}$$

which equals

$$\frac{(m^2-2)!}{(m!)^{m-1}(m-2)!(m-1)!}.$$

Note that there is an obvious 1-1 correspondence between minimal good graphs and maximal bad graphs. Namely, if $g \in G_0$, then the complementary graph $b \in B_0$, where for distinct i, j, $\{i, j\}$ is an edge of g exactly when $\{i, j\}$ is not an edge of b.

Suppose that a monotonic real circuit C of size s separates B_0 from G_0, where $C_1, C_2, \ldots, C_s$ is an enumeration of the inputs and gates of C in a *topologically sorted* manner; i.e., if there is a directed path from gate C_i to gate C_j, then $i < j$. Suppose that we had an injection $\phi : (G_0 \cup B_0) \to \{1, \ldots, s\}$. Then clearly, $s \geq |G_0| + |B_0| = 2|G_0|$. The idea of the Haken–Cook lower bound for BMS_s is to define a map $\phi : A \subseteq (G_0 \cup B_0) \to \{1, \ldots, s\}$, such that A is a large subset of the collection of minimal good and maximal bad graphs, and though ϕ is not a 1-1 mapping, it is the case that few graphs are mapped to the same gate. If ϕ maps at most K graphs to a gate, then clearly $s \geq \frac{|A|}{K}$. This type of argument has been termed *bottleneck counting* by A. Haken, and is similar to Haken's earlier exponential lower bound on resolution proofs of the pigeonhole principle, which we present in a later chapter. In defining the map ϕ, a graph is mapped to the first gate, where *substantial* progress is made in *classifying* the graph as good or bad. The amount of progress is measured in terms of number of input bits which matter at the gate, a kind of *sensitivity* measure.

Definition 2.4.3 (Fence). *Suppose that g is a good graph, $G_t \subseteq G_0$, and $B_t \subseteq B_0$. A fence F around g at gate C at time t is the conjunction*

$$x_{i_1,j_1}, \wedge \cdots \wedge x_{i_q j_q}$$

of edges in g such that

1. $F(g) = 1$
2. $(\forall b \in B_t)[C(b) < C(g) \to F(b) = 0]$.

Suppose that b is a bad graph. A fence F around b at gate C at time t is a disjunction

$$x_{i_1,j_1}, \vee \cdots \vee x_{i_q,j_q}$$

such that

1. $F(b) = 0$,
2. $(\forall g \in G_t)[C(b) < C(g) \to F(g) = 1]$.

Since boolean variable $x_{i,j} = 1$ exactly if there is an edge $\{i,j\}$ in graph g, a fence for a good graph g is a collection of edges from g which distinguish g from bad graphs at time t just as well as the gate C does. Note that a fence F separates g from *all* graphs in B_t, the collection of bad graphs at time t. Though the fence F is good for g, it may not be good for a different good graph g'. Since $x_{i,j} = 1$ exactly if there is an edge $\{i,j\}$ in graph b, a fence is a collection of edges, none of which are in b, such that F distinguishes b from good graphs at time t just as well as gate C does.

Remark 2.4.1. Note that in either case in the above definition, a fence consists of variables, i.e., positive literals. By monotonicity of the gates, if g is a good graph, then the conjunction of two fences for g at gate C at time t is also a fence, and similarly if b is a bad graph, then the disjunction of two fences for b at gate C at time t is also a fence. This is the only place where monotonicity of the gates is used.

Let $k = \frac{m}{2}$. A fence is said to be *long* if it contains at least $\frac{k}{2}$ literals; otherwise the fence is called *short.*

Example 2.4.1.

1. Suppose that $\{i,j\}$ is an edge of g. Then $x_{i,j}$ is a fence around g at input gate $x_{i,j}$ at all times.
2. Suppose that $\{i,j\}$ is not an edge of g. Then $1 \equiv \bigwedge$ is a fence around g at gate x_{ij} at all times.
3. Suppose that $\{i,j\}$ is an edge of bad graph b. Then $0 \equiv \bigvee$ is a fence around b at gate $x_{i,j}$ at all times.
4. Suppose that $\{i,j\}$ is not an edge of b. Then $x_{i,j}$ is a fence around b at the gate x_{ij} at all times.
5. The conjunction of all edges of good graph g is a fence around g at the output gate at all times. The disjunction of all nonedges of b is a fence around b at the output gate at all times.

Progress in distinguishing between good and bad graphs is made at gates having long minimal fences. Note that a gate C may simultaneously have a fence F for good graph g and a fence F' for bad graph b.

Definition 2.4.4 (Map ϕ). *The mapping $\phi :\subseteq G_0 \cup B_0 \to \{1, \ldots, s\}$, whose domain is a (sufficiently large) subset A of the collection of all minimal good and maximal bad graphs, is defined as follows. Suppose that G_t and B_t are defined. Find the first gate C for which there is a graph $h \in G_t \cup B_t$ having a long minimal fence around h at time t. Map h to gate C and set*

$$G_{t+1} = G_t - \{h\}$$
$$B_{t+1} = B_t - \{h\} \ .$$

If no gate has a long minimal fence at time t around a graph in $G_t \cup B_t$, then terminate the construction of ϕ, and letting $T = t$, define

$$A = (G_0 \cup B_0) - (G_T \cup B_T).$$

Lemma 2.4.3. $|A| \geq |G_0|$.

Proof. If ϕ maps all of G_0, or all of B_0, then the assertion of the lemma holds. Otherwise, $A = (G_0 \cup B_0) - (G_T \cup B_T)$ and choose $g \in G_T$, $b \in B_T$. Since the construction of ϕ could not be continued at time T, every fence around b at time T at any gate, in particular output gate C_s, has length strictly less than $k/2$. Consider the fence F around b at the output gate C_s:

$$x_{i_1,j_1} \vee \ldots \vee x_{i_q,j_q}$$

where $q < k/2$. By definition of the fence F

$$(\forall g \in G_T)[C_s(b) < C_s(g) \to F(g) = 1].$$

As C_s is the output gate, $C_s(b) = 0$ and $C_s(g) = 1$, so that $F(g) = 1$. Thus all graphs in G_T have an edge from the fence $x_{i_1,j_1} \vee \ldots \vee x_{i_q,j_q}$. How many minimal good graphs in G_0 can contain an edge among $\{i_1, j_1\}, \ldots, \{i_q, j_q\}$?
CLAIM.

$$\Pr[\{i_1, j_1\} \text{ is an edge of } g | g \in G_0] < \frac{1}{m}.$$

Proof of Claim. If i_1 is in a particular row, then there are $m - 1$ positions from which to choose in placing j_1 in the same row, as opposed to $m^2 - 3$ possible positions for an arbitrary graph $g \in G_0$. Thus the number of ways of constructing minimal good graphs containing a fixed edge $\{i_1, j_1\}$ divided by the number of minimal good graphs is

$$\frac{m-1}{m^2-3} < \frac{1}{m}.$$

This concludes the proof of the claim.

Thus

$$\Pr[(x_{i_1,j_1} \vee \ldots \vee x_{i_q,j_q}) \restriction_g = 1 | g \in G_0] < \frac{q}{m} < \frac{m}{4m} = \frac{1}{4}.$$

It follows that $|G_T| \leq \frac{|G_0|}{4}$, and so the fraction $\frac{3}{4}$ of all minimal good graphs in G_0 are mapped by ϕ. By a symmetrical argument, $|B_T| \leq \frac{|B_0|}{4}$, and we have already noted that $|G_0| = |B_0|$. Thus ϕ maps

$$\frac{3}{4}|G_0| + \frac{3}{4}|B_0| \geq |G_0|$$

elements before time T is reached.

Lemma 2.4.4 (Main Lemma for Broken Mosquito Screen). *Let r be the greatest even number not exceeding $\sqrt{\frac{m}{2}}$. Then the number of graphs mapped by ϕ to any one gate is at most*

$$\frac{2(km)^{r/2}(m^2-m)^{r/2}(m^2-r-2)!}{(m!)^{m-1}(m-2)!(m-1)!}.$$

Proof. Let C be a gate of a minimal circuit $C_1, \ldots, C_s$ which solves BMS_m. We show that the number of ordered partitions leading to good graphs which ϕ maps to C is bounded by $(km)^{r/2}(m^2-m)^{r/2}(m^2-r-2)!$. The number of good graphs is then obtained by quotienting out by the factor $(m!)^{m-1}(m-2)!(m-1)!$ to pass from ordered to unordered partitions. By symmetry, one has the same upper bound for bad graphs which map to the same gate, hence providing the bound stated in the lemma.

Suppose that $g \in G_t$, and that ϕ maps g to gate C for the first time at time t_0. List the maximal bad graphs of B_{t_0} as $\{b_1, b_2, \ldots, b_z\}$ where $C(b_1) \leq C(b_2) \leq \ldots \leq C(b_z)$. Let L be the least index for which $C(b_{L+1}) = 1$, so for $1 \leq i \leq L$, $C(b_i) < 1$. Input gates have short fences, and graphs are mapped to gates by ϕ only if the corresponding fence is long. Thus C cannot be an input gate. We discuss the case that C is a binary gate with inputs from gates D, E; the case where C is a unary gate is similar and simpler. Now, each $b_1, \ldots, b_L$ is not yet mapped, so each has a short minimal fence at D, E. Taking the disjunction of these fences[3] each b_ℓ has fence F_ℓ

$$x^{(\ell)}_{i_1,j_1} \vee \ldots \vee x^{(\ell)}_{i_k,j_k}$$

about C for bad graph b_ℓ at time t_0 for $\ell = 1, \ldots, L$. Suppose that ϕ maps h at time $t > t_0$ to gate C. By the definition of fence F_ℓ for bad b_ℓ, $C(b_\ell) = 0$ and

$$(\forall g\ G_t)[C(b_\ell) < C(g) \rightarrow F_\ell(g) = 1].$$

Thus any $h \in G_{t_0}$ later mapped to C *must* contain an edge from each fence $F_1, \ldots, F_L$, and the size of each F_ℓ is at most k.

[3] As mentioned in Remark 2.4.1, this is the only point in the proof requiring that gates are monotonic.

How many ways can one choose $\frac{k}{2}$ distinct edges, the ℓ-th edge from

$$\{i_1^{(\ell)}, j_1^{(\ell)}\}, \ldots, \{i_k^{(\ell)}, j_k^{(\ell)}\}$$

and produce a good graph? An upper bound for this quantity is an upper bound for the number of good graphs mapped to C.

Having chosen ℓ edges, pick a distinct edge from

$$\{i_1^{(\ell+1)}, j_1^{(\ell+1)}\}, \ldots, \{i_k^{(\ell+1)}, j_k^{(\ell+1)}\}. \tag{2.3}$$

Case 1. There is a vertex i belonging to one of the ℓ edges already chosen, and one can find a vertex j for which $\{i, j\}$ occurs among (2.3).

In this case, there are at most k edges in (2.3) among which to find vertex j, and there are at most $m-1$ positions in the row of i in which to place vertex j. Thus the number of possibilities is bounded above by $k \cdot (m-1)$.

Case 2. There is an edge $\{i, j\}$ occurring in (2.3), where neither i nor j occurs in an edge already chosen.

In this case, there are at most k edges in (2.3) among which to choose the pair i, j, once chosen there are m rows in which to place i, j, and at most $m(m-1)$ ordering positions for i, j in each respective row. This produces an upper bound of $(mk) \cdot (m(m-1))$ possibilities.

Note that the quantity $k(m-1)$ from case 1 is bounded by mk, hence by $m(m-1)$. Thus a larger upper bound in always obtained by applying case 2.

Recall that r is defined to be the largest even integer bounded by $\sqrt{\frac{m}{2}}$. Let N be the smallest number of vertices for which the complete graph or clique on N vertices has at least $\frac{k}{2}$ edges, so

$$\begin{aligned} \binom{N}{2} &\geq \frac{k}{2} \\ N(N-1) &\geq k \\ N^2 &\geq k + N \geq \frac{m}{2} \end{aligned}$$

and hence $N \geq \sqrt{\frac{m}{2}}$. Thus, to produce $\frac{k}{2}$ distinct edges it suffices to choose $r/2$ pairs of edges (all with distinct endpoints) according to the provisions of Case 2. Now complete the ordered partition by drawing from an urn. This gives $(m^2 - r - 2)!$ remaining possibilities. Applying case 2 a total of $r/2$ times yields upper bound of

$$(km)^{r/2} \cdot (m(m-1))^{r/2}$$

and hence overall at most

$$(km)^{r/2}(m^2 - m)^{r/2} \cdot (m^2 - r - z)!$$

ways of choosing a good graph mapped to gate C.

As mentioned at the beginning of the proof, by symmetry we have the same bound for bad graphs mapped to gate C, yielding the factor 2 in the statement of the lemma. Dividing out by a factor to pass from ordered partitions to unordered partitions, we obtain the desired upper bound of

$$\frac{2(km)^{r/2}(m^2-m)^{r/2}(m^2-r-2)!}{(m!)^{m-1}(m-2)!(m-1)!}$$

many graphs mapped to same gate.

Theorem 2.4.1 ([HC99]). *A fan-in 2 monotonic real circuit separating good from bad instances of* BMS_m, *for* $m \geq 5$, *must have at least*

$$\frac{1.8^{\lfloor\sqrt{m/2}\rfloor}}{2}$$

many gates.

Proof. We defined the mapping

$$\phi : A \subseteq (G_0 \cup B_0) \to \{1,\ldots,s\}$$

and have shown an upper bound of

$$M = \frac{2(km)^{r/2}(m^2-m)^{r/2}(m^2-r-2)!}{(m!)^{m-1}(m-2)!(m-1)!}$$

many graphs mapped to the same gate, and so

$$\begin{aligned} s &\geq \frac{|A|}{M} \geq \frac{|G_0|}{M} \\ &\geq \frac{(m^2-2)!}{2(km)^{r/2}(m^2-m)^{r/2}(m^2-r-2)!}. \end{aligned}$$

Note that

- The last m^2-r-2 factors cancel.
- The remaining last $r/2$ factors are greater than or equal to 1 since $r \leq \sqrt{m/2}$, so $2r^2 \leq m$, hence $r-1 \leq m$, and so $\frac{m^2-2-i}{m^2-m} \geq 1$, for $r/2 \leq i \leq r-1$. Thus

  $$\frac{(m^2-2-(\frac{r}{2}))\cdots(m^2-2-(r-1))}{(m^2-m)\cdots(m^2-m)} \geq 1$$

  and $r \leq \sqrt{m/2}, 2r^2 \leq m$ so $r+1 \leq m$.
- Consider first $r/2$ terms

  $$\frac{(m^2-2)(m^2-3)\cdots(m^2-2-(\frac{r}{2}-1))}{2(km)\cdot(km)\cdots(km)}.$$

  The product of these terms is greater than or equal to

  $$\frac{(m^2-2-(\frac{r}{2}-1))^{r/2}}{2(km)^{r/2}}. \tag{2.4}$$

Finally note that for $m \geq 5$, the numerator of (2.4) is at least $0.9m^2$. Thus

$$s \geq \frac{1}{2}\left(\frac{0.9m^2}{0.5m^2}\right)^{r/2} = \frac{1.8^{\lfloor\sqrt{m/8}\rfloor}}{2}.$$

Corollary 2.4.1. *There is a polynomial time computable language $L \subseteq \{0,1\}^*$ with monotonic fan-in 2 real circuit size lower bound of $2^{\Omega(N^{1/8})}$.*

Proof. Define L_n to be the collection of graphs on a vertex set of size $n = m^2 - 2$ having at least $\binom{m^2-2}{2} - (m-1)\cdot\binom{m}{2} - \binom{m-2}{2}$ distinct edges, and let $L = \cup_{n\geq 0} L_n$. For fixed $n = m^2 - 2$ with m sufficiently large, $B_0 \subseteq L_n$, and $L_n \cap G_0 = \emptyset$. If a monotonic real circuit correctly computes L_n, then it solves the broken mosquito screen problem BMS_m. Now $n = m^2 - 2$ and the bit vector encoding g has length $N = \binom{n}{2} = \Theta(m^4)$ and so $m = \Theta(N^{1/4})$. The circuit size lower bound now follows.

This particular circuit lower bound will later yield a lower bound for proof size of cutting plane proofs.

In [Juk99, Juk97], S. Jukna combined the bottleneck argument just given with M. Sipser's notion of finite limit, in order extend lower bounds to large, but not arbitrary fan-in circuits. Following [Juk99], a *norm* $\mu : 2^n \to \mathbf{N}$ is an integer-valued mapping from the power set 2^n of $\{1,\ldots,n\}$, which is monotonic under set-theoretic inclusion; i.e., for $S \subseteq T \subseteq \{1,\ldots,n\}$ we have $\mu(S) \leq \mu(T)$. The *Deviation* $\lambda : \mathbf{N} \to \{0,\ldots,n\}$ is defined by $\lambda(t) = \max\{|S| : \mu(S) \leq t\}$. The *defect* c_μ of μ is $\max\{\mu(\{e\}) : e \in \{1,\ldots,n\}\}$; i.e., the largest norm of a single bit position. Note that $\mu(S) \leq c_\mu \cdot |S|$ and $|S| \leq \lambda(\mu(S))$. For boolean function $f : 2^n \to \{0,1\}$, input $x \in \{0,1\}^n$, value $\epsilon \in \{0,1\}$, and set $A \subseteq f^{-1}(\epsilon)$ define $\min_b(x, A, \mu)$ to be

$$\min\{\Pr[x \in A, (\forall i \in S)(x_i \equiv 1 \oplus \epsilon)] : S \subseteq \{1,\ldots,n\}, \mu(S) \leq b\}$$

and $\max_b(x, A, \mu)$ to be

$$\max\{\Pr[x \in A, (\forall i \in S)(x_i \equiv \epsilon)] : S \subseteq \{1,\ldots,n\}, \mu(S) \geq a\}.$$

The ϵ-degree will not be defined here, but depends on an associated monotonic graph related to the given boolean function f. In an application, $\mu(S)$ might be the number of vertices incident to at least one edge from S. For a pair (μ_0, μ_1) of (not necessarily distinct) norms, and $\epsilon \in \{0,1\}$, define

$$F_F^\epsilon(x, a, b, d) = \frac{\min_b(x, X^\epsilon, \mu_{1\oplus\epsilon}]}{(d \cdot \lambda(bc))^a \cdot \max_a(x, X^\epsilon, \mu_\epsilon)}$$

where X^ϵ is the set of all inputs from $f^{-1}(\epsilon)$, such that c, λ are the defect and deviation of $\mu_{1\oplus\epsilon}$.

The main result of [Juk99] is the following. Let f be a monotonic boolean function on n variables, and let C be a monotonic real circuit computing

f. Then for any random inputs x, y, any norms μ_0, μ_1, and any integers $1 \leq a, b \leq n$,

$$size(C) \geq \min\left\{F_f^0(x, a, b, d_1), F_f^1(y, b, a, d_0)\right\}$$

where d_ϵ is the maximum ϵ-degree of a gate in C.

From this result, Jukna is able to obtain lower bounds for unbounded fan-in monotonic boolean circuits, and for monotonic real circuits, of large fan-in.

S. Jukna's proof uses M. Sipser's notion of *finite limit* [Sip85b] defined as follows. An input $x = x_1 \cdots x_n \in \{0,1\}^n$ is a *k-limit* for a set A of inputs, if for every subset $S \subseteq \{1, \ldots, n\}$ of cardinality k, x coincides with at least one vector from A; i.e., $(\exists a \in A)(\forall i \in S)(x_i = a_i)$. If f is a boolean function on n variables, and $f(x) = 0$, where x is a k-limit of $f^{-1}(1)$, then the input x is a *hard* instance for any circuit computing f, since the computation cannot depend just on k bits of x. See [Juk99, Juk97] for details.

2.4.2 Monotonic Real Circuits Are Powerful

In the case of broken mosquito screen from the last section and *st-connectivity*, a lower bound argument for monotonic, fan-in 2, boolean circuits was later extended to the same lower bound for monotonic, fan-in 2, real circuits. How do monotonic real circuits compare in strength to (non-monotonic) boolean circuits? Could both models be equivalent? These questions are answered by the following results of A. Rosenbloom [Ros97].

Define the *k-th slice* S_k^n of $\{0,1\}^n$ to be the set of length n bit vectors of weight k, and say that boolean function $f : 2^n \to \{0,1\}$ is a *k-slice function* for $0 \leq k \leq n$, denoted $f \in \mathcal{S}_k^n$, if $f(x)$ equals 0 (1) for boolean inputs x of weight less than k (more than k). A *slice family* $\mathcal{F} = \{f_n : n \in \mathbf{N}\}$ is a set of slice functions, where $f_n \in \mathcal{B}_n$.

Lemma 2.4.5 ([Ros97]). *Let $\mathcal{F} = \{f_n : n \in \mathbf{N}\}$ be a slice family. Then there exist logarithmic depth, linear size, monotonic real circuits computing $\mathcal{F}$.*

Proof. Suppose the function $f_n \in \mathcal{F}$ is a k-slice function. Define two monotonic functions p, m[4] (for *plus*, *minus*) by

$$p(x) = 2^n \cdot \sum_{i=0}^{n-1} x_i + \sum_{i=0}^{n-1} x_i \cdot 2^i = |x|_1 \cdot 2^n + x$$

$$m(x) = 2^n \cdot \sum_{i=0}^{n-1} x_i - \sum_{i=0}^{n-1} x_i \cdot 2^i = |x|_1 \cdot 2^n - x$$

[4] In [Ros97], these functions were called $Order_+$, $Order_-$; moreover, 2^{n+1} was used in place of 2^n, where the latter occurs in our definition.

where we recall that the *weight* $|x|_1$ of $x = x_{n-1} \cdots x_0$ is $\sum_{i=0}^{n-1} x_i$, and we identify $x \in \{0,1\}^*$ with the integer having binary representation x. It is clear that p, m can both be computed by logarithmic depth, linear size monotonic real circuits. Define the partial ordering on $\{0,1\}^n$ by $x \preceq y \Leftrightarrow p(x) \leq p(y) \wedge m(x) \leq m(y)$.

CLAIM 1. If $x, y \in S_k^n$ and $x \neq y$, then $p(x) \leq p(y) \Leftrightarrow m(x) \geq m(y)$.

Proof of Claim 1. Since x, y have the same weight,

$$\begin{aligned} p(x) \leq p(y) &\Leftrightarrow 2^n \cdot |x|_1 + x \leq 2^n \cdot |y|_1 + y \\ &\Leftrightarrow x \leq y \\ &\Leftrightarrow 2^n \cdot |y|_1 - y \leq 2^n \cdot |x|_1 - x \\ &\Leftrightarrow m(x) \geq m(y). \end{aligned}$$

CLAIM 2. If $x \in S_k^n$ and $y \in S_\ell^n$, with $k < \ell$, then $p(x) < p(y)$ and $m(x) < m(y)$.

Proof of Claim 2. Since $x, y \in \{0,1\}^*$, the corresponding integers x, y are less than 2^n; thus if $|x|_1 = k < \ell = |y|_1$, then $p(x) = 2^n \cdot k + x < 2^n \cdot \ell + y$. Similarly, $m(x) < m(y)$.

From the previous two claims, it follows that the partial ordering $\preceq$ orders the slices by $S_0^n \prec S_1^n \prec \cdots \prec S_n^n$, but that elements x, y in the same slice are incomparable. Recall that f_n is the n-th slice function in the family $\mathcal{F}$. For arguments $a, b < 2^n \cdot n + 2^n = (n+1) \cdot 2^n < 2^{2n}$, if

$$(\exists x < 2^n)(a = 2^n \cdot |x|_1 + x \wedge b = 2^n \cdot |y|_1 + y)$$

then define

$$\psi(a,b) = \begin{cases} 0 & \text{if } x \in S_i^n \text{ and } i < k \\ 1 & \text{if } x \in S_i^n \text{ and } i > k \\ f_n(x) & \text{if } x \in S_k^n. \end{cases}$$

It follows that ψ is a monotonic real *partial* gate with the property that $\psi(p(x), m(x))$ computes the slice function $f_n(x)$. Extending ψ to a monotonic real gate $g \in \mathcal{B}_{2n}$ then establishes the lemma.

Corollary 2.2.1 together with Lemma 2.4.5 immediately establish the following.

Corollary 2.4.2 ([Ros97]). *Monotonic real circuits of fan-in* 2 *are exponentially more powerful than boolean circuits of fan-in* 2 *on slice families.*

2.4.3 *st*-Connectivity

In this subsection, we present a lower bound result of M. Karchmer and A. Wigderson [KW90], which separates monotonic NC^1 from monotonic AC^1.

The problem considered is *st-connectivity*: Given an undirected[5] graph G with two distinguished vertices s, t, determine whether there exists a path from s to t. This can be formulated as a problem st-CONN$_n$, where the (adjacency matrix of the) graph is encoded by boolean variables $x_{i,j}$, $1 \leq i, j \leq n$, with $x_{i,j} = 1$ iff there is an edge from i to j. Identify vertex s (t) with 0 ($n+1$), and assume that s, t are connected to every internal node $1, \ldots, n$. Then an undirected graph G is st-connected iff there is a path from s to t passing through more than one internal node.

Following [KW90], in this section we'll show that:

1. Monotonic boolean formula depth for st-CONN is $\Omega(\log^2 n)$. (Note that monotonic boolean formula depth, in the absence of any simultaneous size constraint, equals depth for monotonic circuits of bounded fan-in.)
2. Monotonic boolean formula size for st-CONN$_n$ is $n^{\Omega(\log n)}$.

By I. Wegener's monotonic version of Spira's theorem (see Theorem 1.10.2), both lower bounds are equivalent. The upper and lower bound coincide, since by repeated squaring, st-CONN$_n$ belongs to monotonic AC^1, hence to monotonic NC^2, and the upper bound for monotonic formula depth (size) is $O(\log^2 n)$ ($O(n^{\log n})$). In a later chapter, we'll see that the Karchmer–Wigderson monotonic circuit lower bound translates into a proof length lower bound for (tree-like) resolution proofs, first observed by P. Clote and A. Setzer [CS98]. This result was generalized to (tree-like) cutting plane proofs by J. Johannsen [Joh98], who used the work of Krajíček [Kra98] relating 2-person real games, real communication complexity and monotonic real circuit depth (plus the proof from [BS90]) to extend the lower bound for st-connectivity to monotonic real circuits. The proof given below redefines the notion of approximator from [BS90], uses Spira's $1/3 - 2/3$-trick, and incorporates Johannsen's idea of sorting (but without using 2-person games), and then follow the exposition in [BS90].

The lower bound for parity in Theorem 2.6.2 is obtained by repeatedly using random restrictions, beginning near the inputs, and working towards the output (bottom-up, when circuits are displayed with output at top). In contrast, the argument of this section begins near the output and works towards the inputs (top-down).

For a leveled circuit, the *level* (*depth*) of a gate g in a circuit is the path length (maximum path length) from g to an output (input) gate. An *ℓ-path* graph P is a path of length $\ell + 1$ between two distinguished nodes s and t. A *cut* or *cut graph* C is a graph, whose edges are those from two disjoint complete subgraphs, one containing distinguished vertex s and the other containing distinguished vertex t. Identify ℓ-paths with ℓ-tuples $v_1, \ldots, v_\ell$ of in-

[5] Undirected st-connectivity is trivially reducible to directed st-connectivity. The lower bound of [KW90] is $\Omega(\log^2 n)$ for undirected connectivity, while the upper bound for directed connectivity is $O(\log^2 n)$, thus for *monotonic* bounded fan-in boolean circuits, the undirected and directed versions of st-connectivity have the same complexity.

ternal nodes, where $v_i \in \{1, \ldots, n\}$. Similarly, identify cuts with subsets of the set $\{1, \ldots, n\}$ of internal nodes. Let $Path_n^\ell$ Cut_n denote the collection of all ℓ-paths (cuts) with vertex set $V = \{1, \ldots, n\}$; thus $|Path_n^\ell| = n^\ell$ and $|Cut_n| = 2^n$.

Fix integers $2 \leq \ell \leq n$, $\mathcal{P} \subseteq Path_n^\ell$ and $\mathcal{C} \subseteq Cut_n$. A monotonic real function $f : 2^{n^2} \to \mathbf{R}$ is an *$(n, \ell, \mathcal{C}, \alpha)$-path acceptor* if

$$|\{p \in Path_n^\ell : (\forall c \in \mathcal{C})(f(p) > f(c))\}| \geq \alpha \cdot n^\ell$$

i.e., there exists $\mathcal{P}' \subseteq Path_n^\ell$ of path density α such that $f(p) > f(c)$ holds for all $p \in \mathcal{P}'$, $c \in \mathcal{C}$. Similarly, f is an *$(n, \mathcal{P}, \beta)$-cut rejector* if

$$|\{c \in Cut_n : (\forall p \in \mathcal{P})(f(p) > f(c))\}| \geq \beta \cdot 2^n$$

i.e., there exists $\mathcal{C}' \subseteq Cut_n$ of cut density β such that $f(p) > f(c)$ holds for all $p \in \mathcal{P}$, $c \in \mathcal{C}'$. Finally, f is an *$(n, \ell, \mathcal{P}, \mathcal{C}, \alpha, \beta)$-approximator* of st-CONN$_n$, if there exist $\mathcal{P}' \subseteq \mathcal{P}$, $\mathcal{C}' \subseteq \mathcal{C}$ with $|\mathcal{P}'| \geq \alpha n^\ell$, $|\mathcal{C}'| \geq \beta 2^n$ and $(\forall p \in \mathcal{P}', c \in \mathcal{C}')[f(p) > f(c)]$. When $\mathcal{P} = Path_n^\ell$ and $\mathcal{C} = Cut_n$, then we call f simply an *(n, ℓ, α, β)-approximator*; i.e., f separates a set of paths of density α from a set of cuts of density β.

In the argument below, we identify a gate of a circuit with the subcircuit determined by that gate, as well as with the function computed by the gate. From context, there should be no confusion. Note that if f is an $(n, \ell, \mathcal{P}, \mathcal{C}, \alpha, \beta)$-approximator, and $\alpha' \leq \alpha$, $\beta' \leq \beta$, then f is an $(n, \ell, \mathcal{P}, \mathcal{C}, \alpha', \beta')$-approximator.

Let f be a monotonic boolean (or real) function acting on graphs $G = (V, E)$ having vertex set V and edge set $E \subseteq V \times V$. For onto map $\rho : V \to V'$ and $G = (V, E)$, define $G_\rho = (V', E')$, where $\{i, j\} \in E'$ iff $(\exists u, v \in V)(\rho(u) = i \wedge \rho(v) = j \wedge \{i, j\} \in E)$. For $H = (V, E_H)$, define the induced monotonic function f_ρ acting on graphs with vertex set V' by

$$f_\rho(H) = \max_G \{f(G) : \text{ vertex set of } G \text{ is } V \text{ and } G_\rho = H\}.$$

By monotonicity of f, $f_\rho(H) = f(G)$, where $G = (V, E_G)$ is uniquely specified by

$$\{i, j\} \in E_G \Leftrightarrow \rho(i) = \rho(j) \vee \{\rho(i), \rho(j)\} \in E_H.$$

If $\mathcal{G}$ a collection of graphs whose vertex set is V and ρ maps V onto V', then we write $\mathcal{G}_\rho$ for $\{G_\rho : G \in \mathcal{G}\}$. Later in this section, we'll use this notation for sets of paths and cuts, i.e., $\mathcal{P}_\rho$, $\mathcal{C}_\rho$.

The probability distribution $\mathcal{R}_n^k$ is defined by randomly choosing a k element subset $W \subseteq V - \{s, t\}$, randomly partitioning W into disjoint S, T, and defining $\rho : V \to V - W$ by sending $S \mapsto s$, $T \mapsto t$, with ρ the identity on elements of $V - W$.

Lemma 2.4.6. *Suppose that monotonic real gate g is an $(n, \ell, \mathcal{P}, \mathcal{C}, \alpha, \beta)$-approximator. If g_1, g_2 are the input gates to g, then either g_1 or g_2 is an $(n, \ell, \mathcal{P}, \mathcal{C}, \alpha/2, \beta/2)$-approximator.*

Proof. By hypothesis, there exists $\mathcal{P}' \subseteq \mathcal{P}$ ($\mathcal{C}' \subseteq \mathcal{C}$) of density α (β) such that $g(p) > g(c)$, for all $p \in \mathcal{P}'$, $c \in \mathcal{C}'$. Sort the paths in $\mathcal{P}'$ by $p \prec p'$ iff $g_1(p) < g_1(p')$ or ($g_1(p) = g_1(p')$ and p precedes p' in lexicographic order), and similarly sort the cuts in $\mathcal{C}'$. Let p_0 (c_0) be the median of $\mathcal{P}'$ ($\mathcal{C}'$) when sorted.

Case 1. $g_1(p_0) > g_1(c_0)$.
Define $\mathcal{P}'' = \{p \in \mathcal{P}' : g_1(p) \geq g_1(p_0)\}$ and $\mathcal{C}'' = \{c \in \mathcal{C}' : g_1(c) \leq g_1(c_0)\}$. Then

$$(\forall p \in \mathcal{P}'', c \in \mathcal{C}'')[g_1(p) \geq g_1(p_0) > g_1(c_0) \geq g_1(c)]$$

and

$$|\mathcal{P}''| \geq |\mathcal{P}'|/2 \geq \alpha/2 \cdot n^\ell$$

and

$$|\mathcal{C}''| \geq |\mathcal{C}'|/2 \geq \beta/2 \cdot 2^n.$$

Thus g_1 is an $(n, \ell, \mathcal{P}, \mathcal{C}, \alpha/2, \beta/2)$-approximator.
Case 2. $g_1(p_0) \leq g_1(c_0)$.
Define $\mathcal{P}'' = \{p \in \mathcal{P}' : g_1(p) \leq g_1(p_0)\}$ and $\mathcal{C}'' = \{c \in \mathcal{C}' : g_1(c) \geq g_1(c_0)\}$. Then

$$(\forall p \in \mathcal{P}'', c \in \mathcal{C}'')[g_1(p) \leq g_1(p_0) \leq g_1(c_0) \leq g_1(c)]$$

and $|\mathcal{P}''| \geq \frac{\alpha}{2} n^\ell$, $|\mathcal{C}''| \geq \frac{\beta}{2} 2^n$. It must be that $g_2(p) > g_2(c)$ for all $p \in \mathcal{P}''$ and $c \in \mathcal{C}''$; otherwise, for some $p \in \mathcal{P}''$, $c \in \mathcal{C}''$, we have $g_2(p) \leq g_2(c)$, and $g_1(p) \leq g_1(c)$, so by monotonicity of g, $g(p) \leq g(c)$, contradicting our hypothesis. Thus g_2 is an $(n, \ell, \mathcal{P}, \mathcal{C}, \alpha/2, \beta/2)$-approximator.[6]

Lemma 2.4.7. *Fix $2 \leq \ell \leq n$, $\mathcal{P} \subseteq Path_n^\ell$, $\mathcal{C} \subseteq Cut_n$, and let C be a monotonic real circuit computing st-*CONN$_n$. *Suppose that f is a gate of C which is an $(n, \ell, \mathcal{P}, \mathcal{C}, \alpha, \beta)$-approximator. Then for $k \geq 0$, there exists a gate g of C at k levels below f, which is an $(n, \ell, \mathcal{P}, \mathcal{C}, \alpha \cdot 2^{-k}, \beta \cdot 2^{-k})$-approximator.*

Proof. By induction on k using the previous lemma.

Corollary 2.4.3. *Fix $2 \leq \ell \leq n$, and let C be a monotonic real circuit computing st-*CONN$_n$. *Then for $k \geq 0$, there exists a gate g of C at level k, which is an $(n, \ell, 2^{-k}, 2^{-k})$-approximator.*

Proof. The output gate of C at level 0 is an $(n, \ell, 1, 1)$-approximator, thus establishing the base case. Now apply the previous lemma.

Lemma 2.4.8. *Let $2 \leq \ell \leq n$, $\mathcal{P} \subseteq Path_n^\ell$, $\mathcal{C} \subseteq Cut_n$, and f be an $(n, \ell, \mathcal{P}, \mathcal{C}, \alpha, \beta)$-approximator, where*

$$\frac{100\ell}{\alpha} \leq k \leq \frac{n}{100\ell},$$

[6] By considering the definition of median when $|\mathcal{P}'|$, $|\mathcal{C}'|$ are even and odd, we actually have $|\mathcal{P}''| \geq \lceil |\mathcal{P}'|/2 \rceil$, and $|\mathcal{C}''| \geq \lceil |\mathcal{C}'|/2 \rceil$.

$\beta \geq 2^{-\frac{n}{100k}}$ and $k \leq n/4$. Then with non-zero probability there exists a restriction $\rho \in \mathcal{R}_n^k$ such that f_ρ is an $(n-k, \frac{\ell}{2}, \mathcal{P}_\rho, \mathcal{C}_\rho, \frac{\sqrt{\alpha}}{2}, \frac{\beta k}{2n})$-approximator.

The lemma follows immediately from the Lemmas 2.4.9 and 2.4.10, proved later. With this lemma, we can prove the lower bound.

Theorem 2.4.2 ([KW90], [Joh98]). *st-CONN$_n$ has depth $\Omega(\log^2 n)$ for monotonic real circuits with fan-in 2.*

Proof. To obtain a contradiction, assume that there exists a monotonic real circuit C of depth at most $\frac{1}{100}\log^2 n$ which computes st-CONN$_n$. By adding artificial gates, without loss of generality we can assume that C has the structure of a complete binary tree with depth $\frac{1}{100}\log^2 n$. Fix $\ell = n^{1/4}$. Since C computes st-CONN$_n$, it is an $(n, \ell, 1, 1)$-approximator, hence *a fortiori* an $(n, \ell, \frac{1}{4}n^{-1/10}, 1)$-approximator. Divide C into $\frac{\log n}{10}$ many levels, each of depth $\frac{\log n}{10}$. By Lemma 2.4.7, with $k = \frac{\log n}{10}$, there exists a gate g at depth $\frac{\log n}{10}$ from the output gate, which is an $(n, \ell, \frac{1}{4}n^{-1/10} \cdot 2^{-\log n/10}, 2^{-\log n/10})$-approximator, hence an $(n, \ell, \frac{1}{4}n^{-1/5}, n^{-1/10})$-approximator. Apply Lemma 2.4.8 with $k = \sqrt{n}$. The hypotheses of this lemma are satisfied, since for sufficiently large n,

$$\frac{100n^{1/4}}{\frac{1}{4}n^{-1/10}} \leq \sqrt{n} \leq \frac{n}{100n^{1/4}},$$

$n^{-1/10} \geq 2^{\frac{-n}{100\sqrt{n}}}$ and $\sqrt{n} \leq n/4$. Thus there exists a restriction $\rho \in \mathcal{R}_n^k$ for which g_ρ is an $(n - \sqrt{n}, \ell/2, \frac{\sqrt{\frac{1}{4}n^{-1/5}}}{2}, \frac{n^{-1/10}\sqrt{n}}{2n})$-approximator, i.e., an $(n - \sqrt{n}, \ell/2, \frac{1}{4}n^{-1/10}, n^{-2})$-approximator. Note that the path density has been restored to the original density of $\frac{1}{4}n^{-1/10}$. Repeat this procedure $\frac{\log n}{10}$ times. Finally, we obtain a monotonic real circuit of depth 0 which is an $(n - \frac{\log n}{10}\sqrt{n}, \frac{\ell}{2^{\log n/10}}, \frac{1}{4}n^{-1/10}, (n^{-2})^{\log n/10})$-approximator, i.e., an $(n - \frac{\log n}{10}\sqrt{n}, n^{3/20}, \frac{1}{4}n^{-1/10}, n^{-\log n/5})$-approximator. A gate at level 0 must be a constant 0, 1 or variable $x_{i,j}$. The constant 0 (1) accepts no paths (rejects no cuts), so cannot separate paths from cuts. As well, there can be at most $\ell n^{\ell-2} \leq n^{\ell-1}$ many ℓ-path graphs containing a given edge $\{i,j\}$ (ℓ choices for coordinate set equal to i, once fixed, the following coordinate must be j, and then there remain $\ell - 2$ free choices), so the path density of the gate $x_{i,j}$ is at most $1/n < \frac{1}{4}n^{-1/10}$, which is a contradiction.

Lemma 2.4.9. *If f is an $(n, \ell, \mathcal{C}, \alpha)$-path acceptor, and $\rho \in \mathcal{R}_n^k$ is chosen randomly, where $\frac{100\ell}{\alpha} \leq k \leq \frac{n}{100\ell}$, then*

$$\Pr[f_\rho \textit{ is } (n-k, \ell/2, \mathcal{C}_\rho, \sqrt{\alpha}/2)\textit{-path acceptor } |\rho \in \mathcal{R}_n^k] \geq \frac{3}{4}.$$

Proof. Let f be an $(n, \ell, \mathcal{C}, \alpha)$-path acceptor. Let V be the set of n internal nodes (different than s, t) and

$$\mathcal{P} = \{p \in Path_n^\ell : (\forall c \in \mathcal{C})(f(p) > f(c))\}.$$

Identify an ℓ-path graph with an element in V^ℓ. Let $L = \{1, \ldots, \frac{\ell}{2}\}$ and $R = \{\frac{\ell}{2}+1, \ldots, \ell\}$. If $I \subseteq \{1, \ldots, \ell\}$, $\rho \in V^I$ and $\rho' \in V^\ell$, then ρ' is an extension of ρ if $\rho' \upharpoonright I = \rho$. Define

$$A_L = \{\rho \in V^L : \ \rho \text{ has at least } \tfrac{\alpha}{4} n^{\ell/2} \text{ extensions in } \mathcal{P}\}$$

and

$$A_R = \{\rho \in V^R : \ \rho \text{ has at least } \tfrac{\alpha}{4} n^{\ell/2} \text{ extensions in } \mathcal{P}\}.$$

Fact 2.4.1. Either A_L or A_R has at least $\sqrt{\frac{\alpha}{2}} n^{\ell/2}$ elements.

Proof of Fact. Every path in $\mathcal{P}$ is either an element of $A_L \times A_R$, an extension of a subpath in $\overline{A}_L = V^L - A_L$, or an extension of a subpath in $\overline{A}_R$. There are $|A_L| \cdot |A_R|$ many paths in $A_L \times A_R$. There are at most $|V^L| = n^{\ell/2}$ subpaths in $\overline{A}_L$, and by definition of A_L, each one has less than $\frac{\alpha}{4} n^{\ell/2}$ extensions. Thus the number of extensions of subpaths in $\overline{A}_L$ is less than $n^{\ell/2} \cdot \frac{\alpha}{4} n^{\ell/2} = \frac{\alpha}{4} n^\ell$. Finally, by analogous reasoning, there are less than $\frac{\alpha}{4} n^\ell$ extensions of subpaths in $\overline{A}_R$. Thus altogether, we have

$$|\mathcal{P}| < |A_L| \cdot |A_R| + \frac{\alpha}{2} n^\ell.$$

On the other hand, because f is an $(n, \ell, \mathcal{C}, \alpha)$-path acceptor, $|\mathcal{P}| \geq \alpha \cdot n^\ell$, so that

$$\alpha n^\ell \leq |\mathcal{P}| < |A_L| \cdot |A_R| + \frac{\alpha}{2} n^\ell$$

hence $|A_L| \cdot |A_R| > \frac{\alpha}{2} n^\ell$, so that either $|A_L|$ or $|A_R|$ must be greater than $\sqrt{\frac{\alpha}{2}} n^{\ell/2}$. This concludes the proof of Fact 2.4.1.

Without loss of generality assume that $|A_L| > \sqrt{\frac{\alpha}{2}} n^{\ell/2}$. For $\rho \in \mathcal{R}_n^k$ random, set

$$\mathcal{P}'_\rho = \{p_L \in A_L : p_L \in (V - W)^L \wedge (\exists p_R \in T^R)(p_L p_R \in \mathcal{P})\}$$

where $p_L p_R$ is the path obtained by concatenating the left subpath p_L with right subpath p_R, $W \subseteq V - \{s, t\}$ and T are given by the choice of ρ.

By definition, $f_\rho(p_L) = f(G)$, where $G = (V, E_G)$ and $\{i, j\} \in E_G$ iff $\rho(i) = \rho(j) \vee \{\rho(i), \rho(j)\}$ is an edge of p_L; similarly for $f_\rho(c)$, for $c \in \mathcal{C}_\rho$.

If $p_L \in V^L$ has extension $p \in \mathcal{P}$, then

$$(\forall c \in \mathcal{C})(f(p) > f(c))$$

so by definition of f_ρ

$$(\forall c \in \mathcal{C}_\rho)(f_\rho(p_L) > f_\rho(c)).$$

In particular, if $p_L \in \mathcal{P}'$, then there exists $p_R \in T^R$, $p_L p_R \in \mathcal{P}$, so $(\forall c \in \mathcal{C}_\rho)(f_\rho(p_L) > f_\rho(c))$, hence it suffices to show that $\mathcal{P}'_\rho$ is large, with high probability. Now

$$\Pr[p_L \notin \mathcal{P}'_\rho] = \Pr[p_L \notin (V-W)^L \vee \neg(\exists p_R \in T^R)(p_L p_R \in \mathcal{P})].$$

Let us consider each of these terms.

Fact 2.4.2. $\Pr[p_L \notin (V-W)^L] \le \frac{\ell}{2} \cdot \frac{k}{n}$.

Proof of Fact. $\Pr[p_L \notin (V-W)^L] = 1 - \Pr[p_L \in (V-W)^L]$, and the latter is the probability that all $\ell/2$ coordinates do not belong to W, which by independence is $(1-\frac{k}{n})^{\ell/2}$. Hence $\Pr[p_L \notin (V-W)^L] = 1-(1-\frac{k}{n})^{\ell/2} \le \frac{\ell}{2} \cdot \frac{k}{n}$.[7]

Fact 2.4.3. $\Pr[\neg(\exists p_R)(p_R \in T^R \wedge p_L p_R \in \mathcal{P})] \le 2^{-k/10} + (1-\frac{\alpha}{4})^{\frac{k}{2\ell}}$.

Proof of Fact. We have that

$$\Pr[\neg(\exists p_R)(p_R \in T^R \wedge p_L p_R \in \mathcal{P})] = (2.5) + (2.6)$$

where

$$\Pr[|T| < k/4 \wedge \neg(\exists p_R)(p_R \in T^R \wedge p_L p_R \in \mathcal{P})] \tag{2.5}$$

and

$$\Pr[|T| \ge k/4 \wedge \neg(\exists p_R)(p_R \in T^R \wedge p_L p_R \in \mathcal{P})] \tag{2.6}$$

Noting that $(2.5) \le \Pr[|T| < k/4]$, we upper bound each of these terms.

Subfact. $\Pr[|T| < k/4] \le 2^{-k/10}$.

Proof of Subfact. Since $|W| = k$ and T is a random subset of W,

$$\Pr[|T| < k/4] = \frac{\sum_{i=0}^{\lfloor k/4-1 \rfloor} \binom{k}{i}}{2^k}.$$

Now $\sum_{i=0}^{\lfloor k/4-1 \rfloor} \binom{k}{i} \le \frac{k}{4}\binom{k}{k/4}$ and Stirling's approximation yields that for all $n \ge 1$, $(\frac{n}{3})^n \le n! \le 3n(\frac{n}{3})^n$, so that

$$\binom{k}{k/4} \le \frac{3k(\frac{k}{3})^k}{(\frac{k}{12})^{k/4} \cdot (\frac{3k}{12})^{3k/4}} \le \frac{3k(\frac{k}{3})^k}{(\frac{k}{12})^k \cdot (3)^{3k/4}} \le 3k(\frac{4}{3^{3/4}})^k \le 3k(1.76)^k$$

and so $\Pr[|T| < k/4] \le \frac{3}{4}k^2(\frac{1.76}{2})^k \le 2^{-k/10}$, for sufficiently large k. □

Subfact. $\Pr[|T| \ge k/4 \wedge \neg(\exists p_R)(p_R \in T^R \wedge p_L p_R \in \mathcal{P})] < (1-\frac{\alpha}{4})^{\frac{k}{2\ell}}$.

Proof of Subfact. Assume that T is a random subset of W having at least $k/4$ elements, and that T' is a random subset of T with exactly $k/4$ elements.

[7] For the latter inequality, use a Taylor expansion, or let $g(x) = \frac{\ell}{2}x - 1 + (1-x)^{\ell/2}$, for $0 \le x \le 1$. Then $g'(x) = \frac{\ell}{2} - \frac{\ell}{2}(1-x)^{\frac{\ell}{2}-1} = \frac{\ell}{2}(1-(1-x)^{\frac{\ell}{2}-1}) \ge 0$.

This is the same as choosing a random $k/4$ element subset T' of V, among the $\binom{n}{k/4}$ many $k/4$ element subsets of V.

One manner of choosing T' is to choose random subpaths $p_R^1, p_R^2, \ldots, p_R^{\frac{k}{2\ell}}$, each having $\ell/2$ coordinates. The union of the vertices in these paths is a set having at most $\frac{\ell}{2} \cdot \frac{k}{2\ell} = \frac{k}{4}$ independently chosen elements. If necessary, add sufficiently many extra vertices to this set so that the size is $k/4$. Thus

$$\Pr[|T| \geq k/4 \wedge \neg(\exists p_R)(p_R \in T^R \wedge p_L p_R \in \mathcal{P})]$$

is at most

$$\Pr[|T| \geq k/4 \wedge \neg(\exists p_R^i)(p_R^i \in T'^R \wedge p_L p_R^i \in \mathcal{P})]$$

which is bounded above by

$$\begin{aligned}\Pr[\neg(\exists p_R^i)(p_R \in T'^R \wedge p_L p_R^i \in \mathcal{P})] &\leq \Pr[(\forall p_R^i \in T'^R)(p_L p_R^i \notin \mathcal{P})] \\ &= \prod_{i=1}^{k/2\ell} \Pr[p_L p_R^i \notin \mathcal{P}]\end{aligned}$$

where the last equality holds by independence. Now $p_L \in A_L$, hence p_L has at least $\frac{\alpha}{4} n^{\ell/2}$ extensions in $\mathcal{P}$ (by definition of A_L), and p_R^i is a random subpath in V^R, so

$$\Pr[p_L p_R \in \mathcal{P}] \geq \frac{\frac{\alpha}{4} n^{\ell/2}}{n^{\ell/2}} = \frac{\alpha}{4}$$

hence

$$\Pr[|T| \geq k/4 \wedge \neg(\exists p_R)(p_R \in T^R \wedge p_L p_R \in \mathcal{P})] \leq \prod_{i=1}^{k/2\ell} (1 - \frac{\alpha}{4}) = \left(1 - \frac{\alpha}{4}\right)^{\frac{k}{2\ell}}.$$

From the previous subfacts, we have that

$$\Pr[\neg(\exists p_R)(p_R \in T^R \wedge p_L p_R \in \mathcal{P})] \leq 2^{-k/10} + \left(1 - \frac{\alpha}{4}\right)^{\frac{k}{2\ell}}.$$

Now $2^{-k/10} \leq 1/400$ for $k \geq 10 \cdot \log_2 400 \approx 86$, and $\frac{k}{2\ell} \geq \frac{50}{\alpha}$, since by hypothesis of the lemma, $\frac{100\ell}{\alpha} \leq k$. Thus $(1 - \frac{\alpha}{4})^{\frac{k}{2\ell}} \leq (1 - \frac{\alpha}{4})^{\frac{50}{\alpha}} \leq 1/400$.[8]

Thus

$$\Pr[\neg(\exists p_R)(p_R \in T^R \wedge p_L p_R \in \mathcal{P})] \leq \frac{1}{400} + \frac{1}{400} = \frac{1}{200}.$$

On the other hand,

$$\Pr[p_L \notin (V - W)^L] \leq \frac{\ell}{2} \cdot \frac{k}{n} \leq \frac{1}{200}$$

[8] Note that this is equivalent to $-\alpha \ln 400 - 50 \ln(1 - \alpha/4) \geq 0$. Set $g(x) = -x \ln 400 - 50 \ln(1 - x/4)$ and note that $g'(x) = -\ln 400 + \frac{50}{4-x} \geq 0$ for $x \geq 0$, so that $g(x) \geq g(0) = 0$.

where the last inequality holds because of the lemma's assumption $k \le \frac{n}{100\ell}$. Thus

$$\Pr[p_L \notin \mathcal{P}'_\rho] \le \frac{1}{200} + \frac{1}{200} = \frac{1}{100}$$

and

$$\Pr[p_L \in \mathcal{P}'_\rho] \ge 1 - \frac{1}{100} = \frac{99}{100}.$$

Consider the random variable $\frac{|\mathcal{P}'_\rho|}{|A_L|}$. The previous inequality implies that $E[\frac{|\mathcal{P}'_\rho|}{|A_L|}] \ge \frac{99}{100}$; moreover we have $|\mathcal{P}'_\rho| \le |A_L|$.

Fact 2.4.4. Let X be a real random variable with $0 \le X \le 1$ and $E[X] \ge \frac{99}{100}$. Then $\Pr[X \ge \frac{24}{25}] \ge \frac{3}{4}$.

Proof of Fact. Let $u = \Pr[X \ge \frac{24}{25}]$. Since $k \le 1$ we have

$$\begin{aligned}\frac{99}{100} \le E[X] &= \sum_{k<24/25} k \cdot \Pr[X = k] + \sum_{k \ge 24/25} k \cdot \Pr[X = k] \\ &\le \frac{24}{25} \sum_{k<24/25} \Pr[X = k] + \sum_{k \ge 24/25} \Pr[X = k] \\ &= \frac{24}{25} \Pr[X < \frac{24}{25}] + \Pr[X \ge \frac{24}{25}] \\ &= \frac{24}{25}(1-u) + u = \frac{24}{25} + \frac{u}{25}.\end{aligned}$$

Thus $\frac{24}{25} + \frac{u}{25} \ge \frac{99}{100}$, and $u \ge 25(\frac{99}{100} - \frac{24}{25}) = \frac{3}{4}$.

Applying this last fact to the random variable $\frac{|\mathcal{P}'_\rho|}{|A_L|}$, we have that with probability at least $\frac{3}{4}$, $\frac{|\mathcal{P}'_\rho|}{|A_L|} \ge \frac{24}{25}$, so

$$|\mathcal{P}'_\rho| \ge \frac{24}{25}|A_L| \ge \frac{24}{25}\sqrt{\frac{\alpha}{2}} n^{\ell/2} > \frac{\sqrt{\alpha}}{2} n^{\ell/2}$$

where we have used the assumption that $|A_L| \ge \sqrt{\frac{\alpha}{2}} n^{\ell/2}$. Thus f_ρ is an $(n-k, \ell/2, \mathcal{C}_\rho, \sqrt{\alpha}/2)$-path acceptor with probability at least 3/4.

Lemma 2.4.10. *If $\mathcal{P} \subseteq Path_n^\ell$ and f is an $(n, \mathcal{P}, \beta)$-cut rejector, and $\rho \in \mathcal{R}_n^k$ is chosen randomly, where $\beta \ge 2^{-\frac{n}{100k}}$ and $k \le n/4$, then*

$$\Pr[f_\rho \textit{ is } (n-k, \mathcal{P}_\rho, \tfrac{\beta k}{2n})\textit{-cut rejector } | \rho \in \mathcal{R}_n^k] \ge \frac{3}{4}.$$

Proof. Let f be an $(n, \mathcal{P}, \beta)$-cut rejector, k such that $\beta \ge 2^{-\frac{n}{100k}}$, $k \le n/4$, and let V be the set of vertices $\{1, \ldots, n\}$ (distinct from s, t). Let

$$\mathcal{C} = \{c \in Cut_n : (\forall p \in \mathcal{P})(f(p) > f(c))\}.$$

A cut can be considered as an element of $\{0,1\}^V$, by mapping to 0 those vertices connected to s and to 1 those connected to t. If $I \subseteq V$ and $c \in \{0,1\}^I$, then define $d \in \{0,1\}^V$ to be an extension of c if $d \upharpoonright I = c$.

For $I \subseteq V$, $|I| = k$, we set

$$A(I) = \{c \in \{0,1\}^I : \text{ } c \text{ admits at least } \tfrac{\beta k}{2n} 2^{n-k} \text{ extensions in } \mathcal{C}\}.$$

Fact 2.4.5. If I is a random k-element subset of V, then the expected number of elements of $A(I)$ is at least $\frac{3}{4}2^k$.

Proof of Fact. Let $I_1, \ldots, I_{n/k}$ be a partition of V in k-element subsets. Every cut of $\mathcal{C}$ must either be an element of $A(I_1) \times \cdots \times A(I_{n/k})$, or an extension of an element of $\overline{A(I_i)} = \{0,1\}^V - A(I_i)$ for some $1 \leq i \leq n/k$. In the first case, there are $|A(I_1)| \cdots |A(I_{n/k})|$ many elements in $A(I_1) \times \cdots \times A(I_{n/k})$. In the second case, there are n/k choices of i, and at most 2^k choices of an element of $\{0,1\}^{I_i}$. Each element of $\overline{A(I_i)}$ has less than $\frac{\beta k}{2n} 2^{n-k}$ many extensions. Thus there are less than $\frac{n}{k}2^k \cdot \frac{\beta k}{2n}2^{n-k} = \frac{\beta}{2}2^n$ extensions of an element of $\overline{A(I_i)}$, for some i. Thus $|\mathcal{C}| < |A(I_i)| \cdots |A(I_{n/k}| + \frac{\beta}{2}2^n$, hence $|A(I_1)| \cdots |A(I_{n/k}| > \frac{\beta}{2}2^n$. Thus we have

$$\begin{aligned}
\frac{|A(I_1)| + \cdots + |A(I_{n/k})|}{n/k} &\geq (|A(I_1)| \cdots |A(I_{n/k}|)^{k/n} \\
&\quad \text{(by the arithmetic geometric inequality)} \\
&\geq \left(\frac{\beta}{2}2^n\right)^{k/n} \\
&\geq \left(\frac{2^{-\frac{n}{100k}}}{2}2^n\right)^{k/n} \qquad (\text{since } \beta \geq 2^{-\frac{n}{100k}}) \\
&= \frac{2^{-1/100}}{2^{n/k}}2^k \\
&\geq \frac{2^{-1/100}}{2^{1/4}}2^k \quad (\text{since } k \leq n/4) \\
&\geq \frac{3}{4}2^k \quad (\text{since } 2^{-1/100-1/4} \approx 0.83).
\end{aligned}$$

For every partition V into n/k many k-element subsets $I_1, \ldots, I_{n/k}$, we have

$$\frac{|A(I_1)| + \cdots + |A(I_{n/k})|}{n/k} \geq \frac{3}{4}2^k.$$

Compute as follows the average of these terms for all partitions of V into n/k many k-element sets. Let N be the number of such partitions, and π such that each k-element set I appears in π many of the N partitions. We have

$$\begin{aligned}\tfrac{N}{N}\cdot\tfrac{3}{4}2^k &\le \tfrac{1}{N}\sum\left\{\tfrac{|A(I_1)|+\cdots+|A(I_{n/k})|}{n/k} : I_1,\ldots,I_{n/k} \text{ partitions } V\right\}\\ &= \tfrac{1}{Nn/k}\sum_{I\subseteq V,|I|=k}\pi\cdot|A(I)|\\ &= \tfrac{\pi}{Nn/k}\sum_{I\subseteq V,|I|=k}|A(I)|.\end{aligned}$$

By counting the k-element subsets appearing in the N partitions in two different manners, we have $N\frac{n}{k}=\binom{n}{k}\pi$ hence $\frac{\pi}{Nn/k}=\frac{1}{\binom{n}{k}}$, and

$$\frac{3}{4}\cdot 2^k \le \frac{1}{\binom{n}{k}}\sum_{I\subseteq V,|I|=k}|A(I)|.$$

This concludes the proof of the fact. □

For restriction $\rho\in\mathcal{R}_n^k$, there is an associated cut $d\in\{0,1\}^W$ with $d\restriction_S = 0$ and $d\restriction_T=1$, so define

$$\mathcal{C'}_\rho = \{c\in\{0,1\}^{V-W} : cd\in\mathcal{C}\}$$

where cd is the cut, whose edges are those from c and d. Then

$$\begin{aligned} c\in\mathcal{C'}_\rho &\Rightarrow cd\in\mathcal{C} \quad \text{(by definition of } \mathcal{C'}_\rho)\\ &\Rightarrow (\forall p\in\mathcal{P})(f(p)>f(cd)) \quad \text{(by definition of } \mathcal{C})\\ &\Rightarrow (\forall p\in\mathcal{P}_\rho)(f_\rho(p)>f_\rho(c)) \quad \text{(by definition of } f_\rho)\end{aligned}$$

so that

$$|\{c\in\{0,1\}^{V-W} : (\forall p\in\mathcal{P}_\rho)(f_\rho(p)>f_\rho(c))\}| \ge |\mathcal{C'}_\rho|$$

and $\Pr[f_\rho$ is an $(n-k,\mathcal{P}_\rho,\frac{\beta k}{2n})$-cut rejector$]$ equals

$$\Pr\left[|\{c\in\{0,1\}^{V-W} : (\forall p\in\mathcal{P}_\rho)(f_\rho(p)>f_\rho(c))\}| \ge \frac{\beta k}{2n}\cdot 2^{n-k}\right]$$

which is at least

$$\Pr\left[|\mathcal{C'}_\rho|\ge\frac{\beta k}{2n}\cdot 2^{n-k}\right].$$

On the other hand, each element of $\mathcal{C'}_\rho$ has distinct extensions d in $\mathcal{C}$, thus $|\mathcal{C'}_\rho|\ge\frac{\beta k}{2n}\cdot 2^{n-k}$ provided that each $c\in\mathcal{C'}_\rho$ has at least $\frac{\beta k}{2n}$ many extensions in $\mathcal{C}$, which occurs if $c\in A(W)$ (by definition of $A(W)$). In Fact 2.4.5, we proved that $\Pr[c\in A(W)]\ge 3/4$, hence the assertion of the lemma holds.

Corollary 2.4.4 ([KW90]).

1. *Monotonic fan-in* 2 *boolean circuit depth for st*-CONN *is* $\Omega(\log^2 n)$.
2. *Monotonic boolean formula size for st*-CONN$_n$ *is* $n^{\Omega(\log n)}$.

It is currently an open problem whether the monotonic real analogue of Spira's Theorem 1.10.2 holds. Nevertheless, the monotonic real analogue of the lower bound in Corollary 2.4.4 (2) was first proved by J. Johannsen [Joh98], using J. Krajíček's [Kra98] 2-person real games and real communication complexity. Our proof simply substitutes Lemma 2.4.7 by the following lemma.

Lemma 2.4.11. *Fix $2 \leq \ell \leq n$, $\mathcal{P} \subseteq Path_n^\ell$, $\mathcal{C} \subseteq Cut_n$, and let C be a monotonic real circuit computing st-*CONN$_n$. *Suppose that f is a gate of C which is an $(n, \ell, \mathcal{P}, \mathcal{C}, \alpha, \beta)$-approximator, and that the number of leaves below f is m. Then there exists a gate g below f, such that:*

1. *the number m' of leaves below g satisfies $\lceil \frac{m}{3} \rceil \leq m' \leq \lfloor \frac{2m}{3} \rfloor$;*
2. *g is an $(n, \ell, \mathcal{P}, \mathcal{C}, \alpha/2, \beta/2)$-approximator.*

Proof. By hypothesis, there exist $\mathcal{P}' \subseteq \mathcal{P}$, $\mathcal{C}' \subseteq \mathcal{C}$ with $|\mathcal{P}'| \geq \alpha \cdot n^\ell$ and $|\mathcal{C}'| \geq \beta \cdot 2^n$, such that $(\forall p \in \mathcal{P}', c \in \mathcal{C}')(f(p) > f(c))$. As in the proof of Lemma 1.6.1, find gate g whose leafsize satisfies (1). Sort $\mathcal{P}$ by $p \prec p'$ iff $g(p) < g(p')$ or ($g(p) = g(p')$ and p precedes p' in lexicographic order), and let p_0 be the median of $\mathcal{P}$.

Case 1. $|\{c \in \mathcal{C}' : g(c) < g(p_0)\}| \geq \frac{1}{2} \cdot |\mathcal{C}'|$.

In this case, let $\mathcal{P}'' = \{p \in \mathcal{P}' : g(p) \geq g(p_0)\}$ and $\mathcal{C}'' = \{c \in \mathcal{C}' : g(c) < g(p_0)\}$. Then $(\forall p \in \mathcal{P}'', c \in \mathcal{C}'')(g(p) > g(c))$, $|\mathcal{P}''| \geq \alpha/2 \cdot n^\ell$ and $|\mathcal{C}''| \geq \beta/2 \cdot 2^n$, so g is an $(n, \ell, \mathcal{P}, \mathcal{C}, \alpha/2, \beta/2)$-approximator.

Case 2. $|\{c \in \mathcal{C}' : g(c) < g(p_0)\}| < \frac{1}{2} \cdot |\mathcal{C}'|$.

In this case, let $\mathcal{P}'' = \{p \in \mathcal{P}' : g(p) \leq g(p_0)\}$ and $\mathcal{C}'' = \{c \in \mathcal{C}' : g(c) \geq g(p_0)\}$. Suppose that $g(p_0) = m$, and define f_m to be the circuit obtained from f by replacing subcircuit g by real value m. We claim that $(\forall p \in \mathcal{P}'', c \in \mathcal{C}'')(f_m(p) > f_m(c))$. If not, then let $p \in \mathcal{P}'', c \in \mathcal{C}''$ and $f_m(p) \leq f_m(c)$. By definition of $\mathcal{C}''$, we have $g(c) \geq g(p_0) = m$, and hence by monotonicity $f(c) \geq f_m(c)$; as well, by definition of $\mathcal{P}''$, we have $m = g(p_0) \geq g(p)$, so by monotonicity of f, $f_m(p) \geq f(p)$. It follows that $f(c) \geq f(p)$, which contradicts our assumption that f separates $\mathcal{P}'$ from $\mathcal{C}'$. This establishes the claim, so that f_m separates $\mathcal{P}''$ from $\mathcal{C}''$. Since $|\mathcal{P}''| \geq \alpha/2 \cdot n^\ell$ and $|\mathcal{C}''| \geq \beta/2 \cdot 2^n$, it follows that f_m is an $(n, \ell, \mathcal{P}, \mathcal{C}, \alpha/2, \beta/2)$-approximator, whose circuit size is between 1/3 and 2/3 of the circuit size of f.

By induction on k using the previous lemma, we have the following.

Lemma 2.4.12. *Fix $2 \leq \ell \leq n$, $\mathcal{P} \subseteq Path_n^\ell$, $\mathcal{C} \subseteq Cut_n$, and let C be a monotonic real circuit computing st-*CONN$_n$. *Suppose that f is a gate of C which is an $(n, \ell, \mathcal{P}, \mathcal{C}, \alpha, \beta)$-approximator. Then for $k \geq 0$, there exists a gate g of C whose leafsize m' satisfies $m \cdot (\frac{1}{3})^k \leq m' \leq m \cdot (\frac{2}{3})^k$, where m is the leafsize of f, and which is an $(n, \ell, \mathcal{P}, \mathcal{C}, \alpha \cdot 2^{-k}, \beta \cdot 2^{-k})$-approximator.*

Define a *monotonic real formula* to be a monotonic real circuit with fan-in 2 and fan-out 1.

Corollary 2.4.5 ([Joh98]). *Monotonic real formula size of st-*CONN$_n$ *is $n^{\Omega(\log n)}$.*

Proof. Redo the depth lower bound, by replacing Lemma 2.4.7 by the previous lemma.

2.5 Parity and the Random Restriction Method

The size of the smallest circuit computing a boolean function depends very much on the type of gates belonging to the basis. For example, the parity function

$$\mathrm{PAR}_n(x_1, \ldots, x_n) = x_1 \oplus \cdots \oplus x_n.$$

can be computed by a circuit of depth 1 and size $O(n)$ with the arbitrary fan-in gate $\bigoplus$. Over a basis containing the fan-in 2 gate $\oplus$, clearly PAR_n can be computed in size n and depth $O(\log n)$. In this section, we investigate the size and depth of unbounded fan-in boolean circuits for the problem of PAR_n.

Concerning the De Morgan basis $\{0, 1, \wedge, \vee, \neg\}$, it is easy to see that every DNF formula for $x_1 \oplus \cdots \oplus x_n$ must be a disjunction of at least 2^{n-1} conjunctive terms. This is because every conjunction must mention each variable $x_1, \ldots, x_n$. Indeed, if a conjunction C does not mention x_{i_0}, then there are truth assignments σ, σ' which differ only on x_{i_0}, so that $\sigma \models C$ iff $\sigma' \models C$; however $\sigma(x_1) \oplus \cdots \oplus \sigma(x_n)$ is clearly different from $\sigma'(x_1) \oplus \cdots \oplus \sigma'(x_n)$, a contradiction. Since every conjunction mentions each variable $x_1, \ldots, x_n$ and there are 2^{n-1} different satisfying assignments for $x_1 \oplus \cdots \oplus x_n$, there must be exactly 2^{n-1} conjunctive terms. Taking complements then yields the dual result that there must be exactly 2^{n-1} disjunctive clauses in every CNF formula for PAR_n. The size of a depth 2 circuit computing f with OR (AND) gate at the output gate is at least the number of terms (clauses) in a minimal DNF (CNF) formula representing f. Hence we have proved the following.

Theorem 2.5.1 ([Lup61b]). 2^{n-1} *conjuncts (respectively, disjuncts) are necessary and sufficient in order to represent* PAR_n *in conjunctive (respectively, disjunctive) normal form; moreover, size*(PAR_n) *equals* $2^{n-1} + 1$ *over the De Morgan basis* $\neg, \vee, \wedge, 0, 1$ *with unbounded fan-in.*

Let MULT be the function which accepts as inputs two sequences of bits each of length n and outputs their product in binary form. The following result shows that multiplication is at least as hard as parity.

Theorem 2.5.2 ([FSS84]). *Parity is* AC^0 *reducible to multiplication.*

Proof. Let n be fixed and put $k = \lceil \log n \rceil$. Suppose that $x_0, \ldots, x_{n-1}$ are the variables for which we want to construct a bounded depth polynomial size circuit, computing PAR_n, using $\vee, \wedge, \neg$, MULT-gates. Define the numbers a, b as follows:

$$a = \sum_{i=0}^{n-1} x_i 2^{ki}, \quad b = \sum_{i=0}^{n-1} 2^{ki}.$$

Clearly the $2kn$ bits in the above binary representation can be computed easily from the variables $x_0, \ldots, x_{n-1}$. The product ab is given by the formula

$$ab = \sum_{i=0}^{2n-2} c_i 2^{ki},$$

where $c_i = \sum_{j=0}^{i} a_j b_{n-1-j} = \sum_{j=0}^{i} x_j$. In particular,

$$c_{n-1} = \sum_{i=0}^{n-1} x_i$$

and the low order bit of c_{n-1} is equal to $\text{PAR}_n(x_0, \ldots, x_{n-1})$.

Lupanov's Theorem 2.5.1 is generalized in the following sections to show that parity requires exponentially large depth d unbounded fan-in formulas as well. This lower bound result is sufficiently important to warrant illustration from different viewpoints.

A *restriction* is a mapping $\rho : \{x_1, \ldots, x_n\} \to \{0, 1, *\}$. A restriction ρ is identified with the *partial truth assignment* which on each x_i takes the value $0, 1, x_i$ depending on whether $\rho(x_i)$ is $0, 1, *$. The restriction ρ is said to be *set* on the elements of the domain of the associated partial truth assignment; i.e., on those x_i for which $\rho \in \{0, 1\}$. Restrictions ρ, γ are said to be disjoint if the domains of the associated partial truth assignments are disjoint, i.e., $\{x_i : \rho(x_i) \neq *\} \cap \{x_i : \gamma(x_i) \neq *\} = \emptyset$. The composition of disjoint restrictions ρ, γ is written $\rho\gamma$. If $f \in \mathcal{B}_n$ and ρ is a restriction, then $f \restriction_\rho$ is the induced boolean function with domain $\{x_i : \rho(x_i) = *\}$ and value $f(\rho(x_1), \ldots, \rho(x_n))$. A boolean function is in $\Sigma_i^{S,t}$ if it is computable by a leveled boolean circuit having at most $i+1$ levels with an OR as the output (top) gate, where the leaves are labeled by the inputs $x_1, \ldots, x_n$ or their negations $\overline{x}_1, \ldots, \overline{x}_n$, where there are at most S internal (non-input) gates, all of which are ANDs or ORs, and the fanout of the bottom gates (next to inputs) is at most t. The definition of $\Pi_i^{S,t}$ is identical with the exception that the output gate is an AND. For the example $f(x_1, x_2, x_3) = x_1 \oplus x_2 \oplus x_3$ from Figure 1.1, it is clear that $f \in \Sigma_1^{5,3}$, or in the terminology of the previous chapter, f can be written in 3-DNF with size 5. More generally, $x_1 \oplus \cdots \oplus x_n \in \Sigma_1^{2^{n-1}+1,n}$. In the literature, a boolean function f is sometimes called t-*open* [resp t-*closed*] if f can be written in t-DNF or $\Sigma_1^{S,t}$ (t-CNF or $\Pi_1^{S,t}$) form for some S.

A *minterm* C of a boolean function $f : \{0,1\}^n \to \{0,1\}$ is a minimal length term or conjunction $\alpha_1 \wedge \cdots \wedge \alpha_m$ of literals among $x_1, \overline{x}_1, \ldots, x_n, \overline{x}_n$ with the property that for every truth assignment σ of the variables $x_1, \ldots, x_n$, it is the case that

$$(\alpha_1 \wedge \cdots \wedge \alpha_m) \restriction_\sigma = 1 \Rightarrow f \restriction_\sigma = 1.$$

Recall that a minterm C of f can be identified with a restriction $\rho : \{x_1, \ldots, x_n\} \to \{0, 1, *\}$, defined by

$$\rho(x_i) = \begin{cases} 0 & \text{if } \overline{x}_i \text{ appears in } C \\ 1 & \text{if } x_i \text{ appears in } C \\ * & \text{otherwise.} \end{cases}$$

Also, the restriction ρ can be identified in a natural manner with the partial truth assignment $\widetilde{\rho}: \{0,1\}^{n-m} \to \{0,1\}$, defined by setting $\widetilde{\rho}(a_1, \ldots, a_{n-m})$ to be ρ applied to the n-tuple obtained by replacing the i-th star '$*$' by a_i. Depending on the context, by minterm we may mean a conjunction of literals, a restriction, or a partial truth assignment. With this identification, a minterm of f can be defined as a partial truth assignment π for which $f|_\pi \equiv 1$ and no partial truth assignment π' properly contained in π satisfies $f|_{\pi'} \equiv 1$. The *size of minterm* π is the number of 0s and 1s assigned. In other words, the size of a minterm is the size of the domain of the minterm, considered as a partial truth assignment.

As an example, let $f(x_1, x_2, x_3)$ be the boolean function with formula $(x_1 \vee x_2) \wedge \overline{x}_3$. Then the disjunctive normal form of f is $(x_1 \wedge x_2 \wedge \overline{x}_3) \vee (x_1 \wedge \overline{x}_2 \wedge \overline{x}_3) \wedge (\overline{x}_1 \wedge x_2 \wedge \overline{x}_3)$. Written as a sum of products, this is $x_1 x_2 \overline{x}_3 + x_1 \overline{x}_2 \overline{x}_3 + \overline{x}_1 x_2 \overline{x}_3$.

x_1	x_2	x_3	$(x_1 \vee x_2) \wedge \overline{x}_3$
0	0	0	0
0	0	1	0
0	1	0	1
0	1	1	0
1	0	0	1
1	0	1	0
1	1	0	1
1	1	1	0

The function f has only two minterms, π_1 and π_2, where

$$\pi_1 = 1 * 0$$
$$\pi_2 = * 1 0$$

or, in other words,

$$\pi_1(x_1) = 1,\ \pi_1(x_2) = x_2,\ \pi_1(x_3) = 0$$
$$\pi_2(x_1) = x_1,\ \pi_2(x_2) = 1,\ \pi_2(x_3) = 0.$$

In this case, both π_1 and π_2 have size 2. Note that there are 2^{n-1} minterms of $x_1 \oplus \cdots \oplus x_n$, each of size n. Let $\min(f)$ denote the size of the largest minterm of f. Clearly, a boolean function f is in t-DNF (i.e., is t-open) iff $\min(f) \leq t$.

Dual to the notion of minterm is that of *maxterm*. A maxterm B of a boolean function $f : \{0,1\}^n \to \{0,1\}$ is a maximal length clause or disjunction $\alpha_1 \vee \cdots \vee \alpha_m$ of literals among $x_1, \overline{x}_1, \ldots, x_n, \overline{x}_n$ with the property that for every truth assignment σ of the variables $x_1, \ldots, x_n$, it is the case that

$$f \restriction_\sigma = 1 \Rightarrow (\alpha_1 \wedge \cdots \wedge \alpha_m) \restriction_\sigma = 1.$$

We leave the proof of the following observation as an exercise.

Fact 2.5.1. Let f be a boolean function on n variables. Then $\alpha_1 \wedge \cdots \wedge \alpha_m$ is a minterm of f iff $\overline{\alpha}_1 \vee \cdots \vee \overline{\alpha}_m$ is a maxterm of $1 - f$.

After these preliminaries, as a warm-up to the parity lower bound, we present a simpler result due to M. Sipser [Sip85a] that no infinite parity function is constant depth computable over infinite boolean circuits. This discussion assumes knowledge of set theory (cardinals, axiom of choice).

We begin by defining the infinite analogues of boolean circuit, restriction, etc. An infinite boolean circuit is built up inductively from the constants $0, 1$ and boolean variables $x_0, x_1, x_2, \ldots$ by negation, and both finite and countably infinite fan-in $\bigvee$ and $\bigwedge$ gates. An infinite fan-in $\bigvee$ ($\bigwedge$) gate outputs 1 iff one (all) of its inputs are 1. An infinite restriction is a mapping

$$\rho : \{x_1, x_1, x_2, \ldots\} \to \{0, 1, *\}.$$

If ρ is an infinite restriction (identified with a partial truth assignment) and C is an infinite boolean circuit, then $C \upharpoonright_\rho$, the restriction of C by ρ, is defined by induction on the formation of C: $x_i \upharpoonright_\rho$ is x_i if $\rho(x_i) = *$, else $x_i \upharpoonright_\rho$ is $\rho(x_i)$; $(\neg F) \upharpoonright_\rho$ is $\neg(F \upharpoonright_\rho)$; $(\bigvee_{i \in I} F_i) \upharpoonright_\rho$ is $\bigvee_{i \in I}(F_i \upharpoonright_\rho)$; $(\bigwedge_{i \in I} F_i) \upharpoonright_\rho$ is $\bigwedge_{i \in I}(F_i \upharpoonright_\rho)$. A function $f : \{0,1\}^\omega \to \{0,1\}$ is an infinite parity function if whenever s, t are ω-sequences of 0s and 1s which differ on exactly one bit, then $f(s) \neq f(t)$; i.e.,

$$(\exists! n)[s(n) \neq t(n)] \to f(s) \neq f(t).$$

Proposition 2.5.1. *Assuming the axiom of choice, there exist infinite parity functions.*

Proof. Using the axiom of choice, well-order the continuum $\{0,1\}^\omega$ in a sequence $\{s_\alpha : \alpha < \kappa\}$. For each $\alpha < \kappa$, we define disjoint subsets G_α, B_α of $\{0,1\}^\omega$ such that $\cup_{\alpha<\kappa} G_\alpha \cap \cup_{\alpha<\kappa} B_\alpha = \emptyset$, $\cup_{\alpha<\kappa}(G_\alpha \cup B_\alpha) = \{0,1\}^\omega$, and for all $s, t \in \{0,1\}^\omega$, if $s \in G_\alpha$ (B_α) and t differs from s in exactly one bit, then $t \in B_\alpha$ (G_α). It then follows that $f : \{0,1\}^\omega \to \{0,1\}$ defined by $f(s) = 1$ iff $s \in G_\alpha$ is an infinite parity function.

Assume that G_α, B_α have already been defined for $\alpha < \beta$. Let G_β (B_β) be $\emptyset$ if $s_\beta \in \cup_{\alpha<\beta}(G_\alpha \cup B_\alpha)$. Otherwise let G_β consist of s_β together with all those t which differ from s_β in an even number of bits, while B_β consists of all t which differ from s_β in an odd number of bits. This concludes the construction.

Theorem 2.5.3 ([Sip85a]). *Let f be an infinite parity function. Then there is no finite depth infinite boolean circuit computing f.*

Proof. Suppose that C is a finite depth infinite boolean circuit computing f. Without loss of generality we may assume that C is a leveled circuit having alternating levels of $\bigwedge$'s and $\bigvee$'s, with $\bigvee$'s at the lowest level (closest to the input), and inputs $x_0, \neg x_0, x_1, \neg x_1, \ldots$. The proof outline is as follows:

1. Define a restriction ρ which removes all infinite fan-in $\bigvee$ gates at the lowest level; $C \upharpoonright_\rho$ has only finite fan-in $\bigvee$'s remaining at the lowest level.
2. Define a restriction π which converts all infinite fan-in $\bigwedge$ gates at the next level into finite fan-in $\bigwedge$'s; $C \upharpoonright_{\rho\pi}$ then has finite fan-in $\bigwedge$'s of $\bigvee$'s at the lowest two levels.
3. Convert the $\bigwedge$'s of $\bigvee$'s into $\bigvee$'s of $\bigwedge$'s, and merge the second and third lowest $\bigvee$ levels resulting in a depth $d-1$ circuit computing $f \upharpoonright_{\rho\pi}$.
4. The function $(1 - f \upharpoonright_{\rho\pi})$ is an infinite parity function and by the previous step is computable by a depth $d-1$ infinite circuit with $\bigvee$'s at the lowest level (moreover, the $\bigvee$'s have finite fan-in). Repeat the previous two steps $d-3$ many times to produce a depth 2 finite fan-in circuit $\widetilde{C}$ computing an infinite parity function g. But $\widetilde{C}$ depends on only finitely many inputs, whereas g depends on infinitely many inputs. Contradiction.

In defining the restrictions ρ, π for steps (1) and (2), care must be taken to ensure that ρ, π have the value $*$ for infinitely many variables.

Step 1. "Kill all infinite fan-in OR gates." Let $D_1, D_2, \ldots$ be a list of all infinite fan-in $\bigvee$ gates at the lowest level of circuit C. Let $Var(D_i)$ denote the set of variables mentioned in D_i. Define ρ by stages, $\emptyset = \rho_0 \subset \rho_1 \subset \rho_2 \subset \cdots$ with $\rho = \cup_s \rho_s$. At each stage, ρ_s is initially set to be ρ_{s-1}, and then extended by setting a new variable to $*$ and another to either 0 or 1 (to make sure that $f \upharpoonright_\rho$ depends on infinitely many variables).

> **Stage s.** Find the least two distinct indices i_0, i_1 of variables x_{i_0}, x_{i_1} from $Var(D_s)$ which are not yet in the domain of ρ_s, and set x_{i_0} to $*$ and x_{i_1} to 1 if x_{i_1} appears in D_s, and to 0 if $\overline{x_{i_1}}$ appears in D_s.

This completes step (1). Now let D be the resulting simplification of circuit $C \upharpoonright_\rho$: remove all $\bigvee$ gates at the lowest level which contain a literal set to 1, and among the remaining finite fan-in $\bigvee$ gates at the lowest level, remove all literals set to 0.

Step 2. "Convert all AND gates to finite fan-in." Let $E_1, E_2, \ldots$ enumerate all infinite fan-in $\bigwedge$ gates at the next lowest level. Define $\pi_0 \subset \pi_1 \subset \pi_2 \subset \cdots$ with $\pi = \cup_s \pi_s$ by stages, where π_0 is set to ρ where all $(x_i, *)$ have been removed. Initialize π_s to be π_{s-1}, and in the course of stage s, finitely many finite extensions to π_s may be added.

> **Stage s.** Let i_0 be the least index of a variable x_{i_0} from $Var(D)$ not in the domain of π_s. Extend π_s by adding $(x_{i_0}, *)$. Unlike step (1), one cannot kill all E_s by setting all the variables in one of its input $\bigvee$ gates so that the $\bigvee$ outputs 0, since it could happen that all the input $\bigvee$ gates to E_s share a particular variable which has already been starred by ρ_s. Let K be the set of variables starred by π_s. Note that K is finite. For each truth assignment σ of the variables in K, there being $2^{||K||}$ many, finitely extend π_s so that E_s is decided. For

different truth assignments, different truth values for E_s might be forced. Note that if $E_s \upharpoonright_{\pi_s}$ already evaluates to 0 or 1, then nothing need be done. Stage s involves substages $1 \leq t \leq 2^{||K||}$.

Substage t. Let σ be the t-th truth assignment to the variables of K. If $E_s \upharpoonright_{\pi_s \sigma} \neq 1$, then extend π_s finitely so as to ensure $E_s \upharpoonright_{\pi_s \sigma} = 0$.

At the end of stage s, $E_s \upharpoonright_{\pi_s}$ is equivalent to a finite boolean function on the variables of K, hence can be expressed in DNF form as an $\bigvee$ of $\bigwedge$ gates.

This completes the description of steps (1) and (2), and thus the proof is now complete.

2.6 Probabilistic Methods

In this section, we assume that all circuits are leveled, that $\bigwedge$ and $\bigvee$ gates alternate by depth, that negations only occur at the inputs $x_1, \overline{x}_1, \ldots, x_n, \overline{x}_n$. The probabilistic technique to be developed in this section was initiated by Furst, Saxe and Sipser [FSS84] and independently (with rather different notation and methods) by Ajtai [Ajt83].

Lower bounds for depth k circuits can be proved by induction on k. The idea is as follows. Take any gate at distance 2 from the inputs, which represents a circuit of depth 2; without loss of generality assume that this last circuit is an $\bigwedge$ of $\bigvee$'s. Now change this circuit (by distributivity) into an $\bigvee$ of $\bigwedge$'s. By doing this substitution we obtain two adjacent levels of $\bigvee$ gates which we can merge into a single level thus reducing the depth by 1. The only problem with this argument is that by reducing the depth by 1 the size of the circuit may become exponential. To overcome this problem [FSS84] introduced the notion of *random restriction*. This is a random assignment ρ of the values $0, 1, *$ to the variables $x_1, \ldots, x_n$ such that each variable x_i is substituted by $0, 1, x_i$ depending on whether $\rho(x_i) = 0, 1, *$, respectively. Let $f \upharpoonright_\rho$ be the boolean function resulting from f restricted to the variables assigned the value $*$ by ρ. Given $p > 0$ let $\mathcal{R}_p$ be the set of random restrictions ρ with distribution

$$\Pr[\rho(x_i) = 0] = \Pr[\rho(x_i) = 1] = \frac{1-p}{2}, \Pr[\rho(x_i) = *] = p.$$

Recall that a *minterm* of a boolean function g is a partial assignment σ of the variables of g which makes g true, but such that no subassignment of σ does. Also define the event $\min(g) \geq s$ to be "$g \upharpoonright_\rho$ has a minterm of size $\geq s$".

In the sequel we present Håstad's lower bound, which is based on the previously described technique. Subsequently we study the difference between depth k and depth-$(k-1)$ circuits [Sip83].

2.6.1 Håstad's Lower Bound for Parity

The main idea of Håstad's proof is to show by induction on k that "small" depth k parity circuits can be converted to "small" depth $k-1$ parity circuits. The main result of his proof is given in the following lemma.

Theorem 2.6.1 (Switching Lemma, [Hås87]). *For $p > 0$, and s, t non-negative integers, let α be the unique solution of the equation*

$$\left(1 + \frac{4p}{(1+p)\alpha}\right)^t = 1 + \left(1 + \frac{2p}{(1+p)\alpha}\right)^t. \tag{2.7}$$

For arbitrary boolean functions f, g if g is an $\bigwedge$ of $\bigvee$'s of fan-in $\leq t$ then

$$\Pr[\min(g) \geq s | f \restriction_\rho \equiv 1] \leq \alpha^s. \tag{2.8}$$

Proof. Assume $g = \bigwedge_{i=1}^{w} g_i$ where each g_i is an $\bigvee$ of fan-in $\leq t$. The proof is by induction on w. If $w = 0$ then $g \equiv 1$ and hence g has no minterm of size at least s. Hence the left-hand side of (2.8) is 0. By induction hypothesis assume the lemma is true for $w - 1$. By standard probability theory we have that one of the following inequalities holds:

$$\Pr[\min(g) \geq s | f \restriction_\rho \equiv 1] \leq \Pr[\min(g) \geq s | f \restriction_\rho \equiv 1, g_1 \restriction_\rho \equiv 1] \tag{2.9}$$

$$\Pr[\min(g) \geq s | f \restriction_\rho \equiv 1] \leq \Pr[\min(g) \geq s | f \restriction_\rho \equiv 1, g_1 \restriction_\rho \not\equiv 1]. \tag{2.10}$$

In inequality (2.9) the condition $f \restriction_\rho \equiv 1, g_1 \restriction_\rho \equiv 1$ is equivalent to $(f \wedge g_1) \restriction_\rho \equiv 1$ and under this condition $g = \bigwedge_{i=2}^{w} g_i$. Hence the induction hypothesis applies. Consequently, it is enough to prove the upper bound for the right-hand side of (2.10). Without loss of generality we may assume that $g_1 = \bigvee_{i \in T} x_i$, i.e., g_1 is disjunction of positive literals. Let $\rho = \rho_1 \rho_2$, where ρ_1 is the restriction of ρ on $\{x_i : i \in T\}$ and ρ_2 is the restriction of ρ on $\{x_i : i \notin T\}$. Then the condition $g_1 \restriction_\rho \not\equiv 1$ is equivalent to $\rho(x_i) \in \{0, *\}$, for all $i \in T$. For $Y \subseteq T$ let $Min(Y)$ be the set of minterms σ of $g \restriction_\rho$ such that σ assigns values only to the variables in Y or to the variables outside T. Let $\min(g)^Y \geq s$ denote the event "$g \restriction_\rho$ has a minterm $\sigma \in Min(Y)$ of size $\geq s$". Then we have that the right-hand side of (2.10) is at most

$$\sum_{\substack{Y \subseteq T \\ Y \neq \emptyset}} \Pr[\min(g)^Y \geq s | f \restriction_\rho \equiv 1, g_1 \restriction_\rho \not\equiv 1].$$

Since $\sigma \in Min(Y)$, among variables in T σ assigns values only to those variables in $Y \subseteq T$, so ρ (and in particular ρ_1) assigns $*$'s to the variables of Y. Thus the previous sum is equal to

$$\sum_{\substack{Y \subseteq T \\ Y \neq \emptyset}} \Pr[\min(g)^Y \geq s, \rho_1(Y) = * \mid f \restriction_\rho \equiv 1, g_1 \restriction_\rho \not\equiv 1].$$

This last sum can be rewritten as

$$\sum_{\substack{Y \subseteq T \\ Y \neq \emptyset}} \left\{ \begin{array}{c} \Pr[\rho_1(Y) = * \mid f \restriction_\rho \equiv 1, g_1 \restriction_\rho \not\equiv 1] \times \\ \Pr[\min(g)^Y \geq s | f \restriction_\rho \equiv 1, g_1 \restriction_\rho \not\equiv 1, \rho_1(Y) = *] \end{array} \right\} \tag{2.11}$$

It remains to majorize (2.11). To do this we prove the following two lemmas.

Lemma 2.6.1. *The top factor of the summand in the right-hand side of (2.11) is at most* $\left(\frac{2p}{1+p}\right)^{|Y|}$.

Proof. of Lemma 2.6.1. Consider the events

$$A = \text{“}\rho_1(Y) = *\text{”}, B = \text{“}f \restriction_\rho \equiv 1\text{”}, C = \text{“}g_1 \restriction_{\rho_1} \not\equiv 1\text{”}.$$

It is easy to show that

$$\Pr[A|B \wedge C] \leq \Pr[A|C] \iff \Pr[B|A \wedge C] \leq \Pr[B|C]. \tag{2.12}$$

Since the right-hand side of (2.12) is true (requiring some variables to be $*$ cannot increase the probability that a function is determined) it is enough to show that the term $\Pr[A|C]$ on the left-hand side of (2.12) is bounded by $\left(\frac{2p}{1+p}\right)^{|Y|}$. By the previous observation the event C is equivalent to the event “$\rho(x_i) \in \{0, *\}$, for all $i \in T$”. Hence the required result follows from the fact that ρ assigns values independently to different x_i's, as well as the identity

$$\Pr[\rho(x_i) = * | \rho(x_i) \in \{0, *\}] = \frac{p}{p + (1-p)/2} = \frac{2p}{1+p}.$$

This completes the proof of Lemma 2.6.1.

Lemma 2.6.2. *The bottom factor of the summand on the right-hand side of (2.11) is at most* $\left(2^{|Y|} - 1\right) \alpha^{s-|Y|}$.

Proof. For minterm σ of $g \restriction_\rho$, if $\sigma \in Min(Y)$, then σ can be decomposed into $\sigma_1\sigma_2$, where σ_1 assigns values to the variables in Y, and σ_2 assigns values to variables outside of T. Clearly σ_2 is a minterm of the function $(g \restriction_\rho) \restriction_{\sigma_1}$. Let $\min(g)^{Y,\sigma_1} \geq s$ denote the event “$g \restriction_\rho$ has a minterm of size $\geq s$ assigning the values σ_1 to the variables in Y and does not assign values to any other variables in T”. By maximizing over all ρ_1 satisfying the condition $g_1 \restriction_{\rho_1} \not\equiv 1$ we obtain that

$$\Pr[\min(g)^Y \geq s | f \restriction_\rho \equiv 1, g_1 \restriction_{\rho_1} \not\equiv 1, \rho_1(Y) = *] \leq$$

$$\sum_{\sigma_1} (\max_{\rho_1} \Pr_{\rho_2}[\min(g)^{Y,\sigma_1} \geq s | (f \restriction_{\rho_1}) \restriction_{\rho_2} \equiv 1]),$$

where σ_1 ranges over $\{0,1\}^{|Y|}, \sigma_1 \neq 0^{|Y|}$ and the maximum over ρ_1 satisfying $\rho_1(Y) = *$, and $\Pr_{\rho_2}$ is the probability taken over ρ_2 with ρ_1 fixed. Clearly, $\min(g)^{Y,\sigma_1} \geq s$ implies that $(g \restriction_{\rho_1\sigma_1}) \restriction_{\rho_2}$ has a minterm of size $s - |Y|$ on the variables outside of T. Hence, by the induction hypothesis, this probability is majorized by $\alpha^{s-|Y|}$. This completes the proof of Lemma 2.6.2.

Assuming these lemmas the proof of Theorem 2.6.1 is now easy. Indeed,

$$\begin{aligned}
(2.11) &\leq \textstyle\sum_{Y \subseteq T} \left(\frac{2p}{p+1}\right)^{|Y|} \cdot (2^{|Y|} - 1) \cdot \alpha^{s-|Y|} \\
&= \alpha^s \cdot \textstyle\sum_{Y \subseteq T} \left(\frac{2p}{p+1} \cdot \frac{2}{\alpha}\right)^{|Y|} - \alpha^s \cdot \textstyle\sum_{Y \subseteq T} \left(\frac{2p}{p+1} \cdot \frac{1}{\alpha}\right)^{|Y|} \\
&= \alpha^s \cdot \left(1 + \frac{4p}{\alpha(p+1)}\right)^{|T|} - \alpha^s \cdot \left(1 + \frac{2p}{\alpha(p+1)}\right)^{|T|} \\
&= \alpha^s \left[\left(1 + \frac{4p}{\alpha(p+1)}\right)^{|T|} - \left(1 + \frac{2p}{\alpha(p+1)}\right)^{|T|}\right] \\
&= \alpha^s \left[\left(1 + \frac{4p}{\alpha(p+1)}\right)^{t} - \left(1 + \frac{2p}{\alpha(p+1)}\right)^{t}\right] \\
&= \alpha^s .
\end{aligned}$$

This completes the proof of the Switching Lemma (Theorem 2.6.1).

Before proceeding with the main theorem observe that the unknown α appearing in the Switching Lemma must satisfy $\alpha < O(pt)$, for t large enough. To see this, substitute the term $2p/\alpha$ by ϵ/t in (2.7), thus obtaining

$$\left(1 + \frac{2\epsilon}{t(1+p)}\right)^t = 1 + \left(1 + \frac{\epsilon}{t(1+p)}\right)^t$$

which for large t is approximately

$$e^{\frac{2\epsilon}{t(1+p)}} = 1 + e^{\frac{\epsilon}{t(1+p)}}$$

whose solution is $e^{\frac{\epsilon}{t(1+p)}} = \phi$, where $\phi = \frac{\sqrt{5}+1}{2}$ is the golden ratio. From $\epsilon = (\ln \phi)(1+p)$, we have $\frac{2p}{\alpha} = \frac{(\ln \phi)(1+p)}{t}$ thus yielding $\alpha \leq \frac{2pt}{(\ln \phi)(1+p)} \leq \frac{2pt}{\ln \phi}$. Taking into account the approximation, we have $\alpha \leq \frac{2pt}{\ln \phi} + O(tp^2)$.

Theorem 2.6.2 ([Hås87]). *There is a constant n_0 such that for all $n \geq n_0$ there are no depth-k parity circuits of size $2^{(1/2c)^{k/(k-1)} \cdot n^{1/(k-1)}}$.*

Proof. Assume otherwise and consider the circuit as a depth-$(k+1)$ circuit with bottom fan-in 1. Apply a random restriction from $\mathcal{R}_p$, where $p = 1/2c$, which assigns $*$ to an expected number $m = n/2c$ of variables. Now use the Switching Lemma with $s = (1/2c)(n/2c)^{1/(k-1)}$ to shorten the circuit to a new circuit of depth-k, having $2^{(1/2c)m^{1/(k-1)}}$ gates at distance at least 2 from the inputs and bottom fan-in $(1/2c)m^{1/(k-1)}$, for m large enough. We prove that such circuits cannot exist by induction on k. For $k = 2$ the result is implied from the fact that depth-2 circuits for PAR_m must have bottom fan-in m. Assume that such circuits do not exist when the parameter is $k-1$. We prove the same holds when the parameter is k. Assume otherwise. Suppose that the depth-k circuits are such that the gates at distance 2 from the inputs are $\bigwedge$ gates. Hence they represent depth-2 circuits with bottom fan-in $\leq (1/2c)m^{1/(k-1)}$. Apply a random restriction from $\mathcal{R}_p$, with $p = m^{-1/(k-1)}$.

Using the Switching Lemma every such depth-2 circuit can be written as an $\bigvee$ of $\bigwedge$'s of size $< s$. In view of the above choice of p we have that $\alpha < cpt = 1/2$. If we choose $s = (1/2c)m^{1/(k-1)}$ the probability that it is not possible to write all depth-2 circuits as $\bigvee$ of $\bigwedge$'s of size $\leq s$ is bounded by $2^{(1/2c)m^{1/(k-1)}} \cdot \alpha^s = (2\alpha)^s$. Consequently with probability $\geq 1-(2\alpha)^s$ we can interchange the order of $\bigwedge$ and $\bigvee$ in all of these depth-2 subcircuits, maintain bottom fan-in bounded by s and reduce the depth to $k-1$. Put $r = m^{\frac{k-2}{k-1}}$. The expected number of variables (i.e., those assigned value $*$) is $pm = r$ and with probability $> 1/3$ we will get at least this number for m large enough. It follows that the resulting circuit which computes either PAR_r or $\neg\text{PAR}_r$ has bottom fan-in $(1/2c)r^{1/(k-2)}$ and $2^{(1/2c)r^{1/(k-2)}}$ gates at distance at least 2 from the inputs. This last contradiction completes our inductive proof and hence also the proof of Theorem 2.6.2.

It will be seen later in Section 2.7 that Theorem 2.6.2 can be significantly improved by using algebraic methods.

2.6.2 Depth-k Versus Depth-$(k-1)$

In this section we give a collection $\{f_k^n\}$ of boolean functions which can be computed in depth-k linear size circuits but which cannot be computed in depth-$(k-1)$ polynomial size circuits. Define the *Sipser function* f_k^n as an alternating, layered boolean circuit of fan-out 1 (i.e., a tree) having

- bottom fan-out $\sqrt{kn \log n/2}$,
- top fan-out $\sqrt{n/\log n}$, and
- all other fan-outs n

and whose root is labeled with $\bigwedge$. Thus, for even k, f_k^n is

$$\bigwedge_{i_1=1}^{\sqrt{n \log n}} \bigvee_{i_2=1}^{n} \bigwedge_{i_3=1}^{n} \cdots \bigwedge_{i_{k-1}=1}^{n} \bigvee_{i_k=1}^{\sqrt{kn \log n/2}} x_{i_1,\ldots,i_k}.$$

Since each variable occurs at a unique leaf, the total number of variables is

$$\sqrt{kn \log n/2} \cdot n^{k-2} \cdot \sqrt{n/\log n} = n^{k-1} \cdot \sqrt{k/2}$$

and the circuit size is linear in the number of variables. The main theorem is the following.

Theorem 2.6.3 ([Sip83], [Hås87]). *There is a constant n_0 such that for all $n \geq n_0$, depth-$(k-1)$ circuits computing f_k^n require size*

$$2^{\frac{1}{12}\sqrt{\frac{n}{2k \log n}}}.$$

Proof (Outline). The proof requires a new notion of random restriction which depends on partitioning the set of variables into blocks. Namely, for $p > 0$, $b \in \{0,1\}$, and $B = (B_i)_{i=1}^r$ a partition of the variables, the probability space $\mathcal{R}_{p,B}^b$ of random restrictions ρ is defined as follows. For every B_i, $1 \le i \le r$, independently let $s_i = *$ with probability p and $s_i = b \oplus 1$, otherwise. Then for every variable $x_k \in B_i$ let the random restriction ρ satisfy $\rho(x_k) = s_i$ with probability p and $\rho(x_k) = b$, otherwise. In addition, for any restriction $\rho \in \mathcal{R}_{p,B}^b$ define the new restriction $g(\rho)$ as follows. For all $1 \le i \le r$ with $s_i = *$, $n(\rho)$ gives the value b to all variables in B_i given the value $*$ by ρ except for one to which it gives the value $*$. To make $n(\rho)$ deterministic we assume that it gives the value $*$ to the variable with highest index given the value $*$ by ρ.

Let "$\bigwedge(g \restriction_{\rho n(\rho)}) \ge s$" denote the event "$g \restriction_{\rho n(\rho)}$ cannot be written as an $\bigvee$ of $\bigwedge$'s of size $< s$", where $\rho n(\rho)$ denotes the composition of ρ and $n(\rho)$. The Switching Lemma corresponding to these distributions can now be stated as follows.

Lemma 2.6.3. *Let g be an $\bigwedge$ of $\bigvee$'s each of fan-in $\le t$ and let f be an arbitrary function. If ρ is a random restriction in $\mathcal{R}_{p,B}^1$ then*

$$\Pr\left[\bigwedge(g \restriction_{\rho n(\rho)}) \ge s | f \restriction_\rho \equiv 1\right] \le \alpha^s,$$

where $\alpha = 4p/(2^{1/t} - 1)$.

We omit the proof of this lemma and refer the reader instead to [Hås87] for details. Next we continue with the proof of Theorem 2.6.3. Since depth-$(k-1)$ circuits can be considered as depth-k circuits with bottom fan-in 1, Theorem 2.6.2 reduces to proving the following statement.

> *There is a constant n_0 such that for all $n \ge n_0$, there are no depth-k circuits computing f_k^n with bottom fan-in $\frac{1}{12}\sqrt{\frac{n}{2k\log n}}$ and $\le 2^{\frac{1}{12}\sqrt{\frac{n}{2k\log n}}}$ gates of depth ≥ 2.*

The proof of this statement is by induction on k. The case $k = 2$ follows immediately from the definition of f_2^n. For k odd (even) let $B^1 = \{B_i^1\}$ ($B^0 = \{B_i^0\}$) be the partition where B_i^1 (B_i^0) is the set of variables leading into the i-th $\bigwedge$ ($\bigvee$) gate. Further let $p = \sqrt{\frac{2k\log n}{n}}$. With these definitions in mind, we can prove the following lemma.

Lemma 2.6.4. *There is a constant n_1 such that for all $n \ge n_1$ and $\frac{n}{\log n} \ge 100k$, if k is odd (even) then the circuit that defines $f_k^n \restriction_{\rho n(\rho)}$ for random $\rho \in \mathcal{R}_{p,B^1}^1$ ($\rho \in \mathcal{R}_{p,B^0}^0$) will contain the circuit that defines f_{k-1}^n with probability $\ge 2/3$.*

Proof. We consider only the case k is odd (the other case being entirely similar). This implies that the k-th level of the circuit consists of $\bigwedge$ gates while the $(k-1)$-th level of $\bigvee$ gates. The $\bigwedge$ gate corresponding to block B_i^1 takes the value s_i precisely when not only 1's are given to the block. The probability of this happening is

$$(1-p)^{|B_i^1|} = \left(1 - \sqrt{\frac{2k\log n}{n}}\right)^{\sqrt{kn/2\log n}} \leq \frac{1}{6}n^{-k},$$

for n large enough. It follows that there is a constant n_0 such that for all $n \geq n_0$ the $\bigwedge$ gate corresponding to the block B_i^1 takes the value s_i for all i with probability at least 5/6.

Next we claim that with probability $\geq 5/6$, for n large enough at least $\sqrt{(k-1)n\log n/2}$ inputs are given the value $*$ by $\rho n(\rho)$ to each $\bigvee$ gate at level $k-1$. Moreover the expected number of such inputs is $\sqrt{2kn\log n}$. To prove this let p_i be the probability that an $\bigvee$ gate has as input exactly i $\bigwedge$ gates which take the value $*$. Then

$$p_i = \binom{n}{i}\left(\frac{2k\log n}{n}\right)^{i/2}\left(1-\sqrt{\frac{2k\log n}{n}}\right)^{n-i}$$

and $p_i/p_{i-1} \geq \sqrt{2}$, for $i < \sqrt{kn\log n}$. Using $p_{\sqrt{nk\log n}} < 1$ we obtain the estimate

$$\begin{aligned}\sum_{i=1}^{\sqrt{nk\log n/2}} p_i &\leq p_{\sqrt{nk\log n}/2}\sum_{i=0}^{\infty} 2^{-i}\\ &\leq 2p_{\sqrt{nk\log n}/2}\\ &\leq 2\sqrt{2}^{-(1-\sqrt{2}^{-1})\sqrt{nk\log n}}\, p_{\sqrt{nk\log n}}\\ &\leq \tfrac{1}{6}n^{-k}.\end{aligned}$$

It follows that with probability at least 2/3 of all $\bigvee$ gates at level $k-2$ will remain undetermined and will have at least $\sqrt{(k-1)n\log n/2}$ variables as inputs. This defines the circuit for f_{k-1}^n and completes the proof of Lemma 2.6.4.

Now we return to Theorem 2.6.3. Clearly, Theorem 2.6.3 follows directly from the following stronger theorem.

Theorem 2.6.4 ([Sip83, Hås87]). *There is a constant n_0 such that for all $n \geq n_0$, depth-k circuits with bottom fan-in $\frac{1}{12}\sqrt{\frac{n}{2k\log n}}$ computing f_k^n have more than $2^{\frac{1}{12}\sqrt{\frac{n}{2k\log n}}}$ gates of depth ≥ 2.*

Proof. We use induction on k. The base case $k=2$ is easy. For the induction step we argue as in the proof of Theorem 2.6.2. Apply a restriction from $\mathcal{R}_{p,B}^1$ to the circuit. Without loss of generality we may assume that $\frac{n}{\log n} \geq 100k$ (otherwise the result of the theorem is easy). In this case Lemma 2.6.4 is applicable and the defining circuit still computes a function as difficult as

f_{k-1}^n. Thus setting some of the remaining variables, the circuit can be made into the defining circuit for f_{k-1}^n.

Now assume to the contrary there existed a circuit of depth k, bottom fan-in $\frac{1}{12}\sqrt{\frac{n}{2k\log n}}$ and size $2^{\frac{1}{12}\sqrt{\frac{n}{2k\log n}}}$ computing f_k^n. Arguing as in Theorem 2.6.2 we can interchange the $\bigvee$'s and $\bigwedge$'s on the last two levels without increasing the bottom fan-in. Now collapse the two adjacent levels of $\bigvee$ gates and obtain a contradiction as in Theorem 2.6.2. The proof of Theorems 2.6.3 and 2.6.4 is now complete.

2.6.3 Razborov's Simplification and Decision Trees

In [Raz94][pp. 380–383], A.A. Razborov developed a new, very elegant combinatorial method for proving the Håstad Switching Lemma.[9] The bounds from Razborov's method are not quite as good as those from Håstad; nevertheless, the method is significantly simpler, and historically led to an immediate simplification of related lower bound results (see for instance P. Beame [Bea94], A. Urquhart [Urq95]). Using Razborov's technique, a decision tree form of the Switching Lemma was derived by P. Beame ([Bea94]) and S. Buss (unpublished).

In this Subsection, following [Bea94], we present the decision tree form of the Switching Lemma. Razborov's form of the Switching Lemma is then derived as Theorem 2.6.6,[10] and the exponential lower bound for PARITY in Corollary 2.6.1.

Definition 2.6.1 ([Bea94]). *Let $f = F_1 \vee \cdots \vee F_m$ be a* DNF *formula. The canonical decision tree $T(F)$ is defined as follows:*

- *If $F \equiv 0$ (1) then $T(F)$ consists of a single node labeled* 0 *[resp* 1*].*
- *Let $K = \{x_{i_1}, \cdots, x_{i_k}\}$ be the set of variables in F_1, and define S to be a complete binary tree of depth k, whose internal nodes at level j are labeled by x_{i_j}, and whose leaves ℓ are labeled by the corresponding restriction $\sigma_\ell : K \to \{0,1\}$, determined by the branch from root to ℓ, where $\sigma_\ell(x_{i_j}) = 0$ (1) if the branch leading to leaf ℓ goes to the left (right) at the node labeled by x_{i_j}. Now T is obtained from S by replacing each leaf ℓ in S by $T(F \lceil_{\sigma_\ell})$.*

If σ_ℓ is the unique restriction with domain K which satisfies F_1, then $T(F \lceil_{\sigma_\ell})$ is a single node labeled by 1.

Example 2.6.1. Assume that f is the DNF formula $(x_1 \wedge \overline{x}_3) \vee (\overline{x}_1 \wedge \overline{x}_4) \vee x_2$. Then f is of the form $F_1 \vee F_2 \vee F_3$, where $F_1 = x_1 \wedge \overline{x}_3$, so $K = \{x_1, x_3\}$, and we have the restrictions $\rho_1, \ldots, \rho_4$ corresponding to the full binary tree on K. Thus

[9] It appears that an unpublished method of A. Woods is similar.
[10] See Exercise 2.14.4.

$\rho_1 = (0 * 0 *)$	$F \upharpoonright_{\rho_1} = \overline{x}_4 \vee x_2$
$\rho_2 = (0 * 1 *)$	$F \upharpoonright_{\rho_2} = \overline{x}_4 \vee x_2$
$\rho_3 = (1 * 0 *)$	$F \upharpoonright_{\rho_3} = 1$
$\rho_4 = (1 * 1 *)$	$F \upharpoonright_{\rho_4} = x_2$

The decision tree $T(F)$ given in Figure 2.1.

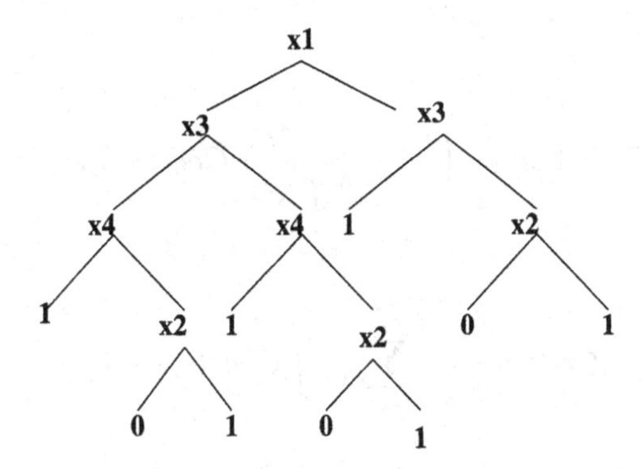

Fig. 2.1. Decision tree $T(f)$, for $f = (x_1 \wedge \overline{x}_3) \vee (\overline{x}_1 \wedge \overline{x}_4) \vee x_2$

Let $Code(r, s)$ be the set of sequences $\beta = (\beta_1, \ldots, \beta_k)$ where $\beta_i \in \{\downarrow, *\}^r - \{*\}^r$ for each $1 \leq i \leq k$, and where there are exactly s occurrences of $\downarrow$ in the entire sequence β. Note that $k \leq s$.

Lemma 2.6.5. *For $0 < s \leq r$, we have $|Code(r, s)| \leq (2r)^s$.*

Proof. For a given sequence $\beta = (\beta_1, \cdots, \beta_k)$ in $Code(r, s)$, define the associated mapping $f_\beta : \{1, \cdots, s\} \to \{1, \ldots, r\} \times \{0, 1\}$ by

$$f_\beta(1) = (j, 1)$$

if the first $\downarrow$ occurs in position j of β, and for $1 < i \leq s$

$$f_\beta(i) = (j, 0) \quad ((j, 1))$$

if the i-th $\downarrow$ occurs in position j of some β_ℓ and the $(i-1)$-st $\downarrow$ occurs in the same β_ℓ (does not occur in the same β_ℓ). Since the association between $\beta \in Code(r, s)$ and f_β is injective, and there are at most $(2r)^s$ maps f_β, the lemma follows.

For completeness, we give a refinement, due to P. Beame [Bea94].

Lemma 2.6.6 ([Bea94]). $|Code(r, s)| < (r/\ln 2)^s$.

Proof. We prove by induction on $s \geq 1$ that $|Code(r,s)| \leq \gamma^s$ where γ satisfies the equation $(1+1/\gamma)^r = 2$. Note that $1+x < e^x$ for $x \neq 0$, so that $(1+\frac{1}{\gamma})^r < e^{r/\gamma}$ which implies that $\gamma < r/\ln 2$. As well $r \leq \gamma$ iff $(1+\frac{1}{r})^r \geq (1+\frac{1}{\gamma})^r = 2$, and the latter clearly holds. When $s = 1$, we have $|Code(r,1)| = r \leq \gamma^1$. Assume now that $s > 1$. If $\beta \in Code(r,s)$, then either $\beta = (\beta_1)$ where β_1 contains s many $\downarrow$ symbols, or $\beta = (\beta_1, \beta')$ where β_1 contains i many $\downarrow$ symbols, for $1 \leq i < s$, and $\beta' \in Code(r, s-i)$. There are $\binom{r}{i}$ many choices for β_1 having i occurrences of $\downarrow$.

Case 1. $s \leq r$

$$|Code(r,s)| = \binom{r}{s} + \sum_{i=1}^{s-1} \binom{r}{i} |Code(r, s-i)|$$

Case 2. $r < s$

$$|Code(r,s)| = \sum_{i=1}^{r} \binom{r}{i} |Code(r, s-i)|$$

Thus in either case, using the induction hypothesis

$$\begin{aligned} |Code(r,s)| &\leq \sum_{i=1}^{r} \binom{r}{i} \cdot \gamma^{s-i} \\ &= \gamma^s \sum_{i=1}^{r} \binom{r}{i} \cdot \left(\frac{1}{\gamma}\right)^i \\ &= \gamma^s \left[\left(1 + \frac{1}{\gamma}\right)^r - 1 \right] \\ &= \gamma^s. \end{aligned}$$

This completes the induction. Since $\gamma < r/\ln 2$ the lemma follows.

This leads to the following dcision tree form of Håstad's Switching Lemma, a result of Beame [Bea94].

Theorem 2.6.5. *Let $f = F_1 \vee \cdots \vee F_m$ be an r-DNF formula in variables $x_1, \ldots, x_n$. Denote the depth of a minimum depth decision tree for restriction $f \upharpoonright_\rho$ by $d(T(f \upharpoonright_\rho))$. For $s \geq 0$ and $p = \ell/n < \frac{1}{7}$*

$$\Pr[\, d(T(f \upharpoonright_\rho)) \geq s \,|\, \rho \in \mathcal{R}_n^\ell] < (7pr)^s.$$

Proof. Define

$$Bad_n^\ell = \{\rho \in \mathcal{R}_n^\ell : d(T(f \upharpoonright_\rho)) \geq s\}.$$

We must show that

$$\frac{|Bad_n^\ell|}{|\mathcal{R}_n^\ell|} < (7pr)^s.$$

To this end, define an injection from Bad_n^ℓ into

$$\mathcal{R}_n^{\ell-s} \times Code(r,s) \times 2^s.$$

Let π be the restriction associated with the leftmost (lexicographically least) node in $T(f\lceil_\rho)$ at level s, so that $|\pi| = s$. Let F_{ν_1} be first term of f for which $F_{\nu_1}\lceil_\rho \not\equiv 0$. There must be such, since otherwise $f\lceil_\rho \equiv 0$ so $d(T(f\lceil_\rho)) = 0$. Let K_1 be the set of variables of $F_{\nu_1}\lceil_\rho$ and σ_1 the unique restriction setting $F_{\nu_1}\lceil_{\sigma_1} \equiv 1$. Let π_1 be the restriction of π to domain K_1.

Case 1. $\pi_1 \neq \pi$.
By construction of the canonical decision tree, $K_1 \subseteq do(\pi_1)$, so $F_{\nu_1}\lceil_{\rho\pi_1} \equiv 0$.

Case 2. $\pi_1 = \pi$.
It is possible that $K_1 \not\subseteq do(\pi_1)$, so it may not be the case that $F_{\nu_1}\lceil_{\rho\pi_1} \equiv 0$. However, in any case , $F_{\nu_1}\lceil_{\rho\pi_1} \not\equiv 1$. Trim σ_1 to have domain $do(\pi_1)$.

Define $\beta_1 \in \{\downarrow,*\}^r$ by setting the j-th component of β_1 to be $\downarrow$ iff the j-th variable of F_{ν_1} is set by σ_1. Since $F_{\nu_1}\lceil_\rho \not\equiv 0$, at least one variable of F_{ν_1} is set by σ_1, so that $\beta_1 \in \{\downarrow,*\}^r - \{*\}^r$. From F_{ν_1} and β_1, one can reconstruct σ_1.

If $\pi_1 \neq \pi$ then repeat the previous construction with $\pi - \pi_1, T(f\lceil_{\rho\pi_1})$ in place of $\pi, T(f\lceil_\rho)$. Determine the first F_{ν_2} in f such that $F_{\nu_2}\lceil_{\rho\pi_1} \not\equiv 0$. Let K_2 be the set of variables in $F_{\nu_2}\lceil_{\rho\pi_1}$ and define the unique restriction σ_2 for which $F_{\nu_2}\lceil_{\rho\pi_1\sigma_2} \equiv 1$. Define $\beta_2 \in \{\downarrow,*\}^r - \{*\}^r$ to have j-th component $\downarrow$ iff the j-th variable of F_{ν_2} is set by σ_2. Let π_2 be the restriction of π to K_2. If $\pi_1\pi_2 \neq \pi$ then $K_2 \subseteq do(\pi_2)$ and $F_{\nu_2}\lceil_{\rho\pi_1\pi_2} \equiv 0$. If $\pi_1\pi_2 = \pi$ then it may be that $K_2 \not\subseteq do(\pi_2)$, in which case σ_2 is restricted to $do(\pi_2)$. In this case $F_{\nu_2}\lceil_{\rho\pi_1\sigma_2} \not\equiv 0$. Continuing in this fashion, we define the following

- $1 \leq \nu_1 < \cdots < \nu_k \leq m$, where each F_{ν_i} is first term of f for which $F_{\nu_i}\lceil_{\rho\pi_1\cdots\pi_{i-1}} \not\equiv 0$,
- $K_1, \ldots, K_k$ where each K_i is the set of variables in the domain of π and in $F_{\nu_i}\lceil_{\rho\pi_1\cdots\pi_{i-1}}$; i.e.,

$$K_i = Var(F_{\nu_i}\lceil_{\rho\pi_1\cdots\pi_{i-1}}) \cap do(\pi).$$

  Note that for $i < k$, we have $K_i = Var(F_{\nu_k}\lceil_{\rho\pi_1\cdots\pi_{i-1}})$.
- $\sigma_1, \ldots, \sigma_k$ where each σ_i has domain K_i and is uniquely defined so that $F_{\nu_i}\lceil_{\rho\pi_1\cdots\pi_{i-1}\sigma_i} \not\equiv 0$. Note that for $i < k$,

$$F_{\nu_i}\lceil_{\rho\pi_1\cdots\pi_{i-1}\sigma_i} \equiv 1.$$

- $\pi_1, \ldots, \pi_k$ where each π_i is the restriction of π to domain K_i.
- $\beta_1, \ldots, \beta_k$ in $\{\downarrow,*\}^r - \{*\}^r$, where the j-th component of each β_i is $\downarrow$ exactly when the j-th variable of F_{ν_i} is set by σ_i.

Now let $\sigma = \sigma_1 \cdots \sigma_k$, $\beta = (\beta_1, \ldots, \beta_k)$ and define $\delta : \{1, \ldots, s\} \to \{0, 1\}$ by setting $\delta(i) = 1$ iff the i-th variable in $do(\pi) = do(\sigma)$ has the same value in π as in σ. Finally, map

$$Bad_n^\ell \to \mathcal{R}_n^{\ell-s} \times Code(r, s) \times 2^s$$

by $\rho \mapsto (\rho\sigma, \beta, \delta)$.

Claim. The mapping $Bad_n^\ell \to \mathcal{R}_n^{\ell-s} \times Code(r, s) \times 2^s$ is injective.

Proof of Claim. To show that this mapping is injective, we indicate how to uniquely determine ρ from its image. Note first the following *crucial observation:*

$$F_{\nu_i} \upharpoonright_{\rho\pi_1\cdots\pi_{i-1}\sigma_i} \equiv F_{\nu_i} \upharpoonright_{\rho\pi_1\cdots\pi_{i-1}\sigma_i\sigma_{i+1}\cdots\sigma_k}.$$

For $i < k$, this value is 1; for $i = k$, the value is perhaps not 1, but nevertheless $\not\equiv 0$. Now ν_1 is the first index for which $F_{\nu_1} \upharpoonright_{\rho\sigma} \not\equiv 0$. From F_{ν_1} and β_1 we can determine K_1, as the set of those variables of F_{ν_1} set by σ_1. Thus σ_1 is the restriction of $\rho\sigma$ to K_1.

From σ_1 and δ we can determine π_1 and hence $\rho\pi_1\sigma_2 \cdots \sigma_k$. Next, ν_2 is the first index for which $F_{\nu_2} \upharpoonright_{\rho\pi_1\sigma_2\cdots\sigma_k} \not\equiv 0$. In a similar manner, we can determine K_2, σ_2, π_2 and $\rho\pi_1\pi_2\sigma_3 \cdots \sigma_k$. Iterating this procedure, we can reconstruct $K_2, \cdots, K_k$ and so determine the domain of ρ from $\rho\pi_1 \ldots \pi_k$. In this manner, ρ can be uniquely reconstructed. This concludes the proof of the claim. □

Definition 2.6.2. *The falling factorial $a^{\underline{m}}$ is defined by the m term product $a(a-1)\cdots(a-(m-1)) = m!\binom{a}{m}$.*

Note that $a^{\underline{m+n}} = a^{\underline{m}}(a-m)^{\underline{n}}$.

Now $|\mathcal{R}_n^\ell| = \binom{n}{\ell} \cdot 2^{n-\ell}$, so

$$\begin{aligned}
\frac{|\mathcal{R}_n^{\ell-s}|}{|\mathcal{R}_n^\ell|} &= \frac{\binom{n}{\ell-s}2^{n-\ell+s}}{\binom{n}{\ell}2^{n-\ell}} \\
&= \frac{\ell(\ell-1)\cdots(\ell-s+1)}{(n-\ell+s)(n-\ell+s-1)\cdots(n-\ell+1)} \cdot 2^s \\
&< \left(\frac{\ell}{n-\ell+1}\right)^s \cdot 2^s < \left(\frac{2\ell}{n-\ell}\right)^s.
\end{aligned}$$

Thus

$$\begin{aligned}
\frac{|Bad_n^\ell|}{|\mathcal{R}_n^\ell|} &\le \frac{|\mathcal{R}_n^{\ell-s}|}{|\mathcal{R}_n^\ell|} \cdot |Code(r, s)| \cdot 2^s \\
&\le \left(\frac{2\ell}{n-\ell}\right)^s \cdot \left(\frac{r}{\ln 2}\right)^s \cdot 2^s \\
&= \left(\frac{4\ell r}{(n-\ell)\ln 2}\right)^s \\
&= \left(\frac{4npr}{n(1-p)\ln 2}\right)^s = \left(\frac{4pr}{(1-p)\ln 2}\right)^s
\end{aligned}$$

since $p = \ell/n$. For $p < \frac{1}{7}$, we have $\frac{1}{1-p} < \frac{7}{6}$ and $\frac{4}{(1-p)\ell n 2} < \frac{14}{3\ln 2} < 7$. Thus

$$\Pr[\, d(T(f \upharpoonright_\rho)) \geq s \mid \rho \in \mathcal{R}_n^\ell \,] < (7pr)^s.$$

We now state Razborov's original form of the Switching Lemma, where a conjunction of small clauses is switched to a disjunction of small terms, modulo a random restriction. The proof is similar to that of the decision tree form in Theorem 2.6.5, and is sketched in Exercise 2.14.4.

Theorem 2.6.6 ([Raz94], pp. 380–383). *Let f be an r-CNF formula over the variables $x_1, \ldots, x_n$. Recall that $\min(g)$ denotes the largest size of a minterm of g. For ρ randomly chosen in $\mathcal{R}_n^\ell$ and $p = \ell/n$,*

$$\Pr[\min(f \upharpoonright_\rho) \geq s] < (7pr)^s.$$

The following is a corollary to Theorem 2.6.6, but can also be derived from Theorem 2.6.5.

Corollary 2.6.1. *For all $n, d > 0$, $\textsc{Parity} \notin \Sigma_d^{2^s,s}$ for $s \leq \frac{n^{1/d}}{28}$.*

Proof. Let $s \leq \frac{n^{1/d}}{28}$, $S = 2^s$ and $C \in \Sigma_d^{S,s}$. For $1 \leq i \leq d-1$, define $\ell_i = n^{1-i/d}$ and $n_i = \ell_i \cdot n^{1/d}$. Inductively define restrictions $\rho_1, \ldots, \rho_{d-1}$ with $\rho_i \in \mathcal{R}_{n_i}^{\ell_i}$, and circuits C_i formed by applying Theorem 2.6.6 to C_{i-1} with n, ℓ, r replaced respectively by n_i, ℓ_i, s. Then C_i can be expressed in $\sum_{d-i}^{S,s}$ form. In going from C_i to C_{i+1}, a restriction $\rho_{i+1} \in \mathcal{R}_{n_i}^{\ell_i}$ is chosen which sets $n^{1/d}$ additional input bits.

Suppose that $\rho_1, \ldots, \rho_{i-1}$ have been constructed so that $C_k \in \Sigma_{d-k}^{S,s}$ and $C_k \equiv C \upharpoonright_{\rho_1 \cdots \rho_k}$, for $k = 1, \ldots, i-1$. By Theorem 2.6.6, for $\rho \in \mathcal{R}_{n_i}^{\ell_i}$

$$\begin{aligned} \Pr[\min(C_{i-1} \upharpoonright_\rho) \geq s] &\leq S \cdot \left(\frac{7\ell_i s}{n_i}\right)^s \leq 2^s \cdot \left(\frac{7\ell_i s}{n^{1/d}\ell_i}\right)^s \\ &\leq 2^s \cdot \left(\frac{7}{n^{1/d}} \cdot \frac{n^{1/d}}{28}\right)^s \leq \left(\frac{1}{2}\right)^s. \end{aligned}$$

Thus a suitable $\rho_i \in \mathcal{R}_{n_i}^{\ell_i}$ can be chosen for which $C_i = C_{i-1} \upharpoonright_{\rho_i}$ belongs to $\Sigma_{d-1}^{S,s}$.

Now $C_{d-1} \in \Sigma_1^{S,s}$ computes $\textsc{Parity}$ (or its negation) and has $n^{1/d}$ unset variables. But any non-constant $\Sigma_1^{S,s}$ function (i.e., any s-DNF formula) can be forced to 1 by setting s input bits. Since $n^{1/d}/28 < n^{1/d}$, the resulting circuit still has unset variables and computes $\textsc{Parity}$ or its negation, yet outputs a constant value, giving a contradiction.

2.6.4 A Hybrid Switching Lemma and *st*-Connectivity

In [BIP98], Beame, Impagliazzo and Pitassi developed a new type of Switching Lemma and gave an application to *st*-connectivity. Their hybrid Switching Lemma involves aspects of the *independent set-style* Switching Lemma of Ajtai [Ajt83], [FSS84] and the Håstad-style Switching Lemma, using the combinatorial argument due to Razborov [Raz94].

To get a flavor of the independent set-style argument, following [Ajt83] and especially p. 37 of [Wil83], define a set $S \subseteq \mathcal{R}_n^\ell$ to be *complete*, provided that S satisfies the following conditions.

1. *Incompatibility*: For distinct $\sigma, \tau \in S$ there exists x on which both σ, τ are defined and different.
2. *Completeness*: For all total truth assignments $G \in 2^n$ there is a $\sigma \in S$ with $\sigma \subseteq G$.

In Theorem 5.7 of [Wil83], Wilkie gave a simplified proof of (a weaker form of) Ajtai's theorem,[11] which asserts, that for any AC^0 language $L \subseteq \{0,1\}^*$, there exists k such that for sufficiently large n, there exists a complete set $S \subseteq \mathcal{R}_n^{n^{1/k}}$ and subset $T \subseteq S$ for which the symmetric difference $L_n \mathrel{\Delta} \cup_{\tau \in T} B_\tau^n$ has cardinality at most $2^{n-n^{1/k}}$. Here, $L_n = L \cap \{0,1\}^n$, Δ denotes symmetric difference, and the basic open sets B_τ^n of the finite Borel hierarchy are defined by

$$B_\sigma^n = \{G \in 2^n : \sigma \subseteq G\}$$

where σ is a partial truth assignment (i.e., a restriction). Dividing the expression bounding the symmetric difference by 2^n, we have that the probability is small (at most $2^{-n^{1/k}}$) that the boolean function f, giving the characteristic function of L_n, *cannot* be written in $n^{1/k}$-DNF as a disjunction of small conjunctions.

Recall from Section 2.4.3, that *st-connectivity* is the problem, given a finite undirected graph G with designated vertices s, t, whether there is a path from s to t. Appropriately formulated, it is well-known that the analogous problem of *st*-connectivity for finite *directed* graphs is complete for nondeterministic logarithmic space. Note that distance bounded connectivity for undirected graphs is just as difficult as that for directed graphs (see Exercise 2.14.25).

Definition 2.6.3. *st-CONN$(k(n))$ is the problem, given undirected graph G on n vertices, with designated vertices s, t, whether there is a path from s to t of length at most $k(n)$.*

Since multiplication of two $n \times n$ boolean matrices can be done by depth 2 polynomial size semi-unbounded fan-in[12] monotonic boolean circuits by

$$c_{i,j} = \bigvee_{k=1}^{n} (a_{i,k} \wedge b_{k,j})$$

it follows by repeated squaring that st-CONN$(k(n)) \in \mathrm{AC}^1$, and is computed by depth $2 \cdot \log k(n)$ polynomial size semi-unbounded fan-in boolean circuits.

[11] The weaker form still implies Theorem [FSS84].

[12] Semi-unbounded means unbounded fan-in $\bigvee$ gates, and fan-in 2 $\bigwedge$ gates.

Replacing the unbounded $\bigvee$ gates by a logarithmic depth tree of fan-in 2 $\bigvee$ gates, it follows that st-CONN$(k(n))$ is computable by depth $O(\log n \cdot \log k)$ polynomial size fan-in 2 boolean circuits, hence belongs to monotonic NC^2.

M. Ajtai [Ajt89] proved that st-CONN$(k(n))$ cannot be computed in constant depth polynomial size unbounded fan-in boolean circuits, i.e., st-CONN$(k(n)) \notin \mathrm{AC}^0$. For $k = \log^{O(1)} n$, Ajtai's work, as analyzed by [BPU92], yields a lower bound of $\Omega(\log^* n)$ for polynomial size circuits. The main result of this section, due to P. Beame, R. Impagliazzo, and T. Pitassi [BIP98], improves this lower bound as follows.

Theorem 2.6.7 ([BIP98]).

1. *For $k = \log^{O(1)} n$, every polynomial size unbounded fan-in boolean circuit computing st-CONN$(k(n))$ has depth $\Omega(\log \log k(n))$.*
2. *There exists $c > 0$ such that for all $k \leq \log n$, every depth d unbounded fan-in boolean circuit computing st-CONN$(k(n))$ has size at least $n^{ck^{\epsilon_d}}$, where $\epsilon_d = \phi^{-2d/3}$, and $\phi = \frac{1+\sqrt{5}}{2}$ is the golden mean.*

Unfortunately, this theorem says nothing about whether st-CONN$(k(n))$ belongs to NC^1, and the best current result is Theorem 2.4.2 of Karchmer–Wigderson: *st-connectivity belongs to monotonic AC^1, but not to monotonic NC^1.* As well, there is still a gap between the $O(\log k)$ upper bound and the $\Omega(\log \log k)$ lower bound for polynomial size circuits, when k is polylogarithmic in n.

The method introduced by [BIP98] is of independent interest, as it develops a *hybrid* Switching Lemma, combining aspects of the *independent set-style* Switching Lemma due to Ajtai [Ajt83] and Furst, Saxe and Sipser [FSS84] with the Håstad-style Switching Lemma, using Razborov's proof technique from Theorems 2.6.5 and 2.6.6.

In the independent set-style form of Switching Lemma, one shows that, given DNF formula f of term size r, with high probability, given a random restriction ρ, there is a small set V of variables, which when arbitrarily set by restriction σ yields a DNF formula $f \upharpoonright_{\rho\sigma}$ of term size at most $r - 1$. Decision trees can then be built in r stages,[13] where the term size is reduced by 1 in each stage. With this method, the *same* restriction ρ can be re-used at each stage of the tree-building process.

With the Håstad-style of Switching Lemma, one shows that, given DNF formula f of term size r, with high probability, given a random restriction ρ, the restriction $f \upharpoonright_{\rho}$ is equivalent to a CNF formulas of small clause size. Informally, this type of lemma asserts, that given an OR of small AND-gates, after applying a random restriction, one obtains an AND of small OR-gates. Rather than "switching" the DNF into CNF form, one can directly construct

[13] In the $(i+1)$-st stage, the leaves of the decision tree from the i-th stage are extended by appending to its leaves truth assignments to the additional set V_i of variables.

the decision tree, generally with smaller tree depth. Nevertheless, [BIP98] contends that independent set-style arguments are more flexible for certain situations.

The plan of this section is as follows. We first present the new hybrid Switching Lemma for the uniform probability distribution on boolean variables, then extend this Switching Lemma to the problem of st-connectivity, involving a nonuniform distribution. Finally, we indicate how the Switching Lemma implies a lower bound for circuit depth of polynomial size circuits computing st-connectivity. Roughly, if st-CONN$(k(n))$ is computed by shallow polynomial size circuits, then one can build a decision tree of depth $< k$ for distance k connectivity on a layered graph $\mathcal{G}(n', k)$, for $n' > 1$ (definition given later). However, this decision tree does not mention enough edge variables to decide the connectivity problem, a contradiction.

2.6.5 Hybrid Switching with the Uniform Distribution

Let f be a DNF formula over the variables $x_1, \ldots, x_n$, having term size at most r. Let ρ be chosen uniformly randomly from the space $\mathcal{R}_n^\ell$ of all restrictions on the variables $x_1, \ldots, x_n$ leaving exactly ℓ unset variables (or $*$'s). A set B of literals is *consistent* if for no variable x, is it the case that $x, \overline{x} \in B$. Sets $B_1, \ldots, B_s$ of literals are *T-consistent* if $T \cup B_1 \cup \cdots \cup B_s$ is consistent, and are *T-independent* if $B_i \cap B_j \subseteq T$ for all $1 \leq i < j \leq n$. Throughout this section, we identify a set T of literals with the minimum restriction σ_T forcing each literal to 1:

$$\sigma_T(x_i) = \begin{cases} 1 \text{ if } x_i \in T \\ 0 \text{ if } \overline{x}_i \in T \\ * \text{ otherwise.} \end{cases}$$

A restriction $\rho \in \mathcal{R}_n^\ell$ is said to be *s-bad for f* if there exists a set T of literals unset by ρ, and s-many terms $C_1, \ldots, C_s$ in $f = F_1 \vee \cdots \vee F_m$ such that:

1. $f \upharpoonright_{\rho \cup T} \not\equiv 0, 1$,
2. $C_1, \ldots, C_s$ are T-consistent,
3. $C_1, \ldots, C_s$ are T-independent.

In the construction below, given s-bad $\rho \in \mathcal{R}_n^\ell$, we will always take the lexicographic first set T of literals and terms $C_1, \ldots, C_s$ appearing in f which witness the s-badness of ρ. As well, it is helpful to picture the $T, C_1, \ldots, C_s$ as a sunflower, where the C_i are the petals and T the kernel. A warning is in order: unlike the Håstad-style approach, terms will not be simplified after a restriction is applied.

Lemma 2.6.7 (Uniform s-bad Lemma, [BIP98]). *Assume that $s \leq \ell \leq \frac{n}{2}$. Let f be a* DNF *formula with term size r over the variables $x_1, \ldots, x_n$, and $\rho \in \mathcal{R}_n^\ell$. Then*

$$\Pr[\rho \text{ is } s\text{-bad for } f \mid \rho \in \mathcal{R}_n^\ell] \leq \left(2r2^r \frac{\ell}{n}\right)^s .$$

Before beginning the proof, it should be mentioned that by applying techniques from the proof of Theorem 2.6.1 or Theorem 2.6.6, one obtains a substantially better bound.

Proof. Let $Bad_n^\ell(f,s)$ designate the set of all $\rho \in \mathcal{R}_n^\ell$, which are s-bad for f. Recall that for $\rho \in Bad_n^\ell(f,s)$, the set T of variables and terms $C_1,\ldots,C_s$ from $f = F_1 \vee \cdots \vee F_m$ have been chosen lexicographically least to witness the s-badness of ρ for f. In this case, define $S_\rho = (C_1 \cup \cdots \cup C_s) - \rho$, and note that $|S_\rho| \leq rs$, because term size of each C_i is at most r. Throughout the proof, we'll use $G \in 2^n$ to vary over total truth assignments for the variables $x_1,\ldots,x_n$.

Given $\rho \in Bad_n^\ell(f,s)$, a total truth assignment $G \in 2^n$ is defined to be an *encoding of* ρ, if $\rho \cup S_\rho \subseteq G$, i.e., G extends ρ and $G \models C_1 \wedge \cdots \wedge C_s$. Let $Encode_n^\ell(\rho)$ denote the set of encodings of s-bad restriction ρ. Why the term *encoding*? Later we'll obtain an upper bound on the number of s-bad ρ's for which G can be an encoding, because one can determine s of the ℓ unset variables of ρ, using the fact that $\rho \cup S_\rho \subseteq G$ together with a small set α of *advice*; i.e., G, together with small additional advice, encodes s of the unset variables of ρ. This leads to an expression of the form $\frac{\binom{n}{s}/\binom{\ell}{s}}{\binom{n}{\ell}}$ which yields an upper bound of s-bad ρ roughly of the form $(\frac{\ell}{n})^s$. Hence we'll have that the probability of choosing s-bad ρ is exponentially small in s.

CLAIM. Fix s-bad restriction $\rho' \in \mathcal{R}_n^\ell$. Then

$$|Encode_n^\ell(\rho')| \geq 2^{\ell - rs}$$

i.e., there are at least $2^{\ell-rs}$ encodings which extend ρ'.

Proof of Claim. G encodes ρ' means that $\rho' \cup S_{\rho'} \subseteq G$. Now ρ' has ℓ unset variables, $|S_{\rho'}| \leq rs$ of which will be set by $S_{\rho'}$. The values of G are those set on at most $\ell - rs$ variables, so there are at least $2^{\ell-rs}$ total truth assignments extending $\rho' \cup S_{\rho'}$. □

CLAIM. Given $\rho \in Bad_n^\ell(f,x)$, encoding G of ρ, and $s\log r$ bits of advice α, one can determine s unset variables of ρ.

Proof of Claim. Let G be a fixed encoding of s-bad ρ, and set $G_1 = G$. Since $\rho \cup S_\rho \subseteq G$, $G \models C_1 \wedge \cdots \wedge C_s$, as well as perhaps other terms from $f = F_1 \vee \cdots \vee F_m$. Let C_1' be the first term from f satisfied by G_1 (note that C_1' might possibly not be among the $C_1,\ldots,C_s$). Now $f \upharpoonright_{\rho \cup T} \not\equiv 1$, so since $f = F_1 \vee \cdots \vee F_m$, no $F_i \upharpoonright_{\rho \cup T}$ evaluates to 1, and hence there is a literal in $C_1' - \rho - T$. Use $\log r$ bits of advice to designate the first such literal α_1 in $C_1' - \rho - T$. Unset the literal α_1 from G_1 to produce G_2.

If C_1' is one of the $C_1,\ldots,C_s$ (say C_{i_0}), then $\alpha_1 \in C_{i_0} - \rho - T$, and by virtue of T-independence, α_1 cannot appear in any other C_j. Since $G \models C_1 \wedge \cdots \wedge C_s$,

G_2 must satisfy at least $s-1$ of the C_is. In the case that C_1' is not one of the $C_1, \ldots, C_s$, either α_1 belongs to none of the C_i, in which case G_2 satisfies all of $C_1, \ldots, C_s$, or α_1 belongs to one of the C_i, hence by T-independence to a unique C_{i_0}. Thus in any case, G_2 satisfies at least $s-1$ of the $C_1, \ldots, C_s$. Let C_2' be the first term of $f = F_1, \ldots, F_m$ satisfied by G_2, and α_2 the first literal in $C_2' - \rho - T$. Unset α_2 from G_2 to produce G_3, and similarly argue, using T-independence, that G_3 satisfies at least $s-2$ of the terms $C_1, \ldots, C_s$. In this manner, one can produce s variables unset by ρ while total advice is $s \log r = \log(r^s)$ bits. □

CLAIM. Fix total truth assignment G'. Then

$$|\{\rho \in \mathcal{R}_n^\ell : \rho \in Bad_n^\ell(f,s), G' \in Encode_n^\ell(\rho)\}| \leq r^s \binom{n-s}{\ell-s}$$

i.e., there are at most $r^s \binom{n-s}{\ell-s}$ many s-bad restrictions ρ, for which G' is an encoding.

Proof of Claim. Given G' and an advice string α of $s \log r$ bits, determine s variables as follows. Let $G_1 = G'$. Find the first term of $f = F_1 \vee \cdots \vee F_m$ satisfied by G_1, and determine the literal α_1 of this term, using $\log r$ bits from the advice string. Unset α_1 from G_1 to form G_2, then find the first term of f satisfied by G_2, and determine the literal α_2 from this term, using an additional $\log r$ bits from the advice string. Continuing, this process produces the variables $x_{i_1}, \ldots, x_{i_s}$ among $x_1, \ldots, x_n$. For each of the $\binom{n-s}{\ell-s}$ sets $Z_j = \{z_1^j, \ldots, z_{\ell-s}^j\}$ of variables chosen from $\{x_1, \ldots, x_n\} - \{x_{i_1}, \ldots, x_{i_s}\}$, define $\rho_{G',\alpha,j} \in \mathcal{R}_n^\ell$ by

$$\rho_{G',\alpha,j}(x_i) = \begin{cases} * & \text{if } x_i \in \{x_{i_1}, \ldots, x_{i_s}\} \cup \{z_1^j, \ldots, z_{\ell-s}^j\} \\ G'(x_i) & \text{otherwise.} \end{cases}$$

Now, if $\rho \in Bad_n^\ell(f,s)$ and G' is an encoding of ρ, then ρ must be $\rho_{G',\alpha,j}$ for some advice string α of $s \log r$ bits, $1 \leq j \leq \binom{n-s}{\ell-s}$. There are at most $2^{s \log r} = r^s$ advice strings, hence at most $r^s \binom{n-s}{\ell-s}$ many $\rho \in Bad_n^\ell(f,s)$, for which G' is an encoding of ρ. □

Define $D = \{(\rho, G) : \rho \in Bad_n^\ell(f,s), G \in Encode_n^\ell(\rho)\}$. We've shown that for each fixed $\rho' \in Bad_n^\ell(f,s)$, there are at least $2^{\ell - rs}$ encodings G of ρ'; i.e., letting $D_{\rho'}$ denote the vertical ρ'-section $\{(\rho', G) : (\rho', G) \in D\}$ (see Figure 2.2)

$$|D_{\rho'}| \geq 2^{\ell - rs}.$$

It follows that

$$|D| \geq |Bad_n^\ell(f,s)| \cdot 2^{\ell - rs}.$$

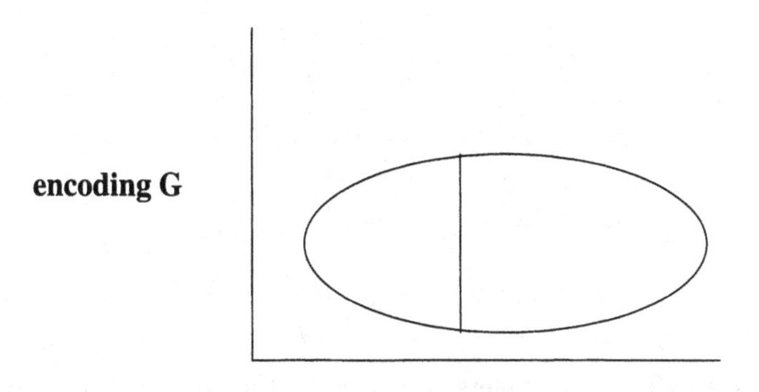

Fig. 2.2. Vertical section of D

As well, we have shown that for each fixed G' encoding some s-bad restriction, there are at most $r^s\binom{n-s}{\ell-s}$ s-bad restrictions, for which G' is an encoding; i.e., letting $D_{G'}$ denote the horizontal G'-section $\{(\rho, G') : (\rho, G') \in D\}$,

$$|D_{G'}| \le r^s \binom{n-s}{\ell-s}$$

and so

$$|D| \le r^s \binom{n-s}{\ell-s} \cdot |Encodings_n^\ell|$$

where $Encodings_n^\ell = \cup_\rho Encode_n^\ell(\rho)$, the union being taken over all s-bad restrictions ρ in $\mathcal{R}_n^\ell$. An encoding is a total truth assignment, so the number $|Encodings_n^\ell|$ of s-bad restrictions is bounded by 2^n, hence

$$|D| \le r^s \binom{n-s}{\ell-s} \cdot 2^n.$$

Thus

$$|Bad_n^\ell(f,s)| \le \frac{r^s\binom{n-s}{\ell-s}2^n}{2^{\ell-rs}}$$

hence

$$\begin{aligned}
\Pr[\,\rho \in Bad_n^\ell(f,s) \,|\, \rho \in \mathcal{R}_n^\ell\,] &= \frac{|Bad_n^\ell(f,s)|}{|\mathcal{R}_n^\ell|} \\
&\le \frac{r^s\binom{n-s}{\ell-s}2^n}{2^{\ell-rs}\binom{n}{\ell}2^{n-\ell}} \\
&= \frac{r^s\binom{n-s}{\ell-s}2^{rs}}{\binom{n}{\ell}}
\end{aligned}$$

$$
\begin{aligned}
&= \frac{r^s (n-s)!\ell!}{n!(\ell-s)!} \cdot 2^{rs} \\
&\leq \frac{r^s \ell^s}{(n-s)^s} \cdot 2^{rs} \\
&\leq \left(2r\frac{\ell}{n}\right)^s \cdot 2^{rs} \\
&= \left(2^r 2r\frac{\ell}{n}\right)^s
\end{aligned}
$$

since $n - s \geq n/2$. This completes the proof of the uniform s-bad lemma.

As a corollary we obtain the following decision tree form of the Uniform hybrid Switching Lemma.

Corollary 2.6.2. *Let $f = F_1 \vee \cdots \vee F_m$ be a* DNF *formula with term size at most r. Suppose that $\rho \in \mathcal{R}_n^\ell$ is not $(s+1)$-bad for f and that $s \leq \ell \leq \frac{n}{2}$. Then $f \upharpoonright_\rho$ has a boolean decision tree of depth at most $r^2 s$.*

Proof. By hypothesis, ρ is not $(s+1)$-bad for f, so $f \upharpoonright_\rho$ has a maximal consistent and independent set of at most s terms. Build a decision tree by querying the at most rs variables from these terms. At each leaf ℓ of the tree so far defined, let σ_ℓ^1 denote the restriction which sets the variables along the branch according to the directions taken by the branch. If $f \upharpoonright_{\rho \cup \sigma_\ell^1} \not\equiv 1$, then since ρ is not $(s+1)$-bad for $f \upharpoonright_{\sigma_\ell^1}$, find a maximal consistent and independent set of at most s terms for $f \upharpoonright_{\rho \cup \sigma_\ell^1}$. Extend the decision tree from each leaf ℓ by continuing to query the at most rs variables from these new terms. Repeat this procedure until $f \upharpoonright_\ell$ is 0 or 1. Each stage shortens the term size by at least 1, so after r stages, the restriction of f along a branch is identically 0 or 1. This results in a decision tree of depth at most $r(rs) = r^2 s$.

Corollary 2.6.3. *Let f be a* DNF *formula over the variables $x_1, \ldots, x_n$ with term size at most r. If $s \leq \ell \leq \frac{n}{2}$ and $\rho \in \mathcal{R}_n^\ell$ is chosen uniformly, then the probability that $f \upharpoonright_\rho$ does not have a decision tree of depth at most $r^2 s$ is at most $(2r2^r p)^{s+1}$, where $p = \frac{\ell}{n}$.*

Nonuniform Connectivity Switching Lemma

Following Ajtai [Ajt89], let $\mathcal{G}(n,k)$ be the set of all undirected graphs on $n(k+1)$ nodes, whose vertices can be partitioned into disjoint sets $V_0, \ldots, V_k$, and whose edges constitute bijections between V_i and V_{i+1}, for $i < k$. Since n is fixed throughout the section, a member of $\mathcal{G}(n,k)$ is called a *k-layered graph.* We represent k-layered graphs using boolean variables

$$
x_{i,j}^{k'} = \begin{cases} 1 \ \exists \text{ edge from } i \in V_{k'} \text{ to } j \in V_{k'+1} \\ 0 \text{ otherwise.} \end{cases}
$$

Note that there are kn^2 boolean variables used to represent a k-layered graph, as opposed to the usual manner of using $\binom{n(k+1)}{2}$ variables for an undirected graph on $n(k+1)$ vertices.

Let $\mathcal{R}^{\ell}_{n,k}$ denote the set of all random restrictions ρ obtained as follows:

1. Choose $U_i \subseteq V_i$, $|U_i| = \ell$, for $0 \leq i \leq k$.
2. Choose bijections from $V_i - U_i$ to $V_{i+1} - U_{i+1}$ for $0 \leq i < k$, thus producing a k-layered graph $G \in \mathcal{G}(n-\ell, k)$.
3. Define the restriction

$$\rho(x^{k'}_{i,j}) = \begin{cases} * \text{ if } i \in U_{k'} \text{ and } j \in U_{k'+1} \\ 1 \text{ if } i \in V_{k'} - U_{k'}, j \in V_{k'+1} - U_{k'+1} \\ \quad \text{and } \{i,j\} \text{ is an edge in } G \\ 0 \text{ otherwise.} \end{cases}$$

Note that $|\mathcal{R}^{\ell}_{n,k}| = \binom{n}{\ell}^{k+1}(n!)^k$, and that the distribution is nonuniform, since the variables are partially dependent:

$$x^{k'}_{i,j} = 1 \Rightarrow x^{k'}_{i',j} = 0, \quad x^{k'}_{i,j'} = 0$$

for $i \neq i'$ and $j \neq j'$. For a set E of edges in a k-layered graph, define the restriction

$$\rho_E(x^{k'}_{i,j}) = \begin{cases} 1 \text{ if edge from } i \in V_{k'} \text{ to } j \in V_{k'+1} \text{ belongs to } E \\ 0 \text{ if another edge of } E \text{ is incident to } i \in V_{k'} \text{ or } j \in V_{k'+1} \\ * \text{ otherwise.} \end{cases}$$

Conversely, given restriction $\rho \in \mathcal{R}^{\ell}_{n,k}$, define E_ρ to consist of those edges from $i \in V_{k'}$ to $j \in V_{k'+1}$ such that $\rho(x^{k'}_{i,j}) = 1$. Thus without the possibility of confusion, edge sets of k-layered graphs will be identified with restrictions.

Let $E, C_1, \ldots, C_s$ be sets of edges. $C_1, \ldots, C_s$ are *E-consistent* if there exists $G \in \mathcal{G}(n,k)$ containing $E \cup C_1 \cup \cdots \cup C_s$. If $C_1, \ldots, C_s$ are E-consistent, then two edges e, e' in the same layer of a term C_i sharing an incident vertex must be identical; formally if $e = \{x_{k'}, y_{k'}\}$ and $e' = \{u_{k'}, v_{k'}\}$ are such that $\{x_{k'}, y_{k'}\} \cap \{u_{k'}, v_{k'}\} \neq \emptyset$, then $e = e'$.

Define $C_1, \ldots, C_s$ to be *E-independent* if

1. $C_i \cap C_j \subseteq E$, for $i \neq j$,
2. there do not exist distinct i, j with edges $e \in C_i - E$, $e' \in C_j - E$ such that e, e' are on the same partial path in $E \cup \{e, e'\}$.

A DNF formula $f = F_1 \vee \cdots \vee F_m$ is called an *r-disjunction*, if term size of the F_i is at most r, and the F_i contain only positive literals. Every DNF of term size r is equivalent over $\mathcal{G}(n,k)$ to an r-disjunction, since term size is not changed by replacing $\neg x^{k'}_{i,j}$ by its equivalent $\bigvee_{i' \neq i} x^{k'}_{i',j}$ or $\bigvee_{j' \neq j} x^{k'}_{i,j'}$.

A restriction $\rho \in \mathcal{R}^{\ell}_{n,k}$ is defined to be s-bad for r-disjunction f if there exist a set T of edges disjoint from ρ[14] and terms $C_1, \ldots, C_s$ from $f = F_1 \vee \cdots \vee F_m$, such that:

1. $f \upharpoonright_{\rho \cup T} \not\equiv 0, 1$ over $\mathcal{G}(n,k)$,[15]
2. $C_1, \ldots, C_s$ are $\rho \cup T$-consistent,
3. $C_1, \ldots, C_s$ are $\rho \cup T$-independent.

Let $Bad^{\ell}_{n,k}(f,s)$ denote the collection of s-bad restrictions $\rho \in \mathcal{R}^{\ell}_{n,k}$ for f.

Lemma 2.6.8 (Connectivity s-bad lemma). *Let f be an r-disjunction over $\mathcal{G}(n,k)$, and let $\rho \in \mathcal{R}^{\ell}_{n,k}$ be a random restriction. Assume that $4s^2r^2k < \ell$. Then*

$$\Pr[\,\rho \in Bad^{\ell}_{n,k}(f,s) \,|\, \rho \in \mathcal{R}^{\ell}_{n,k}\,] \leq (3er(2k\ell)^r p)^s$$

where $p = \ell/n$.

Proof. Given s-bad ρ for r-disjunction f, let $T, C_1, \ldots, C_s$ be the lexicographically least set and terms witnessing the s-badness of ρ. Define $T' = T \cap (\cup_{i=1}^{s} C_i)$. The $\rho \cup T$-consistency and $\rho \cup T$-independence of $C_1, \ldots, C_s$ easily imply $\rho \cup T'$-consistency and $\rho \cup T'$-independence. Define $S_\rho = (C_1 \cup \cdots \cup C_s) - \rho$, and note that $|S_\rho| \leq rs$ because term size of C_i is at most r.

By $(\rho \cup T')$-consistency, S_ρ consists of vertex disjoint paths in the unset variables of ρ. By $(\rho \cup T')$-independence, each path P of S_ρ is contained entirely in some $C_i \cup T'$. Indeed, otherwise, one could choose edges $e \in C_i - \rho - T'$, $e' \in C_j - \rho - T'$ closest together, yielding a path in $T' \cup \{e, e'\}$. This would then contradict $\rho \cup T'$-independence.

Define $G \in \mathcal{G}(n,k)$ to be an *encoding* of s-bad restriction ρ if

1. $\rho \cup S_\rho \subseteq G$,
2. no two paths[16] in S_ρ are part of the same path of G.

For s-bad $\rho \in \mathcal{R}^{\ell}_{n,k}$, let $Encode^{\ell}_{n,k}(\rho)$ denote the set of all encodings $G \in \mathcal{G}(n,k)$ of ρ. The definition of $\rho \cup T'$-independence is so fashioned, that for all paths P in $(C_1 \cup \cdots \cup C_s) - \rho$, P is contained in a unique C_{i_0}.

CLAIM. Fix an s-bad restriction $\rho' \in \mathcal{R}^{\ell}_{n,k}$. Then

$$|Encode^{\ell}_{n,k}(\rho')| \geq \frac{(\ell!)^k}{e\ell^{|S_{\rho'}|}}$$

---

[14] i.e., T consists of certain variables $x^{k'}_{i,j}$, where i (j) is among the unset vertices of $V_{k'}$ ($V_{k'+1}$).

[15] A boolean function or formula f is equivalent to 0 (1) over $\mathcal{G}(n,k)$ if no (every) k-layered graph satisfies f.

[16] Here, "path" means a maximal path in S_ρ; i.e., a path which cannot be extended by edge variables from S_ρ. Of course, such paths may be partial in the sense that they do not go from a vertex in layer 0 to a vertex in layer k.

i.e., there are at least $\frac{(\ell!)^k}{e\ell^{|S_{\rho'}|}}$ encodings G, which extend ρ'.

Proof of Claim. Given s-bad ρ' and $S_{\rho'}$, let $P_1, \ldots, P_p$ be paths in $S_{\rho'}$. Extend P_1 backwards and forwards one edge at a time, avoiding nodes in any other paths. Next, similarly extend P_2 on the remaining nodes, etc. When all paths have been extended, choose a random layered graph for the $\ell - p$ remaining nodes of each layer.

There are $pk - |S_{\rho'}|$ extension phases, and in each phase there are at least $\ell - p$ choices for an edge of that layer. After extending the paths $P_1, \ldots, P_p$, we have $(\ell - p)!^k$ remaining choices to form a layered graph on the $\ell - p$ remaining nodes for each layer. Each s-bad ρ' has at least

$$
\begin{aligned}
(\ell - p)^{pk-|S_{\rho'}|}(\ell - p)!^k &\geq (\ell - p)^{pk} \frac{(\ell - p)!^k}{\ell^{|S_{\rho'}|}} \\
&\geq \left(1 - \frac{p}{\ell}\right)^{pk} \frac{\ell^{pk}}{\ell^{|S_{\rho'}|}} (\ell - p)!^k \\
&\geq \left(1 - \frac{p}{\ell}\right)^{pk} \frac{(\ell!)^k}{\ell^{|S_{\rho'}|}} \\
&\geq e^{-4p^2k/\ell} \frac{(\ell!)^k}{\ell^{|S_{\rho'}|}} \\
&\geq e^{-1} \frac{(\ell!)^k}{\ell^{|S_{\rho'}|}}.
\end{aligned}
$$

encodings. The third inequality follows since

$$\ell^p \geq \ell^{\underline{p}} = \ell(\ell - 1) \cdots (\ell - (p - 1)) = \frac{\ell!}{(\ell - p)!}$$

so that $(\ell^p(\ell - p)!)^k \geq (\ell!)^k$ and $\ell^{pk}(\ell - p)!^k \geq (\ell!)^k$. The fourth inequality follows since

$$\left(1 - \frac{p}{\ell}\right)^{pk} = \left[\left(1 - \frac{1}{\ell/p}\right)^{\ell/p}\right]^{(p/\ell)\cdot pk}$$

and by hypothesis $p \leq rs$ and $4(rs)^2k \leq \ell$, so $4p \leq \ell$ and $(1 - \frac{1}{\ell/p})^{\ell/p} \geq e^{-4}$. The last inequality follows from $4p^2k \leq 4(rs)^2k \leq \ell$. This concludes the proof of the claim.

Decoding Claim. Given $\rho \in Bad^{\ell}_{n,k}(f, s)$, $G \in Encode^{\ell}_{n,k}(\rho)$, and $s(\log r + 1) + rs(\log k + 1)$ bits of advice, one can determine s unset variables of ρ.

Proof of Claim. Fix an encoding G for s-bad ρ. Let $G_1 = G$. Since $\rho \cup S_\rho \subseteq G$, we have $G \models C_1 \wedge \cdots \wedge C_s$. Let C_1' be the first term of $f = F_1 \vee \cdots \vee F_m$ forced to be true by G_1. Now $f \upharpoonright_{\rho \cup T'} \not\equiv 1$ over $\mathcal{G}(n, k)$, so $C_1' \upharpoonright_{\rho \cup T'} \not\equiv 1$, hence there is an edge $e_1 \in C_1 - \rho - T'$ and the accompanying path P_1 determined from G_1. Use $\log r$ bits of advice to specify e_1 among the r literals of C_1'. Use one additional bit to specify whether there are additional edges in $P_1 \cap T'$. Use

$(\log k + 1)$ bits per additional edge to specify all remaining edges of $P_1 \cap T'$, where one bit is used to say whether there is an additional edge, and if so, $\log k$ bits to indicate the layer containing the edge (from e_1, G_1, and the layer, one can find the associated edge of path $P_1 \cap T'$).

Next, obtain the k-layered graph G_2 from G_1 by deleting all edges of $P_1 - T'$ from G_1.

SUBCLAIM. G_2 satisfies at least $s-1$ of the terms $C_1, \ldots, C_s$.

Proof of Subclaim. Suppose that C_1' is one of the terms $C_1, \ldots, C_s$. By the property of being an encoding, path P_1 contains at most one maximal partial path from S_ρ. By $(\rho \cup T')$-independence, there do not exist edges $e \in C_i$, $e' \in C_j$, $i \neq j$, for which there exists a partial path in $T' \cup \{e, e'\}$. It follows that $(P_1 - T') \cap (C_1 \cup \cdots \cup C_s)$ is contained in a unique C_{i_0}. Since G_1 satisfies all the $C_1, \ldots, C_s$, by deleting all edges in $P_1 - T'$ from G_1 to form G_2, it must be that G_2 satisfies at least $s-1$ of the terms $C_1, \ldots, C_s$. This concludes the proof of the subclaim when C_1' is one of the $C_1, \ldots, C_s$.

If C_1' is not one of the $C_1, \ldots, C_s$, then either an edge of P_1 belongs to one of the C_i (and hence a unique C_{i_0} by the previous argument), or not (in which case, G_2 satisfies all the terms $C_1, \ldots, C_s$). Thus G_2 satisfies at least $s-1$ of the terms $C_1, \ldots, C_s$). □

Let C_2' be the first term from $f = F_1 \vee \cdots \vee F_m$ which is satisfied by G_2. Since $f \upharpoonright_{\rho \cup T'} \not\equiv 1$ over $\mathcal{G}(n,k)$, we have $C_2' \upharpoonright_{\rho \cup T'} \not\equiv 1$, so there is an edge $e_2 \in C_2' - \rho - T'$. Using G_2, determine the path P_2 containing edge e_2. Use $\log r$ bits of advice to specify edge e_2, one additional bit to specify if there are other edges in $P_2 \cap T'$, and then $\log k + 1$ edges per additional edge to specify the layer and whether there remaining edges in $P_2 \cap T'$. Form G_3 from G_2 by deleting all edges of $P_2 - T'$; argue as in the preceding subclaim that G_3 must satisfy at least $s-2$ of the terms $C_1, \ldots, C_s$, etc. Since $|T'| \leq rs$, there are at most rs edges in $\cup_{i=1}^{s} (P_i \cap T')$, so repeating this process s stages, total advice is at most

$$s(\log r + 1) + rs(\log k + 1).$$

This concludes the proof of the decoding claim.

CLAIM. Fix $G' \in \mathcal{G}(n,k)$. Then

$$|\{\rho \in \mathcal{R}_{n,k}^{\ell} : \rho \in Bad_{n,k}^{\ell}(f,s), G' \in Encode_{n,k}^{\ell}(\rho)\}| \leq (2r)^s (2k)^{rs} \binom{n-s}{\ell-s}$$

i.e., there are at most $(2r)^s (2k)^{rs} \binom{n-s}{\ell-s}$ many s-bad ρ for which G' is an encoding.

Proof of Claim. Set $G_1 = G'$. Since G_1 is an encoding of s-bad ρ, by the previous claim, let α be an advice string of $s(\log r + 1) + rs(\log k + 1)$ bits. Determine the first C_1' from $f = F_1 \vee \cdots \vee F_m$ satisfied by G_1. Use $\log r$ bits of α to determine the edge e_1 from C_1', and let P_1 be the associated path in

G_1 containing edge e_1. Using $1 + m_1(\log k + 1)$ bits of advice, determine all the m_1 edges in $P_1 \cap T'$. Delete from G_1 all edges in $P_1 - T'$ to form G_2. Find the first term C_2' from $f = F_1 \vee \cdots \vee F_m$ satisfied by G_2. Using $\log r$ bits of α, determine edge e_2 of C_2', and associated path P_2 in G_2 containing edge e_2. Delete all edges of $P_2 - T'$ from G_2 to form G_3, etc. In this manner, we specify s unset edges among the ℓ unset edges of ρ, and hence s vertex disjoint paths $P_1, \ldots, P_s$ (vertex disjoint, with the exception of common vertices incident to edges in $\rho \cup T'$) in the original k-layered graph G'.

Since $2^{s(\log r+1)+rs(\log k+1)} = (2r)^s(2k)^{rs}$, G' can be an encoding of at most

$$(2r)^s(2k)^{rs}\binom{n-s}{\ell-s}$$

many s-bad restrictions. □

Define $D = \{(\rho, G) : \rho \in Bad^{\ell}_{n,k}(f,s), G \in Encode^{\ell}_{n,k}(\rho)\}$. We have shown that for each fixed $\rho' \in Bad^{\ell}_{n,k}(f,s)$, there are at least $\frac{(\ell!)^k}{e\ell^{|S_{\rho'}|}}$ encodings G, which extend ρ'; i.e., letting $D_{\rho'}$ denote the vertical ρ'-section $\{(\rho', G) : (\rho', G) \in D\}$,

$$|D_{\rho'}| \geq \frac{(\ell!)^k}{e\ell^{|S_{\rho'}|}}.$$

It follows that

$$|D| \geq |Bad^{\ell}_{n,k}(f,s)|\frac{(\ell!)^k}{e\ell^{|S_{\rho'}|}}.$$

As well, we have shown that for each fixed G' encoding some s-bad restriction, there are at most $(2r)^s(2k)^{rs}\binom{n-s}{\ell-s}$ many s-bad restrictions, for which G' is an encoding; i.e., letting $D_{G'}$ denote the horizontal G'-section $\{(\rho, G') : (\rho, G') \in D\}$,

$$|D_{G'}| \leq (2r)^s(2k)^{rs}\binom{n-s}{\ell-s}$$

so

$$|D| \leq (2r)^s(2k)^{rs}\binom{n-s}{\ell-s}|Encodings^{\ell}_{n,k}|.$$

The number $|Encodings^{\ell}_{n,k}|$ of s-bad restrictions is clearly bounded by $(n!)^k$, the number of graphs in $\mathcal{G}(n,k)$, so

$$|D| \leq (2r)^s(2k)^{rs}\binom{n-s}{\ell-s}(n!)^k.$$

Thus

$$|Bad^{\ell}_{n,k}(f,s)| \leq \frac{(2r)^s(2k)^{rs}\binom{n-s}{\ell-s}(n!)^k}{(\ell!)^k/e\ell^{|S_{\rho'}|}}$$

so

$$\Pr[\rho \in Bad^{\ell}_{n,k}(f,s) \mid \rho \in \mathcal{R}^{\ell}_{n,k}] = \frac{|Bad^{\ell}_{n,k}(f,s)|}{|\mathcal{R}^{\ell}_{n,k}|}$$
$$\leq \frac{(2r)^s(2k)^{rs}\binom{n-s}{\ell-s}(n!)^k e\ell^{|S_{\rho'}|}}{(\ell!)^k\binom{n}{\ell}^{k+1}(n-\ell)!^k}$$
$$= \frac{(2r)^s(2k)^{rs}\binom{n-s}{\ell-s}}{\binom{n}{\ell}} \cdot \frac{(n!)^k e\ell^{|S_{\rho'}|}}{(\ell!)^k\binom{n}{\ell}^k(n-\ell)!^k}.$$

Now

$$\frac{(2r)^s(2k)^{rs}\binom{n-s}{\ell-s}}{\binom{n}{\ell}} \leq \frac{(2r)^s(2k)^{rs}(n-s)!(\ell)!}{n!(\ell-s)!}$$
$$\leq \frac{(2r)^s(2k)^{rs}(\ell)^s}{(n-s)^s}$$
$$\leq (3r(2k)^r(\ell)/n)^s$$

because $n - s \geq 2n/3$. As well, $\binom{n}{\ell}^k = \left(\frac{n!}{\ell!(n-k)!}\right)^k$ so

$$\frac{(n!)^k e\ell^{|S_{\rho'}|}}{(\ell!)^k\binom{n}{\ell}^k(n-\ell)!^k} = e\ell^{|S_\rho|} \leq e\ell^{rs}.$$

Hence

$$\Pr[\rho \in Bad^{\ell}_{n,k}(f,s) \mid \rho \in \mathcal{R}^{\ell}_{n,k}] \leq (3r(2k)^r p)^s \cdot e\ell^{rs}$$
$$= e(3r(2k\ell)^r p)^s \leq (3er(2k\ell)^r p)^s$$

where $p = \ell/n$. This concludes the proof of the connectivity s-bad lemma.

Decision Trees over $\mathcal{G}(n,k)$

A *decision tree for layered graphs over* $\mathcal{G}(n,k)$ is defined to be a rooted, labeled tree with the following properties. Leaves are labeled by $0, 1$. Internal nodes are labeled by $(v,+)$ for $v \in V_i$, $0 \leq i < k$, or by $(v,-)$ for $v \in V_i$, $0 < i \leq k$, thus indicating a forward or backward query. For $(v,+)$ $((v,-))$, $v \in V_i$, the outedges are labeled by those nodes $u \in V_{i+1}$ $(u \in V_{i-1})$, which preserve the property that every path from root to leaf determines a partial k-layered graph over $V_0 \cup \cdots \cup V_k$.

Definition 2.6.4. *A decision tree T represents a boolean function or formula f over $\mathcal{G}(n,k)$, if every k-layered graph $G \in \mathcal{G}(n,k)$ satisfies f if and only if the path in T from root to leaf determined by G is labeled by* 1.

Lemma 2.6.9 (Connectivity switching lemma (decision tree)). *Let f be an r-disjunction over $\mathcal{G}(n,k)$ and $\rho \in \mathcal{R}^{\ell}_{n,k}$ a random restriction. Assume that s satisfies $4r^2s^2k < \ell$. Then with probability at least*

$$\gamma = 1 - (3er(2k\ell)^r\ell/n)^s$$

$f \upharpoonright_\rho$ can be represented by a decision tree over $\mathcal{G}(n,k)$ of depth at most $4r^2s$.

Proof. By Lemma 2.6.8, with probability at least γ a random restriction from $\mathcal{R}^{\ell}_{n,k}$ satisfies the following property $\mathcal{P}$:

> *For every set T of edges not set by ρ, if $f \upharpoonright_{\rho \cup T} \not\equiv 0, 1$ over $\mathcal{G}(n,k)$, then every maximal collection of $\rho \cup T$-consistent and $\rho \cup T$-independent terms from $f = F_1 \vee \cdots \vee F_m$ has size $\leq s$.*

Let ρ satisfy $\mathcal{P}$. We construct the decision tree $\mathcal{T}$ for $f \upharpoonright_{\rho}$ of depth at most $4r^2 s$, by a construction involving r stages, where unlabeled leaves of the i-th stage decision tree are extended by at most $4rs$ queries. The initial decision tree $\mathcal{T}_0$ consists of a single point. Given decision tree $\mathcal{T}_i$ and restriction σ_ℓ corresponding to the branch from root to leaf ℓ of $\mathcal{T}_i$, if $f \upharpoonright_{\rho\sigma_\ell} \equiv 0$ (1) over $\mathcal{G}(n,k)$, then label leaf ℓ by 0 (1). Otherwise, let $C_1, \ldots, C_{s'}$ be a maximal set of $\rho\sigma_\ell$-consistent and $\rho\sigma_\ell$-independent edges. By property $\mathcal{P}$, $s' \leq s$, so the set S_{i+1} of vertices incident to edges of $C_1 \cup \cdots \cup C_{s'}$ has size at most $2rs$ (term size of each C_i is $\leq r$, since f is an r-disjunction, and there are two vertices per edge). For each $v \in S_{i+1}$, if vertex v is not incident to an edge of T_i, then extend $\mathcal{T}_i$ by the query $(v, +)$ and then $(v, -)$ to determine the predecessor and successor of v in a path of a layered graph. Otherwise, v belongs to a path P of σ_ℓ, so query the predecessor of the first node of P and the successor of the last node of P. In each case, at most two queries have been made per vertex, so the depth of the resulting decision tree $\mathcal{T}_{i+1}$ has increased by at most $2(2rs)$.

By construction, all the nodes incident to $C_1, \ldots, C_{s'}$ are queried, so letting $\pi_{\ell'}$ represent the restriction corresponding to any branch in the extended tree $\mathcal{T}_{i+1}$ from root to leaf ℓ', $C_i \upharpoonright_{\rho\pi_{\ell'}} \equiv 0, 1$ over $\mathcal{G}(n,k)$ for $1 \leq i \leq s'$. For term C from $f = F_1 \vee \cdots \vee F_m$ not among the C_i, by maximality, $C, C_1, \ldots, C_{s'}$ are either $\rho\sigma_\ell$-inconsistent or $\rho\sigma_\ell$-dependent. In either case, a small inductive argument shows that if $\rho \cup \sigma_\ell \cup C$ is consistent over $\mathcal{G}(n,k)$, then at least i edges from C are determined in every branch of decision tree $\mathcal{T}_i$. The argument goes as follows. If $C, C_1, \ldots, C_t$, for $t \leq s$, are $\rho \cup \sigma_\ell$-consistent and $\rho \cup \sigma_\ell$-*dependent*, then there are edges $e = (u,v) \in C - \rho - \sigma_\ell$ and $e' = (u', v') \in C_i - \rho - \sigma_\ell$, for some $1 \leq i \leq t$, such that e is connected to e' by a (possibly empty) path $P \subseteq \rho \cup \sigma_\ell$. Assume that P begins by v and ends by u', the other case being analogous. By construction, the path P is extended to both to the left and to the right by an additional edge, thus producing path P'. If e is an edge of path P', then the corresponding literal is removed from $C \upharpoonright_{\rho\sigma_\ell P'}$, while if e is not on path P', then $C \upharpoonright_{\rho\sigma_\ell P'} \equiv 0$ over $\mathcal{G}(n,k)$. Since C has at most r literals, after r stages, C has been decided by all branches of the resulting decision tree. This is a central point of the entire argument, for which the definition of E-consistency and E-independence was correspondingly tailored.

Lower Bound for Distance k Connectivity

The following technical lemma stipulates appropriate parameters r_i, s_i, n_i for which the previous lemma allows one to iteratively find restrictions ρ_i yielding shallow decision trees for subcircuits of a given shallow circuit.

Lemma 2.6.10 (Connectivity iteration lemma). *Suppose that C is a boolean circuit of depth d and size S in the variables $x_{i,j}^{k'}$ for $0 \leq k' < k$, $1 \leq i,j \leq n$. Let $n_0 = n$, $\ell_0 = 0$, $r_0 = 4$, $s_0 = 4\log_n S$, and for all $i < d$, $r_{i+1} = 4r_i^2 s_i$, $s_{i+1} = 4r_i s_i$, $n_{i+1} = n_i^{1/4r_i}$ and for $1 < i < d$ $\ell_i = n_{i+1}$. Assume that $n_d > (3er_d(2k)^{r_d})^3$. Then for each $0 \leq i \leq d$, there exists a restriction $\rho_i \in \mathcal{R}_{n,k}^{n_i}$ such that for every gate g of C of depth at most i, $g \restriction_{\rho_i}$ is represented by a decision tree of depth at most r_i.*

Proof. Note that

$$n_0^{-s_0/3} = n^{(-4/3)\log_n S} = S^{-4/3} < 1/S$$

and that $n_i^{-s_i/3} = n^{-s_0/3} < 1/S$ for each $i \geq 0$. Moreover, the r_i and s_i increase with increasing i, and the n_i decrease with increasing i.

The lemma is proved by induction on i. Without loss of generality circuit C may be assumed to contain only OR- and NOT-gates, and depth d given by the maximum number of alternations between OR, NOT along a path from input to output node.

Inductively assume that restriction $\rho_i \in \mathcal{R}_{n,k}^{\ell_i}$ is chosen so that for all gates g at depth $i < d$, g_{ρ_i} has a decision tree of depth at most r_i. Let g be a gate at depth i. If $g = \neg h$, then by substituting leaf labels 0 (1) by 1 (0) in the decision tree for $h \restriction_{\rho_i}$, we have a decision tree for $g \restriction_{\rho_i}$ of depth at most r_i. Suppose that $g = \bigvee g_i$, where each $g_i \restriction_{\rho_i}$ has decision tree $\mathcal{T}_i$ of depth at most r_i. Then each $g_i \restriction_{\rho_i}$ is equivalent over $\mathcal{G}(n,k)$ to the r_i-disjunction (each term of which is the conjunction of the edge variables along a branch of $\mathcal{T}_i$, whose leaf is labeled by 1). Thus $g \restriction_{\rho_i}$ is an r_i-disjunction over $\mathcal{G}(n,k)$. Noting that

$$4r_i^2 s_i^2 k \leq 4r_{d-1}^2 s_{d-1}^2 k \leq r_d^2 k < n_d \leq n_{i+1}$$

we apply the previous lemma with $f = g \restriction_{\rho_i}$, $r = r_i$, $s = s_i$, $n = n_i$ and $\ell = n_{i+1}$ to show that all but

$$(3er_i(2kn_{i+1})^{r_i} n_{i+1}/n_i)^{s_i}$$

restrictions $\rho \in \mathcal{R}_{n_i,k}^{n_{i+1}}$, there is a decision tree for $g \restriction_{\rho_i\rho}$ of depth at most $4r_i^2 s_i = r_{i+1}$. From assumptions, $n_i > (3er_i(2k)^{r_i})^3$, so the probability of not choosing such ρ is bounded by

$$(3en_{i+1}^{r_{i+1}} r_i(2k)^{r_i}/n_i)^{s_i} \leq \left(\frac{n_{i+1}^{r_{i+1}}}{n_i^{2/3}}\right)^{s_i}$$

$$\leq \left(\frac{n_{i+1}^{5r_i/4}}{n_i^{2/3}} \right)^{s_i} \quad \text{since } r_i \geq 4$$

$$\leq \left(\frac{n_i^{5/16}}{n_i^{2/3}} \right)^{s_i} < n_i^{-s_i/3} < 1/S.$$

Since there are at most S gates of depth $i+1$, we can find $\rho \in \mathcal{R}_{n_i,k}^{n_{i+1}}$ which works for all gates g of depth $i+1$. Defining $\rho_{i+1} = \rho_i \rho$ then satisfies the conditions of the lemma for $i+1$.

Let DISTCONN(n, k) be the problem, given $G \in \mathcal{G}(n, k)$, of determining, for all n^2 choices of $s \in V_0$, $t \in V_k$, whether there is a path in G from s to t.

Fact 2.6.1. Suppose that C is a circuit of depth d and size S which solves st-connectivity for all $G \in \mathcal{G}(n, k)$. Then there is a circuit D of depth d and size $n^2 S$ solving DISTCONN(n, k).

The fact is easily seen by taking D to be the union of n^2 many circuits $C_{s,t}$ for all choices of $s \in V_0$, $t \in V_k$.

Theorem 2.6.8 (k-bounded st-connectivity lower bound). *Let $F_{-1} = 1$, $F_0 = 0$, and $F_{i+1} = F_i + F_{i-1}$ for $i \geq 0$. Assume $k \leq \log n$. For sufficiently large n, k every depth d unbounded fan-in boolean circuit computing* DISTCONN(n, k) *requires size at least $n^{\delta_d k^{1/(3F_{2d})}}$, where $\delta_d = 4^{-(F_{2d+3}-1)/F_{2d}}$.*

Proof. Suppose that S is the least size of a depth d circuit C computing DISTCONN(n, k). One establishes

$$s_i = 4^{F_{2i+2}} (\log_n S)^{F_{2i-1}}$$
$$r_i = 4^{F_{2i+3}-1} (\log_n S)^{F_{2i}}$$
$$n_i = n^{1/(4^i \prod_{j=0}^{i-1} r_j)}$$
$$4^i \prod_{j=0}^{i-1} r_j = 4^{F_{2i+2}-1} (\log_n S)^{F_{2i-1}-1} < r_i.$$

Suppose that $S < n^{\delta_d k^{1/(3F_{2d})}}$, so $\log_n S < \delta_d k^{1/(3F_{2d})}$, and $r_d < k^{1/3}$. Then $(3er_d(2k)^{r_d})^3 \leq k^{4r_d} \leq k^{4k^{1/3}}$, and $n_d \geq n^{1/r_d} \geq n^{1/k^{1/3}}$. By assumption, $k \leq \log n$, so $k^{4k^{2/3}} < n$, hence $n_d > (er_d k^{r_d})^3$. Apply the previous lemma to find restriction $\rho_d \in \mathcal{R}_{n,k}^{n_d}$ for which $g \upharpoonright_{\rho_d}$ has a decision tree of depth $< k$ over $\mathcal{G}(n_k, k)$, for every output gate g of C. Taking as s (t) one of the n_d vertices of V_0 (V_k) left unset by ρ_d, a decision tree of depth $< k$ cannot possibly determine whether s is connected to t, a contradiction. This concludes the proof of the theorem.

The results stated in Theorem 2.6.7 are now corollaries of the last theorem. The intuitive idea can be summarized as follows: st-connectivity is a *global* property, depending on all k edges of a candidate path from $s \in V_0$ to $t \in V_k$, so clearly cannot be decided by a decision tree of height less than k. By repeated application of the switching lemma, restrictions can be found which transform small depth boolean circuits into decision trees of height strictly less than k, thus the boolean circuit lower bound is proved.

2.7 Algebraic Methods

In contrast to the probabilistic techniques employed in Section 2.6.1 in this section we employ only techniques of an algebraic nature. Razborov's elegant methodology gives an exponential lower bound for majority on constant depth parity circuits (these are circuits with $\wedge, \oplus$-gates) thus significantly improving Håstad's result. It follows in particular that majority is harder than parity (in the AC^0-reduction).

At the heart of Razborov's technique lies an algebraic version of the Switching Lemma: for each ℓ, an $\bigwedge$ of polynomials of degree $\leq d$ lies to within a distance $\leq 2^{n-\ell}$ from a polynomial of degree $\leq d\ell$ (Lemma 2.7.1). As a consequence, we can absorb $\bigwedge$-gates without unnecessarily blowing up the size of the circuit. Smolensky [Smo87] pushed Razborov's idea a step further by proving the generalization of this lemma to finite fields of characteristic $p \neq 0$ (Lemma 2.7.3). As a consequence, if the natural number r is not a power of q then any depth k circuit with $\neg, \vee, \wedge, \text{MOD}_q$-gates computing the function MOD_r must have at least $2^{\Omega(n^{1/2k})}$ $\vee, \wedge$-gates.

2.7.1 Razborov's Lower Bound for Majority over Boolean Circuits with Parity

Before giving an outline of the main ideas of the proof we will need to provide some additional definitions. For each $f, g \in \mathcal{B}_n$ define $|f|$ as the number of n-bit inputs x such that $f(x) = 1$ and $\rho(f, g) = |f \oplus g|$, i.e., the number of inputs x such that $f(x) \neq g(x)$. Notice that ρ is a metric on $\mathcal{B}_n$, where we now consider $\mathcal{B}_n$ as a vector space on the two element field $\mathbf{Z}_2$. The set $P(d)$ of polynomials of degree $\leq d$ is then a vector subspace of $\mathcal{B}_n$. For any set $F \subseteq \mathcal{B}_n$ we also define

$$\rho(f, F) = \min_{g \in F} \rho(f, g).$$

For any set $K \subseteq \{1, 2, \ldots, n\}$ let $V_n(K) = \{x \in 2^n : (\forall i \in K)(x_i = 0)\}$. For each d', d'' with $d' + d'' \leq n$ and any boolean function $f \in \mathcal{B}_n$ define a matrix $A_{d',d''}(f)$ with $\binom{n}{d'}$ rows and $\binom{n}{d''}$ columns as follows: the (I, J)-th entry is given by the formula

$$a_{I,J}(f) = \bigoplus_{x \in V_n(I \cup J)} f(x) \tag{2.13}$$

where I (J) ranges over d' (d'') element subsets of $\{1, 2, \ldots, n\}$.

Theorem 2.7.1 ([Raz87a]).

$$L_k^{\oplus,\wedge}(\text{MAJ}_n) = 2^{\Omega(n^{\frac{1}{k+1}})}. \tag{2.14}$$

Proof. The main ideas in the proof of Theorem 2.7.1 are as follows.

(A) First show that for all $f \in \mathcal{B}_n$ and all integers ℓ,

$$L_k^{\oplus,\wedge}(f) \geq \rho(f, P(\ell^{\lceil k/2 \rceil})) \cdot 2^{\ell-n}. \tag{2.15}$$

This reduces the lower bound on majority to a lower bound involving the distance of the function f from the space $P(\ell^{\lceil k/2 \rceil})$.

(B) Reduce the lower bound on the distance to a lower bound on the rank of a matrix by showing that for all $f \in \mathcal{B}_n$ and all integers $d + d' + d'' < n$,

$$\rho(f, P(d)) \geq \text{rank}(A_{d',d''}(f)). \tag{2.16}$$

(C) Show that there exists a symmetric function f for which

$$(\forall d' < n/2)(\exists d'' \leq d') \left(\text{rank}(A_{d',d''}(f)) \geq \frac{1}{n} \cdot \binom{n}{d'} \right). \tag{2.17}$$

Apply now (A), (B), (C) with $d' = \lfloor n/2 - \sqrt{n} \rfloor$, $d = n - d' - d'' - 1$ and $\ell = d^{2/k}$ to conclude that

$$L_k^{\oplus,\wedge}(f) = 2^{\Omega(n^{\frac{1}{k+1}})}, \tag{2.18}$$

for some symmetric boolean function $f \in \mathcal{B}_n$. Using (2.18) we can show that the same lower bound is also valid for the majority function, thus proving Theorem 2.7.1. Indeed, consider a symmetric function f satisfying inequality (2.18). The value of a symmetric function depends only on the weight of its inputs. If the weight $|x|_1$ of x is j then

$$\text{MAJ}_{2n}(x, 1^i, 0^{n-i}) = 0,$$

for $i + j < n$, and $= 1$, otherwise. Hence from Theorem 1.4.1 there exist constants b_i, $i = 0, \ldots, n$, such that

$$f(x) = \bigoplus_{i=0}^{n} \left(b_i \cdot \text{MAJ}_{2n}(x, 1^i, 0^{n-i}) \right). \tag{2.19}$$

Now take a depth k circuit for MAJ_{2n} and connect n copies of it as indicated by identity (2.19) using $\oplus$ gates and adding a new depth. This proves that

$$L_{k+1}^{\oplus,\wedge}(f) \leq n \cdot L_k^{\oplus,\wedge}(\text{MAJ}_{2n}),$$

which implies the desired exponential lower bound for majority. Now we can concentrate on the proofs of claims (A), (B) and (C).

Proof of (A). The proof is based on the idea of the *regular model* of depth k. This is a tuple $\mathcal{M} = (M_0, M_1, \ldots, M_k, \Pi_1, \ldots, \Pi_k)$ where

- $M_i \subseteq \mathcal{B}_n$, for $i = 0, \ldots, k$,

- $\{x_j, 1 \oplus x_j : j = 1, \ldots, n\} \subseteq M_0$,
- $\Pi_i : \mathcal{P}(M_{i-1}) \longrightarrow M_i$, for $i = 0, \ldots, k$.

For any circuit $C \in \mathcal{C}_k$ of depth $i \leq k$ and any regular model $\mathcal{M}$ of depth k, we define a boolean function $f_C^{\mathcal{M}} \in M_i$ as follows by induction on i: $f_C^{\mathcal{M}} = f_C$ if $i = 0$ and $f_C^{\mathcal{M}} = \Pi_i(\{f_B^{\mathcal{M}} : B \in C\})$ if $i > 0$.

The *defect* $\delta(\mathcal{M})$ of a regular model $\mathcal{M}$ is the $\max_{F \subseteq M_{i-1}, 1 \leq i \leq k} \delta(F, i)$, where $\delta(F, i) = \rho(\Pi_i(F), \circ F)$ for $F \subseteq M_{i-1}$ and $\circ = \oplus$ or $= \wedge$ depending on whether or not i is odd or even, respectively. The significance of regular models lies in the following inequality:

$$L_k(f) \geq \frac{1}{\delta(\mathcal{M})} \cdot \rho(f, M_k). \tag{2.20}$$

This is easy to prove using the fact that for any circuit of depth $i > 0$

$$f_C^{\mathcal{M}} \oplus f_C \leq \bigvee_{D \to C} \{\Pi_j(\{f_B^{\mathcal{M}} : B \in D\}) \oplus \circ_{B \in D} f_B^{\mathcal{M}}\}.$$

Thus, in order to prove (A), we choose a suitable regular model of depth k, which will satisfy $M_{2j} = M_{2j+1} = P(\ell^j)$, for $j \geq 0$, and hence $M_k = P(\ell^{\lfloor k/2 \rfloor})$. In order to guarantee that the defect of $\mathcal{M}$ is $\leq 2^{n-\ell}$ it will be sufficient to ensure that $\delta(F, i) \leq 2^{n-\ell}$ for all $F \subseteq M_{i-1}, 1 \leq i \leq k$. If i is odd, then we can define $\Pi_i(F) = \bigoplus F$, which easily implies that $\delta(F, i) = 0$. But if i is even, then it would be wrong to define $\Pi_i(F) = \bigwedge F$ because $\bigwedge F$ is not necessarily a polynomial of degree $\leq \ell^{\lfloor i/2 \rfloor}$. Instead we assign to $\Pi_i(F)$ the boolean function g given in Lemma 2.7.1.

Lemma 2.7.1. *If $F \subseteq P(d)$ then there exists a $g \in P(d \cdot \ell)$ such that*

$$\rho(\bigwedge F, g) = |(\bigwedge F) \oplus g| \leq 2^{n-\ell}$$

Proof. of Lemma 2.7.1. $Z(h) = \{f \in P(d) : f \wedge h = 0\}$ is a vector subspace of $P(d)$. Put $h = \bigwedge F$. By induction, we will construct a sequence of boolean functions $f_1, f_2, \ldots, f_t, \ldots \in Z(h)$ satisfying $\rho(h, \bigwedge_{i=1}^t (1 \oplus f_i)) \leq 2^{n-t}$, in which case the required function is $g = \bigwedge_{i=1}^{\ell} (1 \oplus f_i)$. Suppose that the first t functions have been constructed. We will construct the $(t+1)$st function f_{t+1}. Put $\Delta_j = \{x \in V_n : h(x) \neq \bigwedge_{i=1}^j (1 \oplus f_i)\}$. Since each $f_i \in Z(h)$ we must have that $h \leq \bigwedge_{i=1}^t (1 \oplus f_i)$. Hence for each $x \in \Delta_t$ there exists a boolean function $g_x \in F$ such that $1 \oplus g_x \in Z(h)$. It follows that for each $x \in \Delta_t$ the linear functional

$$\mathcal{F}_x : Z(h) \to \{0, 1\} : f \to \mathcal{F}_x(f) = f(x)$$

is non-trivial. Moreover if $f' = f \oplus (1 \oplus g_x)$ then the mapping $f \longrightarrow f'$ is 1-1 and onto $Z(h)$ and satisfies $\mathcal{F}_x(f) = 0 \Leftrightarrow \mathcal{F}_x(f') = 1$. So $\mathcal{F}_x$ partitions $Z(h)$ into two sets each of size $\frac{|Z(h)|}{2}$, namely

$$\{f \in Z(h) : f(x) = 0\}, \{f \in Z(h) : f(x) = 1\}.$$

It follows that $\sum_{f \in Z(h)} |\{x \in \Delta_t : f(x) = 1\}| = \frac{|\Delta_t| \cdot |Z(h)|}{2}$, and hence there exists a function $f_{t+1} \in Z(h)$ such that $|\{x \in \Delta_t : f_{t+1}(x) = 1\}| \geq \frac{|\Delta_t|}{2}$. Since $h(x) \neq \bigwedge_{i=1}^{t+1}(1 \oplus f_i(x))$ for at most half the elements of Δ_t it easy to see that $|\Delta_{t+1}| \leq \frac{|\Delta_t|}{2}$, which completes the proof of the lemma.

Proof of (B). In view of (A), to obtain a lower bound on $L_k(f)$ it would be enough to obtain a lower bound on $\rho(f, P(d))$ for a suitably chosen d. Clearly there exists $g \in P(d)$ such that $\rho(f, P(d)) = \rho(f, g) = |f \oplus g|$. Can we find a "matrix transformation" which assigns to every boolean function f a matrix $A(f)$ such that $|f \oplus g| \geq \text{rank}(A(f \oplus g))$? One bothersome point is what to do with the polynomial g. By linearity $A(f \oplus g) = A(f) \oplus A(g)$. Thus in order to avoid mentioning g it would be desirable to have $A(g) = 0$ for all polynomials $g \in P(d)$. But this is achieved by the matrix $A_{d',d''}$ which is given by (2.13), provided that $d + d' + d'' < n$. Indeed, let $g \in P(d)$. By linearity we can assume that g is a monomial of degree $\leq d$, say $g(x_1, \ldots, x_n) = \bigwedge_{i \in K} x_i$, where K is a subset of $\{1, 2, \ldots, n\}$ of size $\leq d$. Then we have that $a_{I,J}(g) = \bigoplus_{x \in V_n(I \cup J)} \bigwedge_{i \in K} x_i$. It is then clear that if $K \cap (I \cup J) \neq \emptyset$ then $a_{I,J}(g) = 0$. So without loss of generality we may assume that $K \cap (I \cup J) = \emptyset$. By assumption we have that $|K \cup I \cup J| \leq d + d' + d'' < n$ and $a_{I,J}(g)$ must be 0 because it is the sum modulo 2 of an even number of 1's. It remains to prove inequality (2.16). In view of the above discussion, it is enough to show that for any f, $\text{rank}(A_{d',d''}(f)) \leq |f|$. If $|f| = 1$ then

$$\left(\bigoplus_{x \in V_n(I)} f(x) \right) \cdot \left(\bigoplus_{x \in V_n(J)} f(x) \right) = \bigoplus_{x \in V_n(J) \cap V_n(J)} f(x)$$

which easily implies that $A_{d',d''}(f) = A_{d',0}(f) \times A_{0,d''}(f)$. This implies that $\text{rank}(A_{d',d''}(f)) \leq 1$. And in general, for any f let $S = \{x \in V_n : f(x) = 1\}$. For each $x \in S$ let $f_x(y) = 1$ if $y = x$ and is 0 otherwise. Clearly $f = \bigoplus_{x \in S} f_x$ and hence

$$\text{rank}(A_{d',d''}(f)) \leq \sum_{x \in S} \text{rank}(A_{d',d''}(f_x)) \leq |S| = |f|,$$

which completes the proof of (B).

Proof of (C). Define the matrices $P_{d_1,d_2} = (p_{I,J})$, with $p_{I,J} = 0$ if $I \cap J \neq \emptyset$ and $= 1$ otherwise. We claim that the rows of the matrix $P_d = (P_{d,0}; P_{d,1}; \cdots; P_{d,d})$ are linearly independent. Indeed, let H be an arbitrary $1 \times \binom{n}{d}$ vector which is $\neq 0$. We must show that $H \cdot P_d \neq 0$. Let $h_{1,I}$ be the I-th element of H and put

$$f_H(x_1, \ldots, x_n) = \bigoplus_{|I|=d} \left(h_{1,I} \cdot \bigwedge_{i \in I} x_i \right).$$

Choose I_0 such that $h_{1,I_0} \neq 0$. After performing the substitution: $x_i \longrightarrow 1$, for $i \notin I_0$ we obtain a non-zero polynomial in the variables $\{x_i : i \in I_0\}$

such that the monomial $\bigwedge_{i\in I_0} x_i$ is unaffected. For this monomial there is an assignment that makes it equal to 1. Let ϵ be the resulting assignment by setting all the remaining variables equal to 1. Clearly we have that

$$1 = f_H(\epsilon) = \bigoplus_{|I|=d} \left(h_{1,I} \cdot \bigwedge_{i\in I} \epsilon_i \right) = \bigoplus_{|I|=d} (h_{1,I} \cdot p_{I,J}) = H \cdot P_d \neq 0,$$

where $J = \{i : \epsilon_i = 0\}$ and $|J| \leq d$. To sum up we have shown that

$$\text{rank}(P_d) = \text{rank}(P_{d,0}; P_{d,1}; \cdots; P_{d,d}) = \binom{n}{d}. \tag{2.21}$$

Now assume that $d' < n/2$ and let $P'_{d''}$ be obtained from $P_{d'}$ by substituting each submatrix $P_{d',i}$, for $i \neq d''$ by the zero matrix. Clearly $P_{d'} = \sum_{d''\leq d'} P'_{d''}$ and

$$\begin{aligned} \text{rank}(P_{d'}) &= \textstyle\sum_{d''\leq d'} \text{rank}(P'_{d''}) \\ &\leq n \cdot \max_{d''\leq d'} \text{rank}(P'_{d''}) \\ &= n \cdot \text{rank}(P'_{d''}), \end{aligned} \tag{2.22}$$

for some $d'' \leq d'$. Using (2.21), (2.22) we obtain that for all $d' < n/2$ there exists $d'' \leq d'$ such that

$$\text{rank}(P_{d',d''}) \geq \frac{1}{n} \cdot \binom{n}{d'}. \tag{2.23}$$

Hence (C) would follow if we could prove the following lemma.

Lemma 2.7.2. . *If $d' + d'' < n$ then there exists a symmetric function f such that $A_{d',d''}(f) = P_{d',d''}$.*

Proof. Write the desired function f as $f = \bigoplus_{d=0}^{d'+d''} b_d \cdot T_{n-d,n}$, where $T_{s,n} = \bigoplus_{|I|=s} \bigwedge_{i\in I} x_i$. Put $|I| = d', |J| = d'', |K| = I \cup J, |K| = u$ and notice that

$$a_{I,J}(f) = \bigoplus_{x\in V_n(K)} \bigoplus_{d=0}^{d'+d''} b_d \cdot T_{n-d,n}(x) = \bigoplus_{d=0}^{d'+d''} b_d \cdot \left(\bigoplus_{x\in V_n(K)} T_{n-d,n}(x) \right).$$

If we call $\phi(n,d,u) = \bigoplus_{x\in V_n(K)} T_{n-d,n}(x)$ then it is easy to see from the definitions that $\phi(n,d,u) = \bigoplus_{i=0}^{n-u} \binom{n-u}{i} \cdot \binom{i}{n-d}$. Since $\binom{p}{q} = 0$ for $p < q$ we obtain that $\phi(n,d,u) = 1$ if $d = u$ and $= 0$ if $u > d$. So the function ϕ is known and the conditions for computing f are $A_{d',d''}(f) = P_{d',d''}$. Hence we obtain the diagonal system (notice that for $I \cap J = \emptyset$, $u = d' + d''$)

$$\left\{ \bigoplus_{d=0}^{d'+d''} b_d \cdot \phi(n,d,u) = \delta_{u,d'+d''} \right\}_{u=0}^{d'+d''}$$

from which we can compute the b_d's.

This concludes the proof of Theorem 2.7.1.

2.7.2 Smolensky's Lower Bound for MOD_p Versus MOD_q

We begin with some basic definitions. Let F be a finite field of characteristic p and let U_F^n be the algebra of functions $f : \{0,1\}^n \to F$ with pointwise addition and multiplication. The boolean variable x_i is considered an element of U_F^n by identifying it with the i-th projection function. For each $a \in \{0,1\}^n$ consider the polynomial

$$P_a(x_1, \ldots, x_n) = \prod_{a_i=1} x_i \cdot \prod_{a_i=0} (1 - x_i).$$

Clearly, $P_a(x) = 1$ if $x = a$ and is 0 otherwise. Each $f \in U_F^n$ is a linear combination of these polynomials since

$$f(x) = \sum_a f(a) \cdot P_a(x).$$

There are 2^n polynomials of the form P_a and the above representation is unique, assuming $x_i^2 = x_i$, $i = 1, \ldots, n$. Hence $\dim(U_F^n) = 2^n$.

For any $E \subseteq \{0,1\}^n$ let $I(E)$ be the ideal of functions $f \in U_F^n$ which are 0 outside of E. $I(E)$ is generated by the set $\{P_a : a \in E\}$ and hence $\dim(I(E)) = |E|$. Moreover by Hilbert's *Nullstellensatz*[17] every ideal of U_F^n is of the form $I(E)$, for some $E \subseteq \{0,1\}^n$ (namely, given an ideal I it is easy to see that the set $E = \{x \in \{0,1\}^n : \forall f \in I(f(x) = 0)\}$ works). Every ideal I determines the quotient algebra $A = U_F^n / I$ and for any $f \in U_F^n$ we define

$$\deg_A(f) = \min\{\deg(f') : f - f' \in I\}.$$

An m-ary boolean function $f(g_1, \ldots, g_m)$ is *F-easy* if it can be represented in the g_is as a polynomial of constant degree $O(1)$. An m-ary boolean function $f(x_1, \ldots, x_m)$ is *nearly F-easy* if for any choice of n-ary boolean functions $g_1, \ldots, g_m$ and any integer l there is a quotient F-algebra A of dimension at least $2^n - 2^{n-l}$ such that $f(g_1, \ldots, g_m)$ can be written in A as a polynomial in the g_is of degree $O(l)$. Clearly, $\neg$ is F-easy and so is $\text{MOD}_p(g_1, \ldots, g_m) = \left(\sum_{i \le m} g_i\right)^{p-1}$ if the field F is of characteristic p. This is not true though for $\bigvee, \bigwedge$ since the degree of the resulting boolean function may depend on the number m of boolean variables of f. However generalizing Razborov's Lemma 2.7.1 (which was proved only for the finite field $F = \mathbf{Z}_2$) we have the following result.

Lemma 2.7.3. *$\bigvee, \bigwedge$ are nearly F-easy for any field of characteristic $p \neq 0$.*

[17] Hilbert's Nullstellensatz states that if K is a field and the polynomial $f \in K[x_1, \ldots, x_n]$ vanishes at all the common zeros of polynomials $f_1, \ldots, f_r$, then there exists natural number m for which $f^m = h_1 \cdot f_1 + \cdots + h_r \cdot f_r$. It follows that if the polynomials $f_1, \ldots, f_r$ have no common zero in field k, then there exist $h_1, \ldots, h_r$ such that $1 = h_1 \cdot f_1 + \cdots + h_r \cdot f_r$.

Proof. Since $\bigwedge_{i=1}^m g_i = \neg \bigvee_{i=1}^m \neg g_i = 1 - \bigvee_{i=1}^m (1-g_i)$ it is enough to prove the lemma for the boolean function $\bigvee$. Hence let $f = \bigvee_{i=1}^m g_i$. Let ℓ be arbitrary but fixed. It is enough to find a polynomial f' of degree $\leq (|F|-1) \cdot \ell$ and differing from f on at most $2^{n-\ell}$ assignments. Then the required algebra would be $A = U_F^n/I$, where I is the ideal generated by $f - f'$. The required function f' will be found among the collection of boolean functions

$$S = \left\{ s : s = \bigvee_{j=1}^{\ell} \left(\sum_{i=1}^{m} c_{i,j} \cdot g_i \right)^{|F|-1}, c_{i,j} \in F \right\}$$

(These are indeed boolean functions, since $x^{|F|-1}$ is always 0 or 1, for all $x \in F$.) Clearly, each $s \in S$ has degree $\leq (|F|-1) \cdot \ell = O(\ell)$ in the g_i. We show that for each $a \in \{0,1\}^n$ the probability that a random $s \in S$ satisfies $s(a) \neq f(a)$ is $\leq |F|^{-\ell}$. Assuming this, the result of the lemma follows easily, since then

$$\begin{aligned} |\{(a,s) : s(a) \neq f(a)\}| &= \sum_a |\{s : s(a) \neq f(a)\}| \\ &= \sum_s |\{a : s(a) \neq f(a)\}| \\ &= 2^n \cdot |S| \cdot |F|^{-\ell}, \end{aligned}$$

which in turn easily implies that there is an $s \in S$ differing from f in at most $2^{n-\ell}$ assignments. To conclude the proof, let a be an arbitrary assignment in $\{0,1\}^n$. If $f(a) = 0$ then each g_i satisfies $g_i(a) = 0$ and hence also $s(a) = 0$ for all $s \in S$. If $f(a) = 1$ then for some g_i, say g_{i_0}, we have that $g_{i_0}(a) = 1$. Hence for any choice of the $c_{i,j}$, for $i \neq i_0$, there is a unique choice of $c_{i_0,1}, \ldots, c_{i_0,\ell}$ such that $s(a) = 0$ for the $s \in S$ defined from these parameters. This proves that for any a the probability that a random s satisfies $s(a) \neq f(a)$ is $|F|^{-\ell}$, which proves the required assertion and hence also the lemma.

Lemma 2.7.4. *If a depth k circuit has $2^{o(n^{1/2k})}$ nearly F-easy gates, then there is a quotient algebra A of dimension $2^n - o(2^n)$ such that all outputs of the circuit have degree $o(\sqrt{n})$ in A.*

Proof. Let $\ell = 2r$, where $r = o(n^{1/2k})$. By definition, each nearly F-easy gate has an ideal I of dimension $\leq 2^{n-\ell}$ such in U_F^n/I the operation performed by the gate can be expressed in terms of its children by a polynomial of degree $O(\ell)$. If I_0 is the sum of these ideals taken over all these nearly F-easy gates, then $\dim(I_0) \leq 2^r 2^{n-\ell} = 2^{n-r} = o(2^n)$. Clearly in $A = U_F^n/I_0$ each gate computes a function that can be expressed in terms of its children by a polynomial of degree $O(\ell) = o(n^{1/2k})$. Since we start with x_i's and the depth is k all outputs have degree $o(\sqrt{n})$ in A. This proves the lemma.

A family $u_1, \ldots, u_s \in U_F$ is called *U_F-complete* if for any F-algebra A and any $v \in U_F$, $\deg_A(v) \leq \frac{n}{2} + \max_i \deg_A(u_i)$. The next lemma proves the existence of U_F-complete families.

Lemma 2.7.5. $\{\text{MOD}_{0,q}, \ldots, \text{MOD}_{q-1,q}\}$ *is* U_F*-complete, provided that* F *contains a* q*-th root of unity.*

Proof. First we show that the function $\prod_{i=1}^{n}(1 + (h-1)x_i)$ is U_F-complete, provided that $h \neq 0, 1$. Indeed, first observe that if $y_i = 1 + (h-1)x_i$ then $y_i^{-1} = 1 + (h^{-1} - 1)x_i$. Given any monomial $\prod_{i \in S} y_i$, where $S \subseteq \{1, \ldots, n\}$, we have that $\prod_{i \in S} y_i = \prod_{i=1}^{n} y_i \cdot \prod_{i \notin S} y_i^{-1}$. Thus regardless of whether or not $|S| \leq n/2$ it is clear that

$$\deg_A(\prod_{i \in S} y_i) \leq \frac{n}{2} + \deg_A(\prod_{i=1}^{n} y_i).$$

Since $x_i = (h-1)^{-1}(y_i - 1)$ it follows that the function $\prod_{i=1}^{n} y_i$ is U_F-complete.

To prove the lemma it is enough to show that the function $\prod_{i=1}^{n} y_i$ has a polynomial expression in terms of the functions

$$\text{MOD}_{0,q}, \ldots, \text{MOD}_{q-1,q}.$$

Indeed, let $h \neq 0, 1$ be a q-th root of unity. For any assignment $a \in \{0,1\}^n$ of weight k if $s < q$ is such that $k \equiv s \bmod q$ then we have that $\prod_{i=1}^{n} y_i(a) = h^k = h^s$. As a consequence we obtain

$$\prod_{i=1}^{n} y_i = \sum_{s=0}^{q-1} h^s \text{MOD}_{s,q}(x_1, \ldots, x_n).$$

Now the lemma follows easily from the above identity.

Lemma 2.7.6. *If the* U_F*-complete element* u *satisfies* $\deg_A(u) = o(\sqrt{n})$ *then* $\dim(A) \leq 2^{n-1} + o(2^n)$.

Proof. By U_F-completeness of u, every element of A can be written as a polynomial of degree $\leq \frac{n}{2} + o(\sqrt{n})$; moreover A is the linear span of all monomials of degree $\leq \frac{n}{2} + o(\sqrt{n})$. Counting all these monomials we obtain

$$\dim(A) \leq \sum_{i=0}^{\frac{n}{2}+o(\sqrt{n})} \binom{n}{i} \leq 2^{n-1} + o(2^n).$$

Lemma 2.7.7. *If a depth* k *circuit uses* $2^{o(n^{1/2k})}$ *nearly* F*-easy gates, then its output* g *will differ from any* U_F*-complete element* f *on* $2^{n-1} - o(2^n)$ *assignments.*

Proof. By Lemma 2.7.4 there is an algebra A of dimension $\geq 2^n - o(2^n)$ such that g has degree $o(\sqrt{n})$ in A. If we ignore the assignments where g differs from f we get an even smaller algebra $\bar{A}$ on which $g = f$. It follows that $\deg_{\bar{A}}(g) = o(\sqrt{n})$. So by Lemma 2.7.6, $\dim(\bar{A}) \leq 2^{n-1} + o(2^n)$. Hence we had to ignore at least $\geq 2^n - (2^{n-1} + o(2^n)) = 2^{n-1} - o(2^n)$ assignments.

We conclude with the following two theorems.

Theorem 2.7.2 ([Smo87]). *The output of any depth k unbounded fan-in circuit of size $2^{o(n^{1/2k})}$ with basic operations $\vee, \wedge, \neg$ differs from* MOD_2 *on $2^{n-1} + o(2^n)$ assignments.*

Proof. The gates $\vee, \wedge$ are nearly F-easy and the family $\text{MOD}_{0,2}, \text{MOD}_{1,2}$ is U_{Z_2}-complete. Hence the theorem follows from Lemma 2.7.7 and the fact that $\text{MOD}_2 = \neg\text{MOD}_{0,2}$.

Theorem 2.7.3 ([Smo87]). *Let p be a prime and suppose that r is not a power of p. Then computing* MOD_r *by depth k unbounded fan-in circuits with basic operations $\vee, \wedge, \neg, \text{MOD}_p$ requires $2^{\Omega(n^{1/2k})}$ many $\vee, \wedge$ gates.*

Proof. Let q be a prime divisor of r, with $r \neq p$. The Galois field $F = GF(p^{q-1})$ has a q-th root of unity. By Lemma 2.7.5 the family

$$\text{MOD}_{0,q}, \ldots, \text{MOD}_{q-1,q}$$

is U_F-complete and the gate MOD_p is F-easy. By Theorem 1.4.1

- $\text{MOD}_{0,q}, \ldots, \text{MOD}_{q-1,q} \in \text{AC}^0(\vee, \wedge, \neg, \text{MOD}_q)$, and
- $\text{MOD}_q \in \text{AC}^0(\vee, \wedge, \neg, \text{MOD}_r)$.

Hence the circuit requires $2^{\Omega(n^{1/2k})}$ many $\vee, \wedge$ gates.

2.8 Polynomial Method

In Section 2.7, we saw how algebraic techniques can be used in the study of the circuit complexity of the MAJ, MOD functions. In this section we represent boolean functions as polynomials modulo an integer and obtain lower bounds on the MOD_m-degree of boolean functions.

2.8.1 On the Strength of MOD_m Gates

Smolensky's lower bound in Theorem 2.7.3 shows that constant depth circuits which allow MOD_p gates, for some fixed prime p, cannot compute the MOD_r function if r is not a power of p. However, what happens if we allow MOD_m gates, for m a composite number? It is conjectured [Smo87] that constant depth circuits with MOD_m gates can not compute the MOD_r function, if there is a prime divisor of m which does not divide r. Nevertheless, it is not even known whether or not constant depth circuits with the MOD_6 gates can compute every function in NP! In this section we study this question in the polynomial model of computation.

A boolean function $f \in \mathcal{B}_n$ is *represented* by a polynomial P modulo m if for all boolean inputs $x \in \{0,1\}^n$,

$$f(x) = 0 \Leftrightarrow P(x) \equiv 0 \bmod m.$$

The MOD_m-degree of f, denoted by $\delta(f, m)$, is the smallest degree of a polynomial representing f modulo m.

Theorem 2.8.1 ([BBR92]). *Let m be a square-free positive integer. Then the OR_n function has MOD_m-degree $O(n^{1/r})$, where r is the number of distinct prime factors of m. In addition, any symmetric polynomial representing OR_n must have MOD_m-degree $\Omega(n^{1/r})$.*

Proof. First we prove the upper bound. Let the k-th *elementary symmetric function* $S_k(x_1, \ldots, x_n)$ be the sum of all monomials of degree k in the input variables $x = (x_1, \ldots, x_n)$.[18] If the weight $|x|_1$ of the input $x = x_1 \cdots x_n$ is j, then it is clear that $S_k(x)$ is $\binom{j}{k}$, independently of n. For simplicity we write this as $S_k(j)$. A *symmetric polynomial* of degree d is simply a linear combination of $S_0, S_1, \ldots, S_d$.

For prime p the function $S_k(j) = \binom{j}{k} \bmod p$ is periodic with period p^e, where e is minimal such that $k \leq p^e$. The polynomials $S_0, \ldots, S_{p^e-1}$ are linearly independent modulo p and hence they form a basis of the vector space of symmetric functions with period p^e. If $p^e > n$ then OR_n is represented modulo p by the function $f(j)$, where $f(j) = 0$ for $j \equiv 0 \bmod p^e$ and $f(j) = 1$ otherwise, which has degree at most $p^e - 1$.

Consider an arbitrary degree d. For each p prime divisor of m let p' be the largest power of p such that $p' - 1 \leq d$. By the previous paragraph there is a degree d polynomial f_p such that $f_p(j) \equiv 0 \bmod p \Leftrightarrow j \equiv 0 \bmod p'$. By the Chinese remainder theorem there is a unique polynomial f modulo m such that $f \equiv f_p \bmod p$, for all p prime divisors of m. It is now clear that f represents the OR of up to $q - 1$ variables, where q is the product of the p's. Since $p' = \Theta(d)$ we have that $q = \Theta(d^r)$. This proves the result when m is square-free.

Theorem 2.8.1 is also true if m is not square-free. For details see Exercise 2.14.12. We are now in a position to study Smolensky's conjecture in the context of the MOD_m-degree.

Theorem 2.8.2 ([BBR92]). *If m is a positive, square-free composite integer and $p_{\max}$ is the largest prime divisor of m, then $\delta(\neg\text{MOD}_m^n, m) \geq n/(2p_{\max})$.*

Proof. Let $Q(x_1, \ldots, x_n)$ be a polynomial representing $\neg\text{MOD}_m^n$, so that for all $x = x_1 \cdots x_n \in \{0,1\}^n$, $\sum_{i=1}^n x_i \not\equiv 0 \bmod m$ if and only if $Q(x_1, \ldots, x_n) = 0 \bmod m$. Thus it certainly is the case that

- $Q(0, \ldots, 0) \not\equiv 0 \bmod m$, and
- $Q(x_1, \ldots, x_n) \equiv 0 \bmod m$, if $\sum_{i=1}^n x_i$ is a power of a prime divisor of m.

[18] By *input variables*, we mean the restriction of variables $x_1, \ldots, x_n$ to $0, 1$ values, rather than to values in $\mathbf{Z}_m$.

We claim that the degree of Q is at least $n/(2p_{\max})$. Assuming that this is not the case, we derive a contradiction, Since $\delta(\neg\text{MOD}_m^n, m) < n/(2p_{\max})$, it follows that for any prime divisor p of m, we have $\delta(\neg\text{MOD}_m^n, m) < n/(2p)$.

Now, let p be a fixed prime divisor of n. Let k be maximal, such that $2p^k - 1 \leq n$ and put $n' = 2p^k - 1$. Also, define $Q'(x_1, \dots, x_{n'}) = Q(x_1, \dots, x_{n'}, 0, \dots, 0)$. For any set $S \subseteq \{x_1, \dots, x_{n'}\}$ define the polynomials $P'_S = \prod_{u \in S} u$ and $P_S = P'_S \cdot \prod_{u \notin S}(1 - u)$. Then we have that $Q'(x_1, \dots, x_{n'}) = \sum_S c_S P_S = \sum_S c'_S P'_S$, where c_S, c'_S satisfy $c_\emptyset \not\equiv 0 \bmod m$ and

$$\begin{aligned} c_S &\equiv 0 \bmod m \text{ if } |S| \text{ is a power of a prime divisor of } m \\ c'_S &= 0 \qquad\qquad \text{if } |S| \geq n/(2p). \end{aligned}$$

If we define $\sigma_i = \sum_{|S|=i} c_S$ and $\sigma'_i = \sum_{|S|=i} c'_S$ then observe that $\sigma_0 = c_\emptyset$ and

$$\begin{aligned} \sigma_i &\equiv 0 \bmod m \text{ if } |S| \text{ is a power of a prime divisor of } m \\ \sigma'_i &= 0 \qquad\qquad \text{if } |S| \geq n/(2p). \end{aligned}$$

If we note that $c'_S = \sum_{T \subseteq S} (-1)^{|S|-|T|} c_T$ then we obtain

$$\begin{aligned} \sigma'_i &= \textstyle\sum_{|S|=i} \sum_{T \subseteq S} (-1)^{|S|-|T|} c_T \\ &= \textstyle\sum_T \sum_{|S|=i, T \subseteq S} (-1)^{|S|-|T|} c_T \\ &= \textstyle\sum_j \sum_{|T|=j} \sum_{|S|=i, T \subseteq S} (-1)^{i-j} c_T \\ &= \textstyle\sum_j \sum_{|T|=j} \binom{n'-j}{i-j} (-1)^{i-j} c_T \\ &= (-1)^i \textstyle\sum_j \binom{n'-j}{i-j} (-1)^j \sum_{|T|=j} c_T \\ &= (-1)^i \textstyle\sum_j \binom{n'-j}{i-j} (-1)^j \sigma_j \\ &= (-1)^i \textstyle\sum_j \binom{n'-j}{n'-i} (-1)^j \sigma_j. \end{aligned}$$

By assumption, $n' = 2p^k - 1$. So, if we let $i = p^k$ then $n - i = p^k - 1$. By *Kummer's theorem*

$$\begin{aligned} &\binom{n'}{n'-i} \not\equiv 0 \bmod p \\ &\binom{n'-j}{n'-i} \equiv 0 \bmod p \text{ if } 0 < j < i. \end{aligned}$$

Since $\sigma_0 = c_\emptyset$ and $\sigma_i \equiv 0 \bmod m$ we obtain

$$\begin{aligned} \sigma'_i &\equiv (-1)^i \left(\binom{n'-i}{n'-i} (-1)^i \sigma_i + \binom{n'-0}{n'-i} (-1)^0 \sigma_0 \right) \\ &\equiv (-1)^i \left((-1)^i \sigma_i + \binom{n'}{n'-i} c_\emptyset \right) \\ &\equiv (-1)^i \binom{n'}{n'-i} c_\emptyset \bmod p. \end{aligned}$$

By definition of k, $2p^{k+1} - 1 > n$ and hence $i = p^k \geq n/(2p)$. So $\sigma'_i = 0$. Since $\binom{n'}{n'-i} \not\equiv 0 \bmod p$, it follows that $c_\emptyset \equiv 0 \bmod p$. Therefore $Q(0, \dots, 0) = Q'(0, \dots, 0) = c_\emptyset \equiv 0 \bmod p$. Since m is square-free we conclude that $Q(0, \dots, 0) \equiv 0 \bmod m$, which is a contradiction. This completes the proof of the theorem.

The MOD_m-degree of the functions $\mathrm{MOD}_p, \neg\mathrm{MOD}_p$ can be computed in a similar manner.

Theorem 2.8.3 ([BBR92]). *Let m be a square-free positive integer, whose largest prime divisor is $p_{\max}$, and let p be any prime not dividing m. Then*

1. $\delta(\neg\mathrm{MOD}_p^n, m) \geq \frac{n}{2p_{\max}}$.
2. $\delta(\mathrm{MOD}_p^n, m) \geq \frac{\lfloor((n-1)/(p-1))^{1/(p-1)}\rfloor}{2p_{\max}(p-1)}$.

Proof. The proof of part (1) is as before. So we only prove part (2). Let Q be a polynomial representing MOD_p^n. Put $t = \lfloor((n-1)/(p-1))^{1/(p-1)}\rfloor$ and let $\ell = (p-1)t^{p-1}$. We can write $(p-1)(x_1+\cdots+x_n)^{p-1}$ as the sum $y_1+\cdots+y_\ell$ of ℓ monomials, each with coefficient 1. Define

$$R(x_1, \ldots, x_t) = Q(y_1, \ldots, y_\ell, 1, 0, \ldots, 0).$$

Put $s = \sum_{i=1}^t x_i$, and use Fermat's theorem to conclude that the following four assertions are equivalent

$$\begin{aligned}
&R(x_1, \ldots, x_t) \equiv 0 \bmod m \\
&(p-1)s^{p-1} + 1 \equiv 0 \bmod p \\
&s^{p-1} \equiv 1 \bmod p \\
&s \not\equiv 0 \bmod p.
\end{aligned}$$

It follows from Theorem 2.8.2 that $\frac{\lfloor((n-1)/(p-1))^{1/(p-1)}\rfloor}{2p_{\max}}$ is a lower bound on the degree of R. Hence the degree of Q must be as in the statement of the theorem.

Corollary 2.8.1. *Assume that m is square-free positive integer, and that p is a prime which does not divide m. Then the MOD_m-degree of $\neg\mathrm{MOD}_p^n$ is $\Omega(n)$, while the MOD_m-degree of MOD_p^n is $\Omega(n^{1/(p-1)})$.*

2.8.2 The MOD_m-Degree of Threshold Functions

By using the periodic property of the combinatorial coefficients modulo m as well as the well known *Möbius inversion formula*, it is possible to give simpler and more elegant proofs of the MOD_m-degree of threshold functions [Tsa93].

Let us define $[n] = \{1, 2, \ldots, n\}$. For $A \subseteq [n]$ define $x_A = \prod_{i\in A} x_i$. It is clear that the set $\{x_A : A \subseteq [n]\}$ ($\{x_A : A \subseteq [n], |A| \leq d\}$) forms a *basis* for the polynomials (of degree at most d) in $\mathbf{Z}_m[x_1, \ldots, x_n]$. Let P be a polynomial representing a boolean function $f \in \mathcal{B}_n$. For $A \subseteq [n]$ let us define $f(A) = f(a_1, \ldots, a_n)$, where $a_i = 1$ if $i \in A$, and is 0 otherwise. $P(A)$ is defined similarly. Write $P(x) = \sum_{A\subseteq[n],|A|\leq d} c_A x_A$. For any $A \subseteq [n]$ it is clear that $x_D(A) = 1$ if $D \subseteq A$ and is 0 otherwise. It follows that $P(A) = \sum_{D\subseteq A,|D|\leq d} c_D$. Using the well-known Möbius inversion formula [Lov79] we can prove the following result.

Lemma 2.8.1.

1. *If* $A \subset [n]$ *and* $|A| \leq d$ *then*

$$c_A = \sum_{D \subseteq A} (-1)^{|A|-|D|} P(D).$$

2. *If* $A \subseteq [n]$ *and* $|A| > d$ *then*

$$P(A) = \sum_{D \subset A, |D| \leq d} (-1)^{d-|D|} \binom{|A| - |D| - 1}{d - |D|} P(D).$$

Proof. Part 1 is proved by induction on $|A|$. The result is trivial if $|A| \leq 1$. Suppose that it is true if $|A| \leq k$. Consider the case $|A| = k + 1 \leq d$. Using the induction hypothesis we obtain

$$P(A) = c_A + \sum_{S \subset A} c_S = c_A + \sum_{S \subset A} \left(P(S) + \sum_{D \subset S} (-1)^{|S|-|D|} P(D) \right),$$

which implies that

$$c_A = P(A) - \sum_{S \subset A} \left(P(S) + \sum_{D \subset S} (-1)^{|S|-|D|} P(D) \right).$$

Restrict the above formula to subsets D_ℓ of A of size $\ell < |A|$ and we have

$$\begin{aligned}
&- \left(\sum_{D_\ell \subset A} P(D_\ell) + \sum_{S \subset A} \sum_{D_\ell \subset S} (-1)^{|S|-\ell} P(D_\ell) \right) \\
&= - \left(\sum_{D_\ell \subset A} P(D_\ell) + \sum_{D_\ell \subset A} \sum_{i=1}^{|A|-\ell-1} (-1)^i \binom{|A| - \ell}{i} P(D_\ell) \right) \\
&= - \sum_{D_\ell \subset A} \sum_{i=1}^{|A|-\ell-1} (-1)^i \binom{|A| - \ell}{i} P(D_\ell) \\
&= - \sum_{D_\ell \subset A} (-1)^{|A|-\ell-1} P(D_\ell) \\
&= \sum_{D_\ell \subset A} (-1)^{|A|-\ell} P(D_\ell).
\end{aligned}$$

This proves part 1. Next we prove part (2).

$$P(A) = \sum_{D \subset A, |D| \leq d} c_D$$

$$= \sum_{D \subset A, |D| \leq d} \left(\sum_{T \subseteq D} (-1)^{|D|-|T|} P(T) \right)$$

$$= \sum_{T \subset A, |T| \leq d} \sum_{i=0}^{d-|T|} (-1)^i \binom{|A| - |T|}{i} P(T)$$

$$= \sum_{T \subset A, |T| \leq d} (-1)^{d-|T|} \binom{|A| - |T| - 1}{d - |T|} P(T).$$

This proves the lemma.

The main theorem of this section is the following.

Theorem 2.8.4 ([Tsa93]). *$\delta(\text{TH}_k^n, m) \geq k$, where $m \geq 1$.*

Proof. Assume on the contrary that there is a polynomial of degree $< k$ representing TH_n^k over $\mathbf{Z}_m$. By definition of the threshold function we must have $P(A) = 0$, for all $A \subseteq [n]$ such that $|A| < k$. Hence by part (1) of Lemma 2.8.1 we must have $c_A = 0$, for all $A \subseteq [n]$ such that $|A| < k$. Hence, by part (2) of Lemma 2.8.1 $P(A) = 0$, for all $A \subseteq [n]$ such that $|A| \geq k$, which contradicts the definition of threshold function.

As an immediate consequence of Theorem 2.8.4, we can obtain strong lower bounds on the MOD_m-degree of the majority MAJ_n and the logical $\bigwedge$ functions (see Exercise 2.14.15).

2.9 Method of Filters

The method of filters was first proposed by Razborov in [Raz89] in order to give lower bounds for the monotone circuit complexity of boolean functions. The main idea of the method is as follows. Suppose that C is a small circuit for computing a hard function f. Further assume that C rejects every vector x such that $f(x) = 0$. In order to show that C does not compute f correctly, we combine rejecting computations for zeros of f to get rejecting computations for the ones of f. Combining computations is achieved by constructing filters over the set $f^{-1}(\{0\})$. An interesting consequence of this method is that it gives a framework for comparing deterministic and nondeterministic computations.

Let U be a subset of $\{0,1\}^n$. A *filter* $\mathcal{F}$ over U is a set of subsets of U such that $\emptyset \notin \mathcal{F}$ and if $A \in \mathcal{F}$ and $A \subseteq B$ then $B \in \mathcal{F}$. An *ultrafilter* on U is a filter $\mathcal{U}$ such that if $A \notin \mathcal{U}$ then $U \setminus A \in \mathcal{U}$. For any function $g \in \mathcal{B}_n$ let $\| g \| = \{u \in U : g(u) = 1\}$. It is easy to see that $\| \cdot \|$ is a homomorphism of the boolean algebra $\mathcal{B}_n$ into the boolean algebra of subsets of U, in the sense that for any boolean functions $g, h \in \mathcal{B}_n$,

- $\| g \wedge h \| = \| g \| \cap \| h \|$,
- $\| g \vee h \| = \| g \| \cup \| h \|$,
- $\| \neg g \| = U \setminus \| g \|$.

A filter $\mathcal{F}$ *preserves a pair* (A, B) of subsets of U if $A, B \in \mathcal{F}$ implies that $A \cap B \in \mathcal{F}$. A filter *preserves a gate* $g \wedge h$ if it preserves the pair $(\| g \|, \| h \|)$. A filter preserves a set of gates if it preserves every gate in the set. $\mathcal{F}$ is *above* a vector $v \in \{0,1\}^n$ if for all $i = 1, 2, \ldots, n$, $v_i = 1 \Rightarrow \| x_i \| \in \mathcal{F}$ and $v_i = 0 \Rightarrow \| \neg x_i \| \in \mathcal{F}$. Let C be a circuit with $\vee, \wedge$ gates, where all the negations are at the input level. Let $\Lambda(C)$ denote the set of $\wedge$ gates of G. A filter $\mathcal{F}$ *majorizes* the computation of a vector v if for every subfunction g of the circuit C, $g(v) = 1 \Rightarrow \| g \| \in \mathcal{F}$.

A *nondeterministic circuit* with m nondeterministic variables is a circuit with $2n + 2m$ inputs labeled

$$x_1, \ldots, x_n, \neg x_1, \ldots, \neg x_n \text{ and } y_1, \ldots, y_m, \neg y_1, \ldots, \neg y_m.$$

A nondeterministic circuit computes a function $f \in \mathcal{B}_n$ in the following way: $f)x) = 1$ if and only if there exists an assignment of the nondeterministic variables $y_1, \ldots, y_m$ which makes the circuit output 1.

Let $s_\wedge(f)$ $(\bar{s}_\wedge(f))$ be the number of $\wedge$-gates of an optimal deterministic (nondeterministic) circuit computing f. For any boolean function $f \in \mathcal{B}_n$ let $\rho(f)$ $(\bar{\rho}(f))$ be the minimum size of a collection Λ of pairs of subsets of $f^{-1}(\{0\})$ such that there is no filter (ultrafilter) above a vector in $f^{-1}(\{1\})$ which preserves Λ. We have the following result.

Theorem 2.9.1. *For any* $f \in \mathcal{B}_n$,

1. *(*[**Raz89**]*)* $\rho(f) \leq s_\wedge(f) = O(\rho(f)^2)$.
2. *(*[**Kar93**]*)* $\bar{\rho}(f) \leq \bar{s}_\wedge(f) = O(\bar{\rho}(f))$.

Proof. We give only an outline of the proof. 1. First we prove the lower bound. Let C be a circuit with less than $\rho(f)$ $\wedge$-gates. Then there is a filter $\mathcal{F}$ preserving $\Lambda(C)$ and which is above a vector $v \in f^{-1}(\{1\})$. We claim that $\mathcal{F}$ majorizes the computation of v. If not, then consider the first node of C which is not majorized. Say this node computes the function g. Since $\mathcal{F}$ is above v it is clear that g cannot be an input literal. Since $\mathcal{F}$ preserves $\Lambda(C)$, g cannot be the output of an $\wedge$-gate. Since $\mathcal{F}$ is a filter, g cannot be the output of an $\vee$-gate. This gives the desired contradiction.

Now we prove the upper bound. Let Λ be an optimal collection of $\wedge$-gates such that there is no filter preserving Λ and which is above a vector in $f^{-1}(\{1\})$. In this case for any x, $f(x) = 0$ if and only if there is a filter preserving Λ and which is above x. Indeed, if $f(x) = 0$ then the filter generated by $\{x\}$ (i.e., $A \in \mathcal{F} \Leftrightarrow x \in A$) preserves Λ and is above x. Thus we can test whether $f(x) = 0$ by trying to construct such a filter. Given x, we put in the filter the necessary sets $\| x_i \|, \| \neg x_i \|$ in order to guarantee that the filter is above x. To make sure that the filter preserves Λ we put $A \cap B$ in the filter if both A, B are in the filter and in addition $(A, B) \in \Lambda$. If at some point we are forced to put the empty set in the filter then we conclude that $f(x) = 1$. Using Razborov's technique [Raz89] it is not hard to design a circuit with $O(\rho(f)^2)$ $\wedge$-gates that checks whether such a filter exists.

2. First we prove the upper bound. Suppose that C is an optimal nondeterministic circuit for f with m nondeterministic input variables. For each $u \in U$ let $w^u \in \{0,1\}^m$ be a rejecting witness. For each subfunction of C define $\| g \| = \{u : g(u, w^u) = 1\}$. Now assume on the contrary that $\bar{s}_\wedge(f) < \bar{\rho}(f)$. Then there exists an ultrafilter preserving $\Lambda(C)$ and which is above a vector $v \in f^{-1}(\{1\})$. The ultrafilter gives values to every edge of circuit C which majorizes the computation of v for some setting of the nondeterministic variables. Since this computation is rejecting, C does not compute f correctly.

Now we prove the upper bound. Let Λ be an optimal collection of $\wedge$-gates such that there is no ultrafilter preserving Λ and which is above a vector in $f^{-1}(\{1\})$. As before we can test whether a given vector x is in $f^{-1}(\{0\})$ by trying to construct an ultrafilter above x which preserves Λ. Nondeterminism makes this task easier.

An easy application of Theorem 2.9.1 is the following result which provides an elegant lower bound technique for the complexity of monotone boolean functions. Let $f \in \mathcal{B}_n$ be a monotone boolean function. Call a vector $u \in \{0,1\}^n$ a *maximal zero* of f if $f(u) = 0$, but for any v obtained from u by flipping a bit, we have $f(v) = 1$.

Theorem 2.9.2 ([Kar93]). *If M is the set of maximal zeros of the monotone boolean function $f \in \mathcal{B}_n$ then $\bar{s}_\wedge(f) \geq \lceil \log |M| \rceil$.*

Proof. Take $U = M$. In view of Theorem 2.9.1 it is enough to show that $\bar{\rho}(f) \geq \lceil \log |M| \rceil$. For any $u, w \in M$ define the ultrafilter $\mathcal{U}_{(u,w)}$ by $A \in \mathcal{U}_{(u,w)} \Leftrightarrow A \cap \{u, w\} \neq \emptyset$. Clearly, $\mathcal{U}_{(u,w)}$ is above the vector $u \vee w$ and $f(u \vee w) = 1$. Let Λ be an optimal set of $\wedge$-gates. The ultrafilter $\mathcal{U}_{(u,w)}$ does not preserve the $\wedge$-gate (g, h) if and only if $u \in \| g \| \setminus \| h \|$ and $w \in \| h \| \setminus \| g \|$ (in which case we say that the gate *separates the pair* $\{u, w\}$). Thus if a pair is not separated by any $\wedge$-gate in Λ then $\mathcal{U}_{u,w}$ preserves Λ. Now enumerate the set Λ as $(g_1, h_1), \ldots, (g_t, h_t)$. An $\wedge$-gate (g, h) will separate the greatest number of pairs if $\| g \|$ and $\| h \|$ form a partition of M. Assuming this is the case for every member of Λ, consider the function $r : M \to \{g, h\}^t$ defined by $r(u)_i = g$ if and only if $u \in \| g \|$. Since every pair from M is separated, the function is injective, which completes the proof of the theorem.

A simple application of Theorem 2.9.2 is given in Exercise 2.14.16. The method of filters is also applicable to non-monotone computation (for more details see Exercise 2.14.17).

2.10 Eliminating Majority Gates

The main result of this section is that if a boolean function can be computed by a constant depth circuit having 2^m $\bigwedge, \bigvee, \neg$ gates, and m majority gates then it can also be computed by a constant depth circuit having $2^{m^{O(1)}}$ $\bigwedge, \bigvee, \neg$ gates, and a single majority gate.

Real function $g(x_1, \ldots, x_n)$ *approximates* boolean function $f(x_1, \ldots, x_n)$ with error ϵ if for all $x_1, \ldots, x_n \in \{0,1\}$, $|f(x_1, \ldots, x_n) - g(x_1, \ldots, x_n)| \leq \epsilon$. The *norm* of a polynomial is the sum of the absolute value of its coefficients; the norm of a rational function is the norm of its numerator plus the norm of its denominator. We mention without proof the following lemma.

Lemma 2.10.1 ([Bei92]). *The function* $\text{MAJ}_n(x_1, \ldots, x_n)$ *can be approximated with error* ϵ *by a rational function with norm* $2^{O(\log^2 n \log(1/\epsilon))}$.

Now we can prove the main theorem.

Theorem 2.10.1 ([Bei92]). *Every boolean function computable by a depth* d *circuit of size* s *with* m *majority gates is also computable by a depth* $d+2$ *circuit of size* $2^{m(O(\log s))^{2d+1}}$ *with a single majority gate at the output gate.*

Proof. We give only an outline. For this proof it will be convenient to assume that the boolean function $f \in \mathcal{B}_n$ has its arguments in $\{0,1\}^n$ but the result is in $\{-1,1\}$, where -1 denotes false and 1 denotes true. Let $\mathcal{T}_k$ be the class of boolean functions computable by depth d circuits of size s with m majority gates all occurring in levels 0 to k. We will show that if $f \in \mathcal{T}_k$ then $f = \text{sgn}(P(f_1, \ldots, f_\ell))$, where P is a polynomial having norm $N_P(k) = 2^{m(O(\log s))^{2d+1}}$ and $f_1, \ldots, f_\ell$ are boolean functions computable with depth d, size s circuits having no majority gates. The theorem follows by taking $k = d$. Since products can be computed by $\bigwedge$ gates the proof of the theorem would be complete. Moreover, we reduce the number of majority gates in the circuit representing f from m to 1, and in fact this majority gate is at the root.

Let k be fixed and $f \in \mathcal{T}_k$. We compute f by summing over all sequences of possible outputs for the majority gates on level k, (a) the value of f given those outputs (here we use -1 for false, and 1 for true) multiplied by (b) the $\bigwedge$ of the corresponding majorities or their complement (here we use 0 for false, and 1 for true). Each term in (a) is the sign of a polynomial P of functions computable with depth d, size s circuits having no majority gates and the norm of P is bounded by $N_P(k-1)$. Suppose there are t majority gates. Clearly, $t \leq m$. The terms in (b) are products of exactly t factors each of which is either a majority or its complement. Let $\epsilon = 1/(m2^m N_P(k-1)+m)$. Each majority has at most s inputs and can be approximated within error ϵ by a rational function R whose norm is $2^{(\log^2 s)(m+\log N_P(k-1))}$, by Lemma 2.10.1. However, if a majority gate is approximated by a rational function R then its complement is approximated by the rational function $1-R$ within the same error, namely $2^{(\log^2 s)(m+\log N_P(k-1))}$. We approximate each term in (b) by the product of the rational functions that approximate the corresponding majorities or their complements. The resulting error is at most $(1+\epsilon)^m - 1$. Now the function f is approximated by taking the sum of the 2^m terms (a) times (b). Since each term in (a) is majorized by $N_P(k-1)$, the error in approximating f is majorized by $2^m N_P(k-1)((1+\epsilon)^m - 1)$, which is easily shown to be less than 1 (use Exercise 2.14.11 with $N = N_P(k-1)$), hence the approximation has the same sign as f.

Since all the rational functions used above have the same denominator, we obtain a polynomial that has the same sign as f by multiplying by the square of that common denominator. Now, if N_R bounds the norm of the rational functions and $N_P(k-1)$ the norm of the polynomials P used for (a) then $2^m N_P(k-1)N_R^2$ bounds the norm of the resulting polynomial, hence

$$N_P(k) \leq 2^m N_P(k-1) 2^{(\log^2 s)(m+\log N_P(k-1))}.$$

Since, $N_P(0) < s$, an easy induction shows that

$$N_P(k) \leq 2^{(m+\log s)(O(\log s))^{2k}} = 2^{m(O(\log s))^{2k+1}}.$$

2.11 Circuits for Symmetric Functions

Section 2.11 provides a complexity result, due to Fagin, Klawe, Pippenger and Stockmeyer [FKPS85], for sequences $f = \{f_n : \{0,1\}^n \rightarrow \{0,1\}\}$ of boolean functions. If $\mu_f(n)$ is "the least number of variables that must be set to constants in order that the resulting function f_n is constant" then it is shown that the growth of $\mu_f(n)$ completely determines whether or not f can be realized by a family of constant depth polynomial size circuits.

The *spectrum* of a symmetric function $f \in \mathcal{B}_n$ is a sequence $w \in \{0,1\}^{n+1}$ whose i-th bit w_i is the output of f on inputs of weight i, where $0 \leq i \leq n$. Since a symmetric function is uniquely determined by its spectrum, we will often identify the function with its spectrum. For each word $w \in \{0,1\}^{n+1}$ let $m(w) = n + 1 -$ "length of largest constant subword of w". (If f has spectrum w then we also use the notation $m(f) = m(w)$.) If w is the spectrum of f then it is clear that $m(w)$ is "the least number of variables that must be set to constants in order that the resulting function f is constant". For sequences $f = \{f_n : \{0,1\}^n \rightarrow \{0,1\}\}$ of boolean functions with corresponding spectra w^n let $\mu_f(n) = m(w^n)$. A (p,d)-*circuit* for a sequence of symmetric functions $f = \{f_n\}$ is a sequence $C = \{C_n\}$ of circuits such that $size(C_n) \leq p(n)$ and $depth(C_n) \leq d$; moreover, in this case the sequence $\{w^n\}$ of spectra of the functions $\{f_n\}$ is called a (p,d)-*spectrum*. We have the following simple results on spectra whose proofs we leave as exercises to the reader.

Lemma 2.11.1. *The complement* $\bar{w}$ *(i.e., taking the complement of each bit) as well as the reverse* w^R *(i.e., the result of writing* w *backwards) of a* (p,d)*-spectrum* w *is a* (p,d)*-spectrum.*

Lemma 2.11.2. *Let $w \in \{0,1\}^{n+1}$ be a (p,d)-spectrum. If p, g are monotone increasing functions with $g(1) = 1$ and $g^{-1}(i) =$ "the greatest integer j for which $g(j) \leq i$" then each subword of w of length $\geq g(n)+1$ is a $(p \circ g^{-1}, d)$-spectrum.*

Lemma 2.11.3. *If $w^i \in \{0,1\}^{n+1}$ is a (p,d)-spectrum, for $i \leq q(n)$, where p, q are polynomials, then both $w^1 \wedge \cdots \wedge w^{q(n)}$, $w^1 \vee \cdots \vee w^{q(n)}$ (i.e., the result of taking the $\wedge$ and $\vee$ of the corresponding bits, respectively) are $((p(n)+1)q(n), d+1)$-spectra.*

To prove the next lemma we require some definitions. For H a subset of a permutation group G (acting on the set X) and $S \subseteq X$ define $H(S) = \{h(s) : h \in H, s \in S\}$.

Lemma 2.11.4. *If a finite group G acts transitively on the set X and $\emptyset \neq S \subseteq X$ then $X = H(S)$, for some $H \subseteq G$ of size $\leq \frac{|X|}{|S|}(1 + \ln|S|)$. (Clearly, the size of H must be at least $\frac{|X|}{|S|}$.)*

Proof. We will define a random $H \subseteq G$ and show that its expected size is $\leq \frac{|X|}{|S|}(1 + \ln|S|)$. Let $0 < p < 1$ be arbitrary but fixed and let H_1 be a random subset of G obtained by taking each element of G independently with probability p. It follows that the expected size of H_1 is $p|G|$. For each $x \in X \setminus H(S)$ there is a $g \in G$ such that $x \in g(S)$. Let H_2 be a set of such g's. If $H = H_1 \cup H_2$ then it is clear that $H(S) = X$.

Next we study the expected size of H. Put $G(x,y) = \{g \in G : x = g(y)\}$. By transitivity, $G(x,y)$ is a left coset of $G(y,y)$ and for each $y \in X$, the family $\{G(x,y)\}_{x \in X}$ is a partition of G. Hence, $|G| = \sum_{x \in X} |G(x,y)|$ and $|G(x,y)| = |G(y,y)| = |G|/|X|$. If we define $G(x,S) = \{g \in G : x \in g(S)\}$, then as before the family $\{G(x,y)\}_{y \in S}$ is a partition of $G(x,S)$, and hence $|G(x,S)| = |G||S|/|X|$. Since for an element $x \in X$, $x \notin H_1(S) \iff H_1 \cap G(x,S) = \emptyset$ it follows that

$$\Pr[x \notin H_1(S)] \leq (1-p)^{|G(x,S)|} = (1-p)^{|G||S|/|X|}.$$

This implies that the expected size of H_2 is $\leq |X|(1-p)^{|G||S|/|X|}$ and hence the expected size of $H = H_1 \cup H_2$ is $\leq p|G| + |X|(1-p)^{|G||S|/|X|}$. Setting $p = (|X| \ln|S|)/(|G||S|)$ and using the inequality $1 - p \leq e^{-p}$ we obtain the desired result.

Now we use Lemma 2.11.4 to give an upper bound for the circuit complexity of threshold functions.

Lemma 2.11.5. *For all $m \leq n$ and all ℓ,*

$$L_{k+2}^{\neg,\wedge,\vee}(\mathrm{TH}_{\ell m}^{\ell n}) \leq (8m)^{\ell/2}(\ell n + 1) \cdot \ell \cdot L_k^{\neg,\wedge,\vee}(\mathrm{TH}_m^n).$$

Proof. Partition the variables $x_1, \ldots, x_{\ell n}$ into ℓ blocks $B_1, \ldots, B_\ell$ each of size n and let C be a depth k circuit for TH^n_m. Let C_j be the circuit obtained from C by substituting the variables $x_1, \ldots, x_n$ by the variables of B_j. The circuit D obtained by taking the $\wedge$ of $C_1, \ldots, C_\ell$ has size $\ell \cdot L_k(\mathrm{TH}^n_m)$ and depth $k+1$.

Anticipating an application of Lemma 2.11.4 take X to be the set of conjunctions $\bigwedge V$, where V is a subset of the ℓn variables of size ℓm, $S \subseteq X$ is the set of elements of X accepted by D, and let G be the symmetric group on the variables $x_1, \ldots, x_{\ell n}$. Clearly, $|X| = \binom{\ell n}{\ell m}, |S| = \binom{n}{m}^\ell$, the group G acts transitively on X, and the circuit $g(D)$ accepts each member of $g(S)$. It follows from Lemma 2.11.4 that there is a set $H \subseteq G$ such that $X = H(S)$ and $|H| \leq \left(1 + \ln \binom{n}{m}\right) \cdot \binom{\ell n}{\ell m} \cdot \binom{n}{m}^{-\ell}$. Using the inequalities

$$\frac{2^{nH(m/n)}}{(8m)^{1/2}} \leq \binom{n}{m} \leq 2^{nH(m/n)},$$

(see [Pet61]) where $H(m/n) = -\frac{m}{n} \log \frac{m}{n} - \frac{n-m}{n} \log \frac{n-m}{n} \leq 1$ we obtain that $|H| \leq (8m)^{\ell/2}(\ell n + 1)$. Now if we take the $\vee$ of the family of circuits $\{g(D)\}_{g \in H}$ we obtain a circuit for the threshold $\mathrm{TH}^{\ell n}_{\ell m}$ of the asserted size. The proof of the lemma is now complete.

Lemma 2.11.5 has two immediate corollaries which we leave as exercises. From now on we assume that p is a polynomial.

Lemma 2.11.6. *If TH^n_m has a (p, d)-circuit and $m + s \leq n$ then TH^{n+s}_m has a $(p', d+2)$-circuit, where p' is a polynomial depending only on p.*

Lemma 2.11.7. *If TH^n_m has a (p, d)-circuit then TH^n_{cm} has a $(p', d+2)$-circuit, where p' is a polynomial depending only on p and c.*

2.11.1 Negative Results

We are now in a position to prove a sufficient condition on the nonexistence of constant depth polynomial size circuits.

Theorem 2.11.1 ([FKPS85]). *If $f = \{f_n\}$ is a family of symmetric boolean functions, and*

$$\mu_f(n) \geq n^{\Omega(1)} \text{ for infinitely many } n$$

then f is not computable by a polynomial size constant depth family of unbounded fan-in boolean circuits; i.e., $f \notin \mathrm{AC}^0$.

Proof. We will need the following lemma.

Lemma 2.11.8. *If $w = w_0 w_1 \cdots w_n \in \{0,1\}^{n+1}$ is a $(p, 2)$-spectrum and p a polynomial of degree k then $w_{k+1} = w_{k+2} = \cdots = w_{n-k-1}$, for n sufficiently large.*

Proof. Assume that the lemma does not hold, and let r be minimal such that $w_r \neq w_{r+1}$ and $k+1 \leq r < n-k-1$. Using Lemma 2.11.1, we can assume without loss of generality that the output gate is an $\vee$. There are two cases to consider depending on whether or not $w_r = 0$ or 1.

Case 1. $w_r = 1$ (in which case $w_{r+1} = 0$). Let X be a subset of the n variables of size r and let X' be the set of $n-r$ remaining variables. Since $w_r = 1$ we know that the output node takes on the value 1 when all the variables in X take on the value 1 and all the variables in X' take on the value 0. Moreover for this assignment there is an $\wedge$-gate, denoted by v_X, that takes on the value 1. But then, for each variable $x' \in X'$ the literal $\neg x'$ is an input node connecting to v_X (if not, $w_{r+1} = 1$). It follows that if $X_1 \neq X_2$ then $v_{X_1} \neq v_{X_2}$, for X_1, X_2 as above. Thus there are at least as many $\wedge$-gates as subsets of size r of a set of n elements. Hence, $\binom{n}{r} \leq p(n)$. Since $p(n)$ is of degree k and $r \geq k+1$ this is a contradiction.

Case 2. $w_r = 0$ (in which case $w_{r+1} = 1$). Argue exactly as in Case 1 but with sets of variables X of size $r+1$ to conclude that $\binom{n}{r+1} \leq p(n)$. As before this gives a contradiction. This completes the proof of Lemma 2.11.8.

Now we concentrate on the proof of the theorem. Assume on the contrary that f is computable by a (p, d)-circuit, where p is a polynomial. The following theorem is a corollary of Theorem 2.6.2.

Theorem 2.11.2. *If $w \in \{0,1\}^{n+1}$ is (p, d)-spectrum, then there is a subword w' of w of length $\geq n^{1/4}/4+1$ which is a $(p', d-1)$-spectrum, for some polynomial p' depending only on p, provided that n is sufficiently large.*

Applying Theorem 2.11.2 $d-2$ times we find a polynomial q such that for n large enough the spectrum w of f_n contains a subword w' of length $\geq 4^{-(d-2)}n^{4^{-(d-2)}} + 1$ which is a $(q, 2)$-spectrum. If the degree of q is k then by Lemma 2.11.8 all bits of w' but the first $k+1$ and last $k+1$ are the same. It follows that the spectrum of f_n contains a constant subword of length $n^{\Omega(1)}$, for n sufficiently large. Now fix n and write $w = sut$ (i.e., the concatenation of s, u, t), where u is the longest constant subword of w. Hence, $|u| = n^{\Omega(1)}$ and $\mu_f(n) = |s| + |t| = n^{\Omega(1)}$. Using Lemma 2.11.1 and reversing and/or complementing the bits of w, we can assume that $|s| = n^{\Omega(1)}$ and u consists only of 1's. Hence $w = s'01^{|u|}t$, where $|s'| = |s| - 1$. Now let j be an integer such that $|s'|, |u| \geq n^{\Omega(1)} \geq N = \lfloor n^{1/j} \rfloor$. However the spectrum of the majority function on N variables is $0^{1+\lfloor N/2 \rfloor}1^{1+\lceil N/2 \rceil}$. Hence by taking an $\wedge$ of appropriate subwords of w of length N and using Lemmas 2.11.2 and 2.11.3 it is easy to see that we can define majority on N variables by a $(p', d+1)$-circuit, where p' is a polynomial depending only on p. This contradicts the fact that majority is not computable by a constant depth polynomial size circuit (e.g., see Theorem 2.7.1).

2.11.2 Positive Results

Theorem 2.11.3 ([FKPS85]). *If $f = \{f_n\}$, $f' = \{f'_n\}$ are families of symmetric boolean functions,*

$$\mu_{f'}(n) = O(\mu_f(n))$$

and if f is computable by a polynomial size constant depth family of boolean circuits of unbounded fan-in, then so is f'; i.e.,

$$(\mu_{f'}(n) = O(\mu_f(n)) \wedge f \in \mathrm{AC}^0) \Rightarrow f' \in \mathrm{AC}^0.$$

Proof. The proof requires the following three lemmas.

Lemma 2.11.9. *Assume that TH_m^n has a (p,d)-circuit and $m < n/2$. If either $0 \le i < m$ or $n - m < i \le n$, then EXACT_i^n has $(p', d+3)$-circuit, where p' is a polynomial depending only on p.*

Proof. Since $\mathrm{EXACT}_i^n = \mathrm{TH}_i^n \wedge \neg\mathrm{TH}_{i+1}^n$, it is enough to show that both $\mathrm{TH}_i^n, \neg\mathrm{TH}_{i+1}^n$ are computable by circuits of the appropriate size and depth. Let $u' = 0^i 1^{n+1-i}$ and $u'' = 0^{i+1}1^{n-i}$ be the corresponding spectra of $\mathrm{TH}_i^n, \mathrm{TH}_{i+1}^n$. By Lemma 2.11.6 the spectrum of TH_m^{n+m}, which is $u = 0^m 1^{n+1}$, has a $(p', d+2)$-circuit, for some polynomial p' depending only on p. Clearly, both u', u'' are subwords of u each of length $> g(m)+1$, where $g(i) = \lceil i/2 \rceil$. Hence for the case where $i < m$ the result follows from Lemma 2.11.2. The case $n - m < i \le n$ follows by reversing the spectrum of EXACT_i^n and applying Lemma 2.11.1.

Lemma 2.11.10. *Assume that TH_m^n has a (p,d)-circuit and $f \in \mathcal{B}_n$ is symmetric. If $m(f) \le m < \frac{n}{2}$ then f has a $(p', d+4)$-circuit, where p' is a polynomial depending only on p.*

Proof. If $w = 0^m 1^{n+1-m}$ is the spectrum of TH_m^n then $v = 0^m 1^{n+1-2m} 0^m = w \wedge w^R$ is a $(2p+2, d+1)$-spectrum. Let u be the spectrum of f. Then by definition $m(u) = n+1-|u_0|$, where u_0 is a constant subword of u of maximal length. Hence $|u_0| \ge n+1-m$. Without loss of generality we may assume that $u_0 = 1^{|u_0|}$. If w^i is the spectrum of EXACT_i^n, then it is easy to see that u can be obtained as the $\vee$ of u with appropriate choices of the w^is. Hence the result follows from Lemmas 2.11.2, 2.11.3, and 2.11.9.

Lemma 2.11.11. *If $w \in \{0,1\}^{n+1}$ has a (p,d)-spectrum then $\mathrm{TH}_{m(w)/2}^{n-m(w)}$ has a $(p', d+1)$-circuit, where p' is a polynomial depending only on p.*

Proof. Without loss of generality we may assume $m(w) > 0$. By either reversing or complementing w (Lemma 2.11.1), we may assume that $w = s01^j t$, where $|u| \ge \lceil m(w)/2 \rceil$ and $j \ge n - m(w)$. Hence the spectrum of $\mathrm{TH}_{m(w)/2}^{n-m(w)}$ can be obtained by taking the $\wedge$ of appropriate subwords of w of length $n - m(w) + 1$, as in the proof of Theorem 2.11.1.

Now we return to the proof of Theorem 2.11.3. Let $f = \{f_n\}$ and $f' = \{f'_n\}$. Assume each f_n has a (p, d)-circuit, for some polynomial p and let c be a constant (sufficiently large) such that $\mu_{f'}(n) \leq c\mu_f(n)$, for all n. By Theorem 2.11.1, $\mu_f(n) < n/(2c)$, for n sufficiently large. By Lemma 2.11.11, $\text{TH}^{n-\mu_f(n)}_{\mu_f(n)/2}$ has a $(p_1, d+1)$-circuit, where p_1 is a polynomial depending only on p. By Lemma 2.11.6, $\text{TH}^n_{\mu_f(n)/2}$ has a $(p_2, d+3)$-circuit, where p_2 is a polynomial depending only on p and c. By Lemma 2.11.7, TH^n_j has $(p_3, d+5)$-circuit, where $j = 2c\lceil \mu_f(n)/2 \rceil$, and p_3 is a polynomial depending only on p. Since $j \geq c\mu_f(n) \geq \mu_{f'}(n)$ and $\mu_{f'}(n) \leq c\mu_f(n) < \frac{n}{2}$, the result follows by applying Lemma 2.11.10.

As a corollary, we also obtain the following theorem:

Theorem 2.11.4 ([FKPS85]). *If $\mu_f(n) \leq (\log n)^{O(1)}$ then $f = \{f_n\}$ is computable by a polynomial size constant depth family of boolean circuit of unbounded fan-in; i.e., $f \in \text{AC}^0$.*

Proof. In view of Theorem 2.11.3 it is enough to show that TH^n_m has $(p, 2k+1)$-circuit, where $m = O((\log n)^k/(\log\log n)^{k-1})$. The proof is by induction on k. Define $\ell = m$, if $k = 1$, and $\ell = \lceil \log n/\log\log n \rceil$, otherwise. Put $m' = \lceil m/\ell \rceil$ and $n' = \lceil (n + \ell m' - m)/\ell \rceil$. Now observe that if $k \geq 2$, then $m' = O((\log n)^{k-1}/(\log\log n)^{k-2})$, and hence by the induction hypothesis $\text{TH}^{n'}_{m'}$ has a $(p', 2k-1)$-circuit. On the other hand, if $k = 1$, then $m' = 1$ and by taking the $\vee$ of variables, we obtain again that $\text{TH}^{n'}_{m'}$ has a $(p', 2k-1)$-circuit, for some polynomial p'. Now apply Lemma 2.11.5 to show that $L^{\neg,\wedge,\vee}_{2k+1}(\text{TH}^{\ell n'}_{\ell m'}) \leq p''(\ell n')$, for some polynomial p'', depending only on p'. Substituting 1's for $\ell m' - m$ variables and 0's for $(\ell n' - n) - (\ell m' - m)$ variables we obtain the desired result.

2.12 Probabilistic Circuits

In the previous sections, we studied the computational limitations of deterministic constant depth polynomial size families of boolean circuits with unbounded fan-in. A natural question to ask is whether anything is to be gained by considering *probabilistic* constant depth circuits. These are circuits which have deterministic as well as probabilistic inputs. To evaluate C on a given input $x = (x_1, \ldots, x_n)$, we set the probabilistic variables $y_1, \ldots, y_m$ to 0 or 1 each with probability 1/2 and then compute the unique output. A family $\{C_n\}$ of probabilistic circuits $\{\epsilon_n\}$-*computes* the family $\{f_n : \{0,1\}^n \to \{0,1\}\}$ of boolean functions if for all n, $x \in \{0,1\}^n$,

$$f_n(x) = 1 \Rightarrow \Pr[C_n(x) = 1] \geq \frac{1}{2} + \epsilon_n,$$

$$f_n(x) = 0 \Rightarrow \Pr[C_n(x) = 1] \leq \frac{1}{2},$$

i.e., the circuit C_n has an ϵ_n-*advantage* in producing the correct output. We can prove the following theorem.

Theorem 2.12.1 ([ABO84]). *Let $r \geq 1$. If $\{f_n\}$ is $\{\frac{1}{(\log n)^r}\}$-computable by a polynomial size constant depth probabilistic circuit then it is also computable by a polynomial size constant depth deterministic circuit.*

Proof. For $1 \geq p \geq q \geq 0$ a probabilistic circuit C (p,q)-*separates* A from B, and we abbreviate this with $[C, A, B, p, q]$, if

$$x \in A \Rightarrow \Pr[C(x) = 1] \geq p,$$
$$x \in B \Rightarrow \Pr[C(x) = 1] \leq q,$$

where $A, B \subseteq \{0,1\}^n$ and C has n deterministic inputs. We prove a series of claims which will imply the result of the theorem.
Claim C1. If $[C, A, B, p, q]$ and $p \geq p_1, q \leq q_1$, then $[C, A, B, p_1, q_1]$.
Proof of Claim C1. Trivial from the definitions.
Claim C2. If $[C, A, B, p, q]$, then there is a circuit C' such that $size(C') = size(C)$, $depth(C') = depth(C)$ and $[C', B, A, 1-q, 1-p]$.
Proof of Claim C2. C' is the negation of C.
Claim C3. If $[C, A, B, p, q]$ and $\ell \geq 1$, then there is a circuit C^ℓ such that $size(C^\ell) = \ell size(C) + 1$, $depth(C^\ell) = depth(C) + 1$ and $[C^\ell, A, B, p^\ell, q^\ell]$.
Proof of Claim C3. Take ℓ independent copies of C and connect their outputs with a single $\bigwedge$-gate. It is clear that $[C^\ell, A, B, p^\ell, q^\ell]$.
Claim C4. If $[C, A, B, p, q]$ and $\ell - p + q < 2^{-n}$, then there is a circuit C', such that $size(C') = size(C)$, $depth(C') = depth(C)$ and $[C', A, B, 1, 0]$.
Proof of Claim C4. For $x \in A$ ($x \in B$) let R_x be the set of random assignments to random variables which produce the output 0 (1). By hypothesis $\Pr[R_x] < 2^{-n}$ and hence also $\Pr[\bigcup_{x \in A \cup B} R_x] < 1$. It follows that there exists an assignment $y \notin \bigcup_{x \in A \cup B} R_x$ to the random variables which evaluates 1 for all $x \in A$ and 0 for all $x \in B$. Using this y it is easy to construct the required deterministic circuit.

We will see that (C1) can be used to eliminate error terms, while (C2) in conjunction with (C3) can be used to amplify an advantage. Finally (C4) is used to convert probabilistic to deterministic circuits. We prove two more claims.
Claim C5. If $[C, A, B, \frac{1}{2}(1 + (\log n)^{-r}), \frac{1}{2}]$, $r \geq 2$, then there is a circuit C' such that $size(C') = O(n^2 \log n \cdot size(C))$, $depth(C') = depth(C) + 2$ and $[C, A, B, \frac{1}{2}(1 + (\log n)^{-r+1}), \frac{1}{2}]$.
Proof of Claim C5. Using the inequality $(1+x)^a > 1 + ax$ we see that $(1 + (\log n)^{-r})^{2\log n} > 1 + 2/(\log n)^{r-1}$. Hence applying (C3) with $\ell = 2\log n$ and then (C1) we obtain a circuit C_1 such that $[C_1, A, B, \frac{1}{n^2}(1+1/(\log n)^{r-1}), \frac{1}{n^2}]$. Next apply (C2), use the inequality

$$\left(1 - \frac{1}{n^2}(1 + 2/(\log n)^{r-1})\right)^{n^2 \log_e 2} < e^{\log_e 2(1+2/(\log n)^{r-1})}$$

$$< 1 - \frac{1}{(\log n)^{r-1}},$$

apply (C3) with $\ell = n^2 \log_e 2$ and apply (C2) once again in order to get the desired circuit.

The sole effect of (C5) is that it amplifies a $\frac{1}{(\log n)^r}$-advantage into a $\frac{1}{\log n}$-advantage.

CLAIM C6. If $[C, A, B, \frac{1}{2}(1 + (\log n)^{-1}), \frac{1}{2}]$, $r \geq 2$, then there is a circuit C' such that $size(C') = O(n^8 size(C))$, $depth(C') = depth(C) + 4$ and $[C, A, B, 1, 0]$.

Proof of claim C6. Apply (C3) with $\ell = 2 \log n$ and (C1) to conclude that for some circuit C_1, $[C_1, A, B, \frac{2}{n^2}, \frac{1}{n^2}]$ (this follows from the trivial inequality $(1 + 1/\log n)^{2 \log n} > 2$). Now apply (C2), then (C3) with $\ell = 2n^2 \log n$ and finally (C1) to conclude that for some circuit C_2, $[C_2, B, A, \frac{1}{n^2}, \frac{1}{n^4}]$. Now apply (C2) and then (C3) with $\ell = n^2$ to obtain $[C_3, A, B, 1 - \frac{2}{n}, e^{-n}]$ for some circuit C_3. Again apply (C2) and then (C3) with $\ell = n$ to show that $[C_4, B, A, 1 - 2ne^{-n}, (2/n)^n]$ for some circuit C_4. Now use the fact that $2ne^{-n} + (2/n)^n < 2^{-n}$, for n large enough, and apply (C2) and finally (C4) to obtain the desired circuit.

The proof of the theorem is now immediate by using the above claims on the sets $A_n = f_n^{-1}\{1\}, B_n = f_n^{-1}\{0\} \subseteq \{0, 1\}^n$.

2.13 Historical and Bibliographical Remarks

The analysis of polynomial size, constant depth circuits has given rise to a variety of very sophisticated mathematical techniques which are applicable to the study of the complexity of several important boolean functions, such as parity, majority, MOD_p, etc. Nevertheless, and despite the fact that almost all boolean functions on n variables have circuit size $\frac{2^n}{n}$ (see Theorem 2.2.1) no NP function is known whose circuit size is super-polynomial (a positive answer to this question would prove P $\neq$ NP). As a matter of fact, the best known lower bound for such a function is $3n - o(n)$ [Blu84, Pau77].

The exponential lower bound for monotonic, bounded fan-in real circuits which solve the *Broken Mosquito Screen* problem in Section 2.4 follows [HC99]. The success of lifting lower bounds, originally proved for monotonic boolean circuits, to monotonic real circuits, suggested the question of whether monotonic real circuits are not any more powerful. The negative answer to this question was given by A. Rosenbloom [Ros97].

Our presentation of the *st*-connectivity lower bound for (due to Karchmer and Wigderson) follows that of [BS90] and notes of P. Clote and P. Michel, from a seminar we gave at the Université Paris VII. Our proof of Johannsen's extension of the Karchmer–Wigderson lower bound to monotonic real circuits is new, produced by an appropriate modification of [BS90] by small definitional changes, and using Johannsen's idea of sorting.

Our treatment of monotonic circuits is restricted to the recent lower bounds for monotonic real circuits, which can be applied to proof size lower bounds for resolution and cutting plane refutations, covered in a later chapter. Important papers on monotonic circuits not covered in this text include Razborov's $n^{\Omega(\log n)}$ bound for the clique function [Raz87b], Andreev's exponential lower bound for an artificially constructed boolean function [And85], and Alon and Boppana's exponential lower bound for the clique function [AB87]. The treatment of infinite parity circuits follows the argument given in unpublished lecture notes of M. Sipser [Sip85a]. J. Håstad's argument comes from his published dissertation [Hås87], while A.A. Razborov's remarkable combinatorial simplification of the Switching Lemma first appeared in [Raz94]. A valuable compendium of lower bound arguments, using the Razborov simplification as applied to decision trees, appears in the unpublished *Switching Lemma Primer* of P. Beame [Bea94]. The hybrid Switching Lemma argument for a lower bound to *st*-connectivity for (non-monotonic) boolean circuits is due to Beame, Impagliazzo, and Pitassi [BIP98].

The idea of representing boolean functions as polynomials over the two-element field $\mathbf{Z}_2 = GF(2)$ dates back to Shannon [Sha38]. A similar idea of considering the depth of $\bigwedge$ gates as the degree of the circuit was also introduced by Skyum and Valiant [SV81].

Polynomials have also been used extensively in order to prove complexity bounds. Minsky and Papert used them in their study of perceptrons [MP68]. More recently, Razborov [Raz87a, Raz93] and Smolensky [Smo87] used them to obtain the lower bounds given in Sections 2.7. Razborov's main idea is based on the fact that the class of boolean functions computable by constant depth, polynomial size circuits with unbounded fan-in $\bigwedge, \bigvee, \bigoplus$ gates can be approximated by polynomials of low degree, while the majority function can not be so approximated. Smolensky extended Razborov's ideas by introducing MOD_p gates, for p prime, and extending the algebraic setting from $\mathbf{Z}_2$ to the algebra of polynomials over a field of characteristic p in the variables $x_1, \ldots, x_n$, satisfying the identities $x_i = x_i^2$. A beautiful survey article describing the polynomial method in circuit complexity is [Bei93]. We also recommend the survey articles by Boppana and Sipser [BS90] and Sipser [Sip92].

The algebraic techniques employed in this chapter have also been used to obtain lower bounds in the "Programs-over-Monoid" model. Since the computation of a finite state machine can be viewed as an iterated multiplication over a particular finite monoid, every regular language has an associated *syntactic semigroup*.[19] There is a well-known structure theory for such semigroups due to Krohn–Rhodes [Arb68], which techniques have found application to the "Programs-over-Monoid" model, as developed by Mix Barrington,

[19] Given a minimal finite state automaton M with state set Q, which accepts a regular language L, the syntactic semigroup of L is the set $\{f_w : w \in L\}$, where $f_w : Q \to Q$ is defined by $f_w(q) = \widehat{\delta}(q, w)$.

Straubing, Thérien and others (see for instance [BT88a]). For a nice survey on this important topic, the reader is referred to [Bar92].

The presentation of Section 2.9 follows the paper of Karchmer [Kar93] which also applies the method of ultrafilters in order to give a new proof for the exponential monotone size lower bound for the clique function. Karchmer's presentation also makes explicit the analogy of the method with the method of ultraproducts in model theory [CK73]. For more information on probabilistic techniques, the reader should consult [ASE92].

2.14 Exercises

Exercise 2.14.1. There are several ways to represent the inputs of boolean functions.

1. *Standard representation:* FALSE is 0, TRUE is 1. Here the logical $\wedge$ is equivalent to multiplication.
2. *Dual representation:* FALSE is 1, TRUE is 0. Here the logical $\vee$ is equivalent to multiplication.
3. *Fourier representation:* FALSE is 1, TRUE is -1. Here the logical $\oplus$ is equivalent to multiplication.
4. *Sign representation:* FALSE is -1, TRUE is 1. Here the logical $\equiv$ is equivalent to multiplication.

Give algorithms to convert from one representation to another. Note that if we use the latter representation -1 for FALSE and 1 for TRUE, then $\text{MAJ}_n(x_1, \ldots, x_n) = \text{sgn}(\sum_{i=1}^n x_i)$. Express MAJ_n in terms of the other representations.

Exercise 2.14.2 (M. Sipser). Prove that a function $f : \{0,1\}^\omega \to \{0,1\}$ is Borel if and only if f is computed by an infinite boolean circuit.

Exercise 2.14.3. Prove Fact 2.5.1, which states the following. Let f be a boolean function on n variables. Then $\alpha_1 \wedge \cdots \wedge \alpha_m$ is a minterm of f iff $\overline{\alpha}_1 \vee \cdots \vee \overline{\alpha}_m$ is a maxterm of $1 - f$.

Exercise 2.14.4 ([Raz94], pp. 380–383). Let f be an r-CNF formula over the variables $x_1, \ldots, x_n$. Using the technique of the proof of Theorem 2.6.5 as applied to CNF formulas rather than decision trees, prove directly that for ρ randomly chosen in $\mathcal{R}_n^\ell$ and $p = \ell/n$, $\Pr[\min(f \upharpoonright_\rho) \geq s] < (7pr)^s$.
HINT. Let $Bad_n^\ell(f,s) = \{\rho \in \mathcal{R}_n^\ell : \min(f \upharpoonright_\rho) \geq s\}$. Given $\rho \in Bad_n^\ell(f,s)$, let π be a minterm of $f \upharpoonright_\rho$ of size at least s. Decompose π into disjoint restrictions $\pi_1, \ldots, \pi_k$ as follows. Since f is an r-CNF formula, $f \upharpoonright_\rho$ can be written as a conjunction $C_1 \wedge \cdots \wedge C_m$, where each clause C_i has size at most r. Suppose that $\pi_1, \ldots, \pi_{i-1}$ have been defined and that $\pi \neq \pi_1 \cdots \pi_{i-1}$. Let ν_i be the least index for which $C_{\nu_i} \upharpoonright_{\rho\pi_1\cdots\pi_{i-1}} \not\equiv 1$. Such must exist, since $f \upharpoonright_{\rho\pi} \equiv 1$ (recall that π is a minterm of $f \upharpoonright_\rho$) and $f \upharpoonright_{\rho\pi_1\cdots\pi_{i-1}} \not\equiv 1$ (as

minterm $\pi \neq \pi_1 \cdots \pi_{i-1}$). Let T_i denote the set of variables of C_{ν_i}, and let Y_i be the set of those variables in T_i set by π but not by $\pi_1 \cdots \pi_{i-1}$. Define π_i to be $\pi \upharpoonright_{Y_i}$. Note that $Y_i \neq \emptyset$ since $f \upharpoonright_{\rho\pi} \equiv 1$. Let k be the least integer for which the composition $\pi_1 \cdots \pi_k$ sets s variables. Since $C_{\nu_k}|_{\rho\pi_1\cdots\pi_k} \equiv 1$ and $C_{\nu_k}|_{\rho\pi_1\cdots\pi_{k-1}} \not\equiv 1$, and C_{ν_k} is a disjunction, if necessary we can trim π_k so that $\pi_1 \cdots \pi_k$ sets exactly s variables and $C_{\nu_k} \upharpoonright_{\rho\pi_1\cdots\pi_k}$ is still set to 1. Now proceed in a similar fashion as in the proof of Theorem 2.6.5 to bound the number of bad restrictions.

Exercise 2.14.5. Let p be a polynomial of degree d in n variables such that p is symmetric when restricted to $\{0,1\}^n$. Show that there exist integers $c_0, c_1, \ldots, c_d$ such that

$$p(x_1, \ldots, x_d) = \sum_{k=0}^{d} c_k \binom{x_1 + \cdots + x_n}{k}.$$

HINT. Since $x_i^2 = x_i$ we can rewrite p so that every monomial in p is of the form $\prod_{i \in S} x_i$, where $S \subseteq \{1, 2, \ldots, n\}$. Since p is symmetric, an easy induction on $|S|$ shows that the coefficient of this monomial must be a constant depending only on the size of the set S; if $|S| = k$ then call this constant c_k. Now notice that $c_k = 0$, for $k > d$.

Exercise 2.14.6 ([MP68]). The *symmetrization* of a boolean function $f \in \mathcal{B}_n$ is the function

$$f^{sym}(x_1, \ldots, x_n) = \frac{\sum_{\sigma \in S_n} f(x_{\sigma(1)}, \ldots, x_{\sigma(n)})}{n!}$$

for $x_1, \ldots, x_n \in \{0,1\}$. Show that f^{sym} depends only on the sum $x_1 + \cdots + x_n$ of the boolean variables $x_1, \ldots, x_n$.

Exercise 2.14.7. Prove that any polynomial over $\mathbf{Z}_m$ representing the $\bigwedge$ of n variables must have degree n.

Exercise 2.14.8. If g, h are represented by degree d polynomials over $\mathbf{Z}_p$, where p is prime, then prove

1. $\neg g$ is represented by a degree $(p-1)d$ polynomial over $\mathbf{Z}_p$,
2. $g \wedge h$ is represented by a degree $2d$ polynomial over $\mathbf{Z}_p$,
3. $g \vee h$ is represented by a degree $2(p-1)d$ polynomial over $\mathbf{Z}_p$,
4. for any $m \geq 1$, g is represented by a degree d polynomial over $\mathbf{Z}_{mp}$.

Exercise 2.14.9 ([NS92]). In this exercise we view boolean functions in $\mathcal{B}_n$ as real functions $\{-1,1\}^n \to \{-1,1\}$. Prove the following:

1. Every boolean function can be represented as a real multivariate polynomial in which every variable appears with degree at most 1.

2. ($\star$) A boolean function $f \in \mathcal{B}_n$ *depends* on a variable x_i if there is an input to f such that changing only the i-th bit changes the value of the function. Show that the degree of the multivariate polynomial representing a boolean function $f \in \mathcal{B}_n$ (which depends on all its n variables) is $\geq \log n - O(\log \log n)$.

HINT. For any string $x \in \{0,1\}^n$, let $x^{(i)}$ be the string x with the i-th bit flipped. For any boolean function $f \in \mathcal{B}_n$ let the *influence* of the variable x_i on f, denoted by $INF_i(f)$, be $\Pr[f(x) \neq f(x^{(i)})]$. Results of [KKL88] show that a lower bound on the above degree is $\sum_{i=1}^n INF_i(f)$. Show that for any nonzero multilinear polynomial P of degree d, $\Pr[P(x) \neq 0] \geq 2^{-d}$. Now for each i define a function

$$f^i(x_1, \ldots, x_{i-1}, x_{i+1}, \ldots, x_n) =$$
$$f(x_1, \ldots, x_{i-1}, -1, x_{i+1}, \ldots, x_n) - f(x_1, \ldots, x_{i-1}, 1, x_{i+1}, \ldots, x_n)$$

and use the previous observations to derive the desired lower bound.

Exercise 2.14.10 ([Pat92]).

1. Show that for any non-constant symmetric function, the degree of the multivariate polynomial representing it is at least $n/2$.
2. ($\star\star$) For any symmetric function f let f_k be the value of f on inputs of weight k. The *jump* of the symmetric function f is defined by $\Gamma(f) = \min\{|2k-n+1| : f_k \neq f_{k+1} \text{ and } 0 \leq k \leq n-1\}$. The approximate degree of a boolean function $f \in \mathcal{B}_n$ is the minimal degree of a multivariate polynomial P such that for all $x \in \{0,1\}^n$, $|f(x) - P(x)| < 1/3$. Show that the approximate degree of a symmetric boolean function is precisely $\Theta(\sqrt{n(n-\Gamma(f))})$.

Exercise 2.14.11. Show that $2^m N((1+\epsilon)^m - 1) < 1$, where $\epsilon = 1/(m2^m N + m)$.
HINT. Use the inequality $1 + y \leq e^y$.

Exercise 2.14.12 ([BBR92]). Show that Theorem 2.8.1 holds also if m is neither square-free nor a prime power.
HINT. Using the notation in the proof of Theorem 2.8.1 show by induction on $i < p^z$ that $S_i(j + p^{e+z-1}) \equiv S_i(j) \bmod p^e$. Moreover, the functions S_i, for $i < p^z$, generate a function g satisfying $g(j) \equiv 0 \bmod p^e \Leftrightarrow j \equiv 0 \bmod p^z$.

Exercise 2.14.13 ([BBR92]). Extend Theorems 2.8.1 and 2.8.2 to arbitrary (not necessarily square-free) integers m and $p \not| m$.

Exercise 2.14.14 ([BBR92]). Let m, r be integers such that the set of prime divisors of r is not contained in the set of prime divisors of m. Then the MOD_m-degree of the functions MOD_m^n, $\neg\text{MOD}_m^n$ is in both cases $n^{\Omega(1)}$.

Exercise 2.14.15 ([Tsa93]). Use Theorem 2.8.4 to prove the following lower bounds on the MOD_m-degree:

$$\delta(\text{MAJ}_n, m) \geq n/2, \delta(\bigwedge_n, m) \geq n,$$
$$\delta(\text{MB}_n, m) = \Omega(\sqrt{n}),$$

where the MID-bit function $\text{MB}_n(x_1, \ldots, x_n)$ is defined to be the $\lfloor \frac{\log n}{2} \rfloor$-th bit of the binary representation of $\sum_{i=1}^{n} x_i$.

Exercise 2.14.16 ([Kar93]). Use Theorem 2.9.2 to conclude that $\bar{s}_{\wedge}(f) \geq n$, where f is the boolean function given by the formula $(x_1 \wedge y_1) \vee \cdots \vee (x_n \wedge y_n)$.

Exercise 2.14.17 ([Kar93]). The machinery of filters works for computation with monotone circuits as well. Prove the analogue of part (1) of Theorem 2.9.1 for monotone boolean functions.
HINT. Give appropriate definitions for a filter to be "weakly" above a vector, and define the appropriate parameters $s^+(f), \rho^+(f)$ for monotone boolean functions $f \in \mathcal{B}_n$.

Exercise 2.14.18. Prove Lemma 2.11.1.
HINT. The first part follows easily by replacing $\vee, \wedge, \neg, 0, 1$ by their duals. For the second part replace every literal by its negation.

Exercise 2.14.19. Prove Lemma 2.11.2.
HINT. Let C be a (p, d)-circuit with spectrum w and let $w' \in \{0,1\}^{m+1}$ be a subword of w with $m \geq g(n)$. By setting $n - m$ variables of C to appropriate constants we get a circuit C' on m input variables, depth $\leq d$ and spectrum w'. By monotonicity we have that $g^{-1}(m) \geq g^{-1}(g(n)) \geq n$ and hence $p(g^{-1}(m)) \geq p(n)$, which proves the desired result.

Exercise 2.14.20. Prove Lemma 2.11.3.

Exercise 2.14.21. Prove the assertion $|H| \geq \frac{|X|}{|S|}$ in Lemma 2.11.4.
HINT. Consider the mapping $(h, s) \mapsto h(s)$.

Exercise 2.14.22. Prove Lemma 2.11.6.
HINT. By Lemma 2.11.5, TH^{2n}_{2m} has a $(q, d+2)$-circuit, for some polynomial q. Now substitute 1s for m variables and 0s for $n - m - s$ variables.

Exercise 2.14.23. Prove Lemma 2.11.7.
HINT. By Lemma 2.11.5, TH^{cn}_{cm} has a $(q, d+2)$-circuit, for some polynomial q. Now substitute $(c-1)m$ of the variables with 0s.

Exercise 2.14.24. Prove Lemma 2.6.1 under the simplifying assumption that no variables of distinct terms are shared.

Exercise 2.14.25 ([BIP98]). Show that distance bounded *st*-connectivity for undirected graphs is as hard as that for directed graphs.
HINT. Convert a directed graph into a layered undirected graph.

Exercise 2.14.26 ([ABO84]). ($\star$) Show that for $r \geq 1$, $\{\text{TH}^n_{(\log n)^r}\}$ is computable by a polynomial size constant depth deterministic circuit.
HINT. Put $k = (\log n)^r$. With $\log n$ random variables we can pick a random input x_r. (Indeed, let $i_1 \cdots i_{\log n}$ be the binary representation of $i \leq n$; if $y_1, \ldots, y_{\log n}$ are $\log n$ random variables then the disjunction

$$\bigvee_{i=1}^{n} \left(x_i \wedge y_1^{i_1} \wedge \cdots \wedge y_{\log n}^{i_{\log n}} \right)$$

picks a random element from the input.) Hence using $[n/k]$ blocks, each of length $\log n$, we can pick $[n/k]$ random elements from the input. Taking the $\bigvee$ of these components we obtain a size $O(n\lfloor n/k \rfloor)$, depth 2 probabilistic circuit C^n_k. If s is an input with exactly s 1s let $p_k(s) = \Pr[C^k_n(x) = 1]$. Show that $p_k(s) = 1 - (1 - s/n)^{[s/k]} \approx 1 - e^{[s/k]}$ and conclude that C^n_k $\{\frac{1}{k^2}\}$-computes the function $\{\text{TH}^n_{(\log n)^r}\}$. A different proof of this result follows from the main result of Section 2.11.

Exercise 2.14.27 ([Hås87]). ($\star$) There is a uniform family of NC^0 permutations which are P-complete to invert under LOGSPACE reductions.

3. Circuit Upper Bounds

> The originality of mathematics consists in the fact that in mathematical science connections between things are exhibited which, apart from the agency of human reason, are extremely unobvious. *A. N. Whitehead [Whi25]*

3.1 Introduction

In Chapter 2, we investigated techniques for proving size lower bounds for restricted classes of circuits (monotonic or constant depth). Returning to the circuit synthesis problem of Chapter 1, recall that in Section 1.8.4, we showed an $O(n)$ upper bound for circuit size for *symmetric* boolean functions $f \in \mathcal{B}_n$. In this chapter, using methods from finite permutation group theory, we extend this result to "almost symmetric" boolean functions, and more generally study the notion of *invariance* or *automorphism* group of a boolean function. In [CK91], Clote and Kranakis defined the invariance group $\mathrm{AUT}(f)$ of a function $f \in \mathcal{B}_n$ to be the set of permutations in S_n which leave f invariant under all inputs. Is there a relation between the algebraic structure and/or size of $\mathrm{AUT}(f)$ and the circuit size $C(f)$? For how many boolean functions $f \in \mathcal{B}_n$ is $\mathrm{AUT}(f)$ equal to a given subgroup G of the full symmetric group S_n? These and other questions are treated in the following pages.

The results of this chapter have a very distinct group-theoretic flavor in the methods used. After building intuition by presenting several examples which suggest relations between algebraic properties of groups and computational complexity of languages, we give sufficient conditions via the *Pólya cycle index* (i.e., the number of *orbits* of the group $G \leq S_n$ acting on 2^n) for an arbitrary finite permutation group to be of the form $\mathrm{AUT}(f)$, for some $f \in \mathcal{B}_n$. We show that asymptotically "almost all" boolean functions have trivial invariance groups. For cyclic groups $G \leq S_n$, we give a logspace algorithm for determining whether the given group is of the form $\mathrm{AUT}(f)$, for some $f \in \mathcal{B}_n$. Throughout this chapter we use standard terminology and notation from permutation group theory as found in Wielandt's classic treatise [Wie64].

Invariance groups demonstrate (for the first time) the applicability of group theoretic techniques in the study of upper bounds concerning the circuit size of languages. For any language L, let L_n be the characteristic function of the set of all strings in L of length exactly n, and let $\text{AUT}_n(L)$ be the invariance group of L_n. We consider the *index* $|S_n : \text{AUT}_n(L)| = n!/|\text{AUT}_n(L)|$ as a function of n and study the class of languages whose index is polynomial in n. We use well-known lower bound results on the index of primitive permutation groups together with the O'Nan-Scott theorem, a deep result in the classification of finite simple groups, to show that any language with polynomial index is in (non-uniform) TC^0 and hence in (non-uniform) NC^1. Next, we present the beautiful result of Babai, Beals, and Takácsi-Nagy [BBTN92], which states that if a language $L \subseteq \{0,1\}^*$ has transitive invariance groups $\text{AUT}(L_n)$ and only a polynomial number of orbits, then $L \in \text{TC}^0$ (this establishes a conjecture of [CK91]).

In Section 3.9, we explore several applications of the theory of invariance groups to the problem of computing boolean functions on anonymous, unlabeled networks. This leads to interesting efficient algorithms for computing boolean functions on rings [ASW88], tori [BB89], hypercubes [KK97] and Cayley networks [KK92].

3.2 Definitions and Elementary Properties

Given a function $f : \{0,\ldots,m-1\}^n \to \{0,\ldots,k-1\}$, the *invariance* or *automorphism group* of f, denoted by $\text{AUT}(f)$, is the set of permutations on $\{1,\ldots,n\}$ which "respect" f, i.e., the set of $\sigma \in S_n$ such that for all $x_1,\ldots,x_n \in \{0,\ldots,m-1\}$,

$$f(x_1,\ldots,x_n) = f(x_{\sigma(1)},\ldots,x_{\sigma(n)}). \tag{3.1}$$

Definition 3.2.1. *For any permutation $\sigma \in S_n$, any n-tuple $x = (x_1,\ldots,x_n)$ of elements from the set $\{0,\ldots,m-1\}$, and any function $f : \{0,\ldots,m-1\} \to \{0,\ldots,k-1\}$, define*

$$x^\sigma = (x_{\sigma(1)},\ldots,x_{\sigma(n)})$$

and define $f^\sigma : \{0,\ldots,m-1\} \to \{0,\ldots,k-1\}$ by

$$f^\sigma(x) = f(x^\sigma),\ \textit{for all } x \in \{0,1\}^n.$$

The invariance group of f indicates how symmetric f is, in the sense that the larger the group $\text{AUT}(f)$, the more symmetric the function f is. If for an input $x = (x_1,\ldots,x_n) \in \{0,1\}^n$ and a permutation σ, Equation (3.1) holds, then we also say that σ *fixes* f on input x. In what follows, it will be seen that there is a rich class of permutation groups which are representable as the invariance groups of boolean functions. For any language $L \subseteq \{0,1\}^*$ let L_n be the characteristic function of the set $L \cap \{0,1\}^n$ and let $\text{AUT}_n(L)$ denote the invariance group of L_n.

A language L is said to *realize* a sequence $\mathbf{G} = \langle G_n : n \geq 1\rangle$ of permutation groups $G_n \leq S_n$, if it is true that $\text{AUT}_n(L) = G_n$, for all n. To build intuition, as an example, we consider the following groups.

- **Identity.** I_n is generated by the identity permutation.
- **Reflection.** $R_n = \langle \rho \rangle$, where $\rho(i) = n+1-i$ is the reflection permutation

$$\begin{pmatrix} 1 & 2 & \cdots & n \\ n & n-1 & \cdots & 1 \end{pmatrix}.$$

- **Cyclic.** $C_n = \langle (1, 2, \ldots, n) \rangle$.
- **Dihedral.** $D_n = C_n \times R_n$.
- **Hyperoctahedral.** $O_n = \langle (i, i+1) : i \text{ is even } \leq n \rangle$.

For the groups above we determine regular, as well as non-regular languages which realize them. We summarize the corresponding representability results in the following theorem. The details of the proof are left as Exercise 3.11.1.

Theorem 3.2.1 ([CK91]). *Each of the identity, reflection, cyclic (in the cyclic case only if $n \neq 3, 4, 5$), and hyperoctahedral groups can be realized by regular languages.*

Not every permutation group is representable as the invariance group of a boolean function.

Theorem 3.2.2 ([CK91]). *The alternating group A_n is not the invariance group of any boolean function $f \in \mathcal{B}_n$, provided that $n \geq 3$.*

Proof. Although this follows directly from our representability results given later, it is instructive to give a direct proof. Suppose that the invariance group of $f \in \mathcal{B}_n$ contains the alternating group A_n. Given $x \in 2^n$, for $3 \leq n$ there exist $1 \leq i < j \leq n$, such that $x_i = x_j$. It follows that the alternating group A_n, as well as the transposition (i, j) fix f on the input x. Consequently, every permutation in S_n must also fix f on x. As this holds for every $x \in 2^n$, it follows that $\text{AUT}(f) = S_n$.

Before we proceed with the general representability results, we will prove several simple observations that will be used frequently in the sequel. We begin with a few useful definitions.

Definition 3.2.2.

1. *For any $f \in \mathcal{B}_n$, define $\text{AUT}^-(f)$ to be the set*

$$\{\sigma \in S_n : (\forall x \in 2^n)(f(x) = 0 \Rightarrow f(x^\sigma) = 0)\}.$$

2. *For any $f \in \mathcal{B}_n$, define $\text{AUT}^+(f)$ to be the set*

$$\{\sigma \in S_n : (\forall x \in 2^n)(f(x) = 1 \Rightarrow f(x^\sigma) = 1)\}.$$

3. *For any permutation group $G \leq S_n$ and any $\Delta \subseteq \{1, 2, \ldots, n\}$, let G_Δ be the set of permutations $\sigma \in G$ such that $(\forall i \in \Delta)(\sigma(i) = i)$. The group G_Δ is called the pointwise stabilizer[1] of G on Δ (see [Wie64]).*

[1] We will not in general consider the *setwise* stabilizer of G with respect to Δ, defined as the set of permutations $\sigma \in G$ such that $(\forall i \in \Delta)(\sigma(i) \in \Delta)$.

4. *For any permutation σ and permutation group G, let $G^\sigma = \sigma^{-1}G\sigma$, also called the conjugate of G by σ.*
5. *For any $f \in \mathcal{B}_n$, let $1 \oplus f \in \mathcal{B}_n$ be defined by $(1 \oplus f)(x) = 1 \oplus f(x)$, for $x \in 2^n$.*
6. *If $f_1, \ldots, f_k \in \mathcal{B}_n$ and $f \in \mathcal{B}_k$, then $g = f(f_1, \ldots, f_k) \in \mathcal{B}_n$ is defined by $g(x) = f(f_1(x), \ldots, f_k(x))$.*

Define the natural isomorphism $\phi : S_n \to (S_{n+m})_{n+1,\ldots,n+m}$ by

$$\phi(\sigma)(i) = \begin{cases} \sigma(i) \text{ if } 1 \leq i \leq n \\ i \quad\ \text{ if } n+1 \leq i \leq n+m. \end{cases}$$

For $X \subseteq S_n$, let $\phi(X)$ denote the image of ϕ on $X \subseteq S_n$. Now if $G \leq S_{n+m}$, and $H = G_{n+1,\ldots,n+m}$ is the pointwise stabilizer of G on $\{n+1, \ldots, n+m\}$, then we may at times identify $H \leq S_{n+m}$ with its isomorphic image $\phi^{-1}(H) \leq S_n$, and indeed write statements like $G_{n+1,\ldots,n+m} \leq S_n$. From the context, the meaning should be clear, and so cause no confusion.

Theorem 3.2.3 ([CK91]).

1. *If $f \in \mathcal{B}_n$ is symmetric, then $\text{AUT}(f) = S_n$.*
2. *Let $0 \leq m \leq n$. Given $f \in \mathcal{B}_n$, define flip(f, m) to be that $g \in \mathcal{B}_n$ satisfying*

$$g(x_1, \ldots, x_n) = \begin{cases} f(x_1, \ldots, x_n) & \text{if the weight } |x_1 \cdots x_n|_1 \neq m \\ 1 - f(x_1, \ldots, x_n) & \text{otherwise.} \end{cases}$$

   Then $\text{AUT}(g) = \text{AUT}(f)$. This observation can be iterated, and so clearly $\text{AUT}(f) = \text{AUT}(1 \oplus f)$, for all $f \in \mathcal{B}_n$.
3. *For any permutation σ, $\text{AUT}(f^\sigma) = \text{AUT}(f)^\sigma$.*
4. *For each $f \in \mathcal{B}_n$, $\text{AUT}(f) = \text{AUT}^-(f) = \text{AUT}^+(f)$.*
5. *If $f_1, \ldots, f_k \in \mathcal{B}_n$ and $f \in \mathcal{B}_k$ and $g = f(f_1, \ldots, f_k) \in \mathcal{B}_n$ then $\text{AUT}(f_1) \cap \cdots \cap \text{AUT}(f_k) \subseteq \text{AUT}(g)$.*
6. *$(\forall k \leq n)(\exists f \in \mathcal{B}_n)(\text{AUT}(f) = S_k)$.*

Proof. The proofs of (1) - (3), (5) are easy and are left as an exercise to the reader. We only prove the assertion of (4) for $\text{AUT}^+(f)$, since the proof for $\text{AUT}^-(f)$ is similar. Note that $\text{AUT}^+(f)$ is finite and closed under the group operation of composition, hence is a group. Trivially $\text{AUT}(f) \subseteq \text{AUT}^+(f)$. If $\sigma \in \text{AUT}^+(f)$, and $f(x) = 1$, then by hypothesis $f(x^\sigma) = 1$. If $f(x^\sigma) = 0$, then since $\sigma^{-1} \in \text{AUT}^+(f)$, we have that $f(x) = f((x^\sigma)^{\sigma^{-1}}) = 0$. It follows that $\text{AUT}^+(f) \subseteq \text{AUT}(f)$, as desired. To prove (6) we consider two cases. If $k + 2 \leq n$, then define f by

$$f(x) = \begin{cases} 1 \text{ if } x_{k+1} \leq x_{k+2} \leq \cdots \leq x_n \\ 0 \text{ otherwise.} \end{cases}$$

Let $\sigma \in \text{AUT}(f)$. First notice that $(\forall i > k)(\sigma(i) > k)$. Next, it is easy to show that if σ is a nontrivial permutation, then there can be no $k \leq i < j \leq n$ such that $\sigma(j) < \sigma(i)$. This proves the desired result. If $k = n - 1$, then define the function f as follows.

$$f(x) = \begin{cases} 1 \text{ if } x_1, \dots, x_{n-1} \leq x_n \\ 0 \text{ otherwise.} \end{cases}$$

A similar proof will show that $\text{AUT}(f) = S_{n-1}$. This completes the proof of the theorem.

Representability will play an important role throughout the chapter.

Definition 3.2.3. *For $k \geq 2$, let $\mathcal{B}_{n,k}$ be the set of functions $f : \{0,1\}^n \to \{0, \dots, k-1\}$. A permutation group $G \leq S_n$ is called k-representable if there exists a function $f \in \mathcal{B}_{n,k}$ such that $G = \text{AUT}(f)$. A 2-representable group is also called strongly representable. $G \leq S_n$ is called representable if it is k-representable for some k.*

We will also consider a variant of the previous definition, by considering functions $f : \{0, \dots, m-1\}^n \to \{0, \dots, k-1\}$, in place of functions in $\mathcal{B}_{n,k}$.

Definition 3.2.4. *A permutation group $G \leq S_n$ is called weakly representable if there exists an integer $k \geq 2$, an integer $2 \leq m < n$ and a function $f : m^n \to k$, such that $G = \text{AUT}(f)$.*

In our definition of representable and weakly representable, we required that an n-variable boolean function represent a subgroup $G \leq S_n$, where $m = n$. This is an important definitional point, as illustrated by the next result.

Theorem 3.2.4 (Isomorphism Theorem, [CK91]). *Every permutation group $G \leq S_n$ is isomorphic to the invariance group of a boolean function $f \in \mathcal{B}_{n(\lfloor \log n + 1 \rfloor)}$.*

Proof. First, some notation. Let $w = w_1 \cdots w_n$ be a word in $\{0,1\}^*$. Recall that the weight $|w|_1$ of w is the number of occurrences of 1 in w, and that $|w|$ denotes the length n of w. The word w is *monotone* if for all $1 \leq i < j \leq |w|$, $w_i = 1 \Rightarrow w_j = 1$. The *complement* of w, denoted by $\overline{w}$ is the word which is obtained from w by "flipping" each bit w_i, i.e., $|w| = |\overline{w}|$ and $\overline{w}_i = 1 \oplus w_i$, for all $1 \leq i \leq |w|$. Fix n and let $s = \lfloor \log n + 1 \rfloor$. View each word $w \in \{0,1\}^{ns}$ (of length ns) as consisting of n blocks, each of length s, and let $w(i) = w_{(i-1)s+1} \cdots w_{is}$ denote the i-th such block. For a given permutation group $G \leq S_n$, let L_G be the set of all words $w \in \{0,1\}^{ns}$ such that one of the following holds: either

1. $|w|_1 = s$ and if w is divided into n blocks

$$w(1), w(2), \dots, w(n)$$

each of length s, then exactly one of these blocks consists entirely of 1s, while the other blocks consist entirely of 0s, or

2. $|w|_1 \leq s-1$ and for each $1 \leq i \leq n$, the complement $\overline{w}$ of the i-th block of w is monotone (thus each $w(i)$ consists of a sequence of 1s concatenated with a sequence of 0s), or
3. a) $|w|_1 \geq n$
   b) for each $1 \leq i \leq n$, the first bit of $w(i)$ is 0,
   c) the integers $bin(w, i)$, whose binary representations are given by the words $w(i)$ for $1 \leq i \leq n$, are mutually distinct
   d) $\sigma_w \in G$, where $\sigma_w : \{1, \ldots, n\} \to \{1, \ldots, n\}$ is the permutation defined by $\sigma_w(i) = bin(w, i)$.

The intuition for items (3a) and (3b) above is the following. The words with exactly s many 1s have all these 1s in exactly one block. This guarantees that any permutation respecting the language L_G must map blocks to blocks. By considering words with a single 1 (which by monotonicity must be located at the first position of a block), we guarantee that each permutation which respects L_G must map the first bit of a block to the first bit of some other block. Inductively, by considering the word with exactly $(r-1)$ many 1s, all located at the beginning of a single block, while all other bits of the word are 0s, we guarantee that each permutation which respects L_G must map the $(r-1)$-th bit of each block to the $(r-1)$-st bit of some other block. It follows that any permutation which respects L_G must respect blocks as well as the order of elements in the blocks; i.e., for every permutation $\tau \in \text{AUT}_{ns}(L_G)$,

$$(\forall k \in \{0, \ldots, n-1\})(\exists m \in \{0, \ldots, n-1\})(\forall i \in \{1, \ldots, n\})(\tau(ks+i) = ms+i).$$

Call such a permutation *s-block invariant.* Given a permutation τ in the invariance group $\text{AUT}_{ns}(L_G)$, let $\overline{\tau} \in S_n$ be the induced permutation defined by

$$\overline{\tau}(k) = m \Leftrightarrow (\forall 1 \leq i \leq n) \left[\tau(ks + i) = ms + i\right].$$

CLAIM. $G = \{\overline{\tau} : \tau \in \text{AUT}^+_{ns}(L_G)\}$.

Proof of Claim. ($\subseteq$) Notice that every element $\overline{\tau}$ of $G \leq S_n$ gives rise to a unique s-block invariant permutation $\tau \in S_{ns}$. If $w \in L_G \subseteq \{0,1\}^{ns}$, then considering separately the cases $|w|_1 \leq s$ and $|w|_1 \geq n$, by s-block invariance of τ, $w^\tau \in L_G$.

($\supseteq$) First, notice that if $w \in L_G \subseteq \{0,1\}^{ns}$ and the associated permutation $\sigma_w \in G \leq S_n$, then $\sigma_{(w^\tau)} = \overline{\tau} \circ \sigma_w \in G$. Now, let $w \in L_G$ be such that the associated σ_w is the identity on S_n. Then for any $\tau \in \text{AUT}_{ns}(L_G)$, $w^\tau \in L_G$, so $\sigma_{(w^\tau)} = \overline{\tau} \circ \sigma_w = \overline{\tau} \in G$. This establishes the claim, which completes the proof of the theorem.

We conclude this section by comparing the different definitions of representability given above.

Theorem 3.2.5 ([CK91]). *For any permutation group $G \le S_n$ the following statements are equivalent:*

1. *G is representable.*
2. *G is the intersection of a finite family of strongly representable permutation groups.*
3. *For some m, G is the pointwise stabilizer of a strongly representable group over S_{n+m}, i.e., $G = (\text{AUT}_{n+m}(f))_{\{n+1,\ldots,n+m\}}$, for some $f \in \mathcal{B}_{n+m}$ and $m \le n$.*

Proof. First we prove that $1 \Rightarrow 2$. Indeed, let $f \in \mathcal{B}_{n,k}$ such that $G = \text{AUT}(f)$. For each $b < k$ define as follows a 2-valued function $f_b : 2^n \to \{b, k\}$:

$$f_b(x) = \begin{cases} b \text{ if } f(x) = b \\ k \text{ if } f(x) \neq b \end{cases}$$

It is straightforward to show that $\text{AUT}(f) = \text{AUT}(f_0) \cap \cdots \cap \text{AUT}(f_{k-1})$. But also conversely we can prove that $2 \Rightarrow 1$. Indeed, assume that $f_b \in \mathcal{B}_n$, $b < k$, is a given family of boolean valued functions such that G is the intersection of the strongly representable groups $\text{AUT}(f_b)$. Define $f \in \mathcal{B}_{n,2^k}$ as follows

$$f(x) = \langle f_0(x), \ldots, f_{k-1}(x) \rangle,$$

where for any integers $n_0, \ldots, n_{k-1}$, the symbol $\langle n_0, \ldots, n_{k-1} \rangle$ represents a standard encoding of the k-tuple $(n_0, \ldots, n_{k-1})$ as an integer. It is then clear that $\text{AUT}(f) = \text{AUT}(f_0) \cap \cdots \cap \text{AUT}(f_{k-1})$, as desired.

We now prove that $2 \Rightarrow 3$. Suppose that $G = \text{AUT}(f_0) \cap \cdots \cap \text{AUT}(f_k) \le S_n$, where $f_0, \ldots, f_k \in \mathcal{B}_n$, and let $m = |k|$. Define $f \in \mathcal{B}_{n+m}$ by

$$f(x_1, \ldots, x_n, b_1, \ldots, b_m) = \begin{cases} f_r(x_1, \ldots, x_n) & \text{if } r = \sum_{i=1}^m b_i \cdot 2^{m-i} \le k \\ 0 & \text{otherwise.} \end{cases}$$

Define the isomorphism $\phi : S_n \to (S_{n+m})_{n+1,\ldots,n+m}$ by

$$\phi(\sigma)(i) = \begin{cases} \sigma(i) & \text{if } 1 \le i \le n \\ i & \text{if } n+1 \le i \le n+m \end{cases}$$

and let $\psi : (S_{n+m})_{n+1,\ldots,n+m} \to S_n$ denote the inverse ϕ^{-1} of ϕ.

CLAIM. $\text{AUT}(f_0) \cap \cdots \cap \text{AUT}(f_k) = \psi(\text{AUT}(f)_{n+1,\ldots,n+m})$.

Proof of Claim. ($\subseteq$) Let $\sigma \in \text{AUT}(f_0) \cap \cdots \cap \text{AUT}(f_k) \subseteq S_n$, and let $\widetilde{\sigma} = \phi(\sigma) \in S_{n+m}$. Given $x \in \{0,1\}^n$ and $b \in \{0,1\}^m$, if $r = \sum_{i=1}^m b_i \cdot 2^{m-i} \le k$, then $f_r(x) = f_r(x^\sigma)$ and so $f(x,b) = f((x,b)^{\widetilde{\sigma}}$. As well, if $r = \sum_{i=1}^m b_i \cdot 2^{m-i} > k$, then $f(x,b) = 0 = f((x,b)^{\widetilde{\sigma}}$. It follows that $\widetilde{\sigma} \in \text{AUT}(f)$, so $\sigma \in \psi(\text{AUT}(f)_{n+1,\ldots,n+m})$.

($\subseteq$) Let $\widetilde{\sigma} \in \text{AUT}(f)_{n+1,\ldots,n+m}$. Given $x \in \{0,1\}^n$ and $b \in \{0,1\}^m$, we have $f(x,b) = f((x,b)^{\widetilde{\sigma}}$. If $r = \sum_{i=1}^m b_i \cdot 2^{m-i} \le k$, then

$$f_r(x) = f(x,b) = f((x,b)^{\widetilde{\sigma}}) = f(x^\sigma, b) = f_r(x^\sigma)$$

and so $\sigma \in \text{AUT}(f_r)$. Since this holds for all $r \leq k$, $\sigma \in \text{AUT}(f_0) \cap \cdots \cap \text{AUT}(f_k)$.

Finally, we prove that $3 \Rightarrow 2$. Let $G \leq S_n$ denote $\psi(\text{AUT}(f)_{n+1,\ldots,n+m})$, and let $\sigma \in G$, and $\widetilde{\sigma} = \phi(\sigma) \in \text{AUT}(f)_{n+1,\ldots,n+m}$. Then for any $x \in \{0,1\}^n$ and $b \in \{0,1\}^m$, if $r = \sum_{i=1}^m b_i \cdot 2^{m-i} \leq k$, then $f(x,b) = f_r(x) = f_r(x^\sigma) = f(x^\sigma, b) = f((x,b)^{\widetilde{\sigma}})$, while if $r = \sum_{i=1}^m b_i \cdot 2^{m-i} > k$, then $f(x,b) = 0 = f(x^\sigma, b) = f((x,b)^{\widetilde{\sigma}})$. Thus $\widetilde{\sigma} \in \text{AUT}(f_0) \cap \cdots \cap \text{AUT}(f_k)$, so $\sigma \in \psi(\text{AUT}(f)_{n+1,\ldots,n+m})$. This concludes the proof of the theorem.

3.3 Pólya's Enumeration Theory

In the section, we present the rudiments of Pólya's enumeration theory. Our goal here is to emphasize the relevant elements of the theory without providing any complete proofs. The interested reader is advised to consult [Ber71] and [PR87] or better yet complete details of the proofs on her own.

Let G be a permutation group on n elements. Define an equivalence relation on integers as follows: $i \sim j \bmod G$ if and only if for some $\sigma \in G$, $\sigma(i) = j$. The equivalence classes under this equivalence relation are called *orbits*. Let $G_i = \{\sigma \in G : \sigma(i) = i\}$ be the *stabilizer* of i, and let i^G be the orbit of i. An elementary theorem [Wie64] asserts that $|G : G_i| = |i^G|$. Using this, we can obtain the well-known theorem of Burnside and Frobenius [Com70].

Theorem 3.3.1. *For any permutation group G on n elements, the number of orbits of G is equal to the average number of fixed points of a permutation $\sigma \in G$; i.e.,*

$$\omega_n(G) = \frac{1}{|G|} \sum_{\sigma \in G} |\{i : \sigma(i) = i\}|, \tag{3.2}$$

where $\omega_n(G)$ is the number of orbits of G.

A group G *acts* on a set X, if there is a map $\phi : G \times X \to X$, such that

1. $\phi(\sigma, x) = x$,
2. $\phi(\sigma \circ \tau, x) = \phi(\sigma, \phi(\tau, x))$

where e is the identity element of G, and $\circ$ is the group multiplication. The group G acts *transitively* on X if additionally

$$(\forall x, y \in X)(\exists \sigma \in G)(\phi(\sigma, x) = y).$$

Note that any group $G \leq S_n$ acts on $\{0,1\}^n$ by the group action

$$\phi(\sigma, x) = \phi(\sigma, x_1 \cdots x_n) = (x_{\sigma(1)} \cdots x_{\sigma(n)}) = x^\sigma.$$

Moreover, any permutation $\sigma \in S_n$ can be identified with a permutation on 2^n defined as follows:

$$x = (x_1, \dots, x_n) \to x^\sigma = (x_{\sigma(1)}, \dots, x_{\sigma(n)}).$$

Hence, any permutation group G on n elements can also be thought of as a permutation group on the set 2^n. It follows from (3.2) that

$$|\{x^G : x \in 2^n\}| = \frac{1}{|G|} \sum_{\sigma \in G} |\{x \in 2^n : x^\sigma = x\}|,$$

where $x^G = \{x^\sigma : \sigma \in G\}$ is the orbit of x. We would like to find a more explicit formula for the right-hand side of the above equation. To do this notice that $x^\sigma = x$ if and only if x is invariant on the orbits of σ. It follows that $|\{x \in 2^n : x^\sigma = x\}| = 2^{o(\sigma)}$, where $o(\sigma)$ is the number of orbits of (the group generated by) σ acting on 2^n. Using the fact that $o(\sigma) = c_1(\sigma) + \cdots + c_n(\sigma)$, where $c_i(\sigma)$ is the number of i-cycles in σ (i.e., in the cycle decomposition of σ), we obtain Pólya's formula:

$$|\{x^G : x \in 2^n\}| = \frac{1}{|G|} \sum_{\sigma \in G} 2^{o(\sigma)} = \frac{1}{|G|} \sum_{\sigma \in G} 2^{c_1(\sigma) + \cdots + c_n(\sigma)}. \tag{3.3}$$

The number $|\{x^G : x \in 2^n\}|$ is called the *cycle index* of the permutation group G and will be denoted by $\Theta(G)$. If we want to stress the fact that G is a permutation group on n letters then we write $\Theta_n(G)$, instead of $\Theta(G)$. For more information on Pólya's enumeration theory, the reader should consult [Ber71] and [PR87].

Since the invariance group $\text{AUT}(f)$ of a function $f \in \mathcal{B}_n$ contains G if and only if it is invariant on each of the different orbits x^G, $x \in 2^n$, we obtain that

$$|\{f \in \mathcal{B}_n : \text{AUT}(f) \geq G\}| = 2^{\Theta(G)}.$$

It is also not difficult to compare the size of $\Theta(G)$ and $|S_n : G|$. Indeed, let $H \leq G \leq S_n$. If

$$Hg_1, Hg_2, \dots, Hg_k$$

are the distinct right cosets of G modulo H then for any $x \in 2^n$ we have that

$$x^G = x^{Hg_1} \cup x^{Hg_2} \cup \cdots \cup x^{Hg_k}.$$

It follows that $\Theta_n(H) \leq \Theta_n(G) \cdot |G : H|$. Using the fact that $\Theta_n(S_n) = n+1$ we obtain as a special case that $\Theta_n(G) \leq (n+1)|S_n : G|$. In addition, using a simple argument concerning the size of the orbits of a permutation group we obtain that if $\Delta_1, \dots, \Delta_\omega$ are different orbits of the group $G \leq S_n$ acting on $\{1, 2, \dots, n\}$ then $(|\Delta_1| + 1) \cdots (|\Delta_\omega| + 1) \leq \Theta_n(G)$. We summarize these results in the following theorem.

Theorem 3.3.2. *For any permutation groups $H \leq G \leq S_n$, we have*

1. $\Theta_n(G) \leq \Theta_n(H) \leq \Theta_n(G) \cdot |G : H|$.
2. $\Theta_n(G) \leq (n+1) \cdot |S_n : G|$.
3. $n+1 \leq \Theta_n(G) \leq 2^n$.
4. *If* $\Delta_1, \ldots, \Delta_\omega$ *are different orbits of* G *acting on* $\{1, \ldots, n\}$ *then* $(|\Delta_1|+1)\cdots(|\Delta_\omega|+1) \leq \Theta_n(G)$.

3.4 Representability of Permutation Groups

Next we study the representability problem for permutation groups and give sufficient conditions via Pólya's cycle index for a permutation group to be representable. In addition we consider the effect on representability of several well-known group operations, like product, wreath product, etc.

A simple observation due to Kisielewicz [Kis99] relates representable groups with the automorphism groups of undirected graphs.

Theorem 3.4.1 ([Kis99]). *The automorphism group of an undirected graph is 2-representable.*

Proof. For each two-element set $e = \{i,j\}$, consider the n-tuple $x^e = (x_1^e, \ldots, x_n^e) \in \{0,1\}^n$ such that $x_i^e = x_j^e = 1$ and $x_k^e = 0$, for all $k \neq i,j$. Let the graph $G = (V, E)$ with vertex set $V = \{1, 2, \ldots, n\}$ and edge set E. Define the boolean function

$$f(x) = \begin{cases} 1 \text{ if } x = x^e, \text{ for some } e \in E \\ 0 \text{ otherwise.} \end{cases}$$

It is a simple exercise to show that $\text{AUT}(f)$ is precisely the automorphism group of the given graph.

In order to state the first general representation theorem we define for any $n+1 \leq \theta \leq 2^n$ and any permutation group $G \leq S_n$ the set $\mathbf{G}_\theta^{(n)} = \{M \leq G : \Theta_n(M) = \theta\}$. Also, for any $H \subseteq S_n$, and any $g \in S_n$, the notation $\langle H, g\rangle$ denotes the smallest subgroup of S_n containing the set $H \cup \{g\}$.

Theorem 3.4.2 (Representation Theorem, [CK91]). *For any permutation groups* $H < G \leq S_n$ *if* $H = G \cap K$*, for some representable permutation group* $K \leq S_n$*, then* $(\forall g \in G - H)(\Theta_n(\langle H, g\rangle) < \Theta_n(H))$*. Moreover, this last statement is equivalent to* H *being maximal in* $\mathbf{G}_\theta^{(n)}$*, where* $\Theta_n(H) = \theta$.

Proof. By Theorem 3.2.5, K is the intersection of a family strongly representable groups. Hence let $f_1, \ldots, f_k \in \mathcal{B}_n$ be such that $K = \cap_{i=1}^k \text{AUT}(f_i)$. Then

$$H = \bigcap_{i=1}^{k} \text{AUT}(f_i) \cap G.$$

Assume, to the contrary, that there exists a subgroup $K \leq G$ such that $H < K$ and $\Theta(K) = \Theta(H)$. This implies

$$\forall x \in 2^n (x^K = x^H).$$

We claim, however, that

$$K \subseteq \bigcap_{i=1}^{k} \text{AUT}(f_i) \cap G = H$$

which then contradicts the assumption that $H < K$. Indeed, let $\sigma \in K$ and $x \in \{0,1\}^n$. Then

$$x^K = (x^\sigma)^K = (x^\sigma)^H$$

so it follows that $x = (x^\sigma)^\tau$, for some $\tau \in H$. Consequently, $f_i(x) = f_i((x^\sigma)^\tau) = f_i(x^\sigma)$, for all $1 \leq i \leq k$, and so $K \subseteq \bigcap_{i=1}^{k} \text{AUT}(f_i) \cap G$, which establishes our claim.

It remains to prove the equivalence of the last statement in the theorem. Assume that H is a maximal element of $\mathbf{G}_\theta^{(n)}$, but that for some $g \in G - H$, we have that $\Theta_n(\langle H, g\rangle) = \Theta_n(H)$. But then $H < \langle H, g\rangle \leq G$, contradicting the maximality of H. To prove the other direction we argue as follows. Assume, to the contrary, that the hypothesis is true but that H is not maximal in $\mathbf{G}_\theta^{(n)}$. This means there exists $H < K \leq G$ such that $\Theta_n(K) = \Theta_n(H)$. Take any $g \in K - H$ and notice that

$$\Theta_n(\langle H, g\rangle) \geq \Theta_n(K) = \theta = \Theta_n(H) \geq \Theta_n(\langle H, g\rangle).$$

Hence, $\Theta_n(H) = \Theta_n(\langle H, g\rangle)$, contradicting our assumption.

Let $O(G)$ denote the set $\{x^G : x \in \{0,1\}^n\}$ of orbits of G acting on $\{0,1\}^n$. If the group G is of the form $\text{AUT}(f)$ for some function $f \in \mathcal{B}_{n,k}$, then (1) all n-tuples in every orbit of $O(G)$ must have the same value under f, and (2) for every permutation $\tau \notin G$, there must be an element x of $\{0,1\}^n$, such that x and x^τ belong to different orbits of $O(G)$, and these orbits have different values under f. Hence, in order to find a k-valued boolean function f whose invariance group is G, it is necessary and sufficient to find a function $F : O(G) \to \{0, \ldots, k-1\}$, such that for every permutation $\tau \notin G$, there exists $x \in \{0,1\}^n$, such that x and x^τ belong to different orbits of $O(G)$ and these orbits have different values under F.

Using deliberations issuing from the previous observation, we prove the following result concerning the representation of *maximal* permutation groups.

Theorem 3.4.3 (Maximality Theorem, [CK91, Kis99]).

1. *A permutation group $G \leq S_n$ is representable if and only if it is a maximal subgroup of S_n among those having the same number $\Theta(G)$ of orbits in $\{0,1\}^n$. In such a case G is representable by a function $f \in \mathcal{B}_{n,k}$ with $k \leq \binom{n}{\lfloor (n)/2 \rfloor}$.*

2. *All maximal subgroups of S_n are strongly representable, the only exceptions being: (a) the alternating group A_n, for all $n \geq 3$, and the conjugates of the following three types of groups: (b) the 1-dimensional, linear, affine group $AGL_1(5)$ over the field of 5 elements, for $n = 5$; (c) the group of linear transformations $PGL_2(5)$ of the projective line over the field of 5 elements, for $n = 6$; (d) the group of semi-linear transformations $P\Gamma L_2(8)$ of the projective line over the field of 8 elements, for $n = 9$.*

Proof. To prove 1, we argue as follows. If G is representable, then it must be maximal among the subgroups of S_n with the same set $O(G)$ of orbits of $\{0,1\}^n$. Indeed, if there were a permutation $\tau \notin G$ such that for every $x \in \{0,1\}^n$, x and x^τ always belong to the same orbit of $O(G)$, then the group $G' = \langle G, \tau \rangle$ has precisely the same orbits in $\{0,1\}^n$ as G itself. Therefore every function f on $\{0,1\}^n$ invariant under G is invariant under G'. Hence, G cannot be representable at all.

Conversely, suppose that $G \leq S_n$ is maximal among those having the same number $k = \Theta_n(G)$ of orbits in $\{0,1\}^n$. Define $f \in \mathcal{B}_{n,k}$ by $f(x) = i$ iff x is in the i-th orbit in some canonical listing of orbits.[2] Clearly $G = \text{AUT}(f)$. To achieve the upper bound $k \leq \binom{n}{\lfloor (n)/2 \rfloor}$, note that the orbits of $O(G)$ can be partitioned into $n+1$ natural levels, according to the weight (number of 1s) $|x|_1$ of an element $x \in \{0,1\}^n$ belonging to an orbit. It thus suffices to assign different values to those orbits of elements having median weight, and clearly there are at most $\binom{n}{\lfloor (n)/2 \rfloor}$ of these.

To prove 2, let M be a maximal subgroup of S_n. We distinguish two cases.

Case 1. $\Theta_n(M) > n+1$.
In this case, there is a level $O_i(M)$ consisting of more than one orbit. If $T \in O_i(M)$ then the boolean function f assigning 1 to all n-tuples in T, and 0 otherwise, strongly represents M.
Case 2. $\Theta_n(M) = n+1$.
In this case, M is not representable at all. Moreover, for any two subsets S, S' of $\{1, 2, \ldots, n\}$ of the same cardinality there is a permutation $\pi \in M$ such that $\pi(S) = S'$. We know from the main theorem of [BP55] that M is of one of the forms in the statement of the theorem.

Our previous study focused on representability results for maximal permutation groups. The following refinement appears to be very natural.

Definition 3.4.1. *Let $\mathcal{R}_k^n$ denote the class of k-representable permutation groups on n letters.*

[2] For $x, y \in \{0,1\}^n$, a possible canonical ordering is given by $x^G < y^G$ iff the lexicographic least element in the orbit of x is less the lexicographic least element in the orbit of y).

Clearly $\mathcal{R}_k^n \subseteq \mathcal{R}_{k+1}^n$. It is interesting to note that it is not known whether or not $\mathcal{R}_k^n$ forms a proper hierarchy. However, the following can be proved.

Theorem 3.4.4 ([Kis99]). $\mathcal{R}_2^n \neq \mathcal{R}_3^n$, *i.e., there exist 3-representable groups which are not 2-representable.*

Proof. The desired group D consists of the identity permutation, as well as the permutations

$$(1,2)(3,4), (1,3)(3,4), \text{ and } (1,4)(2,3).$$

It is easily checked that $\Theta(D) = 7$. Indeed, the orbits are the following:
weight 0: $\{0000\}$,
weight 1: $\{1000, 0100, 0010, 0001\}$,
weight 2: $\{1100, 0011\}, \{1010, 0101\}, \{1001, 0110\}$,
weight 3: $\{0111, 1011, 1101, 1110\}$,
weight 4: $\{1111\}$.

To show that D is 3-representable, we define a function $f : \{0,1\}^4 \to \{0,1,2\}$, which assigns different values to the weight 2 orbits. Inspection of these orbits shows that $\text{AUT}(f)$ cannot contain a transposition and it follows easily that $D = \text{AUT}(f)$.

However, D is not 2-representable. Assume, on the contrary, that there is a boolean function $g \in \mathcal{B}_2$, which represents (i.e., 2-represents) D. Two of the weight 2 orbits must be assigned the same value, say the first and the second one. It follows easily that the transposition $(2,3) \in \text{AUT}(g) = D$. However, this is a contradiction.

As noted above, all maximal permutation groups with the exception of A_n are of the form $\text{AUT}(f)$, provided that $n \geq 10$. Such maximal permutation groups include: the cartesian products $S_k \times S_{n-k}$ $(k \leq n/2)$, the *wreath products* $S_k \wr S_l$ $(n = kl, k, l > 1)$,[3] the affine groups $AGL_d(p)$, for $n = p^d$, etc. The interested reader will find a complete survey of classification results for maximal permutation groups in [KL88]. As well, it should be pointed out that there are many (nonmaximal) permutation groups which are not representable – for example wreath products $G \wr A_n$. For additional representation results, we refer the reader to Exercise 3.11.11.

Theorem 3.4.5. ([Kis99]) *If $G \leq S_n, H \leq S_m$ are k-representable for some $k \geq 2$ then $G \times H \leq S_{n+m}$ is r-representable for every r satisfying $r(r-1) \geq k$. In particular, $G \times H$ is k-representable.*

Proof. We follow the proof of Kisielewicz [Kis99]. Without loss of generality, assume $m \leq n$. Let $g, h \in \mathcal{B}_{n,k}$ be such that $G = \text{AUT}(g), H = \text{AUT}(h)$.

[3] The *wreath* product $G \wr H$ of $G \leq S_n$ with $H \leq S_m$ is a subgroup of $S_{n \cdot m}$, defined as $\{(\sigma_1, \ldots, \sigma_n; \tau) : \sigma_1, \ldots, \sigma_n \in G, \tau \in H\}$. Here, for $\sigma_1, \ldots, \sigma_m \in G \leq S_n, \tau \in H \leq S_m$, define $(\sigma_1, \ldots, \sigma_n; \tau)$ to be that permutation $\rho \in S_{n \cdot m}$ such that for $1 \leq i \leq n, 1 \leq j \leq m$, $\rho(i,j) = (\sigma_j(i), \tau(j))$.

Since $r(r-1) \geq k$, we may assume that g, h take values from the cartesian product $P = \{0, \ldots, r-1\} \times \{0, \ldots, r-2\}$. Let π_1, π_2 be the first and second projection operations on the set P. We define an r-valued boolean function $f : \{0,1\}^{m+n} \to \{0, \ldots, r-1\}$ as follows:

$$f(z) = \begin{cases} \pi_1(g(x)) & \text{if } z = x0^m, \text{ for some } x \in 2^n, x \neq 0^n, 1^n \\ \pi_2(g(x)) & \text{if } z = x1^m, \text{ for some } x \in 2^n, x \neq 0^n, 1^n \\ \pi_1(h(y)) & \text{if } z = 0^n y, \text{ for some } y \in 2^m, y \neq 0^m, 1^m \\ \pi_2(h(y)) & \text{if } z = 1^n y, \text{ for some } y \in 2^m, y \neq 0^m, 1^m \\ r-1 & \text{if } z = 1^n 0^m \\ 0 & \text{otherwise.} \end{cases}$$

CLAIM. $G \times H = \text{AUT}(f)$.
Proof of Claim. ($\subseteq$) Let $\sigma \in G, \tau \in H$ and let $z \in 2^{m+n}$ such that $z = xy$, with $x \in \{0,1\}^n, y \in \{0,1\}^m$. Then by the above definition,

$$f(x^\sigma y) = f(xy) = f(xy^\tau)$$

since $g(x^\sigma) = g(x)$ and $h(y^\tau) = h(y)$.
($\supseteq$) It is easily checked that by definition of f, for all $z \in \{0,1\}^{n+m}$, $|z|_1 = n$, and $z \neq 1^n 0^m$, we have $f(1^n 0^m) = r - 1 > f(z)$. Thus it easily follows that $G \times H \subseteq S_n \times S_m$. Now let $\rho = (\sigma, \tau) \in (S_n \times S_m - G \times H)$, and for specificity, assume that $\sigma \notin G$ (a similar argument works when $\tau \notin H$). There is an $x \in \{0,1\}^n$ such that $g(x) \neq g(x^\sigma)$ and $x \notin \{0^n, 1^n\}$. It follows that $\pi_i(g(x^\sigma)) \neq \pi_i(g(x))$ for $i = 1$ or $i = 2$. Consequently, $f(z) \neq f(z^\sigma)$ for either $z = x0^m$ or $z = x1^m$. This proves the desired assertion.

3.5 Algorithm for Representing Cyclic Groups

In this section we prove the following represention theorem for cyclic groups.

Theorem 3.5.1 ([CK91]). *There is a logspace algorithm, which, when given as input a cyclic group $G \leq S_n$, decides whether the group is 2-representable, in which case it outputs a function $f \in \mathcal{B}_n$ such that $G = \text{AUT}(f)$.*

Proof. We establish the correctness of the following algorithm:

Input
$G = \langle \sigma \rangle$ cyclic group.
Step 1
Decompose $\sigma = \sigma_1 \sigma_2 \cdots \sigma_k$, where $\sigma_1, \sigma_2, \ldots, \sigma_k$ are disjoint cycles of lengths $l_1, l_2, \ldots, l_k \geq 2$, respectively.
Step 2
if for all $1 \leq i \leq k$,
$l_i = 3 \Rightarrow (\exists j \neq i)(3 | l_j)$ and

$l_i = 4 \Rightarrow (\exists j \neq i)(\gcd(4, l_j) \neq 1)$ and
$l_i = 5 \Rightarrow (\exists j \neq i)(5|l_j)$
then output G is 2-representable.
else output G is not 2-representable.
end

Before proceeding with the main proof we introduce some definitions.

Definition 3.5.1.

1. *A boolean function $f \in \mathcal{B}_n$ is called special if for all words w of length n, $|w|_1 = 1 \Rightarrow f(w) = 1$.*
2. *The support of a permutation σ, denoted by Supp(σ), is the set of i such that $\sigma(i) \neq i$. The support of a permutation group G, denoted Supp(G), is the union of the supports of the elements of G.*
3. *Let $\sigma_1, \ldots, \sigma_k$ be a collection of cycles. We say that the group $G = \langle \sigma_1, \ldots, \sigma_k \rangle$ generated by the permutations $\sigma_1, \ldots, \sigma_k$ is specially representable if there exists a special boolean function $f : \{0,1\}^\Omega \to \{0,1\}$ (where Ω is the union of the supports of the permutations $\sigma_1, \ldots, \sigma_k$), such that $G = \mathrm{AUT}(f)$. Note that by definition every specially representable group is strongly representable.*

We now turn our attention to the proof of correctness of the above algorithm. The proof is in a series of lemmas.

Lemma 3.5.1. *Suppose that $\sigma_1, \ldots, \sigma_{n+1}$ is a collection of cycles such that both $\langle \sigma_1, \ldots, \sigma_n \rangle$ and $\langle \sigma_{n+1} \rangle$ are specially representable and have disjoint supports. Then $\langle \sigma_1, \ldots, \sigma_{n+1} \rangle$ is specially representable.*

Proof. Put $\Omega_0 = \cup_{i=1}^n Supp(\sigma_i), \Omega_1 = Supp(\sigma_{n+1})$ and let $|\Omega_0| = m$, $|\Omega_1| = k$. Suppose that $f_0 : 2^{\Omega_0} \to 2$ and $f_1 : 2^{\Omega_1} \to 2$ are special boolean functions representing the groups $\langle \sigma_1, \ldots, \sigma_n \rangle$ and $\langle \sigma_{n+1} \rangle$, respectively. By Theorem 3.2.3, without loss of generality we may assume that $1 = f_0(0^m) \neq f_1(0^k) = 0$, and for $u \in \{0,1\}^m$, $v \in \{0,1\}^k$, $|u|_1 = 1 = |v|_1$ we have $f_0(u) = 1 = f_1(v)$. Let $\Omega = \Omega_0 \cup \Omega_1$ and define $f : \{0,1\}^\Omega \to \{0,1\}$ by $f(w) = f_0(w \upharpoonright \Omega_0) \cdot f_1(w \upharpoonright \Omega_1)$.

CLAIM. $\langle \sigma_1, \ldots, \sigma_{n+1} \rangle = \mathrm{AUT}_\Omega(f)$.

Proof of Claim. The containment from left to right is clear, so it remains to prove that $\mathrm{AUT}_\Omega(f) \subseteq \langle \sigma_1, \ldots, \sigma_{n+1} \rangle$. Assume, on the contrary, that there is a permutation $\tau \in \mathrm{AUT}_\Omega(f) - \langle \sigma_1, \ldots, \sigma_{n+1} \rangle$. We distinguish two cases.

Case 1. $(\exists i \in \Omega_0)(\exists j \in \Omega_1)(\tau(i) = j)$.

Let $w \in \{0,1\}^\Omega$ be defined by $w \upharpoonright \Omega_1 = 0^k$, and

$$(w \upharpoonright \Omega_0)(\ell) = \begin{cases} 0 \text{ if } \ell \neq i \\ 1 \text{ if } \ell = i \end{cases}$$

for $\ell \in \Omega_0$. Since f is a special boolean function, by using the fact that $1 = f_0(0^m) \neq f_1(0^k) = 0$, we obtain that $f(w) = 0 \neq f(w^\tau) = 1$, which is a contradiction.

Case 2. $(\forall i \in \Omega_0)(\tau(i) \in \Omega_0)$.

Put $\tau_0 = (\tau \restriction \Omega_0) \in \text{AUT}_{\Omega_0}$ and $\tau_1 = (\tau \restriction \Omega_1) \in \text{AUT}_{\Omega_1}$. By hypothesis, for all $w \in 2^\Omega$, we have that

$$f(w) = f_0(w \restriction \Omega_0) \cdot f_1(w \restriction \Omega_1) = f(w^\tau) = f_0((w \restriction \Omega_0)^{\tau_0}) \cdot f_1((w \restriction \Omega_1)^{\tau_1}),$$

which implies $\tau_0 \in \text{AUT}_{\Omega_0}(f_0)$ and $\tau_1 \in \text{AUT}_{\Omega_1}(f_1)$.

This completes the proof of the lemma.

An immediate consequence of the previous lemma is the following.

Lemma 3.5.2. *If G, H have disjoint support and are specially representable, then $G \times H$ is specially representable.*

In view of Theorem 3.2.1, we know that the cyclic group $\langle(1, 2, \ldots, n)\rangle$ is 2-representable exactly when $n \neq 3, 4, 5$. In particular, the groups $\langle(1, 2, 3)\rangle$, $\langle(1, 2, 3, 4)\rangle$, $\langle(1, 2, 3, 4, 5)\rangle$ are not representable. The following lemma may be somewhat surprising, since it implies that the group $\langle(1, 2, 3)(4, 5, 6)\rangle$, though isomorphic to $\langle(1, 2, 3)\rangle$, *is* strongly representable.

Lemma 3.5.3. *Let the cyclic group G be generated by a permutation σ, which is the product of two disjoint cycles of lengths ℓ_1, ℓ_2, respectively. Then G is specially representable exactly when the following conditions are satisfied:*
$(\ell_1 = 3 \Rightarrow 3|\ell_2)$ *and* $(\ell_2 = 3 \Rightarrow 3|\ell_1)$,$(\ell_1 = 4 \Rightarrow \gcd(4, \ell_2) \neq 1)$ *and* $(\ell_2 = 4 \Rightarrow \gcd(4, \ell_1) \neq 1)$, $(\ell_1 = 5 \Rightarrow 5|\ell_2)$ *and* $(\ell_2 = 5 \Rightarrow 5|\ell_1)$.

Proof. It is clear that the assertion of the lemma will follow if we can prove that the three assertions below are true.

1. The groups $\langle(1, 2, \ldots, n)(n+1, n+2, \ldots, kn)\rangle$ are specially representable when $n = 3, 4, 5$.
2. The groups $\langle(1, 2, 3, 4)(5, \ldots, m + 4)\rangle$ are specially representable when $\gcd(4, m) \neq 1$.
3. Let m, n be given integers, such that either $m = n = 2$, or $m = 2$ and $n \geq 6$, or $n = 2$ and $m \geq 6$, or $m, n \geq 6$. Then $\langle(1, 2, \ldots, m)(m + 1, m + 2, \ldots, m + n)\rangle$ is specially representable.

Proof of (1). We give the proof only for the case $n = 5$ and $k = 2$. The other cases $n = 3$, $n = 4$ and $k \geq 3$ are treated similarly. Let $\sigma = \sigma_0\sigma_1$, where $\sigma_0 = (1, 2, 3, 4, 5)$ and $\sigma_1 = (6, 7, 8, 9, 10)$. From the proof of Theorem 3.2.3, we know that

$$D_5 = \text{AUT}_5(L') = \text{AUT}_5(L''),$$

where $L' = 0^*1^*0^* \cup 1^*0^*1^*$ and $L'' = \{w \in L' : |w|_0 \geq 1\}$. Let L consist of all words w of length 10 such that

- either $|w|_1 = 1$,
- or $|w|_1 = 2$ and $(\exists 1 \le i \le 5)(w_i = w_{5+i}$ and $(\forall j \ne i, 5+i)(w_j = 0))$,
- or $|w|_1 = 3$ and $(\exists 0 \le i \le 4)(w = (1000011000)^{\sigma^i}$ or $w = (1100010000)^{\sigma^i})$,
- or $|w|_1 = 3$ and $w_1 \cdots w_5 \in L'$ and $w_6 \ldots w_{10} \in L''$.

CLAIM. $\langle(1,2,3,4,5)(6,7,8,9,10)\rangle = \text{AUT}_{10}(L)$.

Proof of Claim. The containment from left to right is clear. For the containment from right to left, i.e., $\text{AUT}_{10}(L) \subseteq \langle(1,2,3,4,5)(6,7,8,9,10)\rangle$, suppose that $\tau \in \text{AUT}_{10}(L)$, but that, on the contrary, there exists an $1 \le i \le 5$ and a $6 \le j \le 10$ such that $\tau(i) = j$. Let the word w be defined such that $w_\ell = 0$, if $\ell = j$, and $= 1$ otherwise. From the fact that $0^5 \notin L''$, and the last clause in the definition of L, it follows that $w \notin L$ and $w^\tau \in L$, contradicting the assumption $\tau \in \text{AUT}_{10}(L)$. Thus τ is the product of two disjoint permutations τ_0 and τ_1 acting on $1, 2, \ldots, 5$ and $6, 7, \ldots, 10$, respectively. Hence from the last clause in the definition of L we have that $\tau_0 \in D_5$ and $\tau_1 \in \pi^{-1} D_5 \pi$, where $\pi(i) = 5 + i$, for $i = 1, \ldots, 5$. Let $\rho_0 = (1,5)(2,4)$ and $\rho_1 = (6,10)(7,9)$ be the reflection permutations on $1, 2, \ldots, 5$ and $6, 7, \ldots, 10$, respectively. To complete the proof of (1), it is enough to show that none of the permutations $\rho_0, \rho_1, \rho_0\rho_1, \rho_0\sigma_1^i, \sigma_0^i\rho_1, \sigma_0^i\sigma_1^j$, for $i \ne j$, belongs to $\text{AUT}_{10}(L)$. To see this let $x = 1000011000 \in L$. Then if $\tau = \rho_0, \rho_1, \rho_0\rho_1, \rho_0\sigma_1^i$, for any $i = 1, 2, 3, 5$ or $\tau = \sigma_0^i\rho_1$ for $i = 1, 2, 4, 5$, then it is easily seen that $x^\tau \notin L$. Now, let $x = 110001000$. Then for $\tau = \rho_0\sigma_1^4$ and $\tau = \sigma_0^3\rho_1$ it is easy to check that $x^\tau \notin L$. Finally, for $x = 1000010000 \in L$ and $\sigma_0^i\sigma_1^j$, where $i \ne j$, we have that $x^\tau \notin L$. This completes the proof of part (1) of the lemma.

Proof of (2). Put $\sigma_0 = (1,2,3,4)$, $\sigma_1 = (5, 6, \ldots, m+4)$, $\sigma = \sigma_0\sigma_1$. Let L be the set of words of length $m + 4$ such that

- either $|w|_1 = 1$,
- or $|w|_1 = 2$ and $(\exists 0 \le i \le \text{lcm}(4, m) - 1)(w = (100010^{m-1})^{\sigma^i})$,
- or $|w|_1 = 3$ and $(\exists 0 \le i \le \text{lcm}(4, m) - 1)(w = (110010^{m-1})^{\sigma^i})$,
- or $|w|_1 > 3$ and $w_1 \cdots w_4 \in L'$ and $w_5 \cdots w_{m+5} \in L''$,

where $L' = 0^*1^*0^* \cup 1^*0^*1^*$ and L'', as given by Theorem 3.2.1, satisfies $\text{AUT}_m(L'') = C_m$, and moreover, for all $i \ge 1$, $0^i \notin L''$. Clearly, $\langle(1,2,3,4)(5,6,\ldots,m+4)\rangle \subseteq \text{AUT}_{m+4}(L)$. It remains to prove that

$$\text{AUT}_{m+4}(L) \subseteq \langle(1,2,3,4)(5,6,\ldots,m+4)\rangle.$$

Let $\tau \in \langle(1,2,3,4)(5,6,\ldots,m+4)\rangle$. As before, τ can be decomposed into $\tau = \tau_0\tau_1$, where $\tau_0 \in D_4$, $\tau_1 \in \pi^{-1} D_m \pi$, and $\pi(i) = 4 + i$ for $i = 1, 2, \ldots, m$. Let $\rho = (1,4)(2,3)$ be the reflection on $1, 2, 3, 4$. It suffices to show that none of the permutations $\rho\sigma_1^i, \sigma_0^i\sigma_1^j$, for $i \not\equiv$ mod4 are in $\text{AUT}_{m+4}(L)$. Indeed, if $\tau = \sigma_0^i\sigma_1^j$, then let $x = 100010^{m-1}$. It is clear that $x \in L$, but $x^\tau \notin L$. Next assume that $\tau = \rho\sigma_1^i$. We distinguish the following two cases.

Case 1. $m = 4k$, i.e., a multiple of 4.

Let $x = 100010^{m-1}$. Then $x \in L$, but $x^\tau \notin L$ unless $x^\tau = x^{\sigma^j}$ for some j. In this case $j \equiv 3 \bmod 4$ and $j \equiv i \bmod 4k$. So it follows that $i = 3, 7, 11, \ldots, 4k-1$. Now let $y = 110010^{m-1}$. Then $y \in L$, but $y^\tau \notin L$ for the above values of i, unless $y^\tau = y^{\sigma^\ell}$ for some ℓ. In that case we have that $\ell \equiv 2 \bmod 4$ and $\ell \equiv i \bmod 4k$. So it follows that $i = 2, 6, 10, \ldots, 4k-2$. Consequently, $\tau \notin \text{AUT}_{m+4}(L)$.

Case 2. $\gcd(4, m) = 2$.
Let $x = 100010^{m-1}$. Then $x \in L$, but $x^\tau \notin L$ unless $x^\tau = x^{\sigma^j}$ for some j. In this case $j \equiv 3 \bmod 4$ and $j \equiv i \bmod 4k$. So it follows that for even values of i, $\tau \notin \text{AUT}_{m+4}(L)$. Let $y = 110010^{m-1}$. Then $y \in L$, but $y^\tau \notin L$ unless $y^\tau = y^{\sigma^\ell}$ for some ℓ. In that case we have that $\ell \equiv 2 \bmod 4$ and $\ell \equiv i \bmod m$. So it follows that for odd values of i, $\tau \notin \text{AUT}_{m+4}(L)$. This completes the proof of (2).
Proof of (3). A similar technique can be used to generalize the representability result to more general types of cycles.

A straightforward generalization of Lemma 3.5.3 is given without proof in the next lemma.

Lemma 3.5.4. *Let G be a permutation group generated by a permutation σ which can be decomposed into k-many disjoint cycles of lengths $\ell_1, \ell_2, \ldots, \ell_k$, respectively. The group G is specially representable exactly when the following conditions are satisfied for all $1 \leq i \leq k$,*
$\ell_i = 3 \Rightarrow (\exists j \neq i)(3|\ell_j)$ **and**
$\ell_i = 4 \Rightarrow (\exists j \neq i)(\gcd(4, \ell_j) \neq 1)$ **and**
$\ell_i = 5 \Rightarrow (\exists j \neq i)(5|\ell_j)$.

The correctness of the algorithm is an immediate consequence of the previous lemmas. This completes the proof of Theorem 3.5.1.
A slightly modified proof of Theorem 3.5.1 can also be found in [Kis99].

3.6 Asymptotics for Invariance Groups

Shannon's theorem from Section 2.2 states that almost all boolean functions require exponential size boolean circuits, and so are as difficult to compute as the hardest boolean function. Since any symmetric language $L \subseteq \{0,1\}^*$ (i.e., for which $\text{AUT}(L_n) = S_n$) can be computed with logdepth fan-in-2 boolean circuits, one might conjecture an inverse relationship between the size (or possibly algebraic structure) of the invariance group $\text{AUT}(f)$ of an n-ary boolean function f, and its boolean complexity. Indeed, we show below that almost all boolean functions have trivial invariance group (i.e., $\text{AUT}(f) = \{id_n\}$, where id_n is the identity permutation in S_n). This yields a type of 0-1 *law* , where for any sequence $\langle G_n \leq S_n n \geq 1\rangle$ of permutation groups, we prove that the limit $\lim_{n\to\infty} |\{f \in \mathcal{B}_n : \text{AUT}(f) = G_n\}| 2^{-2^n}$ is either 0 or 1.

Theorem 3.6.1. *For any family $\langle G_n : n \geq 1\rangle$ of permutations groups such that each $G_n \leq S_n$*

$$\lim_{n\to\infty} \frac{|\{f \in \mathcal{B}_n : \text{AUT}(f) = \{id_n\}\}|}{2^{2^n}} = \lim_{n\to\infty} \frac{|\{f \in \mathcal{B}_n : \text{AUT}(f) \leq G_n\}|}{2^{2^n}} = 1.$$

Moreover, if $\liminf |G_n| > 1$ *then*

$$\lim_{n\to\infty} \frac{|\{f \in \mathcal{B}_n : \text{AUT}(f) \geq G_n\}|}{2^{2^n}} = \lim_{n\to\infty} \frac{|\{f \in \mathcal{B}_n : \text{AUT}(f) = G_n\}|}{2^{2^n}} = 0.$$

Proof. During the course of this proof we use the abbreviation $\Theta(m) := \Theta_m(\langle(1,2,\ldots,m)\rangle)$. First, we prove the second part of the theorem. By assumption, there exists an n_0, such that for all $n \geq n_0$, $|G_n| > 1$. Hence, for each $n \geq n_0$, G_n contains a permutation of order $k(n) \geq 2$, say σ_n. Without loss of generality we can assume that each $k(n)$ is a prime number. Since $k(n)$ is prime, σ_n is a product of $k(n)$-cycles. If $(i_1, \ldots, i_{k(n)})$ is the first $k(n)$-cycle in this product, then it is easy to see that

$$\Theta_n(\langle\sigma_n\rangle) \leq \Theta_n(\langle(i_1, \ldots, i_{k(n)})\rangle).$$

It follows that

$$\begin{aligned} |\{f \in \mathcal{B}_n : \text{AUT}(f) \geq G_n\}| &\leq |\{f \in \mathcal{B}_n : \sigma_n \in \text{AUT}(f)\}| \\ &= 2^{\Theta_n(\sigma_n)} \\ &\leq 2^{\Theta(k(n))\cdot 2^{n-k(n)}}. \end{aligned}$$

Pólya's cycle index formulas have been worked out for particular permutation groups, including the cyclic groups. In particular from [Ber71], we have the formula

$$\Theta(m) = \frac{1}{m} \cdot \sum_{k|m} \phi(k) \cdot 2^{m/k}$$

which gives the Pólya cycle index of the group $\langle(1,2,\ldots,m)\rangle$ acting on the set $\{1,2,\ldots,m\}$, where $\phi(k)$ is Euler's totient function.

However, it is easy to see that for k prime

$$\frac{\Theta(k)}{2^k} = \frac{1}{k} + \frac{2}{2^k} - \frac{2}{k2^k}.$$

In fact, the function on the right-hand side of the above equation is decreasing in k. Hence, for k prime,

$$\frac{\Theta(k)}{2^k} \leq \frac{\Theta(2)}{2^2} = \frac{3}{4}.$$

It follows that

$$\frac{|\{f \in \mathcal{B}_n : \text{AUT}(f) \geq G_n\}|}{2^{2^n}} \leq 2^{2^n \cdot [\Theta(k(n)) \cdot 2^{-k(n)} - 1]} \leq 2^{-2^{n-2}}.$$

Since the right-hand side of the above inequality converges to 0, the proof of the second part of the theorem is complete. To prove the first part notice, that

$$\{f \in \mathcal{B}_n : \text{AUT}(f) \neq id_n\} \subseteq \bigcup_{\sigma \neq id_n} \{f \in \mathcal{B}_n : \sigma \in \text{AUT}(f)\},$$

where σ ranges over cyclic permutations of order a prime number $\leq n$. Since there are at most $n!$ permutations on n letters we obtain from the last inequality that

$$\frac{|\{f \in \mathcal{B}_n : \text{AUT}(f) \neq \{id_n\}\}|}{2^{2^n}} \leq n! \cdot 2^{-2^{n-2}} = 2^{O(n \log n)} \cdot 2^{-2^{n-2}} \to 0,$$

as desired.

As a consequence of the above theorem we obtain that asymptotically almost all boolean functions have trivial invariance group.

An interesting generalization of Theorem 3.6.1 has been given by M. Clausen [Cla91]. Consider the group $GL(n,2)$ of invertible $n \times n$ matrices with entries 0, 1. Let $K_n \leq GL(n,2)$ and for any boolean function $f \in \mathcal{B}_n$ define $K_n(f) = \{A \in K_n : (\forall x \in \{0,1\}^n) [f(A^{-1}x) = f(x)\}]$. We mention without proof (see Exercise 3.11.20) the following result.

Theorem 3.6.2 ([Cla91]). *Let $H_n \leq K_n \leq G_n(n,2)$ be such that $H_n > 1$ for all but a finite number of n. Then*

$$\lim_{n\to\infty} \frac{|\{f \in \mathcal{B}_n : K_n(f) = \{id_n\}\}|}{2^{2^n}} = \lim_{n\to\infty} \frac{|\{f \in \mathcal{B}_n : K_n(f) \leq H_n\}|}{2^{2^n}} = 1,$$

and

$$\lim_{n\to\infty} \frac{|\{f \in \mathcal{B}_n : K_n(f) \geq H_n\}|}{2^{2^n}} = \lim_{n\to\infty} \frac{|\{f \in \mathcal{B}_n : K_n(f) = H_n\}|}{2^{2^n}} = 0.$$

Note that 0-1 laws of the type described in Theorem 3.6.1 have been studied extensively in many branches of mathematical logic. For example, in Exercise 3.11.21, we state Fagin's 0-1 law for graphs.

3.7 Almost Symmetric Languages

In this section, we study the complexity of languages $L \in L(\mathbf{P})$. These are languages whose invariance groups have polynomial index; i.e., $|S_n : \text{AUT}_n(L)| = n^{O(1)}$. Using the classification results on finite simple groups, we will prove that languages in $L(\mathbf{P})$ are precisely the *almost symmetric* languages. The following result is proved by applying the intricate NC algorithm of [BLS87] for permutation group membership. By delving into a deep result in classification theory of finite simple groups, we later improve the conclusion to that of Theorem 3.7.3. For clarity however, we present the following theorem.

Theorem 3.7.1 ([CK91]). *For any language $L \subseteq \{0,1\}^*$, if $L \in L(\mathbf{P})$ then L is in non-uniform* NC.

Proof. As a first step in the proof we will need the following claim.
CLAIM. There is an NC^1 algorithm which, when given $x \in \{0,1\}^n$, outputs $\sigma \in S_n$ such that $x^\sigma = 1^m 0^{n-m}$, for some m.
Proof of Claim. We first illustrate the idea of proof by an example. Suppose that $x = 101100111$. By simultaneously going from left to right and from right to left, we swap an "out-of-place" 0 with an "out-of-place" 1, keeping track of the respective positions. (This is a well-known trick for improving the efficiency of the "partition" or "split" algorithm used in quick-sort.) This gives rise to the desired permutation σ. In the case at hand we find $\sigma = (2,9)(5,8)(6,7)$ and $x^\sigma = 1^6 0^3$.

Now we proceed with the proof of the claim. For $b \in \{0,1\}$, define the predicate $E_{k,b}(u)$, to hold when there are exactly k occurrences of b in the word u. The predicates $E_{k,b}$ are obviously computable in constant depth, polynomial size threshold circuits, i.e., in TC^0. By work of Ajtai, Komlós, and Szemerédi [AKS83], we have $\text{TC}^0 \subseteq \text{NC}^1$. For $k = 1, \ldots, \lfloor n/2 \rfloor$ and $1 \leq i < j \leq n$, let $\alpha_{i,j,k}$ be a log depth circuit which outputs 1 exactly when the k-th "out-of-place" 0 is in position i and the k-th "out-of-place" 1 is in position j. It follows that $\alpha_{i,j,k}(x) = 1$ if and only if "there exist $k-1$ zeros to the left of position i, the i-th bit of x is zero and there exist k ones to the right of position i" and "there exist $k-1$ ones to the right of position j, the j-th bit of x is one and there exist k zeros to the left of position j". This in turn is equivalent to

$$E_{k-1,0}(x_1, \ldots, x_{i-1}) \text{ and } x_i = 0 \text{ and } E_{k,1}(x_{i+1}, \ldots, x_n) \text{ and}$$

$$E_{k-1,1}(x_{j+1}, \ldots, x_n) \text{ and } x_j = 1 \text{ and } E_{k,0}(x_1 \ldots x_{j-1}).$$

This implies that the required permutation can be defined by

$$\sigma = \prod \left\{ (i,j) : i < j \text{ and } \bigvee_{k=1}^{\lfloor n/2 \rfloor} \alpha_{i,j.k} \right\}.$$

Converting the $\vee$-gate of fan-in $\lfloor n/2 \rfloor$ into a $\log(\lfloor n/2 \rfloor)$ depth tree of $\vee$-gates of fan-in 2, we obtain an NC^1 circuit to compute σ. This completes the proof of the claim.

Next we continue with the proof of the main theorem. Put $G_n = S_n(L)$ and let $R_n = \{h_1, \ldots, h_q\}$ be a complete set of representatives for the left cosets of G_n, where $q \leq p(n)$ and $p(n)$ is a polynomial such that $|S_n : G_n| \leq p(n)$. Fix $x \in \{0,1\}^n$. By the previous claim there is a permutation σ which is the product of disjoint transpositions and an integer $0 \leq k \leq n$ such that $x^\sigma = 1^k 0^{n-k}$. Since σ is its own inverse, $x = (1^k 0^{n-k})^\sigma$. In parallel for $i = 1, \ldots, q$ test whether $h_i^{-1}\sigma \in G_n$ by using the principal result of [BLS87], thus determining i such that $\sigma = h_i g$, for some $g \in G_n$. Then we obtain that

$$L_n(x) = L_n((1^k0^{n-k})^\sigma) = L_n((1^k0^{n-k})^{h_i g}) = L_n((1^k0^{n-k})^{h_i}).$$

By hardwiring the polynomially many values $L_n(1^k0^{n-k})^{h_i})$, for $0 \leq k \leq n$ and $1 \leq i \leq q$, we produce a family of polynomial size, polylogarithmic depth boolean circuits for L.

Theorem 3.7.1 involves a straightforward application of the beautiful NC algorithm of Babai, Luks and Seress [BLS87] for testing membership in a finite permutation group. By using the deep structure consequences of the O'Nan-Scott theorem below, together with Bochert's result on the size of the index of primitive permutation groups we can improve the NC algorithm of Theorem 3.7.1 to an optimal TC^0 (and hence NC^1) algorithm. First, we take the following discussion and statement of the O'Nan-Scott theorem from [KL88], page 376.

Let $I = \{1, 2, \ldots, n\}$ and let S_n act naturally on I. Consider all subgroups of the following five classes of subgroups of S_n.

α_1: $S_k \times S_{n-k}$, where $1 \leq k \leq n/2$,
α_2: $S_a \wr S_b$, where either ($n = ab$ and $a, b \geq 1$) or ($n = a^b$ and $a \geq 5, b \geq 2$),
α_3: the affine groups $AGL_d(p)$, where $n = p^d$,
α_4: $T^k \cdot (Out(T) \times S_k)$, where T is a non-abelian simple group, $k \geq 2$ and $n = |T|^{k-1}$, as well as all groups in the class
α_5: almost simple groups acting primitively on I.[4]

Theorem 3.7.2 (O'Nan-Scott). *Every subgroup of S_n not containing A_n is a member of $\alpha_1 \cup \cdots \cup \alpha_5$.*

Now we can improve the result of Theorem 3.7.1 in the following way.

Theorem 3.7.3 ([CK91]). *For any language $L \subseteq \{0,1\}^*$, if $L \in L(\mathbf{P})$ then $L \in \text{TC}^0$, hence $L \in \text{NC}^1$.*

Proof. The proof requires the following consequence of the O'Nan-Scott theorem.

Lemma 3.7.1 ([CK91]). *Suppose that $\langle G_n \leq S_n : n \geq 1 \rangle$ is a family of permutation groups, such that for all n, $|S_n : G_n| \leq n^k$, for some k. Then for sufficiently large N, there exists an $i_n \leq k$ for which $G_n = U_n \times V_n$ with the supports of U_n, V_n disjoint and $U_n \leq S_{i_n}, V_n = S_{n-i_n}$.*

Before proving the lemma, we complete the details of the proof of Theorem 3.7.3. Apply the lemma to $G_n = \text{AUT}_n(L)$ and notice that given $x \in \{0,1\}^n$, the question of whether x belongs to L is decided completely by the number

[4] Consider a permutation group G acting on a nonempty set X. A subset B of X is called a block if for all $g \in G$ the sets B and B^g are either equal or disjoint. The empty set, X itself, and all singletons of X are blocks (also called trivial blocks). A transitive permutation group G with no non-trivial blocks is called primitive.

of 1s in the support of $K_n = S_{n-i_n}$ together with information about the action of a finite group $H_n \leq S_{i_n}$, for $i_n \leq k$. Using the counting predicates as in the proof of Theorem 3.7.1, it is clear that appropriate TC^0 circuits can be built. This completes the proof of Theorem 3.7.3, assuming Lemma 3.7.1.

Proof. We have already observed that $G_n \neq A_n$. By the O'Nan-Scott theorem, G_n is a member of $\alpha_1 \cup \cdots \cup \alpha_5$. Using Bochert's theorem on the size of the index of primitive permutation groups (if a primitive permutation group $H \leq S_n$ does not contain the alternating group A_n, then $|S_n : H| \geq \lfloor (n+1)/2 \rfloor!$ [Wie64]), the observations of [LPS88] concerning the primitivity of the maximal groups in $\alpha_3 \cup \alpha_4 \cup \alpha_5$ and the fact that G_n has polynomial index with respect to S_n, we conclude that the subgroup G_n cannot be a member of the class $\alpha_3 \cup \alpha_4 \cup \alpha_5$. It follows that $G_n \in \alpha_1 \cup \alpha_2$. We show that in fact $G_n \notin \alpha_2$. Assume on the contrary that $G_n \leq H_n = S_a \wr S_b$. It follows that $|H_n| = a!(b!)^a$. We distinguish the following two cases.

Case 1. $n = ab$, for $a, b > 1$.

In this case it is easy to verify using Stirling's formula

$$(n/e)^n \sqrt{n} < n! < (n/e)^n 3\sqrt{n}$$

that

$$|S_n : H_n| = \frac{n!}{a!(b!)^a} \sim \frac{a^{n-a}}{3b^{a/2}(3/a)^a \sqrt{a}}.$$

Moreover, it is clear that the right-hand side of this last inequality cannot be asymptotically polynomial in n, since $a \leq n$ is a proper divisor of n, which is a contradiction.

Case 2. $n = a^b$, for $a \geq 5, b \geq 2$.

A similar calculation shows that asymptotically

$$|S_n : H_n| = \frac{n!}{a!(b!)^a} = \frac{n!}{a!(b'!)^a},$$

where $b' = a^{b-1}$. It follows from the argument of case 1 that this last quantity cannot be asymptotically polynomial in n, which is a contradiction. It follows that $G_n \in \alpha_1$. Let $G_n \leq S_{i_n} \times S_{n-i_n}$, for some $1 \leq i_n \leq n/2$.

We claim that there exists a constant k, for which $i_n \leq k$, for all but a finite number of ns. Indeed, notice that

$$|S_n : S_i \times S_{n-i}| = \frac{n!}{i!(n-i)!} = \Omega(n^i) \leq |S_n : G_n| \leq n^k,$$

which proves that $i_n \leq k$. It follows that $G_n = U_n \times V_n$, where $U_n \leq S_{i_n}$ and $V_n \leq S_{n-i_n}$. Since $i_n \leq k$ and $|S_n : G_n| \leq n^k$, we have that for n large enough, $V_n = S_{n-i_n}$. This completes the proof of the claim. Now let $L \subseteq \{0,1\}^*$ have polynomial index. Given a word $x \in \{0,1\}^n$, in TC^0, one can test whether the number of 1s occurring in the $n - i_n$ positions (where

$V_n = S_{n-i_n}$) is equal to a fixed value, hardwired into the n-th circuit. This, together with a finite look-up table corresponding to the U_n part, furnishes a TC^0 algorithm for testing membership in L.

3.8 Symmetry and Complexity

In [CK91], by adapting the counting argument of [Lup61a], it was shown that for any superpolynomial function f, there exist languages $L \subseteq \{0,1\}^*$ whose invariance groups G_n have at most $f(n)$ orbits when acting on $\{0,1\}^n$ and yet L is not computable in polynomial size circuits. Against this negative result it was there conjectured that if $L \subseteq \{0,1\}^*$ is a language whose invariance groups have polynomially many orbits ($\Theta_n(L_n) \leq n^{O(1)}$) then L is computable in non-uniform NC. Babai, Beals and Takácsi-Nagy [BBTN92] proved this conjecture by developing some very elegant structure theory for groups having polynomially many orbits. As an additional corollary, they obtained an NC solution of a specific case of the bounded valency graph isomorphism problem.

For group $G \leq S_n$ and words $x, y \in \{0,1\}^n$, recall the group action If $G \leq S_n$ is a permutation group, then recall the action of G on the collection of n-length words; namely, for $x, y \in \{0,1\}^n$, we write $x \sim y \bmod G$ to assert the existence of σ in G for which $x^\sigma = y$. The *orbit* of x is $\{y \in \{0,1\}^n : x \sim y\}$. We define the ORBIT PROBLEM for group $G \leq S_n$ as follows.

Input: $x, y \in \{0,1\}^n$
Output: Whether $x \sim y \bmod G$.

For families $\mathcal{G} = \langle G_n : G_n \leq S_n \rangle$ and $\mathcal{H} = \langle H_n : H_n \leq S_n \rangle$, we write $\mathcal{H} \leq \mathcal{G}$ to indicate $H_n \leq G_n$ for all $n \in \mathbf{N}$. Let $\Theta(G_n)$ be the number of orbits of G_n acting on $\{0,1\}^n$. For simplicity, we write G instead of $\mathcal{G}$ and suppress indices n in G_n. We also use the notation $Sym(\Omega)$ for the of permutations on the set Ω.

Proposition 3.8.1. *If $\mathcal{H} \leq \mathcal{G}$ and $\Theta(H) \leq n^{O(1)}$, then the orbit problem for $\mathcal{G}$ is* AC^0 *reducible to the orbit problem for $\mathcal{H}$.*

Proof. Since H_n is a subgroup of G_n, every H_n orbit is contained in a G_n orbit. There are at most $p(n)$ many orbits of H_n acting on 2^n, so

$$x \sim y \bmod G_n \iff \bigvee_{i=1}^{p(n)} x \sim y_i \bmod H_n$$

where $y_1, \ldots, y_{p(n)}$ are fixed representatives for those H_n orbits contained in the G_n orbit of y.

The following proposition lists some elementary facts about the number of orbits of a group with permutation domain Ω, when acting on the power set of Ω.

Proposition 3.8.2. *Let G, H be permutation groups.*

1. *If $H \leq G$ then $\Theta(H) \geq \Theta(G)$.*
2. *Assuming that G, H are have disjoint supports, $\Theta(G \times H) = \Theta(G) \cdot \Theta(H)$.*
3. $\Theta(H \wr S_k) = \binom{\Theta(H)+k-1}{k} = \binom{\Theta(H)+k-1}{\Theta(H)-1}$
4. *For $k \geq 3$, $\Theta(A_k) = \Theta(S_k)$ and $\Theta(H \wr A_k) = \Theta(H \wr S_k)$.*

Proof. of (1). Clear since every H orbit is contained in a G orbit.
Proof of (2). Straightforward.
Proof of (3). If the degree of H is m, then recall that the *wreath product* $H \wr S_k$ is given by the collection of permutations $\pi \in Sym(A \times B)$, where $|A| = m$, $|B| = k$ and $\pi = \langle \sigma_1, \ldots, \sigma_k; \tau \rangle$ for $\sigma_1, \ldots, \sigma_k$ independent permutations in H and τ in S_k. The action of π on the permutation domain $A \times B$ is given by $(i,j)^\pi = (i^{\sigma_j}, j^\tau)$.

CLAIM. There is a 1-1 correspondence between $\Theta(H \wr S_k)$ and the collection of all non-decreasing maps from $\{1, \ldots, k\}$ into $\{1, \ldots, \Theta(H)\}$.

Proof of Claim. Temporarily define a *canonical ordering* on $\{0,1\}^m$ as follows. For $x, y \in \{0,1\}^m$, let $x \prec y$ iff the weight $|x|_1$ of x is less than the weight $|y|_1$ of y or x, y have equal weights and x precedes y in the lexicographic ordering. Define $x \in \{0,1\}^m$ to be a *canonical representative* of an orbit of H if for all lexicographically smaller $y \in \{0,1\}^m$, $y \not\sim x \bmod H$. Let $\phi : \{0,1\}^m \to \{0,1\}^m$ by setting $\phi(u)$ to be that canonical representative lying in the same H-orbit as u. Let $\{x_1, \ldots, x_{\Theta(H)}\}_\prec$ be a listing of the canonical representatives of the orbits of H acting on $\{0,1\}^m$. Now given $u \in \{0,1\}^{mk}$, where $u = u_1 \cdots u_k$, and each $u_i \in \{0,1\}^m$, determine a permutation $\sigma \in S_k$ for which

$$\phi(u_{\sigma(1)}) \preceq \phi(u_{\sigma(2)}) \preceq \cdots \preceq \phi(u_{\sigma(k)}).$$

The claim now readily follows.

It is well-known (see for instance [Ber71]), that the number of non-decreasing maps from k into m is equal to the number of ways of choosing k objects from a collection of m objects, *allowing repetitions*, given by

$$\frac{(m+k-1)\cdots(m+1)(m)}{k!} = \binom{m+k-1}{k}.$$

Since we have established a 1-1 correspondence between $\Theta(H \wr S_k)$ and the collection of all non-decreasing maps from $\{1, \ldots, k\}$ into $\{1, \ldots, \Theta(H)\}$, it follows that $\Theta(H \wr S_k) = \binom{\Theta(H)+k-1}{k}$. Using the symmetry of the binomial coefficients, i.e., that $\binom{n}{k} = \binom{n}{n-k}$, the equality $\Theta(H \wr S_k) = \binom{\Theta(H)+k-1}{\Theta(H)-1}$ is immediate.

Proof of (4). Suppose that $x, y \in \{0,1\}^k$ and $x^\sigma = y$ for some $\sigma \in S_k$. If $\sigma \in A_k$, then let $\overline{\sigma} = \sigma$, otherwise, since $k \geq 3$, let τ be the transposition interchanging i, j, where $x_i = x_j$ and set $\overline{\sigma} = \sigma \circ \tau$. Then $\overline{\sigma} \in A_k$ and $x^{\overline{\sigma}} = y$. It follows that $x, y \in \{0,1\}^k$ are in the same S_k orbit iff they are in the same A_k orbit. The assertion for $H \wr A_k$ and $H \wr S_k$ is similarly proved. This concludes the proof of the Proposition.

Lemma 3.8.1 ([BBTN92]). *If $G \leq H \wr S_k$ and $\Theta(G) \leq n^c$, then*

$$\min(\Theta(H) - 1, k) \leq 2c.$$

Proof. Since $G \leq H \wr S_k$, Proposition 3.8.2 implies that $\Theta(G) \geq \Theta(H \wr S_k)$.
Case 1. $k \leq \Theta(H) - 1$.
Noting that for $a, b \geq 1$, and $i \geq 0$

$$\frac{a+b-i}{a-i} \geq \frac{a+b}{a}$$

so that

$$\binom{a+b}{a} \geq \left(\frac{a+b}{a}\right)^a$$

it follows that

$$\Theta(H \wr S_k) = \binom{\Theta(H)+k-1}{k} \geq \binom{2k}{k} \geq 2^k.$$

Thus $k \leq \log \Theta(G)$. For sufficiently large n, $n/(c \cdot \log(n))^2 \geq \sqrt{n}$, so

$$\begin{aligned}
n^c &\geq \Theta(G) \\
&\geq \binom{\Theta(H)+k-1}{k} \\
&\geq \binom{\Theta(H)}{k} \\
&\geq \binom{n/k}{k} \\
&\geq \left(\frac{n}{k^2}\right)^k \\
&\geq n^{k/2}
\end{aligned}$$

Hence $k \leq 2c$.
Case 2. $k > \Theta(H) - 1$.

$$\begin{aligned}
\Theta(H \wr S_k) &= \binom{\Theta(H)+k-1}{\Theta(H)-1} \\
&\geq \binom{2 \cdot (\Theta(H)-1)}{\Theta(H)-1} \\
&\geq 2^{\Theta(H)-1}
\end{aligned}$$

so $\Theta(H) - 1 \leq \log \Theta(G)$. Thus

$$\begin{aligned}
n^c &\geq \Theta(G) \\
&\geq \binom{\Theta(H)+k-1}{\Theta(H)-1} \\
&\geq \binom{k}{\Theta(H)-1} \\
&\geq \binom{n/(\Theta(H)-1)}{\Theta(H)-1} \\
&\geq \left(\frac{n}{(\Theta(H)-1)^2}\right)^{\Theta(H)-1} \\
&\geq n^{(\Theta(H)-1)/2}
\end{aligned}$$

Hence $\Theta(H) - 1 \leq 2c$.

We require some definitions in order to establish structure results for groups having polynomially many orbits.

Definition 3.8.1. *A subset $\Delta \subseteq \Omega$ is a block of imprimitivity of group $G \leq Sym(\Omega)$ if for every $\sigma \in G$, $\Delta^\sigma = \Delta$ or $\Delta^\sigma \cap \Delta = \emptyset$. The group G is primitive if the only blocks of imprimitivity of G are Ω and the singleton subsets of Ω. The group $G \leq Sym(\Omega)$ is transitive if for every $x, y \in \Omega$, there is $\sigma \in G$ such that $x^\sigma = y$.*

It is clear that if G is transitive and $\Delta_1, \ldots, \Delta_m$ is a system of blocks of imprimitivity, then all blocks have the same number of elements. Notice that for $G \leq S_n$, we distinguish between G acting on its permutation domain $\{1, \ldots, n\}$, G acting on the set 2^n of all n-length binary words, and G acting on the set 2^{2^n} of all boolean functions on n variables. A *structure forest* $\mathcal{F}$ for permutation group $G \leq Sym(\Omega)$ is a forest on which G acts as automorphisms such that the leaves form the permutation domain Ω and the roots correspond to orbits. Each node $v \in \mathcal{F}$ is identified with a block $B(v)$ of imprimitivity of G acting on Ω, where $B(v)$ consists of the leaves of $\mathcal{F}$ below v. Let

$$\mathcal{B}(v) = \{B(u) : u \text{ is a child of } v\}$$

Let $L(v) \leq Sym(B(v)$ denote the action of G_v on $B(v)$, and let $H(v) \leq Sym(\mathcal{B}(v))$ denote the action of G_v on $\mathcal{B}(v)$. A node $v \in \mathcal{F}$ is *primitive* if $H(v)$ is primitive, while v is a *giant* if $H(v)$ is the alternating or symmetric group. If G is transitive, then the structure forest is a tree and we write $k_i = |\mathcal{B}(v)|$ for $v \in \mathcal{L}_i$. In the general case where $\mathcal{F}$ is not a tree, we write $k_{i,j} = \mathcal{B}(v)$ where $v \in \mathcal{L}_i$ on tree T_j. The group K_i is the pointwise stabilizer of $\mathcal{L}_i$. Note that K_i is a normal subgroup of G, denoted by $K_i \triangleleft G$, since K_i is the kernel of the action of G on $\mathcal{L}_i$. If $v \in \mathcal{L}_i$ then $K_i \leq L(v)^{|\mathcal{L}_i|}$.

Theorem 3.8.1 ([Bab81]). *Suppose that $G \leq S_n$ is a primitive permutation group of degree n not containing A_n. Then*

$$|G| < \exp\{4\sqrt{n}\log^2(n)\}$$

The proof of this estimate will not be given, but we note that the proof does not use classification theory.

Theorem 3.8.2 (Babai–Pyber). *Suppose that $G \leq Sym(\Omega)$, $|\Omega| = n$, $\mathcal{F}$ is a primitive structure forest for G. For any $t > 1$, if $\mathcal{F}$ has no giant node of degree strictly greater than t, then*

$$\Theta(G) \geq 2^{n/c_1 t}$$

for some absolute constant c_1.

Proof. Let $\{\Delta_1, \ldots, \Delta_m\}$ be the orbits of G acting on Ω. Then $G \leq \Pi_{i=1}^m G^{\Delta_i}$, so $\Theta(G) \geq \Pi_{i=1}^m \Theta(G^{\Delta_i})$. Thus it suffices to prove the theorem for transitive groups G. We may suppose t is sufficiently large to satisfy

$$t^{x-1} \geq \exp\{4\sqrt{x}\log^2(x)\}$$

for all $x \geq 2$. Set $c_2 \geq 8$ and $c_3 = 4c_2$. For t given, let $\Theta_t(n)$ be the minimum value of $\Theta(G)$ as G ranges over all transitive permutation groups of degree n having a primitive structure tree T with no giant node of degree strictly greater than t. For $1 \leq n \leq c_2 t$, it is clear that

$$\Theta_t(n) \geq n + 1 \geq 2 \geq 2^{c_2/c_3} \geq 2^{n/c_3 t}.$$

By induction on n, we show that following claim which immediately implies the statement of the theorem.

CLAIM. For $n \geq c_2 t$, $\Theta_t(n) \geq t 2^{n/c_3 t}$.

Proof of Claim. Suppose that G is a transitive permutation group of degree $n \geq c_2 t$ and T is a primitive structure tree for G with no giant nodes of degree $> t$. Assume the claim holds for values less than n. Collapse all levels below $\mathcal{L}_1$ to a single level. Let $H = H(\text{root})$, $L = L(u)$ for some $u \in \mathcal{L}_1$.

Case 1. $k_1 \geq c_2 t$.

H is of degree k_0, so $|H_0| \leq k_0!$ and for $k_0 > t$, since H is primitive, by Theorem 3.8.1, $|H| \leq \exp\{4\sqrt{k_0}\log^2(k_0)\}$, so $|H| \leq t^{k_0 - 1}$. By the induction hypothesis, as L is of degree $k_1 < n$, $\Theta(L) \geq t \cdot 2^{k_1/c_3 t}$, so

$$\begin{aligned}\Theta(G) &\geq \Theta(K_1)/|H| \geq \Theta(L^{k_0})/|H| = \Theta(L)^{k_0}/|H| \\ &\geq (t2^{k_1/c_3 t})^{k_0}/t^{k_0-1} = t2^{k_1 k_0/c3t} = t2^{n/c3t}.\end{aligned}$$

Case 2. $k_1 < c_2 t \leq k_0$.

By Theorem 3.8.1,

$$\begin{aligned}|H| &\leq e^{4\sqrt{k_0}\log^2(k_0)} \leq 2^{4\log(e)\sqrt{k_0}\log^2(k_0)} \\ &\leq 2^{8\sqrt{k_0}\log^2(k_0)} \leq 2^{k_0/2}.\end{aligned}$$

Also,

$$2^{n/c_3 t} = 2^{k_0 k_1/c_3 t} < 2^{k_0 c_2 t/c_3 t} = 2^{k_0/4}.$$

Thus

$$\begin{aligned}\Theta(G) &\geq \Theta(s)^{k_0}/|H| \geq 2^{k_0}/|H| \geq 2^{k_2/2}\\ &\geq 2^{k_0/4} \cdot 2^{n/c_3 t} \geq t \cdot 2^{n/c_3 t}.\end{aligned}$$

Case 3. $k_0, k_1 < c_2 t$.
Then $G \leq S_{k_1} \wr S_{k_0}$ so

$$\Theta(G) \geq \Theta(S_{k_1} \wr S_{k_0}) = \binom{k_1 + k_0}{k_0} = \binom{k_0 + k_1}{k_1}.$$

By symmetry, we can assume that $k_0 \leq k_1$. As $n = k_0 k_1 \geq c_2 t$ and $k_0, k_1 < c_2 t$, it follows that $2 \leq k_0 \leq k_1$. Hence

$$\begin{aligned}\Theta(G) &\geq \binom{k_0 + k_1}{k_0} = (\frac{k_1 + k_0}{k_0})(\frac{k_1 + k_0 - 1}{k_0 - 1}) \cdots (\frac{k_1 + 1}{1})\\ &\geq 2^{k_0 - 3} \cdot (\frac{k_1 + 3}{3})(\frac{k_1 + 2}{2})(\frac{k_1 + 1}{1})\\ &\geq 2^{k_0 - 3} k_1^2 \geq 2^{k_0 - 3} k_1 k_0 \geq 2^{k_0 - 3} c_2 t \geq 2^{k_0} t \geq t 2^{(\frac{c_2}{c_3}) k_0}\\ &= t 2^{(\frac{c_2 t}{c_3 t}) k_0} \geq t 2^{n k_0 / c_3 t} \geq t 2^{n/c_3 t}.\end{aligned}$$

This completes the proof of the claim and hence of the theorem.

Corollary 3.8.1. *For transitive group $G \leq Sym(\Omega)$, there exists a depth 3 structure tree T, such that $k_0 k_2 \leq c_1 \log(\Theta(G))$, and the nodes on level 1 of T are giants.*

Proof. Let T' be the primitive structure tree for G and let t be the largest degree of giant nodes in T'. The level of these nodes is called the *explosion level.* Contract all levels above and below the explosion level to one level, keeping the root separate. This produces a depth 3 structure tree T. By Theorem 3.8.2, we have $k_1 \geq \frac{n}{c_1 \log(\Theta(G))}$. Now $n = k_0 k_1 k_2$, so

$$k_0 k_2 = n/k_1 \leq \frac{n}{n/c_1 \log(\Theta(G))} \leq c_1 \log(\Theta(G)).$$

We introduce some definitions. For subgroups H, K of group G, H is said to be a complement of K in G if $H \cap K = 1$ and $HK = G$. Let $B_1, \ldots, B_k$ be a system of blocks of imprimitivity for G. An element $\sigma \in G$ is clean if for all $1 \leq i \leq k$ either $B_i^\sigma \neq B_i$ or σ acts trivially on B_i ($\sigma(x) = x$ for all $x \in B_i$). A subgroup H is clean if it consists only of clean elements; H is a clean complement of K if it is clean and is a complement to K.

Lemma 3.8.2. *If $G \leq Sym(\Omega)$ is a transitive permutation group having a depth 2 structure tree T such that $H(root) = A_{k_0}$ and $k_0 \geq 4k_1$, then K_1 has a clean complement.*

Proof. Let $\mathcal{L}_1 = \{v_1, \ldots, v_{k_0}\}$ be the collection of nodes on the first level of T. For $\tau \in G$, let $\overline{\tau}$ denote the action of τ on $\mathcal{L}_1$. By Bertrand's postulate, there is a prime p satisfying $k_1 < p < k_0/2$. Take $\pi \in G$ such that $\overline{\pi}$ is a p-cycle. Since $k_1 < p$ and the order of an element divides the order of the group to which it belongs, there is an integer m not divisible by p for which $\overline{\pi}^m$ is the identity on $\mathcal{L}_1$, hence π is clean. Without loss of generality, suppose that $m = 1$ and that π permutes $v_1, \ldots, v_p$ cyclically and fixes each of $v_{p+1}, \ldots, v_{k_0}$ and their children. Similarly, there is an element $\pi' \in G$ such that π' permutes $v_p, \ldots, v_{2p-1}$ cyclically and fixes each v_i and its children for i different from $p, \ldots, 2p-1$. By abuse of language, we temporarily call a permutation $\sigma \in G$ a clean 3-cycle if $\overline{\sigma}$ is a 3-cycle permuting cyclically $v_{i_1}, v_{i_2}, v_{i_3}$ while fixing v_i and all its children for i different from i_1, i_2, i_3. It follows that the commutator $\sigma = [\pi, pi'] = \pi\pi'\pi^{-1}\pi'^{-1}$ is a clean 3- cycle and $\overline{\sigma} = (v_{p+1}, v_p, v_1)$. We leave it to the reader to verify that the conjugate $\theta\sigma\theta^{-1}$ of a clean 3-cycle is a clean 3- cycle and that a group generated by clean elements is a clean group. For $1 \leq i < k_0$, let $\sigma_i \in G$ be a clean 3- cycle with $\overline{\sigma_i} = (v_i, v_{i+1}, v_{k_0})$.
Case 1. k_0 is odd.
Then H is generated by $\sigma_1, \sigma_3, \sigma_5, \ldots, \sigma_{k_0-2}$.
Case 2. k_0 is even.
Let $A = \langle \sigma_1, \sigma_3 \rangle_{v_{k_0}}$, consisting of those $\sigma \in G$ generated by σ_1, σ_3 where σ fixes v_{k_0} and its children. Then A is clean. Let H be generated by $A, \sigma_4, \sigma_6, \ldots, \sigma_{k_0-2}$. It follows that H is a clean complement to K_1.

Theorem 3.8.3 ([BBTN92]). *Every language L with transitive automorphism groups* $\mathrm{AUT}(L_n)$ *and polynomially many cycles, i.e.,* $\Theta(\mathrm{AUT}(L_n)) \leq n^{O(1)}$, *is in* TC^0.

Proof. By Corollary 3.8.1, let T be a depth 3 structure tree where $H(u)$ is a giant for each $u \in \mathcal{L}_1$. Applying the clean complement Lemma 3.8.2 to each $B(u)$ for $u \in \mathcal{L}_1$, there is a clean complement $H_u = \langle 1_{B(u)} \rangle \wr A_{k_1}$ of K_2 with respect to $H(u)$. Thus $H_u K_2 = H(u)$ and $\Pi_{u \in \mathcal{L}_1} H_u \leq G$, so

$$\Theta(\Pi_{u \in \mathcal{L}_1} H_u) = \binom{k_1 + 2^{k_2} - 1}{2^{k_2} - 1}^{k_0} \geq \Theta(G).$$

By Lemma 3.8.1 $k_0, k_2 \leq 2c$ for an absolute constant c, so $\binom{k_1 + 2^{k_2} - 1}{2^{k_2} - 1}$ is polynomial in k_1 and hence polynomial in n. The orbit problem for $\Pi_{u \in \mathcal{L}_1} H_u$ is solved essentially by counting, and hence belongs to TC^0.

3.9 Applications to Anonymous Networks

The anonymous network was introduced in Section 1.11.8. In this section we concentrate on the study of the bit complexity of computing boolean functions on Rings and Hypercubes.

3.9.1 Rings

Recall that C_N is the cyclic group generated by the cycle $(1, 2, \ldots, N)$ and D_N is the dihedral group generated by the cycle $(1, 2, \ldots, N)$ and the reflection

$$\rho_N = \begin{pmatrix} 1 & 2 & \cdots & N \\ N & N-1 & \cdots & 1 \end{pmatrix}.$$

Let R_N denoted the ring of N processors.

Theorem 3.9.1 ([ASW88]). *Let f be a boolean function in $\mathcal{B}_N$. Then*

1. *f is computable in the oriented ring R_N if and only if* $\mathrm{AUT}(f) \geq C_N$.
2. *f is computable in the unoriented ring R_N if and only if* $\mathrm{AUT}(f) \geq D_N$.

Proof. The if part follows easily from the fact that if a boolean function is computable in the network then it must be invariant under its group of automorphisms. So we concentrate on the proof of the other direction.

For the case of oriented rings we have the following algorithm.

```
Algorithm for processor p:
send your bit left;
for N steps do
     send the bit you receive from
     the right to the left;
     od
endfor
```

For the case of unoriented rings we have the following algorithm.

```
Algorithm for processor p:
send your bit both left and right;
for ⌊N/2⌋ steps do
     send the bit you receive in the direction
     opposite to the one you got it from;
     od
endfor
```

It is easy to see that these algorithms are correct.

3.9.2 Hypercubes

A natural labeling of the hypercube is the following, $\mathcal{L}$: the edge connecting nodes $x = (x_1, \ldots, x_n)$ and $y = (y_1, \ldots, y_n)$ is labeled by i if and only if $x_i \neq y_i$, i.e., $\mathcal{L}(x, y) = \mathcal{L}(y, x) = i$. In this subsection we will refer to a hypercube with this labeling as an canonically labeled hypercube and we will reserve the symbol $\mathcal{L}$ to denote this canonical labeling.

Of particular interest in the case of the canonically labeled hypercube are the *bit-complement* automorphisms that complement the bits of certain components, i.e., for any set $S \subseteq \{1, \ldots, n\}$ let $\phi_S(x_1, \ldots, x_n) = (y_1, \ldots, y_n)$, where $y_i = x_i + 1$, if $i \in S$, and $y_i = x_i$ otherwise (here addition is modulo 2). Let F_n denote the group of bit-complement automorphisms of Q_n.

Theorem 3.9.2. *The group of automorphisms of the canonically labeled hypercube $Q_n[\mathcal{L}]$ is exactly the group F_n of bit-complement automorphisms.*

Proof. Let $\phi \in Aut(Q_n[\mathcal{L}])$. We claim that for all $x, y \in Q_n$,

$$\phi(x) + \phi(y) = x + y. \tag{3.4}$$

Indeed, given x, y there is a path $x_0 := x, x_1, \ldots, x_k := y$ joining x to y. By definition, ϕ must preserve labels, i.e., for all $i < k$, $\phi(x_i) + \phi(x_{i+1}) = x_i + x_{i+1}$. Adding these congruences we obtain $\phi(x_0) + \phi(x_k) = x_0 + x_k$, which proves (3.4). Using (3.4) it is now easy to show that ϕ is uniquely determined from the value of $\phi(0^n)$, say $\phi(0^n) = (p_1, \ldots, p_n)$. It follows easily that $\phi = \phi_S$, where $S = \{1 \leq i \leq n : p_i \neq 0\}$.

The automorphism group of the unlabeled hypercube Q_n is larger than F_n. For any permutation $\sigma \in S_n$ let $\phi_\sigma(x_1, \ldots, x_n) = (x_{\sigma(1)}, \ldots, x_{\sigma(n)})$ and let P_n denote the group of these automorphisms. We mention without proof that it can be shown easily that P_n is a normal subgroup of $Aut(Q_n)$ and in fact $Aut(Q_n) = F_n \cdot P_n$.

First we characterize the class of boolean functions which are computable in the canonically labeled hypercube in terms of its group of automorphisms and provide an algorithm with bit complexity $O(N^2)$ for computing all such functions.

Theorem 3.9.3. *On the canonically labeled hypercube Q_n of degree n and for any boolean function $f \in \mathcal{B}_N$, $N = 2^n$, f is computable on the hypercube Q_n if and only if f is invariant under the bit-complement automorphisms of Q_n. Moreover, the bit complexity of any such computable function is $O(N^2)$.*

Proof. The "if" part is straightforward so we need only prove the "only if" part. Let $f \in \mathcal{B}_N$ be invariant under all bit-complement automorphisms of the hypercube. The algorithm proceeds by induction on the dimension n of the hypercube. Intuitively, it splits the hypercube into two $n-1$ dimensional hypercubes. The first hypercube consists of all nodes with $x_n = 0$ and the second of all nodes with $x_n = 1$. By the induction hypothesis the nodes of these hypercubes know the entire input configuration of their corresponding hypercubes. Every node in the hypercube with $x_n = 0$ is adjacent to unique node in the hypercube with $x_n = 1$. By exchanging their information all processors will know the entire input configuration and hence they can all compute the value of f on the given input.

More formally, the algorithm is as follows. For any sequences of bits I, J let IJ denote the concatenation of I and J. Let I_p^i denote the input to processor

p at the i-th step of the computation. Initially I_p^0 is the input bit to processor p.

Algorithm for processor p:
initialize: I_p^0 is the input bit to processor p;
for $i := 0, \ldots, n-1$ **do**
  send message I_p^i to ps neighbor q along the i-th link
  let I_q^i be the message received by p from p's neighbor
  q along the i-th link and **put** $I_p^{i+1} := I_p^i I_q^i$;
od;
output $f(I_p^n)$

The algorithm is depicted in Figure 3.3 for a given input. To prove the correctness of the algorithm it must be shown that all processors output the same correct bit, i.e., for all processors p, q, $f(I_p^n) = f(I_q^n)$. Let $I_p = I_p^n$ be the sequence obtained by processor p at the n-th stage of the above algorithm. We call I_p the view of processor p on input I. Let p, q be any two processors of the hypercube. Clearly, there is a unique bit-complement automorphism ϕ satisfying $\phi(p) = q$, namely $\phi = \phi_S$, where $i \in S$ if and only if $p_i \neq q_i$. Now it can be shown that this automorphism will map processor p's view, namely I_p, to the view of processor q, namely I_q. For any sequence $b_x b_{x'} \cdots$ of bits indexed by elements $x, x', \ldots \in Q_n$ define $\phi(b_x b_{x'} \cdots) = b_{\phi(x)} b_{\phi(x')} \cdots$. The proof of correctness is based on the identity

$$\phi(I_p) = I_{\phi(p)}. \tag{3.5}$$

To prove (3.5) it is sufficient to show that for all $i \leq n = \log N$, $\phi(I_p^i) = I_{\phi(p)}^i$. The proof is by induction on $i \leq n$. The result is clear for $i = 0$. Assume the result true for i. Let p', q' be p's and q's neighbors along the i-th edge, respectively. Then by definition we have

$$I_p^{i+1} = I_p^i I_{p'}^i \text{ and } I_q^{i+1} = I_q^i I_{q'}^i.$$

Since ϕ is a bit-complement automorphism and p, p' are connected via the i-th edge it follows that $\phi(p) = q$ and $\phi(p') = q'$. Using the induction hypothesis $\phi(I_p^i) = I_{\phi(p)}^i$ we obtain

$$\phi(I_p^{i+1}) = \phi(I_p^i)\phi(I_{p'}^i) = I_q^i I_{\phi(p')}^i = I_q^{i+1} = I_{\phi(p)}^{i+1}$$

This completes the inductive proof. It follows now that $\phi(I_p) = I_q$ which implies that $f(I_p) = f(I_q)$, since f is invariant under the bit-complement automorphisms of Q_n.

To study the bit complexity of the above algorithm, let $T(N)$ be the number of bits transmitted in order that at the end of the computation all the processors in the hypercube know the input of the entire hypercube. By performing a computation on each of the two $n-1$-dimensional hypercubes we obtain that their nodes will know the entire input corresponding to their nodes in $T(N/2)$ bits. The total number of bits transmitted in this case is $2 \cdot T(N/2)$. The final exchange transmission consists of $N/2$ bits being transmitted by $N/2$ nodes to their $N/2$ corresponding other nodes, for a total of $2 \cdot N/2 \cdot N/2 = N^2/2$. Hence we have proved that $T(N) \leq 2 \cdot T(N/2) + N^2/2$. It follows that $T(N) \leq N^2$, as desired.

Next we make several alterations to the previous algorithm and show how to improve the complexity bound to $O(N \cdot \log^4 N)$, for each boolean function $f \in \mathcal{B}_N$ which is computable in the hypercube. In all our subsequent discussions we use the notation and terminology established in the previous discussion. As before the new algorithm is also executed in $n = \log N$ steps, one step per dimension. However, now we take advantage of the information provided to p about the hypercube from its i-th view I_p^i. The main ingredients of the new algorithm are the following.

- We introduce a leader election mechanism which for each $i \leq \log N$ elects leaders among the processors with lexicographically maximal view at the i-th step of the algorithm.
- We use elementary results from the theory of finite permutation groups [Wie64] in order to introduce a coding mechanism of the views; leaders at the $(i-1)$st step exchange the encoded versions of their views I_p^{i-1}; upon receipt of the encoded view they recover the original view sent and elect new leaders for the i-th step.
- The leader election and coding mechanisms help keep low the number of bits transmitted during the i-th step of the algorithm to $O(N \cdot i^3)$ bits.

The technical details of the above description will appear in the sequel. We begin with some preliminary lemmas that will be essential in the proof of the main theorem.

Lemma 3.9.1. *If $I_p = I_q$ then the hypercube as viewed from p is identical to the hypercube as viewed from q. More formally, for each p let $I_p = \langle b_x : x \in N \rangle$. If $I_p = I_q$ and $\phi = \phi_S$, where $S = \{i \leq n : p_i \neq q_i\}$, then $\forall x \in Q_n (b_x = b_{\phi(x)})$.*

Proof. Indeed, notice that since $q = \phi(p)$

$$I_p = I_q = I_{\phi(p)} = \phi(I_p),$$

where the right-most equality follows from (3.5). This proves the lemma.

Lemma 3.9.2. *Let I be a fixed sequence of bits of length 2^n. Then the set of processors p such that $I_p = I$ can be identified in a natural way with a group of bit-complement automorphisms. Moreover, the number of processors p such that $I_p = I$ is either 0 or a power of 2.*

Proof. Let $\mathcal{G}$ be the following set of automorphisms

$$\mathcal{G} = \{\phi \in F_n : \forall p \in Q_n (I_p = I \Rightarrow I_{\phi(p)} = I)\}. \tag{3.6}$$

The identity element is in $\mathcal{G}$. In addition the identity

$$I_{\phi(\psi(p))} = \phi(\psi(I_p))$$

implies that $\mathcal{G}$ is also closed under multiplication. Since $\mathcal{G}$ is finite it is a group.

Next consider the set $\mathcal{J}$ of processors q satisfying $I_q = I$ and assume that $\mathcal{J} \neq \emptyset$. Let p_0 be an arbitrary but fixed element of $\mathcal{J}$. Without loss of generality we may assume that $p_0 = 0^n$. We claim that the sets $\mathcal{J}$ and $\mathcal{G}$ are equipotent. First we prove $|\mathcal{J}| \leq |\mathcal{G}|$. Indeed, for each $p \in \mathcal{J}$ there is a unique bit-complement automorphism, say ϕ_p, such that $\phi_p(0^n) = p$. We show that in fact $\phi_p \in \mathcal{G}$. To see this let q be an arbitrary element of $\mathcal{J}$. By assumption we have

$$I_{0^n} = I_p = I_q = I.$$

Thus using identity (3.5) we obtain

$$I = I_{\phi_p(0^n)} = \phi_p(I_{0^n}) = \phi_p(I_q) = I_{\phi_p(q)}.$$

In turn, this implies the desired inequality $|\mathcal{J}| \leq |\mathcal{G}|$. To complete the proof of the claim it remains to prove that $|\mathcal{G}| \leq |\mathcal{J}|$. But this is obvious since the mapping $\phi \to \phi(0^n)$ is 1-1. The above considerations complete the proof of the first part of the lemma.

To prove the second assertion we note that F_n can be identified with an n-dimensional vector space over the finite field $Z_2 = \{0, 1\}$ of two elements. The standard basis of this vector space consists of the bit-complement automorphisms

$$\phi_{\{1\}}, \phi_{\{2\}}, \ldots, \phi_{\{n\}}.$$

Any other bit-complement automorphism ϕ_S can be written as the sum (which in this case is the regular composition of functions) of the automorphisms $\phi_{\{i\}}$, where $i \in S$. As a vector subspace $\mathcal{G}$ has a basis consisting of a fixed number of bit-complement automorphisms. Moreover, $|\mathcal{G}|$ is a power of 2. It follows that if $|\mathcal{J}|$ is nonempty it must be a power of 2.

The group $\mathcal{G}$ defined in Lemma 3.9.2 is called the automorphism group of the string I. Clearly, it depends on the string I. However we do not mention it explicitly in $\mathcal{G}$ in order to avoid unnecessary notational complications.

Lemma 3.9.3. *Let $\mathcal{G}$ be the automorphism group of the string I. If $|\mathcal{G}| = 2^l$ then I can be coded with a string of length 2^{n-l} and l bit-complement automorphisms.*

Proof. We continue using the notation of Lemma 3.9.2. The group $\mathcal{G}$ defined above has a natural action on the hypercube Q_n. For each $x \in Q_n$ let $x^{\mathcal{G}}$ be the orbit of x under $\mathcal{G}$, i.e.,

$$x^{\mathcal{G}} = \{\phi(x) : \phi \in \mathcal{G}\}.$$

For each x the stabilizer $\mathcal{G}_x$ of $\mathcal{G}$ under x is the identity group, where the stabilizer group [Wie64] is defined by

$$\mathcal{G}_x = \{\phi \in \mathcal{G} : \phi(x) = x\}.$$

By the well-known stabilizer theorem [Wie64]

$$|\mathcal{G}_x| \cdot |x^{\mathcal{G}}| = |\mathcal{G}|.$$

Since $|\mathcal{G}_x| = 1$ we obtain that all the orbits of $\mathcal{G}$ have exactly the same size, namely $|\mathcal{G}| = 2^l$, and since $|Q_n| = 2^n$, there are exactly

$$\frac{2^n}{|\mathcal{G}|} = 2^{n-l}$$

pairwise disjoint orbits.

The above discussion gives rise to the following "coding" algorithm which can be applied by the processors concerned in order to code the given configuration I with a new (generally shorter) string. Each processor that knows I can execute the following "coding algorithm" (i.e., processor p applies this algorithm to the string $I = I_p^n$).

Coding Algorithm:
Input: $I = \langle b_x : x \in Q_n \rangle$ is the given configuration, where b_x is the bit corresponding to processor x.

1. Compute the group $\mathcal{G}$ of bit-complement automorphisms ϕ such that

$$\forall p \in Q_n (I_p = I \Rightarrow I_{\phi(p)} = I).$$

   Assume that l is such that $|\mathcal{G}| = 2^l$.
2. Compute a set of l generators, i.e., a set $\phi_1, \ldots, \phi_l$ of bit-complement automorphisms which generate the group $\mathcal{G}$.
3. Compute the set of orbits of $\mathcal{G}$ in its natural action on Q_n. There are 2^{n-l} such orbits. For each orbit the processors choose a representative of the orbit in some canonical way, say lexicographically minimal; let $x(1), x(2), \ldots, x(2^{n-l})$ be the representatives chosen. Next the processor arranges them in increasing order according to the lexicographic order $\prec$, i.e., $x(1) \prec x(2) \prec \ldots \prec x(2^{n-l})$.
4. The code of I is defined to be the sequence $\langle I'; \phi_1, \phi_2, \ldots, \phi_l \rangle$, where I' is the sequence of bits of length 2^l given by

$$I' := b_{x(1)} b_{x(2)} \cdots b_{x(2^{n-l})}$$

   and

$$\phi_1, \phi_2, \ldots, \phi_l$$

   is a sequence of bit-complement automorphisms generating the group $\mathcal{G}$.

Output: $\langle I'; \phi_1, \phi_2, \ldots, \phi_l \rangle$.

It remains to prove that a processor can reconstruct I from its encoding. To do this it executes the following decoding algorithm.

Decoding Algorithm:
Input: $\langle I'; \phi_1, \phi_2, \ldots, \phi_l \rangle$, where I' is a string of length 2^{n-l} and $\phi_1, \phi_2, \ldots, \phi_l$ are bit-complement automorphisms.

1. Let $\mathcal{G}$ be the group generated by these automorphisms. Compute the set of orbits of $\mathcal{G}$ in its natural action on Q_n. There are 2^{n-l} such orbits. For each orbit choose as representative of the orbit the lexicographically minimal string in the orbit. Let $x(1), x(2), \ldots, x(2^{n-l})$ be the representatives chosen. Next the processor arranges them in increasing order according to the lexicographic order $\prec$, i.e., $x(1) \prec x(2) \prec \ldots \prec x(2^{n-l})$.
2. The coding algorithm guarantees that $I' = b_{x(1)} b_{x(2)} \cdots b_{x(2^{n-l})}$. Hence we can "fill-in" the remaining bits to form the string I since $b_x = b_y$ for x, y in the same orbit.

Output: I.

Indeed, by definition of the group $\mathcal{G}$ we have that for all $\phi \in \mathcal{G}$, $\phi(I) = I$. Hence by Lemma 3.9.1

$$\forall x \in Q_n \forall \phi \in \mathcal{G} (b_x = b_{\phi(x)}),$$

where $I = \langle b_x : x \in Q_n \rangle$. This explains why the decoding algorithm works.

Now we can prove the following theorem which significantly improves the upper bound of Theorem 3.9.3.

Theorem 3.9.4 ([KK97]). *There is an algorithm computing every boolean function $f \in \mathcal{B}_N$ (which is invariant under all bit-complement automorphisms) on the canonically labeled hypercube Q_n, $N = 2^n$, with bit complexity $O(N \cdot \log^4 N)$.*

Proof. For each fixed string $x = x_{i+1} \cdots x_n$ of bits of length $n - i$ let

$$Q_i(x) = \{u_1 \cdots u_i x : u_1, \ldots, u_i \in \{0, 1\}\}.$$

For each processor p represented by the sequence $p_1 \cdots p_n$ of bits the i-th hypercube of p is defined to be $Q_i(p_{i+1} \cdots p_n)$. Clearly we have that

$$Q_i(x) = Q_{i-1}(0x) \cup Q_{i-1}(1x).$$

Initially, I_p^0 = "input bit to processor p" and each processor declares itself leader of its 0-dimension hypercube $Q_0(p) = \{p\}$. The leaders at the i-th step of the algorithm are among those processors whose "view" I_p^i of their i-th hypercube is lexicographically maximal among the set of strings I_q^i. Assume by induction that we have elected leaders for the $(i-1)$-th stage of the algorithm and that each processor has a path to such a leader along its

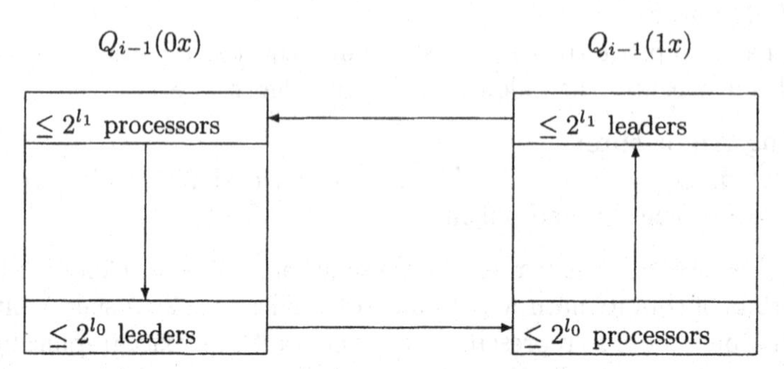

Fig. 3.1. Exchange of views among leaders in hypercube $Q_i(x)$

hypercube with edges $\leq i-1$. We show how to extend these assumptions to the i-th stage of the algorithm. The i-th stage of the new algorithm consists of the following steps.

1. The leader processors (elected at the $(i-1)$st stage) send their encoded views of their hypercube to their neighbors along the i-th dimension.
2. The processors of the opposite hypercube receiving the views route them to their leaders. (By induction hypothesis, all the processors know routes to their leaders along their hypercube; hence they can transmit the view received along such a route, for example the lexicographically minimal one.) Leaders that receive such encoded views decode the messages received as in Lemma 3.9.3, compute the corresponding views of their neighbors along their i-th edge and append it to their own view thus forming views at step i. To compute the view of their neighbors along their i-th edge each leader ℓ executes the following algorithm.
   a) Let ℓ's neighbor along the i-th edge be p and let $1 \leq k_1, \dots, k_r \leq i-1$ be a path along p's subcube leading to a leader ℓ' in this subcube (by the induction hypothesis we can assume that such a path is known to p). By the previous argument the view $I_{\ell'}^{i-1}$ of ℓ' is known to ℓ (see Figure 3.2). Now ℓ requests this path from its neighbor p.
   b) Since $\phi_{\{k_1,\dots,k_r\}}(\ell') = p$ it is clear that ℓ can compute p's view via the identity
   $$\phi_{\{k_1,\dots,k_r\}}(I_{\ell'}^{i-1}) = I_p^{i-1}.$$
3. If I_0 is the leader view in hypercube $Q_{i-1}(0x)$ and I_1 is the leader view in hypercube $Q_{i-1}(1x)$ then the leader view in hypercube $Q_i(x)$ will be
   $$\begin{cases} I_0 I & \text{if } I_0 \succ I_1 \\ I_1 I & \text{if } I_1 \succ I_0 \\ I_0 I_1 & \text{if } I_0 = I_1 \end{cases}$$
   for some string of length 2^{i-1} ($\succ$ denotes the lexicographic ordering). If L_0 is the set of leaders in hypercube $Q_{i-1}(0x)$ and L_1 is the set of

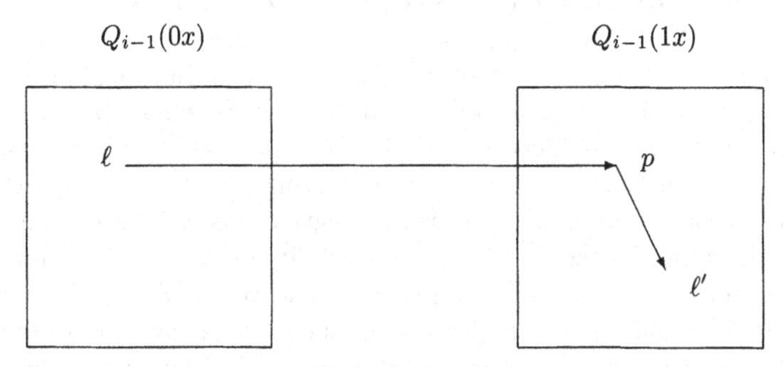

Fig. 3.2. In subcase 2(a), ℓ sends its view to p which routes it to ℓ'.

leaders in hypercube $Q_{i-1}(1x)$ then the set of leaders of the i-th stage will be among either L_0 or L_1 or $L_0 \cup L_1$, depending on the lexicographic comparison of I_0 and I_1. It follows that all the processors of $Q_i(x)$ will know paths to these new leaders. (Indeed, if $p \in Q_i(x)$ then either $p \in Q_{i-1}(0x)$ or $p \in Q_{i-1}(1x)$. Say, $p \in Q_{i-1}(0x)$. By induction p knows a path to a leader at the $i-1$st stage. But by the previous argument this leader knows a path to a leader at the i-th stage.)

4. Return to 1 and iterate, for $i = 1, 2, \ldots, \log N$.

The mechanism for exchanging views at the i-th iteration of the above algorithm is depicted in Figure 3.1.

Now we estimate the bit complexity of the algorithm. The coding and decoding algorithms are "internal" and do not contribute anything to the total bit complexity. Suppose there are $\leq 2^l$ leaders elected at the i-th step of the algorithm. There exists a message w of length 2^{i-l} and a sequence of $l \leq i$ bit-complement automorphisms of the hypercube Q_i which "code" the view I_p^i. Since only the leaders transmit messages at the i-th step while the rest of the processors are "routing" messages to the leaders (processors are always at a distance $\leq i$ from a leader, since the diameter of the i-th hypercube is i), the total bit complexity at the i-th step of the algorithm is $O(2^i \cdot i^3)$ (since each encoded view consists of at most i bit-complement automorphisms and each bit-complement automorphism can be coded with i bits). Clearly, for each $i \leq \log N$ this algorithm is applied to 2^{n-i} subcubes simultaneously. Since the algorithm is iterated $\log N$ times it follows that the bit complexity of the new algorithm is

$$\sum_{i=1}^{\log N} 2^{n-i} \cdot O(2^i \cdot i^3) = O(N \cdot \log^4 N).$$

3.10 Historical and Bibliographical Remarks

The most important application of invariance groups is in providing a precise upper bound on the circuit complexity of boolean functions based on their degree of symmetry (see Theorem 3.8.3). There are several interesting open problems. Two such problems concern improving the $2^{O(\log n)}$ algorithm for testing the representability of an arbitrary group, as well as extending the logspace algorithm for testing the representability of cyclic groups to a larger class of permutation groups (Theorem 3.5.1). The work of Furst, Hopcroft and Luks [FHL80] should play a major role in such an endeavor. For additional information and results on the representation problem for boolean functions the reader should consult the papers [Kis99] and [Xia00].

The computability problems studied in Section 3.9 are a special case of the problem of collecting input data in a deterministic, distributed environment. There are several papers on trade-offs for input collection on anonymous networks as well as studies for randomized evaluation of boolean functions on anonymous rings [AAHK88]. For more details and references on anonymous networks the reader is referred to the survey article [Kra97b].

3.11 Exercises

Exercise 3.11.1 ([CK91]). ($\star$) Prove Theorem 3.2.1.
HINT: For the identity group take $L = 0^*1^*$, for the dihedral group $L = 0^*1^*0^* \cup 1^*0^*1^*$, for the reflection group $L = 0^*1^*0^*$. For the cyclic groups, if $n = 2$ take $L = (01 \cup 10)0^*1^*$, and if $n \geq 6$ then take $L = (L_0^1 \cup L_1^1) \cap L^2$, where

$$L_0^1 = 1^*0^*1^* \cup 0^*1^*0^* \cup 101000^*1 \cup 0^*1101000^*$$

$$L_1^1 = 0^*011010 \cup 0^*001101 \cup 10^*00110 \cup 010^*0011$$

and L^2 is the language $\overline{10^*00101}$. Notice that for $3 \leq n \leq 5$, if $C_n \subseteq \text{AUT}(f) \subseteq D_n$ then $\text{AUT}(f) = D_n$. For the hyperoctahedral group let L consist of the set of all finite strings $x = (x_1, \ldots, x_n)$ such that for some $i \leq n/2$, $x_{2i-1} = x_{2i}$.

Exercise 3.11.2. Use the fact that for any permutation group G not containing A_n, $|S_n : G| \geq n$ [Wie64] to conclude that A_n is not isomorphic to the invariance group $\text{AUT}(f)$ of any $f \in \mathcal{B}_n$. However, A_n is isomorphic to the invariance group $\text{AUT}(f)$ for some boolean function $f \in \mathcal{B}_{n(\log n+1)}$ (see Theorem 3.2.4).

Exercise 3.11.3. One can generalize the notion of invariance group for any language $L \subseteq (k+1)^*$ by setting $L_n = L \cap \{0, \ldots, k\}^n$ and $\text{AUT}(L_n)$ to be the set of permutations $\sigma \in S_n$ such that

$$\forall x_1, \ldots, x_n \in \{0, 1, \ldots, k\}(x_1, \ldots, x_k \in L_n \iff x_{\sigma(1)}, \ldots, x_{\sigma(n)} \in L_n).$$

Show that for all n, there exist groups $G_n \leq S_n$ which are strongly representable as $G_n = \text{AUT}(L_n)$ for some $L \subseteq \{0, 1, \ldots, n-1\}^n$ but which are not so representable for any language $L' \subseteq \{0, 1, \ldots, n-2\}^n$.
HINT: The alternating group $A_n = \text{AUT}(L_n)$, where $L_n = \{w \in \{0, \ldots, n-1\}^n : \sigma_w \in A_n\}$, where $\sigma_w : i \mapsto w(i-1)+1$. By a variant of the previous argument, A_n is not so representable by any language $L' \subseteq \{0, 1, \ldots, n-2\}^n$.

Exercise 3.11.4. Compared to the difficulties regarding the question of representing permutation groups $G \leq S_n$ in the form $G = \text{AUT}(f)$, for some $f \in \mathcal{B}_n$, it is interesting to note that a similar representation theorem for the groups $S(x) = \{\sigma \in S_n : x^\sigma = x\}$, where $x \in 2^n$, is relatively easy. It turns out that these last groups are exactly the permutation groups which are isomorphic to $S_k \times S_{n-k}$ for some k.
HINT: Given $x \in 2^n$ let

$$X = \{i : 1 \leq i \leq n \text{ and } x_i = 0\}, Y = \{i : 1 \leq i \leq n \text{ and } x_i = 1\}.$$

It is then easy to see that $S(x)$ is isomorphic to $S_X \times S_Y$. In fact, $\sigma \in S(x)$ if and only if $X^\sigma = X$ and $Y^\sigma = Y$.

Exercise 3.11.5. Notice the importance of assuming $m < n$ in the definition of weak representability. If $m = n$ were allowed then every permutation group would be weakly representable.
HINT: Given any permutation group $G \leq S_n$ define the function f as follows:

$$f(x_1, \ldots, x_n) = \begin{cases} 0 \text{ if } (x_1, \ldots, x_n) \in G \\ 1 \text{ otherwise} \end{cases}$$

(here, we think of $(x_1, \ldots, x_n)$ as the function $i \to x_i$ in n^n) and notice that for all $\sigma \in S_n$, $\sigma \in \text{AUT}(f)$ if and only if $\forall \tau \in S_n(\tau \in G \Leftrightarrow \tau\sigma \in G)$. Hence $G = \text{AUT}(f)$, as desired.

Exercise 3.11.6. Incidentally, it is not known if the $n(1+\log n)$ upper bound of Theorem 3.2.4 can be improved. However the idea of the proof of Theorem 3.2.4 can also be used to show that for any alphabet Σ, if $L \subseteq \Sigma^n$ then $\text{AUT}_n(L)$ (the set of permutations in S_n "respecting" the language L) is isomorphic to $\text{AUT}_{ns}(L')$, for some $L' \subseteq \{0,1\}^{ns}$, where $s = 1 + \log|\Sigma|$.

Exercise 3.11.7. The well-known graph non-isomorphism problem (NGIP) is related to the above group representation problem. Indeed, let

$$G = (\{v_1, \ldots, v_n\}, E_G), H = (\{u_1, \ldots, u_n\}, E_H)$$

be two graphs each on vertices and let $\text{ISO}(G, H) \leq S_{n+3}$ have generators σ satisfying:

$$\forall 1 \leq i, j \leq n(E_G(v_i, v_j) \Leftrightarrow E_H(u_{\sigma(i)}, u_{\sigma(j)})),$$

and in addition the permutation $n + i \to \sigma(n + i)$, $i = 1, 2, 3$, belongs to the group $C_3 = (n + 1, n + 2, n + 3)$. It is easy to show that if G, H are isomorphic then there exists a group $K \leq S_n$ such that $\mathrm{ISO}(G, H) = K \times C_3$. On the other hand, if G, H are not isomorphic then $\mathrm{ISO}(G, H) = \langle id_{n+3} \rangle$. As a consequence of the non-representability of C_3, and the representability theorem of direct products, it follows that G, H are not isomorphic if and only if $\mathrm{ISO}(G, H) = \langle id_{n+3} \rangle$.

Exercise 3.11.8. An idea similar to that used in the proof of the representation theorem can also be used to show that for any representable permutation groups $G < H \leq S_n$,

$$2 \cdot |\{h \in \mathcal{B}_n : H = \mathrm{AUT}(h)\}| \leq |\{g \in \mathcal{B}_n : G = \mathrm{AUT}(g)\}|.$$

HINT: Assume that G, H are as above. Without loss of generality we may assume that there is no representable group K such that $G < K < H$. As in the proof of the representation theorem there exist $x, y \in 2^n$ such that $x = y \bmod H, x \neq y \bmod G$. Define two boolean functions $h_b \in \mathcal{B}_n$, $b = 0, 1$, as follows for $w \in 2^n$,

$$h_b(w) = \begin{cases} h(w) & \text{if } w \neq x \bmod G, w \neq y \bmod G \\ b & \text{if } w = x \bmod G \\ \bar{b} & \text{if } w = y \bmod G \end{cases}$$

Since $G \leq \mathrm{AUT}(h_b) < S(h)$, it follows from the above definition that each $h \in \mathcal{B}_n$ with $H = \mathrm{AUT}(h)$ gives rise to two distinct $h_b \in \mathcal{B}_n$, $b = 0, 1$, such that $G = \mathrm{AUT}(h_b)$. Moreover it is not difficult to check that the mapping $h \to \{h_0, h_1\}$, where $H = \mathrm{AUT}(h)$, is 1-1. It is now easy to complete the proof of the assertion.

Exercise 3.11.9. ($\star$) Prove all the assertions made in Section 3.3.

Exercise 3.11.10. The automorphism group of a directed graph may not be 2-representable.
HINT: Look at the cyclic groups C_3, C_4, C_5 from Section 3.3.

Exercise 3.11.11. In this exercise we develop representability theorems for wreath products of permutation groups. For details on proofs the reader may consult [Kis99], [CK91]. Let $G \leq S_m, H \leq S_n$. Then

1. ($\star$) G and H are k-representable $\Rightarrow G \wr H$ is k-representable.
2. $G \wr H$ is 2-representable $\Rightarrow H$ is representable.
3. $G \wr H$ is 2-representable and $2^n < m \Rightarrow G$ is weakly representable.
4. For p prime, a p-Sylow subgroup P of S_n is representable $\Leftrightarrow p \neq 3, 4, 5$.

Exercise 3.11.12. It is easy to see that in general $|S_n : G|$ and $\Theta_n(G)$ can diverge widely. For example, let $f(n) = n - \log n$ and let G be the group $\{\sigma \in S_n : \forall i > f(n)(\sigma(i) = i)\}$. It is then clear that $\Theta_n(G) = (f(n) + 1) \cdot 2^{\log n}$ is of order n^2, while $|S_n : G|$ is of order $n^{\log n}$. Another simpler example is obtained when G is the identity subgroup of S_n.

Exercise 3.11.13. The converse of part (1) of the Theorem 3.11.11 is not necessarily true. This is easy to see from the wreath product $A_3 \wr S_2$ which is representable, but that A_3 is not.
HINT: Consider the language

$$L = \{001101,\ 010011,\ 110100,\ 001110,\ 100011,\ 111000\} \subseteq 2^6.$$

We already proved that A_3 is not representable. We claim that $A_3 \wr S_2 = \text{AUT}_6(L)$. Consider the 3-cycle $\tau = (\{1,2\}, \{3,4\}, \{5,6\})$. It is easy to see $A_3 \wr S_2$ consists of the 24 permutations σ in S_6 which permute the two-element sets $\{1,2\}$, $\{3,4\}$, $\{5,6\}$ like in the three-cycles τ, τ^2, τ^3. A straightforward (but tedious) computation shows that $\text{AUT}_6(L)$ also consists of exactly the above 24 permutations.

Exercise 3.11.14. Another class of examples of nonrepresentable groups is given by the direct products of the form $A_m \times G$, $G \times A_m$, where G is any permutation group acting on a set which is disjoint from $\{1, 2, \ldots, m\}$, $m \geq 3$.

Exercise 3.11.15 (Open Problem). At present, we do not know how to efficiently test the representability of arbitrary abelian groups (or other natural classes of groups such as solvable, nilpotent, etc.)

Exercise 3.11.16. If a given abelian group K can be decomposed into disjoint cyclic factors, then we have the following NC algorithm for testing representability: (1) use an NC algorithm [BLS87], [MC85], [Mul86] to "factor" K into its cyclic factors and then (2) apply the "cyclic-group" algorithm to each of the cyclic factors of K. In view of the result below the group K is representable exactly when each of its disjoint, cyclic factors is.

1. Let $G \leq S_m, H \leq S_n$ be permutation groups. Then $G \times H$ is representable $\Leftrightarrow$ both G, H are representable.

Exercise 3.11.17. ($\star$) Show that

1. there is no regular language L such that for all but a finite number of n we have that $\text{AUT}_{2n}(L) = (S_{2n})_{\{1,2,\ldots,n\}}$.
2. there is a regular language L such that for all n we have that $\text{AUT}_{2n}(L) = (S_{2n})_{\{2i:i \leq n/2\}}$.

Exercise 3.11.18. The group S_n is generated by the cyclic permutation $c_n = (1, 2, \ldots, n)$ and the transposition $\tau = (1, 2)$ (in fact any transposition will do) [Wie64].

Exercise 3.11.19 ([CK91]). ($\star$) Consider a term $t(x,y)$ built up from the variables x, y by concatenation The number of occurrences of x and y in the term $t(x,y)$ is called the length of t and is denoted by $|t|$. For any permutations σ, τ let the permutation $t(\sigma,\tau)$ be obtained from the term $t(x,y)$ by substituting each occurrence of x, y by σ, τ, respectively, and interpreting concatenation as product of permutations. A sequence $\sigma = \langle \sigma_n : n \geq 1 \rangle$ of permutations is term-generated by the permutations c_n, τ if there is a term $t(x,y)$ such that for all $n \geq 2$, $\sigma_n = t(c_n, \tau)$. Show that

1. Let $\sigma = \langle \sigma_n : n \geq 1 \rangle$ be a sequence of permutations which is term-generated by the permutations $c_n = (1,2,\ldots,n), \tau = (1,2)$. Then for any regular language L, L^σ is also regular.
2. For any term t of length $|t|$ the problem of testing whether for a regular language L, $L = L^\sigma$, where $\sigma = \langle \sigma_n : n \geq 1 \rangle$ is a sequence of permutations generated by the term t via the permutations $c_n = (1,2,\ldots,n), \tau = (1,2)$, is decidable; in fact it has complexity $O(2^{|t|})$.

Exercise 3.11.20 ([Cla91]). Prove Theorem 3.6.2.
HINT: Show that if A is not the identity matrix then the number of orbits of A in $GL(n,2)$ is at most $2^{n-1} + 2^{n-2} = 3 \cdot 2^{n-2}$.

Exercise 3.11.21 ([FKPS85]). ($\star$) Besides equality, the language of graph theory has a single binary relation I. The axioms of the theory of loopless, undirected graphs LUG are: $\forall x \neg I(x,x)$ and $\forall x, y(I(x,y) \leftrightarrow I(y,x))$. For arbitrary but fixed $0 < p < 1$, let $G_n = (V,E)$ run over random graphs of n nodes such that $\Pr[(i,j) \in E] = p$. Let $\phi_{r,s}$ denote the sentence: for any distinct $x_1, \ldots, x_r, y_1, \ldots, y_s$ there is an x adjacent to all the x_is and none of the y_is. Show that

1. any two models of LUG satisfying all sentences $\phi_{r,s}$ are isomorphic,
2. the set $\{\phi_{r,s}\}$ is complete,
3. $\lim_{n\to\infty} \Pr[G_n \models \phi_{r,s}] = 1$,
4. for any sentence ϕ of LUG, $\lim_{n\to\infty} \Pr[G_n \models \phi = 0$ or 1.

Exercise 3.11.22. For any language L and any sequence $\sigma = \langle \sigma_n : n \geq 1 \rangle$ of permutations such that each $\sigma_n \in S_n$ we define the language $L_n^\sigma = \{x \in 2^n : x^{\sigma_n} \in L_n\}$. For each n let $G_n \leq S_n$ and put $\mathbf{G} = \langle G_n : n \geq 1 \rangle$. Define $L^{\mathbf{G}} = \bigcup_{\sigma_n \in G_n} L_n^{\sigma_n}$. For each $1 \leq k \leq \infty$ let $\mathbf{F}_k$ be the class of functions $n^{c \log^{(k)} n}$, $c > 0$, where $\log^{(1)} n = \log n$, $\log^{(k+1)} n = \log \log^{(k)} n$, and $\log^{(\infty)} n = 1$. Clearly, $\mathbf{F}_\infty$ is the class $\mathbf{P}$ of polynomial functions. We also define $\mathbf{F}_0$ as the class of functions 2^{cn}, $c > 0$. Let $\mathbf{L}(\mathbf{F}_k)$ be the set languages $L \subseteq \{0,1\}^*$ such that there exists a function $f \in \mathbf{F}_k$ satisfying $\forall n(|S_n : \mathrm{AUT}_n(L)| \leq f(n))$. We will also use the notation $L(\mathbf{EXP})$ and $L(\mathbf{P})$ for the classes $\mathbf{L}(\mathbf{F}_0)$ and $\mathbf{L}(\mathbf{F}_\infty)$, respectively. Show that

1. for any $0 \leq k \leq \infty$ and any language $L \in \mathbf{L}(\mathbf{F}_k)$,
2. $\mathbf{L}(\mathbf{F}_k)$ is closed under boolean operations and homomorphisms,

3. $(L \cdot \Sigma) \in \mathbf{L}(\mathbf{F}_k)$,
4. $L^\sigma \in \mathbf{L}(\mathbf{F}_k)$, where $\sigma = \langle \sigma_n : n \geq 1 \rangle$, with each $\sigma_n \in S_n$,
5. if $|S_n : N_{S_n}(G_n)| \leq f(n)$ and $f \in \mathbf{F}_k$ then $L^{\mathbf{G}} \in \mathbf{L}(\mathbf{F}_k)$, where $\mathbf{G} = \langle G_n : n \geq 1 \rangle$.
6. $L \in L(\mathbf{P})$ and $p \in \mathbf{P} \Rightarrow |S_{p(n)} : S_{p(n)}(L)| = n^{O(1)}$.
7. $L^1, L^2 \in L(\mathbf{EXP}) \Rightarrow L = \{xy : x \in L^1, y \in L^2, l(x) = l(y)\} \in L(\mathbf{EXP})$.
8. $\mathbf{L}(\mathbf{F}_\infty) = L(\mathbf{P}) \subset \ldots \subset \mathbf{L}(\mathbf{F}_{k+1}) \subset \mathbf{L}(\mathbf{F}_k) \subset \cdots \subset L(\mathbf{EXP}) = \mathbf{L}(\mathbf{F}_0)$,
9. $\mathbf{REG} \cap L(\mathbf{P}) \neq \emptyset$, $\mathbf{REG} - L(\mathbf{EXP}) \neq \emptyset$, $L(\mathbf{P}) - \mathbf{REG} \neq \emptyset$.

Exercise 3.11.23. A family $\langle p_n : n \geq 1 \rangle$ of multivariate polynomials in the ring $Z_2[x_1, \ldots, x_n]$ is of polynomial index if $|S_n : \mathrm{AUT}(p_n)| = n^{O(1)}$. Show that for such a family of multivariate polynomials there is a family $\langle q_n : n \geq 1 \rangle$ of multivariate polynomials (in $Z_2[x_1, \ldots, x_n]$) of polynomial length such that $p_n = q_n$.

Exercise 3.11.24. Because of the limitations of families of groups of polynomial index proved in the claim above, we obtain a generalization of the principal results of [FKPS85]. Namely, for $L \subseteq \{0,1\}^*$ let $\mu_L(n)$ be the least number of input bits which must be set to a constant in order for the resulting language $L_n = L \cap \{0,1\}^n$ to be constant (see [FKPS85] for more details). Then we can prove the following result. If $L \in L(\mathbf{P})$ then $\mu_L(n) \leq (\log n)^{O(1)} \iff L \in \mathrm{AC}^0$.

Exercise 3.11.25. Our characterization of permutation groups of polynomial index given during the proof of Theorem 3.7.3 can also be used to determine the parallel complexity of the following problem concerning "weight-swapping". Let $\mathbf{G} = \langle G_n : n \in \mathbf{N} \rangle$ denote a sequence of permutation groups such that $G_n \leq S_n$, for all n. By SWAP($\mathbf{G}$) we understand the following problem:

Input. $n \in \mathbf{N}$, $a_1, \ldots, a_n$ positive rationals, each of whose (binary) representations is of length at most n.
Output. A permutation $\sigma \in G_n$ such that for all $1 \leq i \leq n$, $a_{\sigma(i)} + a_{\sigma(i+1)} \leq 2$, if such a permutation exists, and the response "NO" otherwise.

Show that for any sequence $\mathbf{G}$ of permutation groups of polynomial index, the problem SWAP($\mathbf{G}$) is in non-uniform NC^1.

Exercise 3.11.26. Deduce from the proof of Theorem 2.2.1 that the number of boolean functions $f \in \mathcal{B}_n$ which can be computed by a circuit of size s with n input gates is $O(s^{2s})$.

Exercise 3.11.27. For any sequence $\mathbf{G} = \langle G_n : n \geq 1 \rangle$ of permutation groups $G_n \leq S_n$ it is possible to find a language L such that

$$L \notin SIZE(\sqrt{\Theta(G_n)}), \text{ and } \forall n(\mathrm{AUT}(L_n) \supseteq G_n).$$

HINT: By Exercise 3.11.26 $|\{f \in \mathcal{B}_n : c(f) \leq q\}| = O(q^{2q}) = 2^{O(q \log q)}$, where $c(f)$ is the size of a circuit with minimal number of gates computing

f. Hence, if $q_n \to \infty$ then $|\{f \in \mathcal{B}_n : c(f) \leq q_n\}| < 2^{q_n^2}$. In particular, setting $q_n = \sqrt{\Theta(G_n)}$ we obtain

$$|\{f \in \mathcal{B}_n : c(f) \leq \sqrt{\Theta(G_n)}\}| < 2^{\Theta(G_n)} = |\{f \in \mathcal{B}_n : \text{AUT}(f) \supseteq G_n\}|.$$

It follows that for n big enough there exists an $f_n \in \mathcal{B}_n$ such that $\text{AUT}(f_n) \supseteq G_n$ and $c(f_n) > \sqrt{\Theta(G_n)}$.

Exercise 3.11.28. In this exercise we develop the notion of structure forest used extensively in Section 3.7

1. A structure tree for a transitive permutation group G acting on Ω can be constructed as follows. Take a strictly decreasing sequence
$$B_0 := \Omega \supset B_1 \supset \cdots \supset B_{r-1} \supset B_r = \{x\},$$
of blocks of G with $B_0 := \Omega$ and B_r a singleton. Then the blocks $\{B_i^\sigma : i = 0, \ldots, r, \sigma \in G\}$ form a tree with respect to inclusion whose root is Ω and leaves are the singletons $\{x\}$, where $x \in \Omega$. Each element of the i-th level, denoted by $\mathcal{L}_i$, of this tree, can be written as the disjoint union of elements of the $i+1$st level. The number of elements of this union is a constant k_i which is independent of the level i. Moreover, $|\{\mathcal{L}_i\}| = k_0 k_1 \cdots k_{i-1}$. In particular, $n = |\{\mathcal{L}_{r+1}\}| = k_0 k_1 \cdots k_r$.
2. For each block B let $G_{\{B\}} = \{\sigma \in G : B^\sigma = B\}$ be the stabilizer of B. Let $L(B)$ denote the action of $G_{\{B\}}$ on B. Let $\mathcal{B}$ be the set of blocks which are sons of B in the above structure tree. Denote by $H(B)$ the action of $G_{\{B\}}$ on $\mathcal{B}$. Then we have that $L(B) \leq Sym(B), H(B) \leq Sym(\mathcal{B})$. Show that the groups $G_{\{B\}}$ of each level are conjugate of each other.
3. For each i let K_i stand for the pointwise stabilizer of level i, i.e.,
$$K_i = \{\sigma \in G : \forall B \in \mathcal{L}_i (B^\sigma = B)\}$$
Show that K_i is a normal subgroup of G and $\frac{G}{K_i} \leq Sym(\mathcal{L}_i)$.
HINT: For any groups $N \leq M$ consider the set $\mathcal{C}$ of left cosets of N with respect to M. Show that the kernel of the homomorphism $m \to mN$ is $\bigcap_{m \in M} m^{-1}Nm$ which is also the largest normal subgroup of M contained in N.
4. Show that for $B \in \mathcal{L}_i$, $K_i \leq L(B)^{|\mathcal{L}_i|}, G \leq L(B) \wr (G/K_i)$.
5. If the group is not transitive break-up Ω into disjoint orbits. The action of the group on each of these orbits gives rise to a structure tree. The totality of these trees is called a structure forest.

Exercise 3.11.29 (Open Problem). For every permutation group $G \leq S_n$ let $k_n(G)$ denote the smallest integer k such that G is isomorphic to the invariance group of f. By Theorem 3.2.4, $k(G)n(1+\log n)$. Determine tighter bounds.

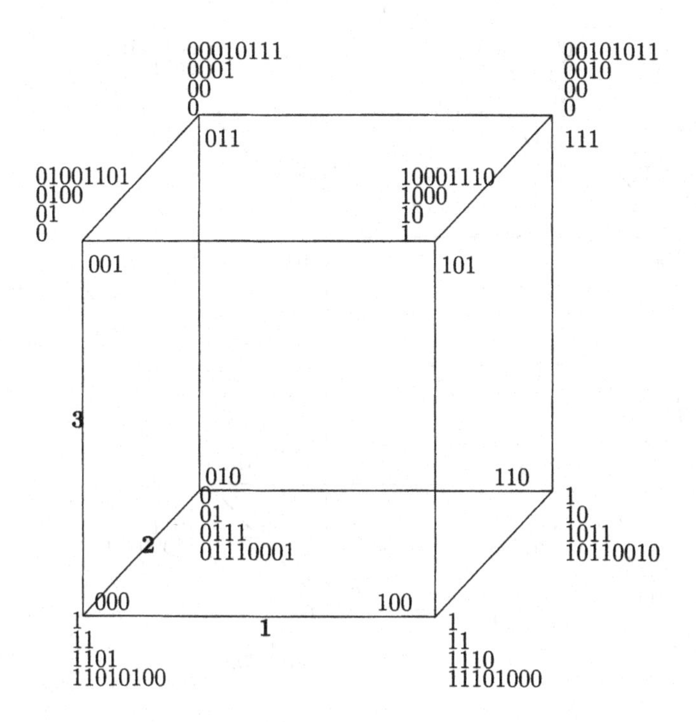

Fig. 3.3. Executing the $O(N^2)$ algorithm on a three-dimensional hypercube

Exercise 3.11.30 (Open Problem). $\mathcal{R}_k^n$ is the class of k-representable permutation groups on n letters. It is clear that $\mathcal{R}_k^n \subseteq \mathcal{R}_{k+1}^n$. Is $\mathcal{R}_k^n$ a proper hierarchy?

Exercise 3.11.31.
(1) Consider the three-dimensional hypercube depicted in Figure 3.3 with the input indicated. Let us trace the behavior of the algorithm on the given input for the bottom-left processor, say $p = 000$. Let $p_1 = 100, p_2 = 010, p_3 = 001$ be the neighbors of p along dimensions $1, 2, 3$, respectively. Following the algorithm the successive views of processor p are

$$\begin{aligned} I_p^0 &= 1 \\ I_p^1 &= I_p^0 I_{p_1}^0 = 11 \\ I_p^2 &= I_p^1 I_{p_2}^1 = 1101 \\ I_p^3 &= I_p^2 I_{p_3}^2 = 11010100. \end{aligned}$$

(2) Let b_p denote the input bit to processor p. A similar reasoning shows that $I_{111}^3 = 00101011$. We can show that I_{000}^3, I_{111}^3 are identical up to automorphism. Indeed, take the unique automorphism which maps 000 into 111,

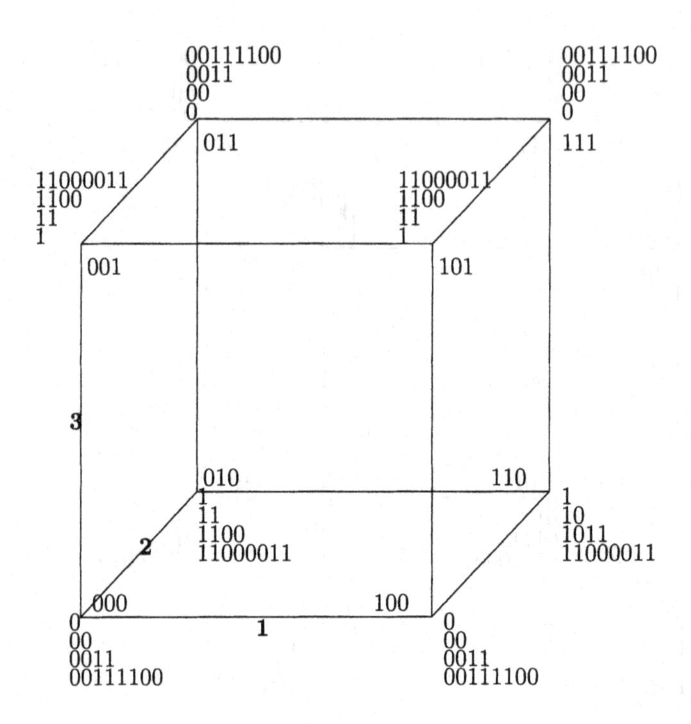

Fig. 3.4. Example illustrating the coding of views

namely $\phi = \phi_{\{1,2,3\}}$. Then we have

$$\begin{aligned} I^3_{000} &= b_{000}b_{100}b_{010}b_{110}b_{001}b_{101}b_{011}b_{111} \\ &= 11010100 \end{aligned}$$

and

$$\begin{aligned} I^3_{111} &= b_{\phi(000)}b_{\phi(100)}b_{\phi(010)}b_{\phi(110)}b_{\phi(001)}b_{\phi(101)}b_{\phi(011)}b_{\phi(111)} \\ &= b_{111}b_{011}b_{101}b_{001}b_{110}b_{010}b_{100}b_{000} \\ &= 00101011. \end{aligned}$$

Exercise 3.11.32. (1) Consider a three-dimensional hypercube with input as depicted in Figure 3.3. After the third iteration of the algorithm the view of processor 000 is $I^3_{000} = 11010100$. From its view I^3_{000} the processor 000 can reconstruct the views of all other processors. 000 is the only processor with this view. The group $\mathcal{G} = \{\phi : I^3_{000} = I^3_{\phi(000)}\}$ defined by Equation (3.6) is easily seen to be the trivial identity group generated by the identity automorphism e. The group has a natural action on the set of processors and gives rise to eight orbits:

$$\{000\}, \{100\}, \{011\}, \{111\}, \{010\}, \{110\}, \{101\}, \{100\}$$

Now the code for processor 000 is $\langle 11010100, e \rangle$.
(2) Consider a three-dimensional hypercube with input as depicted in Figure 3.4. After the third iteration of the algorithm the view of processor 000 is $I^3_{000} = 00111100$. From its view I^3_{000} the processor 000 can reconstruct the views of all other processors. There are four processors with this view, namely $000, 100, 011, 111$. The group $\mathcal{G} = \{\phi : I^3_{000} = I^3_{\phi(000)}\}$ defined by Equation (3.6) is easily seen to be generated by the automorphisms $\phi_{\{1\}}, \phi_{\{2,3\}}$ and has size exactly $4 = 2^2$. The group has a natural action on the set of processors and gives rise to two orbits: $\{000, 100, 011, 111\}, \{001, 101, 010, 110\}$. Now the code for processor 000 is $\langle 01, \phi_{\{1\}}, \phi_{\{2,3\}} \rangle$, where 0 is the input bit of processor 000 and 1 is the input bit of processor 010.
(3) Here is how the decoding algorithm works. Suppose that a processor receives the code $\langle 01, \phi_{\{1\}}, \phi_{\{2,3\}} \rangle$. The processor constructs the orbit of the lexicographically minimal processor, namely 000. This is the orbit $\{000, 100, 011, 111\}$. Since $b_{000} = 0$ we know that

$$b_{000} = b_{100} = b_{011} = b_{111} = 0.$$

The remaining processors also form an orbit and the lexicographically minimal processor among them is 001. Since $b_{001} = 1$ we know that

$$b_{001} = b_{101} = b_{010} = b_{110} = 1.$$

Hence the decoded view of the processor is 00111100, as desired.

Exercise 3.11.33. The input configuration depicted in Figure 3.3 has a single leader, namely processor 100 with view 11101000. The input configuration depicted in Figure 3.4 has four leaders, namely $100, 110, 011, 101$ with view 11000011. Notice that all processors can check from their view who and where the leaders are with respect to themselves. Now assume that the configuration depicted in Figure 3.3 is in the left-most hypercube in Figure 3.1 and the configuration depicted in Figure 3.4 is in the right-most hypercube in Figure 3.1. It is now clear that if the leaders of the corresponding three-dimensional hypercubes transmit their encoded views along dimension 4 all the processors of the four dimensional hypercube will be able to form views of the entire four dimensional hypercube.

Exercise 3.11.34 ([KK97]). On the canonically labeled hypercube Q_n, every symmetric function can be computed in $O(N \cdot \log^2 N)$ bits. Moreover the threshold function Th_k can be computed in $O(N \cdot \log N \cdot \log k)$ bits, where $k \leq N$.

Exercise 3.11.35 ([KK92]). ($\star$) Theorem 3.9.4 generalizes to arbitrary anonymous Cayley networks.

1. Show that if G is a set of generators for a group $\mathcal{G}$ then a boolean function f is computable on the naturally labeled Cayley network $\mathcal{N}_G[\mathcal{L}_G]$ if and only if f is invariant under all automorphisms of the network.
2. The bit complexity of computing all boolean functions which are computable on $\mathcal{N}_G[\mathcal{L}_G]$ is $O(|\mathcal{G}| \cdot \log^2 |\mathcal{G}| \cdot \delta^2 \cdot \sum_{g \in G} |g|^2)$, where δ is the diameter of the network, and $|g|$ the order of g in $\mathcal{G}$.
3. For any group $\mathcal{G}$ there is a set G of generators of $\mathcal{G}$ such that the above bit-complexity is $O(|\mathcal{G}|^3 \cdot \log^4 |\mathcal{G}|)$.
4. Contrast the classes of boolean functions computable on labeled and unlabeled Cayley networks.

Exercise 3.11.36. In this exercise we refer to the anonymous ring on N processors.

1. Show that OR_N requires $\Omega(N^2)$ bits on the anonymous ring.
2. ($\star$) ([**MW86**]) Non-constant boolean functions on N variables which are computable on an anonymous ring (oriented or not) of size N require $\Omega(N \log N)$ bits.
   HINT: First consider the case of oriented rings. Consider an arbitrary algorithm $\mathcal{A}$ computing a given non-constant boolean function f on N variables. Let S be the set of inputs w accepted by $\mathcal{A}$, i.e., such that $f(w) = 1$. We prove the lower bound for the synchronous ring (in which case it will be valid for the asynchronous case as well). Show that
   (a) if algorithm $\mathcal{A}$ rejects 0^N but accepts $0^n w$, for some word w, then $\mathcal{A}$ requires at least $N\lfloor n/2 \rfloor$ messages in order to reject 0^N, and
   (b) the average length of k pairwise distinct words $w_1, \ldots, w_k$ on an alphabet of size r is $> \frac{\log_r(k/2)}{2}$.
   Without loss of generality assume that 0^N is rejected. Assume that all processors terminate before time t when the input to $\mathcal{A}$ is ω, where ω is a word in S. Let $h_i(s)$ denote the history of processor p_i, i.e., $h_i(s) = m_i(1)\$ \cdots \$m_i(k)$, where \$ is a special symbol and $m_i(1), \ldots, m_i(k)$ are the messages received by p_i before time s in this order. Then $H_i = h_i(t)$ is the total history of p_i (on this computation). Since the length of H_i is less than twice the number of bits received by p_i a lower bound on the sum of the lengths of the histories of the processors implies a lower bound on the bit complexity of algorithm $\mathcal{A}$. Now to obtain the desired lower bound $\Omega(N \log N)$ we construct either an input with $\log N$ consecutive 0s (in which case part (a) applies) or else an input under whose execution the algorithm gives rise to at least $N \log N$ processors with distinct histories (in which case part (b) applies).
3. ($\star$) [[**ASW88**]] Assume N is odd and $N = 2n + 1$. Define $f_N(x) = 1$ if x is either $0(01)^n$ or a cyclical shift of it, and 0 otherwise. Show that f_N can be computed in $O(N)$ messages.
4. ($\star$) [[**MW86**]] There is a family $\{f_N\}$ of boolean functions computable with message complexity $O(N \log^* N)$. Use this to construct a family

of boolean functions computable with bit complexity $O(N \log N)$. This shows that the lower bound of part (2) is optimal.

4. Randomness and Satisfiability

> However the formulas (of mathematical probability) may be derived, they frequently prove remarkably trustworthy in practice. The proper attitude is not to reject laws of doubtful origin, but to scrutinize them with care, with a view to reaching the true principle underneath. *J. L. Coolidge [Coo25]*

4.1 Introduction

Let a literal be a propositional variable or its negation, and a clause a conjunction of literals. Consider formulas in conjunctive normal form on n variables, $x_1, x_2, \ldots, x_n$, and k literals per clause (these are known as instances of k-SAT), where k is an integer ≥ 1. Given an instance of k-SAT the problem is to determine whether there is an assignment to the variables such that all clauses evaluate to true, in which case the formula is called satisfiable. The satisfiability problem for 3-SAT lies at the root of theoretical computer science. Aside from the fact that it was the first problem discovered to belong to the class NP, it is of practical interest as well, in fields ranging from theorem provers in Artificial Intelligence to scheduling in computer software.

The idea of classifying problems as NP-complete is based entirely on worst-case analysis. This does not take into account other interesting cases of the problem, such as average and probabilistic. A remarkable idea, which was developed in the context of graph theory by Erdős and Rényi [ER60] (see also [Spe94, ASE92]) has found many applications: by interpreting events in a mathematical system as phase transitions they identified sharp thresholds in the context of graphs. This has proved useful in the study of satisfiability as well.

In recent years satisfiability has attracted attention for a similar reason. Although the hardest instances of satisfiability are very hard indeed, some instances seem to be easy to solve. In particular, investigators have looked at the pattern of "hard" versus "easy" instances of 3-SAT. Let m be the number of clauses and let $r = m/n$ be the clause to variable ratio. It has been observed experimentally that for a random instance ϕ of 3-SAT, if r is "small" then ϕ is almost certainly satisfiable, while if r is "big" then ϕ is almost certainly unsatisfiable. Moreover, as the ratio $r = m/n$ varies from 0 to infinity the satisfiability problem goes from easy to hard and then back to easy again (see Figure 4.1), in the sense that if r is “below“ (“above”) a certain value then it is easy to find a satisfying truth assignment (prove the formula is unsatisfiable). This is known as a phase transition phenomenon and experimental evidence suggests that for 3-SAT this “easy-hard-easy” transition occurs at $r \approx 4.2$ [KS94, CM97]. To facilitate our subsequent discussion we

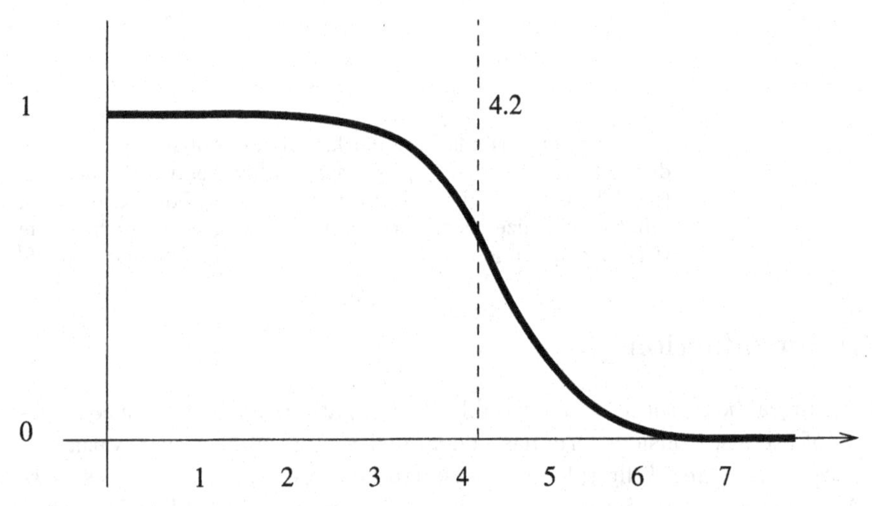

Fig. 4.1. Phase transition phenomenon for 3-SAT. The horizontal axis represents the value of the clause to variable ratio of a random instance of 3-SAT, while the vertical axis represents the probability that a random instance is satisfiable (value 1) or unsatisfiable (value 0). The threshold value 4.2 depicted in the graph is predicted by experimental results.

consider the following two “intuitive” definitions. Let r^k_{sat} (r^k_{unsat}) denote the supremum (infimum) of all real numbers r such that if the clause to variable ratio is less (bigger) than r then almost surely a random instance of k-SAT is satisfiable (unsatisfiable). It follows that $r^k_{sat} \leq r^k_{unsat}$ and if a threshold exists then it has to lie in the interval $[r^k_{sat}, r^k_{unsat}]$.

The present chapter is devoted to the study of the threshold phenomenon for 3-SAT. An outline of the presentation is as follows. In Section 4.2 we prove that 2-SAT has a threshold, namely $r^2_{sat} = r^2_{unsat} = 1$. Section 4.3 is devoted to upper bounds on the unsatisfiability threshold for 3-SAT, and we prove that $r^3_{unsat} \leq 4.601$. Section 4.4 considers lower bounds for the satisfiability threshold for 3-SAT: we prove that $r^3_{unsat} \geq 2/3$; however a deeper analysis of this lower bound technique can be used to prove that $r^3_{unsat} \geq 3.003$. Section 4.5 considers the same threshold phenomenon for $(2+p)$-SAT (a variant of satisfiability that has a smoother transition from 2-SAT to 3-SAT) and Section 4.6 the same problem for Constraint programming. We substantiate mathematically the experimental easy-hard-easy transition on the difficulty of finding a satisfying truth assignment for a random instance of k-SAT. We consider the length of resolution proofs of a randomly chosen instance of k-SAT. In particular, we show that for any $\epsilon > 0$ random instances of 3-SAT on n variables and m clauses require an exponential size resolution refutation, provided that $m/n \leq n^{1/7+\epsilon}$.

4.2 Threshold for 2-SAT

Throughout this section and for the rest of this chapter we will also use the following additional notation.

Definition 4.2.1. *CNF is used as an abbreviation for Conjunctive Normal Form. An event which depends on the parameter n holds almost surely if the probability that the event holds tends to 1 as n tends to infinity.*

For the case of 2-SAT we can prove a sharp threshold. We have the following theorem, which was first proved by V. Chvátal and B. Reed [CR92] and independently by A. Goerdt and W. F. de la Vega.

Theorem 4.2.1 ([CR92]).

1. *If the clause to variable ratio is less than* 1 *then almost surely a random instance of* 2*-SAT is satisfiable.*
2. *If the clause to variable ratio exceeds* 1 *then almost surely a random instance of* 2*-SAT is unsatisfiable.*

Proof. We follow closely the proof of Chvátal and Reed [CR92].

(1) First we consider the satisfiability threshold. The proof is based on the idea of "bicycle" . This is an instance

$$(u \vee u_1) \wedge (\overline{u}_1 \vee u_2) \wedge (\overline{u}_2 \vee u_3) \wedge \cdots \wedge (\overline{u}_{s-1} \vee u_s) \wedge (\overline{u}_s \vee v)$$

of 2-SAT with at least two clauses and literals $u, u_1, u_2, \ldots, u_s, v$ chosen from $x_1, x_2, \ldots, x_s, \overline{x}_1, \overline{x}_2, \ldots, \overline{x}_s$, where $x_1, x_2, \ldots, x_s$ are distinct variables. The following two lemmas will be used in the proof.

Lemma 4.2.1. *A random instance of* 2*-SAT with $(1 + o(1))cn$ clauses over n variables such that $c < 1$ contains a bicycle with probability $o(1)$.*

Proof. Let p denote the probability that a random instance of 2-SAT with m clauses over n variables contains a bicycle. A simple argument counting the number of bicycles shows that

$$\begin{aligned} p &\leq \sum_{s=2}^{n} n^s 2^s (2s)^2 m^{s+1} \left(1 / \left(4\binom{n}{2}\right)\right)^{s+1} \\ &= \frac{2m}{n(n-1)} \sum_{s=2}^{n} s^2 \left(\frac{m}{n-1}\right) \\ &= O(1/n), \end{aligned}$$

since $m = (1 + o(1))cn$ with $c < 1$. This proves Lemma 4.2.1.

Lemma 4.2.2. *Every unsatisfiable instance of 2-SAT contains a bicycle.*

Proof. For any instance ϕ of 2-SAT on n variables consider the directed graph $G(\phi)$: it has $2n$ vertices (one for each literal arising from the n variables) and for each clause $u \vee v$ of ϕ the directed edges $\overline{u} \to v$, $\overline{v} \to u$. This graph was defined by Aspvall, Plass and Tarjan [APT79] where they prove ϕ is unsatisfiable if and only if some variable and its complement belong to the same strongly connected component of $G(\phi)$. In particular, if ϕ is unsatisfiable then $G(\phi)$ contains a directed walk $u_0, u_1, \ldots, u_r$ such that $u_0 = u_r$ and $u_t = \overline{u}_0$ for some t. Now choose a walk that minimizes t and observe that all the literals $u_1, u_2, \ldots, u_t$ are distinct and none of them is the complement of another. This easily gives rise to a bicycle. The proof of Lemma 4.2.2 is complete.

The first part of the theorem is now an immediate consequence of Lemmas 4.2.1, 4.2.2. Indeed, if ϕ is a random instance of 2-SAT with $m = (1+o(1))cn$ clauses where $c < 1$ then $\Pr[\phi \text{ is unsatisfiable}] \leq \Pr[\phi \text{ contains a bicycle}] = o(1)$.

(2) In the second part we consider the unsatisfiability threshold. First we select an integer $t := t(n)$ (which depends on n) such that

$$\frac{t}{\log n} \to \infty \text{ and } \frac{t}{n^{1/9}} \to 0. \tag{4.1}$$

Next we define the notion of "snake": this is an odd sequence of distinct literals $u_1, u_2, \ldots, u_s$, $s = 2t - 1$, none of which is a complement of another. With each snake A we associate the set F_A of $s+1$ clauses $\overline{u}_r \vee u_{r+1}$, $0 \leq r \leq s$, such that $u_0 = u_{s+1} = \overline{u}_t$. Let ϕ_A be the formula corresponding to F_A. Every truth assignment that satisfies ϕ_A must satisfy $u_i \leq u_j$, whenever $i \leq j$, and hence ϕ_A is unsatisfiable. We can view F_A as a graph with vertices $x_1, x_2, \ldots, x_s$ (where x_i is the variable such that u_i is either x_i or $\overline{x}_i$) and edges $\{x_r, x_{r+1}\}$, $0 \leq r \leq s$, such that $x_0 = x_{s+1} = x_t$.

Let $p_t(n)$ denote the probability that for a fixed snake A and a snake B chosen at random from the uniform distribution the sets F_A and F_B share precisely t clauses. View F_A, F_B as graphs and let $F_{A,B}$ denote their intersection with all isolated vertices removed; for a fixed snake A let $N(i,j)$ denote the number of snakes B such that $F_{A,B}$ has i edges and j vertices. We want to obtain an upper bound on $N(i,j)$. To obtain all the snakes counted by $N(i,j)$,

1. choose j terms of A for membership in $F_{A,B}$,
2. choose j terms of B for membership in $F_{A,B}$,
3. assign values to the j terms in $B \cap F_{A,B}$, and
4. assign values to the $s - j$ terms in $B \setminus F_{A,B}$.

In item (1) members of $F_{A,B}$ can be selected from A by first deciding whether $\{x_0, x_1\}$ belongs to $F_{A,B}$ or not and then placing a member at each x_r with $1 \leq r \leq s$ such that precisely one of $\{x_{r-1}, x_r\}$ and $\{x_r, x_{r+1}\}$ belongs to

$F_{A,B}$. Since there are at least $2(j-i)-1$ and at most $2(j-i)+2$ markers we conclude that there are at most $2\binom{s+3}{2j-2i+2}$ choices for item (1) and at most that many for item (2). For the analysis of item (3) let k denote the number of connected components in $F_{A,B}$. Components that are paths may be mixed and matched with their counterparts in up to $k!$ ways and each of these components may flip. In addition there may be a unique component that is not a path and this may be mapped onto its counterpart in at most $2t$ ways. It follows that there are at most 2^k choices in item (3) and trivially at most $(2n)^{s-j}$ choices in item (4). If $1 \le i \le t-1$ then $k = j-i$; hence

$$\begin{aligned} N(i,j) &\le 4\binom{2t+2}{2j-2i+2}^2 t(j-i)!2^{j-i}(2n)^{s-j} \\ &\le 4t(2t+2)^{4(j-i)+4}2^{j-i}(2n)^{s-j}. \end{aligned}$$

It follows that for sufficiently large n the following holds whenever $1 \le i \le t-1$:

$$\begin{aligned} p_i(n) &\le \tfrac{1}{\binom{n}{s}s!2^s}\textstyle\sum_{j\ge i+1} N(i,j) \\ &< \tfrac{5t(2t+2)^4}{(2n)!}\textstyle\sum_{j\ge i+1}\left(\tfrac{(2t+2)^4}{n}\right)^{j-i} \\ &< \tfrac{1{,}300t^9}{n(2n)^i}. \end{aligned} \tag{4.2}$$

If only $1 \le i \le 2t$ then we have $k \le j-i+2$ and hence

$$\begin{aligned} N(i,j) &\le 4\binom{2t+2}{2j-2i+2}^2 t(j-i+2)!2^{j-i+2}(2n)^{s-j} \\ &\le 16t(2t+j)^{4(j-i)+4}2^{j-i}(2n)^{s-j}. \end{aligned}$$

It follows that for sufficiently large n the following holds whenever $1 \le i \le 2t$:

$$\begin{aligned} p_i(n) &\le \tfrac{1}{\binom{n}{s}s!2^s}\textstyle\sum_{j\ge i+1} N(i,j) \\ &< \tfrac{17t(2t+2)^4}{(2n)!}\textstyle\sum_{j\ge i+1}\left(\tfrac{(2t+2)^4}{n}\right)^{j-i} \\ &< \tfrac{18t^9}{(2n)^i}. \end{aligned} \tag{4.3}$$

In the sequel it will be shown that with probability $1-o(1)$ a random instance ϕ of 2-SAT (with $(1+o(n))cn$ clauses over n variables $c > 1$) includes all the clauses of some ϕ_A. Write $X = \sum_A X_A$, with $X_A = 1$ if each clause of ϕ_A appears precisely once in ϕ and $X_A = 0$ otherwise.

Consider arbitrary snakes A, B. We have $\mathbf{E}[X_A] = \mathbf{E}[X_B] = f(2t)$ with

$$f(x) = \binom{m}{x}x!\left(\frac{1}{4\binom{n}{2}}\right)^x\left(1-\frac{x}{4\binom{n}{2}}\right)^{m-x};$$

in the case when F_A and F_B have exactly i clauses in common then $\mathbf{E}[X_A X_B] = f(4t-i)$. Since $m = O(n)$ we have that

$$f(x) = (1 + o(1)) \left(\frac{m}{2n(n-1)} \right)^x$$

uniformly in every range $x = O(n^\alpha)$ with $\alpha < 1/2$; hence the limit conditions (4.1) guarantee that

$$\mathbf{E}[X_A X_B] = (1 + o(1))\mathbf{E}[X_A]\mathbf{E}[X_B] \left(\frac{2n(n-1)}{m} \right)^i$$

uniformly in the range $0 \le i \le 2t$. It follows that

$$\begin{aligned}
\mathbf{E}[X^2] &= \sum_{A,B} \mathbf{E}[X_A X_B] \\
&= (1 + o(1)) \left(\sum_{A,B} \mathbf{E}[X_A]\mathbf{E}[X_B] \right) \sum_{i=0}^{2t} p_t(n) \left(\frac{2n(n-1)}{m} \right)^i \\
&= (1 + o(1))\mathbf{E}[X]^2 \sum_{i=1}^{2t} p_i(n) \left(\frac{2n(n-1)}{m} \right)^i .
\end{aligned}$$

Using (4.1) and (4.2) we obtain

$$\sum_{i=1}^{t-1} p_i(n) \left(\frac{2n(n-1)}{m} \right)^i < \frac{1,300t^9}{n} \sum_{i=1}^{t-1} \left(\frac{n-1}{m} \right)^i = o(1),$$

and using (4.1) and (4.3) we obtain

$$\sum_{i=t}^{2t} p_i(n) \left(\frac{2n(n-1)}{m} \right)^i < 18tn \sum_{i=t}^{2t} \left(\frac{n-1}{m} \right)^i = o(1),$$

It follows that

$$\mathbf{E}[X^2] \le (1 + o(1))(\mathbf{E}[X])^2$$

and by Chebyshev's inequality

$$\Pr[X > 0] \le \Pr[|X - \mathbf{E}[X]| > \mathbf{E}[X]] \le \frac{\mathbf{E}[X^2] - \mathbf{E}[X]^2}{\mathbf{E}[X]^2} = o(1).$$

This completes the proof of the theorem.

4.3 Unsatisfiability Threshold for 3-SAT

In this section we study a new technique leading to new and improved upper bounds on the unsatisfiability threshold. We begin by discussing a general methodology based on the first moment method and indicate its potential applicability. Then we consider a detailed analysis of the method of single- and double-flips.

4.3.1 A General Method and Local Maxima

The following simple theorem is an important observation that was made by several researchers, including Chvátal et al. [CS88, CR92], Franco et al. [FP83, CF86], and Simon et al. [SCDM86].

Theorem 4.3.1. *If the clause to variable ratio exceeds* 5.19 *then almost surely a random instance of* 3-*SAT is unsatisfiable.*

Proof. Let ϕ be a random formula on n variables $x_1, x_2, \ldots, x_n$ which is an instance of 3-SAT and let X be the random variable denoting the number of truth assignments satisfying ϕ. For a truth assignment A, consider the indicator function X_A defined as follows: $X_A = 1$ if A satisfies ϕ and $X_A = 0$, otherwise. Using Markov's inequality we have

$$\begin{aligned}\Pr[X > 0] &\leq \mathbf{E}[X] \\ &= \mathbf{E}[\sum_A X_A] \\ &= \sum_A \mathbf{E}[X_A] \\ &= \sum_A \Pr[X_A = 1] \\ &= \sum_A \Pr[A \text{ satisfies } \phi] \\ &= 2^n (1 - 2^{-3})^m \\ &= 2^n (7/8)^m,\end{aligned}$$

where A ranges over truth assignments on the x variables $x_1, x_2, \ldots, x_n$. To guarantee that this last term converges to 0 it is enough to assume that $(2^n(7/8))^m)^{1/n} < 1$. From this it follows that the desired threshold value is obtained by solving the equation

$$\mathbf{E}[X] = \left(2^n \left(\frac{7}{8}\right)^m\right)^{1/n} = 1, \tag{4.4}$$

which easily implies the value

$$r = \frac{m}{n} = -\frac{\ln 2}{\ln(7/8)} \approx 5.19.$$

This completes the proof of the theorem.

An important question is how to improve on the above threshold value. We can follow one of the following two approaches

1. Use techniques that give more precise closed formulas approximating $\Pr[X > 0]$ (i.e., the probability that the random formula is satisfiable).

One such approach is provided by the second moment method as employed by Kamath et al. [KMPS95] which leads to the unsatisfiability threshold value 4.78.

2. Use the first moment method but on new random variables. To accomplish this we would like to have a new random variable X_{new} which has the following relation to the random variable X above.
   - $X_{new} \leq X$, and
   - $X > 0 \Rightarrow X_{new} > 0$.

   It would then follow that

$$\begin{aligned}\Pr[X > 0] &\leq \Pr[X_{new} > 0] \\ &\leq \mathbf{E}[X_{new}] \\ &\leq \mathbf{E}[X].\end{aligned} \tag{4.5}$$

   In general it is not obvious what random variable X_{new} we should select. Nevertheless, the previous approach that reduced to equation (4.4) and the fact that $\mathbf{E}[X_{new}] \leq \mathbf{E}[X]$ would seem to indicate that the ratio r_{new} obtained by solving the new equation $\mathbf{E}[X_{new}] = 1$ should satisfy $r_{new} \leq r$, i.e., a value which is closer to the real threshold.

It is this second approach which is followed in [KKK97]. For a random formula ϕ, call a satisfying truth assignment A local maximum for single flips, if A satisfies ϕ but if we flip a false value into a true value then the resulting assignment does not satisfy A. Let X_{new} be the random variable of the number of satisfying truth assignments which are local maxima for single flips. It is easy to see that this random variable satisfies the conditions above. Indeed it can also be used to improve the unsatisfiability threshold [KKK96, KKKS98].

4.3.2 Method of Single Flips

In the remainder of this section we follow closely the details of the proof in [KKKS98] (see also [KKK96] and [KKK97]).

Definition 4.3.1. *Let $\mathcal{A}_n$ be the set of all truth assignments on the n variables $x_1, \ldots x_n$, and let $\mathcal{S}_n$ be the set of truth assignments that satisfy the random formula ϕ. Thus the cardinality $|\mathcal{S}_n|$ is a random variable. Also, for an instantiation ϕ of the random formula, let $|\mathcal{S}_n(\phi)|$ denote the number of truth assignments that satisfy ϕ.*

We now define a class even smaller than $\mathcal{S}_n$.

Definition 4.3.2. *For a random formula ϕ, $\mathcal{S}_n^\sharp$ is defined to be the random class of truth assignments A such that (i) $A \models \phi$, and (ii) any assignment obtained from A by changing exactly one* FALSE *value of A to* TRUE *does not satisfy ϕ.*

Notice that the truth assignment with all its values equal to TRUE vacuously satisfies condition (ii) of the previous definition. Consider the lexicographic ordering among truth assignments, where the value FALSE is considered smaller than TRUE and the values of variables with higher index are of lower priority in establishing the way two assignments compare. It is not hard to see that $\mathcal{S}_n^\sharp$ is the set of elements of $\mathcal{S}_n$ that are local maxima in the lexicographic ordering of assignments, where the neighborhood of determination of local maximality is the set of assignments that differ from A in at most one position. Just like the proof of inequality (4.5) we can now prove:

Lemma 4.3.1. *The following Markov type inequality holds for* $\mathcal{S}_n^\sharp$*:*

$$\Pr[\text{the random formula is satisfiable}] \leq \mathbf{E}[|\mathcal{S}_n^\sharp|]. \tag{4.6}$$

Proof. From the previous definition we easily infer that if an instantiation ϕ of the random formula is satisfiable, then $\mathcal{S}_n^\sharp(\phi) \neq \emptyset$. (Recall that $\mathcal{S}_n^\sharp(\phi)$ is the instantiation of the random class $\mathcal{S}_n^\sharp$ at the instantiation ϕ.) We also have that Pr[the random formula is satisfiable] $= \sum_\phi (\Pr[\phi] \cdot I_\phi)$, where

$$I_\phi = \begin{cases} 1 \text{ if } \phi \text{ is satisfiable,} \\ 0 \text{ otherwise.} \end{cases} \tag{4.7}$$

On the other hand,

$$\mathbf{E}[|\mathcal{S}_n^\sharp|] = \sum_\phi \left(\Pr[\phi] \cdot |\mathcal{S}_n^\sharp(\phi)|\right).$$

The lemma now immediately follows from the above.

We also have the following:

Lemma 4.3.2. *The expected value of the random variable* $|\mathcal{S}_n^\sharp|$ *is given by the formula*

$$\mathbf{E}[|\mathcal{S}_n^\sharp|] = (7/8)^{rn} \sum\nolimits_{A \in \mathcal{A}_n} \Pr[A \in \mathcal{S}_n^\sharp \mid A \in \mathcal{S}_n]. \tag{4.8}$$

Proof. First observe that the random variable $|\mathcal{S}_n^\sharp|$ is the sum of indicator variables and then condition on $A \models \phi$ (recall, r is the number of clause-to-variable ratio of ϕ, so $m = nr$).

Definition 4.3.3. *We call a change of* exactly one FALSE *value of a truth assignment A to* TRUE *a* single flip. *The number of possible single flips, which is of course equal to the number of* FALSE *values of A, is denoted by $sf(A)$. The assignment obtained by applying a single flip sf on A is denoted by A^{sf}.*

We now prove the following:

Theorem 4.3.2 ([KKK97, KKKS98]). *If the clause to variable ratio exceeds* 4.667 *then almost surely a random instance of* 3-*SAT is unsatisfiable. More formally, the expected value* $\mathbf{E}[|\mathcal{S}_n^\sharp|]$ *is at most* $(7/8)^{rn}(2 - e^{-3r/7} + o(1))^n$. *It follows that the unique positive solution of the equation*

$$(7/8)^r(2 - e^{-3r/7}) = 1,$$

is an upper bound for κ *(this solution is less than 4.667).*

Proof. Fix a single flip sf_0 on A and assume that $A \models \phi$. Observe that the assumption that $A \models \phi$ excludes $\binom{n}{3}$ clauses from the conjuncts of ϕ, i.e., there remain $7\binom{n}{3}$ clauses from which to choose the conjuncts of ϕ. Now consider the clauses that are not satisfied by A^{sf_0} and contain the flipped variable. There are $\binom{n-1}{2}$ of them. Under the assumption that $A \models \phi$, in order to have that $A^{sf_0} \not\models \phi$, it is necessary and sufficient that at least one of these $\binom{n-1}{2}$ clauses be a conjunct of ϕ. Therefore, for each of the m clause selections for ϕ, the probability of being one that guarantees that $A^{sf_0} \not\models \phi$ is $\binom{n-1}{2}/7\binom{n}{3} = 3/(7n)$. Therefore, the probability that $A^{sf_0} \not\models \phi$ (given that $A \models \phi$) is equal to $1 - (1 - 3/(7n))^m$. Now, there are $sf(A)$ possible flips for A. The events that ϕ is not satisfied by the assignment A^{sf} for *each* single flip sf (under the assumption that $A \models \phi$) refer to disjoint sets of clauses. Therefore, the dependencies among them are such that:

$$\begin{aligned} \Pr[A \in \mathcal{S}_n^\sharp \mid A \models \phi] &\le \left(1 - \left(1 - \tfrac{3}{7n}\right)^m\right)^{sf(A)} \\ &= \left(1 - e^{-3r/7} + o(1)\right)^{sf(A)}. \end{aligned} \tag{4.9}$$

Petr Savický has supplied us with a formal proof of the above inequality. In addition, a result that implies it is presented in [McD92]. Indeed, in the notation of the main theorem in [McD92], it is enough, in order to obtain the above inequality, to let (i) $V = \{1, \dots, m\}$, (ii) $I = \{1, \dots, sf(A)\}$, (iii)$X_v = i$ if and only if the v-th clause of ϕ is satisfied by A but not satisfied by A^{sf_i}, where A^{sf_i} is the truth assignment obtained from A by flipping the i-th FALSE value of A, and (iv) for all i, $\mathcal{F}_i$ be the "increasing" collection of non-empty subsets of V.

Now recall that $sf(A)$ is equal to the number of FALSE values of A. Therefore, by equation (4.8) and by Newton's binomial formula, $\mathbf{E}[|\mathcal{S}_n^\sharp|]$ is bounded above by $(7/8)^{rn}(2 - (1 - 3/(7n))^{rn})^n$, which proves the first statement of the theorem.

It also follows that $\mathbf{E}[|\mathcal{S}_n^\sharp|]$ converges to zero for values of r that strictly exceed the unique positive solution of the equation $(7/8)^r(2 - e^{-3r/7}) = 1$. By Lemma 4.3.1, this solution is an upper bound for κ. As it can be seen by any program that computes roots of equations with accuracy of at least four decimal digits (we used Maple [Red94]), this solution is less than 4.667.

At this point it is interesting to note that Dubois and Boufkhad [DB97] independently also analyzed the single flip method and obtained the upper bound 4.642. However, their analysis is more complicated than the one presented here and also does not lead to the idea of multiple flips.

4.3.3 Approximating the Threshold

The method of single flips can be generalized to arbitrary range of locality when selecting the subset of $\mathcal{S}_n$. We start with a definition:

Definition 4.3.4. *Given a random formula ϕ and a non-negative integer l, $\mathcal{A}_n^l (l \leq n)$ is defined to be the random class of truth assignments A such that (i) $A \models \phi$, and (ii) any assignment that differs from A in at most l variables and is lexicographically strictly larger than A does not satisfy ϕ.*

Observe that $\mathcal{S}_n$ of the previous section, i.e., the class of truth assignments satisfying the random formula is now redefined as $\mathcal{A}_n^0$ and $\mathcal{S}_n^\sharp$ is redefined as $\mathcal{A}_n^1$. $\mathcal{A}_n^l$ is the subset of $\mathcal{S}_n$ that consists of the lexicographic local maxima of $\mathcal{S}$ where the neighborhood of locality for an assignment A is the set of assignments that differ from A in at most l places. Moreover, $\mathcal{A}_n^l$ is a sequence of classes which is non increasing (with respect to set inclusion).

Now, exactly as in Lemma 4.3.1, we have the following result.

Lemma 4.3.3. *The random variables $|\mathcal{A}_n^1|$ satisfy:*

1. $\Pr[\phi \text{ is satisfiable}] = \mathbf{E}[|\mathcal{A}_n^n|]$, *and*
2. $$\mathbf{E}[|\mathcal{A}_n^n|] \leq \mathbf{E}[|\mathcal{A}_n^{n-1}|] \leq \cdots \leq \mathbf{E}[|\mathcal{A}_n^1|] \leq \mathbf{E}[|\mathcal{A}_n^0|]. \tag{4.10}$$

In principle, for a fixed l, by letting $\lim_n \mathbf{E}[|\mathcal{A}_n^l|] = 0$, we obtain upper bounds for the unsatisfiability threshold which decrease as l increases. In practice, the mathematical analysis required when $l \geq 2$ s nontrivial and is left as an open problem. In the sequel we will study the case of double flips in detail following the outline of Kirousis, Kranakis, Krizanc, and Stamatiou [KKKS98].

4.3.4 Method of Double Flips

We concentrate below on the case $l = 2$, which corresponds to the case of double flips.. A change of exactly two values of a truth assignment A that gives a truth assignment which is lexicographically greater than A must be of one of the following types:

1. a change of the value FALSE of a variable to TRUE and a change of the value TRUE of a higher indexed variable to FALSE, or
2. a change of two variables both of value FALSE to TRUE.

Definition 4.3.5. *Of these two possible types of changes, we consider only the first, since the calculations become easier, while the final result remains the same. We call such changes* double flips. *Define A^{df} and $df(A)$ in a way analogous to the single flip case.*

Notice that if A is considered as a sequence of the boolean values 0 and 1, then $df(A)$ is equal to the number of order inversions as we move along A from high-indexed variables to low-indexed ones, i.e., from right to left.

Definition 4.3.6. *Let $\mathcal{A}_n^{2\sharp}$ be the set of assignments A such that $A \models \phi$ and for all single (double) flips sf (df), $A^{sf} \not\models \phi$ ($A^{df} \not\models \phi$).*

It can be easily seen that $\mathcal{A}_n^2$ is a subset of $\mathcal{A}_n^{2\sharp}$ (in general a proper one, because in the definition of $\mathcal{A}_n^{2\sharp}$ we did not take into account the changes of type (2)).

Therefore a value of r that makes the expected value $\mathbf{E}[|\mathcal{A}_n^{2\sharp}|]$ converge to zero is, by Lemma 4.3.3, an upper bound for κ. Actually, it can be proved that both $\mathbf{E}[|\mathcal{A}_n^{2\sharp}|]$ and $\mathbf{E}[|\mathcal{A}_n^2|]$ converge to zero for the same values of r (see Exercise 4.8.16).

Now in analogy to Lemma 4.3.2 we have

Lemma 4.3.4. *The expected value of the random variable $|\mathcal{A}_n^{2\sharp}|$ is equal to*

$$(7/8)^{rn} \cdot \sum_{A \in \mathcal{A}_n} \Pr[A \in \mathcal{A}_n^1 \mid A \models \phi] \cdot \Pr[A \in \mathcal{A}_n^{2\sharp} \mid A \in \mathcal{A}_n^1]. \quad (4.11)$$

Therefore, by the remarks in the beginning of the current section, an upper bound for κ can be found by computing a value (the smaller the better) for r for which the right-hand side of the equation above converges to zero. We will do this in two steps. First we will compute an upper bound for each term of the second sum in the equation above; then we will find an upper bound for $\mathbf{E}[|\mathcal{A}_n^{2\sharp}|]$ which will be a closed expression involving r and n. Letting this closed expression converge to zero with n, we will get an equation in terms of r that gives the required bound for κ.

To compute an upper bound for the terms of the sum, we will make use of an inequality that appears in [Jan98] as Theorem 7, which gives an estimate for the probability of the intersection of dependent events. We give the details in the sequel.

4.3.5 Probability Calculations

Given a fixed assignment A, we will now find an upper bound for $\Pr[A \in \mathcal{A}_n^1 \mid A \models \phi] \cdot \Pr[A \in \mathcal{A}_n^{2\sharp} \mid A \in \mathcal{A}_n^1]$. We assume for the rest of this subsection that the condition $A \models \phi$ holds. This is equivalent to assuming that the space of all clauses from which we uniformly, independently, and with replacement choose the ones that form ϕ is equal to the set of all clauses satisfied by A. This set of clauses has cardinality $7\binom{n}{3}$. Also notice that under the condition $A \models \phi$, the event $A \in \mathcal{A}_n^1$ is equivalent to the statement that for any single flip sf, $A^{sf} \not\models \phi$. In the sequel, all computations of probabilities, analysis of events, etc., will be carried out assuming that $A \models \phi$, usually without explicit mention of it.

To compute $\Pr[A \in \mathcal{A}_n^{2\sharp}]$, it is more convenient to work in another model for random formulas. In the next paragraphs, we give the necessary definitions and notations.

Consistent with the standard notation of the theory of random graphs [Bol85], let $\mathcal{G}_p$ be the model for random formulas where each clause has an independent probability p of appearing in the formula, let $\mathcal{G}_m$ be the model where the random formula is obtained by uniformly and independently selecting m clauses *without* replacement, and, finally, let $\mathcal{G}_{mm}$ be the model that we use in this book, where the formula is obtained by uniformly and independently selecting m clauses *with* replacement (recall that according to our assumption, we only refer to clauses that are satisfied by A).

The probability of an event E in $\mathcal{G}_p$ ($\mathcal{G}_m$) will be denoted by $\Pr_p[E]$ ($\Pr_m[E]$). In order not to change our notation, we continue to denote the probability of E in the model $\mathcal{G}_{mm}$ by $\Pr[E]$. Set $p = m/(7\binom{n}{3}) \sim 6r/(7n^2)$. By [Bol85][Theorem 2 (iii), Chapter 3, page 55], we have that for any property Q of formulas, $\Pr_m[Q] \leq 3m^{1/2}\Pr_p[Q]$. Additionally, if Q is monotonically increasing (i.e., if it holds for a formula, it also holds for any formula containing more clauses) and *reducible* (i.e., it holds for a formula if and only if it holds for the formula where multiple occurrences of clauses have been omitted), then $\Pr[Q] \leq \Pr_m[Q]$. Intuitively, this is so because by the assumptions of increasing monotonicity and reducibility for Q, when selecting the clauses to be included in ϕ, we increase the probability to satisfy Q by selecting a "new" clause, rather than by selecting one that has already been selected. A formal proof of this property can be found in [KS96]. Therefore, as non-satisfiability is both monotonically increasing and reducible, we conclude that

$$\begin{aligned}\Pr[A \in \mathcal{A}_n^{2\sharp}] &\leq 3m^{1/2}\Pr_p[A \in \mathcal{A}_n^{2\sharp}] \\ &= 3m^{1/2}\Pr_p[A \in \mathcal{A}_n^{2\sharp} \wedge A \in \mathcal{A}_n^{1}] \\ &\quad \text{(because } A \in \mathcal{A}_n^1 \text{ is implied by } A \in \mathcal{A}_n^{2\sharp}\text{)} \\ &= 3m^{1/2}\Pr_p[A \in \mathcal{A}_n^1] \cdot \Pr_p[A \in \mathcal{A}_n^{2\sharp} \mid A \in \mathcal{A}_n^1].\end{aligned} \tag{4.12}$$

It is easy to see, carrying the corresponding argument in the proof of Theorem 4.3.2 within the model $\mathcal{G}_p$, that

$$\Pr_p[A \in \mathcal{A}_n^1] = \left(1-(1-p)^{\binom{n-1}{2}}\right)^{sf(A)} = (1-e^{-3r/7}+o(1))^{sf(A)}. \tag{4.13}$$

So, by Equations (4.11), (4.12), and (4.13) to find an upper bound for

$$\Pr[A \in \mathcal{A}_n^1] \cdot \Pr[A \in \mathcal{A}_n^{2\sharp} \mid A \in \mathcal{A}_n^1],$$

it is enough to find an upper bound for

$$\Pr_p[A \in \mathcal{A}_n^{2\sharp} \mid A \in \mathcal{A}_n^1].$$

Computing this last probability is equivalent to computing the probability that for all double flips df, $A^{df} \not\models \phi$, under the condition that for all single flips sf, $A^{sf} \not\models \phi$. In the next lemma, given a fixed double flip df_0, we will compute the probability that $A^{df_0} \not\models \phi$, under the same condition. We will then compute the joint probability for all double flips.

At this point it is convenient to introduce the following notation to be used below: for a variable x_i, x_i^A is the literal x_i if the value of x_i in A is TRUE, and it is the literal $\neg x_i$, otherwise. Also let $q = 1 - p$.

First, fix a double flip df_0. Then we have:

Lemma 4.3.5. *The following holds:*

$$\Pr_p[A^{df_0} \not\models \phi \mid A \in \mathcal{A}_n^1] = 1 - \frac{q^{(n-2)^2}(1-q^{n-2})}{1-q^{\binom{n-1}{2}}} = 1 - \frac{6e^{-6r/7}r}{7(1-e^{-3r/7})}\frac{1}{n} + o\left(\frac{1}{n}\right). \tag{4.14}$$

Proof. Assume without loss of generality that df_0 changes the values of x_1 and x_2 and that these values are originally FALSE and TRUE, respectively. Also let sf_0 be the *unique* single flip that changes a value which is also changed by df_0. In this case, sf_0 is the flip that changes the value of x_1 from FALSE to TRUE.

Notice that because all single flips that are distinct from sf_0 change values which are not changed by df_0, $\Pr_p[A^{df_0} \not\models \phi \mid A \in \mathcal{A}_n^1] = \Pr_p[A^{df_0} \not\models \phi \mid A^{sf_0} \not\models \phi]$. To compute the "negated" probability in the right-hand side of the above inequality, we proceed as follows:

It is easy to see, carrying the corresponding argument in the proof of Theorem 4.3.2 within the model $\mathcal{G}_p$, that $\Pr_p[A^{sf_0} \not\models \phi] = 1-q^{\binom{n-1}{2}}$. We now first compute the "positive" (with respect to A^{df_0}) probability: $\Pr_p[A^{df_0} \models \phi \wedge A^{sf_0} \not\models \phi]$. Observe that in order to have that $A^{df_0} \models \phi$, any clause that contains *at least one* of the literals $\neg x_1, x_2$ and whose remaining literals belong to $\neg x_i^A$, $i > 2$, *must not be* among the conjuncts of ϕ. The number of these clauses is equal to $2\binom{n-2}{2} + n - 2 = (n-2)^2$. However the additional requirement that $A^{sf_0} \not\models \phi$, in conjunction with the requirement that $A^{df_0} \models \phi$, makes necessary that at least one clause that contains *both* $\neg x_1, \neg x_2$ and one of $\neg x_i^A$, $i > 2$, *is* among the conjuncts of ϕ (the number of such clauses is $n-2$). The probability these events occurring simultaneously is equal to $q^{(n-2)^2}(1-q^{n-2})$. This last expression gives the probability $\Pr_p[A^{df_0} \models \phi \wedge A^{sf_0} \not\models \phi]$.

From the above, it follows that

$$\Pr_p[A^{df_0} \not\models \phi \mid A^{sf_0} \not\models \phi] = 1 - \frac{q^{(n-2)^2}(1-q^{n-2})}{1-q^{\binom{n-1}{2}}}.$$

This concludes the proof.

Unfortunately, we cannot just multiply the probabilities in the previous lemma to compute $\Pr_p[A \in \mathcal{A}_n^{2\sharp} \mid A \in \mathcal{A}_n^1]$, because these probabilities are not independent. This is so because two double flips may have variables in common. Fortunately, we can apply a variant of Suen's inequality that was proved by S. Janson (this inequality appears as Theorem 7 in [Jan98]; for the original version of Suen's inequality see [Sue90]) and gives an estimate for the probability of the intersection of dependent events. In what follows, we will first present the inequality as well as the assumptions under which it is applicable, and then apply it in the context of our problem.

Let $\{I_i\}_{i\in\mathcal{I}}$ be a finite family of indicator random variables defined on a probability space. Let also Γ be a dependency graph for $\{I_i\}_{i\in\mathcal{I}}$, i.e., a graph with vertex set $\mathcal{I}$ such that if A and B are two disjoint subsets of $\mathcal{I}$, and Γ contains no edge between A and B, then the families $\{I_i\}_{i\in A}$ and $\{I_i\}_{i\in B}$ are *independent*. If $p_i = \Pr[I_i = 1]$, $\Delta = \frac{1}{2}\sum_{(i,j):i\sim j} \mathbf{E}[I_i I_j]$ (summing over ordered pairs (i,j)), $\delta = max_{i\in\mathcal{I}} \sum_{j\sim i} p_j$, $\mu = \sum_{i\in\mathcal{I}} p_i$ and $\epsilon = max_{i\in\mathcal{I}} p_i$ and, in addition, $\delta + \epsilon \leq e^{-1}$, then

$$\Pr\left[\sum_{i\in\mathcal{I}} I_i = 0\right] \leq e^{\Delta\phi_2(\delta+\epsilon)} \prod_{k\in\mathcal{I}} (1 - p_k) \tag{4.15}$$

where $\phi_2(x)$ is the smallest root of the equation $\phi_2(x) = e^{x\phi_2(x)}$, given x such that $0 \leq x \leq e^{-1}$ (ϕ_2 is increasing in this range).

Now, in our context, given a truth assignment, let DF (the index set $\mathcal{I}$ above) be the class of all double flips. For an element df of $\mathcal{I}$, let $I_{df} = 1$ if and only if $A^{df} \models \phi$, given that $A \in \mathcal{A}_n^1$. Then, $p_{df} = \Pr_p[I_{df} = 1] = \Pr_p[A^{df} \models \phi \mid A \in \mathcal{A}_n^1]$, and, from Lemma 4.3.5 is equal to

$$\frac{6e^{-6r/7}r}{7(1-e^{-3r/7})}\frac{1}{n} + o\left(\frac{1}{n}\right).$$

Also, it holds that:

$$\delta \leq \frac{6e^{-6r/7}r}{7(1-e^{-3r/7})} + o(1)$$

since a double flip may share a flipped variable with at most n other double flips. Also, from Lemma 4.3.5, $\epsilon = o(1)$. Therefore, for the range of r that is of concern to us ($r > 3.003$, which is the best known lower bound), $\delta + \epsilon \leq 6e^{-6r/7}r/(7(1-e^{-3r/7})) + o(1) \leq e^{-1}$ (for sufficiently large n).

For two elements df and df' of DF, let $df \sim df'$ denote the fact that df and df' are distinct double flips sharing a flipped variable. Then

$$\begin{aligned} \Delta &= \tfrac{1}{2}\textstyle\sum_{(df,df'):df\sim df'} \mathbf{E}[I_{df} I_{df'}] \\ &= \tfrac{1}{2}\textstyle\sum_{(df,df'):df\sim df'} \Pr_p[A^{df} \models \phi, A^{df'} \models \phi \mid A \in \mathcal{A}_n^1]. \end{aligned} \tag{4.16}$$

Before calculating the probability that is involved in equation (4.16), we show that the events we are considering, i.e., the events that $A^{df} \models \phi$, df a double flip, conditional on $A \in \mathcal{A}_n^1$, form a dependency graph. In other words, we must check whether for any two sets J_1 and J_2 of double flips such that no flip in J_1 shares a variable with a flip in J_2, any boolean combination of conditional events corresponding to flips in J_1 is independent of any boolean combination of conditional events corresponding to flips in J_2. Suppose that the conditional were not $A \in \mathcal{A}_n^1$, but $A \models \phi$. Then the resulting space is a $\mathcal{G}_p$ space, i.e, each clause satisfied by A has an independent probability of appearing in the random formula. Then the mutual independence required to obtain the above inequality would be obviously satisfied, as the two boolean combinations that must be shown independent correspond to distinct clauses. In our case however, where the conditional is $A \in \mathcal{A}_n^1$, the probability space is not a $\mathcal{G}_p$ space. Nevertheless, the required independence still holds. To prove this let B_1 and B_2 be two boolean combinations of *unconditional* events corresponding to two sets of double flips that do not share a variable. The *conditional* independence that is required to obtain the above inequality is equivalent to:

$$\Pr_p[B_1, B_2, \bigwedge_{sf \in SF} A^{sf} \not\models \phi \mid A \models \phi] \cdot \Pr_p[\bigwedge_{sf \in SF} A^{sf} \not\models \phi \mid A \models \phi] =$$

$$\Pr_p[B_1, \bigwedge_{sf \in SF} A^{sf} \not\models \phi \mid A \models \phi] \cdot \Pr_p[B_2, \bigwedge_{sf \in SF} A^{sf} \not\models \phi \mid A \models \phi].$$

Notice that because the conditional in the probabilities in the above equality is $A \models \phi$, the resulting space is from the model $\mathcal{G}_p$. Now, the above equation is trivial to prove using the fact that in such a space combinations of events corresponding to either single or double flips with no common variables are independent.

We now compute the exponential correlation factor that appears in inequality (4.15). The computation is a bit tedious. In the following lemmas, we give the results of the various steps hoping that the interested (and patient) reader can carry them out by herself. The method to be used is very similar to that of the proof of Lemma 4.3.5. For the sake of further economy, we set

$$u = e^{-r/7}.$$

Notice that then, by Lemma 4.3.5,

$$\Pr_p[A^{df_0} \models \phi \mid A \in \mathcal{A}_n^1] = \frac{6u^6 \ln(1/u)}{1-u^3} \frac{1}{n} + o\left(\frac{1}{n}\right). \tag{4.17}$$

Lemma 4.3.6. *Let df_0 and df_1 be two double flips that share the variable that they change from* FALSE *to* TRUE*. Then*

$$\begin{aligned} \Pr_p[A^{df_0} \models \phi, A^{df_1} \models \phi \mid A \in \mathcal{A}_n^1] &= \frac{q^{2(n-2)} q^{3\binom{n-2}{2}} q^{n-3} p}{1-q^{\binom{n-1}{2}}} \\ &= \frac{6u^9 \ln(1/u)}{1-u^3} \frac{1}{n^2} + o\left(\frac{1}{n^2}\right). \end{aligned} \tag{4.18}$$

Lemma 4.3.7. *Let df_0 and df_1 be two double flips that share the variable that they change from* TRUE *to* FALSE*. Then*

$$\begin{aligned} \Pr_p[A^{df_0} \models \phi, A^{df_1} \models \phi \mid A \in \mathcal{A}_n^1] &= \frac{q^{2(n-2)} q^{3\binom{n-2}{2}} q^{n-3} (1-q^{n-2})^2}{(1-q^{\binom{n-1}{2}})^2} \\ &= \frac{36u^9 \ln^2(1/u)}{(1-u^3)^2} \frac{1}{n^2} + o\left(\frac{1}{n^2}\right). \end{aligned} \tag{4.19}$$

Now observe that the number of intersecting ordered pairs of double flips is at most $df(A) \cdot n$. Finally, it is easy to see that the probability in Lemma 4.3.6 is smaller than the probability in Lemma 4.3.7. From these observations, and by substituting in equation (4.16) the right-hand side of equation (4.19), we get that:

$$\Delta = \frac{1}{2}\sum_{(df,df'):df\sim df'} \mathbf{E}[I_{df} I_{df'}]$$

$$\leq df(A) \cdot \left(\frac{18u^9 \ln^2(1/u)}{(1-u^3)^2}\frac{1}{n} + o\left(\frac{1}{n}\right)\right) \phi_2\left(\frac{6u^6\ln(1/u)}{1-u^3} + o(1)\right).$$

From this and equation (4.17), it follows, by inequality (4.15), that the probability

$$\Pr_p[A \in \mathcal{A}_n^{2\sharp} \mid A \in \mathcal{A}_n^1]$$

is

$$[1 - \frac{6u^6 \ln(1/u)}{1-u^3}\frac{1}{n} + \frac{18u^9 \ln^2(1/u)}{(1-u^3)^2}\frac{1}{n}\phi_2\left(\frac{6u^6\ln(1/u)}{1-u^3} + o(1)\right) + o\left(\frac{1}{n}\right)]^{df(A)}. \tag{4.20}$$

It is easy to see (e.g. by using Maple, or by a bit tedious analytical computations) that the expression at the base of the right-hand side of the above inequality is at most 1, for $3 \leq r \leq 5$. Now, by equations (4.11), (4.12), (4.13), and (4.20), we get that:

$$\mathbf{E}[|\mathcal{A}_n^{2\sharp}|] \leq 3(rn)^{1/2}(7/8)^{rn}\sum_A X^{sf(A)} Y^{df(A)}, \tag{4.21}$$

where

$$X = 1 - u^3 + o(1) \tag{4.22}$$

and

$$Y = 1 - \frac{6u^6\ln(1/u)}{1-u^3} \left(1 - \frac{3u^3\ln(1/u)}{1-u^3} \cdot \phi_2\left(\frac{6u^6\ln(1/u)}{1-u^3} + o(1)\right)\right)\frac{1}{n} + o\left(\frac{1}{n}\right). \tag{4.23}$$

Using estimates for hypergeometric sums [GR90] whose details we omit here (see [KKKS98]) we can give an estimate for the sum in inequality (4.21). These computations lead to a closed expression that is an upper bound for $\mathbf{E}[|\mathcal{A}_n^{2\sharp}|]$ and can be used to derive the following result.

Theorem 4.3.3 ([KKKS98]). *If the clause to variable ratio exceeds* 4.601+ *then almost surely a random instance of* 3-*SAT is unsatisfiable.*

The required calculations of the value 4.601+ use the mathematical package Maple [Red94] and are guaranteed with accuracy of at least four decimal digits.

4.4 Satisfiability Threshold for 3-SAT

In Section 4.3 we considered techniques for estimating upper bounds on the unsatisfiability thresholds for 3-SAT. In this section our goal is to study lower bounds on the satisfiability threshold for 3-SAT. For a random instance ϕ of 3-SAT with a "critical number" of clauses of size 3 we are interested in establishing

$$\Pr[\phi \text{ is satisfiable}] = 1 - o(1) \tag{4.24}$$

Rather than merely establishing inequality (4.24) we will show how to find a truth assignment which satisfies ϕ with probability $1 - o(1)$. We will analyze an algorithm due to Chvátal and Reed [CR92]: given a random instance of 3-SAT with $(1 + o(1))rn$ clauses over n variables such that $r < 2/3$ the algorithm finds a truth assignment that satisfies this formula with probability at least $1 - o(1)$.

4.4.1 Satisfiability Heuristics

Several heuristics have been studied in the literature for finding satisfying truth assignments of a given instance of SAT. We consider the following definition:

Definition 4.4.1. *Setting a literal u in a formula ϕ at* 1 *amounts to removing from ϕ all the clauses that contain u and deleting $\overline{u}$ from all the clauses that contain $\overline{u}$. Let $\phi[u]$ denote the resulting formula.*

We consider the following general heuristic for satisfiability:

$\phi_0 = \phi$;
for $t = 1$ to n do
  choose a literal u_t whose value has not yet been set;
  $u_t = 1$;
  $\phi_t = \phi_{t-1}[u_t]$;
end

For a given formula ϕ the general heuristic above sets selected literals to 1. The heuristic produces a formula ϕ_n and the truth assignment resulting from this scheme satisfies ϕ if and only if ϕ_n includes no null clause (i.e., a clause of size zero).

Selection of the literal u_t (in step t) may be done in several ways. The selection procedure leads to different algorithms which we consider.

- UNIT CLAUSE (abbreviated UC): if ϕ_{t-1} includes a clause of size one then set such a variable at 1, otherwise u_t is chosen at random from the set of literals not yet set.
- GENERALIZED UNIT CLAUSE (abbreviated GUC): u_t is chosen at random from a randomly selected clause of ϕ_{t-1} of smallest size.

- SMALL CLAUSE (abbreviated SC): u_t is chosen at random from a clause of ϕ_{t-1} of size one; if no such clause exists then u_t is chosen at random from a clause of size two; if no clause of size one or two exists then u_t is chosen at random from the set of literals not yet set.

Thus SC differs from GUC in that if there are no clauses of size one or two then u_t is chosen at random from the set of all literals not yet used.

UC and GUC were first analyzed by Chao and Franco [CF86]. Our analysis in Subsection 4.4.2 is based on algorithm SC which was introduced by Chvátal and Reed [CR92].

In what follows we consider the formal description of SC and describe the algorithm in more detail.

- Represent the input as a matrix with m rows (one row for each clause) and k columns: the i-th row lists the literals occurring in the i-th clause. The output of SC is an array of length n that lists the values of the n variables (thus corresponding to a truth assignment).
- Construct $2n$ linked lists (called *incidence lists*), one for each literal, that enumerate all the clauses that contain the given literal u using a pointer to the corresponding row of the input matrix.
- For each $i = 1, 2, \ldots, n$ we associate an integer $size[i]$ (initialized to k) which keeps track of the current size of the i-th input clause, as well as a bit $sat[i]$ (initiated to false) which keeps track of whether or not the i-th input clause has been satisfied.
- We also use two stacks. STACK1 lists all clauses whose current size is one (and also possibly some clauses of size zero). STACK2 lists all the clauses whose current size is two (and possibly also some clauses of size at most one). Each clause in the stacks is represented by a pointer to the corresponding row of the input matrix.

Nest we consider the selection of the literal u_t from ϕ_{t-1} in the t-th iteration of SC. First attempt to find a clause of size one by popping STACK1 until an i with $size[i] = 1$ is found or the stack becomes empty; if there is no such clause then we attempt to find a clause of size two by popping STACK2 until an i with $size[i] = 2$ is found or the stack becomes empty. If no clause of size one or two is found then we let u_t be a randomly chosen entry in the corresponding row of the input matrix whose value has not yet been set. Otherwise we let u_t be the first variable in the output array whose value has not yet been set. Then we replace formula ϕ_t by scanning the incidence lists of literals.

It is important to note that SC maintains randomness in the following sense: if C is any clause of size s in ϕ_{t-1} other than the small clause from which u_t is selected then "u_t occurs in C" ("u_t does not occur in C") with probability $\frac{s}{2(n-t+1)}$ (see Exercise 4.8.6).

4.4.2 Threshold

We now give a precise proof of a general lower bound on the satisfiability threshold for k-SAT. For $k = 3$ this gives the value $2/3$.[1]

Theorem 4.4.1 ([CR92]). *Let k be an integer ≥ 3. If the clause to variable ratio of a random instance of k-SAT on n variables is less than*

$$\left(\frac{k-1}{k-3}\right)^{k-3} \cdot \frac{k-1}{k-2} \cdot \frac{2^k}{8k} \tag{4.25}$$

then heuristic SC, given a random formula, finds a truth assignment that satisfies this formula with probability $1 - o(1)$.

Proof. Let c be a positive constant less than the quantity in (4.25). There exists $\epsilon > 0$ such that

$$\frac{1+\epsilon}{1-2\epsilon} \cdot c < \left(\frac{k-1}{k-3}\right)^{k-3} \cdot \frac{k-1}{k-2} \cdot \frac{2^k}{8k}.$$

Let p_k denote the probability that a fixed input clause shrinks to size two after precisely t iterations. Then we have that

$$p_k = \frac{\binom{t-1}{k-3} \cdot \binom{n-t}{2}}{\binom{n}{k}} \cdot \frac{1}{2^{k-2}} = \frac{\binom{t-1}{k-3} \cdot \binom{n-t}{2}}{\binom{n-1}{k-1}} \cdot \frac{1}{n2^{k-2}}$$

and p_k is maximized when

$$\frac{k-3}{k-1} n \leq t \leq \frac{k-3}{k-1} n + 1.$$

This implies that

$$p_k \leq (1+\epsilon) \binom{k-1}{2} \left(\frac{k-3}{k-1}\right)^{k-3} \left(\frac{2}{k-1}\right)^2 \cdot \frac{k}{n2^{k-2}} < \frac{1-\epsilon}{m},$$

for all sufficiently large n.

Let a_t be the number of input clauses that shrink to size two after precisely t iterations. The number of clauses of size one or two in formula ϕ_t is at most b_t where it is easy to see that b_t satisfies the following recursive definition:

$$b_0 = 0 \text{ and } b_s = \max\{b_{s-1} - 1, 0\}, \text{ for } s \geq 1.$$

We have the following lemma.

Lemma 4.4.1. *We can prove*

1. $\Pr[\sum_{t=r}^{s} a_t \leq s - r] = 1 - o(1)$ *whenever* $s - r \geq 5 \ln n / \epsilon^2$.

---

[1] We use the convention $\left(\frac{2}{0}\right)^0 = 1$ in interpreting Formula (4.25).

2. $\Pr[a_t \le 1] = 1 - o(1)$ *whenever* $n - n^{3/4} < t \le n$.
3. $\Pr[b_t \le 5 \ln n/\epsilon^2] = 1 - o(1)$, *for all* t.
4. $\Pr[b_t \le 1] = 1 - o(1)$ *whenever* $n - n^{2/3} \le t \le n$.

Proof. The well-known inequality

$$\sum_{i \ge x} \binom{m}{i} p^i (1-p)^{m-i} \le \left(\frac{pm}{x}\right)^x \left(\frac{m - pm}{m - x}\right)^{m-x} \tag{4.26}$$

implies that

$$\sum_{i \ge x} \binom{m}{i} p^i (1-p)^{m-i} \le \left(\frac{pm}{x}\right)^x e^{x - pm} \le \exp(-\epsilon^2 x/2),$$

whenever $pm \le (1-\epsilon)x$. Thus if $1 \le r \le s \le n$ then

$$\Pr\left[\sum_{t=r}^{s} a_t \ge s - r + 1\right] \le \exp(-\epsilon^2 (s - r + 1)/2) = o(1/n^2),$$

whenever $s - r \ge 5 \ln n/\epsilon^2$. Since there are $\binom{n+1}{2}$ choices for r, s the proof of part (1) is complete. To prove part (2) note that inequality (4.26) implies that

$$\sum_{i \ge 2} \binom{m}{i} p^i (1-p)^{m-i} \le \left(\frac{\epsilon pm}{2}\right)^2.$$

Also note that

$$p_t m = O\left(\left(\frac{n-t}{n}\right)^2\right).$$

It follows that $\Pr[a_t \ge 2] = O(((n-t)/n)^4) = o(1)$, which proves part (2) of the lemma since there are at most $1 + n^{3/4}$ choices of t with $n - n^{3/4} < t \le n$. Parts (3) and (4) follow easily using inequality

$$b_t \le 1 + \max_r \sum_{t=r}^{s} (a_t - 1),$$

which is easily proved by induction on s. The proof of Lemma 4.4.1 is now complete.

To conclude the proof it will be necessary to prove that with probability $1 - o(1)$ the clauses of size two in ϕ_t remain distributed in a special way throughout the run of the algorithm. To capture this we will need the definition below.

Definition 4.4.2. *The formula* ϕ_t *is called well-behaved if*

1. the multigraph whose vertices are the variables in ϕ_t *and whose edges are the clauses of size two in* ϕ_t *contains no cycles (including cycles of length two),*

2. *for each connected component (including single-vertex components) there is at most one clause of size one in ϕ_t whose variable is a vertex of the component, and*
3. *there are no clauses of size zero in ϕ_t.*

The next lemma asserts that ϕ_t remains well-behaved with probability $1-o(1)$ throughout the run of SC.

Lemma 4.4.2. $\Pr[\phi_t$ *is well-behaved*$] = 1 - o(1)$, *for all* t.

Proof. Let ψ_t be the formula consisting of all the clauses of size at most two in ϕ_t. Let N_t be the clauses of size two that appear in ψ_t but not in ψ_{t-1}, and V_t the set of variables involved in the clauses of $\psi_{t-1}[u_t]$. Consider the events

- C_t: the multigraph formed by N_t contains a cycle.
- P_t: the multigraph formed by N_t contains a path with both endpoints in V_t.
- B_t: $|V_t| \leq 2b$ and $|N_t| \leq b$, with $b = 5\ln n/\epsilon^2$.

If ψ_t is well-behaved so is $\psi_{t-1}[u_t]$ and hence ϕ_t ceases to be well-behaved if and only if at least one of the events C_t, P_t takes place. It follows that

$$\Pr[\phi_j \text{ is not well-behaved}] \leq \sum_{t=1}^{j}(\Pr[C_t] + \Pr[P_t]).$$

Using part (3) of Lemma 4.4.1 we obtain $\Pr[B_1 B_2 \cdots B_n) = 1 - o(1)$. Hence

$$\Pr[C_t|B_t] \leq \sum_{i=2}^{n-t}(n-t)^i b^i \binom{n-t}{2}^{-i} < \frac{8b^2}{(n-t-1)^2}$$

and

$$\Pr[P_t|B_t] \leq \sum_{i=2}^{n-t-2}(2b)^2(n-t)^{i-1} b^i \binom{n-t}{2}^{-i} < \frac{16b^3}{(n-t)(n-t-1)}$$

whenever $n-t-1 \geq 4b$. Setting $j = n - \lfloor n^{2/3} \rfloor$ we conclude that

$$\Pr[\phi_t \text{ is well-behaved}] = 1 - o(1).$$

If $n - n^{2/3} \leq t \leq n$ holds then part (4) of Lemma 4.4.1 is true and ϕ_t includes at most one clause of size at most two and at most one new clause of size at most two appears in each subsequent iteration of SC. Hence formula ϕ_t remains well-behaved throughout $t = j+1, j+2, \ldots, n$ with probability $1-o(1)$. This completes the proof of Lemma 4.4.2.

The proof of Theorem 4.25 is now complete.

Theorem 4.4.1 is based on the study of the behavior of algorithm SC which is easier to analyze than algorithm GUC. Since at least one clause is satisfied each time GUC assigns a value to a variable it is intuitively clear that GUC is likely to perform better than SC. Frieze and Suen [FS96] give the precise limiting probability that GUC succeeds when applied to random instances of 3-SAT.

Theorem 4.4.2 ([FS96]). *If the clause to variable ratio is less than* 3.003 *then almost surely a random instance of* 3-*SAT is satisfiable.*

The precise analysis of this theorem represents intermediate states of GUC as a Markov chain. We will not prove the theorem here. Instead we refer the reader to [FS96] for details. We also note the recent paper of A.C. Kaporis, L.M. Kirousis, and E.G. Lalas that uses the technique of diffrential equations to prove that if the clause to variable ratio is less than 3.42 then almost surely a random instance of 3-SAT is satisfiable.[2] Theorem 4.3.3 in conjunction with Theorem 4.4.2 implies that a "threshold" for 3-SAT, if it exists, must lie between 3.003 and 4.601, a gap which is still rather large.

4.5 $(2+p)$-SAT

The $(2+p)$-SAT model was introduced by Monasson et al. [MZK$^+$] in order to illuminate the transition from 2-SAT to 3-SAT and is defined as follows.

Definition 4.5.1. *Consider n boolean variables and a random CNF formula $\phi = \phi_2 \wedge \phi_3$ consisting of m clauses such that ϕ_2 (ϕ_3) is a random* 2-*CNF* (3-*CNF*) *with $(1-p)m$ (pm) clauses. ϕ_2 and ϕ_3 are drawn independently from each other and p is a real number $0 \leq p \leq 1$.*

Notice that the case $p = 0$ corresponds to 2-SAT while the case $p = 1$ corresponds to 3-SAT.

Monasson et al. [MZK$^+$] observe that the known bounds for 2-SAT and 3-SAT imply that the critical value for $(2+p)$-SAT is bounded above by $\min\{1/(1-p), r_3/p\}$, where r_3 is an upper bound on the unsatisfiability threshold for 3-SAT (see Exercise 4.8.11). They also use methods from statistical mechanics in order to derive predictions for the value of the threshold for $(2+p)$-SAT. For example, using the "replica method" (see [MPV87, MZK$^+$]) they predict that for $p < 0.413$, $1/(1-p)$ is also a lower bound, i.e., for values of p in the range 0 to .413, $(2+p)$-SAT behaves in a manner similar to 2-SAT.

It can be shown (see [AKKK97]) that for $p \leq 2/5$ the threshold is indeed $1/(1-p)$, i.e., the clauses of length 3 are "irrelevant" to the critical behavior of $(2+p)$-SAT in this range. They also provide improved upper and lower

[2] "The Probabilistic Analysis of a Greedy Satisfiability Algorithm", in Proceedings of the European Symposium of Algorithms, 2002, to appear.

bounds for the threshold of $(2+p)$-SAT for values of $p > 2/5$. Their main theorem is the following.

Theorem 4.5.1 ([AKKK97]). *Let ϕ be a random instance of $(2+p)$-SAT.*

1. *If $p \leq 2/5$ then ϕ is almost surely satisfiable if $r < 1/(1-p)$ and almost surely unsatisfiable if $r > 1/(1-p)$.*
2. *If $p > 2/5$ then ϕ is almost surely satisfiable if $r < 24p/(p+2)^2$ and almost surely unsatisfiable if*

$$(7/8)^{rp}(3/4)^{r(1-p)}(2-e^{-2/3(1-5p/14)}) < 1.$$

We will not give the complete proof of this theorem here. Instead we will highlight the applicability of the single flip method and outline how Friedgut's technique in [Fri97] can be used to obtain the precise phase transition stated in Theorem 4.5.1. First we study an upper bound for the unsatisfiability threshold for $(2+p)$-SAT and then we study the behavior of $(2+p)$-SAT in the transition from 2-SAT to 3-SAT. Our presentation follows [AKKK97].

4.5.1 Unsatisfiability Threshold

We begin with the following definition.

Definition 4.5.2. *For a set of n variables V, let $C_i(V)$ denote the set of all $2^i\binom{n}{i}$ possible clauses of length i, on the variables of V. Unless, otherwise stated we will consider $V = \{x_1, \ldots, x_n\}$ and write C_i for $C_i(V)$. For a literal ℓ, $v(\ell)$ will denote the corresponding variable.*

Using [CR92, KKK96] we know that a random instance of $(2+p)$-SAT is almost surely unsatisfiable for $r > \min\{1/(1-p), 4.601/p\}$ simply by combining the bounds for 2-SAT and 3-SAT. It is easily seen that for $p \leq 0.821$ we have that $\min\{1/(1-p), 4.601/p\} = 1/(1-p)$. Extending the simple argument of Theorem 4.3.1 we can show that (see Exercise 4.8.11) a random instance of $(2+p)$-SAT is almost surely unsatisfiable for

$$r > \frac{\ln 2}{\ln(4/3) - p\ln(7/6)}.$$

Moreover it is easy to see that the right hand side above is less than $1/(1-p)$ for $p \geq 0.752$. In what follows we extend the single flip method to devise a bound that is tighter for values of $p \geq 0.6945$.

Lemma 4.5.1 ([AKKK97]). *Let r be the clause to variable ratio of a random instance ϕ of $(2+p)$-SAT with n variables. If*

$$(7/8)^{rp}(3/4)^{r(1-p)}(2-e^{-2/3(1-5p/14)}) < 1$$

then ϕ is almost surely unsatisfiable. Moreover, for $p \geq 0.6945$ the solution of

$$(7/8)^{rp}(3/4)^{r(1-p)}(2-e^{-2/3r(1-5p/14)}) = 1$$

is less than $1/(1-p)$.

Proof. Let $\mathcal{S}_n$ be the set of satisfying truth assignments of the random instance of $(2+p)$-SAT. As before we can apply the first moment method to the above random variable of the number of satisfying truth assignments (see Exercise 4.8.11). We use the refinement of the first moment method introduced in [KKK97, KKK96] which "compresses" $\mathcal{S}_n$ by requiring truth assignments not only to be solutions of ϕ but also to satisfy a certain "local maximality" condition. We need the following definition.

Definition 4.5.3. *The set $\mathcal{S}_n^\sharp$ is defined as the random set of truth assignments A such that*

1. *A satisfies ϕ, and*
2. *any truth assignment obtained from A by changing the value of exactly one variable from 0 to 1, does not satisfy ϕ.*

For a truth assignment A, and a variable x assigned value 0 by A, $A(x)$ will denote the truth assignment obtained from A, by changing the value of x from 0 to 1. The number of possible such changes for a truth assignment A (i.e., the number of 0s A contains) will be denoted by $sf(A)$ (for "single flips"). From [KKK96] we have that

$$\Pr[A \models \phi\,] \leq E[|\mathcal{S}_n^\sharp|] \tag{4.27}$$

and

$$E[|\mathcal{S}_n^\sharp|] = \Pr[A \in \mathcal{S}_n] \sum_{A \in \mathcal{A}_n} \Pr[A \in \mathcal{S}_n^\sharp \mid A \in \mathcal{S}_n] \ . \tag{4.28}$$

As we saw in the application of the first moment method (see Exercise 4.8.11),

$$\Pr[A \in \mathcal{S}_n] = (7/8)^{rpn}(3/4)^{r(1-p)n} \ . \tag{4.29}$$

We will now compute an upper bound on $\Pr[A \in \mathcal{S}_n^\sharp \mid A \in \mathcal{S}_n]$.

Fix a truth assignment $A \in \mathcal{S}_n$. Conditioning on $A \in \mathcal{S}_n$ exclude $\binom{n}{3}$ clauses from C_3 and $\binom{n}{2}$ clauses from C_2 from the conjuncts of ϕ. Now, consider changing the value of variable x from 0 to 1. The event $A(x) \notin \mathcal{S}_n$ occurs if and only if among the rn clauses in ϕ, there is a clause that contains $\overline{x}$ and is not satisfied by $A(x)$. In particular,

$$\begin{aligned}
\Pr[A(x) \notin \mathcal{S}_n] &= 1 - \left(1 - \frac{\binom{n-1}{2}}{7\binom{n}{3}}\right)^{rpn} \left(1 - \frac{\binom{n-1}{1}}{3\binom{n}{2}}\right)^{r(1-p)n} \\
&= 1 - \left(1 - \frac{3}{7n}\right)^{rpn} \left(1 - \frac{2}{3n}\right)^{r(1-p)n} \\
&= 1 - e^{-2/3r(1-5p/14)} + o(1) \ .
\end{aligned}$$

Let $\mathcal{A}_i$ denote the event $A(x_i) \notin \mathcal{S}_n$ and let $\mathcal{D}(x_i)$ denote the set of clauses associated with (that can cause) $\mathcal{A}_i$. The events $\mathcal{A}_i$ are not independent but $\mathcal{D}(x_i) \cap \mathcal{D}(x_j) = \emptyset$ for $i \neq j$. Intuitively, the occurrence of $\mathcal{A}_i$ "exposes" only clauses that cannot contribute to any $\mathcal{A}_j$, $i \neq j$. Moreover, it exposes the fact that at least one of the rn clauses was "used" to cause $\mathcal{A}_i$. Hence, the events $\mathcal{A}_i$ are in fact negatively correlated. Formally, this fact follows from the main Theorem of [McD92]. This implies that

$$\Pr[A \in \mathcal{S}_n^\sharp \mid A \in \mathcal{S}_n] \leq \left(1 - e^{-2/3(1-5p/14)} + o(1)\right)^{sf(A)} . \tag{4.30}$$

Hence, from (4.27),(4.28),(4.29),(4.30) we have that $\Pr[\phi$ is satisfiable$]$ is bounded above by

$$\begin{aligned}
&\leq (7/8)^{rpn}(3/4)^{r(1-p)n} \sum_{A \in \mathcal{A}_n} \Pr[A \in \mathcal{S}_n^\sharp \mid A \in \mathcal{S}_n] \\
&\leq (7/8)^{rpn}(3/4)^{r(1-p)n} \sum_{k=0}^{n} \binom{n}{k} \left(1 - e^{-2/3r(1-5p/14)} + o(1)\right)^k \\
&= \left((7/8)^{rp}(3/4)^{r(1-p)}(2 - e^{-2/3r(1-5p/14)} + o(1))\right)^n .
\end{aligned}$$

Thus, if

$$(7/8)^{rp}(3/4)^{r(1-p)}(2 - e^{-2/3r(1-5p/14)}) < 1 ,$$

then ϕ is almost surely unsatisfiable. This completes the proof of Lemma 4.5.1.

4.5.2 Transition from 2-SAT to 3-SAT

Let $f_p(n,r)$ denote the probability that a random instance of $(2+p)$-SAT with n variables and rn clauses has a satisfying truth assignment.

Lemma 4.5.2. *For all $p \in [0,1]$, there exists a function $r_p(n)$ such that for any $\epsilon > 0$,*

$$\lim_{n\to\infty} f_p(n, r_p(n) - \epsilon) = 1 \quad \textit{and} \quad \lim_{n\to\infty} f_p(n, r_p(n) + \epsilon) = 0.$$

Hence, for all $r \geq 0, p \in [0,1]$ and $\epsilon > 0$,

$$\textit{if} \ \lim_{n\to\infty} f_p(n,r) \geq \epsilon \ \textit{then} \ \lim_{n\to\infty} f_p(n,r) = 1 - o(1).$$

Proof. The proof is essentially identical to the proof of the existence of a sharp threshold for random k-SAT [Fri97]. In what follows we present an outline, highlighting the differences.

Let U_p denote the property of unsatisfiability for $(2+p)$-SAT, i.e., the set of all $(2+p)$-SAT instances that are unsatisfiable. For $\mathcal{L} = \phi_1, \ldots, \phi_l$ a fixed list of $(2+p)$-SAT formulas, let $F_{\mathcal{L}}$ denote the set of all $(2+p)$-SAT formulas that contain as a subformula (a copy of) some formula from $\mathcal{L}$. Using Theorem 5.1 of [Fri97], it follows that the lemma is false if and only if: for every $\epsilon > 0$ there exists a constant $\ell = \ell(\epsilon)$ and a list $\phi_1, \ldots, \phi_\ell$ such that for a random formula ϕ of $(2+p)$-SAT

$$\Pr[\phi \in U_p \mid \phi \in F_{\mathcal{L}}] \geq 1 - \epsilon \ .$$

In other words, unsatisfiability can be "approximated" by the property of containing some formula from a short, fixed list of formulas. To prove that this last statement leads to a contradiction it suffices to observe that all analogous steps of the proof in [Fri97] can be repeated verbatim, up to the final Lemma 5.7. For this final step, by analogy, we should take the codimension to be 3 with probability p and 2 with probability $1-p$, in each repetition. Taking the codimension to be 3 in all repetitions allows the lemma to be applied verbatim. Since by doing this, the expected number of clauses needed for coverage is increased only by a constant factor, the lemma still holds.

By Lemma 4.5.2, in order to prove that for some fixed r, p, random instances of $(2+p)$-SAT with rn clauses are almost surely satisfiable, it suffices to show that they are satisfiable with probability $\epsilon > 0$, where ϵ is independent of n. To show the latter, we will in fact show how to find a satisfying truth assignment, in polynomial time, with probability $\epsilon > 0$.

The algorithm used, UNIT CLAUSE (UC), was introduced and analyzed for random 3-SAT by Chao and Franco [CF86]. The algorithm makes n iterations, setting permanently one variable in each clause. In the following, "at time t" will mean after t such iterations have been performed (i.e., after t variables have been set). We will let $\mathcal{C}_i(t)$ denote the set of clauses of length i at time t. Unit clause without majority is the following algorithm.

- $V \leftarrow \{x_1, ..., x_n\}$.
- `For` $t = 0, \ldots, n-1$
  1. `If` $\mathcal{C}_1(t) \neq \emptyset$ `then` choose a random clause (literal) $\ell \in \mathcal{C}_1(t)$
     `else` choose a random variable $v \in V$ and
     take $\ell = v, \ell = \overline{v}$ with equal probability.
  2. Set $v(\ell)$ so as to satisfy ℓ.
  3. $V \leftarrow V - v(\ell)$.
  4. Remove all clauses containing ℓ.
  5. Remove $\overline{\ell}$ from all clauses (i.e., "shrink" all clauses containing $\overline{\ell}$).

At each substep of type (4) UC might generate a clause of length 0 and, clearly, such a clause will never be removed. On the other hand, if this never happens, i.e., $\mathcal{C}_0(n) = \emptyset$, then UC finds a satisfying truth assignment, in which case we say that it *succeeds*.

Let $C_i(t) = |\mathcal{C}_i(t)|$, not to be confused with our earlier notation $\mathcal{C}_i(V)$ from Definition 4.5.2. Lemmas 4.5.3, 4.5.5, below, were proven in [CF86] and concern the behavior of UC while processing a random instance ϕ of 3-SAT.

Lemma 4.5.3 (citeChFr86). *Let $V(t)$ denote the set of unset variables at time t. For all $0 \leq t \leq n$, $\mathcal{C}_i(t)$ is distributed as a set of $C_i(t)$ clauses chosen uniformly and independently from $\mathcal{C}_i(V(t))$.*

Proof. We give an outline of the proof. The lemma is true at $t = 0$ by the construction of ϕ. At each step, every clause c either becomes satisfied and is removed, or precisely one literal is removed from c, or c is not affected. Hence, after a step where variable v is set, the only new thing exposed about the remaining clauses is that they do not contain $v, \overline{v}$. The lemma follows inductively.

Lemma 4.5.4. *Lemma 4.5.3 remains true as long as each $C_i(0)$ is distributed as a set of $C_i(0)$ clauses chosen uniformly and independently from $C_i(V)$. In particular, it is true if ϕ is a random instance of $(2+p)$-SAT.*

Using Lemma 4.5.3 it can be shown that

Lemma 4.5.5 ([CF86]). *Let $\Delta C_i(t) = C_i(t+1) - C_i(t)$. For all $0 \leq t \leq n-3$,*

$$\mathbf{E}[\Delta C_3(t)] = -\frac{3}{n-t} C_3(t)$$
$$\mathbf{E}[\Delta C_2(t)] = \frac{3}{2(n-t)} C_3(t) - \frac{2}{n-t} C_2(t)$$

Proof. We give an outline of the proof. For each term we apply linearity of expectation to multiply a probability (that of containing the variable being set) with the number of choices (clauses of length i). Each negative term represents the expected number of clauses leaving $C_i(t)$ during step t, either as satisfied or as "shrunk". The positive term denotes the fact that half the clauses leaving $C_3(t)$, leave it as "shrunk" and hence end up in C_2. The negative terms follow by applying Lemma 4.5.3 to C_1 and noting that any choices made regarding which variable to set are random ones. Hence, in every step a random variable is set. For the positive term it suffices to add that to the previous argument that the choice of which "way" to set a variable, if it exists, is also random.

The next lemma is based on Lemma 4.5.5 and provides us with two functions that "approximate" $C_3(t), C_2(t)$ very well and almost surely.

Lemma 4.5.6. *Assume that functions $c_3(x), c_2(x)$ satisfy the following differential equations and boundary conditions.*

$$c_3'(x) = -\frac{3}{1-x} c_3(x) \ , \ c_3(0) = rp. \tag{4.31}$$

$$c_2'(x) = \frac{3}{2(1-x)} c_3(x) - \frac{2}{1-x} c_2(x) \ , \ c_2(0) = r(1-p). \tag{4.32}$$

Let ϕ be a random instance of $(2+p)$-SAT with n variables and rn clauses. If UC is applied to ϕ then almost surely for all $0 \leq t \leq n$, $i = 2, 3$

$$C_i(t) = c_i(t/n)\, n + o(n). \tag{4.33}$$

The following lemma states that if the rate at which "forced" variables are generated is low enough (less than one per step) then UC has a positive probability of success.

Lemma 4.5.7 ([CF86]). *Let $U(t)$ be the random variable defined as the number of clauses that shrink to unit length, during step t. Let $g(t) = \mathbf{E}[U(t)]$.*
(i) For all $0 \leq t \leq n-2$,

$$g(t) = \frac{1}{n-t} C_2(t) \ .$$

(ii) If there is $\delta > 0$, such that for all $0 \leq t \leq n-2$,

$$g(t) \leq 1 - \delta$$

then UC succeeds with probability $\rho = \rho(\delta) > 0$.

The unique solution of the differential equations (and boundary conditions) (4.31), (4.32) of Lemma 4.5.6 is

$$\begin{aligned} c_3(x) &= rp(1-x)^3 \\ c_2(x) &= 1/2r(3px - 2p + 2)(1-x)^2 \ . \end{aligned} \tag{4.34}$$

Combining (4.33),(4.34) and Lemma 4.5.7 we get that for a random instance ϕ of $(2+p)$-SAT with rn clauses, UC succeeds with positive probability if for all $x \in [0,1]$,

$$1/2r(3px - 2p + 2)(1-x) < 1. \tag{4.35}$$

For $p \leq 2/5$, the left hand side of (4.35) is non-increasing with x and hence (4.35) holds if and only if

$$r < \frac{1}{1-p}. \tag{4.36}$$

For $p > 2/5$, the left hand side of (4.35) is unimodal for $x \in [0,1]$, the unique maximum occuring for $x = (5p-2)/6$. Substituting this value of x in (4.35) we get

$$r < \frac{24p}{(p+2)^2} \ .$$

The discussion above completes the outline of the proof of Theorem 4.5.1.

The reader interested in more details of the proof should consult Friedgut's original paper [Fri97].

4.6 Constraint Programming

In this section we discuss random models of constraint programming and consider associated phase transitions.

4.6.1 Models of CSP

Definition 4.6.1. *A* constraint network *consists of a set $X_1, \ldots, X_n$ of variables with domain D, and a set of constraints C. For $2 \leq k \leq n$ a constraint $R_{i_1,i_2,\ldots,i_k} \in C$ is a subset of D^k, where the $i_1, i_2, \ldots, i_k$ are distinct. We say that $R_{i_1,i_2,\ldots,i_k}$ has arity k and that it bounds the variables $X_{i_1}, \ldots, X_{i_k}$.*

Definition 4.6.2. *For a given constraint network, the Constraint Satisfaction Problem (CSP), asks for all the n-tuples $(d_1, \ldots, d_n)$ of values such that $d_i \in D$, $i = 1, \ldots, n$, and for every $R_{i_1,i_2,\ldots,i_k} \in C$, $(d_{i_1}, d_{i_2}, \ldots, d_{i_k}) \notin R_{i_1,i_2,\ldots,i_k}$. Such an n-tuple is called a solution of the CSP. The decision version of the CSP is determining if a solution exists.*

Definition 4.6.3. *For an instance ϕ of CSP with n variables, its* constraint hypergraph *G^ϕ (or just G when no confusion may arise) has n vertices $v_1, v_2, \ldots, v_n$, which correspond to the variables of ϕ and it contains a hyperedge $\{v_{i_1}, v_{i_2}, \ldots, v_{i_k}\}$ if and only if there exists a constraint of arity k that bounds the variables $X_{i_1}, X_{i_2}, \ldots, X_{i_k}$.*

We will use a convenient graph-theoretic representation of a CSP instance ϕ, defined as follows:

Definition 4.6.4. *The incompatibility hypergraph of ϕ, C^ϕ (or just C, when no confusion may arise) is an n-partite hypergraph. The i-th part of C^ϕ corresponds to variable X_i of ϕ and has exactly D vertices, one for each value in D.[3] In C^ϕ there exists a hyperedge $\{v_{i_1}, v_{i_2}, \ldots, v_{i_k}\}$, if and only if the corresponding vertices $d_{i_1}, d_{i_2}, \ldots, d_{i_k} \in D$ are in (not allowed by) some constraint that bounds the corresponding variables.*

Hence, the decision version of CSP is equivalent to asking if there is a set of vertices in C containing exactly one vertex from each part while not "containing" any hyperedge, i.e., whether there exists an independent set with one vertex from each part.

It is common practice for generating random CSP to make the simplifying assumption that all the constraints are of the same arity, k. Hence, fixing n, k and D the generation of a random CSP is usually done in two parts. First, the constraint hypergraph G is chosen as a random k-uniform hypergraph; then for each hyperedge (constraint) in G a set of hyperedges (forbidden k-tuples of values) is "filled in" in C. A general framework for doing this has been introduced in the papers presented in [BB96]. More precisely,

1. Either each one of the $\binom{n}{k}$ hyperedges is selected to be in G independently with probability p_1, or we uniformly select a random set of hyperedges of size $p_1 \binom{n}{k}$.

[3] To simplify notation we will use the same symbol to denote the domain D and its number of elements. The precise meaning will always be clear from the context.

2. Either for every hyperedge of G each one of the D^k k-tuples is selected with probability p_2, or for every hyperedge of G we uniformly select a random set of k-tuples of size p_2D^k.

The value p_1 determines how many constraints exist in a CSP instance and is called *constraint density*. The value p_2 determines how restrictive the constraints are and is called *constraint tightness*. Here $p_2 \in (0, 1)$ is taken to be a constant independent of any other model parameters. Since, p_2 is a constant, p_1 is varied so that there are $\Theta(n)$ constraints in total. It is not hard to see that unless the total number of constraints is $\Theta(n)$ the resulting instances are trivially underconstrained or overconstrained. Combining the options for the two steps we get four slightly different models for generating random CSP which have received various names in the past literature. In particular, in the terminology of Smith and Dyer in [BB96], if both Step 1 and Step 2 are done using the first option (the $G_{n,p}$ fashion, in random graph terminology) then we get Model A while if they are both done using the second option (the $G_{n,m}$ fashion, in random graph terminology) we get Model B.

Theorem 4.6.1. *Let ϕ be a random CSP with D^{k-1} or more forbidden k-tuples in each constraint (assume that we use the second option for Step 2). Then ϕ almost surely has no solution.*

Proof. We give an outline of the proof. Note that since p_2 is a constant the parameterization implies $\Theta(D^k)$ forbidden k-tuples in each constraint and hence requiring D^{k-1} or more forbidden k-tuples in each constraint is a very weak condition.

Specifically, if $p_2D^k \geq D^{k-1}$ then for every constraint there is a *constant* probability, i.e., depending only on k, D, that some vertex (value) in the domain of one of the variables bound by it, has degree D^{k-1}. We will call such a vertex *universal* since it participates in hyperedges containing all possible D^{k-1} combinations of values for the other $k - 1$ variables in the constraint. It is clear that no solution of ϕ assigns a universal value to any variable and hence if for some variable all the values in its domain are universal then ϕ has no solution.

Now consider the constraint hypergraph G. Recall, that the degree of a variable in G is the number of constraints by which the variable is bound. It's not hard to see that the probability that *all* the values in the domain of a particular variable are universal depends only on the variable's degree, k and D and if the degree is at least D then this probability is a strictly positive constant. Since the total number of hyperedges (constraints) in G is $\Theta(n)$, using methods from Chapter 8 of [ASE92], its not hard to verify that the degree of each vertex (variable) is, asymptotically, a Poisson-distributed random variable with mean a constant for both options for Step 1. This implies that for any constant d there exist $\Omega(n)$ vertices that have degree at least d. Hence, if we take $D = d$, as n tends to infinity, the probability that there exists a variable with all the values in its domain universal, tends to 1 since we are making n experiments each one with a probability of "success" that is positive and independent of n.

Finally, if we use the first option for step 2 and take p_2 so that the expected number of forbidden k-tuples is D^{k-1} or more, then each constraint has a constant probability of having at least D^{k-1} forbidden k-tuples and the same arguments can be applied.

4.6.2 A New Model for Random CSP

Note that fixing the old models by explicitly disallowing universal values, say by requiring each value to have degree less than D^{k-1}, will probably not lead to any interesting models. This is because the old models generate many other such *local* flaws similar to the one in the previous section. For example we could have a pair of variables where neither is impossible to set by itself but where every setting of one makes it impossible to set the other (again for some local reasons as in the problem we point out).

In what follows we discuss a new model for random CSP first proposed by Achlioptas et al. [AKK+97] which does not seem to have any such trivial local inconsistencies. It will be more convenient to describe how to generate C^ϕ for a random CSP instance ϕ.

Definition 4.6.5 (Model E). *C^ϕ is as a random n-partite k-uniform hypergraph with D nodes in each part. That is, either each one of the $\binom{n}{k}D^k$ hyperedges is selected to be in C independently with probability p, or we uniformly select a random set of hyperedges of size $m = p\binom{n}{k}D^k$.*

To generate an instance ϕ of random CSP according to model E we can, alternatively, do the following: We first generate a random k-regular hypergraph G on n vertices (variables), as in Step 1. Then, to construct C^ϕ we expand each vertex v in G to an independent set of D vertices and we substitute each vertex in each hyperedge e of G with a randomly picked vertex from the corresponding independent set. This generates the same distribution of random n-partite hypergraphs with D nodes in each part, as Model E. We will now show that if for a random instance ϕ generated using Model E, C^ϕ has fewer than $\frac{1}{k(k-1)}n$ hyperedges then ϕ almost surely has a solution. This fact follows by using the alternative way to generate C^ϕ, a well-known fact related to the birth of a giant component in a random hypergraph, and a simple lemma. It seems possible that substantially stronger bounds can be derived by more sophisticated analysis.

Lemma 4.6.1. *If a k-regular hypergraph G is formed by selecting each of the $\binom{n}{k}$ hyperedges with probability $p < \frac{(k-2)!}{n^{k-1}}$ then, almost surely, all the components of G will contain* at most one *cycle. This is also true if G is formed by selecting uniformly a set of hyperedges of size $p\binom{n}{k} < \frac{n}{k(k-1)}$.*

Lemma 4.6.2. *If in an instance ϕ of CSP every variable belongs to at most one cycle then ϕ has at least one solution.*

Proof. Assume that a hyperedge e disappears from C^{ϕ} if any of the variables it bounds is set to a value other than that forbidden by e. We take care of cycles first.

If $k \geq 3$, since there is only one cycle in each component, in each hyperedge of the cycle there is a vertex (value) v that is not connected to any other vertex in the cycle. Setting the corresponding variable of each such v to a value other than v eliminates the cycle. If $k = 2$ the lemma is a well known fact when $D = 2$ [CR92] (described in terms of 2-SAT) and increasing D only makes things easier.

Note that after we take care of the cycles, since there is only one cycle in each component, each remaining hyperedge contains at least $k - 1$ unset variables. Hence, any acyclic components remaining are easy to deal with by repeatedly picking an arbitrary hyperedge and an unset variable in it and setting the corresponding variable to a value other than that forbidden by the hyperedge. This completes the proof of Lemma 4.6.2.

4.6.3 The Method of Local Maxima

Let ϕ be a random CSP with n variables, generated in any of the ways defined so far. By $\mathcal{A}_n$ we denote the set of all value assignments for ϕ, and by $\mathcal{S}_n$ we denote the random set of solutions of ϕ. We are interested in establishing a condition subject to which the probability that ϕ has a solution decreases exponentially with n. Such a condition is readily provided by the *first moment* method (for excellent expositions see [ASE92] and [Spe94]). That is, by first noting that

$$E[|\mathcal{S}_n|] = \sum_{\phi} \left(\Pr[\phi] \cdot |\mathcal{S}_n(\phi)|\right) \tag{4.37}$$

and then noting that

$$\Pr[\phi \text{ has a solution}] = \sum_{\phi} \left(\Pr[\phi] \cdot I_{\phi}\right), \tag{4.38}$$

where for an instantiation of the random variable ϕ the indicator variable I_{ϕ} is defined as

$$I_{\phi} = \begin{cases} 1 \text{ if } \phi \text{ has a solution}, \\ 0 \text{ otherwise}. \end{cases}$$

Hence, from (4.37) and (4.38) we get

$$\Pr[\phi \text{ has a solution}] \leq E[|\mathcal{S}_n|]. \tag{4.39}$$

Calculating $E[|\mathcal{S}_n|]$ appears to be much easier than calculating the probability that ϕ has a solution. As an illustration we apply the first moment method to the model we suggest with $m = rn$. There are D^n possible value assignments and each one of the $\binom{n}{k} D^k$ possible constraints has a probability of $1/D^k$ of being violated by a random value assignment. Since, we have m constraints,

$$\begin{aligned}\Pr[\phi \text{ has a solution}] &\le E[|\mathcal{S}_n|] \\ &= D^n \left(1 - \tfrac{1}{D^k}\right)^m \\ &= \left(D\left(1 - \tfrac{1}{D^k}\right)^r\right)^n .\end{aligned} \tag{4.40}$$

Hence, if $r > \ln(1/D)/\ln(1-1/D^k) \approx D^k \ln D$ then $D(1-\frac{1}{D^k})^r < 1$ and the probability that ϕ has a solution drops exponentially with n, asymptotically tending to zero.

The price paid for the simplicity of the first moment method is that instances with a very large number of solutions, although they may occur with very small probability, contribute substantially to $E[|\mathcal{S}_n|]$. Hence, by substituting $|\mathcal{S}_n(\phi)|$ for I_ϕ we might be giving away "a lot". The technique introduced in [KKK96], when applied to CSP amounts to "compressing" $\mathcal{S}_n$ by requiring value assignments not only to be solutions of ϕ but to also satisfy a certain "local maximality" condition. The underlying intuition is that for a random solution of ϕ, if we choose a variable at random and change its value the probability that we will end up with a non-solution is small. Consequently, solutions (when they exist) tend to appear in large "clusters" and instead of counting all solutions in a cluster we need only count a representative one (locally maximum). It appears that this clustering is not specific to the model we put forward and that it is closely related to the notion of *influence* introduced in [KKL88]. The analysis following is similar to that used in Subsection 4.3.2.

Definition 4.6.6. *For each variable in ϕ fix an arbitrary ordering of the values in its domain. The set $\mathcal{S}_n^\sharp$ is defined as the random set of value assignments A such that*

1. *A is a solution of ϕ (written $A \models \phi$), and*
2. *any value assignment obtained from A by changing the value of exactly one variable to some greater value, is not a solution of ϕ.*

For a value assignment A we will denote the change of the value assigned to the variable X to a value v that is greater than the currently assigned value by $A(X,v)$. The number of possible such changes denoted by $sf(A)$. The assignment obtained by applying one such change is denoted by $A(X,v)$. By merely repeating the proof of Lemma 4.3.1 we derive the following result.

Lemma 4.6.3. $\Pr[\phi$ has a solution$] \le E[|\mathcal{S}_n^\sharp|]$.

Since $\mathcal{S}_n$ can be written as the sum of D^n indicator variables, one for each possible value assignment A and nonzero if and only if A is a solution of ϕ, we obtain the following lemma by conditioning on $A \models \phi$:

Lemma 4.6.4.

$$E[|\mathcal{S}_n^\sharp|] = \Pr[A \in \mathcal{S}_n] \sum_{A \in \mathcal{A}_n} \Pr[A \in \mathcal{S}_n^\sharp \mid A \in \mathcal{S}_n]. \tag{4.41}$$

In what follows we show how to calculate an upper bound for $E[|\mathcal{S}_n^\sharp|]$ for model E.

4.6.4 Threshold for Model E

We apply the single flips technique to random CSP instances generated using Model E where we select uniformly a set of $m = rn$ incompatibility hyperedges. It is straightforward to show, using for example the techniques in [Bol85], that if there exists a sharp threshold for having a solution in Model E then its location is the same for both variations of the model. With respect to upper bounds, the model we analyze tends to give somewhat better results.

Theorem 4.6.2 ([AKK+97]). *For a random CSP generated according to Model E*

$$E[|\mathcal{S}_n^\sharp|] \leq \left(\left(1 - \frac{1}{D^k}\right)^r \frac{1}{\zeta}\left(1 - (1-\zeta)^D\right) + o(1)\right)^n, \text{ where } \zeta = e^{-\frac{kr}{D^k-1}}.$$

Proof. As discussed earlier,

$$\Pr[A \in \mathcal{S}_n] = (1 - \frac{1}{D^k})^m.$$

In view of identity (4.41) it is enough to compute an upper bound on $\Pr[A \in \mathcal{S}_n^\sharp \mid A \in \mathcal{S}_n]$. Fix a value assignment $A \in \mathcal{S}_n$. Since $A \in \mathcal{S}_n$, every k-subset of A is *not* a hyperedge in C. Consequently, conditioning on $A \in \mathcal{S}_n$ implies that the set of possible hyperedges has cardinality $D^k\binom{n}{k} - \binom{n}{k} = (D^k - 1)\binom{n}{k}$. Now consider a fixed $A(X, v)$. The event $A^{sf_0} \not\models \phi$ occurs if and only if among the m hyperedges in C, *at least* one hyperedge contains v and $k-1$ other values that remained the same. Hence,

$$\begin{aligned}\Pr[A^{sf_0} \not\models \phi] &= 1 - \left(1 - \frac{\binom{n-1}{k-1}}{(D^k - 1)\binom{n}{k}}\right)^m \\ &= 1 - \left(1 - \frac{k}{(D^k - 1)n}\right)^{rn} \\ &= 1 - e^{-\frac{kr}{D^k-1}} + o(1).\end{aligned}$$

The events $A(X_i, v_j) \not\models \phi$, for each $A(X_i, v_j)$ are not independent. On the other hand, as we saw above, the set of hyperedges associated with each such event is disjoint from all other such sets. Intuitively, any such event $A(X_i, v_j) \not\models \phi$ "exposes" only hyperedges that can be of no harm to any other flip. Moreover, it exposes that at least one of the m hyperedges was "used" to cause $A(X_i, v_j) \not\models \phi$.

Hence, the events $A(X_i, v_j) \not\models \phi$ are in fact negatively correlated. Formally, this fact follows from the main theorem in [McD92]. This implies,

$$\begin{aligned}\Pr[A \in \mathcal{S}_n^\sharp \mid A \in \mathcal{S}_n] &\leq \left(1 - \left(1 - \frac{k}{n(D^k-1)}\right)^{rn}\right)^{sf(A)} \\ &= \left(1 - e^{-\frac{kr}{D^k-1}} + o(1)\right)^{sf(A)}\end{aligned}$$

If for each value assignment A we let k_i denote the number of variables assigned the i-th smallest value in their domain then

$$\begin{aligned}
\mathbf{E}[|\mathcal{S}_n^\sharp|] &\leq \left(1 - \tfrac{1}{D^k}\right)^m \sum_{A \in \mathcal{A}_n} \Pr[A \in \mathcal{S}_n^\sharp \mid A \in \mathcal{S}_n] \\
&\leq \left(1 - \tfrac{1}{D^k}\right)^m \sum_{k_1, k_2, \ldots, k_D} \binom{n}{k_1, k_2, \ldots, k_D} \\
&\quad \cdot \prod_{j=1}^{D} \left(1 - e^{-\frac{kr}{D^k - 1}} + o(1)\right)^{(D-j)\, k_j} \\
&= \left(\left(1 - \tfrac{1}{D^k}\right)^r \sum_{j=0}^{D-1} (1 - e^{-\frac{kr}{D^k-1}})^j + o(1)\right)^n \\
&= \left(\left(1 - \tfrac{1}{D^k}\right)^r \tfrac{1}{\zeta}\left(1 - (1 - \zeta)^D\right) + o(1)\right)^n,
\end{aligned} \tag{4.42}$$

where $\zeta = e^{-\frac{kr}{D^k-1}}$. This completes the proof of Theorem 4.6.2.

Therefore, if $\left(1 - \frac{1}{D^k}\right)^r \sum_{j=0}^{D-1} (1 - e^{-\frac{kr}{D^k-1}})^j < 1$ then $\Pr[\phi$ has a solution$]$ tends to zero with n. This is a more relaxed condition than $\left(1 - \frac{1}{D^k}\right)^r D < 1$, derived earlier by the first moment method, since $\sum_{j=0}^{D-1} (1 - e^{-\frac{kr}{D^k-1}})^j < D$. For example, when $k = 2, D = 3$ the first moment method gives $r < 9.32$ while Theorem 4.6.2 gives $r < 8.21$.

4.7 Historical and Bibliographical Remarks

This chapter was devoted entirely to the study of the threshold phenomenon for satisfiability. We stressed mainly the threshold for 3-SAT but similar patterns hold for k-SAT, $k \geq 3$. For the case $k = 3$ important techniques leading to upper bounds were reported by [KMPS95]. The method of single flips was first introduced by Kirousis, Kranakis, and Krizanc in [KKK97] and independently by Dubois, and Boufkhad in [DB97]. Both papers analyze the single-flip method: the first paper reports a 4.667 upper bound while the second paper 4.642. However the general method of flipping variables was not introduced and studied until [KKK96, KKKS98] which also gives the upper bound 4.601+. Recently, Dubois, Boufkhard and Mandler announced the upper bound 4.506 in [DBM00]. We mentioned in Theorem 4.4.2 the 3.003 lower bound for the satisfiability threshold of 3-SAT. Recently, D. Achlioptas [Ach00] improved this to 3.145 by introducing a new heuristic that sets two variables at a time. For a general analysis of lower bounds cia differential equations we refer the reader to [AF00]. In view of the NP-completeness of the satisfiability problem for 3-SAT and the fundamental importance of this problem in many areas of computer science, this threshold phenomenon has given rise to numerous studies. A recent survey paper concerning the problem of finding hard instances of satisfiability is [CM97].

As reported in this chapter the technique of single flips has proved fruitful for the study of thresholds of related problems, e.g., $(2+p)$-SAT and CSP. It has also been applied successfully to the problem of 3-colorability [AM97a] (see also [AM99]). Phase transitions have been observed in many other areas as well. Influential studies concerning phase transitions in computational systems appeared in [KP83, HH87].

To date the existence of such a threshold is not known. However, Friedgut [Fri97] has recently shown that there is a function $r^k(n)$ such that the phase transition of k-SAT occurs within an ϵ neighborhood of $r^k(n)$: thus it is still possible that although a sharp transition may exist the critical value does not converge to any given constant. Friedgut's technique uses differential equations to describe the "mean path" of Markov chains (in this case defined implicitly), a technique which has found many applications [AM97b, CF86, CF90, FS96, KS81, MR95, PSW96, Wor95]. A rigorous, unifying framework regarding the applicability of this technique was developed by Wormald in [Wor95].

4.8 Exercises

Exercise 4.8.1. Let ϕ be an arbitrary boolean formula with m clauses. Let c be the smallest clause size in ϕ. If $m < 2^c$ then the formula ϕ is satisfiable. Conclude that any collection of $m < 2^k$ clauses of k-SAT must be satisfiable.

Exercise 4.8.2. Imitate the proof of Theorem 4.3.1 in order to prove the following result. If the clause to variable ratio exceeds $2^k \ln 2$ then almost surely a random instance of k-SAT is unsatisfiable.

Exercise 4.8.3. Show that with probability $1-o(1)$ a random instance of 1-SAT with $m(n)$ clauses over n variables is unsatisfiable whenever $\frac{m(n)}{m} \to \infty$ and satisfiable whenever $\frac{m(n)}{m} \to 0$.

Exercise 4.8.4 ([Coo71]). Show that for every satisfiable instance of 2-SAT a satisfying truth assignment can be found in time polynomial in the size of the input. (There are linear time algorithms for this problem [EIS76], [APT79].)

Exercise 4.8.5 ([APT79]). Let ϕ be an instance of 2-SAT on n variables. Consider the directed graph $G(\phi)$: it has $2n$ vertices (one for each literal arising from the n variables) and for each clause $u \vee v$ of ϕ the directed edges $\overline{u} \to v$, $\overline{v} \to u$. The following is used in the proof of Lemma 4.2.2: ϕ is unsatisfiable if and only if some variable and its complement belong to the same strongly connected component of $G(\phi)$.

Exercise 4.8.6. Prove the assertion made in Subsection 4.4.1: Algorithm SC maintains randomness in the following sense: if C is any clause of size s in ϕ_{t-1} other than the small clause from which u_t is selected then "u_t occurs in C" (respectively, "u_t does not occur in C") with probability $\frac{s}{2(n-t+1)}$.

Exercise 4.8.7 ([Ach97]). ($\star$) Let d be the Hamming distance among n-tuples in the set $\{0,1\}^n$. For $x \in \{0,1\}^n$ and S a non-empty subset of $\{0,1\}^n$ define $N_S(x) = |\{y \in S : d(x,y) = 1\}|$ and $val_S^n(x) = 2^{-N_S(x)}$.

1. Prove that
$$val(S) := \sum_{x \in S} val_S^n(x) \geq 1. \tag{4.43}$$
HINT. Use induction on n.
2. For a random instance ϕ of 3-SAT with m clauses let S be the set of truth assignments satisfying ϕ. Let p be the probability that a satisfying truth assignment has a neighbor (i.e., of Hamming distance 1) which is also a satisfying truth assignment of ϕ. Show that
$$\mathbf{E}[val(S)] = (7/8)^m (2-p)^n.$$
HINT. Use identity (4.43).
3. Use this to derive a new proof of Theorem 4.3.2.
HINT. Observe that $\Pr[S \neq \emptyset] \leq \mathbf{E}[val(S)]$.

Exercise 4.8.8. For every $l \geq 0$, the probability that a random formula is satisfiable is given by the following harmonic mean formula due to Aldous [Ald89]
$$\sum_A \left(\Pr[A \in \mathcal{A}_n^l] \cdot \mathbf{E}\left[\frac{1}{|\mathcal{A}_n^l|} \middle| A \in \mathcal{A}_n^l \right] \right).$$
HINT. Use indicator functions (for more details see [KKKS98]).

Exercise 4.8.9 ([KKK97]). Use the technique of Theorem 4.3.2 to the case of arbitrary k-SAT in order to prove the following result.

1. For the case of k-SAT ($k \geq 3$), the expected value $\mathbf{E}[|S_n^\sharp|]$ is at most $((2^k-1)/2^k)^{rn}(2-e^{-kr/(2^k-1)}+o(1))^n$. It follows that the unique positive solution of the equation
$$\left(\frac{2^k-1}{2^k}\right)^r (2-e^{-kr/(2^k-1)}) = 1,$$
is an upper bound for the unsatisfiability threshold for k-SAT).
2. Use the mathematical package Maple to compute the solutions of the corresponding equations for $k = 4, 5, 6, 8$.

Exercise 4.8.10. Consider the general heuristic for satisfiability in Subsection 4.4.1 with a given input given formula ϕ. The heuristic outputs a formula ϕ_n and the truth assignment resulting from this scheme satisfies ϕ if and only if ϕ_n includes no null clause.

Exercise 4.8.11. In this exercise we consider elementary threshold bounds for $(2+p)$-SAT.

1. Prove that the critical value for $(2+p)$-SAT is bounded above by $\min\{1/(1-p), r_3/p\}$, where r_3 is an upper bound on the unsatisfiability threshold for 3-SAT.
2. Use Theorem 4.3.3 to conclude that the value above is $\min\{1/(1-p), 4.601/p\}$. Conclude that for $p \leq 0.821$ we have that $\min\{1/(1-p), 4.598/p\} = 1/(1-p)$.
3. Let ϕ be a random instance of $(2+p)$-SAT with n variables and rn clauses and let $\mathcal{S}_n$ denote the random set of solutions of ϕ. Use the first moment method (see Theorem 4.3.1) to show that

$$\mathbf{E}[|\mathcal{S}_n|] = 2^n (7/8)^{rpn} (3/4)^{r(1-p)n}.$$

Conclude that ϕ is unsatisfiable if

$$r > \frac{\ln 2}{\ln(4/3) - p\ln(7/6)}.$$

The right hand side above is less than $1/(1-p)$ for $p \geq 0.752$.

Exercise 4.8.12. Prove Lemma 4.6.3.

Exercise 4.8.13 ([AKK⁺97]). Apply the method of local maxima to the models described in Subsection 4.6.1 and derive upper bounds on the corresponding thresholds.

Exercise 4.8.14. Formulate k-SAT as a constrained satisfaction problem.

Exercise 4.8.15 (Open Problem). Extend the technique of single- and double-flips to f-flips, where $f \geq 3$ variables are being flipped.

Exercise 4.8.16. Prove that both $\mathbf{E}[|\mathcal{A}_n^{2\sharp}|]$ and $\mathbf{E}[|\mathcal{A}_n^{2}|]$ (defined in Subsection 4.3.4 converge to zero for the same values of r.

5. Propositional Proof Systems

... we might imagine a machine where we should put in axioms at one end and take out theorems at the other, like that legendary machine in Chicago where pigs go in alive and come out transformed into hams and sausages. *H. Poincaré [Poi52]*

5.1 Introduction

Mathematicians have always been interested in finding the shortest, most elegant proof of a given theorem, using the minimum number of hypotheses. For example, the prime number theorem was originally proved by de la Vallée Poussin using contour integration over complex numbers. Only much later did Erdős and Selberg give a proof of the this theorem, using only "elementary" properties of the integers. Beginning with the work of Peano and especially Frege [Fre67, Fre93], proofs of various mathematical theorems have been formalized in certain formal proof systems. Hilbert later extensively developed proof theory, attempting to prove the consistency of mathematics by finite combinatorial means; however, Hilbert's program was shown to be unrealizable by Gödel in his 1931 Incompleteness Theorem.[1]

In this chapter, we consider propositional proof systems, because of their relation to boolean circuits and open problems in complexity theory, such as the NP =? *co*-NP. Certain lower bound techniques developed in Chapter 2 can be adapted to prove lower bounds for propositional proof length, though the proofs for propositional proof systems are often substantially more difficult than the case for boolean circuits.

Propositional formulas, built up from propositional variables and connectives, are called *tautologies*, provided that they are true under every truth assignment to the variables. Clearly, the length of proof of a tautology depends on the axiom system and rules of inference, in which the proof is carried out. The present chapter is concerned with the complexity (size, length, width, depth, degree, etc.) of derivation of propositional tautologies in proof systems as a function of the size (number of symbols) of the tautology. Here, we

[1] Less widely known is that in 1956, long before the theory of NP-completeness, Gödel asked J. von Neumann a question concerning proof length, whose affirmative answer would imply P = NP (see [CK93] for explanation and a copy of Gödel's letter).

study Gentzen propositional sequent calculus (LK), resolution (R), Nullstellensatz systems (NS), polynomial calculus (PC), cutting planes (CP), propositional threshold calculus (PTK), Frege systems (F), extended Frege systems (EF), and substitution Frege systems (SF). Proof systems will be compared via polynomial simulations, and combinatorial principles such as the pigeonhole principle are introduced, in order to separate proof systems in terms of strength.

Several axiom systems of propositional logic are known. One of the first is due to Frege, as later simplified by Lukasiewicz [HA50], and uses only the connectives $\rightarrow, \neg$. Formulas are inductively defined from the propositional variables $x_1, x_2, \ldots$ using these connectives, and the following axioms are admitted:

$$\begin{aligned} &p \rightarrow (q \rightarrow p) \\ &(p \rightarrow (q \rightarrow r)) \rightarrow ((p \rightarrow q) \rightarrow (p \rightarrow r)) \\ &(\neg q \rightarrow \neg p) \rightarrow ((\neg q \rightarrow p) \rightarrow q) \end{aligned} \tag{5.1}$$

where p, q, r are propositional variables. The only rule of inference is that of *modus ponens*

$$\frac{p, p \rightarrow q}{q}. \tag{5.2}$$

Instantiations of the axioms, obtained by simultaneous and uniform substitution of arbitrary formulas for the propositional variables, yield

$$\begin{aligned} &F \rightarrow (G \rightarrow F) \\ &(F \rightarrow (G \rightarrow H)) \rightarrow ((F \rightarrow G) \rightarrow (F \rightarrow H)) \\ &(\neg F \rightarrow \neg G) \rightarrow ((\neg F \rightarrow G) \rightarrow F) \end{aligned}$$

and a simular instantiation of the rule of inference yields

$$\frac{F, F \rightarrow G}{G}.$$

where F, G, H are arbitrary formulas. (In this chapter, propositional formulas may be denoted by F,G, etc. or alternately by ϕ, ψ, etc.) This axiom system together with modus ponens is an example of a *Frege system* (see Section 5.7).

If T is a set of propositional formulas then $T \models F$ means that F is valid in every truth assignment in which all the formulas in T are valid. We abbreviate $\emptyset \models F$ by $\models F$ and in this case we call F a *tautology.* We may also use the abbreviation $\models T$ for $\forall F \in T(\models F)$.

For each proof system, the notion of *derivation* of F from a set T of formulas will be defined. The existence of a derivation of F from T is denoted $T \vdash F$ of formula F. For most proof systems which we discuss, $T \vdash F$ will mean that there is a sequence $P = (F_1, \ldots, F_n)$, such that $F_n = F$, and for $1 \leq i \leq n$, each F_i either belongs to T, or is inferred from previous formulas

F_j, $j < i$, using a rule of inference. We will abbreviate $\emptyset \vdash F$ by $\vdash F$. In this case, we call F a *theorem* and say that the derivation P of F from $\emptyset$ is a *proof* of F. When it is important to indicate the proof system $\mathcal{P}$, then we may write $T \vdash_{\mathcal{P}} F$.

A proof system $\mathcal{P}$ is *sound* if every theorem F of $\mathcal{P}$ is valid, i.e., $\models F$. A proof system $\mathcal{P}$ is *implicationally complete* if for any propositional formulas $F_1, \ldots, F_k, G$, it is the case that $F_1, \ldots, F_k \models G$ implies $F_1, \ldots, F_k \vdash_{\mathcal{P}} G$.

The *length* of proof $P = (F_1, \ldots, F_n)$, also called the *number of inferences* or *number of steps*, is n, and the *size* of proof P, denoted $|P|$, is $\sum_{i=1}^{n} |F_i|$, where $|F_i|$ is the number of symbols in F_i.

5.2 Complexity of Proofs

A proof system is stipulated by a collection of axioms together with certain rules of inference. This can be generalized as follows.

Let Σ_1, Σ_2 be finite alphabets, both of cardinality at least 2, and let $L \subseteq \Sigma_2^*$. Following S. Cook and R. Reckhow [CR77], define a *propositional proof system* for L to be a polynomial time computable *surjection* $f : \Sigma_1^* \to L$; i.e., the range of f is all of L. Generally, one is interested in proof systems for the collection TAUT of tautologies in the De Morgan basis $0, 1, \neg, \vee, \wedge$ (where TRUE is represented by 1 and FALSE by 0), considered as a subset of Σ^*, where $\Sigma = \{0, 1, \neg, \wedge, x, `(', `)'\}$. Here, formulas are defined in the usual manner over the De Morgan basis, allowing parentheses, while propositional variables are considered as strings consisting of x followed by an integer in binary.

A propositional proof system $f : \Sigma^* \to L$, is called *polynomially bounded* if there is a polynomial p such that

$$(\forall x \in L)(\exists y \in \Sigma^*)(f(y) = x \wedge |y| \leq p(|x|)) \tag{5.3}$$

where $|x|$ denotes the size (i.e., number of symbols) of the word $x \in \Sigma^*$. This definition encompasses extremely strong propositional proof systems, including those derived from Peano arithmetic PA and Zermelo-Fraenkel set theory ZF, as in[2]

$$f(P) = \begin{cases} F & \text{if } P \text{ encodes a proof of } F \text{ in } PA \text{ (or } ZF) \\ p \vee \neg p & \text{otherwise.} \end{cases}$$

The following theorem relates the existence of polynomially bounded propositional proof systems to a question related to P =?NP, and gives rise to the program of establishing lower bounds for proof size of combinatorial principles in natural proof systems.

[2] It is assumed that every propositional formula F has a natural translation into the language of PA (ZF).

Theorem 5.2.1 ([CR77]). *The following statements are equivalent:*

1. NP $=$ *co*-NP.
2. *There is a polynomially bounded propositional proof system for* TAUT.

Proof. First notice that since SAT is NP-complete and for all x,

$$\neg x \notin \text{TAUT} \Leftrightarrow x \in \text{SAT},$$

TAUT must be *co*-NP-complete. Consider the direction (1) $\Rightarrow$ (2). Let $\Sigma = \{0, 1, \neg, \wedge, x, \text{`('}, \text{`)'}\}$. By the previous observation, TAUT $\in$ *co*-NP, so by hypothesis TAUT $\in$ NP. Hence there exists a polynomial p and a polynomial time computable relation R such that for all x,

$$x \in \text{TAUT} \Leftrightarrow (\exists y \in \Sigma^*)(R(x, y) \wedge |y| \leq p(|x|)).$$

Define the propositional proof system $f : (\Sigma \cup \{\text{`}\langle\text{'}, \text{`},\text{'}, \text{`}\rangle\text{'}\})^* \to \text{TAUT}$ by

$$f(w) = \begin{cases} x & \text{if } \exists y(R(x, y) \wedge w = \langle x, y \rangle) \\ p \vee \neg p & \text{otherwise.} \end{cases}$$

It is clear that f is polynomially bounded.

Now consider the direction (2) $\Rightarrow$ (1). Let $f : \Gamma^* \to \text{TAUT}$ be a polynomially bounded propositional proof system for TAUT, and let p be a polynomial satisfying (5.3). Since for all x,

$$x \in \text{TAUT} \Leftrightarrow (\exists y \in \Gamma^*)(f(y) = x \wedge |y| \leq p(|x|))$$

we obtain that TAUT $\in$ NP. Let $R \in$ *co*-NP. By *co*-NP-completeness of TAUT, R is polynomially reducible to TAUT. Since TAUT $\in$ NP so is R. This shows that *co*-NP $\subseteq$ NP and consequently also *co*-NP $=$ NP.

One motivation for proving lower bounds for propositional logic proof systems is to develop new combinatorial tools pertinent to the problem of NP $=?$*co*-NP. A newer concept which has emerged from recent research trends is that of *automatizability*.

Definition 5.2.1. *A propositional proof system T is automatizable, if there is an algorithm A_T, which given any propositional formula A yields a proof in T of A in time polynomial in the size of A, provided that such exists.*

Later, we will consider refutation systems, such as resolution, cutting planes, polynomial calculus, etc. For such systems, the definition of *automatizable* refutation system is obtained by replacing "proof" by "refutation" in the previous definition. Clearly, whether certain proof systems are automatizable has important consequences for practical work on theorem provers.

Propositional logic can be formulated using alternative connectives, such as $\neg, \rightarrow$, or $\neg, \vee, \oplus$, or even with arbitrary k-ary connectives. Following Reckhow [Rec75], we show that the previous theorem holds for TAUT_κ, where κ is any finite, adequate set of connectives. First, some definitions:

Definition 5.2.2. *A propositional connective is a function symbol of a given arity. A formula in the set κ of connectives is a finite, rooted, ordered, labeled tree,*[3] *which is either a single node labeled by a variable, or whose root is labeled by a connective of arity n from κ, and whose children $F_1, \ldots, F_n$ are formulas. The size of formula F, denoted by $|F|$, is the total number of symbols in F. The formula size (circuit size) of formula F, denoted by $f(F)$ $(c(F))$, is the total number of connectives in F (number of distinct subformulas), and the leaf size of F, denoted $||F||$, is the number of occurrences of variables in F.*

The root label is called the *principal connective.* Subformulas of formula F are defined inductively: if F is the variable x, then the only subformula of F is x; if F is $k(F_1, \ldots, F_n)$, where k is an n-ary connective, then the subformulas of F are F itself, together with all subformulas of $F_1, \ldots, F_n$.

In the previous definition, several different notions of size were introduced. These are related as follows. For clarity when discussing other notions of size, we call the number of symbols occurring in F, denoted $|F|$, the *symbol size* of F. For the Frege system given in (5.1,5.2), we can take alphabet $\Sigma = \{x, 0, 1, \neg, \rightarrow\}$ and define $|x_i| = 1 + |i|$ to be one more than the length of the binary representation of i, $|\neg F| = 1 + |F|$ and $|F \rightarrow G| = 1 + |F| + |G|$. In contrast, the *leaf size* $||F||$ of propositional formula F corresponds to the number of leaves in the formula tree of F, and has already been introduced in the proof of Theorem 1.10.3. Continuing with the example of the Frege system just introduced, $||x_i|| = 1$, $||\neg F|| = ||F||$, and $||F \rightarrow G|| = ||F|| + ||G||$. The *formula size* $f(F)$ of propositional formula F, defined in Chapter 1, is the number of gates in the formula tree (considered as a fan-out 1 circuit) corresponding to F. Finally, the *circuit size* $c(F)$ of propositional formula F, defined in Chapter 1, is the minimum number of gates in a circuit which represents F. Assume that all connectives of formula F have arity at most k and that along every path from root to leaf of the formula tree corresponding to F, there are never two successive occurrences of a unary connective (e.g. in the De Morgan basis, all double negations have been eliminated). Moreover, assume that the variables appearing in F are $x_1, \ldots, x_m$, where $m = ||F||$. Under these hypotheses, $f(F) + ||F||$ is the number of nodes in the formula tree corresponding to F, and we clearly have $c(F) \leq f(F) \leq 2 \cdot ||F||$ and $||F|| \leq (k-1) \cdot f(F) + 1$, hence $||F|| \leq |F| = O(||F|| \cdot \log_2 ||F||)$.

The size (symbol size, variable size, formula size or circuit size) of a proof $P = (F_1, \ldots, F_r)$ is the sum of the sizes of the formulas F_i, for $1 \leq i \leq r$. For propositional proof system T and tautology F, $\text{SIZE}_T(F)$ is defined to be the minimum size of a proof P of F in system T. For certain systems T, it may be more convenient to study the formula size or circuit size of proofs of formula F in T – however, by previous remarks, this can then be related

[3] As discerned from the context, we identify formulas, considered as trees, with the corresponding fully parenthesized expressions and/or the corresponding Polish normal form (prefix traversal of formula tree).

to symbol size. Thus for instance the lower bound for constant-depth Frege proofs of the pigeonhole principle from Section 5.7.1 is proved for circuit size. From the context, the appropriate notion of size will always be clear.

A *total truth assignment*, or *valuation*, for variables $x_1, \ldots, x_n$ is a mapping $\sigma : \{x_1, \ldots, x_n\} \to \{0, 1\}$. Total truth assignments are identified in the obvious manner with elements of $\{0,1\}^n$. A *partial truth assignment* or *restriction* for variables $x_1, \ldots, x_n$ is a mapping $\rho : \{x_1, \ldots, x_n\} \to \{*, 0, 1\}$, where $\rho(x_i) = *$ is an abuse of notation, meaning that $\rho(x_i) = x_i$.

The truth value $F \restriction_\sigma$ of a formula F with respect to total truth assignment σ of the variables of F is defined in the obvious manner, using the semantic interpretation of the connectives. If σ is a partial truth assignment, or restriction, then in the obvious manner, $F \restriction_\sigma$ simplifies to a formula in the unset variables. A boolean function $f \in \mathcal{B}_n$ is represented by the formula F, if $f(\sigma) = F \restriction_\sigma$ for all total truth assignments $\sigma \in \{0,1\}^n$. A set κ of connectives is *adequate* (or *complete*) if every boolean function can be represented by a formula in κ. A *tautology* $F \in \text{TAUT}_\kappa$ is a formula in the connective set κ, whose negation is not satisfiable; i.e., the truth value of F under every truth assignment to its variables is 1 or T.

Let κ be any finite, adequate set of connectives of arity at most k. Denote the set of formulas in connective set κ by $Form_\kappa$, the set of tautologies in connective set κ by TAUT_κ, and note that $Form_\kappa$ is a set of trees, whose branching degree is at most k. Let $Form$ denote the set of formulas over the De Morgan set $\{0, 1, \neg, \vee, \wedge\}$ of connectives,[4] and let TAUT be the corresponding set of tautologies.

Theorem 5.2.2 ([Rec75]). *There is a polynomial time computable translation $tr : Form_\kappa \to Form$ satisfying $tr(F) \equiv F$, for all $F \in Form_\kappa$, and which is surjective in the sense that for every $G \in Form$ there exists $F \in Form_\kappa$ such that $tr(F) \equiv G$.*

Note that the size of the translation $tr(F)$ is polynomial in the size of F, since tr is polynomial time computable.

Proof. If $F \in Form_\kappa$ contains less than k occurrences of variables, then let G be an equivalent formula from $Form$ (for each $\ell \leq k$, preassign 2^{2^ℓ} many formulas in $Form$ for the 2^{2^ℓ} many possible boolean functions on ℓ variables; given formula $F \in Form_\kappa$ having less than k occurrences of variables, determine the corresponding boolean function and by table look-up find the corresponding G). Otherwise, let n be the number of distinct variables appearing in F. By Exercise 1.13.8, a slight generalization of Spira's Theorem 1.10.1, if T is a k-ary tree with leaf size $n \geq k$, then there exists a node x of T, such that the subtree

$$T_x = \{y \in T : \ x \text{ is an ancestor of } y\}$$

[4] Note that 0, 1 are the only possible 0-ary connectives.

of T rooted at x contains m leaves, where $\lceil \frac{n}{k+1} \rceil \leq m \leq \lfloor \frac{kn}{k+1} \rfloor$. Applying this, we determine a subformula H of F, such that the number of occurrences of variables both in H and the *scar* $F[H/p]$ of F (obtained by replacing subformula H by new variable p) is between $\frac{n}{k+1}$ and $\frac{k \cdot n}{k+1}$, not counting occurrences of p. Note that F is equivalent to

$$(F[H/1] \wedge H) \vee (F[H/0] \wedge \neg H)$$

and iterate this construction on $F[H/1] \wedge H$ and $F[H/0] \wedge \neg H$. Let tr : $Form_\kappa \to Form$ be the corresponding translation.

Let $g(n) = \max\{||f(F)|| : F \in Form_\kappa, ||F|| \leq n\}$, where, as in Definition 5.2.2, $||F||$ denotes the number of occurrences of variables in F. Clearly for $n \geq k$, $g(n) \leq 4 \cdot g(\frac{k \cdot n}{k+1})$. Let r satisfy $(\frac{k}{k+1})^r \cdot n < k$, so

$$n < \left(\frac{k+1}{k}\right)^r \cdot k$$

$$\log_{\frac{k+1}{k}}(n) < r + \log_{\frac{k+1}{k}}(k).$$

Thus $\log_{\frac{k+1}{k}}(n)$ stages of the above construction suffice so that all remaining formulas in $Form_\kappa$ have less than k occurrences of variables. It follows that $g(n) \leq 4^{\log_2(n)/\log_2(\frac{k+1}{k})} = (2^{\log_2 n})^{2/\log_2(\frac{k+1}{k})} = O(n^{2k})$ and so the circuit size of the translation $c(tr(F))$ is polynomial in the circuit size $c(F)$ of formula F. By depth first search, it is clear that tr is polynomial time computable. Clearly $tr(F) \equiv F$, for all $F \in Form_\kappa$, and by adequacy of the De Morgan basis, tr is surjective in the sense that for every $G \in Form$ there exists $F \in Form_\kappa$ such that $tr(F)$ is equivalent with G.

Applying the construction of the previous proof to the De Morgan basis (in place of κ), we obtain a translation $tr_1 : Form \to Form$ which balances formulas, in the sense that the circuit depth of $tr_1(F)$ is logarithmic in the circuit size of F, for all $F \in Form$. (see Theorem 1.10.3 for a more precise statement). From Theorem 5.2.2, it follows that Theorem 5.2.1 holds with TAUT$_\kappa$ in place of TAUT, for an arbitrary finite, adequate set κ of connectives.

Theorem 5.2.1 implies that if no propositional proof system is polynomially bounded, then P $\neq$ NP. An interesting general program, attributed to S.A. Cook,[5] consists of establishing superpolynomial (preferably exponential) lower bounds for proof size of proofs of tautologies in specific proof systems, such as analytic tableaux, Gentzen without cut, resolution, cutting planes, etc. While it is not expected that P $\neq$ NP can be established by proving lower bounds for infinitely many proof systems, it is the case that deep new combinatorial techniques must be introduced in proving lower bounds for specific

[5] The first significant contribution to this program was by Tseitin, who proved an exponential lower bound for refutation size of certain graph theoretic formulas (*odd-charged graphs*, to be explained later) in the system of *regular* refutation, a restricted form of resolution. It should be noted that Tseitin's work was in 1968, several years before Theorem 5.2.1 was proved.

proof systems. As in the case of boolean circuits, where it is generally felt that *counting* is difficult to compute (e.g. Theorem 2.6.2), it seems that tautologies, which concern notions from finite combinatorics, require large proof size. The Dirichlet *pigeonhole principle* states that if $n+1$ pigeons occupy n holes then at least one hole must be occupied by at least two pigeons; in other words, there is no injection from a set of size $n+1$ into a set of size n. This principle implies that finite cardinality is a well-defined notion; i.e., a finite set cannot have both cardinality $n+1$ and n.

The pigeonhole principle can be formalized in propositional logic as follows. For each $1 \leq i \leq n+1, 1 \leq j \leq n$, introduce the propositional variables $p_{i,j}$, whose interpretation is that the i-th pigeon sits in the j-th hole. Let PHP_n^{n+1} be the formula

$$\bigvee_{i=1}^{n+1} \bigwedge_{j=1}^{n} \overline{p}_{i,j} \vee \bigvee_{1 \leq i < i' \leq n+1} \bigvee_{j=1}^{n} (p_{i,j} \wedge p_{i',j}). \tag{5.4}$$

PHP_n^{n+1} expresses the *relational* pigeonhole principle, which states that there is no total injective relation from $\{1, \ldots, n+1\}$ into $\{1, \ldots, n\}$. The *functional* pigeonhole principle, $\text{fun} - \text{PHP}_n^{n+1}$

$$\text{PHP}_n^{n+1} \vee \bigvee_{i=1}^{n+1} \bigvee_{1 \leq j < j' \leq n} (p_{i,j} \wedge p_{i,j'}) \tag{5.5}$$

states that there is no total injective function from $\{1, \ldots, n+1\}$ into $\{1, \ldots, n\}$. Finally, the onto version $\text{onto} - \text{PHP}_n^{n+1}$ of the functional pigeonhole principle is expressed by

$$\text{fun} - \text{PHP}_n^{n+1} \vee \bigvee_{j=1}^{n} \bigwedge_{i=1}^{n+1} \overline{p}_{i,j}. \tag{5.6}$$

Clearly, all the above versions of the pigeonhole principle are tautologies, and have $O(n^3)$ symbols.

A slight generalization of the pigeonhole principle, denoted by PHP_n^m, is given by

$$\neg \bigwedge_{i=1}^{m} \bigvee_{j=1}^{n} p_{i,j} \vee \bigvee_{1 \leq i < i' \leq m} \bigvee_{j=1}^{n} (p_{i,j} \wedge p_{i',j}). \tag{5.7}$$

This formula has size $O(m^2 \cdot n)$ and expresses that there is no injective relation from a set of size m into a set of size n. The *weak pigeonhole principle* is generally taken to be PHP_n^m, for $m \geq n^2$.

In the sequel, the size of the smallest proof of combinatorial statements such as the pigeonhole principle will be investigated in different propositional proof systems. The following notion of *polynomial simulation* of one proof system by another is useful in ordering the strength of various systems.

Definition 5.2.3 ([CR74], [CR77]). *Let f, g be proof systems such that $f : \Sigma_1^* \rightarrow$ TAUT and $g : \Sigma_2^* \rightarrow$ TAUT. Then g p-simulates f if there is a polynomial time computable function $h : \Sigma_1^* \rightarrow \Sigma_2^*$ such that $g(h(x)) = f(x)$ for all $x \in \Sigma_1^*$.*

If the associated function h is polynomially bounded but not necessarily polynomial time computable, then we speak of *polynomial simulation.* More formally, we have the following.

Definition 5.2.4. *Let $\mathcal{P}_1$, $\mathcal{P}_2$ be arbitrary proof systems for propositional logic. The system $\mathcal{P}_1$* simulates *system $\mathcal{P}_2$, iff there is a polynomial $p(x)$ such that for any proof Q of formula A in $\mathcal{P}_2$, there is a proof P of (the formula corresponding to) A in $\mathcal{P}_1$ and the size$(P) \leq p($size$(Q))$.*

Note that if proof systems $\mathcal{P}_1, \mathcal{P}_2$ happen to have the same language, then it is *not* required by the previous definitions that the translation of a formula be itself. When formulas are translated by themselves, then we speak of a *strong* p-simulation resp. *strong* polynomial simulation; otherwise the simulations are said to be *weak*. In all cases where the underlying language is the same, simulations given in this chapter are strong (the only exception is Theorem 5.4.4). As well, we often write 'polynomially simulation' when what is proved is actually a 'p-simulation'.

5.3 Gentzen Sequent Calculus LK

In this section, we study the Gentzen propositional sequent calculus LK for propositional logic with the usual connectives $\neg, \vee, \wedge$.[6] Finite sets of propositional formulas are called *cedents* , and will be denoted by upper case Greek letters.[7] If Γ, Δ are cedents, then $\Gamma \mapsto \Delta$ is a *sequent*, and Γ the *antecedent* and Δ the *succedent.*[8] A rule is an expression of the form

$$\frac{S_1 \dots S_m}{S_1' \dots S_n'}$$

where $S_1, \dots, S_m, S_1', \dots, S_n'$ are sequents. $S_1, \dots, S_m$ are called *upper sequents* while $S_1', \dots, S_n'$ *lower sequents* of the corresponding rule. As is usual

[6] Gentzen's system for predicate logic is called *Logischer Kalkül*, denoted by LK. In this text we are concerned with the propositional fragment of LK. Since we never consider predicate logic, there is no risk of confusion in denoting the propositional fragment by LK.

[7] In Gentzen sequent calculus, a cedent is usually taken to mean a *sequence*, rather than a *set* of formulas. In this form, structural rules allow one to permute the order of formulas in a cedent. In Section 5.6.6, we define PTK as a Gentzen sequent calculus, where cedents are taken to be sequences. In the current section, the set formulation streamlines presentation of Statman's lower bound.

[8] The new symbol $\mapsto$ should not be confused with implication $\rightarrow$.

practice, we use the abbreviation Γ, Δ for $\Gamma \cup \Delta$ and Γ, ϕ for $\Gamma \cup \{\phi\}$. Following [Tak75] we have the following logical rules of inference of propositional sequent calculus

$$\neg\text{-left}\ \frac{\Gamma \mapsto \phi, \Delta}{\neg\phi, \Gamma \mapsto \Delta} \qquad \neg\text{-right}\ \frac{\phi, \Gamma \mapsto \Delta}{\Gamma \mapsto \neg\phi, \Delta}$$

$$\vee\text{-left}\ \frac{\phi, \Gamma \mapsto \Delta \qquad \psi, \Gamma \mapsto \Delta}{\phi \vee \psi, \Gamma \mapsto \Delta}$$

$$\vee\text{-right}\ \frac{\Gamma \mapsto \phi, \Delta}{\Gamma \mapsto \phi \vee \psi, \Delta} \qquad \vee\text{-right}\ \frac{\Gamma \mapsto \phi, \Delta}{\Gamma \mapsto \psi \vee \phi, \Delta}$$

$$\wedge\text{-left}\ \frac{\phi, \Gamma \mapsto \Delta}{\phi \wedge \psi, \Gamma \mapsto \Delta} \qquad \wedge\text{-left}\ \frac{\phi, \Gamma \mapsto \Delta}{\psi \wedge \phi, \Gamma \mapsto \Delta}$$

$$\wedge\text{-right}\ \frac{\Gamma \mapsto \phi, \Delta \qquad \Gamma \mapsto \psi, \Delta}{\Gamma \mapsto \phi \wedge \psi, \Delta}$$

The next two rules are the *cut* rule

$$\text{cut}\ \frac{\Gamma \mapsto \phi, \Delta \qquad \phi, \Gamma \mapsto \Delta}{\Gamma \mapsto \Delta}$$

and a *structural* rule, which encompasses the weakening, contraction and permutation rules when cedents are considered as sequences of formulas, rather than sets of formulas, given by

$$\text{structural}\ \frac{\Gamma \mapsto \Delta}{\Gamma' \mapsto \Delta'}$$

where we assume $\Gamma \subseteq \Gamma', \Delta \subseteq \Delta'$. The only *axioms* (also called *initial sequents*) of propositional sequent calculus are of the form $p \mapsto p$, where p is a propositional variable. A *proof* of $\Gamma \mapsto \Delta$ is a sequence P of sequents $S_1, \ldots, S_n$, such that S_n is the *end sequent* $\Gamma \mapsto \Delta$, and for every $1 \leq i \leq n$, either S_i is an initial sequent, or there exists $1 \leq j < i$ such that S_i is obtained by a rule of inference from S_j using a unary rule of inference (i.e., negation, $\vee$-right, $\wedge$-left, or a structural rule), or there exist $1 \leq j, k < i$ such that S_i is obtained by a binary rule of inference from the sequents S_j, S_k (i.e., $\vee$-left, $\wedge$-right, cut).[9]

If each sequent is used at most once as the hypothesis of a rule of inference in a proof, then the proof is said to be *tree-like.*[10] A tree-like proof of $\Gamma \mapsto \Delta$ is thus a tree[11] consisting of sequents, such that the following conditions are satisfied:

[9] Rules for implication $\supset$ will shortly be considered.

[10] Without explicitly making a new definition, for each proof system considered in this chapter, proofs may be considered either as *dag*-like (i.e., a sequence of formulas) or as *tree*-like (i.e., a tree of formulas).

[11] Unlike circuits and formulas, proof trees are envisioned as botanical trees with the root at the bottom.

- $\Gamma \mapsto \Delta$ is the root.
- The leaves are axioms.
- Every node other than the root is an upper sequent of a rule, and every node other than a leaf is a lower sequent of a rule.

A proof without the cut rule is called *cut-free.*

Definition 5.3.1. *The size $S(\Pi)$ of derivation $\Pi = (\phi_1, \ldots, \phi_n)$ is the total number of symbols in Π. The length $L(\Pi)$ is n. If ϕ is a tautology, then $S(\phi)$ $(S_T(\phi))$ is $S(\Pi)$, where Π is the smallest proof (tree-like proof) of ϕ. Similarly $L(\phi)$ $(L_T(\phi))$ is $L(\Pi)$, where Π is the smallest proof (tree-like proof) of ϕ.*

For proof systems we later encounter, we will similarly speak of size and length of derivations within those systems, referring to Definition 5.3.1, as appropriately modified for the proof system under discussion.

The reader may find it convenient to think of $\Gamma \mapsto \Delta$ as $\Gamma \vdash \Delta$, where $\vdash$ is the symbol used in deductions. As such the meaning of $\Gamma \mapsto \Delta$ is $\bigwedge \Gamma \to \bigvee \Delta$. However it should be clear that $\mapsto$ is a primitive symbol used to construct the "formulas" in sequent calculus, while $\vdash$ is a symbol in the metalanguage.

As an example, we give polynomial-size, tree-like, cut-free proofs in LK of the sequent

$$p_1, \neg p_1 \vee p_2, \ldots, \neg p_{n-1} \vee p_n \mapsto p_n. \tag{5.8}$$

When $n = 2$, we have the following proof.

$$\dfrac{\dfrac{\dfrac{p_1 \mapsto p_1}{p_1, \neg p_1 \mapsto}}{p_1, \neg p_1 \mapsto p_2} \qquad \dfrac{p_2 \mapsto p_2}{p_1, p_2 \mapsto p_2}}{p_1, \neg p_1 \vee p_2 \mapsto p_2}$$

Assuming that P_k is a proof of (5.8), when $n = k$, we have the following proof for (5.8), when $n = k + 1$.

$$\dfrac{\dfrac{\dfrac{\dfrac{\vdots}{p_1, \neg p_1 \vee p_2, \ldots, \neg p_{k-1} \vee p_k \mapsto p_k}}{p_1, \neg p_1 \vee p_2, \ldots, \neg p_{k-1} \vee p_k, \neg p_k \mapsto}}{p_1, \neg p_1 \vee p_2, \ldots, \neg p_{k-1} \vee p_k, \neg p_k \mapsto p_{k+1}} \qquad \dfrac{\begin{matrix}\dfrac{\dfrac{p_{k+1} \mapsto p_{k+1}}{p_1, p_{k+1} \mapsto p_{k+1}}}{p_1, \neg p_1 \vee p_2, p_{k+1} \mapsto p_{k+1}} \\ \vdots \end{matrix}}{p_1, \neg p_1 \vee p_2, \ldots, \neg p_{k-1} \vee p_k, p_{k+1} \mapsto p_{k+1}}}{p_1, \neg p_1 \vee p_2, \ldots, \neg p_k \vee p_{k+1} \mapsto p_{k+1}}$$

5.3.1 Completeness

A straightforward proof by induction on the number of inferences in a proof shows that LK is a sound proof system, in the sense that $\Gamma \mapsto \Delta$ implies that $\models \bigvee \Gamma \to \bigvee \Delta$. However, it may come as a surprise that the cut rule does not add to the provability of the sequent calculus. This is made precise in the following completeness theorem, whose proof is elementary. (It should be mentioned that Gentzen's cut elimination theorem for the first order logic is called the Gentzen *Hauptsatz*, and is much more difficult to prove than the following theorem – see, for instance, [Tak75].)

Theorem 5.3.1 ([Gen34]). *The following statements are equivalent for Γ, Δ finite sets of propositional formulas.*

1. $\models \bigwedge \Gamma \to \bigvee \Delta$
2. *There is a sequent proof of* $\Gamma \mapsto \Delta$.
3. *There is a cut-free sequent proof of* $\Gamma \mapsto \Delta$.

Proof. (3) $\Rightarrow$ (2) is trivial. (2) $\Rightarrow$ (1) expresses the soundness of the propositional sequent calculus. The proof is by induction on the number of proof inferences, and is left to the reader. It remains to prove that (1) $\Rightarrow$ (3). Assume $\models \bigwedge \Gamma \to \bigvee \Delta$. Construct a proof tree by induction on the height of a node. At height 0 (root) there is a unique node labeled $\Gamma \mapsto \Delta$. Consider a node labeled by $\Phi \mapsto \Psi$ at height $h \geq 0$ in the tree thus far constructed. We show how to append the children of $\Phi \mapsto \Psi$ to the tree. Consider the first non-atomic formula ϕ occurring in this sequent and extend the proof tree as indicated below depending on the form of ϕ.

First, assume that $\phi \in \Psi$, and in this case, let Ψ be ϕ, Ψ'.

1. If $\phi = \neg\psi$ then add to the proof-tree the sequent $\psi, \Phi \mapsto \Psi'$.
2. If $\phi = \psi \wedge \psi'$ then add to the proof-tree the sequents $\Phi \mapsto \psi, \Psi'$ and $\Phi \mapsto \psi', \Psi'$ (bifurcation step).
3. If $\phi = \psi \vee \psi'$ then add to the proof-tree the following sequents one above the other $\Phi \mapsto \psi \vee \psi', \psi \vee \phi', \Psi'$; $\Phi \mapsto \psi', \psi \vee \psi', \Psi'$; $\Phi \mapsto \psi \vee \psi', \psi', \Psi'$; $\Phi \mapsto \psi, \psi', \Psi'$.

Second, assume $\phi \in \Phi$, and in this case, let Φ be ϕ, Φ'.

1. If $\phi = \neg\psi$ then add to the proof-tree the sequent $\Phi' \mapsto \psi, \Psi$.
2. If $\phi = \psi \wedge \psi'$ then add to the proof-tree the following sequents one above the other $\psi \wedge \psi', \psi' \wedge \psi, \Phi' \mapsto \Psi$; $\psi', \psi \wedge \psi', \Phi' \mapsto \Psi$; $\psi \wedge \psi', \Phi' \mapsto \Psi$; $\psi, \psi', \Phi' \mapsto \Psi$.
3. If $\phi = \psi \vee \psi'$ then add to the proof-tree the sequents $\psi, \Phi' \mapsto \Psi$ and $\psi', \Phi' \mapsto \Psi$ (bifurcation step).

We terminate this construction when both Φ, Ψ consist only of propositional variables. Next, we claim that $\Phi \cap \Psi \neq \emptyset$. Indeed, otherwise consider the truth assignment which assigns the value TRUE to every element of Φ and the value FALSE to every element of Ψ. Extend this truth assignment to the remaining propositional variables in the language. It is then obvious that this truth assignment assigns the value FALSE to every sequent on the branch which begins at the sequent $\Phi \mapsto \Psi$ and ends at the root $\Gamma \mapsto \Delta$, which is a contradiction. To conclude, it remains to convert the above tree into a cut-free proof. For each leaf $\Phi \mapsto \Psi$ choose $p \in \Phi \cap \Psi$ and add the axiom $p \mapsto p$. This is legal in view of the structural rule.

Note that the cut-free proof constructed in the proof of the previous theorem is tree-like, rather than *dag*-like. *Analytical tableaux*, a popular method in the machine theorem prover community, is equivalent to the tree-like, cut-free Gentzen sequent calculus.

Definition 5.3.2. *A proof has the subformula property, if every formula appearing in every sequent of the proof is a subformula of a formula appearing in the end sequent.*

It follows that a cut-free proof has the subformula property.

5.3.2 Lower Bound for Cut-Free Gentzen

A cut-free proof of the sequent $\Gamma \mapsto \Delta$ is preferable to a proof with cuts because of the subformula property, a feature important for computer implementations of theorem provers. At what cost can a proof with cuts be converted into a cut-free proof? If n is the size (i.e., number of symbols) of a valid sequent, then analysis of the proof of Theorem 5.3.1 reveals that the size of the cut-free proof tree is $2^{O(n)}$.

We now investigate the length and size of proofs in the Gentzen sequent calculus with regard to the properties of tree-like vs. dag-like, and cut-free vs. with cut. Theorem 5.3.3 and Theorem 5.3.4 show that tree-like cut-free Gentzen proofs can be exponentially longer than dag-like cut-free proofs, while Theorem 5.3.2 and Theorem 5.3.4 show that tree-like cut-free proofs can be exponentially longer than tree-like proofs with cut. Concerning the efficiency of cut for dag-like proofs in the Gentzen sequent calculus, Theorem 5.3.5 shows that dag-like cut-free proofs can be exponentially longer than dag-like proofs with cut. First, we extend LK by adding new rules for implication.[12]

$$\supset -\text{left}\ \frac{\phi, \Gamma \mapsto \psi, \Delta}{\Gamma \mapsto \phi \supset \psi, \Delta} \qquad\qquad \supset -\text{right}\ \frac{\Gamma \mapsto \phi, \Delta \qquad \Gamma, \psi \mapsto \Delta}{\Gamma, \phi \supset \psi \mapsto \Delta}$$

Without risk of confusion, the new system will be denoted as well by LK (from the context, it will be clear whether the system is intended to include implication or not).

In Theorems 5.3.2, 5.3.3, 5.3.4, we are concerned with the size of Gentzen sequent calculus proofs of $\Gamma_n \mapsto \Delta_n$, which are defined as follows. Let $p_1, \ldots, p_n, q_1, \ldots, q_n$ be propositional variables. For $1 \leq i \leq n$, define ϕ_i as

$$\bigwedge_{j=1}^{i} (p_j \vee q_j).$$

Define α_1 to be p_1 and β_1 to be q_1, and for $2 \leq i \leq n$, define

[12] When working in sequent calculus, to avoid confusion with the sequent primitive $\mapsto$, implication is often denoted by $\supset$ rather than $\rightarrow$.

$$\alpha_i \text{ to be } \left(\bigwedge_{j=1}^{i-1}(p_j \vee q_j)\right) \supset p_i$$
$$\beta_i \text{ to be } \left(\bigwedge_{j=1}^{i-1}(p_j \vee q_j)\right) \supset q_i.$$

For $1 \leq i \leq n$, define

$$\Gamma_i \text{ to be } \{\alpha_1 \vee \beta_1, \ldots, \alpha_i \vee \beta_i\}$$
$$\Delta_i \text{ to be } \{p_i, q_i\}.$$

Let's unravel the definition of $\Gamma_n \mapsto \Delta_n$ for $n = 1, 2, 3$. $\Gamma_1 \mapsto \Delta_1$ is just

$$p_1 \vee q_1 \mapsto p_1, q_1$$

and has the following tree-like, cut-free proof

$$\dfrac{\dfrac{p_1 \mapsto p_1}{p_1 \mapsto p_1, q_1} \qquad \dfrac{q_1 \mapsto q_1}{q_1 \mapsto p_1, q_1}}{p_1 \vee q_1 \mapsto p_1, q_1}$$

The sequent $\Gamma_2 \mapsto \Delta_2$ is

$$p_1 \vee q_1, (p_1 \vee q_1 \supset p_2) \vee (p_1 \vee q_1 \supset q_2) \mapsto p_2, q_2$$

In order to give a tree-like, cut-free proof of $\Gamma_2 \mapsto \Delta_2$, we first give a tree-like, cut-free proof of

$$p_1 \vee q_1, (p_1 \vee q_1) \supset p_2 \mapsto p_2 \tag{5.9}$$

as follows.

$$\dfrac{\dfrac{\dfrac{\dfrac{p_1 \mapsto p_1}{p_1 \mapsto p_1 \vee q_1} \qquad \dfrac{q_1 \mapsto q_1}{q_1 \mapsto p_1 \vee q_1}}{p_1 \vee q_1 \mapsto p_1 \vee q_1}}{p_1 \vee q_1 \mapsto p_1 \vee q_1, p_2} \qquad \dfrac{p_2 \mapsto p_2}{p_1 \vee q_1, p_2 \mapsto p_2}}{p_1 \vee q_1, ((p_1 \vee q_1) \supset p_2) \mapsto p_2}$$

Similarly we have a derivation of

$$p_1 \vee q_1, ((p_1 \vee q_1) \supset q_2) \mapsto q_2. \tag{5.10}$$

By weakening applied to (5.9) and (5.10), followed by applying $\vee$-left, we have a tree-like, cut-free proof of

$$p_1 \vee q_1, (p_1 \vee q_1 \supset p_2) \vee (p_1 \vee q_1 \supset q_2) \mapsto p_2, q_2$$

which is the sequent $\Gamma_2 \mapsto \Delta_2$. Finally, $\Gamma_3 \mapsto \Delta_3$ is

$$p_1 \vee q_1, (p_1 \vee q_1 \supset p_2) \vee (p_1 \vee q_1 \supset q_2),$$
$$(p_1 \vee q_1) \wedge (p_2 \vee q_2) \supset p_3) \vee (p_1 \vee q_1) \wedge (p_2 \vee q_2) \supset q_3) \mapsto p_3, q_3$$

and we might cringe at the effort involved in giving a tree-like, cut-free proof. Nevertheless, after unravelling the definitions, one is easily convinced that $\Gamma_n \mapsto \Delta_n$ is a valid sequent, by using an intuitive argument with modus ponens.

We now turn to estimating the length (i.e., number of steps or lines or sequents) and the size (i.e., number of symbols) of Gentzen proofs, depending on the parameters of tree-like vs. dag-like, and cut-free vs. with cut. Note before we get started that for $1 \le k \le n$, $|\phi_k| = O(k)$, $|\alpha_k| = O(k)$, $|\beta_k| = O(k)$, $|\Gamma_k| = O(k^2)$, $|\Delta_k| = O(1)$, hence $|\Gamma_n \mapsto \Delta_n|$ is of size quadratic in n.

Theorem 5.3.2 ([Sta78]). *There are tree-like proofs of $\Gamma_n \mapsto \Delta_n$ with cut of length $O(n^2)$ and size $O(n^3)$.*

Proof. We begin by proving several claims.

CLAIM 1. For $1 \le i < n$, there exist tree-like, cut-free proofs of

$$\phi_i, \alpha_{i+1} \vee \beta_{i+1} \mapsto \phi_{i+1} \tag{5.11}$$

of length $O(n)$ and size $O(n)$.

Proof of Claim 1. Note that (5.11) is just

$$\phi_i, (\phi_i \supset p_{i+1}) \vee (\phi_i \supset q_{i+1}) \mapsto \phi_i \wedge (p_{i+1} \vee q_{i+1}). \tag{5.12}$$

By Exercise 5.10.4 there is a tree-like, cut-free proof of linear size of $\phi_i \mapsto \phi_i$, i.e., of length $O(|\phi_i|)$ and size $O(|\phi_i|)$. From this, by weakening, we get

$$\phi_i, (\phi_i \supset p_{i+1}) \vee (\phi_i \supset q_{i+1}) \mapsto \phi_i. \tag{5.13}$$

Now

$$\dfrac{\dfrac{\dfrac{p_{i+1} \mapsto p_{i+1}}{\phi_i, p_{i+1} \mapsto p_{i+1}}}{\phi_i, p_{i+1} \mapsto p_{i+1} \vee q_{i+1}} \qquad \dfrac{\phi_i \mapsto \phi_i}{\phi_i \mapsto \phi_i, p_{i+1} \vee q_{i+1}}}{\phi_i, \phi_i \supset p_{i+1} \mapsto p_{i+1} \vee q_{i+1}}$$

Similarly, we can give a tree-like, cut-free proof of

$$\phi_i, \phi_i \supset q_{i+1} \mapsto p_{i+1} \vee q_{i+1}$$

and thus by $\vee$-left we have

$$\phi_i, (\phi_i \supset p_{i+1}) \vee (\phi_i \supset q_{i+1}) \mapsto p_{i+1} \vee q_{i+1}. \tag{5.14}$$

By applying $\wedge$-right to (5.13) and (5.14), we have

$$\phi_i, (\phi_i \supset p_{i+1}) \vee (\phi_i \supset q_{i+1}) \mapsto \phi_i \wedge (p_{i+1} \vee q_{i+1})$$

which completes the proof of Claim 1.

CLAIM 2. There exist tree-like proofs of $\Gamma_n \mapsto \phi_n$ with cut having length $O(n^2)$ and size $O(n^3)$.

Proof of Claim 2. Recall that ϕ_1 is $p_1 \vee q_1$, hence also $\alpha_1 \vee \beta_1$. Recall as well that Γ_i is $\{\alpha_1 \vee \beta_1, \ldots, \alpha_i \vee \beta_i\}$. From repeated applications of Claim 1, for $i = 1, 2, \ldots, n-1$, we have

$$\cfrac{\cfrac{\alpha_1 \vee \beta_1, \alpha_2 \vee \beta_2 \mapsto \phi_2 \qquad \phi_2, \alpha_3 \vee \beta_3 \mapsto \phi_3}{\alpha_1 \vee \beta_1, \alpha_2 \vee \beta_2, \alpha_3 \vee \beta_3 \mapsto \phi_3} \qquad \phi_3, \alpha_4 \vee \beta_4 \mapsto \phi_4}{\cfrac{\Gamma_4 \mapsto \phi_4}{\vdots}}$$

and continuing in this fashion, we have

$$\cfrac{\cfrac{\vdots}{\Gamma_{n-1} \mapsto \phi_{n-1}} \qquad \phi_{n-1}, \alpha_n \vee \beta_n \mapsto \phi_n}{\Gamma_n \mapsto \phi_n}$$

Since $|\Gamma_n| = O(n^2)$ and $|\phi_n| = O(n)$, we have given a tree-like proof of $\Gamma_n \mapsto \phi_n$ with the cut rule, having length $O(n^2)$ and size $O(n^3)$. This proof may seem to have length only $O(n)$, but we must additionally append proof of each $\phi_i, \alpha_{i+1} \vee \beta_{i+1} \mapsto \phi_{i+1}$, to ensure the proof is tree-like. By Claim 1, each of these proofs has length $O(n)$ and size $O(n)$, thus justifying our assertion concerning the length and size of our proof of $\Gamma_n \mapsto \phi_n$.

CLAIM 3. There exist tree-like, cut-free proofs of $\phi_n \mapsto p_n, q_n$ of length $O(n)$ and size $O(n)$.

Proof of Claim 3.

$$\cfrac{\cfrac{\cfrac{\cfrac{\cfrac{p_n \mapsto p_n}{p_n \mapsto p_n, q_n} \qquad \cfrac{q_n \mapsto q_n}{q_n \mapsto p_n, q_n}}{(p_n \vee q_n) \mapsto p_n, q_n}}{(p_{n-1} \vee q_{n-1}) \wedge (p_n \vee q_n) \mapsto p_n, q_n}}{(p_{n-2} \vee q_{n-2}) \wedge (p_{n-1} \vee q_{n-1}) \wedge (p_n \vee q_n) \mapsto p_n, q_n}}{\cfrac{\vdots}{(p_1 \vee q_1) \wedge \ldots \wedge (p_n \vee q_n) \mapsto p_n, q_n}}$$

This completes the proof of Claim 3.

It follows from Claims 2 and 3 that by applying cut to $\Gamma_n \mapsto \phi_n$ and $\phi_n \mapsto p_n, q_n$ that we have tree-like proofs of $\Gamma_n \mapsto \Delta_n$ with the cut rule, having length $O(n^2)$ and size $O(n^3)$. This completes the proof of the theorem.

Theorem 5.3.3. *There exist dag-like, cut-free proofs of $\Gamma_n \mapsto \Delta_n$ in the Gentzen sequent calculus with length $O(n^3)$ and size $O(n^5)$.*

Proof. By induction on $1 \le i \le n$, we first prove that there exists a dag-like, cut-free derivation of $\Gamma_i \mapsto p_i, q_i$ and of $\Gamma_i \mapsto p_i \vee q_i$ of length $O(i^2)$ and size $O(i^4)$, taking the sequents $\Gamma_j \mapsto p_j, q_j$ and $\Gamma_j \mapsto p_j \vee q_j$, for $1 \le j < i$, as hypotheses. For the base case,

$$\cfrac{\cfrac{p_1 \mapsto p_1}{p_1 \mapsto p_1, q_1} \qquad \cfrac{q_1 \mapsto q_1}{q_1 \mapsto p_1, q_1}}{p_1 \vee q_1 \mapsto p_1, q_1} \qquad \cfrac{\cfrac{p_1 \mapsto p_1}{p_1 \mapsto p_1 \vee q_1} \qquad \cfrac{q_1 \mapsto q_1}{q_1 \mapsto p_1 \vee q_1}}{p_1 \vee q_1 \mapsto p_1 \vee q_1}$$

Now, inductively assume that for $1 \le j \le i$ there is a dag-like, cut-free derivation of $\Gamma_j \mapsto \Delta_j$ and of $\Gamma_j \mapsto p_j \vee q_j$ of length $O(j^2)$ and size $O(j^4)$, taking the sequents $\Gamma_k \mapsto p_k, q_k$ and $\Gamma_k \mapsto p_k \vee q_k$, for $1 \leq k < j$, as hypotheses. We give dag-like, cut-free Gentzen derivations of $\Gamma_{i+1} \mapsto p_{i+1}, q_{i+1}$ and $\Gamma_{i+1} \mapsto p_{i+1} \vee q_{i+1}$ of length $O((i+1)^2)$ and size $O((i+1)^4)$, taking the sequents $\Gamma_j \mapsto p_j, q_j$ and $\Gamma_j \mapsto p_j \vee q_j$, for $1 \leq j \leq i$, as hypotheses.

CLAIM. For each $1 \leq j \leq i$, there exists a dag-like Gentzen derivation of $\Gamma_i \mapsto p_j \vee q_j, p_{i+1}, q_{i+1}$ and of $\Gamma_i \mapsto p_j \vee q_j, p_{i+1} \vee q_{i+1}$ from $\Gamma_j \mapsto p_j \vee q_j$ of length $O(1)$ and size $O(i^2)$

Proof of Claim. Consider first the case where $j = i$.

$$\dfrac{\dfrac{\dfrac{\dfrac{\Gamma_i \mapsto p_i, q_i}{\Gamma_i \mapsto p_i \vee q_i, q_i}}{\Gamma_i \mapsto p_i \vee q_i, p_i \vee q_i}}{\Gamma_i \mapsto p_i \vee q_i}}{\Gamma_i \mapsto p_i \vee q_i, p_{i+1}, q_{i+1}} \qquad \dfrac{\dfrac{\dfrac{\dfrac{\Gamma_i \mapsto p_i, q_i}{\Gamma_i \mapsto p_i \vee q_i, q_i}}{\Gamma_i \mapsto p_i \vee q_i, p_i \vee q_i}}{\Gamma_i \mapsto p_i \vee q_i}}{\Gamma_i \mapsto p_i \vee q_i, p_{i+1} \vee q_{i+1}}$$

Now fix $1 \leq j < i$, and assume by the induction hypothesis that we have a proof of $\Gamma_j \mapsto p_j, q_j$ and of $\Gamma_j \mapsto p_j \vee q_j$. Now

$$\dfrac{\dfrac{\dfrac{\Gamma_j \mapsto p_j \vee q_j}{\Gamma_j \mapsto p_j \vee q_j, p_{i+1}, q_{i+1}}}{\Gamma_j, \alpha_{j+1} \vee \beta_{j+1}, \ldots, \alpha_i \vee \beta_i \mapsto p_j \vee q_j, p_{i+1}, q_{i+1}}}{\Gamma_i \mapsto p_j \vee q_j, p_{i+1}, q_{i+1}}$$

and

$$\dfrac{\dfrac{\dfrac{\Gamma_j \mapsto p_j \vee q_j}{\Gamma_j \mapsto p_j \vee q_j, p_{i+1} \vee q_{i+1}}}{\Gamma_j, \alpha_{j+1} \vee \beta_{j+1}, \ldots, \alpha_i \vee \beta_i \mapsto p_j \vee q_j, p_{i+1} \vee q_{i+1}}}{\Gamma_i \mapsto p_j \vee q_j, p_{i+1} \vee q_{i+1}}$$

This completes the proof of the claim.

From the claim, by $i - 1$ applications of $\wedge$-right, we have proofs of

$$\Gamma_i \mapsto \left(\bigwedge_{j=1}^{i} (p_j \vee q_j) \right), p_{i+1}, q_{i+1}$$

and

$$\Gamma_i \mapsto \left(\bigwedge_{j=1}^{i} (p_j \vee q_j) \right), p_{i+1} \vee q_{i+1}$$

of length $O(i)$ and size $O(i^3)$. Recalling that ϕ_i is $\bigwedge_{j=1}^{i} (p_j \vee q_j)$, this yields

$$\Gamma_i \mapsto \phi_i, p_{i+1}, q_{i+1} \text{ and } \Gamma_i \mapsto \phi_i, p_{i+1} \vee q_{i+1}. \tag{5.15}$$

We have

$$\dfrac{\dfrac{p_{i+1} \mapsto p_{i+1}}{\Gamma_i, p_{i+1} \mapsto p_{i+1}}}{\Gamma_i, p_{i+1} \mapsto p_{i+1}, q_{i+1}} \qquad \dfrac{\dfrac{p_{i+1} \mapsto p_{i+1}}{\Gamma_i, p_{i+1} \mapsto p_{i+1}}}{\Gamma_i, p_{i+1} \mapsto p_{i+1} \vee q_{i+1}}$$

so by applying $\supset$-left to this and (5.15), we obtain

$$\Gamma_i, \phi_i \supset p_{i+1} \mapsto p_{i+1}, q_{i+1} \text{ and } \Gamma_i, \phi_i \supset p_{i+1} \mapsto p_{i+1} \vee q_{i+1} \tag{5.16}$$

In a similar fashion, we have a proof of

$$\Gamma_i, \phi_i \supset q_{i+1} \mapsto p_{i+1}, q_{i+1} \text{ and } \Gamma_i, \phi_i \supset q_{i+1} \mapsto p_{i+1} \vee q_{i+1}. \tag{5.17}$$

Recalling that α_{i+1} (β_{i+1}) is the formula $\phi_i \supset p_{i+1}$ ($\phi_i \supset q_{i+1}$), by applying $\vee$-left to (5.16) and (5.17), we have a proof of

$$\Gamma_i, \alpha_{i+1} \vee \beta_{i+1} \mapsto p_{i+1}, q_{i+1} \qquad \Gamma_i, \alpha_{i+1} \vee \beta_{i+1} \mapsto p_{i+1} \vee q_{i+1}. \tag{5.18}$$

Recalling that Γ_{i+1} is $\Gamma_i \cup \{\alpha_{i+1} \vee \beta_{i+1}\}$ and that Δ_{i+1} is $\{p_{i+1}, q_{i+1}\}$, this yields a derivation of

$$\Gamma_{i+1} \mapsto \Delta_{i+1} \text{ and } \Gamma_{i+1} \mapsto p_{i+1} \vee q_{i+1}$$

from the sequents $\Gamma_j \mapsto \Delta_j$ and $\Gamma_j \mapsto p_j \vee q_j$, for $1 \leq j \leq i$. The length of this derivation is $\sum_{j=1}^{i} O(j) = O((i+1)^2)$ and its size is $\sum_{j=1}^{i} O(j^3) = O((i+1)^4)$. This completes the proof by induction.

To prove $\Gamma_n \mapsto \Delta_n$, we give a proof of $\Gamma_1 \mapsto \Delta_1$ and $\Gamma_1 \mapsto p_1 \vee q_1$, then, for each $2 \leq i \leq n$, we give a derivation of $\Gamma_i \mapsto \Delta_i$ and of $\Gamma_i \mapsto p_i \vee q_i$ from the hypotheses $\Gamma_j \mapsto \Delta_j$ and $\Gamma_j \mapsto p_j \vee q_j$, for $1 \leq j < i$. This proof is clearly cut-free and dag-like (where the sequents $\Gamma_j \mapsto \Delta_j$ and $\Gamma_j \mapsto p_j \vee q_j$ appear as antecedents of a proof rule $O(n-j)$ times), and its length is $\sum_{i=1}^{n} O(i^2) = O(n^3)$ and its size is $\sum_{i=1}^{n} O(i^4) = O(n^5)$. This completes the proof of the theorem.

We now present Statman's exponential lower bound for tree-like, cut-free Gentzen proofs of $\Gamma_n \mapsto \Delta_n$.

Theorem 5.3.4 ([Sta78]). *Assuming n is arbitrarily large, every tree-like, cut-free proof of $\Gamma_n \mapsto \Delta_n$ in the Gentzen sequent calculus has at least 2^n sequents.*

Proof. If $\Gamma \mapsto \Delta$ is a sequent, then let $shc(\Gamma \mapsto \Delta)$ denote the number of sequents in the shortest tree-like, cut-free proof $\Gamma \mapsto \Delta$ in the Gentzen sequent calculus. Note that for any sets of formulas Γ, Δ, the $\vee$-left rule connects the proof trees of $\phi, \Gamma \mapsto \Delta$ and $\psi, \Gamma \mapsto \Delta$ in yielding a proof tree for $\phi \vee \psi, \Gamma \mapsto \Delta$. If we could choose the sets of formulas in such a way that the branches of the upper sequents in the above $\vee$-left rule could not be canceled, then it should be true that

$$shc(\phi \vee \psi, \Gamma \mapsto \Delta) \geq shc(\phi, \Gamma \mapsto \Delta) + shc(\psi, \Gamma \mapsto \Delta).$$

Now let us a consider a tree-like, cut-free proof of $\Gamma_n \mapsto \Delta_n$ having the minimum number of sequents. By minimality, the last non-structural rule applied in this derivation must a $\vee$-left rule of the form

$$\frac{(\Gamma_n - \{\alpha_i \vee \beta_i\}) \cup \{\alpha_i\} \mapsto \Delta_n \quad (\Gamma_n - \{\alpha_i \vee \beta_i\}) \cup \{\beta_i\} \mapsto \Delta_n}{\Gamma_n \mapsto \Delta_n} \tag{5.19}$$

for some $i \leq n$, where each upper sequent has a tree-like, cut-free proof. Now consider the case $i = n$ and assume that $\Delta_n = \{p_n, q_n\}$. Since α_n is the formula $\phi_{n-1} \supset p_n$, by Exercise 5.10.4 the tree-like, cut-free proof of the upper sequent on the left side of (5.19) can be shortened to give a tree-like, cut-free proof of

$$\Gamma_n - \{\alpha_n \vee \beta_n\} \mapsto \phi_{n-1}, p_n, q_n.$$

Now ϕ_{n-1} is equal to $\phi_{n-2} \wedge (p_{n-1} \vee q_{n-1})$, and so by Exercise 5.10.4, we get a shorter tree-like, cut-free proof of

$$\Gamma_n - \{\alpha_n \vee \beta_n\} \mapsto p_{n-1} \vee q_{n-1}, p_n, q_n$$

which in turn can be shortened to a tree-like, cut-free proof of either

$$\Gamma_n - \{\alpha_n \vee \beta_n\} \mapsto p_{n-1}, p_n, q_n \text{ or } \Gamma_n - \{\alpha_n \vee \beta_n\} \mapsto q_{n-1}, p_n, q_n.$$

Since p_n, q_n do not occur in $\Gamma_n - \{\alpha_n \vee \beta_n\}$, they must have been inferred by the weakening rule, and so we have tree-like, cut-free proofs of

$$\Gamma_n - \{\alpha_n \vee \beta_n\} \mapsto p_{n-1} \text{ and } \Gamma_n - \{\alpha_n \vee \beta_n\} \mapsto q_{n-1}$$

and hence in either case we have a shorter tree-like, cut-free proof

$$\Gamma_n - \{\alpha_n \vee \beta_n\} \mapsto p_{n-1}, q_{n-1}. \tag{5.20}$$

This last step involves an application of the weakening rule to an already shortened tree-like, cut-free proof, hence the length and size of the proof of (5.20) is at most that of the original proof of $\Gamma_n \mapsto \Delta_n$.

From this discussion, it is clear that using (5.19), the above argument, combined with a similar assertion for the sequent in the right-hand side of (5.19), shows that for the case $i = n$,

$$shc(\Gamma_n \mapsto \Delta_n) \geq 2 \cdot shc(\Gamma_{n-1} \mapsto \Delta_{n-1}). \tag{5.21}$$

We would like to prove that inequality (5.21) holds for the case $i < n$ as well, since then by induction on n, it follows that $shc(\Gamma_n \mapsto \Delta_n) \geq 2^n$.

When $i < n$, we claim that the tree-like, cut-free proof of the upper sequent on the left side of (5.19) can be shortened to give a tree-like, cut-free proof of $\Gamma_{n-1} \mapsto \Delta_{n-1}$. Combining this with a similar assertion for the upper sequent on the right side of (5.19), we obtain the inequality (5.21) when $i < n$, thus yielding the desired exponential lower bound on number of sequents in a tree-like, cut-free proof of $\Gamma_n \mapsto \Delta_n$.

To this end, consider a tree-like, cut-free proof P of

$$(\Gamma_n - \{\alpha_i \vee \beta_i\}) \cup \{\alpha_i\} \mapsto \Delta_n.$$

We show how to shorten P by removing all occurrences of p_i, q_i. This suffices, because if P' were the shorter tree-like, cut-free proof resulting from P after removing the occurrences of p_i, q_i, then in P' we can rename all variables p_j, q_j to p_{j-1}, q_{j-1}, for $i < j \leq n$, thus yielding an even shorter tree-like, cut-free of $\Gamma_{n-1} \mapsto \Delta_{n-1}$.

Given a fixed $1 \leq i < n$ and P, combine the variables

$$p_1, \dots, p_{i-1}, p_{i+1}, \dots, p_n, q_1, \dots, q_{i-1}, q_{i+1}, \dots, q_n$$

to define the formulas $\phi'_j, \alpha'_j, \beta'_j$, for $j \neq i-1$, in a similar fashion as $\phi_j, \alpha_j, \beta_j$, for $j \leq n$. Specifically, let

$$\phi'_j \equiv \bigwedge_{1 \leq k \leq j, k \neq i} (p_k \vee q_k)$$

$$\alpha'_1 \equiv p_1$$

$$\alpha'_{j+1} \equiv \phi'_j \supset p_{j+1}, \text{ for } j+1 \neq i$$

$$\beta'_1 \equiv q_1$$

$$\beta'_{j+1} \equiv \phi'_j \supset q_{j+1}, \text{ for } j+1 \neq i.$$

We show how to shorten P to the desired tree-like, cut-free proof P' for the sequent

$$\alpha'_1 \vee \beta'_1, \dots, \alpha'_{i-1} \vee \beta'_{i-1}, \alpha'_{i+1} \vee \beta'_{i+1}, \dots, \alpha'_n \vee \beta'_n \mapsto p_n, q_n.$$

By hypothesis, P is a tree-like, cut-free proof of the sequent

$$\alpha_1 \vee \beta_1, \dots, \alpha_{i-1} \vee \beta_{i-1}, \alpha_i, \alpha_{i+1} \vee \beta_{i+1}, \dots, \alpha_n \vee \beta_n \mapsto p_n, q_n.$$

In P, replace each occurrence of ϕ_i by ϕ_{i-1} to obtain a new tree P_1; this transformation changes each $\phi_j, \alpha_j, \beta_j$ into $\phi'_j, \alpha'_j, \beta'_j$, respectively. By making only slight modifications, this last tree P_1 can be converted into the desired tree-like, cut-free proof P'. This completes the proof of the lower bound for every tree-like, cut-free proof of $\Gamma_n \mapsto \Delta_n$.

It will follow from later results in this chapter, that the pigeonhole principle does not have polynomial-size cut-free Gentzen sequent calculus proofs. Specifically, we have the following result.

Theorem 5.3.5. *There are polynomial-size proofs of* PHP_n^{n+1} *in Gentzen's system LK (i.e., using the cut rule), but for any* $0 < \delta < 1/5^4$, *every cut-free proof of* PHP_n^{n+1} *in LK asymptotically has size at least* 2^{n^δ}.

Proof. (Outline) Theorem 5.7.15 states that there are polynomial-size Frege proofs of PHP_n^{n+1}. Theorem 5.7.2 by R. Reckhow states that Frege systems and the Gentzen system LK (with cut) are polynomially equivalent. This establishes the first assertion of the theorem.

In the language of Section 5.7.1, the pigeonhole principle is formulated by the formula

$$\bigvee_{i=1}^{n+1} \neg \bigvee_{j=1}^{n} p_{i,j} \vee \bigvee_{1 \le i < i' \le n+1} \bigvee_{j=1}^{n} (\neg p_{i,j} \vee \neg p_{i',j})$$

of depth 4 and size $O(n^3)$. If there were a cut-free LK proof of PHP_n^{n+1} of size less than 2^{n^δ}, for $0 < \delta < 1/5^4$, then by the subformula property, every formula of that proof would have depth at most 4. But this contradicts Theorem 5.7.6, due to Beame, Impagliazzo, Krajíček, Pitassi, Pudlák, Woods.

What about the situation of dag-like Gentzen proofs with cut versus tree-like Gentzen proof with cut? Surprisingly, we have the following.

Theorem 5.3.6 ([Kra95]). *Tree-like LK p-simulates dag-like LK.*

Proof. (Outline) Lemma 4.4.8 of [Kra95] states that if propositional formula ϕ has a dag-like Frege proof $\pi = \phi_1, \ldots, \phi_k$ of k steps, depth d and size m, then there is a tree-like Frege proof π' of ϕ of $ck \log_2 k$ steps, depth $d+c$ and size $cmk \log_2 k$ for some constant c, which depends only on the Frege system, but not on θ. This is done as follows. For $1 \le i \le k$ let ϕ'_i be $\phi_1 \wedge \cdots \wedge \phi_i$, where the parentheses are balanced, so that ϕ'_i corresponds to a formula tree of depth $\lceil \log_2 i \rceil$. Show that for every $1 \le j < i$, ϕ_j has a tree-like Frege proof from ϕ'_i with $O(\log_2 i)$ steps, depth $\text{depth}(\phi'_i) + O(1)$ and size $O(\log_2 i \cdot |\phi'_i|)$. Thus the proof π' has $\sum_{i=1}^k O(\log_2 i) = O(k \log_2 k)$ steps, depth $d+O(1)$ and size $\sum_{i=1}^k O(\log_2 i \cdot |\phi'_i|) = O(mk \log_2 k)$. Now, by Theorem 5.7.2, a similar assertion is holds for LK.

5.3.3 Monotonic Sequent Calculus

Let MLK[13] denote the monotonic version of Gentzen's sequent calculus for propositional formulas, where the only logical connectives are $\wedge$, $\vee$ (no negations), and the rules of inference are the usual rules, without the rules for introducing the negation on the left and right. By monotonic formula, we mean a sequent $\Gamma \mapsto \Delta$, where Γ, Δ are cedents of formulas not containing negation, and $\mapsto$ is the Gentzen sequent arrow. The pigeonhole principle PHP_n^{n+1} can be so represented, as follows:

$$\bigwedge_{i=1}^{n+1} \bigvee_{j=1}^{n} p_{i,j} \mapsto \bigvee_{1 \le i < i' \le n+1} \bigvee_{1 \le j \le n} (p_{i,j} \wedge p_{i',j}).$$

The proof of completeness of LK for all propositional tautologies easily yields the completeness of MLK for monotonic tautologies. In boolean circuit complexity theory, it is well-known that there are monotonic problems having

[13] MLK is the Gentzen-style propositional logic fragment of a monotonic form of first order logic, known in the literature as *Geometric Logic* – see [Pud98].

polynomial-size circuits, but requiring exponential size monotonic circuits – indeed, the Broken Mosquito Screen problem is such an example (see Theorem 2.4.1). In analogy to this, it is natural to ask whether there exists a family of monotonic formulas, having polynomial-size proofs in LK, but requiring superpolynomial-size MLK proofs. Though this question is still open, a partial result along these lines is proved in [AGG00], by formally proving properties of the threshold formulas in Section 5.6.6.

Theorem 5.3.7 ([AGG00]). *There are MLK proofs of the pigeonhole principle* PHP_n^{n+1} *of quasipolynomial-size* $2^{\log^{O(1)} n}$.

5.4 Resolution

Though implicit in J. Herbrand's early work, explicit in A. Blake's dissertation [Bla37], and immediately implied by the work of M. Davis and H. Putnam [Dav58], resolution is popularly credited to J.A. Robinson [Rob65a]. Resolution is a refutation system for formulas in conjunctive normal form, and is an important tool used in automated theorem proving and in the execution of the programming language Prolog. A *clause* $C = \{\ell_1, \ldots, \ell_n\}$ is a set of literals, and is interpreted as the disjunction $\bigvee_{i=1}^n \ell_i$. A CNF formula ϕ is written as the collection $S = \{C_1, \ldots, C_m\}$ of clauses, each C_i corresponding to a conjunct of ϕ. The resolution rule is given by

$$\frac{C \cup \{p\} \qquad D \cup \{\overline{p}\}}{C \cup D}$$

where p is a propositional variable, and C, D are clauses not containing either p or its negation $\overline{p}$. The clauses $C \cup \{p\}, D \cup \{\overline{p}\}$ are said to *clash* on the variable p, and $C \cup D$ is said to be the *resolvent*. The empty clause is usually denoted by $\square$. Note that the resolvent is a logical consequence of the conjunction of the clauses which clash. It follows that resolution is a sound refutation system; i.e., if there is a resolution refutation (derivation of $\square$) from clauses $C_1, \ldots, C_m$, then $C_1, \ldots, C_m$ are not satisfiable. Completeness, the converse, follows from Theorem 5.4.1 and yields the following Davis–Putnam algorithm for testing satisfiability of CNF formulas.

Algorithm 5.4.1 (Davis–Putnam procedure).
Input. Set $S = \{C_1, \ldots, C_m\}$ of clauses.
Output. Whether S is satisfiable.

repeat
  choose $p \in Var(S)$
  $S = S - \{C \in S : p \in C \wedge \overline{p} \in C\}$
  $T = \{C \cup D : C \cup \{p\} \in S, D \cup \{\overline{p}\} \in S\}$
  $S = T \cup \{C \in S : p \notin C \wedge \overline{p} \notin C\}$
until $S = \emptyset$ **or** $\square \in S$

```
if □ ∈ S then
   return "inconsistent"
else
   return "consistent"
endif
```

Theorem 5.4.1 ([DP60]). *If $C_1, \ldots, C_m$ are not satisfiable, then there is a resolution derivation of $\Box$ from $C_1, \ldots, C_m$.*

Proof. Let $S = \{C_1, \ldots, C_m\}$ be an unsatisfiable collection of clauses. First, any clause $C \in S$ which contains both variable p and its negation $\overline{p}$ is trivially satisfied by every truth assignment, so S is satisfiable if and only if $S - \{C \in S : p \in C \wedge \overline{p} \in C\}$ is satisfiable. Next, let T consist of all resolvents of all clauses from S which clash on variable p and define S^* to consist of those clauses from S which contain neither p nor its negation, along with clauses from T; i.e., $S^* = T \cup \{C \in S : p \notin C \wedge \overline{p} \notin C\}$. Note that $p \notin Var(S^*)$. We claim that S^* is unsatisfiable. If not, then let $\sigma : Var(S^*) \to \{0,1\}$ be a satisfying assignment for all clauses of S^*.

CLAIM. Either $\sigma \models C$, for all clauses C such that $C \cup \{p\} \in S^*$, or $\sigma \models D$, for all clauses D such that $D \cup \{\overline{p}\} \in S^*$.

Proof of Claim. If not, then for some resolvent $C \cup D$ of $C \cup \{p\}$ and $D \cup \{\overline{p}\}$, we have $\sigma \not\models C \cup D$. Thus σ is not a satisfying assignment for all clauses in S^*, which contradicts our assumption.

The set S^* is unsatisfiable and $Var(S^*) \subseteq Var(S) - \{p\}$. Now repeat this argument, successively selecting one of the remaining variables, and forming all resolvents of clauses which clash on that variable, together with clauses which mention neither the variable nor its negation. After finitely many steps, the empty resolvent $\Box$ is derived.

To prove using resolution that DNF formula ϕ is a tautology, we attempt to refute $\neg\phi$; if successful, we write $\vdash_R \phi$ and say that ϕ has a resolution proof. Since resolution is a sound and complete refutation system for CNF formulas, it follows that $\models \phi \Leftrightarrow \vdash_R \phi$ for all DNF formulas ϕ.

Much research has been done in finding more efficient theorem provers than that of Davis–Putnam. The *subsumption rule* may be added to the procedure of Davis–Putnam. If clause C is a subset of clause D, then C is said to *subsume* D; in particular D is a logical consequence of C. The subsumption rule states that if D is subsumed by C, then D may be removed without affecting satisfiability of S. This may lead to shorter proofs. Other research has been done in criteria for variable selection and resolution selection strategy (see [WCR64, WRC65, Lov70] and J.A. Robinson's hyper-resolution [Rob65b]).

Given arbitrary propositional formula ϕ, one can always put $\neg\phi$ into CNF and apply resolution; however, the conjunctive normal form for $\neg\phi$ may then have size exponential in the size of ϕ. Following Tseitin [Tse68], we extend resolution in an innocuous manner by adding clauses which define new propositional variables to be equivalent to subformulas of $\neg\phi$. The resulting system, *resolution with limited extension* (denoted RLE), is a sound and complete proof system for propositional logic.

Definition 5.4.1. *Suppose that A is a propositional formula in the connectives $\neg, \wedge, \vee$. The collection $LE[A]$ of clauses added by limited extension is defined by induction on A.*

- *If A is the propositional variable p, then*
  $LE[A] = \{\{\overline{q}_A, p\}, \{\overline{p}, q_A\}\}$.
- *If A is $\neg B$, then*
  $LE[A] = \{\{\overline{q}_A, \overline{q}_B\}, \{q_B, q_A\}\} \cup LE[B]$.
- *If A is $B_1 \vee \cdots \vee B_n$, then*
  $LE[A] = \{\{\overline{q}_A, q_{B_1}, \cdots, q_{B_n}\}, \{\overline{q}_{B_1}, q_A\}, \ldots, \{\overline{q}_{B_n}, q_A\}\} \cup \bigcup_{i=1}^n LE[B_i]$.
- *If A is $B_1 \wedge \cdots \wedge B_n$, then*
  $LE[A] = \{\{\overline{q}_A, q_{B_1}\}, \ldots, \{\overline{q}_A, q_{B_n}\} \cup \{\overline{q}_{B_1}, \ldots, \overline{q}_{B_n}, q_A\} \cup \bigcup_{i=1}^n LE[B_i]$.

For an arbitrary propositional formula A in connectives $\neg, \wedge, \vee$, it is not difficult to see that A is satisfiable if and only if there is a truth assignment which satisfies q_A and all the clauses of $LE[A]$. From completeness of resolution it follows that B is a tautology if and only if there is a resolution refutation of $\{q_{\neg B}\} \cup LE[\neg B]$. Summarizing, we have the following.

Theorem 5.4.2 ([Tse68]). *Resolution with limited extension is a sound and complete proof system for propositional logic, as formulated in the connectives $\neg, \vee, \wedge$; i.e., for any formula A, A is unsatisfiable if and only if there is a resolution refutation of $\{\{p_A\}\} \cup LE[A]$.*

In [Tse68], Tseitin introduced as well the system ER of *extended resolution*, which allows abbreviations of arbitrary formulas.

Definition 5.4.2 ([Tse68]). *Suppose that P is a derivation, p is a propositional variable which has not already appeared in P, and that ℓ_1, ℓ_2 are literals, neither of which is equal to p or $\neg p$. The extension rule allows the inference of the following clauses,*

$$p \vee \neg\ell_1, p \vee \neg\ell_2, \neg p \vee \ell_1 \vee \ell_2.$$

which assert that p is equivalent to $\ell_1 \vee \ell_2$. The system ER of extended resolution is the system R of resolution augmented by the extension rule.

By repeated application of the extension rule, we can introduce new propositional variables q with clauses which assert that q is equivalent to any *arbitrary* propositional formula. The difference between the system ER of extended resolution and the system RLE of resolution with limited extension is that in the former, we allow abbreviations of arbitrary formulas by new propositional variables, while in the latter, we allow only abbreviations of subformulas of the given formula to be refuted. The system ER obviously extends RLE, and hence is complete, in that given any unsatisfiable formula A, there exists a derivation of the empty clause $\Box$ from $\{p_A\} \cup LE[A]$. It is easy to establish that ER is sound, in that if there is a derivation of $\Box$ from $\{p_A\} \cup LE[A]$, then A is not satisfiable. Often we may speak of a proof of A in ER, meaning that $\neg A$ has a refutation in ER. Summarizing this discussion, we have the following.

Theorem 5.4.3 ([Tse68]). *Extended resolution is sound and complete.*

Tseitin introduced the notion of *tree-like regular resolution*, a restriction of tree-like resolution, where no literal is annihilated twice on any path from a leaf to the root in a refutation tree. *Dag-like regular resolution* was similarly defined by Tseitin, where in every subsequence $C_1, \ldots, C_r$ of the refutation sequence P, which satisfies the condition that for $1 \leq i < r$, C_{i+1} is obtained by resolution from premiss C_i and another clause in P, it is not the case that there is a literal ℓ appearing in C_1 and in C_r, but in no intermediate C_i for $1 < i < r$. Clearly, the Davis–Putnam Algorithm 5.4.1 produces dag-like regular resolution refutations. In Theorem 9 of [Tse68], Tseitin gave examples of unsatisfiable sets $\mathcal{C}_n$ of clauses based on graphs constructed from $n \times n$ grids (see Lemma 5.4.4), where $\mathcal{C}_n$ contains $O(n^2)$ clauses, each clause containing 4 literals, and such that every dag-like regular resolution refutation of $\mathcal{C}_n$ contains at least $2^{\Omega(n)}$ clauses. This was later improved by Z. Galil in [Gal77b] to a true exponential lower bound. In the next section, we will present a stronger result by A. Haken for (dag-like) resolution refutation length, without the regularity restriction.

We close this section by stating a recent result of N. Arai, who proved that cut-free LK simulates resolution. First, recall here the distinction between strong and weak simulation from the discussion after Definition 5.2.4.

Theorem 5.4.4 ([Ara00]). *Dag-like, cut-free LK on* CNF *formulas strongly p-simulates regular resolution. Dag-like, cut-free LK on* CNF *formulas weakly p-simulates resolution.*

Tree-like, cut-free LK is well-known to be equivalent to the refutation system of analytic tableaux, so Arai's result should be contrasted with the following.

Theorem 5.4.5 ([Urq95]). *Tree-like resolution strongly p-simulates analytic tableaux, but the method of analytic tableaux does not strongly p-simulate tree-like resolution.*

5.4.1 Resolution and the PHP

In our study of the complexity of resolution, we first consider the pigeonhole principle. The negation $\neg\mathrm{PHP}_n^{n+1}$ of (the relational form) of the pigeonhole principle can be formulated in clausal form by

$$\{p_{i,1}, \ldots, p_{i,n}\}$$

for $1 \leq i \leq n+1$, together with

$$\{\overline{p}_{i,j}, \overline{p}_{i',j}\}$$

for $1 \leq i < i' \leq n+1$, $1 \leq j \leq n$.

As an illustration, we present a resolution proof of PHP^3_2. Assuming $\neg\text{PHP}^3_2$, i.e., the existence of an injection $f : \{1,2,3\} \to \{1,2\}$, we first give resolution derivations of three partial results to the effect that

$$2 \in \{f(1), f(2)\}, 2 \in \{f(2), f(3)\}, 2 \in \{f(1), f(3)\}.$$

It will then follow that $f(1) = 2$ and $f(2) = 2$, violating the assumption that f is an injection.

Fact 5.4.1. $p_{1,2}, p_{2,2}$ has a resolution derivation from $\neg\text{PHP}^3_2$.

$$\dfrac{\dfrac{p_{1,1}, p_{1,2} \qquad \overline{p}_{1,1}, \overline{p}_{2,1}}{p_{1,2}, \overline{p}_{2,1}} \qquad p_{2,1}, p_{2,2}}{p_{1,2}, p_{2,2}}$$

Fact 5.4.2. $p_{2,2}, p_{3,2}$ has a resolution derivation from $\neg\text{PHP}^3_2$.

$$\dfrac{\dfrac{p_{2,1}, p_{2,2} \qquad \overline{p}_{2,1}, \overline{p}_{3,1}}{p_{2,2}, \overline{p}_{3,1}} \qquad p_{3,1}, p_{3,2}}{p_{2,2}, p_{3,2}}$$

Fact 5.4.3. $p_{3,2}, p_{1,2}$ has a resolution derivation from $\neg\text{PHP}^3_2$.

$$\dfrac{\dfrac{p_{3,1}, p_{3,2} \qquad \overline{p}_{1,1}, \overline{p}_{3,1}}{p_{3,2}, \overline{p}_{1,1}} \qquad p_{1,1}, p_{1,2}}{p_{3,2}, p_{1,2}}$$

We now combine these to derive $p_{1,2}$ and $p_{2,2}$, showing that the function f is not an injection.

Fact 5.4.4. $p_{1,2}$ has a resolution derivation from $\neg\text{PHP}^3_2$.

$$\dfrac{\dfrac{p_{1,2}, p_{2,2} (\text{Fact 5.4.1}) \qquad \overline{p}_{2,2}, \overline{p}_{3,2}}{p_{1,2}, \overline{p}_{3,2}} \qquad p_{3,2}, p_{1,2} (\text{Fact 5.4.3})}{p_{1,2}}$$

Fact 5.4.5. $p_{2,2}$ has a resolution derivation from $\neg\text{PHP}^3_2$.

$$\dfrac{\dfrac{p_{2,2}, p_{3,2} (\text{Fact 5.4.2}) \qquad \overline{p}_{1,2}, \overline{p}_{3,2}}{p_{2,2}, \overline{p}_{1,2}} \qquad p_{1,2}, p_{2,2} (\text{Fact 5.4.1})}{p_{2,2}}$$

It now follows that

$$\dfrac{\dfrac{p_{1,2} (\text{Fact 5.4.4}) \qquad \overline{p}_{1,2}, \overline{p}_{2,2}}{\overline{p}_{2,2}} \qquad p_{2,2} (\text{Fact 5.4.5})}{\square}$$

This concludes the rather lengthy resolution refutation of $\neg\text{PHP}_2^3$. We will shortly present A. Haken's result that there are essentially no shorter resolution proofs of the pigeonhole principle than that obtained by generalizing the previous example, or that which follows from the proof of completeness of resolution. Nevertheless, there are a number of "symmetries" in the above proof (as in Facts 5.4.1, 5.4.2, 5.4.3 and in Facts 5.4.4, 5.4.5). Propositional proof systems with some kind of symmetry rule have been introduced by Krishnamurthy [Kri85], A. Urquhart [Urq99] and most recently by N. Arai [Ara96]. Details of such systems differ, but the pigeonhole principle PHP_n^{n+1} does have polynomial-size proofs within these systems.

The best upper bound on the number of clauses of the shortest resolution refutation of PHP_n^{n+1} is $2^{O(n)}$. With respect to lower bounds we have the following result.

Theorem 5.4.6 ([Hak85]). *Any resolution refutation of* $\neg\text{PHP}_n^{n+1}$ *must have* $2^{\Omega(n)}$ *clauses.*

As a matter of fact, a more general theorem regarding the generalized pigeonhole principle can be proved. Let PHP_n^m denote the following generalized pigeonhole principle:

$$\bigwedge_{i=1}^{m} \bigvee_{j=1}^{n} p_{i,j} \rightarrow \bigvee_{1 \leq i < i' \leq m} \bigvee_{j=1}^{n} (p_{i,j} \wedge p_{i',j}). \tag{5.22}$$

This is a formula of size $O(m^2 \cdot n)$, and it is clear that Theorem 5.4.6 follows immediately from Theorem 5.4.7.

Theorem 5.4.7 ([BT88b]). *Any resolution refutation of* $\neg\text{PHP}_n^m$ *must have* $2^{\Omega(n^2/m)}$ *clauses, for* $m > n$.

Proof. The proof is based on the adaptations, due to S. Buss and G. Turán, of A. Haken's original ideas. Let $\rho = C_1, \ldots, C_\ell$ be a resolution derivation of $\Box$ from $\neg\text{PHP}_n^m$. We want to show that $\ell = 2^{\Omega(n^2/m)}$. Any total truth assignment σ of the variables $p_{i,j}$ can be pictured as a 0-1 matrix with n rows and m columns, where the entry in column i and row j is 1 if and only if $\sigma(p_{i,j}) = \text{T}$, for $1 \leq i \leq m$ and $1 \leq j \leq n$. A set of exactly $\lfloor n/4 \rfloor$ many 1's lying in distinct rows and columns of this matrix is called *quarter critical.* A (total) truth assignment σ is called *critical* or *maximal* if its matrix has exactly n many 1's, no two of which lie in the same row or column. Thus the critical assignment σ codes a partial one-to-one function $f_\sigma : \{1, \ldots, m\} \longrightarrow \{1, \ldots, n\}$, namely $f_\sigma(i) = j$ if and only if $\sigma(p_{i,j}) = \text{T}$. The 0-*columns* of a critical truth assignment are those columns, none of whose entries is 1. Our goal is to define a mapping Φ, which to each quarter critical set S associates a clause $\Phi(S) = C^S$ occurring in the refutation $\rho = C_1, \ldots, C_\ell$. Clauses of the form C^S will be called *complex,* and will be shown to contain many variables in a 0-column of σ; in particular, complex clauses cannot be small. The mapping Φ is not 1-to-1, but will be shown to be at most $r(n)$-to-1, and so it will follow that the number ℓ of clauses in refutation ρ is at least the number $q(n)$ of quarter critical sets divided by $r(n)$. Estimation of the latter two parameters leads to a lower bound for ℓ, which is of the form $2^{\Omega(n^2/m)}$. Haken calls this type of argument a *bottleneck* counting argument.

A clause C will be viewed as a matrix with n rows and m columns, where the entry in column i and row j is either $+, -$ or blank; this entry is $+$ $(-)$ if $p_{i,j}$ $(\overline{p}_{i,j})$ occurs in C, and blank otherwise. Clearly, the initial clauses of $\neg\text{PHP}^m_n$ consist either of one column of n many $+$'s or of two $-$'s in one row. Now we can prove the following lemma.

Lemma 5.4.1. *Let $\rho = C_1, \ldots, C_\ell$ be a resolution refutation of $\neg PHP^n_{n-1}$. For any critical truth assignment σ and any clause C_i in ρ having $< \lfloor n/2 \rfloor$ $+$'s in a 0-column of σ there is a $j < i$ such that $\sigma \not\models C_j$ and C_j has $\leq \lfloor n/2 \rfloor$ $+$'s in every 0-column of σ; moreover exactly one 0-column in σ has $\lfloor n/2 \rfloor$ many $+$'s.*

Proof. Let σ be a critical truth assignment. By soundness of resolution, if clause $C \cup D$ is inferred by the resolution rule from $C \cup \{p_{i,j}\}$ and $D \cup \{\overline{p}_{i,j}\}$, and $\sigma \not\models C \cup D$, then either $\sigma \not\models C \cup \{p_{i,j}\}$ or $\sigma \not\models D \cup \{\overline{p}_{i,j}\}$. Starting from clause C_i, trace backwards in this fashion through the resolution refutation ρ, thus obtaining a sequence of clauses to which σ assigns the truth value F, until an initial clause C of PHP^m_n is obtained. Since σ is critical, $\sigma \models \overline{p}_{i,j} \vee \overline{p}_{i',j}$ for all $1 \leq i < i' \leq m$ and $1 \leq j \leq n$, and $\sigma \not\models C$, so this initial clause C must be one with n $+$'s in a 0-column of σ. Since in each application of resolution one $+$ is eliminated, there must exist a clause with exactly $\lfloor n/2 \rfloor$ many $+$'s in some 0-column of σ. Now consider the first among these clauses, say C_j, occurring in ρ. It follows easily that C_j has all the required properties. This completes the proof of the lemma.

For any critical truth assignment σ let C^σ be the first clause in ρ which satisfies the conclusion of Lemma 5.4.1. An assignment σ is *compatible* with a set S of ordered pairs of indices if $\sigma(p_{i,j}) = \text{T}$, for all $(i,j) \in S$. Also, define C^S to be the first clause in ρ, which is of the form C^σ for some critical truth assignment σ compatible with S (recall that such clauses were earlier called *complex*). Next we need the following lemma.

Lemma 5.4.2. *A complex clause has at least $(\lfloor n/4 \rfloor + 1)$ many columns, each containing either a $-$ or at least $\lfloor n/2 \rfloor$ many $+$'s.*

Proof. Let C be a complex clause of the form $C = C^S = C^\sigma$, for some quarter critical set S and some critical assignment σ compatible with S. Then σ has a unique 0-column, say i_0, with exactly $\lfloor n/2 \rfloor$ many $+$'s. It remains to show that there exist at least $\lfloor n/4 \rfloor$ many more such columns, i.e., either with a $-$ or at least $\lfloor n/2 \rfloor$ $+$'s. Divide these columns into two groups G^-, G^+, respectively, the first having a $-$ and the second having $\geq \lfloor n/2 \rfloor$ $+$'s but no $-$'s. Then it is enough to show that $|G^-| + |G^+| \geq \lfloor n/4 \rfloor$. Assume on the contrary that $|G^-| + |G^+| < \lfloor n/4 \rfloor$.

By assumption S is quarter critical, $|S| = \lfloor n/4 \rfloor$, $|G| < \lfloor n/4 \rfloor$, and $|\{p_{i,j} : \sigma \models p_{i,j}\}| = n$. Thus there are $> n - 2 \cdot \lfloor n/4 \rfloor$ hence $> \lfloor n/2 \rfloor$ many column indices i, such that $i \notin G^- \cup G^+$, $(i,j) \notin S$ for all $1 \leq j \leq n$, and $\sigma \models p_{i,j}$ for some $1 \leq j \leq n$. By choice of 0-column i_0, there are exactly $\lfloor n/2 \rfloor$ many row indices j, such that $p_{i_0,j}$ occurs in C. Thus there exists $(i,j) \notin S$, such that $\sigma \models p_{i,j}$, $i \notin G^- \cup G^+$ and moreover $p_{i_0,j}$ does not occur in C. As well, σ is critical and $\sigma \models p_{i,j}$, so $\sigma \models \overline{p}_{i_0,j}$; since $C = C^\sigma$, and hence by definition $\sigma \not\models C$, it follows that $\overline{p}_{i_0,j}$ does not occur in C.

Define a new maximal truth assignment σ' from σ by setting the truth value of $\sigma(p_{i,j})$ to false and that of $\sigma(p_{i_0,j})$ to true. It follows that $\sigma' \not\models C$, and $\sigma' \models p_{k,\ell}$ for all $(k,\ell) \in S$, and all 0-columns of σ' contain $< \lfloor n/2 \rfloor$ many +'s in C. Lemma 5.4.1 implies that $C^{\sigma'}$ precedes C^σ in ρ, contradicting the definition of $C = C^\sigma = C^S$.

We now conclude the proof of Theorem 5.4.7. If C is a complex clause, then let $r_C(n)$ denote the number of quarter critical S for which $C = C^S$, and $r(n) = \max\{r_C(n) : C \text{ is complex}\}$. Let $q(n)$ denote the number of all quarter critical sets S. As earlier mentioned, it will follow that the number ℓ of clauses in refutation ρ is at least $q(n)/r(n)$.

We now give an upper bound for $r(n)$. Let C be a fixed complex clause, for which $r_C(n)$ achieves a maximum, so that $r(n) = r_C(n)$. Put $k = \lfloor n/4 \rfloor$ and choose $k+1$ columns satisfying the conclusion of Lemma 5.4.2. We can describe a quarter critical S for which $C = C^S$ by first choosing i many of these $k+1$ columns together with $k - i$ many of the remaining $m - (k+1)$ columns and require that S have a 1 entry in these columns. Since critical σ compatible with S must satisfy $\sigma \not\models C$, we cannot place the i many 1's among the $k+1$ columns in the region where C has at least $\lfloor n/2 \rfloor$ many +'s. Thus there are at most $\lceil n/2 \rceil^i$ many choices of rows in the placement of these 1's. For the remaining $k-i$ many 1's occurring in the $m-(k+1)$ other columns, we require placement in distinct rows, there being $\binom{n-i}{k-i}$ many choices of distinct rows, and $(k-i)!$ many possible placements. Note that $\frac{(n-i)!}{(n-k)!} = \binom{n-i}{k-i} \cdot (k-i)!$. Thus

$$r(n) = r_C(n) \leq \sum_{i \leq k} \binom{k+1}{i} \cdot \binom{m-k-1}{k-i} \cdot \lceil n/2 \rceil^i \cdot \frac{(n-i)!}{(n-k)!}. \quad (5.23)$$

We now derive an expression for $q(n)$, the number of quarter critical sets. Of course

$$q(n) = \binom{m}{\lfloor n/4 \rfloor} \cdot \binom{n}{\lfloor n/4 \rfloor} \cdot \lfloor n/4 \rfloor !.$$

However, in order to find a lower bound for $q(n)/r(n)$, we consider the previously fixed complex clause C, for which $r_C(n)$ is a maximum. Recall that $k = \lfloor n/4 \rfloor$ and choose $k+1$ columns satisfying the conclusion of Lemma 5.4.2. We can describe a quarter critical S by first choosing i many of these $k+1$ columns together with $k-i$ many of the remaining $m-(k+1)$ columns and require that S have a 1 entry in these columns. We then choose k of the n rows, in which to place a 1, and for each such choice require the 1's to be placed in distinct rows and columns. This yields $\binom{n}{n-k}$ many rows, and $k!$ many placements of 1's. Noting that $\frac{n!}{(n-k)!} = \binom{n}{n-k} \cdot k!$, we have that

$$q(n) = \sum_{i \le k} \binom{k+1}{i} \cdot \binom{m-k-1}{k-i} \cdot \frac{n!}{(n-k)!}. \tag{5.24}$$

A straightforward calculation using the inequalities (5.23) and (5.24) then yields the required lower bound $2^{\Omega(n^2/m)}$. Indeed, for $i \le k$ we have that

$$\begin{aligned} \frac{n!/(n-k)!}{\lceil n/2 \rceil^i (n-i)!/(n-k)!} &= \frac{n!}{\lceil n/2 \rceil^i (n-i)!} \\ &= \frac{n(n-1)\cdots(n-(i-1))}{\lceil n/2 \rceil^i} \\ &\ge \left(\frac{3}{2}\right)^i \end{aligned}$$

where we have used the inequality

$$\frac{n-j}{\lceil n/2 \rceil} \ge 3/2$$

for $j \le k$ and $n > 40$. Thus

$$q(n)/r(n) \ge \frac{\sum_{i \le k} \binom{k+1}{i} \cdot \binom{m-k-1}{k-i} \cdot (3/2)^i}{\sum_{i \le k} \binom{k+1}{i} \cdot \binom{m-k-1}{k-i}}. \tag{5.25}$$

Let d_i denote the term $\binom{k+1}{i} \cdot \binom{m-k-1}{k-i}$, for $0 \le i \le k$. Let $k_0 = \lfloor n^2/50m \rfloor$. Notice that $d_{i-1} < d_i$ for $i \le \lfloor n^2/25m \rfloor = 2k_0$, since $\lfloor n^2/25m \rfloor < n/8$ (recall that $m \ge n+1$), we obtain that $\sum_{i=0}^{k_0-1} d_i \le \sum_{i=k_0}^{2k_0-1} d_i$ and so

$$\begin{aligned} q(n)/r(n) &\ge \frac{\sum_{i=k_0}^{k} d_i \cdot (3/2)^i}{2 \sum_{i=k_0}^{k} d_i} \\ &\ge \frac{(3/2)^{k_0} \cdot \sum_{i=k_0}^{k} d_i}{2 \sum_{i=k_0}^{k} d_i} \\ &\ge \frac{(3/2)^{k_0}}{2} = \frac{(3/2)^{\lfloor n^2/50m \rfloor}}{2} = 2^{\Omega(n^2/m)}. \end{aligned}$$

This concludes the proof of Theorem 5.4.7.

In [BP96], P. Beame and T. Pitassi gave a dramatically simpler proof of Haken's Theorem 5.4.6, yielding a slightly better exponential bound. A rough outline of their proof goes as follows. Repeatedly apply an appropriate random restriction t many times to a minimal size resolution refutation of $\neg PHP_n$, thus killing off all large clauses (clauses containing many literals).

This results in a resolution refutation of $\neg PHP_{n-t}$ having no large clauses; however, it is then shown that any such resolution refutation must have at least one clause containing many variables, a contradiction.

Theorem 5.4.8 ([BP96]). *For n sufficiently large, any resolution refutation of $\neg PHP_n$ contains at least $2^{n/32}$ many clauses.*

Proof. As in Haken's proof, a total truth assignment σ is said to be *i-critical* if it defines a 1-to-1 mapping $f_\sigma : \{1,\ldots,n\}-\{i\} \to \{1,\ldots,n-1\}$; moreover, σ is *critical* if it is i-critical for some $1 \leq i \leq n$. If C is a clause over the variables $p_{i,j}$ for $1 \leq i \leq n$ and $1 \leq j \leq n-1$, then C^* denotes the clause obtained from C by replacing each negated variable $\overline{p}_{i,j}$ by $\bigvee_{1\leq k\leq n, k\neq i} p_{k,j}$. Though logically inequivalent, C and C^* are equivalent over all critical truth assignments. Moreover, the translation of the resolution rule is sound with respect to critical truth assignments. Indeed, suppose that

$$\frac{C \cup \{p_{i,j}\} \qquad D \cup \{\overline{p}_{i,j}\}}{C \cup D}$$

is an instance of the resolution rule. The translation

$$\frac{C^* \cup \{p_{i,j}\} \qquad D^* \cup \{\bigvee_{1\leq k\leq n, k\neq i} p_{k,j}\}}{(C \cup D)^*}$$

is sound with respect to critical assignments σ, since if $\sigma \models C^* \vee p_{i,j}$ and $\sigma \models D^* \vee \bigvee_{1\leq k\leq n, k\neq i} p_{k,j}$ then maximality of σ implies that $\sigma \models C^* \cup D^*$.

Let $\rho = C_1,\ldots,C_\ell$ be a refutation of $\neg PHP^n_{n-1}$ having a minimal number of clauses, and let $\rho^* = C_1^*,\ldots,C_\ell^*$ be the previously described monotone transformation.[14]

Lemma 5.4.3. *The refutation ρ^* contains a clause containing at least $2n^2/9$ many variables.*

Proof. Define the *complexity* of clause C to be the minimum number of initial clauses in $\neg PHP_n$ of the form $\{p_{i,1},\ldots,p_{i,n-1}\}$, whose conjunction implies C over all critical truth assignments; i.e.,

$$complex(C) = \min\{|S| : (\forall \text{ critical } \sigma)\ (\sigma \models \bigwedge_{i\in S}(p_{i,1} \vee \cdots \vee p_{i,n-1}) \to C)\}.$$

Note the analogy between $complex(C)$ and the notion of *fence* in Haken's Broken Mosquito Screen problem. Initial clauses of $\neg PHP_n$ have complexity 1, and the empty clause $\Box$ has complexity n. Moreover, if E is obtained by (monotone) resolution from C, D, then $complex(E) \leq complex(C) + complex(D)$. Using the idea of the proof of the 2-3 trick from Lemma 1.6.1, find clause C in (monotone) refutation ρ^* such that

$$n/3 < complex(C) \leq 2n/3.$$

[14] This type of monotone transformation of resolution refutations was first described by S.R. Buss.

Namely, let $1 \leq i \leq \ell$ be the least such that $complex(C_i^*) > n/3$. We claim that for this i, $complex(C_i^*) \leq 2n/3$. Note that C_i^* must be obtained by (monotonic) resolution from C_j^*, and C_k^*, for $j, k < i$, since the complexity of the initial clauses from $\neg PHP_n$ is 0. By definition of index i, $complex(C_j^*), complex(C_k^*) \leq n/3$, so by a previous remark $complex(C_i^*) \leq complex(C^*j) + complex(C^*k) \leq 2n/3$.

CLAIM. Let m denote the complexity of C. Then C contains at least $m(n-m)$ many variables.

Proof of Claim. Let $S \subseteq \{1, \ldots, n\}$ be such that $|S| = complex(C) = m$, and for all critical σ, $\sigma \models \bigwedge_{i \in S}(p_{i,1} \vee \cdots \vee p_{i,n-1}) \to C$. Fix $i_0 \in S$, and consider i_0-critical σ with $\sigma \not\models C$ (note that if every i_0-critical assignment σ satisfies C, then $|S|$ is not minimal, since $S - \{i_0\}$ would work). Let $i_1 \in \{1, \ldots, n\} - S$. Define σ' from σ by replacing i_0 by i_1; i.e., $\sigma'(p_{i_0,j}) = \sigma(p_{i_1,j})$ and $\sigma'(p_{i_1,j}) = \sigma(p_{i_0,j})$ for all $1 \leq j \leq n-1$, so it follows that σ' is i_1-critical. We claim that $\sigma' \models C$. Indeed, otherwise by taking the contrapositive of the formula in the definition of $complex(C)$, there must exist $i_2 \in S$ for which $\sigma' \not\models p_{i_2,1} \vee \cdots \vee p_{i_2,n-1}$. This means that $i_2 = i_1$, and so $i_1 \in S$, a contradiction.

Now suppose that $\sigma \models p_{i_0,\ell}$. Then $\sigma' \models p_{i_1,\ell}$. Since C is monotone, $\sigma \not\models C$, $\sigma' \models C$, C must contain the variable $p_{i_1,\ell}$. For i_0 and σ fixed, repeat this argument for all $i_1 \in \{1, \ldots, n\} - S$. Altogether C must contain at least $m(n-m)$ many variables. This completes the proof of the lemma.

Recall that $\rho = C_1, \ldots, C_\ell$ is a resolution refutation of $\neg PHP_n$ containing the minimum number ℓ of clauses, and that $\rho^* = C_1^*, \ldots, C_\ell^*$ is the monotonic translation. A clause C in ρ^* is said to be *large* if it contains at least $n^2/8$ many variables, i.e., $n^2/8$ out of the $n(n-1)$, or $> 1/8$ of all variables. Assume now that $\ell < 2^{n/32}$. Then there are at most ℓ large clauses; denote this number by $\ell_0 \leq \ell$.

CLAIM. There exist i, j such that $p_{i,j}$ appears in at least $\ell_0/8$ many large clauses.

Proof of Claim. If not, then $n(n-1) \cdot \frac{\ell_0}{8} > \frac{n^2}{8} \cdot \ell_0$, since there are $n(n-1)$ variables, each of which appears in less than $\ell_0/8$ many large clauses, while there are ℓ_0 many large clauses, hence which contain at least $n^2/8$ variables.

Now for i, j from the claim, set $p_{i,j} = 1$ in all clauses of ρ^*, and obtain a resolution refutation ρ_1^* of $\neg PHP_{n-1}$, having at most $\ell_0(1-1/8)$ many large clauses. Let t be the smallest integer satisfying

$$\begin{aligned} \ell_0(7/8)^t &< 1 \\ \log_2(\ell_0) + t\log_2(7/8) &< 0 \\ t &< \frac{\log_2 \ell_0}{\log_2 8 - \log_2 7} \\ t &< 8\log_2 \ell_0. \end{aligned}$$

Iterate the previous procedure $t = \log_2 \ell$ many times, each time setting some $p_{i,j} = 1$ and removing 1/8-th of the remaining large clauses. In this manner, we obtain a (monotonic) resolution refutation of $\neg PHP_{n-t}$. Now apply Lemma 5.4.3 to find a clause in the refutation of $\neg PHP_{n-t}$ containing at least $\frac{2(n-t)^2}{9}$ many variables. But

$$\begin{aligned}\frac{2(n-t)^2}{9} &= \frac{2(n-8\log_2 \ell)^2}{9} \\ &\geq \frac{2(n-8\log_2 2^{n/32})^2}{9} \\ &> \frac{2(n-n/4)^2}{9} \\ &= \frac{n^2}{8}\end{aligned}$$

which contradicts the fact that all large clauses had been removed. This concludes the proof of Theorem 5.4.8.

We leave as an exercise the application of this technique to yield a simplified proof of Theorem 5.4.7.

5.4.2 Resolution and Odd-Charged Graphs

If N denotes the number of clauses of the formula ϕ_n under consideration (for example, the number of clauses of PHP_n^{n+1} is $N = \Theta(n^3)$), then Haken's lower bound shows that in fact the optimal resolution derivation of the empty clause from $\neg\text{PHP}_n^{n+1}$ must have $2^{\Theta(N^{1/3})}$ clauses. This raises the question whether there are examples of formulas ϕ_n with shortest resolution of size $2^{\Omega(n)}$, where $|\phi_n| = O(n)$. In [Gal77b] Galil was able to improve Tseitin's earlier mentioned lower bound for regular resolution to a true exponential lower bound. Returning to Tseitin's approach, but armed with Haken's bottleneck counting method, A. Urquhart [Urq87] obtained a true exponential lower bound by using expander graphs in place of Tseitin's $n \times n$ grids.

The Tseitin–Urquhart formulas are based on certain graphs, described as follows. Let $G = (V, E)$ denote a finite, undirected, labeled graph without loops and having without multiple edges between the same vertex. Assign a weight $w(u) \in \{0, 1\}$ to each node u; the weight will hereafter be called a *charge*. The total charge $w(G)$ of G is the sum mod 2 of all the charges $w(u)$ for $u \in V$. The edge labels are literals such that if edges e, e' are labeled with the literals ℓ, ℓ', respectively, then $\{\ell, \neg\ell\} \cap \{\ell', \neg\ell'\} = \emptyset$. Usually we identify edges with their labels. If $p_1, \ldots, p_{deg(u)}$ are the literals attached to u let $E(u)$ denote the equation $p_1 \oplus \cdots \oplus p_{\deg(u)} = w(u)$, where $\deg(u)$ is the number of edges adjacent to u. Let $\mathcal{C}(u)$ be the set of clauses formed by the conjunctive normal form of equation $E(u)$ and let $\mathcal{C}(G)$ be the union over $u \in V$ of the sets $\mathcal{C}(u)$ of clauses. It is clear that $|\mathcal{C}(u)| = 2^{deg(u)-1}$.

A graph G is said to be *odd-charged*, if the sum mod 2 of all vertex charges is 1. Lemma 5.4.4 explains why from now on we will be interested in connected graphs with odd charge.

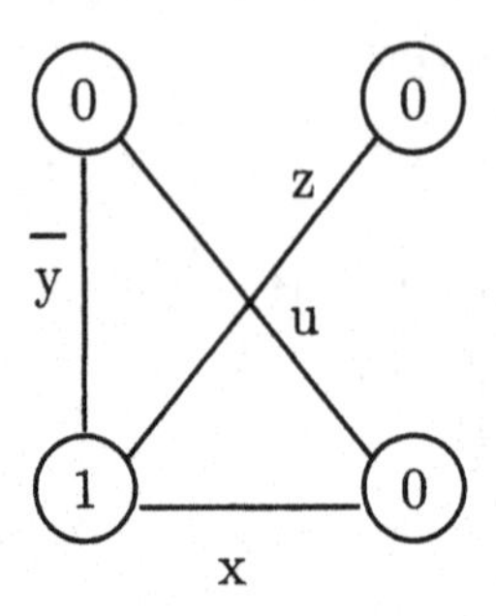

Fig. 5.1. Odd-charged graph with edges labeled by literals

Example 5.4.1. For the graph G depicted in Figure 5.1, the charge equations are given by

1. $\overline{y} \oplus u = 0$
2. $\overline{y} \oplus x \oplus z = 1$
3. $z = 0$
4. $x \oplus u = 0$

with corresponding conjunctive normal form, as expressed by clauses

1. $\{\overline{u}, \overline{y}\}, \{u, y\}$
2. $\{x, y, \overline{z}\}, \{x, \overline{y}, z\}, \{\overline{x}, y, z\}, \{\overline{x}, \overline{y}, \overline{z}\}$
3. $\{\overline{z}\}$
4. $\{x, \overline{u}\}, \{\overline{x}, u\}$

The rule for producing clauses from a charge equation is to place an odd (even) number of negations on the associated literals, if the charge is 0 (1). Clearly, there are 2^{d-1} clauses associated with the charge equation for vertex v if the degree of v is d (note that half of the 2^d truth assignments satisfy the charge equation). When considering proof size, we are thus only interested in graph families of bounded degree.

Now we can prove the following result for connected graphs.

Lemma 5.4.4 ([Tse83]). *For G connected graph,*

$$C(G) \text{ is unsatisfiable} \Leftrightarrow w(G) = 1.$$

Proof. Let $E(G)$ denote the system $\{E(u) : u \in V\}$. First we prove $(\Leftarrow)$. Assume $w(G) = 1$. The modulo 2 sum of the left-hand of the system $E(G)$ is 0 since each literal is attached to two vertices. By assumption the sum in the right-hand side of $E(G)$ is 1. Hence there is no truth assignment satisfying $\mathcal{C}(G)$. Next we prove ($\Rightarrow$). Assume $w(G) = 0$. We show that $\mathcal{C}(G)$ is satisfiable. Let G_p be obtained from G by interchanging p and $\neg p$ and complementing the charges of the vertices incident to p. Clearly the system $E(G)$ and $E(G_p)$ have the same satisfying truth assignments. If u, v are distinct nodes both having charge equal to 1, then there is a sequence of vertices $u = u_1, \ldots, u_r = v$ forming a path from u to v. Applying the above toggling transformation $G \longrightarrow G_p$, we can transfer charges along this path until we obtain a graph, whose total charge is decreased by 2. Since $w(G) = 0$, by repeating this observation, we finally end up with a system of equations whose right-hand side has only 0 charges. A satisfying assignment is now obtained by setting all the literals to 0.

For any truth assignment σ and any vertex u let $w_\sigma(u)$ be the sum mod 2 of $\sigma(p)$ where p runs over the literals adjacent to u. Call σ *u-critical* if $w(v) = w_\sigma(v)$, for all $v \neq u$ and $w(u) \neq w_\sigma(u)$. A partial truth assignment is a truth assignment defined on a subset of the variables; it is called *nonseparating* if the graph resulting from G by deleting all the edges labeled by literals in $dom(\sigma)$ is connected. Then we can prove the following lemma.

Lemma 5.4.5. *For any node u, any partial, nonseparating truth assignment σ can be extended to a u-critical truth assignment.*

Proof. Let T be a spanning tree of the graph resulting from G by deleting all the edges labeled by literals in $dom(\sigma)$. Assign values arbitrarily to any edge not in the spanning tree that has not yet been assigned a value. We extend this to a u-critical truth assignment σ' as follows. Proceeding from the leaves inward toward u assign values to vertices $v \neq u$ such that $w(v) = w_{\sigma'}(v)$. The resulting σ' is uniquely determined from the values given to edges not in T and must be u-critical since $w(G) = 1$.

With these preliminaries, we can now prove Urquhart's exponential lower bound for resolution.

Theorem 5.4.9 ([Tse83], [Urq87]). *There is a sequence ϕ_n of valid formulas consisting of $O(n)$ many constant size clauses such that each $\neg\phi_n$ has a polynomial-size $n^{O(1)}$ Frege refutation proof but every resolution refutation has size $2^{\Omega(n)}$.*

Proof. Tseitin's original construction was based on two dimensional grids, the intuition being that one must remove many edges before the grid is broken into two roughly equal subgraphs (the number of edges required to repeatedly disconnect the graph is related to the regular resolution lower bound). The present construction is based on an idea of M. Ben-Ari [BA80], and provides an example of a bounded-degree family of graphs, which require many edges to be removed before the graph is disconnected. Let H_n be a bipartite graph consisting of two sides, each consisting of $n = m^2$ nodes, such that each node has degree ≤ 5. We construct the graph G_n from the graphs H_n by connecting the nodes of each side into a chain by adding $n-1$ new edges to each side. Each node of the new graph has degree ≤ 7 and hence the clauses of $\mathcal{C}(G_n)$ are of constant size. The formula ϕ_n is the disjunction of the conjuction of the formulas in C, where C is a clause in $\mathcal{C}(G_n)$. Clearly ϕ_n is of size $O(n)$. In view of Lemma 5.4.4, we will assume that the weight function in G_n satisfies $w(G_n) = 1$. Margulis constructed such expander graphs with properties given in the following lemma (consult [Mar73] for details, or see the next section for a probabilistic construction by U. Schöning).

Lemma 5.4.6 ([Mar73]). *There is a constant $d > 0$ such that if V_1 is a set of nodes of size $\leq n/2$ contained in one side of G_n and V_2 is the set of nodes in the opposite side of G_n connected to a node in V_1 by an edge, then $|V_2| \geq (1+d) \cdot |V_1|$.*

Note that $d \leq 4$, since G_n has degree at most 5. First of all we show that ϕ_n has a polynomial-size Frege proof. Letting left$(E(u))$ (right$(E(u))$) denote the left (right) side of the charge equation $E(u)$, use the propositional identities

$$\begin{aligned} p \oplus q &\equiv \neg(p \leftrightarrow q) \\ (\neg p) \leftrightarrow (\neg q) &\equiv p \leftrightarrow q \end{aligned}$$

to convert

$$\bigoplus_{u \in V} \text{left} E(u)) \leftrightarrow \bigoplus_{u \in V} \text{right} E(u))$$

into formulas consisting only of literals and the biconditional $\leftrightarrow$. This takes $O(n)$ steps. Using the associative and commutative laws of the biconditional, we can move double literals to the front and eliminate these double occurrences. Each of these steps takes $O(n^2)$ steps, thus yielding the desired contradiction $0 \leftrightarrow 1$ in a total of $O(n^3)$ steps, each step of length $O(n)$. Hence the size of the Frege proof is $O(n^4)$. See also p. 143 of [Chu56].

Next we prove the lower bound on resolution refutations of $\mathcal{C}(G_n)$. Let ρ be a resolution refutation of $\mathcal{C}(G_n)$. Let us assume that the vertices of the sides of the bipartite graph are numbered in some canonical way (such as $1, \ldots, n$ and $\overline{1}, \ldots, \overline{n}$). Define $\mathcal{R}_n$ to be the set of partial truth assignments (or restrictions) σ specified by choosing $\lfloor dn/16 \rfloor$ vertices from one side together with corresponding vertices in the opposite side and then assigning truth values arbitrarily to the middle edges attached to at least one of the above vertices. For any such restriction $\sigma \in \mathcal{R}_n$, let $V(\sigma)$ be the set of the above $2 \cdot \lfloor dn/16 \rfloor$ vertices chosen. Clearly all such restrictions are non-separating. Further, for any clause C define $Cover(C, \sigma)$ as the set of vertices $u \notin V(\sigma)$ such that for some u-critical truth assignment σ' extending σ, $\sigma'(C) = \text{F}$. For any partial truth assignment σ let C^σ be the first clause in the refutation ρ satisfying $|Cover(C, \sigma)| \geq n/4$. This is well-defined, because $|Cover(\emptyset, \sigma)| \geq n/4$, for every partial truth assignment σ, as $2 \cdot \lfloor \frac{dn}{16} \rfloor \leq \frac{10n}{16} < \frac{3n}{4}$. A clause

C is called *complex* if $C = C^\sigma$ for some partial truth assignment σ. Now we can prove the following lemma.

Lemma 5.4.7. *For any partial truth assignment σ the clause C^σ contains at least $\lfloor dn/16 \rfloor$ literals.*

Proof. Let $C = C^\sigma$ be as above. C must be derived by two earlier clauses in ρ, say D, E. Since $Cover(C, \sigma) \subseteq Cover(D, \sigma) \cup Cover(E, \sigma)$ and C is σ-complex, both sets on the right-hand side have size $< n/4$, so it follows that $Cover(C, \sigma) < n/2$. Now write $Cover(C, \sigma) = W_1 \cup W_2$, where W_1, W_2 are vertices on the opposite sides of G_n and without loss of generality assume that $|W_1| \geq |W_2|$. Let Y_2 be the vertices not in W_2, which are connected to W_1 by a middle edge. Clearly $|W_1| \geq n/8$ and hence $|Y_2| \geq dn/8$ by Lemma 5.4.6. Put $Z_2 = Y_2 - V(\sigma)$. It is clear that $|Z_2| \geq |Y_2| - |V(\sigma)| \geq dn/16$. We show that if $v \in Z_2$ then C mentions a literal incident to v. By definition, there is a literal incident to a middle edge that links v to a vertex $u \in W_1$. Hence there is a u-critical truth assignment σ' extending σ such that $\sigma'(C) = \text{F}$. If the literal ℓ incident to this edge is not mentioned in C, then it is easily seen that σ'' is v critical, where σ'' is the truth assignment identical to σ', except on literal ℓ, in which case $\sigma''(\ell) = 1 - \sigma'(\ell)$. Since $u, v \notin V(\sigma)$ the truth assignment σ'' is also an extension of σ, contradicting $v \notin W_2$. This completes the proof of the lemma.

To complete the proof of the theorem, it is enough to show that for any complex clause C,

$$\Pr[C = C^\sigma] \leq 2^{-O(n)}. \tag{5.26}$$

For any complex clause C, let $E(C)$ be the set of edges mentioned in C, and for any partial truth assignment $\sigma \in \mathcal{R}_n$, let $E(\sigma)$ be the set of vertices contained in a chosen side of G_n with a middle edge attached to them which is mentioned in C. By Lemma 5.4.7, we can choose a side such that $|E(C)| \geq dn/16$. Consider a random variable X representing the overlap between $E(C)$ and $E(\sigma)$, i.e., $X = |E(C) \cap E(\sigma)|$. It is clear that

$$\Pr[C = C^\sigma] = \sum_{i \leq dn/16} \Pr[X = i] \cdot \Pr[C = C^\sigma | X = i]. \tag{5.27}$$

Now, since the edges are set independently, the fraction of restrictions in $\mathcal{R}_n$ with $|E(C) \cap E(\sigma)| = i$, for which C is σ-complex, is at most 2^{-i}, hence $\Pr[X = i] \leq 2^{-i}$. As well, X has the hypergeometric distribution, representing sampling without replacement from a population of size n containing at least $dn/16$ good objects taking samples of size $\geq \lfloor dn/16 \rfloor$. Thus

$$\Pr[C = C^\sigma] = \sum_{i=1}^{s} \binom{M}{i} \binom{N - M}{s - i} \binom{N}{s}^{-1}$$

where $M = |E(C)| \lfloor \frac{dn}{16} \rfloor$, $s = |E(\sigma)| = \lceil \frac{dn}{16} \rceil$, and N is the number of middle edges in H_n.

It is well known that the binomial distribution, representing sampling *with* replacement, approximates the hypergeometric distribution, provided population size is large with respect to sample size (see [Fel68]). Let Y be the random variable with binomial distribution representing sampling with replacement from the same population and with the same sample size as X and denote $\frac{d}{16}$ by f. For appropriately chosen $g \geq f$,

$$\Pr[X \leq \lfloor gn \rfloor] \leq \Pr[Y \leq \lfloor gn \rfloor] \leq \exp(-2nh^2)$$

for $h = f - g$ (see p. 151, 69 of [JK70]), hence the sum of the terms of the sum in (5.27) for $i \leq gn$ is bounded by $2^{-O(n)}$. Since $\Pr[C = C^\sigma | X > gn] \leq 2^{-gn}$, putting both inequalities together we have

$$\Pr[C = C^\sigma] \leq 2^{-O(n)}.$$

In [Urq95], A. Urquhart gave an elementary combinatorial proof of the $\Pr[C = C^\sigma]$, relying only on a tricky, but elementary approximation of the tail of the hypergeometric distribution, due to V. Chvátal. First, some claims:

CLAIM 1.

$$\binom{a}{b}\binom{b}{c} = \binom{a}{c}\binom{a-c}{b-c}.$$

This immediately follows by calculation.

CLAIM 2.

$$\binom{b}{a} = \sum_{i=0}^{a} \binom{c}{i}\binom{b-c}{a-i}$$

This is clear by counting.

Using these, we have the following technical result:

CLAIM 3.

$$\begin{aligned}
&\sum_{i=0}^{s-j} \binom{M}{i}\binom{s-i}{j}\binom{N}{s}^{-1} \\
&= \binom{N-M}{j}\sum_{i=0}^{s-j}\binom{M}{i}\binom{N-M-j}{s-i-j}\binom{N}{s}^{-1} \\
&= \binom{N-M}{j}\binom{N-j}{s-j}\binom{N}{s}^{-1} \\
&= \binom{N-M}{j}\binom{s}{j}\binom{N}{j}^{-1} \\
&\leq \binom{s}{j}\left(\frac{N-M}{N}\right)^{j}.
\end{aligned}$$

The first line follows by setting $a = N - M$, $b = s - i$, $c = j$ and applying Claim 1. The second line follows by setting $a = s - j$, $b = N - j$, $c = M$ and applying Claim 2, and the third line follows by calculation.

From Claim 3, we now have the following

$$\begin{aligned}
&\sum_{i=0}^{s} \binom{M}{i}\binom{N-M}{s-i}\binom{N}{s}^{-1} w^{-i} \\
&= 2^{-s} \sum_{i=0}^{s} \binom{M}{i}\binom{N-M}{s-i}\binom{N}{s}^{-1} \sum_{j=0}^{s-i} \binom{s-i}{j} \\
&= 2^{-s} \sum_{j=0}^{s} \sum_{i=0}^{s-j} \binom{M}{i}\binom{N-M}{s-i}\binom{s-i}{j}\binom{N}{s}^{-1} \\
&\leq 2^{-s} \sum_{j=0}^{s} \binom{s}{j}\left(\frac{N-M}{N}\right)^{j} \\
&= 2^{-s}(2 - \frac{M}{N})^{s} \\
&= (1 - \frac{M}{2N})^{s}.
\end{aligned}$$

Recalling that $f = \frac{d}{16}$, we have that the number N of middle edges is at most $5n$, and $M = \lfloor fn \rfloor$, so $\frac{M}{2N} \geq \frac{f}{11}$, for sufficiently large n. Taking $c = (1 - \frac{f}{11})^{-f}$, it follows that there must be at least c^N complex clauses in the refutation. This completes the proof of Theorem 5.4.9.

5.4.3 Schöning's Expander Graphs and Resolution

In this section, we present U. Schöning's simplification in [Sch97] of the Urquhart exponential lower bound for resolution refutations of Tseitin formulas for a certain class of odd-charged graphs. Schöning's proof uses two basic ideas:

1. By setting certain chosen literals to 0 or 1 appropriately, kill off all large clauses (this is an application of the Beame-Pitassi simplification of Haken's lower bound for the pigeonhole principle within the context of Tseitin's graph formulas).
2. By appropriately toggling certain critical truth assignments, prove that there is a remaining large clause (having many literals). The ability to so toggle certain truth assignments uses the existence of certain expander graphs, whose existence is proved by a new probabilistic construction.[15]

Theorem 5.4.10. *Let $d = 10$, $\beta = \frac{1}{(4 \ln 2)d}$. There exists a family of undirected, degree d graphs $G_n = (V_n, E_n)$, where $V_n = \{1, \ldots, n\}$, such that every*

[15] In [Sch97], the counting argument was couched in terms of Kolmogorov complexity.

resolution refutation of the related CNF *Tseitin formula* ϕ_n *has at least* $2^{\beta n}$ *clauses.*

Proof. Let's begin by an overview of the proof. For $G = (V, E)$, where $|V| = n$ and $S \subseteq V$, define

$$E(S, V - S) = \{e \in E : (\exists x \in S)(\exists y \in V - S)[e = \{x, y\}] \tag{5.28}$$

and let $Exp_n(G, S)$ be the property

$$|E(S, V - S)| > n. \tag{5.29}$$

Let $\gamma = 0.32$. At the end of the proof, we will construct graphs $G = G_n$ with the following expansion property

$$(\forall S \subset \{1, \ldots, n\}) \left[\gamma n < |S| \leq 2\gamma n \to Exp_n(G, S)\right]. \tag{5.30}$$

Suppose now that $P = C_1, \ldots, C_\ell$ is a resolution refutation of the Tseitin formula ϕ_n associated with an odd charged graph $G = G_n$, which satisfies the expansion property (5.30). Without loss of generality, we may assume that the edge labels for G are distinct variables (rather than literals). Assume, in order to obtain a contradiction, that $\ell < 2^{\beta n}$.

Let's call a clause C from refutation P *large* if it contains at least $n/2$ literals. Let $\ell_0 \leq \ell < 2^{\beta n}$ denote the number of large clauses in P. Since G has n vertices and degree d, there are $dn/2$ undirected edges in G, hence $dn/2$ variables labeling edges of G. Thus that there are dn many literals appearing in the Tseitin formula for G.

CLAIM. There exists a literal e appearing in at least $\ell_0 \cdot \frac{n/2}{dn} = \ell_0/2d$ many large clauses of P.

Proof of Claim. If not, then $\frac{\ell_0}{2d} \cdot dn < \ell_0 \cdot n/2$, a contradiction.

Note that if clause C contains literal e, then the clause $C[e/1]$ is equivalent to 1, where $C[e/1]$ is obtained by replacing e by 1 or TRUE and $\overline{e}$ by 0 or FALSE. Similarly if C contains literal $\overline{e}$, then $C[e/1]$ is equivalent to the clause C' obtained from C by removing e. By soundness of resolution, the sequence $P[e/1] = C_1[e/1], \ldots, C_\ell[e/1]$, obtained from P is still a refutation. The clauses where substitution $e/1$ was made are now trivial (i.e., equivalent to TRUE), hence cannot play a role in the refutation. By removing these and possibly other clauses, we obtain a refutation P' involving only non-trivial clauses from $P[e/1]$.

Let G' be obtained from G by removing the edge labeled by $e = \{x, y\}$, and by toggling the charges of the incident vertices x, y. Then G' is still an odd-charged graph, though it is possible that G' is no longer connected. It follows that P' is a resolution refutation of the Tseitin formula related to G'.

Now, from the previous claim, define G' and P' as explained above, for the edge e, such that e appears in fraction $1/2d$ of large clauses of P. The number of large clauses remaining in P' is at most $\ell_0 \cdot (1 - 1/2d)$. Now

$$2^{\beta \cdot n}(1 - 1/2d)^t < 1 \iff \beta n + t \log_2(1 - 1/2d) < 0$$
$$\iff t > \frac{-\beta n}{\log_2(1 - 1/2d)}.$$

Noting that $\ln(1-\epsilon) = -\epsilon+\epsilon^2/2-\epsilon^3/3+\cdots$ for $0 < \epsilon < 1$ and so $\ln(1-\epsilon) < -\epsilon$, hence $\frac{1}{\epsilon} > -\frac{1}{\ln(1-\epsilon)}$ and $\frac{\ln 2}{\epsilon} > -\frac{\ln 2}{\ln(1-\epsilon)} = -\frac{1}{\log_2(1-\epsilon)}$, it follows that after $t := (2\ln 2)\beta nd$ many rounds, each time using the previous claim to remove fraction $1/2d$ of the large clauses, we obtain a resolution refutation P^* of the Tseitin formula for graph G^*, where G^* is obtained from G by removing t edges from G and appropriately toggling the charges of the vertices incident to the edges removed.

A truth assignment σ of the literals labeling G^* is called *x-critical* if for each $y \in \{1,\ldots,n\}$ different from x, $\sigma \models F_y$ and yet $\sigma \not\models F_x$, where F_z is the charge equation of G^* at the vertex z. For C appearing in P^*, define

$$cover(C) = \{x \in \{1,\ldots,n\} : (\exists\sigma)(\sigma \text{ is } x\text{-critical and } \sigma \not\models C)\}.$$

Let C be the first (i.e., leftmost) clause appearing in refutation P^*, for which $|cover(C)| > \gamma n$.

CLAIM. $|cover(C)| \leq 2\gamma n$.

Proof of Claim. Since initial clauses from the Tseitin formula (i.e., clauses from the formulas F_z) have a 1-element cover, C must be derived by an application of the resolution rule from clauses A, B appearing earlier in P^*. Note that in this case, $cover(C) \subseteq cover(A) \cup cover(B)$. If the claim does not hold, then $|cover(C)| > 2\gamma n$, and so $2\gamma n < |cover(A)| + |cover(B)|$, hence either A or B has a cover of size $> \gamma n$. But since A, B occur before C in P^*, this contradicts the choice of C, thus establishing the claim.

Now letting $S = cover(C)$, chosen as above, by the expansion property (5.30), there are n edges $\{x, y\} \in E$ from graph G such that $x \in S$ and $y \notin S$. The graph G^* was obtained from G by removing t edges (and appropriately toggling the charges), so there are $n - t$ edges $\{x, y\}$ in E^* with $x \in S$, and $y \notin S$. For such a pair $\{x, y\}$ labeled by literal e, let σ be an x-critical truth assignment for which $\sigma \not\models C$, and define σ' from σ by $\sigma'(e) = 1-\sigma(e)$. Clearly σ' is now y-critical, but since $y \notin S = cover(C)$, it must be that $\sigma' \models C$. Thus literal e appears in clause C. Since there are $n - t$ edges with $x \in S$, $y \notin S$, the clause C contains $n - t$ literals. Recalling that $t = (2\ln 2)\beta nd$, we have

$$n - t \geq n/2 \iff n/2 \geq t$$
$$\iff n/2 \geq (2\ln 2)\beta nd$$
$$\iff \beta \leq \frac{1}{(4\ln 2)d}.$$

Hence, under the assumption that $\beta = \frac{1}{(4\ln 2)d}$ and that P is a refutation of the Tseitin formula for odd-charged graph G containing less than $2^{\beta n}$ clauses, it follows that the above clause C contains at least $n/2$ literals, and so is large. However, this contradicts the fact that in P^* there are no remaining large clauses. This establishes the lower bound of $2^{\beta n}$.

To conclude the proof of Theorem 5.4.10, we must construct a family of graphs $G_n = (V_n, E_n)$, with $V_n = \{1, \ldots, n\}$, such that $G = G_n$ satisfies the expansion property (5.30). In [Sch97], U. Schöning does this by a probabilistic argument. Consider Figure 5.4.3, where sets $A = \{a_0, \ldots, a_{n-1}\}$, $B = \{b_0, \ldots, b_{nd-1}\}$, $C = \{c_0, \ldots, c_{nd-1}\}$, $D = \{d_0, \ldots, d_{nd/2-1}\}$ are displayed. Each of the n elements in set A is connected to d distinct elements in B, in that $b_i \in B$ is connected to $a_{\lfloor i/d \rfloor} \in A$. The elements of B are connected to those of C by a permutation $\pi \in S_{nd}$. Each of the $nd/2$ elements in D is connected to two distinct elements in C, in that $c_i \in C$ is connected to $d_{\lfloor i/2 \rfloor} \in D$. Define the graph $H_\pi = (V, E)$, where $V = A$ and $\{x, y\} \in E$ if x, y are connected to the same node in D. Let $\mathcal{G}(= \mathcal{G}_n)$ denote the set of degree d undirected graphs $G = (V_n, E_n)$ with $V_n = \{a_0, \ldots, a_{n-1}\}$, and let $g = |\mathcal{G}|$. If x, y are both connected to d_i, either x (y) is connected to c_{2i} (c_{2i+1}) or x (y) is connected to c_{2i+1} (c_{2i}), so we must quotient out by $2^{|D|}$, yielding

$$g = \frac{(nd)!}{2^{(nd/2)}}. \tag{5.31}$$

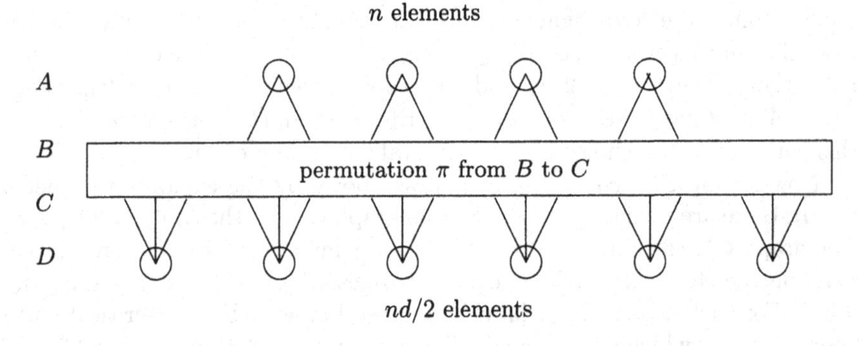

Fig. 5.2. Expander graph construction

Let $\mathcal{K}_\alpha(= \mathcal{K}_{\alpha,n})$ denote the set of degree d graphs $G = (V_n, E_n)$ on vertex set $\{a_0, \ldots, a_{n-1}\}$, which do *not* satisfy the expansion property for sets $S \subseteq V_n$ of cardinality αn. For graphs in $\mathcal{K}_\alpha$, we have

$$(\exists S \subset V_n) [|S| = \lfloor \alpha \cdot n \rfloor \wedge \neg Exp_n(G, S)] \tag{5.32}$$

i.e., there is a set S of size αn, for which there are at most n edges $\{x, y\} \in E_n$ for which $x \in S$ and $y \notin S$.

For $\alpha \in (0, 1)$, every graph $H \in \mathcal{K}_\alpha$ can be obtained by performing the following steps (1) through (5).

1. Choose set R containing αn vertices from A, and let S denote the $d\alpha n$ vertices from B connected with these vertices. This yields

$$\binom{n}{\alpha n}$$

many choices.
2. Choose set T containing n of the $d\alpha n$ many vertices in S, and let $U = S - T$. This yields

$$\binom{d\alpha n}{n}$$

many choices.
3. Choose $\frac{d\alpha n}{2}$ many vertices from D, and let W denote the vertices in C connected to these vertices. This yields

$$\binom{\frac{nd}{2}}{\frac{d\alpha n-n}{2}}$$

many choices.
4. Join the elements of U to those of W in a bijective manner. There are

$$(d\alpha n - n)!$$

many such bijections.
5. Now join the elements of $B - U$ to those of $C - W$ in a bijective manner. There are

$$(nd - [d\alpha n - n])!$$

many such bijections.

It follows that for $\alpha \in (0,1)$, an upper bound on the cardinality $\kappa_\alpha = |\mathcal{K}_\alpha|$ is

$$\frac{\binom{n}{\alpha n} \cdot \binom{d\alpha n}{n} \cdot \binom{\frac{nd}{2}}{\frac{d\alpha n-n}{2}} \cdot (d\alpha n - n)! \cdot (nd - [d\alpha n - n])!}{2^{nd/2}}. \tag{5.33}$$

Lemma 5.4.8. *Let $a_1, \ldots, a_m$ be positive integers, and N be $\sum_{i=1}^m a_i$. Let $p_j = \frac{a_j}{N}$, and $\mathbf{p}$ be the probability distribution $p_1, \ldots, p_m$. Then*

$$\log(a_1! \cdots a_m!) = \log(N!) - h(\mathbf{p})N + O(1)$$

where $h(\mathbf{p}) = -\sum_{i=1}^m \frac{a_i}{N} \log(\frac{a_i}{N})$ is the Shannon entropy of $\mathbf{p}$.

Proof. By Stirling's formula

$$n! \approx \sqrt{2\pi n}\left(\frac{n}{e}\right)^n$$

so

$$\log(n!) \approx n \log n - n \log e + \frac{\log n}{2} + \log(\sqrt{2\pi}).$$

Now the left side is

$$\begin{aligned} LHS &= (\sum_{i=1}^{m} a_i \log a_i) - \sum_{i=1}^{m} a_i + \frac{1}{2}\sum_{i=1}^{m} \log a_i + 2m \\ &= \sum_{i=1}^{m} a_i \log a_i - N + \frac{\log N}{2} + 2m. \end{aligned}$$

Note that

$$\begin{aligned} h(\mathbf{p}) &= -\sum_{i=1}^{m} \frac{a_i}{N} \log(\frac{a_i}{N}) \\ &= -\frac{1}{N}\left[\sum_{i=1}^{m} a_i \log a_i - \sum_{i=1}^{m} a_i \log N\right] \\ &= -\frac{1}{N}\left[\sum_{i=1}^{m} a_i \log a_i - N \log N\right]. \end{aligned}$$

Using this, the right side is

$$\begin{aligned} RHS &= N \log N - N + \frac{\log N}{2} - N\left[(-1/N)(\sum_{i=1}^{m} a_i \log a_i - N \log N)\right] \\ &= N \log N - N + \frac{\log N}{2} + \sum_{i=1}^{m} a_i \log a_i - N \log N \\ &= \sum_{i=1}^{m} a_i \log a_i - N + \frac{\log N}{2} + 2. \end{aligned}$$

This concludes the proof of the lemma.

It is a well-known fact (see [Wel88]) that for $\alpha \leq 1/2$, $\sum_{i=0}^{\alpha n} \binom{n}{i} \leq 2^{h(\alpha)n}$, where h denotes the *entropy*

$$h(\alpha) = -\alpha \log(\alpha) - (1-\alpha)\log(1-\alpha).$$

Since $\binom{n}{\alpha n} = \binom{n}{n(1-\alpha)}$, it follows that $\log(\binom{n}{\alpha n}) \leq h(\alpha)n$. From (5.31) and (5.33), using this fact, it follows that $\log(\frac{\kappa_\alpha}{g})$ is bounded above by

$$\begin{aligned} h(\alpha)n + h\left(\frac{1}{d\alpha}\right) d\alpha n + h\left(\alpha - \frac{1}{d}\right)\frac{dn}{2} + \log((nd)!) \\ -h\left(\frac{d\alpha - 1}{d}\right) \cdot nd + O(1) - \log((nd)!). \end{aligned} \tag{5.34}$$

A small computation establishes that for $0.32 \leq \alpha \leq 0.64$ (i.e., $\alpha \in [\gamma, 2\gamma]$, where $\gamma = 0.32$), the expression (5.34) is asymptotically less than 0, and so

$\mathcal{G} - \mathcal{K}_\alpha \neq \emptyset$; i.e., there are graphs which satisfy the α-expansion property (5.32). The probabilistic construction yields even that *most* degree d graphs satisfy the α-expansion property.

5.4.4 Width-Bounded Resolution Proofs

In this section, the results of which are due to E. Ben-Sasson and A. Wigderson [BSW99], it will be convenient to work with an inessential extension of the system of resolution, so tailored that the restriction of a refutation is a refutation. Recall the rule of *resolution*

$$\frac{C \cup \{x\} \qquad D \cup \{\overline{x}\}}{C \cup D}$$

where C, D are clauses, and $x \notin C$, $\overline{x} \notin D$. If A, B, C are clauses, then the rule of *weakening* is given by

$$\frac{A}{A \cup B}$$

and the rule of *simplification* by

$$\frac{C}{\{1\}}$$

provided that $x, \overline{x} \in C$, for some propositional variable x. The system RWS is a refutation system for CNF formulas whose rules of inference are the rules of resolution, weakening and simplification. The notions of *derivation*, *refutation*, etc. are defined analogously as for the system R. Since it extends R, RWS is clearly a complete and consistent refutation system for all CNF formulas, and that a lower bound for RWI extends *a fortiori* to R.

Recall that we have used the convention that for variable x, x^1 denotes x and x^0 denotes $\overline{x}$. If C is a clause, x is a variable, and $a \in \{0, 1\}$, then the *restriction* $C \upharpoonright_{x=a}$ of C by $x = a$ is defined by

$$C \upharpoonright_{x=a} = \begin{cases} C & \text{if } x, \overline{x} \text{ do not occur in } C \\ \{1\} & \text{if } x^a \text{ occurs in } C \\ C - \{x^{1-a}\} & \text{if } x^{1-a} \text{ occurs in } C. \end{cases}$$

For $\mathcal{C}$ a set of clauses, $\mathcal{C} \upharpoonright_{x=a}$ is $\{C \upharpoonright_{x=a} : C \in \mathcal{C}\}$. If Π is a derivation in RWS from the initial set $\mathcal{C}$ of clauses, then $\Pi \upharpoonright_{x=a}$ is the induced derivation, defined in the obvious manner; i.e., if $\Pi = (C_1, \ldots, C_n)$, then $\Pi \upharpoonright_{x=a} = (C_1', \ldots, C_n')$, where

- C_i' is $C_i \upharpoonright_{x=a}$ if $C_i \in \mathcal{C}$;
- C_i' is $(C_j' - \{x\}) \cup (C_k' - \{\overline{x}\})$, if C_i is obtained by resolution of C_j, C_k where $j, k < i$ and $x \in C_j$, $\overline{x} \in C_k$;
- C_i' is $C_j' \cup A \upharpoonright_{x=a}$, if C_i is obtained from C_j, $j < i$, by the weakening rule;

- C'_i is $\{1\}$ and obtained by the simplification rule from C_j, $j < i$.

Lemma 5.4.9. *If* $\mathcal{C} \upharpoonright_{x=0} \vdash_w A$ *then* $\mathcal{C} \vdash_{w+1} A \vee x$. *If* $\mathcal{C} \upharpoonright_{x=1} \vdash_w A$ *then* $\mathcal{C} \vdash_{w+1} A \vee \overline{x}$.

Proof. Recall that for propositional variable x, x^1 denotes x and x^0 denotes $\overline{x}$. We prove only the first assertion. The second assertion is then proved analogously, by interchanging x and $\overline{x}$.

Let $\Pi = (C_1, \ldots, C_n)$ be a derivation of A from $\mathcal{C} \upharpoonright_{x=0}$, with $C_n = A$. For $1 \leq i \leq n$, let clause D_i be $C_i \cup \{x\}$, and let $\Pi' = (D_1, \ldots, D_n)$. By induction on the number m of inferences, we prove that Π' either is a valid derivation in $\mathcal{C}$, or can be modified to such by the additional application of weakening and simplification rules.

BASE CASE. The number of inferences $m = 0$. Then $A = C \upharpoonright_{x=0}$, for some initial clause $C \in \mathcal{C}$. There are 3 subcases, according to whether C contains $x, \overline{x}$ or neither.

SUBCASE 1. C contains the literal $\overline{x}$. Then $A = C \upharpoonright_{x=0} = \{1\}$. By the simplification rule, $\{1\}$ can be derived from the clause $C \cup \{x\}$, and so $A \cup \{x\} = \{1, x\}$ can be derived by weakening.

SUBCASE 2. C contains x. Then $A = C \upharpoonright_{x=0} = C - \{x\}$, so $A \cup \{x\} = C \in \mathcal{C}$, hence can be derived from $\mathcal{C}$.

SUBCASE 3. C contains neither x nor $\overline{x}$. Then $A = C \upharpoonright_{x=0} = C$, and so $A \cup \{x\}$ is obtained by weakening from an initial clause from $\mathcal{C}$.

INDUCTIVE CASE. The number of inferences $m > 0$.

SUBCASE 1. $A = C_i$ is inferred from C_j, C_k, for $j, k < i$, by resolution on the variable $y \neq x$ in the derivation Π. Then $A \cup \{x\}$ is inferred from $C_j \cup \{x\}$ and $C_k \cup \{\overline{x}\}$ by resolution.

SUBCASE 2. $A = C_i$ is inferred from C_j, for $j < i$, by weakening. Then $A \cup \{x\}$ is inferred from $C_j \cup \{x\}$ by weakening.

SUBCASE 3. $A = C_i = \{1\}$ is inferred from C_j, for $j < i$, by simplification. Then $A \cup \{x\} = \{1, x\}$ is obtained from $C_j \cup \{x\}$ by applying successively simplification and weakening.

This concludes the proof of the lemma.

Lemma 5.4.10. *For* $a \in \{0, 1\}$, *if* $\mathcal{C} \upharpoonright_{x=a} \vdash_{k-1} \square$ *and* $\mathcal{C} \upharpoonright_{x=1-a} \vdash_k \square$, *then* $w(\mathcal{C} \vdash \square) \leq \max\{k, w(\mathcal{C})\}$.

Proof. From $\mathcal{C} \upharpoonright_{x=a} \vdash_{k-1} \square$, by Lemma 5.4.9, we have that $\mathcal{C} \vdash_k \{x^{1-a}\}$. Resolve $\{x^{1-a}\}$ successively with each clause in $\{C : C \in \mathcal{C}, x^a \in C\}$, and note that the width of these resolutions is bounded above by $w(\mathcal{C})$. It follows that each clause D of $\mathcal{C} \upharpoonright_{x=1-a}$, which is different from $\{1\}$, has a derivation from $\mathcal{C}$ of width at most $w(\mathcal{C})$. By hypothesis, there is a derivation of the empty clause $\square$ from $\mathcal{C} \upharpoonright_{x=1-a}$ with width bounded by k, so putting both derivations together, $\square$ has a derivation from $\mathcal{C}$ with width bounded by $\max\{k, w(\mathcal{C})\}$.

Theorem 5.4.11. *Let n denote the number of distinct variables in the unsatisfiable set $\mathcal{C}$ of clauses. Suppose that Π is a tree-like derivation of the empty clause $\Box$ from $\mathcal{C}$, consisting of at most 2^d lines. Then $w(\mathcal{C} \vdash \Box) \leq w(\mathcal{C}) + d$.*

Proof. By double induction on n, d. If $n = 0$ or if $n > 0$ and $d = 0$, then the empty clause $\Box$ belongs to $\mathcal{C}$, so $w(\mathcal{C} \vdash \Box) = 0$ and the assertion of the theorem holds. Consider now the case that $n > 0$, $d > 0$, and let Π be a minimal length tree-like derivation of $\Box$ from $\mathcal{C}$. The last inference of Π must be of the form

$$\frac{\{x\} \qquad \{\overline{x}\}}{\Box}$$

so let Π_x ($\Pi_{\overline{x}}$) be the subderivation of Π whose last clause is $\{x\}$ ($\{\overline{x}\}$). Since the tree-like derivation Π has at most 2^d lines, either Π_x or $\Pi_{\overline{x}}$ has at most 2^{d-1} lines. Without loss of generality, assume the former. It follows that $\Pi_x \upharpoonright_{x=0}$ is a refutation of $\mathcal{C} \upharpoonright_{x=0}$ with length at most 2^{d-1}. The number of distinct variables appearing in $\mathcal{C} \upharpoonright_{x=0}$ is at most $n-1$, so by the induction hypothesis

$$w(\mathcal{C} \upharpoonright_{x=0} \vdash \Box) \leq w(\mathcal{C} \upharpoonright_{x=0}) + d - 1.$$

As well, $\Pi_{\overline{x}} \upharpoonright_{\overline{x}=0}$ is a refutation of $\mathcal{C} \upharpoonright_{x=1}$ with length at most 2^d, and the number of distinct variables in $\mathcal{C}_{x=1}$ is at most $n-1$. By the induction hypothesis,

$$w(\mathcal{C} \upharpoonright_{x=1} \vdash \Box) \leq w(\mathcal{C} \upharpoonright_{x=1}) + d,$$

so by Lemma 5.4.10

$$w(\mathcal{C} \vdash \Box) \leq w(\mathcal{C}) + d.$$

Corollary 5.4.1.

1. $w(\mathcal{C} \vdash \Box) \leq w(\mathcal{C}) + \log_2 L_T(\mathcal{C})$.
2. $L_T(\mathcal{C}) \geq 2^{(w(\mathcal{C} \vdash \Box) - w(\mathcal{C}))}$.

Theorem 5.4.12. *Let $\mathcal{C}$ be an unsatisfiable set of clauses, having at most n distinct variables. Then $w(\mathcal{C} \vdash \Box) \leq w(\mathcal{C}) + O(\sqrt{n \log_2 L(\mathcal{C})})$.*

Proof. By double induction on n, d. If $n = 0$, then $\Box \in \mathcal{C}$, so $w(\mathcal{C} \vdash \Box) = 0$, $L(\mathcal{C}) = 0$ and the assertion of the theorem holds. Suppose now that $n > 0$. Let Π be a refutation of $\mathcal{C}$ of minimum length L. Let $d = \lceil \sqrt{2n \log_2 L(\mathcal{C})} \rceil$ and $a = (1 - \frac{d}{2n})^{-1}$. Let Π^* denote the set of *fat* clauses in Π, i.e., those clauses of width greater than d. By induction on b, we show that if $|\Pi^*| < a^b$, then $w(\mathcal{C} \vdash \Box) \leq w(\mathcal{C}) + d + b$.

BASE CASE. $b = 0$.
Then $|\Pi^*| = 0$, so $w(\Pi) \leq d$, and hence the claim holds.
INDUCTIVE CASE. $b > 0$.
Since $\mathcal{C}$ has at most n variables, there are at most $2n$ literals, and so some literal ℓ appears in at least $\frac{d \cdot |\Pi^*|}{2n}$ fat clauses. Setting $\ell = 1$ *kills* these clauses;

i.e., after restriction, such clauses are set to $\{1\}$, and hence can be removed from the derivation $\Pi \upharpoonright_{\ell=1}$ of the empty clause $\Box$. It follows that there are at most

$$|\Pi^*|\left(1-\frac{d}{2n}\right)=\frac{|\Pi^*|}{a}\leq\frac{a^b}{a}=a^{b-1}$$

remaining fat clauses in the refutation $\Pi \upharpoonright_{\ell=1}$ of $\mathcal{C} \upharpoonright_{\ell=1}$. By the induction hypothesis applied to $b-1$,

$$w(\mathcal{C} \upharpoonright_{\ell=1} \vdash \Box) \leq w(\mathcal{C})+d+b-1.$$

On the other hand, by setting $\ell=0$, the literal ℓ is removed from all clauses in Π, resulting in a refutation $\Pi \upharpoonright_{\ell=0}$ of $\mathcal{C} \upharpoonright_{\ell=0}$, where there are at most a^b fat clauses and at most $n-1$ variables. By applying the induction hypothesis to $n-1$,

$$w(\mathcal{C} \upharpoonright_{\ell=0} \vdash \Box) \leq w(\mathcal{C} \upharpoonright_{\ell=0})+d+b.$$

It now follows by Lemma 5.4.10 that $w(\mathcal{C} \vdash \Box) \leq w(\mathcal{C})+d+b$.

Corollary 5.4.2. $L(\mathcal{C})=\exp\left(\Omega(\frac{[(\mathcal{C}\vdash\Box)-w(\mathcal{C})]^2}{n}\right)$.

The previous results on width bounds for resolution proofs lead to the dynamic programming algorithm, given in Algorithm 5.4.2, to search for resolution proofs. The existence of such an algorithm was first noted by M. Clegg, J. Edmonds and R. Impagliazzo in [CEI96] for the related case of polynomial calculus refutations, and later explicitly for the case of resolution refutations by E. Ben-Sasson and A. Wigderson in [BSW99]. In our presentation below, we attempt to spell out the algorithm in a bit more detail, indicating necessary data structures.

Fix a canonical enumeration of all width w clauses in the literals

$$x_1,\ldots,x_n,\overline{x_1},\ldots,\overline{x_n}$$

and for

$$1\leq i\leq\binom{n}{w}\cdot 2^w$$

let $C_{n,w}(i)$ denote the i-th clause in this enumeration.

Let $\mathcal{C}$ be an unsatisfiable set of clauses consisting of literals among the variables $x_1,\ldots,x_n$ and their negations. For width w, let T_w be a boolean valued table indexed by all width w clauses having literals among $x_1,\ldots,x_n,\overline{x}_1,\ldots,\overline{x}_n$; i.e., for indices $1\leq i\leq\binom{n}{w}\cdot 2^w$ it is the case that $T_w[i]\in\{\text{TRUE},\text{FALSE}\}$ (by abuse of notation, we may sometimes write $T_w[E]$, for clause E, rather than the formally correct $T_w[i]$, where $E=C_{n,w}(i)$).

Initially, each $T_w[i]$ is set to FALSE. During the execution of Algorithm A below (Algorithm 5.4.2), the value of $T_w[i]$ is set to TRUE if the i-th width w clause is derivable from $\mathcal{C}$.

Algorithm 5.4.2 (Algorithm A [BSW99]).
Input. Collection $\mathcal{C}$ of clauses.
Output. Determination whether $\mathcal{C}$ is unsatisfiable.

```
w = 0
while  w ≤ n {
    INITIAL(w)
    DERIVATION(w)
   if T_w[□] = TRUE
      return TRUE
   }
return FALSE
```

The procedure INITIAL(w) sets $T_w[i] = $ TRUE for all clauses of width w which belong to $\mathcal{C}$.

Algorithm 5.4.3 (INITIAL(w)).

```
for i = 1 to (n choose w) · 2^w {
    if C_{n,w}(i) ∈ C
        T_w[i] = TRUE
    }
```

The runtime for INITIAL(w) is clearly $O(n^{2w})$.

The procedure DERIVATION(w) sets $T_v[i] = $ TRUE to all clauses of width $v \leq w$ which can be derived during the w-th pass of the algorithm. At the start of DERIVATION(w), the queue Q is initialized to hold all clauses having a derivation of width strictly less than w, along with the width w clauses of $\mathcal{C}$. The set R, initially empty, consists of all clauses C which are moved from Q, after being "processed"; i.e., all resolutions between C and clauses D, of width at most w which have been so far derived, have been performed.

Algorithm 5.4.4 (DERIVATION(w)).

```
Q = ∅;  R = ∅;
for v = 1 to w
   for i = 1 to (n choose w) · 2^w
      if T_v[i] = TRUE
          place C_i(v) at tail of queue Q;
while Q ≠ ∅ {
   remove head C from queue Q;
   R = R ∪ {C};
   for all D ∈ Q
       RESOLVE_w(C, D, Q, R);
}
```

Algorithm 5.4.5 (RESOLVE$_w(C, D, Q, R)$).

```
if (∃ℓ)(ℓ ∈ C ∧ ℓ̄ ∈ D) {
  E = (C − {ℓ}) ∪ (D − {ℓ̄});
  u = |E|;
  if u ≤ w and T_u[E] = FALSE {
    T_u[E] = TRUE;
    place E at tail of queue Q;
    for all F ∈ R
        RESOLVE_w(E, F, Q, R);
  }
}
```

The runtime for DERIVATION(w) is $O(n^{4w})$, since one must possibly consider all resolutions of clauses C, D, where C, D range over width w clauses. The implementation would use an auxilliary table R_w, where $R_w(A, B) \in \{\text{TRUE}, \text{FALSE}, \text{DONE}\}$. Given clauses C, D, determine whether their resolution has already been accounted for, and if so, set $R_w(C, D) = \text{DONE}$. If not, then determine whether there is a literal $\ell \in C$, whose negation $\bar{\ell} \in D$ and if so, whether the width u of the resolvent E of C, D is at most w. In that case, $R_w(C, D)$ is set to $T_w[E]$, otherwise $R_w(C, D)$ is set to FALSE. Without loss of generality, we can assume that the literal ℓ resolved upon (if it exists) is uniquely determined from C, D, since if there were a distinct literal $\ell' \neq \ell$, for which $\ell' \in C$ and $\overline{\ell'} \in D$, then the resolvent E of C, D would contain a literal and its negation, and hence play no role in a derivation of $\Box$.

5.4.5 Interpolation and *st*-Connectivity

Suppose that $\mathbf{p}$, $\mathbf{q}$, and $\mathbf{r}$ denote sequences $p_1, \ldots, p_{\ell_1}, q_1, \ldots, q_{\ell_2}, r_1, \ldots, r_{\ell_3}$ of distinct propositional variables, and that $A(\mathbf{p}, \mathbf{q})$, $B(\mathbf{p}, \mathbf{r})$ are propositional formulas, such that $A(\mathbf{p}, \mathbf{q}) \to B(\mathbf{p}, \mathbf{q})$ is a tautology. An *interpolant* is a propositional formula $C(\mathbf{p})$ containing only the shared variables $\mathbf{p}$ of A, B, for which $A(\mathbf{p}, \mathbf{q}) \to C(\mathbf{p})$ and $C(\mathbf{p}) \to B(\mathbf{p}, \mathbf{r})$ are tautologies. It's easy to see that an interpolant always exists, for either $A(\mathbf{p}, \mathbf{q})$ is unsatisfiable, in which case we can take $C(\mathbf{p}) = p_1 \wedge \neg p_1$, or $A(\mathbf{p}, \mathbf{q})$ is satisfiable, in which case we can take $C(\mathbf{p})$ to be the disjunction over all conjunctions $p_1^{a_1} \wedge \ldots \wedge p_n^{a_n}$ of signed variables $\mathbf{p}$, for which $A(\mathbf{a}, \mathbf{q})$ is satisfiable; i.e., $C(\mathbf{p})$ is

$$\bigvee \left\{ p_1^{a_1} \wedge \ldots \wedge p_n^{a_n} : \mathbf{a} \in \{0,1\}^n, (\exists \mathbf{q})[A(\mathbf{a}, \mathbf{q}) \equiv 1] \right\} \tag{5.35}$$

Noting that the empty disjunction is defined to be the boolean constant 0, the interpolant is then given in all cases by (5.35). W. Craig actually showed how to construct the interpolant $C(\mathbf{p})$ from a cut-free proof of $A(\mathbf{p}, \mathbf{q}) \to B(\mathbf{p}, \mathbf{r})$ in the propositional sequent calculus LK. In particular, it follows from the subformula property of cut-free proofs that $C(\mathbf{p})$ can be taken to be a subformula of A, B. Unfortunately, the size of the cut-free proof P', obtained from given proof P by applying Gentzen's cut elimination theorem, may be exponentially larger than the original proof size of P. Thus, in general, Craig's technique yields no better size bound for the interpolant than that obtained by disjunctive normal form leading to (5.35). D. Mundici [Mun82, Mun83] noticed connections between interpolation in the propositional logic and computational complexity. For instance, if an interpolant can always be found, whose formula size (circuit size) is polynomial in the size of the implicant, then $\text{NP} \cap co\text{-NP} \subseteq \text{NC}^1/poly$ ($\text{NP} \cap co\text{-NP} \subseteq \text{P}/poly$).

A novel idea, due to J. Krajíček [Kra94a], is to define the interpolant $C(\mathbf{p})$ to be a *circuit*, and to estimate the size of the interpolating circuit in terms of the length of proof of the implication. Let $\mathbf{a}$ be an assignment for $\mathbf{p}$. The interpolating circuit is defined by

$$C(\mathbf{a}) = \begin{cases} 0 \text{ if } A(\mathbf{p}/\mathbf{a}, \mathbf{q}) \text{ is unsatisfiable} \\ 1 \text{ else} \end{cases}$$

where $A(\mathbf{p}/\mathbf{a}, \mathbf{q})$ is the formula obtained from $A(\mathbf{p}, \mathbf{q})$ by instantiating $p_1, \ldots, p_{\ell_1}$ by the boolean values $a_1, \ldots, a_{\ell_1}$. Suppose $A(\mathbf{p}, \mathbf{q}) \to B(\mathbf{p}, \mathbf{r})$ is a tautology. Then $\neg A(\mathbf{p}, \mathbf{q}) \vee B(\mathbf{p}, \mathbf{r})$ is a tautology, so $A(\mathbf{p}, \mathbf{q}) \wedge \neg B(\mathbf{p}, \mathbf{r})$ is unsatisfiable. From the definition of C, it follows that $C(\mathbf{a}) = 0$ (1) if $A(\mathbf{a}, \mathbf{q})$ is unsatisfiable ($\neg B(\mathbf{a}, \mathbf{r})$ is unsatisfiable). Since we'll be applying interpolation to refutation systems of resolution (and later to cutting planes), it makes more sense to define an interpolating circuit $C(\mathbf{p})$ for the *unsatisfiable* conjunction $A(\mathbf{p}, \mathbf{q}) \wedge B(\mathbf{p}, \mathbf{r})$ by

$$C(\mathbf{a}) = \begin{cases} 0 \text{ if } A(\mathbf{a}, \mathbf{q}) \text{ is unsatisfiable} \\ 1 \text{ else (hence } B(\mathbf{a}, \mathbf{r}) \text{ is unsatisfiable.} \end{cases}$$

Theorem 5.4.13 ([Kra97a]). *Let P be a resolution refutation of length k from initial clauses $A_1, \ldots, A_m, B_1, \ldots, B_\ell$, where each A_i has literals among*

$$p_1, \overline{p_1}, \ldots, p_n, \overline{p_n}, q_1, \overline{q_1}, \ldots, q_s, \overline{q_s}$$

and each B_i has literals among

$$p_1, \overline{p_1}, \ldots, p_n, \overline{p_n}, r_1, \overline{q_1}, \ldots, r_t, \overline{r_t}.$$

Then there exists a boolean circuit $C(\mathbf{p})$, which, for every truth assignment $\mathbf{a}$ of variables $\mathbf{p}$, satisfies

$$C(\mathbf{a}) = \begin{cases} 0 \textit{ if } \bigwedge_{i=1}^m A_i(\mathbf{a}, \mathbf{q}), \textit{ is unsatisfiable} \\ 1 \textit{ else (hence } \bigwedge_{i=1}^\ell B_i(\mathbf{a}, \mathbf{r}), \textit{ is unsatisfiable).} \end{cases}$$

Moreover, the circuit C has size bounded by $kn^{O(1)}$. Additionally, if the variables $p_1, \ldots, p_n$ all occur positively in the A_i, or all occur negatively in the B_i, then the circuit is a monotonic boolean circuit.

Rather than prove Theorem 5.4.13 (see Exercise 5.10.10), we will later prove Pudlák's stronger interpolation Theorem 5.6.7 for cutting planes. A proof system $\mathcal{P}$ is said to have the *feasible interpolation property* if whenever the unsatisfiable formula $A(\mathbf{p}, \mathbf{q}) \wedge B(\mathbf{p}, \mathbf{r})$ has a polynomial-size refutation in $\mathcal{P}$, then there is an interpolating boolean circuit of polynomial-size. The *monotone feasible interpolation property* states that if additionally the common variables $\mathbf{p}$ occur only positively in A or only negatively in B, then the circuit can be taken to be monotonic. In [BDG+99] it is observed that if a proof system is automatizable, then it has the feasible interpolation property. We now apply the monotonic feasible interpolation property of resolution.

There are various possible formulations of *st*-connectivity for undirected graphs in propositional logic. Our first formulation, $wSTC_n$, is quite weak, and states that either there exists a path from s to t, or there exists a cut separating s from t. It turns out that this formulation has polynomial-size resolution proofs, yet no polynomial-size *tree-like* resolution proofs.

Definition 5.4.3 ([CS98]). *Assume that G is a finite undirected graph with two distinct, designated vertices s, t. Then either there is a path from s to t, or there is a partition of the vertices of G into two classes, where s and t lie in different classes and no edge goes between vertices lying in different classes (i.e., a cut separating s from t).*

The weak form of *st*-connectivity is formulated as $wSTC_n$, where $\neg wSTC_n$ is the conjunction $A(\mathbf{p}, \mathbf{q}) \wedge B(\mathbf{p}, \mathbf{r})$, where A asserts that either the graph is not undirected, or there exists a path from s to t, and where B asserts the existence of a cut separating s from t. Let $A(\mathbf{p}, \mathbf{q})$ be the conjunction of the following clauses:

1. $\{q_{0,0}\}$
2. $\{q_{n+1,n+1}\}$
3. $\{\overline{q}_{i,j}, \overline{q}_{i,k}\}$, for all $j \neq k$ in $\{0, \ldots, n+1\}$.
4. $\{q_{i,0}, \ldots, q_{i,n+1}\}$, for all $i \in \{1, \ldots, n\}$.
5. $\{\overline{q}_{i,j}, \overline{q}_{i+1,k}, p_{j,k}\}$, for all $j \neq k$ in $\{0, \ldots, n+1\}$.
6. $\{\overline{p}_{i,j}, p_{j,i}\}$, for all $i \neq j$ in $\{0, \ldots, n+1\}$.

Here the p's express the edge relation of G (i.e., $p_{i,j} = 1$ iff there is a directed edge from i to j), and the q's define a path from $s = 0$ to $t = n+1$, where multiple occurrences of the same vertex are allowed along a path (i.e., $q_{i,x}$ asserts that vertex x is the i-th vertex in a path from s to t). Thus $A(\mathbf{p}, \mathbf{q})$ expresses that either G is not an undirected graph, or there is a path from s to t.

The formula $B(\mathbf{p}, \mathbf{r})$ is the conjunction of the following clauses:

1. $\{\overline{r}_0\}$
2. $\{r_{n+1}\}$
3. $\{\overline{r}_i, \overline{p}_{i,j}, r_j\}$, for all $i \neq j$ in $\{0, \ldots, n+1\}$.

Here the p's express the edge relation of G, and the r's express the cut: those vertices i in the same partition class as s (we identify s with 0) satisfy $\overline{r}_i$, while those in the same class as t (we identify t with $n+1$) satisfy r_i.

The resolution formulation of weak st-connectivity is the conjunction of both $A(\mathbf{p},\mathbf{q})$, which expresses that either graph G is not an undirected graph, or there is a path from s to t, and $B(\mathbf{p},\mathbf{r})$, which states that there is a partition of G's vertices, with s,t in different classes, and for which no edge of G goes between vertices in different classes. Note that all occurrences of $\mathbf{p}$ in the clauses B are negative.

Theorem 5.4.14 ([CS98]). *There are polynomial-size resolution proofs of weak st-connectivity.*

Proof. We begin by the following claim.

Claim. For $1 \leq i \leq n+1$, there is a resolution proof of $\{\overline{q}_{i,j}, \overline{r}_j\}$

The proof of the claim is by induction on i. For the base case of $i = 1$, note that

$$\dfrac{\dfrac{\dfrac{\{\overline{q}_{0,0}, \overline{q}_{1,k}, p_{0,k}\} \qquad \dfrac{\{\overline{r}_k, \overline{p}_{k,0}, r_0\} \qquad \{\overline{p}_{0,k}, p_{k,0}\}}{\{\overline{r}_k, \overline{p}_{0,k}, r_0\}}}{\{\overline{q}_{0,0}, \overline{q}_{1,k}, \overline{r}_k, r_0\}} \qquad \{q_{0,0}\}}{\{\overline{q}_{1,k}, \overline{r}_k, r_0\}} \qquad \{\overline{r}_0\}}{\{\overline{q}_{1,k}, \overline{r}_k\}}$$

The resolution proof for the base case is $O(n)$ size. Now, the induction hypothesis is

$$\{\overline{q}_{i,j}, \overline{r}_j\}.$$

We have the following auxiliary result.

$$\dfrac{\{\overline{q}_{i,j}, \overline{r}_j\} \qquad \dfrac{\{\overline{q}_{i,j}, \overline{q}_{i+1,k}, p_{j,k}\} \qquad \dfrac{\{\overline{p}_{j,k}, p_{k,j}\} \qquad \{\overline{r}_k, \overline{p}_{k,j}, r_j\}}{\{\overline{r}_k, \overline{p}_{j,k}, r_j\}}}{\{\overline{q}_{i,j}, \overline{q}_{i+1,k}, \overline{r}_k, r_j\}}}{\{\overline{q}_{i,j}, \overline{q}_{i+1,k}, \overline{r}_k\}}$$

Now

$$\dfrac{\dfrac{\{q_{i,0}, q_{i,1}, \ldots, q_{i,n+1}\} \qquad \{\overline{q}_{i,0}, \overline{q}_{i+1,k}, \overline{r}_k\}}{\{q_{i,1}, q_{i,2}, \ldots, q_{i,n+1}, \overline{q}_{i+1,k}, \overline{r}_k\}} \qquad \{\overline{q}_{i,1}, \overline{q}_{i+1,k}, \overline{r}_k\}}{\{q_{i,2}, \ldots, q_{i,n+1}, \overline{q}_{i+1,k}, \overline{r}_k\}}$$

Inductively continuing in this manner, we obtain

$$\{\overline{q}_{i+1,k}, \overline{r}_k\}.$$

This completes the inductive case. For i,k fixed, there are $O(n)$ additional resolution steps, with overall size $O(n^2)$.

Taking $i = n+1$, it follows that

$$\{\overline{q}_{n+1,k}, \overline{r}_k\}$$

for all k, so that

$$\cfrac{\cfrac{\{\overline{q}_{n+1,n+1}, \overline{r}_{n+1}\} \qquad \{q_{n+1,n+1}\}}{\{\overline{r}_{n+1}\}} \qquad \{r_{n+1}\}}{\Box}$$

We have thus derived the empty clause by a proof of size $O(n^4)$ from the assumptions.

In [CS98], P. Clote and A. Setzer first defined the propositional form of weak st-connectivity, gave the previous polynomial-size resolution proofs of $wSTC_n$, and using the Karchmer-Wigderson lower bound for monotonic boolean circuits for st-connectivity, given in Theorem 2.4.2, along with Krajíček's resolution interpolation theorem, showed an $n^{\Omega \log n}$ size lower bound for tree-like resolution proofs. For a CNF formula F, let $S(F)$ ($S_T(F)$) denote the number of lines (i.e., *length* or number of clauses) in a minimal dag-like (tree-like) resolution refutation of F.

Theorem 5.4.15 ([CS98]). *There exists an infinite family of explicitly constructible unsatisfiable* CNF *formulas F_n, such that $|F_n| = O(n^3)$, $S(F_n) = O(n^4)$, and $S_T(F_n) = 2^{\Omega(\log^2 n)}$.*

Much earlier, Tseitin proved a $2^{\Omega(\log^2 n)}$ lower bound for tree-like resolution refutations of the odd-charged graph formulas for graphs G_n, consisting of n vertices $v_1, \ldots, v_n$, with adjacent vertices v_i and v_{i+1} joined by $\log_2 n$ edges.

Recently essentially optimal exponential separation between tree-like and dag-like resolution was recently given by Ben-Sasson, Impagliazzo and Wigderson [BSIW99], who work with resolution width and pebbling arguments. Recall that $w(F \vdash \Box)$ is the width of a minimal width resolution refutation of F.

Theorem 5.4.16 ([BSIW99]). *There exists an infinite family of explicitly constructible unsatisfiable* CNF *formulas F_n, such that $|F_n| = O(n)$, $S(F_n) = O(n)$, $w(F_n \vdash \Box) = O(1)$ and $S_T(F_n) = 2^{\Omega(n/\log n)}$.*

It is not hard to see that $S_T(F) \leq 2^{S(F)}$. The following result shows that the previous theorem is almost optimal.

Theorem 5.4.17 ([BSIW99]). *For every unsatisfiable* CNF *formula F,*

$$S(F) = 2^{O(\frac{S(F) \log \log S(F)}{\log S(F)})}.$$

5.4.6 Phase Transition and Length of Resolution Proofs

In the introduction Section 4.1 of the previous chapter, we alluded to the fact that experimental evidence indicates that the threshold value for k-SAT points to an easy-hard-easy transition on the difficulty of finding a satisfying truth assignment for a random instance of k-SAT. In this section we substantiate this claim by determining bounds on the length of resolution refutations of k-CNF formulas. The main theorem of this section is the following.

Theorem 5.4.18 ([BP96]). *For $k \geq 6$ almost surely k-CNF formulas in n variables and at most $n^{(k+2)/4-\epsilon}$ clauses do not have subexponential size resolution refutations, where $\epsilon > 0$. . The same result holds for $k \leq 5$ provided that the number of clauses is at most (1) $n^{8/7-\epsilon}$, for $k = 3$, (2) $n^{7/5-\epsilon}$, for $k = 4$, and (3) $n^{22/13-\epsilon}$, for $k = 5$.*

Proof. The idea of the proof is based on the existence of a restriction ρ such that

- most unsatisfiable formulas with short resolution refutations have no long clauses in these refutations after ρ is applied to them, and
- almost no random formulas can be unsatisfiable and have a proof without long clauses after ρ is applied.

It follows that no random formula can be unsatisfiable and have short resolution refutations.

The precise proof of the theorem will follow from a sequence of lemmas which we prove in the sequel. Theorem 5.4.18 will follow immediately from Lemma 5.4.17 by choosing $w(n) = n^{\epsilon/(k+3)}$. Our outline of the proof follows closely the paper [BP96].

Definition 5.4.4. *Let ϕ be a CNF formula.*

1. *ϕ is n'-sparse if every set of $s \leq n'$ variables contains at most s clauses of ϕ.*
2. *Let $n' < n''$. Then ϕ is (n', n'', y)-sparse if every set of s variables, $n' < s \leq n''$, contains at most ys clauses.*

Definition 5.4.5. *The boundary ∂S of a set S of clauses is the set of variables that appear in only one clause of S.*

Lemma 5.4.11. *If a CNF formula ϕ is n'-sparse then every subset of up to n' of its clauses is satisfiable.*

Proof. Let T be a set of clauses of ϕ of size n'. By definition of n'-sparsity every subset S of T contains at least $|S|$ different variables. Hence, by Hall's theorem [BM76] we can choose a system of distinct representative variables one for each clause of T. Thus we can satisfy the clauses of T by setting the representative variable of each clause. This completes the proof of Lemma 5.4.11.

Lemma 5.4.12. *If a k-CNF formula ϕ is $(n'(k+\epsilon)/2, n''(k+\epsilon)/2, 2/(k+\epsilon))$-sparse then every subset of ℓ clauses, with $n' < \ell \leq n''$, has a boundary of size at least $\epsilon\ell$.*

Proof. Let S be a set of ℓ clauses of ϕ and let $n' < \ell \leq n''$. Assume on the contrary that $b(S) < \epsilon\ell$. We have at most $k\ell$ occurrences of variables among the clauses of ϕ and the maximum number of different variables appearing in S is less than

$$\epsilon\ell + (k\ell - \epsilon\ell)/2 \le (k+\epsilon)\ell/2 \le (k+\epsilon)n''/2,$$

since each boundary variable occurs once and every one of the remaining variables occurs at least twice. However this contradicts the assumption that ϕ is $(n'(k+\epsilon)/2, n''(k+\epsilon)/2, 2/(k+\epsilon))$-sparse. This completes the proof of Lemma 5.4.12.

Lemma 5.4.13 (Complex Clause Lemma). *Let $n' \le n''$ and ϕ be an unsatisfiable* CNF *formula on n variables and clauses of size at most k. If ϕ is n'-sparse and $(n'(k+\epsilon)/4, n'(k+\epsilon)/2, 2/(k+\epsilon))$-sparse then every resolution refutation of ϕ must include a clause of length at least $\epsilon n'/2$.*

Proof. Consider a resolution refutation P of a CNF formula ϕ. For any clause C in P let the complexity of C be the smallest number of clauses of ϕ whose conjunction implies C. Since ϕ is n'-sparse we can use Lemma 5.4.11 to conclude that any set of at most n' clauses of ϕ is satisfiable. In particular, the complexity of the empty clause must be $> n'$. Since the complexity of the resolvent is at most the sum of the complexities of the clauses from which it is derived, there must exist a clause C in the proof whose complexity is bigger than $n'/2$ and at most n'.

Let S be a set of clauses of ϕ witnessing the complexity of C whose size is bigger than $n'/2$ and at most n'. It follows from Lemma 5.4.12 and the fact that ϕ is $(n'(k+\epsilon)/4, n'(k+\epsilon)/2, 2/(k+\epsilon))$-sparse that $|\partial S| > \epsilon n'/2$. It suffices to prove that C contains all the variables in $b(S)$. Let x be a variable in $b(S)$ and let C' be the unique clause of S containing x. By definition of S the conjunction of the clauses in $S \setminus \{C'\}$ does not imply C, but S implies C. Therefore there is an assignment of variables of S and C such that all clauses in $S \setminus \{C'\}$ are true but both C and C' are false. If we modify this assignment by toggling the truth value of x in order to satisfy C' then we obtain an assignment that satisfies all clauses of S and therefore satisfies C by definition. Thus we have only modified the truth value of x and have changed the truth value of C. Therefore C contains x. This completes the proof of Lemma 5.4.13.

Lemma 5.4.14. *Let P be a resolution refutation of ϕ of size S. With probability greater than $1 - 2^{1-\alpha t/4}S$ a random restriction of size t sets all large clauses (i.e., clauses with more than αn distinct variables) in S to* 1.

Proof. Let C be a large clause of P. The expected number of variables of C assigned values by a randomly chosen restriction of size t is $\alpha t n/n = \alpha n$. Let D be the random variable representing the domain of ρ. By Chernoff-Hoeffding bounds on the tail of the hypergeometric distribution we have

$$\Pr[|C \cap D| \le \alpha t/4] \le \left(\frac{\sqrt{2}}{e^{3/4}}\right)^{\alpha t} \le 2^{-\alpha t/2}.$$

Given that $|C \cap D| = s$, the probability that $C' \upharpoonright_\rho$ is not set to 1 is 2^{-s}. Therefore the probability that $C' \upharpoonright_\rho$ is not 1 is at most $2^{-\alpha t/2} + 2^{-\alpha t/4} < 2^{1-\alpha t/4}$. Thus the probability that same large clause of P is not set to 1 is less than $2^{1-\alpha t/4} S$. This completes the proof of Lemma 5.4.14.

Lemma 5.4.15. *Let x, y, z be such that $x \leq 1, 1/(k-1) < y \leq 1, 2^{1/k} \leq z$ and let ρ be any restriction on $t \leq \min\{xn/2, x^{1-(1+1/y)/k} n^{1-2/k}/z\}$ variables. If ϕ is chosen as a random k-CNF formula on n variables and $m \leq \frac{y}{e^{1+1/y} 2^{k+1/y}} x^{1/y-(k-1)} n$ clauses then*

$$\Pr[\phi \upharpoonright_\rho \textit{ is both } xn\textit{- and } (xn/2, xn, y)\textit{-sparse}] \geq 1 - 2^{-t} - 2z^{-k} - n^{-1}.$$

Proof. Let S be a fixed subset of variables of size s. Let p' be the probability that a randomly chosen C of size k is such that $C \upharpoonright_\rho \neq 1$ and all variables in $C \upharpoonright_\rho$ are contained in S. This happens when all of the variables of C are either in S or in the domain D of ρ. In particular,

$$p' \leq \frac{\binom{s+t}{k}}{\binom{n}{k}} \leq (s+t)^k / n^k.$$

Define $p = (s+t)^k/n^k$. Since the clauses of ϕ are chosen independently, the distribution of the number of clauses of $\phi \upharpoonright_\rho$ lying in S is the binomial distribution $B(m, p')$ and the probability that more than ys clauses of $\phi \upharpoonright_\rho$ lie in S is $\Pr[B(m,p') \geq ys] \leq \Pr[B(m,p) \geq ys]$. Using Chernoff bounds on the tail of the binomial distribution this last probability is bounded above by

$$\left(\frac{epm}{ys}\right)^{ys} \leq \left(\frac{e(s+t)^k m}{ysn^k}\right)^{ys}.$$

Since there are $\binom{n}{s} \leq (ne/s)^s$ sets of size s the probability that some set of size s contains more than ys clauses is at most

$$\left(\frac{en}{s}\right)^s \left(\frac{e(s+t)^k m}{ysn^k}\right)^{ys} = \left(\frac{e^{1+y}(s+t)^{ky} m^y}{y^y s^{1+y} n^{ky-1} m^{ky-1}}\right)^s. \tag{5.36}$$

For $t < s, s+t \leq 2s$ and since $s \leq xn$ the right-hand side of (5.36) is at most

$$\left(\frac{e^{1+y} 2^{ky} s^{ky-y-1} m^y}{y^y n^{ky-1}}\right)^s \leq \left(\frac{e^{1+y} 2^{ky} x^{ky-y-1} m^y}{y^y n^y}\right)^s \leq 2^{-s},$$

for $m \leq y e^{-1-1/y} 2^{-k} x^{1-k+1/y} n$. It follows that the probability that some set S of size s, $t < s \leq xn$, has more than ys clauses is less than $\sum_{s=t+1}^{xn} 2^{-s} < 2^{-t}$. Therefore

$$\Pr[\phi \text{ is } (xn/2, xn, y)\text{-sparse }] \geq \Pr[\phi \text{ is } (t, xn, y)\text{-sparse }] \geq 1 - 2^{-t}.$$

Next we consider xn-sparsity of ϕ. If ϕ is (t, xn, y)-sparse and $y \leq 1$ then no set of size s with $t < s \leq xn$ can contain more than s clauses. So now we consider only sets of size $s \le t$. Using (5.36) with $y = 1$ we see that the probability that a set S of size s has more than s clauses of $\phi \restriction_\rho$ is at most

$$\left(\frac{e^2(s+t)^k m}{s^2 n^{k-1}}\right)^s \le \left(\frac{e^2 2^k t^k m}{s^2 n^{k-1}}\right)^s \le \left(\frac{e^2 2^k t^k m}{n^{k-1}}\right)^s,$$

since $s \ge 1$. Now the term inside the parenthesis at the right-hand side of the above inequality is at most

$$\left(\frac{t}{x^{k-1-1/y)/k} n^{1-2/k}}\right)^k.$$

Hence the bound on t implies that the total failure probability for sets of size s, $1 \le s \le t$ is at most

$$\sum_{s=1}^{k} \left[\frac{1}{z}\right]^{ks} \le \frac{2}{z^k}$$

since $z \ge 2^{1/k}$. It follows that the probability that $\phi \restriction_\rho$ contains the empty clause is less than the probability that some clause of ϕ lies entirely in the domain of D of ρ, which by the previous calculation is at most

$$\frac{m\binom{t}{k}}{\binom{n}{k}} \le \frac{mt^k}{n^k} < \frac{1}{n}.$$

This completes the proof of Lemma 5.4.15.

Lemma 5.4.16. *Assume that $k \ge 3, 1 > \epsilon > 0, y = 2/(k+\epsilon)$ and x, t, z are functions of n such that t, z are $\omega(1)$ and t satisfies the conditions of Lemma 5.4.15 for all sufficiently large n. Then almost surely a randomly chosen k-CNF formula on n variables with $m \le 2^{-7k/2} x^{-(k-2-\epsilon)/2} n$ clauses does not have a resolution refutation of size at most $2^{\epsilon xt/(4(k+\epsilon))}/8$.*

Proof. Put $S = 2^{\epsilon xt/(4(k+\epsilon))}/8$ and let U be the set of unsatisfiable k-CNF formulas with n variables and m clauses. For each $\phi \in U$ let P_ϕ be some shortest resolution refutation of ϕ. Let B be the subset of those formulas $\phi \in U$ such that the size of P_ϕ is at most S.

By Lemma 5.4.14 for $\phi \in B$ the fraction of restrictions ρ which set t variables such that $P_\phi \restriction_\phi$ contains a clause of length at least $\epsilon xn/(k+\epsilon)$ is at most $\alpha = 2^{1-\epsilon xt/4(k+\epsilon)} S \le 1/4$. For $\phi \in U$ call the pair (ρ, ϕ) bad if $P_\phi \restriction_\phi$ contains a large clause (i.e., a clause of size at least $\epsilon xt/(k+\epsilon)$). By Markov's inequality and since the total fraction of bad pairs (ρ, ϕ) with $\phi \in B$ is at most $1/4$ we have that the fraction of ρ such that (ρ, ϕ) is bad for at least half of the formulas ϕ in B is at most $1/2$. Fix some ρ for which less than half of the $\phi \in B$ have a clause of length $\ge \epsilon xn/(k+\epsilon)$ in $P_\phi \restriction_\rho$. Now observe that x is $\omega(1)$ as a function of n and in fact it is $\ge 2^{1/k}$ for sufficiently large n. Moreover, $k - 1 - 1/y = (k-2-\epsilon)/2$ and $2^{-7k/2} \le y/(e^{1+1/y} 2^{k+1/y})$.

It follows that all the conditions of Lemma 5.4.15 are satisfied for y, z, t and for $m \leq 2^{-7k/2} x^{-(k-2-\epsilon)/2} n$. Since both t, x are $\omega(1)$ it follows that with probability tending to 0 as n tends to infinity, for random k-CNF formulas ϕ with m clauses and n variables either ϕ is satisfiable (i.e., not in U) or else the restriction $\phi \upharpoonright_\rho$ fails to be both $(xn/2, xn, 2/(k+\epsilon)))$-sparse and xn-sparse. By Lemma 5.4.13 and since xn-sparsity implies $2xn/(k+\epsilon)$-sparsity we conclude that almost surely all ϕ are either satisfiable or else have a clause of length at least $\epsilon xn/(k+\epsilon)$ in $P_\phi \upharpoonright_\rho$. Since $B \subseteq U$ at least half the formulas ϕ in B do not have such a large clause in $P_\phi \upharpoonright_\rho$ the set B is negligibly small. It follows that almost all k-CNF formulas in n variables with m clauses do not have resolution refutations of size at most S. This completes the proof of Lemma 5.4.16.

Lemma 5.4.17. *Assume that $k \geq 3, 1 > \epsilon > 0$. If the function $w(n)$ satisfies*

$$(a)\ w(n) = o(n^{(k-6+\epsilon)/(k+2+\epsilon)}),\ \textit{resp.}\ (b)\ w(n) = \Omega(n^{(k-6+\epsilon)/(k+2+\epsilon)})$$

then a negligible fraction of all k-CNF formulas in n variables with

$$(a) \leq n^{(k+2-\epsilon)/4} 2^{-4k} w(n)^{-(k-2-\epsilon)/4},\ \textit{resp.}\ (b)\ o(n^{(k^2-k+2)/(3k-2)} w(n)^{-k/3})$$

clauses have resolution refutations of size at most $2^{\epsilon w(n)/(4(k+\epsilon))}/8$.

Proof. First consider part (a). If $y = 2/(k+\epsilon)$ then $1 - (1+1/y)/k = (k-2-\epsilon)/(2k)$. For $w(n)$ as above define $x(n) = \sqrt{2w(n)/n}$ and $t(n) = xn/2 = \sqrt{w(n)n/2}$. It is easy to verify that

$$t(n) = o(x(n)^{(k-2-\epsilon)/(2k)} n^{1-2/k}) = o(x(n)^{1-(1+1/y)/k} n^{1-2/k}),$$

which shows that the conditions of Lemma 5.4.15 are met. Using Lemma 5.4.16 we conclude that a random k-CNF formula having

$$m \leq 2^{-7k/2} x^{-(k-2-\epsilon)/2} n$$

clauses almost surely does not have a resolution refutation of size at most $2^{\epsilon w(n)/(4(k+\epsilon))}/8$. This proves part (a).

To prove part (b) let $w(n)$ and m be as above. Observe that m is also $o(g)$ where

$$g(n) = n^{\frac{k^2-k+2-(k-1)\epsilon}{3k-2-\epsilon}} w(n)^{-\frac{k(k-2-\epsilon)}{3k-2-\epsilon}}.$$

Define

$$\begin{aligned} z(n) &= (2^{-7k/2} q(n)/m)^{3/k} \\ x(n) &= n^{-2(k-2)/(3k-2-\epsilon)} (w(n) z(n))^{2k/(3k-2-\epsilon)} \\ t(n) &= x(n)^{(k-2-\epsilon)/(2k)} n^{1-2/k} / z(n). \end{aligned}$$

Since m is $o(q)$ we observe that $z(n)$ is $\omega(1)$. Also, note that $x(n)t(n) = w(n)$ and the condition on $w(n)$ implies that $t(n) \leq x(n)n/2$. It follows that $t(n)$ satisfies the conditions of Lemma 5.4.15. Moreover,

$$\begin{aligned} 2^{-7k/2}x^{-(k-2-\epsilon)/2}n &= q(n)2^{-7k/2}z(n)^{-\frac{k(k-2-\epsilon)}{3k-2-\epsilon}} \\ &= q(n)2^{-7k/2}z(n)^{-k/3} \\ &= m. \end{aligned}$$

By Lemma 5.4.16 a random k-CNF formula with m clauses almost surely does not have a resolution refutation of size at most $2^{\epsilon x(n)t(n)/(4(k+\epsilon))}/8 = 2^{\epsilon w(n)/(4(k+\epsilon))}/8$. This completes the proof of part (b) and the proof of Lemma 5.4.17.

This also completes the proof of Theorem 5.4.18.

We also note the recent result of Ran Raz that any Resolution proof for the weak pigeonhole principle, with n holes and any number of pigeons, is of length $\Omega(2^{n^\epsilon})$, (for some global constant $\epsilon > 0$). One corollary is that certain propositional formulations of the statement P $\neq$ NP do not have short Resolution proofs.[16]

5.5 Algebraic Refutation Systems

In this section, we survey work on algebraic refutation systems, where the equality $1 = 0$ is derived by performing certain operations on polynomials over a field. In particular, the *Nullstellensatz system* (NS), *polynomial calculus* (PC), *Gaussian calculus* (GC), and *binomial calculus* (BC) will be considered. Each of these depends on a fixed field F, so these systems actually constitute families of refutation systems, one for each fixed field F. Before defining algebraic refutation systems, we begin by reviewing some elementary concepts and notation from algebra.

A finite field must be of characteristic p, where p is a prime. Moreover, for every power prime power p^n, $n \geq 1$, there exists up to isomorphism a unique finite field with p^n elements, called the *Galois field of order* p^n, and denoted by $GF(p^n)$. For p prime, $GF(p)$ is often denoted by $\mathbf{Z}_p$, the field consisting of elements $\{0, 1, \ldots, p-1\}$ with the operations of addition and multiplication modulo p.

Let F be a finite field of characteristic p. The *order* of an element $a \in F$ is the smallest positive exponent m for which $a^m = 1$ in F. If a is an element of F such that $a^m = 1$ then a is called an *m-th root of unity* in F. The multiplicative group of the field F is cyclic of order $p-1$. Let $q \neq p$ be prime. There are at most q, q-th roots of unity in F and they obviously form a group which is cyclic. A generator for the group of qth roots of unity is called a *primitive* qth root of unity. If a primitive q-th root of unity exists in F then the set of q-th roots of unity in F forms a group whose order q divides $p-1$.

[16] "Resolution Lower Bounds for the Weak Pigeon Hole Principle", in Proceedings of 34th Annual ACM Symposium on Theory of Computing, 2002.

If R is a ring, then $I \subseteq R$ is an *ideal* if I is an additive subgroup of R, which is closed under the operation of multiplication by elements of R. We'll be interested in ideals I of $F[x_1, \ldots, x_n]$, the ring of polynomials in indeterminates $x_1, \ldots, x_n$ with coefficients from field F. The notation $I = \langle f_1, \ldots, f_m \rangle$ means that ideal I is generated by the polynomials $f_1, \ldots, f_m \in F[x_1, \ldots, x_n]$; i.e., I is the smallest additive subgroup of $F[x_1, \ldots, x_n]$ containing $f_1, \ldots, f_m$ and closed under multiplication of arbitrary polynomials from $F[x_1, \ldots, x_n]$.

Let $PP[x_1, \ldots, x_n]$ denote the set of *power products* $\prod_{i \in I} x_i^{\alpha_i}$ where $I \subseteq \{1, \ldots, n\}$ and $\alpha_i \in \mathbf{N}^+$; i.e., *monomials* in $F[x_1, \ldots, x_n]$ with coefficient 1. A *multilinear monomial*, is a monomial of the form $c \prod_{i \in I} x_i$, where $c \in F$; i.e., the exponent of every variable appearing in the monomial is 1. A *multilinear* polynomial, or *multinomial*, in indeterminates $x_1, \ldots, x_n$ is an element of $F[x_1, \ldots, x_n]/\langle x_1^2 - x_1, \ldots, x_n^2 - x_n \rangle$; e.g. $x_1^4 x_2^2 x_4^3 + 2x_1^2 x_3^2$ is equivalent over $F[x_1, \ldots, x_4]/\langle x_1^2 - x_1, \ldots, x_4^2 - x_4 \rangle$ to the multinomial $x_1 x_2 x_4 + 2x_1 x_3$. The multinomial of p is denoted by $\text{ML}(p)$. Sometimes multinomial power products will be identified with subwords w of the word $x_1 \cdots x_n$, denoted by $w \subseteq x_1 \cdots x_n$. Thus $\sum_{w \subseteq x_1 \cdots x_n} a_w \cdot w$ is the general form for a multinomial, where $a_w \in F$.

The fact that polynomial f divides polynomial g with 0 remainder is denoted by $f|g$. The least common multiple of f, g is denoted by $LCM(f, g)$. If $t \in PP[x_1, \ldots, x_n]$ and $f \in F[x_1, \ldots, x_n]$, then $C(f, t)$ is the coefficient of t in f. The monomial at t in f is denoted by $Mon(f, t)$, and satisfies $Mon(f, t) = C(f, t) \cdot t$. The *support* of polynomial $f \in F[x_1, \ldots, x_n]$, denoted $supp(f)$, is defined by $\{t \in PP[x_1, \ldots, x_n] : C(f, t) \neq 0\}$.

Crucial to the Gröbner Basis Algorithm, not presented in this monograph, but which yields an alternative proof of the automatizability of PC (explained later), is the notion of *admissible ordering*. An admissible ordering $\prec$ is a total ordering on the set of power products, which satisfies the properties

- $1 \prec t$, for all $t \in PP[x_1, \ldots, x_n]$.
- $\prec$ is monotonic; i.e., for all $u, v, t \in PP[x_1, \ldots, x_n]$, if $u \prec v$ then $t \cdot u \prec t \cdot v$.

It can be shown that if $\prec$ is admissible and t, u are power products satisfying $t|u$, then $t \prec u$. Moreover, $\prec$ is well-founded; i.e., there exist no infinite descending chains in the total order $\prec$ (this property is also called *Noetherian*). An example of an admissible ordering for the set of power products in variables $x_1, \ldots, x_n$ is the lexicographic ordering. Another example is the *total degree ordering*, where $t \prec u$ if $deg(t) < deg(u)$, or $deg(t) = deg(u)$ and t precedes u in the lexicographic ordering. With this notation, $LPP_\prec(p)$ denotes the leading power product in polynomial p; i.e., $\max_\prec supp(p)$. The *leading coefficient* $LeadC_\prec(p)$ is defined by $C(p, LPP_\prec(p))$. The *leading monomial* $LeadMon_\prec(p)$ is defined by $LeadC_\prec(p) \cdot LPP_\prec(p)$, while the *remaining part* $R_\prec(p)$ is defined by $p - LeadMon_\prec(p)$. A *monic polynomial* is a polynomial whose leading coefficient is 1. When the ordering $\prec$ is understood, we may suppress $\prec$ from the notation introduced. Finally, when the ordering $\prec$ is fixed, we may drop reference to $\prec$, as in $LPP(p)$ and $LeadMon(p)$, rather than $LPP_{\prec}(p)$ and $LeadMon_{\prec}(p)$.

5.5.1 Nullstellensatz

The *Nullstellensatz system* (NS) is a refutation system for unsatisfiable propositional formulas, based on showing that 1 belongs to the ideal generated by a certain set of multivariate polynomials. The completeness of this propositional proof system depends on the following weak form of Hilbert's Nullstellensatz Theorem, whose presentation is taken from [BW93].

Theorem 5.5.1 (D. Hilbert). *Let F be a field, and*

$$g(x_1,\ldots,x_n), f_1(x_1,\ldots,x_n),\ldots,f_m(x_1,\ldots,x_n)$$

be polynomials over F. Then the following are equivalent.

1. *The assertion*

$$(\forall x_1,\ldots,x_n)\left[\bigwedge_{i=1}^{m} f_i(x_1,\ldots,x_n)=0 \rightarrow g(x_1,\ldots,x_n)=0\right]$$

   holds in all extension rings of F.
2. *$g \in I$, where $I = \langle f_1,\ldots,f_m\rangle$ is the ideal generated by $f_1,\ldots,f_m$.*

Proof. We prove the direction (1) $\Rightarrow$ (2) by contradiction. Assume that (2) fails, so $g \notin I$ and hence $I \subset F[x_1,\ldots,x_n]$ is a proper ideal. Define the mapping $\phi : F[x_1,\ldots,x_n] \rightarrow F[x_1,\ldots,x_n]/I$ by $\phi : f \mapsto f+I$. We claim that the restriction of ϕ to F is injective. If not, then for distinct $a,b \in F$, $\phi(a)=\phi(b)$, hence $\phi(a-b)=0$ in $F[x_1,\ldots,x_n]/I$, and so $a-b \in I$. Since I is closed under multiplication by elements of $F[x_1,\ldots,x_n]$, $1=(a-b)^{-1}\cdot(a-b) \in I$, and hence $I = F[x_1,\ldots,x_n]$. This contradicts the assumption that I is proper, so it follows that the restriction of ϕ to F is injective, and hence $F[x_1,\ldots,x_n]/I$ is isomorphic to a ring extension of F. The mapping ϕ is a ring homomorphism, and $I=\langle f_1,\ldots,f_m\rangle$, so for each $1 \le i \le m$,

$$f_i(\phi(x_1),\ldots,\phi(x_n)) = \phi(f_i(x_1,\ldots,x_n)) = 0$$

holds in $F[x_1,\ldots,x_n]/I$, and yet

$$g(\phi(x_1),\ldots,\phi(x_n)) = \phi(g(x_1,\ldots,x_n)) \neq 0$$

holds in $F[x_1,\ldots,x_n]/I$, since $\phi(g(x_1,\ldots,x_n)) \notin I$. This contradicts assumption (1), and hence establishes (1) $\Rightarrow$ (2).

The direction (2) $\Rightarrow$ (1) easily follows from definitions.

The analogous theorem, which states that (1) holds in all algebraically closed extension fields of F if and only if $g^r \in I$, for some $r \geq 1$ is known as *Hilbert's Nullstellensatz.*

To present the Nullstellensatz system (NS), we begin by translating propositional formulas A into multivariate polynomials p_A, in the following canonical manner.

Definition 5.5.1 (Canonical polynomial p_A). *Define* $p_0 = 0$ *(i.e.,* FALSE *is represented by* 0*),* $p_1 = 1$ *(i.e.,* TRUE *is represented by* 1*),* $p_{x_i} = x_i$*,* $p_{\neg A} = 1 - p_A$*,* $p_{A \wedge B} = p_A \cdot p_B$*, and* $p_{A \vee B} = p_A + p_B - p_A p_B$*.*

Let F be an arbitrary field, let A be a propositional formula in variables $x_1, \ldots, x_n$, and let $a_1, \ldots, a_n$ be arbitrary elements of F. Then it is clear that $a_1, \ldots, a_n$ is a common zero of

$$1 - p_A(x_1, \ldots, x_n), x_1^2 - x_1, \ldots, x_n^2 - x_n$$

if and only if σ is a satisfying truth assignment for A, where $\sigma : \{x_1, \ldots, x_n\} \to \{0, 1\}$ is defined by $\sigma(x_i) = a_i$.

Definition 5.5.2 ([BIK+96]). *Let F be a fixed field. A Nullstellensatz refutation of propositional formula A, using canonical translation p_A, is given by*

$$1 = (1 - p_A(x_1, \ldots, x_n)) \cdot g + \sum_{i=1}^{n} (x_i^2 - x_i) \cdot h_i \tag{5.37}$$

for $g, h_1, \ldots, h_n \in F[x_1, \ldots, x_n]$*. More generally, given polynomials*

$$f_1, \ldots, f_m \in F[x_1, \ldots, x_n],$$

a Nullstellensatz refutation of $f_1, \ldots, f_m$ *is given by*

$$1 = \sum_{i=1}^{m} f_i \cdot g_i + \sum_{i=1}^{n} (x_i^2 - x_i) \cdot h_i \tag{5.38}$$

where $g_1, \ldots, g_m, h_1, \ldots, h_n \in F[x_1, \ldots, x_n]$*. The degree of Nullstellensatz refutation (5.37) is* $deg(p_A \cdot g)$*, and the degree of Nullstellensatz refutation (5.38) is* $\max\{deg(f_i \cdot g_i) : 1 \leq i \leq m\}$*.*

The degree and size (i.e., number of symbols) of a Nullstellensatz refutation are related. Indeed, if F is a finite field, then there are $\sum_{m=0}^{d} |F| \cdot \binom{n}{m} = n^{O(d)}$ many monomials of degree at most d, so it follows that Nullstellensatz refutations have constant degree if and only if their size is polynomial. The definition of *Nullstellensatz degree* does not involve the coefficients h_i of $x_i^2 - x_i$ in the previous definition for the following reason.

Proposition 5.5.1 ([BIK+97]). *Suppose that* $W = \prod_{i \in I} x_i^{\alpha_i}$ *and* $U = \prod_{i \in I} x_i^{\beta_i}$*, where* $\emptyset \neq I \subseteq \{1, \ldots, n\}$*, and* $1 \leq \beta_i \leq \alpha_i$ *for* $i \in I$*. Then there exist* $h_i \in F[x_1, \ldots, x_n]$ *of degree at most* $deg(W) - 2$ *such that* $W - U = \sum_{i \in I} (x_i^2 - x_i) \cdot h_i$*.*

Proof. By induction on the degree $d = \deg(W) - \deg(U)$. The base case $d = 0$ is trivial, since then $W = U$, so assume that $d > 0$ and that the proposition has been established for smaller degrees. Let $i_0 \in I$ be such that $\alpha_{i_0} > 1$, and define $W' = \frac{W}{x_{i_0}^2}$ and $W'' = \frac{W}{x_{i_0}}$. Then $W - W'' = W'(x_{i_0}^2 - x_{i_0})$, so using the induction hypothesis we have

$$W - U = (W - W'') + (W'' - U) = W'(x_{i_0}^2 - x_{i_0}) + \sum_{i \in I}(x_i^2 - x_i) \cdot h_i'.$$

If we define

$$h_i = \begin{cases} h_{i_0}' + W'' & \text{if } i = i_0 \\ h_i' & \text{else} \end{cases}$$

then $W - U$ is written in the required form.

Corollary 5.5.1. *Let $p \in F[x_1, \ldots, x_n]$ be of degree d. Then there exist $h_1, \ldots, h_n \in F[x_1, \ldots, x_n]$ be of degree at most $d - 2$ such that*

$$p - \mathrm{ML}(p) = \sum_{i=1}^{n}(x_i^2 - x_2) \cdot h_i.$$

The following two theorems give the soundness and completeness of the Nullstellensatz system.

Theorem 5.5.2 (Soundness, [BIK+96]). *The Nullstellensatz system is sound, in that if propositional formula A has a Nullstellensatz refutation, using canonical translation p_A, then A is unsatisfiable.*

Proof. Suppose that A has a Nullstellensatz refutation, hence

$$1 = (1 - p_A(x_1, \ldots, x_n)) \cdot g + \sum_{i=1}^{n}(x_i^2 - x_i) \cdot h_i. \tag{5.39}$$

Suppose, in order to obtain a contradiction, that A is satisfiable. Let $\sigma : \{x_1, \ldots, x_n\} \to \{0, 1\}$ be such that $\sigma \models A$. Then substituting x_i by $\sigma(x_i)$ in (5.39) we obtain the contradiction $1 = 0$, hence A is unsatisfiable.

Theorem 5.5.3 (Completeness, Beame et al. [BIK+96]). *The Nullstellensatz system is complete, using the canonical representation p_A, in that if propositional formula A is unsatisfiable, then A has a Nullstellensatz refutation.*

Proof. Suppose that A is unsatisfiable, so there is no common zero of $1 - p_A, x_1^2, -x_1, \ldots, x_n^2 - x_n$. Assume, in order to obtain a contradiction, that A has no Nullstellensatz refutation. Then, since the antecedent of the following implication

$$(\forall x_1, \ldots, x_n) \left[\left(1 - p_A = 0 \wedge \bigwedge_{i=1}^{n} x_i^2 - x_i = 0 \right) \to 1 = 0 \right]$$

holds, the implication must vacuously hold. From Theorem 5.5.1, it now follows that $1 \in I$, where $I = \langle 1 - p_A, x_1^2 - x_1, \ldots, x_n^2 - x_n \rangle$.

The proof yields a bit more – namely, that the Nullstellensatz system is implicationally complete.

Corollary 5.5.2. *NS is implicationally complete; i.e., whenever*

$$q, p_1, \ldots, p_m \in F[x_1, \ldots, x_n]$$

and

$$(\forall x_1, \ldots, x_n \in F) \left[\bigwedge_{i=1}^{m} p_i(x_1, \ldots, x_n) = 0 \to q(x_1, \ldots, x_n) = 0 \right]$$

then there exist $g_1, \ldots, g_m \in F[x_1, \ldots, x_n]$ such that $q = \sum_{i=1}^{m} p_i \cdot g_i$.

Summarizing, we have that propositional formula A is unsatisfiable if and only if $1 \in I$, where I is the ideal generated by $1 - p_A(x_1, \ldots, x_n), x_1^2 - x_1, \ldots, x_n^2 - x_n$. Since $p_{\neg A} = 1 - p_A$, it follows that A is a tautology if and only if $\neg A$ is unsatisfiable if and only if

$$1 \in \langle p_A, x_1^2 - x_i, \ldots, x_n^2 - x_n \rangle.$$

To fix ideas, here is a small example. Recall that the pigeonhole principle PHP_n^{n+1} is expressed by

$$\bigvee_{i=1}^{n+1} \bigwedge_{j=1}^{n} \neg p_{i,j} \vee \bigvee_{1 \le i < i' \le n+1} \bigvee_{j=1}^{n} (p_{i,j} \wedge p_{i',j}).$$

In particular PHP_1^2 is given by

$$(\overline{p}_{1,1} \vee \overline{p}_{2,1}) \vee (p_{1,1} \wedge p_{2,1}).$$

Let q abbreviate

$$(1 - x_{1,1}) + (1 - x_{2,1}) - (1 - x_{1,1})(1 - x_{2,1}) = 1 - x_{1,1}x_{2,1} + x_{1,1}^2 x_{2,1}^2.$$

The representation of PHP_1^2 using canonical polynomial p_A is

$$q + x_{1,1}x_{2,1} - q \cdot x_{1,1}x_{2,1}. \tag{5.40}$$

which simplifies to

$$1 - x_{1,1}x_{2,1} + x_{1,1}^2 x_{2,1}^2. \tag{5.41}$$

Denote the polynomial (5.41) by $p(x_{1,1}, x_{2,1})$. A Nullstellensatz refutation of $\neg PHP_1^2$ is

$$1 = p(x_{1,1}, x_{2,1}) + (x_{1,1}^2 - x_{1,1})(-x_{2,1}^2) + (x_{2,1}^2 - x_{2,1})(-x_{1,1})$$

with degree 4.

The canonical polynomial p_A, where A is PHP_n^{n+1}, is tedious to give explicitly. In contrast, an alternative, more elegant approach in providing Nullstellensatz refutations of unsatisfiable CNF formulas uses a different translation of formulas into polynomials.

Definition 5.5.3 (Canonical polynomial q_A). *Let $q_0 = 1$ (i.e., FALSE is represented by 1), $q_1 = 0$ (i.e., TRUE is represented by 0), $q_{x_i} = 1 - x_i$ (i.e., the propositional variable x_i is represented by the expression $1 - x_i$), $q_{\neg A} = 1 - q_A$, $q_{A \wedge B} = q_A + q_B - q_A q_B$, and $q_{A \vee B} = q_A \cdot q_B$.*

To prove a DNF formula A of the form $\bigvee_{i=1}^{r} \bigwedge_{j=1}^{s_i} \ell_{i,j}$ consider the CNF formula $\neg A$ of the form $\bigwedge_{i=1}^{r} \bigvee_{j=1}^{s_i} \overline{\ell_{i,j}}$, and represent each conjunct $C_i \equiv \bigvee_{j=1}^{s_i} \overline{\ell_{i,j}}$ by the polynomial q_{C_i}. Clearly $\neg A$ is satisfiable if and only if the polynomials $q_{C_1}, \ldots, q_{C_r}$ have a common zero in $F[x_1, \ldots, x_n]/\langle x_1^2 - x_1, \ldots, x_n^2 - x_n \rangle$. Thus A is a tautology if and only if $q_{C_1}, \ldots, q_{C_r}$ have no common zero in $F[x_1, \ldots, x_n]/\langle x_1^2 - x_1, \ldots, x_n^2 - x_n \rangle$, which holds if and only if $1 \in I$, where I is the ideal generated by

$$q_{C_1}, \ldots, q_{C_r}, x_1^2 - x_1, \ldots, x_n^2 - x_n.$$

For instance, representing each conjunct C_i of the CNF formula $\neg PHP_n^{n+1}$ by q_{C_i}, we have the polynomials

$$f_i(x_{i,1}, \ldots, x_{i,n}), f_{i,i',j}(x_{i,j}, x_{i',j}) \tag{5.42}$$

for $1 \le i < i' \le n+1, 1 \le j \le n$, where $f_i(x_{i,1}, \ldots, x_{i,n}) = \prod_{j=1}^{n}(1 - x_{i,j})$ and $f_{i,i',j}(x_{i,j}, x_{i',j}) = x_{i,j} x_{i',j}$. It is clear that $\neg PHP_n^{n+1}$ is satisfiable if and only if the polynomials in (5.42) have a common zero, and thus PHP_n^{n+1} is a tautology if and only if there exist polynomials $g_i, g_{i,i',j}, h_i$ such that

$$1 = \sum_{i=1}^{n} f_i g_i + \sum_{1 \le i < i' \le n+1, 1 \le j \le n} f_{i,i',j} g_{i,i',j} + \sum_{i=1}^{n} (x_i^2 - x_i) h_i.$$

Resuming our earlier example of PHP_1^2, the CNF formula $\neg PHP_1^2$ is given by

$$p_{1,1} \wedge p_{2,1} \wedge (\overline{p}_{1,1} \vee \overline{p}_{2,1}).$$

The latter is represented by the system of polynomials q_1, q_2, q_3, where q_1 is $1 - x_{1,1}$, q_2 is $1 - x_{2,1}$, and q_3 is $x_{1,1} x_{2,1}$. A Nullstellensatz refutation is given by

$$1 = q_1 \cdot x_{2,1} + q_2 \cdot 1 + q_3 \cdot 1$$

of degree 2.

The unsatisfiable CNF formula $\neg PHP_2^3$, is given by the conjunction of $p_{1,1} \vee p_{1,2}$, $p_{2,1} \vee p_{2,2}$, $p_{3,1} \vee p_{3,2}$, $\overline{p}_{1,1} \vee \overline{p}_{2,1}$, $\overline{p}_{1,1} \vee \overline{p}_{3,1}$, $\overline{p}_{2,1} \vee \overline{p}_{3,1}$, $\overline{p}_{1,2} \vee \overline{p}_{2,2}$, $\overline{p}_{1,2} \vee \overline{p}_{3,2}$, $\overline{p}_{2,2} \vee \overline{p}_{3,2}$. This yields the system of polynomials $f_{i,j,k}$, for $1 \le i \le 3$, $1 \le j < k \le 2$, and $g_{i,j,k}$, for $1 \le i < j \le 3$, $1 \le k \le 2$, where $f_{i,j,k}$

is $(1 - x_{i,j})(1 - x_{j,k})$, and $g_{i,j,k}$ is $x_{i,j}x_{j,k}$. This is a much more tractable representation than the canonical polynomial p_A, for $A \equiv PHP_2^3$, and hence is the preferred representation when refuting CNF formulas. For unsatisfiable formulas not in CNF form, one can either use the canonical polynomial p_A, or instead introduce new extension variables, paralleling in an obvious way resolution with limited extension.

Definition 5.5.4 ([BIK+96]). *Let F be a fixed field. Let $A \equiv \bigwedge_{i=1}^{r} C_i$ be an unsatisfiable* CNF *formula. A Nullstellensatz refutation of A, using canonical representation q_A, is given by*

$$1 = \sum_{i=1}^{m} q_{C_i} \cdot g_i + \sum_{i=1}^{n} (x_i^2 - x_i) \cdot h_i \tag{5.43}$$

for polynomials $g_1, \ldots, g_m, h_1, \ldots, h_n \in F[x_1, \ldots, x_n]$. The degree of the refutation (5.43) is $\max\{deg(q_{C_i} \cdot g_i) : 1 \leq i \leq m\}$.

As before, by Corollary 5.5.1, the definition of Nullstellensatz degree does not consider the auxilliary polynomials h_i.

What is an upper bound on the degree of a Nullstellensatz refutation for an arbitrary propositional formula? The proof of Theorem 5.5.3 depends on the weak version of Hilbert's Nullstellensatz Theorem, which yields no obvious, immediate upper bound on the degree of a Nullstellensatz refutation. In contrast, for unsatisfiable CNF formulas, using canonical representation q_A, the following constructive proof yields a linear upper bound for the degree.

To simplify presentation, we introduce the following notation. For truth assignment $\sigma : \{x_1, \ldots, x_n\} \to \{0, 1\}$, let f_σ denote

$$\prod_{\sigma(x_i)=0} x_i \cdot \prod_{\sigma(x_i)=1} (1 - x_i).$$

Denote the collection of all total truth assignments, or *valuations*, on variables $x_1, \ldots, x_n$ by Val_n.

The following theorem gives the completeness of the Nullstellensatz system for CNF.

Theorem 5.5.4 ([BIK+97], [Pit97]). *Let $A \equiv \bigwedge_{i=1}^{r} C_i$ be an unsatisfiable* CNF *formula in variables $x_1, \ldots, x_n$. Then there is a Nullstellensatz refutation of A*

$$1 = \sum_{i=1}^{r} q_{C_i} \cdot g_i$$

of degree at most n; i.e., $\max\{deg(q_{C_i} \cdot g_i) : 1 \leq i \leq r\} \leq n$.

Proof. Note well that the auxilliary polynomials $x_i^2 - x_i$ for $1 \leq i \leq n$, are not needed in the case of refutations of CNF formulas. As well, the following proof requires that the polynomials q_{C_i} be multilinear, since it is shown that

q_{C_i} divides the multilinear polynomial f_σ – of course the multilinearity of q_{C_i} is guaranteed by definition, since C_i is a disjunction of literals.

CLAIM. $1 = \sum_\sigma f_\sigma$, where the sum is taken over all $\sigma \in Val_n$.

Proof of Claim. By induction on n. When $n = 1$, we have $1 = x_1 + (1 - x_1)$. Assume now that $n > 1$. Then

$$\sum_\sigma f_\sigma = x_n \cdot \sum_{\sigma \in A} f_\sigma + (1 - x_n) \cdot \sum_{\sigma \in B} f_\sigma$$

where A (B) is the set of truth assignments $\sigma \in Val_n$ such that $\sigma(x_n) = 0$ ($\sigma(x_n) = 1$). Every truth assignment $\tau \in Val_{n-1}$ can be uniquely extended to a truth assignment $\sigma \in A$ ($\sigma \in B$), thus by the induction hypothesis,

$$\sum_\sigma f_\sigma = x_n \cdot 1 + (1 - x_n) \cdot 1 = 1.$$

This completes the proof of the claim.

Now, given any DNF tautology $A \equiv \bigvee_{i=1}^r D_i \equiv \bigvee_i \bigwedge_j \ell_{i,j}$, the CNF formula $\neg A$ is of the form $\bigwedge_{i=1}^r C_i \equiv \bigwedge_i \bigvee_j \overline{\ell_{i,j}}$.

CLAIM. For every $\sigma \in Val_n$, there exists $1 \leq i \leq r$ such that f_σ is divisible by q_{C_i}.

Proof of Claim. Suppose that $\sigma \in Val_n$. Since $\bigvee_{i=1}^r D_i$ is a tautology, let i_0 be such that $\sigma \models D_{i_0}$. $D_{i_0} \equiv \ell_{i_0,1} \wedge \cdots \wedge \ell_{i_0,s}$ and its negation is equivalent to $C_{i_0} \equiv \overline{\ell_{i_0,1}} \vee \cdots \vee \overline{\ell_{i_0,s}}$. From definitions, we have that $q_{C_{i_0}} = p_{D_{i_0}}$, and that $p_{D_{i_0}}$ divides $\prod_{\sigma(x_i)=0} x_i \cdot \prod_{\sigma(x_i)=1} (1 - x_i) = f_\sigma$. This establishes the claim.

A Nullstellensatz refutation of unsatisfiable CNF formula A can be given as follows. Partition Val_n into disjoint sets $V_1, \ldots, V_r$, where $\sigma \in V_i$ implies that $\sigma \models D_i$. Then

$$\begin{aligned} 1 &= \sum_{\sigma \in Val_n} f_\sigma \\ &= \sum_{i=1}^r \left(q_{C_i} \cdot \sum_{\sigma \in V_i} \frac{f_\sigma}{q_{C_i}} \right) \end{aligned}$$

yields a Nullstellensatz refutation of A of degree n.

The previous theorem is due to T. Pitassi [Pit97]. Since a linear degree upper bound is always possible, an obvious question is to establish a matching linear lower bound for an explicitly given family of unsatisfiable CNF formulas, a task to which we will soon turn.

In passing, note that it follows from the NP-completeness of satisfiability that testing whether a set of polynomials of degree at most 2 over a finite field F is also NP-complete, an observation due to L. Valiant. This can be established as follows. For $A \equiv \bigwedge_{i=1}^r C_i$ a propositional formula in conjunctive normal form, where each conjunct is a disjunction of at most 3 literals over $x_1, \ldots, x_n, \overline{x_1}, \ldots, \overline{x_n}$, A is satisfiable if and only if

$$q_{C_1}, \ldots, q_{C_r}, x_1^2 - x_1, \ldots, x_n^2 - x_n$$

has a common root in F. Since each C_i has at most 3 literals, the degree of q_{C_i} is at most 3. To reduce the degree to 2, introduce a new variable w for each monomial xyz, along with the auxilliary polynomials xw, $w - yz$.

We turn now to the question of automatizability of the Nullstellensatz system, first proved by Beame et al. [BIK+96].

Theorem 5.5.5 ([BIK+96]). *Let F be a field. The degree d bounded Nullstellensatz system over F is automatizable; i.e., there is a polynomial time algorithm $\mathcal{A}_d$, which when given polynomials $p_1, \ldots, p_k \in F[x_1, \ldots, x_n]$, outputs polynomials $g_1, \ldots, g_m, h_1, \ldots, h_n \in F[x_1, \ldots, x_n]$, such that*

$$1 = \sum_{i=1}^{m} p_i \cdot g_i + \sum_{i=1}^{n} (x_i^2 - x_i) \cdot h_i$$

and

$$\max\{deg(p_i \cdot g_i), deg((x_j^2 - x_j) \cdot h_j) : 1 \le i \le m, 1 \le j \le n\} \le d$$

provided that there exists a degree d Nullstellensatz refutation of $p_1, \ldots, p_k$.

Proof. For each subset r of $\{1, \ldots, n\}$, let x_r denote the multilinear power product $\prod_{i \in r} x_i$, where if $r = \emptyset$, then $x_r = 1$. Let $\mathcal{P}_{\le d}(\{1, \ldots, n\})$ denote the collection of subsets of $\{1, \ldots, n\}$ of size at most d. Assume that there exists a degree d Nullstellensatz refutation of $p_1, \ldots, p_k$ over field F. Then it follows that there exist $a_{i,r} \in F$, for $1 \le i \le m$, and $b_{j,r} \in F$, for $1 \le j \le n$, such that

$$1 = \sum_{i=1}^{m} \left(p_i \cdot \sum_r a_{i,r} x_r \right) + \sum_{i=1}^{n} \left((x_i^2 - x_i) \cdot \sum_r b_{j,r} x_r \right) \tag{5.44}$$

where in the sum, r varies over $\mathcal{P}_{\le d}(\{1, \ldots, n\})$. Formally multiply out the products $p_i \cdot \sum_r a_{i,r} x_r$ and $(x_i^2 - x_i) \cdot \sum_r b_{j,r} x_r$, and collect all terms in the same power product x_r. This gives rise to a system of linear equations, one for each x_r with $r \in \mathcal{P}_{\le d}(\{1, \ldots, n\})$, where a linear combination of the $a_{i,r}$ and $b_{j,r}$ over F equals 0 (1) if $r \ne \emptyset$ ($r = \emptyset$). By the polynomial time procedure of Gaussian elimination over F, we can solve for the $a_{i,r}$ and $b_{j,r}$ and hence determine

$$g_i = \sum_r a_{i,r} x_r$$
$$h_j = \sum_r b_{j,r} x_r.$$

This completes the proof of automatizability of Nullstellensatz.

Assuming P $\neq$ NP, it follows from Theorem 5.5.5 (along with Valiant's earlier observation on the NP-completeness for the problem of determining the existence of roots for systems of fixed degree polynomials) that there must be a non-constant lower bound for the Nullstellensatz degree for unsatisfiable 3-CNF formulas. This remark applies to all algebraic proof systems which we cover.

We now turn to lower bounds for Nullstellensatz systems. The *housesitting principle* (previously called *iteration principle*) is a restricted version of the pigeonhole principle, which is intuitively expressed as follows. Let $I = \{0, \ldots, n\}$ and $J = \{1, \ldots, n\}$. For each $i \in I$, the i-th person either stays at her own home i, or housesits in a home $j > i$ of a person j, who is not at home. Following [Bus98b], to formalize this we define the polynomials $p_i, q_{i,j}, r_{i,j,j'}, s_{i,i',j,j'}$, where

$$\begin{aligned}
p_i &= x_{i,i} + x_{i,i+1} + \cdots + x_{i,n} - 1 \\
q_{i,j} &= x_{i,j} x_{j,j} \\
r_{i,j,j'} &= x_{i,j} x_{i,j'}, \text{ for } j \neq j' \\
s_{i,i',j,j'} &= x_{i,j} x_{i',j'}, \text{ for } i < i' < j, j' \text{ and } j \neq j' \\
t_{i,j} &= x_{i,j}^2 - x_{i,j}.
\end{aligned}$$

Here polynomial p_i says that person i stays in a house $j \geq i$, $q_{i,j}$ says that if person j is at home, then person i cannot housesit there, and $r_{i,j,j'}$ says that person i cannot be in two places at once. Polynomial $s_{i,i',j,j'}$ is an additional technical condition introduced by Buss to lift the original lower bound proof of [CEI96] from the field $GF(2)$ to an arbitrary field. Finally, the familiar polynomials $t_{i,j}$ allow for the multilinearization of any polynomial in variables $x_{i,j}$. Let HS_n be the housesitting principle, formalized in the $O(n^2)$ variables $x_{i,j}$ for $0 \leq i \leq n$, $1 \leq j \leq n$.

Theorem 5.5.6 ([Bus98b], [CEI96]). *Let F be an arbitrary field. There is no Nullstellensatz refutation of the housesitting principle HS_n of degree $d \leq n$.*

Historically, the Nullstellensatz system arose in an attempt to prove exponential lower bounds for constant-depth Frege systems with a modular counting principle – see Theorem 5.7.8. In the next section, we present proofs of linear lower bounds for the polynomial calculus PC, a system which trivially polynomially simulates the Nullstellensatz system NS. Since it can be shown that there are degree 3 polynomial calculus refutations of HS_n, it follows that PC is strictly stronger than NS. For this reason, we do not present proofs of lower bounds for the Nullstellensatz system, but only for the polynomial calculus.

5.5.2 Polynomial Calculus

The *polynomial calculus* (PC) is a refutation system for unsatisfiable sets of polynomial equations over a field. This system was introduced by Clegg, Edmonds and Impagliazzo in [CEI96] under the name Gröbner system, because the well-known Gröbner basis algorithm provides a proof search mechanism for PC refutations – the more accurate name of *polynomial calculus* has since been adopted.

Fix field F, and let $\mathcal{P} \subseteq F[x_1, \ldots, x_n]$ be a finite set of multivariate polynomials over F. An axiom of PC is a polynomial $p \in \mathcal{P}$ or $x_i^2 - x_i$, for $1 \leq i \leq n$. There are two rules of inference of PC.

- *Multiplication by a variable:* From p, infer $x_i \cdot p$, where $1 \leq i \leq n$.
- *Linear combination:* From p, p', infer $a \cdot p + b \cdot p'$, where $a, b \in F$.

A *derivation* of polynomial q from $\mathcal{P}$ is a finite sequence $\Pi = (p_1, \ldots, p_m)$, where $q = p_m$ and for each $1 \leq i \leq m$, either $p_i \in \mathcal{P}$ or there exists $1 \leq j < i$ such that $p_i = x_k \cdot p_j$, for some $1 \leq k \leq n$, or there exist $1 \leq j, k < i$ such that $p_i = a \cdot p_j + b \cdot p_k$. By $\mathcal{P} \vdash q$, we denote that q has a derivation from $\mathcal{P}$. By $\mathcal{P} \vdash_d q$, we denote that q has a derivation $\Pi = (p_1, \ldots, p_m)$ from $\mathcal{P}$ of degree at most d; i.e., $\max\{\deg(p_i) : 1 \leq i \leq m\} \leq d$. Finally, $\mathcal{P} \vdash_{d,m} q$ means that $\mathcal{P} \vdash_d q$ and additionally that the number of *lines* in the derivation $\Pi = (p_1, \ldots, p_m)$ is m.

A *PC refutation* of $\mathcal{P}$ is a derivation of 1 from $\mathcal{P}$. The *degree* of refutation $\Pi = (p_1, \ldots, p_m)$ is $\min\{\deg(p_i) : 1 \leq i \leq m\}$. The PC degree of an unsatisfiable set $\mathcal{P}$ of polynomials, denoted $\deg(\mathcal{P})$, is the minimum degree of a refutation of $\mathcal{P}$.

Sometimes, instead of polynomials in a PC refutation $\Pi = (p_1, \ldots, p_m)$ (where p_m is 1), one instead writes polynomial equations $p_1 = 0, \ldots, p_m = 0$. In this context, the relationship with the Nullstellensatz system is clearer, where in both NS and PC, a refutation of unsatisfiable CNF formula $\wedge_{i=1}^r C_i$ is a formal manifestation that $1 \in I = \langle q_{C_1}, \ldots, q_{C_r}, x_1^2 - x_1, \ldots, x_n^2 - x_n \rangle$. For NS, 1 is explicitly given as a linear combination over $F[x_1, \ldots, x_n]$ of the q_{C_i} and $(x_i^2 - x_i)$, while in PC, a derivation of the fact that 1 belongs to I is given stepwise. It follows that the degree of a PC refutation of formula A is at most the degree of an NS refutation of A. Later, when considering the Gaussian calculus (GC) and the binomial calculus (BC), the equational form of derivations will be exploited. Generally, for the polynomial calculus, we use the canonical polynomial q_A, rather than p_A, in representing CNF formulas to be refuted.

Here is an example derivation. Consider the unsatisfiable CNF formula obtained by taking the conjunction of $x_1, \neg x_1 \vee x_2, \neg x_2 \vee x_3, \ldots, \neg x_{n-1} \vee x_n, \neg x_n$. Using the q_A translation, we have the polynomials $1 - x_1, x_1 - x_1 x_2, x_2 - x_2 x_3, \ldots, x_{n-1} - x_{n-1} x_n, x_n$. Consider the following derivation.

1. $x_1 - x_1 x_2$, axiom.
2. $x_2 - x_2 x_3$, axiom.
3. $x_1 x_2 x_3 - x_1 x_3$, multiplication of (1) by $-x_3$.
4. $x_1 x_2 - x_1 x_2 x_3$, multiplication of (2) by x_1.
5. $x_1 - x_1 x_2 x_3$, addition of (1),(4).

6. $x_1 - x_1x_3$, addition of (3),(5).

The last line represents $\neg x_1 \vee x_3$. By repeating this, we can derive $\neg x_1 \vee x_n$, i.e., $x_1 - x_1x_n$. From the hypotheses x_1 and $\neg x_n$, we have the additional polynomial equations $1 - x_1$, x_n. Thus we have

1. $x_1 - x_1x_n$, derived from above.
2. x_n, axiom.
3. x_1x_n, multiplication of (1) by x_1.
4. x_1, addition of (1),(3).
5. $1 - x_1$, axiom.
6. 1, addition of (4),(5).

An easy proof by induction on the number of inferences proves that if there is a polynomial calculus refutation of CNF formula A, then A is not satisfiable. Given a Nullstellensatz refutation, we can obviously furnish a refutation in the polynomial calculus, of the same degree or less, hence it follows that PC is complete, with degree bound of n for unsatisfiable CNF formulas on n variables. The weak Nullstellensatz Theorem 5.5.1 yields a bit more.

Theorem 5.5.7 (Completeness of polynomial calculus). *If there is no* $0, 1$ *solution of the polynomial equations* $p(x_1, \ldots, x_n)$ *for all* $p \in \mathcal{P} \subseteq F[x_1, \ldots, x_n]$, *then there is a degree* $n+1$ *derivation of* 1 *from* $\mathcal{P} \cup \{x_1^2 - x_1, \ldots, x_n^2 - x_n\}$ *in PC.*

Proof. (Outline) Theorem 5.5.1 yields a PC derivation of 1 from $\mathcal{P} \cup \{x_1^2 - x_1, \ldots, x_n^2 - x_n\}$. In that derivation, by judicious application of the axioms $x_1^2 - x_1, \ldots, x_n^2 - x_n$, we can ensure that the degree is never larger than $n+1$.

The weak Nullstellensatz yields the immediate corollary that PC is implicationally complete.

Corollary 5.5.3 (Folklore). *PC is implicationally complete; i.e.,*

$$(\forall x_1, \ldots, x_n \in F)\left[\bigwedge_{i=1}^{m} p_i(x_1, \ldots, x_n) = 0 \rightarrow q(x_1, \ldots, x_n) = 0\right]$$

implies that $p_1, \ldots, p_m \vdash_{PC} q$.

The following alternate proof of completeness of PC for CNF formulas yields the simple, but important fact that constant width resolution refutations can be polynomially simulated by constant degree polynomial calculus refutations. This is formalized in the following theorem.

Theorem 5.5.8. *If the set* $\mathcal{C}$ *of clauses has a resolution refutation of width* w, *then* $\mathcal{C}$ *has a polynomial calculus refutation of degree at most* $2w$.

Proof. (Outline) Suppose that we have the resolution inference

$$\frac{A \cup B \cup \{x\} \qquad B \cup C \cup \{\overline{x}\}}{A \cup B \cup C}$$

where $A = \{\alpha_1, \ldots, \alpha_r\}$, $B = \{\ell_1, \ldots, \ell_s\}$, and $C = \{\beta_1, \ldots, \beta_t\}$, and literals $\alpha_i, \ell_i, \beta_i$ range among variables $x_1, \ldots, x_n$ and their negations. Recall that $q_A = \prod_{\overline{x} \in A} x \cdot \prod_{x \in A}(1-x)$, and define the polynomials q_B and q_C analogously for clauses B and C.

With these conventions, $A \cup B \cup \{x\}$ is represented by the polynomial $(1-x) \cdot q_A \cdot q_B$, and $B \cup C \cup \{\overline{x}\}$ is represented by $x \cdot q_B \cdot q_C$. By successive multiplications, we obtain

$$\begin{aligned}(1-x) \cdot q_A \cdot q_B \cdot q_C \\ x \cdot q_A \cdot q_B \cdot q_C\end{aligned}$$

so by addition, we have $q_A \cdot q_B \cdot q_C$, which represents the resolvent $A \cup B \cup C$. Clearly the degree of this derivation is at most $1 + \deg(q_A) + \deg(q_B) + \deg(q_C)$, hence at most twice the width of any clause appearing in the resolution derivation.

We now turn to the *automatizability* of the polynomial calculus and give a characterization of degree d polynomial calculus derivations.

Definition 5.5.5 ([CEI96]). *A degree d pseudoideal I in $F[x_1, \ldots, x_n$ is a vector subspace of $F[x_1, \ldots, x_n]$, say V, consisting of polynomials of degree at most d, such that if $p \in I$ and $\deg(p) < d$, then for $1 \leq i \leq n$, $x_i \cdot p \in I$.*

Let $p_1, \ldots, p_k \in F[x_1, \ldots, x_n]$ be multivariate polynomials of degree at most d. Then $I_{d,n}(p_1, \ldots, p_k)$ denotes the smallest degree d pseudo-ideal of $F[x_1, \ldots, x_n]$.

Recall that we defined $p_1, \ldots, p_k \vdash_d q$ to mean that there is a PC derivation of q from $p_1, \ldots, p_k$ of degree at most d. Additionally, we define $p_1, \ldots, p_k \vdash_{d,m} q$ to mean that there is a PC derivation of q from $p_1, \ldots, p_k$ of degree at most d and length at most m.

Theorem 5.5.9 ([CEI96]). *For any multilinear polynomials*

$$p_1, \ldots, p_k, q \in F[x_1, \ldots, x_n]/\langle x_1^2 - x_1, \ldots, x_n^2 - x_n \rangle$$

of degree at most d,

$$p_1, \ldots, p_k \vdash_d q \iff q \in I_{d,n}(p_1, \ldots, p_k).$$

Proof. Let $V = \{q \in F[x_1, \ldots, x_n] : p_1, \ldots, p_k \vdash_d q\}$. We first show the direction from left to right, i.e., that $V \subseteq I_{d,n}(p_1, \ldots, p_k)$, by induction on the number m of inferences in the derivation of q from $p_1, \ldots, p_k$. If $p_1, \ldots, p_k \vdash_{d,1} q$, then $q \in \{p_1, \ldots, p_k\}$, so that $q \in I_{d,n}(p_1, \ldots, p_k)$.

Suppose now that $\Pi = (r_1, \ldots, r_{m+1})$ is a derivation of $q = r_{m+1}$ of degree at most d from $\{p_1, \ldots, p_k\}$.

Case 1. $\deg(p) < d$ and $q = x_i \cdot r_j$, for some $1 \le i \le n$ and $1 \le j \le m$. Then by definition, $q \in I_{d,n}(p_1, \ldots, p_k)$.

Case 2. $q = a \cdot r + b \cdot r'$, for some $a, b \in F$ and $r, r' \in \{r_1, \ldots, r_m\}$. Since $I_{d,n}(p_1, \ldots, p_k)$ is a vector space, and hence closed under the formation of linear combinations, $q \in I_{d,n}(p_1, \ldots, p_k)$.

Now consider the direction from right to left, i.e., $I_{d,n}(p_1, \ldots, p_k) \subseteq V$. By definition, $\{p_1, \ldots, p_k\} \subseteq V$, and V is closed under linear combinations over F and if $q \in V$ is of degree less than d, then for $1 \le i \le n$, $x_i \cdot q \in V$. By definition, $I_{d,n}(p_1, \ldots, p_k)$ is the smallest vector space satisfying these same properties, and so $I_{d,n}(p_1, \ldots, p_k) \subseteq V$.

We now turn to the question of automatizability of polynomial calculus, and begin by presenting the following algorithms
REDUCE and CONSTRUCTBASIS$_d$.

Algorithm 5.5.1 (REDUCE(p,B)).

```
find q ∈ B such that LPP(p) = LPP(q)
if q does not exist
   return p;
else {
   a = LeadMon(p) / LeadMon(q);
   p = p − aq;
   if (p ≠ 0)
      return REDUCE(p, B);
   else
      return 0;
}
```

The algorithm REDUCE works as follows. If there is $q \in B$ having the same leading monomial as that of p, then we find a scalar $a \in F$ for which $p-aq$ has canceled that leading monomial, and, provided that $p-aq \neq 0$, then recursively reduce $p-aq$. On the other hand, if there is no such q, then $B \cup \{p\}$ is linearly independent, so we return p. Provided that REDUCE$(p, B) \neq 0$, it is clear that $Span(B \cup \{\text{REDUCE}(p, B)\}) = Span(B \cup \{p\})$, and that $Span(B \cup \{\text{REDUCE}(p, B)\})$ is linearly independent.

Algorithm 5.5.2 (CONSTRUCTBASIS$_d(\{p_1, \ldots, p_k\})$).

```
B = ∅;
S = {p1, ..., pk};
while (S ≠ ∅) {
   choose q ∈ S;
   S = S − {q};
   q0 = REDUCE(q, B);
   if (q0 ≠ 0) {
      B = B ∪ {q0};
      if (deg(q0) < d)
         for i = 1 to n
```

```
            S = S ∪ { ML(x_i · q_0)};
        }
    }
```

Algorithm $\textsc{ConstructBasis}_d$ works as follows. Initially, set $B = \emptyset$ and $S = \{p_1, \ldots, p_k\}$, so that

1. $p_1, \ldots, p_k \in Span(B \cup S)$, and
2. $B \cup S \subseteq I_{d,n}(p_1, \ldots, p_k)$.

Construct stepwise a set B, consisting of polynomials having distinct leading power products (thus ensuring that B is linearly independent), such that conditions (1) and (2) are inductively met. At the end of the construction, $S = \emptyset$ and B will be a linearly independent set of polynomials, spanning $I_{d,n}(p_1, \ldots, p_k)$, hence a basis. This fact will be proved in Theorem 5.5.10.

The time analysis of $\textsc{Reduce}$ is given as follows. Use a hash table to store the entries of $(LeadMon(q), q)$ for $q \in B$, so that given p, one can determine in constant time whether there exists $q \in B$ with $LPP(q) = LPP(p)$. If such q exists, then $a = LeadMon(p)/LeadMon(q)$ can be determined in constant time, while performing the subtraction $p - aq$ takes $O(\binom{n}{\leq d})$ time. There are at most $\binom{n}{\leq d}$ stages of recursion, so the algorithm $\textsc{Reduce}$ runs in time $O(\binom{n}{\leq d}^2)$.

The time analysis of $\textsc{ConstructBasis}_d$ is given as follows. At any time, S is a collection of multilinear monomials in $x_1, \ldots, x_n$, hence $|S| \leq \binom{n}{\leq d}$, so there are at most $\binom{n}{\leq d}$ passes through the `while`-loop. Each pass calls the algorithm $\textsc{Reduce}$, hence costs time $O(\binom{n}{\leq d}^2)$, plus and additional time $n \cdot \binom{n}{\leq d-1}$ for adding the multilinearization of $x_i \cdot q_0$, for $1 \leq i \leq n$. Thus the overall time of $\textsc{ConstructBasis}_d$ is

$$O(n \cdot \binom{n}{\leq d-1} \cdot \binom{n}{\leq d}^2) = O(n^{3d}).$$

Theorem 5.5.10 ([CEI96]). *Algorithm* $\textsc{ConstructBasis}_d$ *produces a basis* B *of the vector space* $I_{d,n}(p_1, \ldots, p_k)$.

Proof. Let B_i, S_i denote the values of B, S respectively after the i-th pass through the `while`-loop in $\textsc{ConstructBasis}_d$, and let B_∞, S_∞ denote the terminal values of B, S after execution of $\textsc{ConstructBasis}_d$. Recall that $B_0 = \emptyset$, $S_0 = \{p_1, \ldots, p_k\}$, and that $S_\infty = \emptyset$.

For $i \leq \infty$ and all distinct $p, q \in B_i$, the leading power product $LPP(p)$ of p is unequal to the leading power product $LPP(q)$ of q, hence B_i is a linearly independent subset of $I_{d,n}(p_1, \ldots, p_k)$. Let $Span(B)$ denote the span of B over fixed field F, i.e., the collection of all linear combinations over F of elements of B.

CLAIM. $Span(B_\infty) = I_{d,n}(p_1, \ldots, p_k)$.

Proof of Claim. Clearly, $Span(B_\infty) \subseteq I_{d,n}(p_1,\ldots,p_k)$. Consider now the converse inclusion. We prove the following.

1. $p_1,\ldots,p_k \subseteq Span(B_\infty)$.
2. $Span(B_\infty)$ is a vector space.
3. If $q \in Span(B_\infty)$ and $\deg(q) < d$, then $x_i \cdot q \in Span(B_\infty)$ for $1 \leq i \leq n$.

For 1, let q_i denote the element chosen in S_i in the i-th pass of the `while`-loop of ConstructBasis$_d$, and let r_i = Reduce(q_i, B_i). If $r_i \neq 0$, then $B_{i+1} = B_i \cup \{r_i\}$, otherwise $B_{i+1} = B_i$, and hence $B_i \subseteq B_{i+1}$. Note that $Span(B_i \cup \{q_i\}) = Span(B_i \cup \{r_i\})$, hence

$$Span(B_i \cup S_i) \subseteq Span(B_i \cup \{r_i\} \cup (S_i - \{q_i\})) \subseteq Span(B_{i+1} \cup S_{i+1})$$

since

$$S_{i+1} = (S_i - \{q_i\}) \cup \{x_j \cdot r_i : 1 \leq j \leq n\}$$

provided that $r_i \neq 0$ and $\deg(r_i) < d$, and otherwise $S_{i+1} = S_i - \{q_i\}$. It follows that

$$\{p_1,\ldots,p_k\} = S_0 \subseteq Span(B_0 \cup S_0) \subseteq Span(B_\infty \cup S_\infty) = Span(B_\infty)$$

and so 1 is established.

Since B_∞ is a linearly independent subset of $I_{d,n}(p_1,\ldots,p_k)$, it follows that $Span(B_\infty)$ is a vector space, and so 2 is established.

Suppose now that $r \in Span(B_\infty)$ has degree less than d, so $r = \sum_{i \in I} a_i \cdot r_i$, where I is a finite set of indices, $a_i \in F$, and r_i = Reduce(q_i, B_i) is added to B in the i-th pass of the `while`-loop of algorithm ConstructBasis$_d$. For 3, we must show that $x_j \cdot r \in Span(B_\infty)$ for $1 \leq j \leq n$. To this end, it clearly suffices to show that $x_j \cdot r_i \in Span(B_\infty)$ for $1 \leq j \leq n$ and $i \in I$. In ConstructBasis$_d$, when r_i is added to B_{i+1}, $x_j \cdot r_i$ is added to S_{i+1}, so

$$x_j \cdot r_i \in Span(B_{i+1} \cup S_{i+1}) \subseteq Span(B_\infty \cup S_\infty) = Span(B_\infty).$$

This establishes 3.

Finally, properties 1, 2, 3 imply that $I_{d,n}(p_1,\ldots,p_k) \subseteq Span(B_\infty)$, and so equality holds. This concludes the proof of the claim and hence of the theorem.

The previous theorem actually furnishes yet another proof of the completeness of PC, formalized as follows, where we use the notation from the proof of the last theorem.

Theorem 5.5.11 ([CEI96]). *Let $p_1,\ldots,p_k \in F[x_1,\ldots,x_n]$ and fix $d \geq \max\{deg(p_i) : 1 \leq i \leq k\}$. For $0 \leq i \leq \infty$, for any $r \in B_i$ and $q \in S_i$, $p_1,\ldots,p_k \vdash_d r$ and $p_1,\ldots,p_k \vdash_d q$.*

Proof. By induction on i. When $i = 0$, the assertion is clear, since $B_0 = \emptyset$ and $S_0 = \{p_1, \ldots, p_k\}$. Assume now that the assertion has been proved for all values less than i. Now $B_{i+1} - B_i = \{q_0\}$, where $r_i = \textsc{Reduce}(q_i, B_i)$ and q_i is chosen from S_i in the i-th pass of the `while`-loop of $\textsc{ConstructBasis}_d$. Clearly

$$r_i = \textsc{Reduce}(q_i, B_i) = q_i - a_1 t_1 - \cdots - a_\ell t_\ell$$

where $a_1, \ldots, a_\ell \in F$ and $t_1, \ldots, t_\ell \in B_i$, for $0 \leq \ell$ (if $\ell = 0$, then $r_i = q_i$). By the induction hypothesis applied to B_i, for $1 \leq i \leq \ell$, $p_1, \ldots, p_k \vdash_d t_i$ and by the induction hypothesis applied to S_i, $p_1, \ldots, p_k \vdash_d q_i$. Since r_i is a linear combination of $t_1, \ldots, t_\ell, q_i$, it follows that $p_1, \ldots, p_k \vdash_d r_i$. This establishes the inductive case for B_{i+1}.

Now $S_{i+1} - S_i = \emptyset$ if $r_i = 0$ or $\deg(r_i) = d$. Otherwise,

$$S_{i+1} - S_i = \{\text{ML}(x_j \cdot r_i) : 1 \leq j \leq n\}.$$

Corollary 5.5.1 implies that the multilinearization of $x_j \cdot r_i$ is obtained from $x_j \cdot r_i$ by a PC derivation involving $x_1^2 - x_1, \ldots, x_n^2 - x_n$, and so $p_1, \ldots, p_k \vdash_d \text{ML}(x_j \cdot r_i)$. This establishes the inductive case for S_{i+1}, and completes the proof of theorem.

Corollary 5.5.4 ([CEI96]). *Let $B = \textsc{ConstructBasis}_d(\{p_1, \ldots, p_k\})$, and let q be a polynomial in $F[x_1, \ldots, x_n]$ having degree at least $d \geq \max\{deg(p_i) : 1 \leq i \leq k\}$. If $\textsc{Reduce}(q, B) = 0$, then $p_1, \ldots, p_k \vdash_d q$.*

Theorem 5.5.12 ([CEI96]). *The degree d bounded polynomial calculus is automatizable; i.e., there is an algorithm $\mathcal{A}_d$, which when given polynomials $p_1, \ldots, p_k \in F[x_1, \ldots, x_n]$ of degree at most d having no $0, 1$ solution, yields a derivation of*

$$1 \in \langle p_1, \ldots, p_k, x_1^2 - x_1, \ldots, x_n^2 - x_n \rangle$$

in time $O(n^{3d})$. More generally, if $q \in I_{d,n}(p_1, \ldots, p_n)$ then $\mathcal{A}_d$ yields a PC derivation this fact.

Proof. From Theorem 5.5.10 it follows that the set B constructed by algorithm $\textsc{ConstructBasis}_d$ is a basis for $I_{d,n}(p_1, \ldots, p_k)$. Given a polynomial $q \in F[x_1, \ldots, x_n]$, $q \in I_{d,n}(p_1, \ldots, p_n)$ if and only if $\textsc{Reduce}(q, B) = 0$. The runtime for $\textsc{ConstructBasis}_d$ and $\textsc{Reduce}$ is $O(n^{3d})$, and the previous theorem now furnishes the required derivation.

Up to the present, we have translated arbitrary propositional formulas (CNF formulas) using the canonical polynomials p_A (q_A). In establishing lower bounds for the polynomial calculus, it is notationally expedient to change the representation, adopting the *Fourier basis* . This representation is only valid when working over a field F of characteristic different than 2.

Definition 5.5.6 (Fourier basis). *Let $q_0 = 1$ (i.e., FALSE is represented by 1), $q_1 = -1$ (i.e., TRUE is represented by -1), $q_{x_i} = y_i$ (i.e., the propositional variable x_i is represented by the algebraic variable y_i), $q_{\neg A} = -q_A$, $q_{A \vee B} = \frac{q_A \cdot q_B + q_A + q_B - 1}{2}$, $q_{A \wedge B} = \frac{-q_A \cdot q_B + q_A + q_B + 1}{2}$, and $q_{A \oplus B} = q_A \cdot q_B$.*

When working with the Fourier basis, rather than the auxilliary polynomials $x_i^2 - x_i$, we use the auxilliary polynomials $y_i^2 - 1$, which ensure that y_i takes the value ± 1. Later, it will often be the case that the polynomial representation of propositional formula A in variables $x_1, \ldots, x_n$, when using the Fourier basis, will be written in the form $q(y_1, \ldots, y_m) \in F[y_1, \ldots, y_m]$, where $y_i = 1 - 2x_i$.

To obtain a degree lower bounds for PC derivations, we will soon focus on linear equations over $GF(2)$. In particular, note that the Fourier representation of linear equation $\sum_{i=1}^{r} x_i + a = 0$ over $GF(2)$ is

$$(-1)^{1-a} \cdot \prod_{i=1}^{r} \frac{1 - x_i}{2} = 0$$

which will generally be written in the form

$$(-1)^{1-a} \cdot \prod_{i=1}^{r} y_i = 0$$

where $y_i = 1 - 2x_i$. Later, we will introduce the *balanced Fourier representation* of the form

$$\prod_{i=1}^{\lfloor r/2 \rfloor} y_i + (-1)^{1-a} \cdot \prod_{i=\lfloor r/2 \rfloor + 1}^{r} y_i = 0.$$

The Fourier basis, especially in its application to algebraic proof systems, was originally employed by D. Grigoriev [Gri98], and allows for substantial simplification of lower bound arguments for NS and PC. We'll later prove a lemma which allows lower bounds for the Fourier basis representation of propositional formulas to entail lower bounds for the conventional representation using canonical polynomials q_A.

5.5.3 Gaussian Calculus

The *Gaussian calculus* (GC) is a refutation system for unsatisfiable sets of linear equations over a field, first defined by E. Ben-Sasson and R. Impagliazzo in [BSI99]. Fix prime q, and let $\mathcal{L} = \{\ell_i : 1 \le i \le m\}$ be a set of m linear equations over $GF(q)$, where each ℓ_i has the form

$$\sum_{j \in S_i} a_{i,j} x_j + b_i = 0$$

with $a_{i,j}, b_i \in \{0, 1, \ldots, q-1\}$.

An *axiom* is any linear equation in $\mathcal{L}$. The Gaussian calculus has two rules of inference.

- *Scalar multiplication:* From linear equation ℓ of the form

$$\sum_{j \in S} a_j x_j + b = 0$$

infer the linear equation $\alpha \cdot \ell$ of the form

$$\sum_{j \in S} \alpha a_j x_j + \alpha b = 0$$

where $\alpha \in GF(q)$.
- *Addition:* From linear equations ℓ, ℓ' respectively of the form

$$\sum_{j \in S} a_j x_j + c = 0$$

$$\sum_{j \in S'} b_j x_j + d = 0$$

infer the linear equation $\ell + \ell'$ of the form

$$\sum_{j \in S \cup S'} (a_j + b_j) x_j + (c + d) = 0.$$

Here, if $j \in S - S'$, then $b_j = 0$ and if $j \in S' - S$, then $a_j = 0$.

A *GC derivation* of ℓ from $\mathcal{L}$ is a finite sequence $E_1, E_2, \ldots, E_r$ of linear equations, such that ℓ is the equation E_r and for each $1 \leq i \leq r$, E_i is either an axiom (i.e., element of $\mathcal{L}$), or there exists $1 \leq j < i$ such that E_i is obtained by scalar multiplication from E_j, or there exist $1 \leq j, k < i$ such that E_i is obtained by addition of E_j, E_k. Often, we speak of E_i as a *line* in the derivation. A *GC refutation* of $\mathcal{L}$ is a derivation of $1 = 0$ from $\mathcal{L}$. The *width* of a refutation $E_1, \ldots, E_r$ is the maximum number of variables appearing in any E_i, i.e., $\max\{|Vars(E_i)| : 1 \leq i \leq r\}$. The Gaussian width $w_G(\mathcal{L})$ of an unsatisfiable set $\mathcal{L}$ of linear equations is the minimum width of a refutation of $\mathcal{L}$.

Suppose that $\mathcal{L}$ is a set of linear equations over field F in the variables $x_1, \ldots, x_n$. If the assignment $x_1 = a_1, \ldots, x_n = a_n$ for $a_1, \ldots, a_n \in F$ satisfies linear equations ℓ, ℓ', then certainly the same assignment satisfies the linear equations $\alpha \cdot \ell$ for $\alpha \in F$, and $\ell + \ell'$. By induction on the derivation length, it follows that if $\mathcal{L}$ is a refutable set of linear equations over field F, then $\mathcal{L}$ is unsatisfiable, hence the Gaussian calculus is *sound.*

Standard Gaussian elimination proves that the Gaussian calculus is *complete,* in that if $\mathcal{L}$ is an unsatisfiable set of linear equations over field F, then there is a refutation of $\mathcal{L}$. Moreover, Gaussian elimination yields that the number of lines in a refutation of an unsatisfiable set $\mathcal{L} = \{\ell_i : 1 \leq i \leq m\}$ of linear equations in variables $x_1, \ldots, x_n$ in $GF(q)$ is at most mn.

For clarity, here is an example width 2 Gaussian calculus derivation over $GF(5)$.

Example 5.5.1. The linear equations

$$2x + 3y + 1 = 0$$
$$x - 2y + 2 = 0$$
$$y + 4 = 0$$

are unsatisfiable over the field GF_5, and has the following refutation:

1. $2x + 3y + 1 = 0$, axiom.
2. $x - 2y + 2 = 0$, axiom.
3. $3x + 4y + 1 = 0$, multiplication of (2) by -2 (or equivalently by 3).
4. $2y = 2$, addition of (1),(3).
5. $y + 4 = 0$, axiom.
6. $4y + 3 = 0$, multiplication of (4) by $-\frac{1}{2}$ (or equivalently by 2).
7. $3 = 0$, addition of (5),(6).
8. $1 = 0$, multiplication of (7) by $\frac{1}{4}$ (or equivalently by 4).

5.5.4 Binomial Calculus

There exists a close relationship between the Gaussian calculus and the polynomial calculus, when restricted to multilinear *binomials*, i.e., polynomials of the form $a\prod_{i\in I} x_i + b\prod_{j\in J} x_j$. This restriction is called the *binomial calculus* (BC), defined by Buss et al. [BGIP99] and Ben-Sasson, Impagliazzo [BSI99]. The following theorem was first proved in [BGIP99] by using the method of Laurent relations. The proof given below follows the sketch given in Ben-Sasson and Impagliazzo [BSI99].

Theorem 5.5.13 ([BGIP99], [BSI99]). *If $\mathcal{P} = \{p_1, \ldots, p_k\}$ is a set of binomials having a PC refutation of degree d, then there exists a binomial refutation of degree d.*

Proof. (Outline) Modify the proof of Theorem 5.5.11 where now $p_1, \ldots, p_k$ are binomials.

CLAIM. All polynomials in B_i and S_i are binomials.

Proof of Claim. By induction on i. The claim clearly holds when $i = 0$, since $B_0 = \emptyset$ and $S_0 = \{p_1, \ldots, p_k\}$. For the inductive case, $B_{i+1} - B_i = \{r_i\}$, where $r_i = \text{REDUCE}(q_i, B_i)$, and q_i is chosen from S_i (hence by the inductive hypothesis q_i is a binomial).

Case 1. There exists $p \in B_i$, for which $LPP(q_i) = LPP(p)$.

By the inductive hypothesis, q_i is a binomial of the form $c_1m_1 + c_2m_2$, and p is a binomial of the form $c_3m_1 + c_4m_4$, where $c_1, c_2, c_3, c_4 \in F$ and m_1, m_2, m_3 are power products, m_1 the leading power product of both p and q_i. By definition, $\text{REDUCE}(q_i, B_i) = \text{REDUCE}(c_2m_2 - \frac{c_1c_4}{c_3}m_3, B_i)$, and clearly $c_2m_2 - \frac{c_1c_4}{c_3}m_3$ is a binomial. Induction on the number of recursion steps in

REDUCE(q_i, B_i) completes the argument. This establishes the inductive case for B_{i+1}.

Case 2. For all $p \in B_i$, $LPP(r_i) \neq LPP(p)$.

In this case $r_i = q_i \in S_i$, and so $B_{i+1} = B_i \cup \{r_i\}$ consists of binomials. This establishes the inductive case for B_{i+1}.

In both cases, $S_{i+1} - S_i$ is either $\emptyset$ or a set of multilinearizations of products of the form $x_j \cdot r_i$, where r_i is a binomial. This establishes the inductive case for S_{i+1}. The theorem now follows, as in the proof of Theorem 5.5.11.

Definition 5.5.7. *Let p be a binomial in $F[x_1, \ldots, x_n]$, whose monomial terms have coefficients in $\{1, -1\}$. If p is of the form*

$$(-1)^a \cdot \prod_S y_i + (-1)^b \cdot \prod_T y_i$$

where $a, b \in \{0, 1\}$, then the linear equation $\ell(p)$ over $GF(2)$ is defined by

$$1 + a + \sum_S x_i + b + \sum_T x_i = 0.$$

As well, let $\mathcal{L}(\mathcal{P})$ denote the set $\{\ell(p) : p \in \mathcal{P}\}$ of linear equalities over $GF(2)$.

In the following, Theorem 5.5.14 and Theorem 5.5.15 imply a strong *width/degree* preserving polynomial simulation of the binomial calculus by and vice-versa. Using this result, it will be later possible to prove degree lower bounds for polynomial calculus refutations by proving the more manageable width lower bounds for the Gaussian calculus.

Theorem 5.5.14 ([BSI99]). *Let $p_1, \ldots, p_k$ be binomials in $F[x_1, \ldots, x_n]$, whose monomial terms have coefficients in $\{1, -1\}$. If there is a PC refutation of $p_1, \ldots, p_k$ of degree d, then there is a GC refutation of $\mathcal{L}(\{p_1, \ldots, p_k\})$ of width at most $2d$.*

Proof. Temporarily, let us say that a derivation $\Pi = (s_1, \ldots, s_m)$ from $p_1, \ldots, p_k$ is a *special binomial* derivation of degree d, if

- $\max\{\deg(s_1, \ldots, s_m)\} = d$,
- s_m is a monomial,
- for $1 \leq i < m$, s_i is a binomial, whose monomial terms have coefficients in $\{1, -1\}$,
- for all $1 \leq i \leq m$, either $s_i \in p_1, \ldots, p_k$, or there exists $1 \leq j \leq n$, $1 \leq k < i$, such that $s_i = x_j \cdot s_k$, or there exist $1 \leq j, k < i$ and $a, b \in \{1, -1\}$, such that $s_i = a \cdot s_j + b \cdot s_k$.

Suppose that $p_1, \ldots, p_k \vdash_d 1$. As in the proof of the previous Theorem 5.5.13, we will attempt to build a basis B obtained from the algorithm $\text{CONSTRUCTBASIS}_d(p_1, \ldots, p_k)$, and then establish that $\text{REDUCE}(1, B) = 0$.

CLAIM. *Either* $\text{CONSTRUCTBASIS}_d(p_1, \ldots, p_k)$ produces a basis B consisting of binomials of degree at most d, whose monomial terms have coefficients in $\{1, -1\}$, *or* there exists a special binomial derivation $\Pi = (s_1, \ldots, s_m)$ from $p_1, \ldots, p_k$ of degree d, where additionally s_m is a monomial whose coefficient is 2 or -2.

Proof of Claim. (Outline) Using the notation from the proof of the previous theorem, consider the i-th pass in the `while`-loop in the construction of a binomial basis, where $r_i = \text{REDUCE}(q_i, B_i)$. If $r_i = 0$, then $B_{i+1} = B_i$, and we proceed in the $i + 1$-st pass of the `while`-loop. Assuming that $r_i \neq 0$, we have

$$\text{REDUCE}(q_i, B_i) = \text{REDUCE}(c_2 m_2 - \frac{c_1 c_4}{c_3} m_3, B_i)$$

where $q_i \in S_i$ is of the form $c_1 m_1 + c_2 m_2$, and $p \in B_i$ is of the form $c_3 m_1 + c_4 m_3$, for $c_1, c_2, c_3, c_4 \in \{1, -1\}$ and power products m_1, m_2, m_3. It follows that $\frac{c_1 c_4}{c_3} \in \{1, -1\}$.

Case 1. $m_2 \neq m_3$. Then $c_2 m_2 - \frac{c_1 c_3}{c_4} m_3$ is a binomial, whose monomial terms have coefficients in $\{1, -1\}$.

Case 2. $m_2 = m_3$. In this case, we have derived $(c_2 - \frac{c_1 c_3}{c_4}) m_2$. Since we assumed that $r_i \neq 0$, it must be that $c_2 - \frac{c_1 c_3}{c_4} \in \{2, -2\}$. This concludes the proof of the claim.

CLAIM. There is a degree d special binomial derivation from $p_1, \ldots, p_k$.

Proof of Claim. Assume, in order to obtain a contradiction, that the algorithm $\text{CONSTRUCTBASIS}_d(p_1, \ldots, p_k)$ produces a basis B of $I_{d,n}(p_1, \ldots, p_k)$ consisting of binomials, whose monomial terms have coefficients in $\{1, -1\}$. Since $p_1, \ldots, p_k \vdash_d 1$, it must be that $\text{REDUCE}(1, B) = 0$, and so for some i_0, in the i_0-th pass of the `while`-loop, $r_{i_0} \in F$ is added to B. (This is because for $\text{REDUCE}(1, B)$ to equal 0, it must be that some element p of B has leading power product 1; i.e., $p \in F$.) The only manner in which this could have happened is if $r_{i_0} \in \{2, -2\}$, in which case B does not consist of only binomials. Applying the previous claim, it follows that there must exist a special derivation of the required form. This establishes the claim.

Assume now that $\Pi = (s_1, \ldots, s_m)$ is a degree d special binomial derivation from $p_1, \ldots, p_k$, where s_m is a monomial, whose coefficient is in $\{2, -2\}$. By induction on $1 \leq \alpha \leq m$, define linear equation E_α over $GF(2)$, as follows. If $s_\alpha \in \{p_1, \ldots, p_k\}$, then $E_\alpha = \ell(s_\alpha)$. If there exist $1 \leq j \leq n$ and $1 \leq k < \alpha$, such that $s_\alpha = x_j \cdot s_k$, then $E_\alpha = E_k$. If there exist $1 \leq j, k < \alpha$ and $a, b \in \{1, -1\}$ such that $s_\alpha = a \cdot s_j + b \cdot s_k$, then s_j is of the form

$$(-1)^{c_1} \prod_S y_i + (-1)^{c_2} \prod_T y_i$$

and s_k is of the form

$$(-1)^{c_3} \prod_S y_i + (-1)^{c_4} \prod_R y_i.$$

Letting $c = \frac{1-a}{2}$ and $d = \frac{1-b}{2}$, $a \cdot s_j$ is of the form

$$(-1)^{c_1+c} \prod_S y_i + (-1)^{c_2+c} \prod_T y_i$$

and $b \cdot s_k$ is of the form

$$(-1)^{c_3+d} \prod_S y_i + (-1)^{c_4+d} \prod_R y_i$$

where $c_1 + c$ $(c_3 + d)$ is even (odd), or vice-versa. Hence s_i is of the form

$$(-1)^{c_2+c} \prod_T y_i + (-1)^{c_4+d} \prod_R y_i. \tag{5.45}$$

The equation $\ell(a \cdot s_j)$ is of the form

$$1 + c_1 + c + \sum_S x_i + c_2 + c + \sum_T x_i = 0 \tag{5.46}$$

and $\ell(b \cdot s_k)$ is of the form

$$1 + c_3 + d + \sum_S x_i + c_4 + d + \sum_R x_i = 0. \tag{5.47}$$

Since $c_1 + c + c_3 + d$ is odd, the sum of (5.46) and (5.47) over $GF(2)$ is

$$1 + c + d + c_2 + c_4 + \sum_T x_i + \sum_R x_i = 0 \tag{5.48}$$

which is $\ell(s_\alpha)$. This completes the proof by induction.

For $1 \leq \alpha < m$, s_α is a binomial, and so the sets T, R, in (5.48) are unequal. Since s_m is a monomial with coefficient in $\{2, -2\}$, it is of the form $\prod_T y_i + \prod_T y_i$ or $-\prod_T y_i - \prod_T y_i$, and in either case $\ell(s_m)$, which is E_m, is the equation $1 = 0$.

It follows from the above that from the degree d special binomial derivation $\Pi = (s_1, \ldots, s_m)$, where s_m is a monomial, whose coefficient is ± 2, we have constructed a Gaussian calculus refutation $\widetilde{\Pi} = (E_1, \ldots, E_m)$ of $\ell(p_1), \ldots, \ell(p_k)$ with width at most $2d$. This concludes the proof of the theorem.

Corollary 5.5.5. *Let $\mathcal{P}$ be a set of polynomials in $F[x_1, \ldots, x_n]$ having no common root, and $\mathcal{L}(\mathcal{P})$ be the associated system of linear equations over $GF(2)$. Then*

$$w_G(\mathcal{L}(\mathcal{P}) \leq 2 \cdot deg(\mathcal{P}).$$

The following definition is motivated by, but different from that of [BSI99].

Definition 5.5.8. *Let F be a finite field of characteristic $p > 2$, and let ℓ be a linear equation over $GF(2)$ of the form $\sum_{i \in I} x_i + a = 0$, with $a \in \{0, 1\}$. Define the projections π_1, π_2 so that $\pi_1(I)$,$pi_2(I)$ partition I in such a manner so that $|\pi_1(I)| = \lfloor \frac{|I|}{2} \rfloor$, $|\pi_2(I)| = \lceil \frac{|I|}{2} \rceil$, and for all $i \in \pi_1(I)$, $j \in \pi_2(I)$, we have $i < j$. (The idea is to sort I, and let $\pi_1(I)$ consist of the first $\lfloor \frac{|I|}{2} \rfloor$ many elements, and $\pi_2(I)$ consist of the remaining elements.) Define $P_F(\ell) \in F[y_1, \ldots, y_n]$ by*

$$\prod_{i \in \pi_1(I)} y_i + (-1)^{1-a} \prod_{i \in \pi_2(I)} y_i.$$

If $\mathcal{L}$ is a set of linear equations in the variables $x_1, \ldots, x_n$ over $GF(2)$, then $P_F(\mathcal{L}) = \{P_F(\ell) : \ell \in \mathcal{L}\} \cup \{y_i^2 - 1 : 1 \leq i \leq n\}$.

A set $\mathcal{L}$ of linear equations is a *minimal unsatisfiable* set, if $\mathcal{L}$ is unsatisfiable, and every proper subset $\mathcal{L}' \subset \mathcal{L}$ is satisfiable.

Theorem 5.5.15 ([BSI99]). *Let $\mathcal{L}$ be a minimal unsatisfiable set of linear equations in the variables $x_1, \ldots, x_n$ over $GF(2)$ of width at most k, and let F be a finite field of characteristic $p \neq 2$. Then*

$$deg(P_F(\mathcal{L})) \leq \max\{k, \left\lceil \frac{w_G(\mathcal{L})}{2} \right\rceil + 1\}.$$

Proof. Let $d = \lceil \frac{w_G(\mathcal{L})}{2} \rceil + 1$. Suppose that $E_1, \ldots, E_r$ is a GC refutation of $\mathcal{L}$. By induction on $1 \leq \alpha \leq r$, we show that $P_F(E_\alpha)$ has a PC derivation of degree at most d.

Case 1. E_α is an axiom $\ell \in \mathcal{L}$, so $P_F(\ell) \in P_F(\mathcal{L})$, hence is an axiom of the polynomial calculus.

Case 2. E_α is obtained by scalar multiplication from E_j, for some $j < \alpha$. Since the field is $GF(2)$, this means multiplication by 0 or 1, and so the result holds trivially.

Case 3. E_α is obtained by addition of E_j and E_k, for some $1 \leq j, k < \alpha$. Suppose that E_j respectively E_k is the linear equation

$$\sum_{i \in S} x_i + b = 0 \qquad \text{resp.} \sum_{i \in T} x_i + c = 0$$

and that E_α is

$$\sum_{i \in S \oplus T} x_i + b \oplus c = 0 \tag{5.49}$$

where $S \oplus T$ denotes the disjoint union $(S - T) \cup (T - S)$ of S, T. Let P denote the polynomial $P_F(E_j)$ given by

$$\prod_{i \in \pi_1(S)} y_i + (-1)^{1-b} \cdot \prod_{i \in \pi_2(S)} y_i \tag{5.50}$$

and let Q denote the polynomial $P_F(E_k)$ given by

$$\prod_{i \in \pi_1(T)} y_i + (-1)^{1-c} \cdot \prod_{i \in \pi_2(T)} y_i. \tag{5.51}$$

Case 1. $|S \cap T|$, $|S - T|$, $|T - S|$ are all bounded by $\lceil \frac{w}{2} \rceil$.

1. $\prod_{i \in \pi_1(S)} y_i + (-1)^{1-b} \cdot \prod_{i \in \pi_2(S)} y_i$, axiom $P_F(E_j)$.
2. $\prod_{i \in \pi_1(S) \cup (\pi_2(S) \cap T)} y_i + (-1)^{1-b} \cdot \prod_{i \in \pi_2(S) - T} y_i$, by multiplying (1) by y_i, for $i \in \pi_2(S) \cap T$ and reducing by using $y_i^2 - 1$.
3. $\prod_{i \in \pi_1(T)} y_i + (-1)^{1-c} \cdot \prod_{i \in \pi_2(T)} y_i$, axiom $P_F(E_k)$.
4. $\prod_{i \in \pi_1(T) \cup (\pi_2(T) \cap S)} y_i + (-1)^{1-c} \cdot \prod_{i \in \pi_2(T) - S} y_i$, by multiplying (3) by y_i, for $i \in \pi_2(T) \cap S$ and reducing by using $y_i^2 - 1$.
5. $\prod_{i \in (\pi_1(S) \cap T) \cup (\pi_2(S) \cap T)} y_i + (-1)^{1-b} \cdot \prod_{i \in (\pi_1(S) - T) \cup (\pi_2(S) - T)} y_i$, by multiplying (2) by y_i, for $i \in \pi_1(S) - T$ and reducing by using $y_i^2 - 1$.
6. $\prod_{i \in (S \cap T)} y_i + (-1)^{1-b} \cdot \prod_{i \in (S - T)} y_i$, a reformulation of (5).
7. $\prod_{i \in (\pi_1(T) \cap S) \cup (\pi_2(T) \cap S)} y_i + (-1)^{1-c} \cdot \prod_{i \in (\pi_1(T) - S) \cup (\pi_2(T) - S)} y_i$, by multiplying (2) by y_i, for $i \in \pi_1(T) - S$ and reducing by using $y_i^2 - 1$.
8. $\prod_{i \in (S \cap T)} y_i + (-1)^{1-c} \cdot \prod_{i \in (T - S)} y_i$, a reformulation of (7).
9. $(-1)^{1-b} \cdot \prod_{i \in (S - T)} y_i + (-1)^{2-c} \cdot \prod_{i \in (T - S)} y_i$, by adding (6) and $-1 \cdot (8)$.
10. $\prod_{i \in (S - T)} y_i + (-1)^{1-b+c} \cdot \prod_{i \in (T - S)} y_i$, by multiplying the last equation by $(-1)^{1-b}$.
11. $\prod_{i \in \pi_1(S \oplus T)} y_i + (-1)^{b+c-1} \cdot \prod_{i \in \pi_2(S \oplus T)} y_i$, by multiplying by y_i, for each $i \in ((\pi_1(S \oplus T) - (S - T)) \cup (\pi_2(S \oplus T) - (T - S)))$, and reducing by using $y_i^2 - 1$.

Note that this last polynomial is exactly $P_F(E_\alpha)$, as required. Under the case assumption that $|S \cap T|$, $|S - T|$, $|T - S|$ are all bounded by $\lceil \frac{w}{2} \rceil$, the degree bound of the previous PC derivation is at most $d = 1 + \lceil \frac{w}{2} \rceil$.

Case 2. $|S \cap T| > \lceil \frac{w}{2} \rceil$. In this case, $|S - T| < d$ and $|T - S| < d$. Define R to be an arbitrary subset of $S \cap T$ of size at most $\lceil \frac{w}{2} \rceil$, and analogously as in the previous case give a PC derivation of

$$\prod_{i \in (S - R)} y_i + (-1)^{1-b+c} \cdot \prod_{i \in (T - R)} y_i.$$

Now multiply by y_i, for $i \in (S \cap T) - R$, and reduce by using $y_i^2 - 1$ to obtain

$$\prod_{i \in (S - T)} y_i + (-1)^{1-b+c} \cdot \prod_{i \in (T - S)} y_i$$

and proceed as before. The degree bound of this derivation is clearly at most d.

Case 3. $|S-T| \geq d$, $|S \cap T| < d$, $|T-S| < d$. In this case, we proceed in a similar fashion, but only list the results of each step.

1. $\prod_{i \in \pi_1(S)} y_i + (-1)^{1-b} \cdot \prod_{i \in \pi_2(S)} y_i$, axiom $P_F(E_j)$.
2. $\prod_{i \in \pi_1(S-T)} y_i + (-1)^{1-b} \cdot \prod_{i \in (\pi_2(S)-\pi_1(S-T)) \cup (\pi_1(S)-\pi_1(S-T))} y_i$.
3. $\prod_{i \in \pi_1(S-T)} y_i + (-1)^{1-b} \cdot \prod_{i \in (S-\pi_1(S-T)} y_i$, a reformulation of the previous polynomial.
4. $\prod_{i \in \pi_1(T)} y_i + (-1)^{1-c} \cdot \prod_{i \in \pi_2(T)} y_i$, axiom $P_F(E_k)$.
5. $\prod_{i \in \pi_1(T-S)} y_i + (-1)^{1-c} \cdot \prod_{i \in (T-(T-S))} y_i$.
6. $\prod_{i \in \pi_1(T-S)} y_i + (-1)^{1-c} \cdot \prod_{i \in (S \cap T))} y_i$, a reformulation of the previous polynomial.
7. $(-1)^{1-b} \cdot \prod_{i \in \pi_1(S-T)} y_i + \prod_{i \in S-\pi_1(S-T))} y_i$, multiplication of (3) by $(-1)^{1-b}$.
8. $\prod_{i \in (T-S) \cup \pi_2(S-T)} y_i + (-1)^{1-c} \cdot \prod_{i \in (S \cap T) \cup \pi_2(S-T))} y_i$, multiplication of (6) by y_i, for $i \in \pi_2(S-T)$, with reduction by using $y_i^2 - 1$.
9. $(-1)^c \cdot \prod_{i \in (T-S) \cup \pi_2(S-T)} y_i - \prod_{i \in (S \cap T) \cup \pi_2(S-T)} y_i$, multiplication of (8) by -1. Note additionally that $(S \cap T) \cup \pi_2(S-T) = S - \pi_1(S-T)$.
10. $(-1)^{1-b} \cdot \prod_{i \in \pi_1(S-T)} y_i + (-1)^c \cdot \prod_{i \in (T-S) \cup \pi_2(S-T)} y_i$, by adding (9) and (10).
11. $\prod_{i \in \pi_1(S \oplus T)} y_i + (-1)^{b+c-1} \cdot \prod_{i \in \pi_2(S \oplus T)} y_i$, by appropriate multiplications of y_i, with reduction using $y_i^2 - 1$.

This completes the proof of Case 3.
Case 4. $|T-S| \geq d$, $|S \cap T| < d$, $|S-T| < d$. The proof of this case is analogous to Case 3 (interchange S, T and b, c). This completes the proof of the theorem.

5.5.5 Lower Bounds for the Polynomial Calculus

We are now in a position to furnish lower bounds for the degree of polynomial calculus refutations of certain formulas which have large *expansion*. First, we need some preliminary results, which allow lower bounds for the Fourier basis to translate into lower bounds for the conventional representation in PC.

Definition 5.5.9 ([BGIP99]). *Let*

$$\mathcal{P}(x_1, \ldots, x_n)(\mathcal{Q}(y_1, \ldots, y_m))$$

be finite sets of polynomials in

$$F[x_1, \ldots, x_n](F[y_1, \ldots, y_m]).$$

Then $\mathcal{P}$ is (d_1, d_2) reducible to $\mathcal{Q}$ if the following hold.

1. *For $1 \leq i \leq m$, there is a degree d_1 polynomial $r_i(x_1, \ldots, x_n)$, which defines y_i in terms of $x_1, \ldots, x_n$.*
2. *Letting $\mathcal{Q}(r_1(\mathbf{x}), \ldots, r_m(\mathbf{x}))$ abbreviate the set*

$$\{q(r_1(x_1, \ldots, x_n), \ldots, r_m(x_1, \ldots, x_n)) : q \in \mathcal{Q}\}$$

there exists a degree d_2 PC refutation of $\mathcal{Q}(r_1(\mathbf{x}), \ldots, r_m(\mathbf{x}))$ from $\mathcal{P}(\mathbf{x})$.

Lemma 5.5.1 ([BGIP99]). *Suppose that $\mathcal{P}(x_1, \ldots, x_n)$ is (d_1, d_2)-reducible to $\mathcal{Q}(y_1, \ldots, y_m)$. If there is a PC refutation of $\mathcal{Q}(y_1, \ldots, y_m)$ of degree d_3, then there is a PC refutation of $\mathcal{P}(x_1, \ldots, x_n)$ having degree $\max\{d_2, d_1 d_3\}$.*

Proof. Let Π_1 be a degree d_2 derivation of $\mathcal{Q}(r_1(\mathbf{x}), \ldots, r_m(\mathbf{x}))$ from $\mathcal{P}(\mathbf{x})$; i.e., for each $q \in \mathcal{Q}$, $q(r_1(\mathbf{x}), \ldots, r_m(\mathbf{x}))$ appears in the derivation Π_1. Let Π_3 be a degree d_3 derivation of 1 from $\{q(y_1, \ldots, y_m) : q \in \mathcal{Q}\}$. Let Π_2 be obtained from Π_3 by replacing y_j in q by $r_j(x_1, \ldots, x_n)$, for each $1 \leq j \leq m$ and $q \in \mathcal{Q}$, and let Π be the concatenation of Π_1 with Π_2. Clearly Π is a derivation of 1 from $\mathcal{P}(x_1, \ldots, x_n)$ of degree $\max\{d_2, d_1 d_3\}$.

The following definition facilitates the expression of later results.

Definition 5.5.10. *Let ℓ be a width k linear equation $x_{i_1} + \cdots + x_{i_k} + a = 0$ over $GF(2)$, with $a \in \{0,1\}$, where variables $x_{i_1}, \ldots, x_{i_k}$ are among $x_1, \ldots, x_n$, Define $P_F(\ell)$ to be the balanced Fourier representation of ℓ given by*

$$\prod_{r=1}^{\lfloor k/2 \rfloor} y_{i_r} + (-1)^{1-a} \cdot \prod_{r=\lfloor k/2 \rfloor + 1}^{k} y_{i_r}$$

and define $\mathcal{P}_F(\ell)$ to be $P_F(\ell) \cup \{y_1^2 - 1, \ldots, y_n^2 - 1\}$. If L is a set of linear equations in the variables $x_1, \ldots, x_n$, then define $P_F(L) = \{P_F(\ell) : \ell \in L\}$ and $\mathcal{P}_F(L) = \cup_{\ell \in L} \mathcal{P}_F(\ell)$.

While the previous definition relates to the Fourier basis, the following definition introduces notation corresponding to the canonical polynomial representation of clauses, which is really what interests us (see Definition 5.5.3).

Definition 5.5.11. *If C is a clause in the variables $x_{i_1}, \ldots, x_{i_k}$, where $x_{i_1}, \ldots, x_{i_k}$ are among $x_1, \ldots, x_n$, then the canonical polynomial is*

$$q_C(x_{i_1}, \ldots, x_{i_k}) = \prod_{\epsilon_{i_r}=0} x_{i_r} \cdot \prod_{\epsilon_{i_r}=1} (1 - x_{i_r}),$$

where for variable x, x^0 abbreviates $\neg x$, and x^1 abbreviates x. Define $\mathcal{Q}(C)$ to be $\{q_C(i_1, \ldots, i_k)\} \cup \{x_{i_1}^2 - x_{i_1}, \ldots, x_{i_k}^2 - x_{i_k}\}$, and for the set $\mathcal{C}$ of clauses, define $\mathcal{Q}(\mathcal{C}) = \cup_{C \in \mathcal{C}} \mathcal{Q}(C)$.

Note that for linear equation ℓ of the form $x_1 + \cdots + x_k + a = 0$, $\ell \models C$ means that C is a clause of the form $x_1^{\epsilon_1} \vee \cdots \vee x_k^{\epsilon_k}$, where $\sum_{i=1}^{k} \epsilon \equiv a \,(\mathrm{mod} 2)$.

Lemma 5.5.2 ([BSI99]). *Let ℓ be a linear equation $x_1 + \cdots + x_k + a = 0$ in $GF(2)$, for $a \in \{0,1\}$, and let C be a clause over $x_1, \ldots, x_k$ such that $\ell \models C$. Then $\mathcal{P}_F(\ell)(y_1, \ldots, y_k)$ is $(1, k+1)$-reducible to $\mathcal{Q}(C)$.*

Proof. Before beginning the proof, note that the same result holds for the (usual) Fourier representation $(-1)^{1-a} \cdot \prod_{i=1}^k y_i - 1$, representing the polynomial equation $(-1)^{1-a} \cdot \prod_{i=1}^k y_i = 1$. The interest in using the balanced Fourier representation lies in improving the degree lower bound later by a factor of 2.

Define the function r, to go from the Fourier basis to the boolean basis, by $r(y_i) = \frac{1-y_i}{2}$, and similarly define the function s, to go from the boolean basis to the Fourier basis, by $s(x_i) = 1 - 2x_i$. For $1 \leq i \leq k$, let

$$x_i = r(y_i) = \frac{1 - y_i}{2}.$$

The computation

$$r(y_i)^2 - r(y_i) = \left(\frac{1-y_i}{2}\right)^2 - \left(\frac{1-y_i}{2}\right)$$
$$= \frac{1 - 2y_i + y_i^2}{4} - \frac{1-y_i}{2} = \frac{y_i^2 - 1}{4}$$

yields a degree 2 PC derivation of $r(y_i)^2 - r(y_i) = 0$ from $y_i^2 - 1 = 0$. Moreover, it is not difficult to see that

$$(\forall y_1, \ldots, y_k)[P_F(\ell)(y_1, \ldots, y_k) = 0 \wedge \bigwedge_{i=1}^k y_i^2 - 1 = 0 \rightarrow$$
$$q_C(r(y_1), \ldots, r(y_k)) = 0].$$

Since PC is implicationally complete, there is a derivation Π of the polynomial $q_C(r(y_1), \ldots, r(y_k))$ from $\mathcal{P}(y_1, \ldots, y_k)$. A derivation Π' of degree at most $k+1$ can be constructed from Π by applying the axioms $\{y_1^2 - 1, \ldots, y_k^2 - 1\}$ appropriately whenever the exponent of y_i in a term of a polynomial of Π is 2, thus obtaining the multilinearization of the polynomial. This yields a degree $r+1$ derivation of $q_C(\frac{1-y_1}{2}, \ldots, \frac{1-y_r}{2})$.

Given clause C, there is a unique linear equation ℓ_C over $GF(2)$ such that $\ell_C \models C$ – for instance, if C is the clause $\{x, \overline{y}, z\}$, representing $x \vee \neg y \vee z$, then ℓ_C is the equation $x + y + z = 0$, while if C is the clause $\{x, y, z\}$, then ℓ_C is the equation $x + y + z + 1 = 0$. For a formula in CNF expressed as a set $\mathcal{C}$ of clauses, the *linear closure* of $\mathcal{C}$, denoted $\mathcal{L}_\mathcal{C}$, is defined by

$$\mathcal{L}_\mathcal{C} = \{\ell_C : C \in \mathcal{C}\}.$$

The following theorem states that taking the linear closure of an unsatisfiable CNF formula does not reduce the degree of PC refutation. The proof follows easily from the previous Lemma 5.5.2 and the definition of linear closure, hence is left to the reader.

Theorem 5.5.16 ([BSI99]). *Let F be a field of characteristic greater than 2, and let $\mathcal{C}$ be an unsatisfiable k-CNF formula. Then*

$$deg(\mathcal{Q}(\mathcal{C})) \geq \max\{deg(\mathcal{P}_F(\mathcal{L}_{\mathcal{C}})), k+1\}.$$

We now turn to the requisite notion of *expansion* which yields lower bounds for the polynomial calculus. Recall that for $f \in \mathcal{B}_n$ a boolean function on variables $x_1, \ldots, x_n$, we say that f *depends* on x_{i_0} if there is a partial assignment $\rho : \{x_i : 1 \leq i \leq n, i \neq i_0\} \to \{0,1\}$ to all variables except for x_{i_0}, such that $f \upharpoonright_\rho(0) \neq f \upharpoonright_\rho(1)$; i.e., $f \upharpoonright_\rho$ takes on different values, depending on the truth value of x_{i_0}. Let $Vars(f)$ denote the set of variables, upon which f depends. If f is the linear function $\sum_{i \in S} x_i \bmod 2$ and f depends on variable x_{i_0}, then f *strongly depends* on x_{i_0}, in the sense that for *every* partial assignment $\rho : \{x_i : 1 \leq i \leq n, i \neq i_0\} \to \{0,1\}$, $f \upharpoonright_\rho(0) \neq f \upharpoonright_\rho(1)$. This observation is necessary for the development below.

Definition 5.5.12 (Boundary). *Let $\mathcal{F}$ denote a set of boolean functions. The boundary of $\mathcal{F}$, denoted $\partial\mathcal{F}$, is the set of variables x_i, such that there exists a unique $f \in \mathcal{F}$, which depends on x_i.*

Note that in the case of a set $\mathcal{L}$ of linear equations over $GF(2)$, a variable x depends on $\ell \in \mathcal{L}$ if and only if x appears in ℓ, and the boundary $\partial\mathcal{L}$ of $\mathcal{L}$ is just the set of variables which occur only once, i.e., in a single equation of $\mathcal{L}$.

Definition 5.5.13 (Expansion). *Let $\mathcal{F}$ denote a set of boolean functions, and let s denote the least size of an unsatisfiable subset of $\mathcal{F}$. For any $t \leq s$, define $e_t(\mathcal{F})$ to be*

$$\min\left\{|\partial\mathcal{G}| : \mathcal{G} \subseteq \mathcal{F}, \frac{t}{3} < |\mathcal{G}| \leq \frac{2t}{3}\right\}.$$

The expansion *$e(\mathcal{F})$ is defined by $e(\mathcal{F}) = e_s(\mathcal{F})$.*

This definition should recall the well-known 2-3 trick from Lemma 1.6.1.

The previous definition of expansion generalizes the usual graph-theoretic notion. Indeed, let $G = (V, E)$ be an undirected graph, whose edges $e = \{u, v\}$ are labeled by distinct propositional variables x_e, and whose vertices v are labeled by boolean functions f_v, where f_v is the linear equation $\sum_{v \in e \in E} x_e$. For $V' \subseteq V$ let $\mathcal{F}_{V'} = \{f_v : v \in V\}$, and define $E(V', V - V') = \{e \in E : (\exists u \in V')(\exists v \in V - V')[e = \{u, v\}]$, i.e., $E(V', V - V')$ is the set of "cut" edges V' to its complement. Then

$$\partial\mathcal{F}_{V'} = \{x_e : e \in E(V', V - V')\}.$$

Theorem 5.5.17 ([BSI99]). *Let L be a minimal unsatisfiable set of linear equations in the variables $x_1, \ldots, x_n$ over $GF(2)$, such that each $\ell \in L$ has width k. Suppose that $s = |L|$ and $6 \leq t \leq s$. If F is a field of characteristic greater than 2, then*

$$deg(\mathcal{P}_F(L)) \geq \max\left\{k, \frac{e_t(L)}{2} + \Theta(1)\right\}.$$

Proof. Since L is minimal unsatisfiable, each axiom must occur in a refutation of L, and so $\deg(\mathcal{P}_F(L)) \geq k$. Assume that $k < \frac{e(L)}{2}$. By Theorem 5.5.14, it suffices to prove that $w_G(L) \geq e(L)$. For each $\ell \in L$, define the measure $\mu(\ell) = \min\{|L'| : L' \subseteq L, L' \models \ell\}$. It is easy to see that the following hold.

1. For $\ell \in L$, $\mu(\ell) \leq 1$.
2. $\mu(1 = 0) = s$.
3. $\mu(\ell + \ell') \leq \mu(\ell) + \mu(\ell')$.

Item (3) holds, since if $L \models \ell$ and $L' \models \ell'$, then $L \cup L' \models \ell + \ell'$.

CLAIM. If $6 \leq t \leq s$, then in every Gaussian refutation of L, there must be a line ℓ with $\frac{t}{3} < \mu(\ell) \leq \frac{2t}{3}$.

Proof of Claim. Suppose that $\Pi = (\ell_1, \ldots, \ell_r)$ is a Gaussian refutation of L, and let $i_0 \in \{1, \ldots, r\}$ be the least i such that $\mu(\ell_i) > \frac{t}{3}$. We claim that $\mu(\ell_{i_0}) \leq \frac{2t}{3}$. Indeed, ℓ_{i_0} cannot be an axiom or equal to ℓ, since then $\mu(\ell_0)$ would be bounded by 1, hence ℓ_{i_0} is obtained by addition of earlier equations ℓ_{i_1}, ℓ_{i_2}, for $1 \leq i_1, i_2 < i_0$. By definition of i_0, it must be that $\mu(\ell_{i_1}) \leq \frac{t}{3}$ and $\mu(\ell_{i_2}) \leq \frac{t}{3}$, and so $\mu(\ell_{i_0}) \leq \frac{2t}{3}$. This establishes the claim.

Now, let $\widetilde{\ell}$ denote such a line, and let $\widetilde{L}$ be a minimal subset of L, such that $|\widetilde{L}| = \mu(\widetilde{\ell})$ and $\widetilde{L} \models \widetilde{\ell}$.

CLAIM. Every variable in $\partial\widetilde{L}$ occurs in $\widetilde{\ell}$.

Proof of Claim. Assume that there is a variable $x_i \in \partial\widetilde{L}$ which does not occur in $\widetilde{\ell}$. Since $x_i \in \partial\widetilde{L}$, let $\ell' \in \widetilde{L}$ be the unique equation in which x_i appears. Since $\widetilde{L}$ was chosen to be a minimal subset of L for which $\widetilde{L} \models \widetilde{\ell}$, it must be that $\widetilde{L} - \{\ell'\} \not\models \widetilde{\ell}$, and so there exists a truth assignment $\sigma \models \widetilde{L} - \{\ell'\}$, $\sigma \not\models \ell'$, and $\sigma \not\models \widetilde{\ell}$. If we define the truth assignment σ' by flipping σ at x_i, i.e.,

$$\sigma'(x_1, \ldots, x_n) = \sigma(x_1, \ldots, x_{i-1}, 1 - x_i, x_{i+1}, \ldots, x_n)$$

then $\sigma' \models \ell'$, but since x_i does not occur in $\widetilde{\ell}$, we have $\sigma' \not\models \widetilde{\ell}$, and so $\widetilde{L} \not\models \widetilde{\ell}$. This contradiction establishes the claim.

It follows that the minimum width of a Gaussian refutation of L must be at least $|\partial L'|$, for a set $L' \subset L$, whose size is between $t/3$ and $2t/3$, and hence $w_G(L) \geq e_t(L)$.

Taking $t = s$ in the previous theorem yields the following more succinct statement.

Theorem 5.5.18 ([BSI99]). *Let L be an unsatisfiable set of linear equations over $GF(2)$, each equation of width at most k. Let F be a finite field of characteristic p greater than 2. Then $deg(\mathcal{P}_F(L)) \geq \max\{k, \frac{e(L)}{2} + \Theta(1)\}$.*

As a corollary of the degree lower bound of PC refutations in terms of the expansion, we have a lower bound for PC refutations of Tseitin's odd-charged graph formulas, defined in Section 5.4.2.

Theorem 5.5.19 ([BSI99]). *Let $G = (V, E)$ be a finite k-regular, undirected, connected, odd-charged graph, and $T(G)$ be the set of* CNF *formulas representing the linear equations associated with the charge of the vertices of G. Let F be a field of characteristic greater than 2. Then the degree of any PC refutation of $\mathcal{Q}(T(G))$ is at least* $\max\{\frac{e(G)}{2}, k+1\}$.

Proof. Let $L = \{E(u) : u \in V\}$ consist of the charge equations for the vertices of G. Without loss of generality, we can assume that every $\ell \in L$ is of the form $x_{i_1} + \cdots x_{i_k} + a = 0$, where $a \in \{0, 1\}$. For each $\ell \in L$, let $P_F(\ell)$ be the balanced Fourier representation of ℓ and $\mathcal{P}(L) = \{P_F(\ell) : \ell \in L\} \cup \{y_1^2 - 1, \ldots, y_n^2 - 1\}$.

Represent each $\ell \in L$ having k variables $x_{i_1}, \ldots, x_{i_k}$ in conjunctive normal form, as a set of 2^{k-1} clauses $C_1(\ell), \ldots, C_{2^{k-1}}(\ell)$, and let $\mathcal{Q}(\ell)$ be the set of canonical polynomials $q_{C_1(\ell)}, \ldots, q_{C_{2^k-1}(\ell)}$ along with $x_{i_1}^2 - x_{i_1}, \ldots, x_{i_k}^2 - x_{i_k}$. Then $\mathcal{Q}(T(G)) = \cup_{\ell \in L} \mathcal{Q}(\ell)$.

By Lemma 5.5.2, $\mathcal{P}_F(L)$ is $(1, k+1)$-reducible to $\mathcal{Q}(T(G))$, and by Theorem 5.5.17, the minimal degree of a PC refutation of $\mathcal{P}_L(y_1, \ldots, y_n)$ is $\frac{e(L)}{2} = \frac{e(G)}{2}$. It follows by Lemma 5.5.1 that the minimal degree of a PC refutation of $\mathcal{Q}_G(x_1, \ldots, x_n)$ is $\max\{\frac{e(G)}{2}, k+1\}$.

From Equation (5.30) and the previous theorem, we obtain a linear lower bound on the degree of a polynomial calculus refutations of the Tseitin formulas for expander graphs, as constructed in Section 5.4.3. This linear lower bound was first obtained by Buss, Grigoriev, Impagliazzo and Pitassi [BGIP99] using somewhat different techniques involving Laurent relations.

Theorem 5.5.20 ([BGIP99, BSI99]). *Let F be a field of characteristic greater than 2. There exists a family of finite, k-regular, undirected graphs $G_n = (V_n, E_n)$, where $V_n = \{1, \ldots, n\}$ with the following property. If $T(G_n)$ designates the set of* CNF *formulas representing the linear equations associated with the charge of the vertices of G_n, then*

$$deg(T(G_n)) = \Omega(n).$$

5.5.6 Random CNF Formulas and the Polynomial Calculus

Definition 5.5.14 (Random formulas and linear equations). *Let $\mathcal{F} \in \mathcal{F}_\Delta^{k,n}$ denote[17] that $\mathcal{F}$ is a random k-*CNF *formula on n variables, involving $m = \Delta \cdot n$ clauses, constructed as follows. Choose at random $\Delta \cdot n$ many clauses from the collection of all $\binom{n}{k} \cdot 2^k$ clauses, with repetitions.*

[17] In [BSI99], the notation used is $\mathcal{F} \sim \mathcal{F}_k^{n,\Delta}$.

Let $L \in \mathcal{L}_\Delta^{k,n}$ denote that L is a random set of linear equations over $GF(2)$ in n variables, each equation having at most k variables, constructed as follows. Choose at random $\Delta \cdot n$ equations from the collection of all $2 \cdot \binom{n}{k}$ possible equations of the form $x_{i_1} + \cdots x_{i_k} + a = 0$, for $a \in \{0,1\}$, allowing repetitions.

Within the context of CNF formulas, the parameter Δ is called the *clause density.* Note that the probability distributions $\mathcal{F}_\Delta^{k,n}$ and $\mathcal{L}_\Delta^{k,n}$ are related – namely, given $L \in \mathcal{L}_\Delta^{k,n}$, we form $\mathcal{F} \in \mathcal{F}_\Delta^{k,n}$ by choosing, for each $\ell \in L$, one of the 2^{k-1} possible defining clauses for ℓ, with equal probability.

The following lemma in the case for $k = 3$ was announced without proof in [BSI99], and states, loosely speaking, that all small sets of linear equations are satisfiable. The proof in the case $k = 3$ was provided to the authors by R. Impagliazzo [Imp93] (personal communication). The extension below to arbitrary $k \geq 3$ is straightforward from Impagliazzo's proof.

Lemma 5.5.3. *For any $k \geq 3$ and $0 < \epsilon < 1/2$, there is a constant $c > 0$, independent of n, such that if $\Delta \leq n^{\frac{1}{2}-\epsilon}$ and $L \in \mathcal{L}_\Delta^{k,n}$, then almost surely every subset of at most $\frac{cn}{\Delta^{2/(k-2)}}$ equations of L is satisfiable; i.e.,*

$$\lim_{n\to\infty} \Pr_L[(\exists L' \subseteq L)(|L'| = s, 2 \leq s \leq \frac{cn}{\Delta^{2/(k-2)}}, \ L' \ \textit{satisfiable}) \,|\, L \in \mathcal{L}_\Delta^{k,n} = 0.$$

Proof. Fix $k \geq 3$, $0 < \epsilon < 1/2$, and $2 \leq s$. A set $L' \subseteq L$ of linear equations has no solution if and only if the corresponding set of vectors, obtained from L' by dropping the constants to the right of the equality sign, is linearly dependent. In the case of $GF(2)$, this means that each variable occurring in L' appears an even number of times. We now give an upper bound for the number of ways to construct a subset $L' \subseteq L$ of size s having this property, for L random in $\mathcal{L}_\Delta^{k,n}$.

The linear equations of L' are given by

$$v_{1,1} + v_{1,2} + \cdots + v_{1,k} = a_1$$

$$\vdots$$

$$v_{s,1} + v_{s,2} + \cdots + v_{s,k} = a_s$$

where $v_{i,j} \in \{x_1, \ldots, x_n\}$ and $a_i \in \{0,1\}$. There are $\frac{ks}{2}$ steps in the construction of L', where we choose variables to place in the first $\frac{ks}{2}$ positions from the top of a $k \times s$ grid, proceeding left to right, and for each variable so placed, placing a copy of the same variable in a random position in the grid. Thus in step ℓ, for $1 \leq \ell \leq \frac{ks}{2}$, we choose a variable from $\{x_1, \ldots, x_n\}$ to place in position $v_{i,j}$, where $i = \lfloor \frac{\ell-1}{k} \rfloor + 1$ and $j = ((\ell - 1) \bmod k) + 1$, then place a copy of the same variable in one of ks positions in the grid (this is actually overcounting). After $ks/2$ steps, this involves $(ks \cdot n)^{ks/2}$ choices, whereas there would be $\binom{n}{k}^s$ choices, if we did not have to respect the constraint that

L' is satisfiable. To complete L' to a random set $L \in \mathcal{L}_\Delta^{k,n}$ with $L' \subseteq L$, we must choose an additional $\Delta n - s$ equations having no constraint concerning satisfiability, then choose which lines of L will be occupied by lines from L', then finally choose constants in $\{0, 1\}$ for the right side of the equality side. Temporarily let

$$p_s = \Pr_L[(\exists L' \subseteq L)(|L'| = s,\ L' \text{ satisfiable}) \mid L \in \mathcal{L}_\Delta^{k,n}]$$

and let m denote Δn, the number of equations in L. Recalling that $(\frac{n}{k})^k \leq \binom{n}{k} \leq (\frac{en}{k})^k$, we have

$$\begin{aligned} p_s &\leq \frac{(ksn)^{ks/2} \cdot \binom{n}{k}^{m-s} \cdot \binom{m}{s} \cdot 2^s}{\binom{n}{k}^s \cdot \binom{n}{k}^{m-s} \cdot 2^s} \\ &\leq \frac{(ksn)^{ks/2} \cdot (\frac{em}{s})^s}{(\frac{n}{k})^{ks}} \\ &= \left(\frac{k^3 s}{n}\right)^{ks/2} \left(\frac{e^2 m^2}{s^2}\right)^{s/2} \\ &= \left(\frac{e^2 k^{3k} s^{k-2} \Delta^2}{n^{k-2}}\right)^{s/2}. \end{aligned}$$

Define

$$a(k, n, \Delta, s) = \left(\frac{e^2 k^{3k} s^{k-2} \Delta^2}{n^{k-2}}\right)^{s/2}.$$

The previous calculation shows that

$$\Pr_L[(\exists L' \subseteq L)(|L'| = s,\ L' \text{ satisfiable}) \mid L \in \mathcal{L}_\Delta^{k,n}] \leq a(n, k, \Delta, s).$$

Now since $\Delta \leq n^{\frac{1}{2}-\epsilon}$,

$$\begin{aligned} a(k, n, \Delta, s) &\leq \left(\frac{e^2 k^{3k} s^{k-3} \cdot s \cdot n^{1-2\epsilon}}{n^{k-2}}\right)^{s/2} \\ &= \left(e^2 k^{3k} \left(\frac{s}{n}\right)^{k-3} \left(\frac{s}{n^\epsilon}\right)\right)^{s/2} \cdot \left(\frac{1}{n^\epsilon}\right)^{s/2} \end{aligned}$$

so for $2 \leq s \leq \log_2 n$, it follows that

$$a(k, n, \Delta, s) = O\left(\frac{1}{n^\epsilon}\right). \tag{5.52}$$

For $k \geq 3$ fixed, define

$$c = c(k) = \left(\frac{1}{4 \cdot e^2 k^{3k}}\right)^{\frac{1}{k-2}}$$

and let S abbreviate $\frac{cn}{\Delta^{2/(k-2)}}$, i.e., $S = S(k, n, \Delta)$ is the previous expression depending on parameters k, n, Δ. Clearly, it follows from the definition of $a(k, n, \Delta, s)$ in (5.52) that for $\log_2 n \leq s \leq S(k, n, \Delta)$, we have

$$a(k, n, \Delta, s) \leq \frac{1}{4^{s/2}} = \frac{1}{2^s}. \tag{5.53}$$

From (5.52) and (5.53), it follows that for $\Delta \leq n^{\frac{1}{2}-\epsilon}$,

$$\sum_{2 \leq s \leq S(k,n,\Delta)} a(k, n, \Delta, s) = O\left(\frac{\log_2 n}{n^\epsilon}\right) + \sum_{s=\log_2 n}^{S(k,n,\Delta)} \frac{1}{2^s}$$
$$= O\left(\frac{\log_2 n}{n^\epsilon}\right) + \frac{2}{n}.$$

The lemma now follows.

Corollary 5.5.6 ([BSI99]). *For any $0 < \epsilon < 1/2$, there is a constant $c > 0$, independent of n, such that if $\Delta \leq n^{\frac{1}{2}-\epsilon}$ and $L \in \mathcal{L}_\Delta^{3,n}$, then almost surely every subset of at most $\frac{cn}{\Delta^2}$ equations of L is satisfiable.*

The following lemma states, roughly speaking, that medium size random sets of linear equations are expanding. The proof of the following is adapted from [BSW01].

Lemma 5.5.4 ([BKPS]). *For every $0 < \epsilon < \frac{1}{2}$, if $\Delta = n^{\frac{1}{2}-\epsilon}$, $t = \frac{3 \cdot \epsilon n}{(80\Delta)^{2/(1-\epsilon)}}$, and $L \in \mathcal{L}_\Delta^{3,n}$, then with high probability $e_t(L) \geq t/3$; i.e.,*

$$\lim_{n \to \infty} \Pr\left[e_t(L) \geq \frac{\epsilon n}{(80\Delta)^{2/(1-\epsilon)}} \mid L \in \mathcal{L}_\Delta^{3,n}\right] = 1.$$

Proof. Let $L \in \mathcal{L}_\Delta^{3,n}$. For any subset $L' \subseteq L$ of equations, let $V(L')$ be the set of vertices covered by L'; i.e., $V(L') = \{x \in V : (\exists \ell \in L')(x \in \ell)\}$. The *expansion constant* of L', denoted by $e_{L'}$, is defined by

$$e_{L'} = \frac{2|V(L')| - 3|L'|}{|L'|}. \tag{5.54}$$

CLAIM. $e_{L'} \leq \frac{|\partial L'|}{|L'|}$.

Proof of Claim. Clearly we must have that

$$|V(L')| \leq |\partial L'| + \frac{3|L'| - |\partial L'|}{2}$$

since the set $V(L')$ of vertices covered by L' consists of boundary vertices $\partial L'$ together with non-boundary vertices, where each of the latter belongs to at least 2 equations. This establishes the claim.

It follows that if $\epsilon \le e_{L'}$, then $\epsilon \cdot |L'| \le |\partial L'|$. To show that almost surely $e_t(L) \ge \epsilon n(80\Delta)^{-2/(1-\epsilon)}$, it suffices to prove that

$$(\forall L' \subseteq L)\left[\frac{n}{(80\Delta)^{\frac{2}{1-\epsilon}}} < |L'| \le \frac{2n}{(80\Delta)^{\frac{2}{1-\epsilon}}} \to e_{L'} \ge \epsilon\right]. \tag{5.55}$$

To this end, let A_i designate the event that a random subset $L' \subseteq L \in \mathcal{L}_{\Delta}^{3,n}$ of size i has expansion constant $e_{L'}$ less than ϵ; i.e., that

$$|V(L')| < \frac{3+\epsilon}{2} \cdot |L'|.$$

Let $\lambda = \frac{2+\epsilon}{2}$ and $k = n(80\Delta)^{-2/(1-\epsilon)} = 80^{-2/(1-\epsilon)} \cdot n^{\epsilon/(1-\epsilon)}$. We want to establish an upper bound for

$$\Pr[A_i] = \Pr[|V(L')| < \lambda|L'| : |L'| = i, L' \subseteq L, L \in \mathcal{L}_{\Delta}^{3,n}]$$

for all sizes $i \in \{k+1, \ldots, 2k\}$. There are $\binom{\Delta n}{i}$ possible choices of sets L' with $L' \subseteq L$, $|L'| = i$, and there are $\binom{n}{\lambda i}$ possible choices of sets V' of vertices satisfying $|V'| < \lambda|L'|$. Note that for $a \le b \le c$, we have

$$\frac{\binom{b}{a}}{\binom{c}{a}} = \frac{b(b-1)\cdots(b-a+1)}{c(c-1)\cdots(c-a+1)} = \left(\frac{b}{c}\right)^a$$

since $(b-i)/(c-i) \le b/c$ holds when $b \le c$. Since $\lambda i \le n$, we have that the probability that a single equation is contained in such a small vertex set V' is

$$\frac{\binom{\lambda i}{3}}{\binom{n}{3}} \le \frac{(\frac{\lambda i}{3})^3}{(\frac{n}{3})^3} = (\frac{\lambda i}{n})^3.$$

The probability that this happens i times independently is thus bounded by $(\frac{\lambda i}{n})^{3i}$. Thus

$$\begin{aligned}
\Pr[A_i] &\le \binom{\Delta n}{i} \cdot \binom{n}{\lambda i} \cdot \left(\frac{\lambda i}{n}\right)^{3i} \\
&\le \left(\frac{e\Delta n}{i}\right)^i \left(\frac{en}{\lambda i}\right)^{\lambda i} \left(\frac{\lambda i}{n}\right)^{3i} \\
&\le \left(e^{\lambda+1}\lambda^{3-\lambda}\Delta\left(\frac{i}{n}\right)^{2-\lambda}\right)^i.
\end{aligned}$$

For $0 < \epsilon < \frac{1}{2}$, we have $\frac{3}{2} < \lambda < \frac{7}{4}$ and so $20.22 < e^{\lambda+4}\lambda^{3-\lambda} < 36.22$. Since $\lambda = \frac{3+\epsilon}{2}$, $2 - \lambda = \frac{1-\epsilon}{2}$, and recalling that $\Delta = n^{\frac{1}{2}-\epsilon}$, we have

$$\begin{aligned}
\Pr[A_i] &\le \left(40 \cdot n^{\frac{1}{2}-\epsilon} \cdot \left(\frac{i}{n}\right)^{\frac{1-\epsilon}{2}}\right)^i \\
&\le \left(40 \cdot n^{-\frac{\epsilon}{2}} \cdot i^{\frac{1-\epsilon}{2}}\right)^i
\end{aligned}$$

and so $\Pr[\cup_{i=k+1}^{2k} A_i]$ is bounded by a sum of k terms, each of which is bounded by $\left(40 \cdot n^{-\frac{\epsilon}{2}} \cdot (2k)^{\frac{1-\epsilon}{2}}\right)^{2k}$. Thus

$$\begin{aligned}\Pr[\cup_{i=k+1}^{2k} A_i &\leq k\left(40 \cdot n^{-\frac{\epsilon}{2}} \cdot (2k)^{\frac{1-\epsilon}{2}}\right)^{2k} \\ &\leq k\left(40 \cdot n^{-\frac{\epsilon}{2}} \cdot (2 \cdot 80^{-2/(1-\epsilon)} \cdot n^{\epsilon/(1-\epsilon)})^{\frac{1-\epsilon}{2}}\right)^{2k} \\ &\leq k\left(\frac{40 \cdot 2^{\frac{1-\epsilon}{2}}}{80}\right)^{2k} \leq k\left(\frac{\sqrt{2}}{2}\right)^{2k} = \frac{k}{2^k}.\end{aligned}$$

It follows that as n tends to infinity, $\Pr[\bigcup_{i=k+1}^{2k} A_i]$ tends to 0, thus concluding the proof of the theorem.

The previous lemma can be appropriately generalized for k-CNF formulas, by replacing $\Delta^{2/(1-\epsilon)}$ by $\Delta^{2/(k-1-\epsilon)}$ (see [BKPS] or Exercise 5.10.17).

From the work of this section, we immediately deduce the following.

Theorem 5.5.21 ([BSI99]). *Let F be a field of characteristic greater than 2, let $0 < \epsilon < 1/2$ and $\Delta = n^{\frac{1}{2}-\epsilon}$. If $L \in \mathcal{L}_{\Delta}^{3,n}$ then almost surely deg($\mathcal{P}_F(L)$) is*

$$\Omega(\frac{n}{\Delta^{2/(1-\epsilon)}}).$$

Proof. Let s be the minimum size of an unsatisfiable subset $L' \subseteq L$, where $L \in \mathcal{L}_{\Delta}^{3,n}$. By Lemma 5.5.3, almost surely s is not small, i.e., $s \geq \frac{cn}{\Delta^2}$. Define $t = \frac{3 \cdot \epsilon n}{(80\Delta)^{2/(1-\epsilon)}}$. Clearly $t \leq s$ and so by Lemma 5.5.4, almost surely $e_t(L) \geq \frac{\epsilon n}{(80\Delta)^{2/(1-\epsilon)}}$. By Theorem 5.5.17,

$$\deg(\mathcal{P}_F(L)) \geq \max\left\{k, \frac{e_t(L)}{2} + \Theta(1)\right\}$$

so

$$\deg(\mathcal{P}_F(L)) \geq \max\left\{3, \frac{\epsilon n}{2 \cdot (80\Delta)^{2/(1-\epsilon)}} + \Theta(1)\right\}$$

from which the result follows.

Theorem 5.5.22 ([BSI99]). *Let F be a field of characteristic greater than 2. For any $0 < \epsilon < 1/2$, let $\Delta = n^{\frac{1}{2}-\epsilon}$ and $\mathcal{F} \in \mathcal{F}_{\Delta}^{3,n}$. Then almost surely deg($\mathcal{F}$) is*

$$\Omega\left(\frac{n}{\Delta^{2/(1-\epsilon)}}\right).$$

Proof. Recall that we can obtain $\mathcal{F} \in \mathcal{F}_{\Delta}^{3,n}$ by first choosing $L \in \mathcal{L}_{\Delta}^{3,n}$, and then for each $\ell \in L$, choosing one of the 4 possible defining clauses C for ℓ, with equal probability. Let $\mathcal{Q} = \{q_C : C \in \mathcal{F}\} \cup \{x_1^2 - x_1, \ldots, x_n^2 - x_n\}$, and note that for each $C \in \mathcal{F}$, there is $\ell \in L$ for which $\ell \models C$.

By the previous theorem, almost surely $\deg(\mathcal{P}_F(L)) = \Omega(\frac{n}{\Delta^{2/(1-\epsilon)}})$. From Lemma 5.5.2, it follows that $\mathcal{P}_F(L)$ is $(1,4)$-reducible to $\mathcal{Q}$, and so by Lemma 5.5.1, the minimal degree of a PC refutation of $\mathcal{Q}(x_1,\ldots,x_n)$ is also $\Omega(\frac{n}{\Delta^{2/(1-\epsilon)}})$. This completes the proof of the theorem.

Generalizing to k-CNF formulas for arbitrary $k \geq 3$, we have the following (see Exercise 5.10.17).

Theorem 5.5.23 ([BSI99]). *Let F be a field of characteristic greater than 2, let $0 < \epsilon < 1/2$ and $\Delta = n^{\frac{1}{2}-\epsilon}$. For $L \in \mathcal{L}_\Delta^{k,n}$ and $\mathcal{F} \in \mathcal{F}_\Delta^{k,n}$, almost surely deg($\mathcal{P}_F(L)$) is $\Omega(\frac{n}{\Delta^{2/(k-1-\epsilon)}})$ and deg($\mathcal{F}$) is $\Omega(\frac{n}{\Delta^{2/(k-1-\epsilon)}})$.*

In closing this section, we mention the following result.

Theorem 5.5.24 ([Raz98]). *If F is any field, then*

$$deg(\neg\mathrm{PHP}_n^{n+1}) \geq \lceil n/2 \rceil + 1.$$

Impagliazzo, Pudlák and Sgall [IPS] have shown that this bound is tight for any field of sufficiently large characteristic.

5.6 Cutting Planes CP

Originating in operations research, the cutting plane proof system, CP, (often just called cutting planes) is a sound and complete refutation system for CNF formulas, based on showing the non-existence of solutions for a family of linear inequalities. In this section, we give the representation of CNF formulas in CP, explain the rules of inference, prove Chvátal's strong completeness theorem and provide some examples of polynomial-size CP proofs of hard combinatorial properties related to the pigeonhole principle. We then prove some lower bounds for CP proofs based on P. Pudlák's interpolation theorem, a result allowing one to infer lower bounds for CP proof size from monotonic real circuit size lower bounds.

All inequalities in CP are of the form $\sum a_i \cdot x_i \geq k$, where $a_i, k \in \mathbf{Z}$. If C is a clause, and the variables x_i, $i \in P$, occur positively in C, and variables x_i, $i \in N$, occur negatively in C, then C is represented by the linear inequality

$$\sum_{i\in P} x_i - \sum_{i\in N} x_i \geq 1 - |N|.$$

A CNF formula is represented by the family of linear inequalities corresponding to its clauses. For example, $(x \vee \overline{y} \vee z) \wedge (\overline{x} \vee z \vee u \vee \overline{v})$ is represented by

$$x + -y + z \geq 0$$
$$-x + z + u - v \geq -1.$$

The *axioms* of CP are $x_i \geq 0$, $-x_i \geq -1$. The *rules of inference* of CP are

- addition $\dfrac{\sum a_i \cdot x_i \geq A \qquad \sum b_i \cdot x_i \geq B}{\sum (a_i + b_i) \cdot x_i \geq A + B}$

- division $\dfrac{\sum (c \cdot a_i) \cdot x_i \geq A}{\sum a_i \cdot x_i \geq \lceil \frac{A}{c} \rceil}$ where $c > 1$ is an arbitrary integer,

- multiplication $\dfrac{\sum a_i \cdot x_i \geq A}{\sum (c \cdot a_i) \cdot x_i \geq c \cdot A}$ where $c > 1$ is an arbitrary integer.

A derivation D of the inequality I from inequalities $I_1, \ldots, I_m$ is a sequence $D = (D_0, \ldots, D_n)$ such that $I = D_n$, and for all $i \leq n$ either D_i is an axiom, or one of $I_i, \ldots, I_m$ or inferred from D_j, D_k for $j, k < i$ by means of a rule of inference. The *length* or *number of lines* of derivation $D_0, \ldots, D_n$ is $n + 1$. We denote the existence of a derivation of I from $I_1, \ldots, I_m$ by $I_1, \ldots, I_m \vdash_{CP} I$, or just $I_1, \ldots, I_m \vdash I$, when context of CP is clearly intended. By $I_1, \ldots, I_m \vdash I$, we mean that every 0/1-assignment to the boolean variables which satisfies the inequalities $I_1, \ldots, I_m$ also satisfies I.

A CP *refutation* of $I_1, \ldots, I_m$ is a derivation of $0 \geq 1$ from $I_1, \ldots, I_m$. Its *length* is the number of lines in the derivation, while its *size* is the total number of symbols appearing in the derivation; i.e., the sum of $\sum |a_i| + |A|$, over all inequalities $\sum a_i \cdot x_i \geq A$ occurring in the refutation, where $|A|$ indicates the length of the binary representation of A. The system CP^* is identical to CP, except that all integers are written in unary, rather than binary. The *width* of a refutation is the maximum number of variables occurring in any single inequality appearing in the derivation.

As in the case of resolution, by abuse of terminology, we say that a disjunctive normal form formula has a CP proof if its negation has a CP refutation.

To fix ideas, let's look at a CP proof of an instance of the relational version of the pigeonhole principle PHP_n^{n+1} as given by

$$\bigwedge_{0 \leq i \leq n} \bigvee_{0 \leq j < n} p_{i,j} \rightarrow \bigvee_{0 \leq i < i' \leq n} \bigvee_{0 \leq j < n} (p_{i,j} \wedge p_{i',j}).$$

Consider PHP_2^3, whose negation is represented in CP by

$$\begin{array}{lll} p_{0,0} + p_{0,1} \geq 1 & p_{1,0} + p_{1,1} \geq 1 & p_{2,0} + p_{2,1} \geq 1 \\ -p_{0,0} - p_{1,0} \geq -1 & -p_{0,0} - p_{2,0} \geq -1 & -p_{1,0} - p_{2,0} \geq -1 \\ -p_{0,1} - p_{1,1} \geq -1 & -p_{0,1} - p_{2,1} \geq -1 & -p_{1,1} - p_{2,1} \geq -1 \end{array}$$

When pictured as a 2×3 boolean array, where $p_{i,j}$ corresponds to the entry in column i and row j, the first 3 inequalities assert that each pigeon finds a hole, while the last 6 inequalities assert that no hole contains two pigeons. By addition of $-p_{0,0} - p_{1,0} \geq -1$, $-p_{0,0} - p_{2,0} \geq -1$ and $-p_{1,0} - p_{2,0} \geq -1$, we obtain $-2p_{0,0} - 2p_{1,0} - 2p_{2,0} \geq -3$. Dividing by 2 and rounding up leads to $-p_{0,0} - p_{1,0} - p_{2,0} \geq -1$. In a similar fashion, we derive $-p_{0,1} - p_{1,1} - p_{2,1} \geq -1$. Adding both yields $\sum_{i,j} -p_{i,j} \geq -2$. On the other hand, from the addition of $p_{0,0} + p_{0,1} \geq 1$, $p_{1,0} + p_{1,1} \geq 1$, and $p_{2,0} + p_{2,1} \geq 1$, it follows that $\sum_{i,j} p_{i,j} \geq 3$. Adding the latter with $\sum_{i,j} -p_{i,j} \geq -2$ yields the desired contradiction $0 \geq 1$.

In later examples, we'll work more informally. For example, the previous derivation more informally explained gives that the sum of all the rows in the boolean matrix is at most 2, while the sum of all the columns is at least 3, yielding the contradiction that $2 \geq 3$.

5.6.1 Completeness of CP

We've seen that resolution is sound and complete for CNF formulas; i.e., *sound*, in the sense that given any CNF formula ϕ which has a resolution refutation, ϕ is not satisfiable, and *complete*, in the sense that given any unsatisfiable CNF formula ϕ, there is a resolution refutation of ϕ. W. Cook et al. [CCT87] observed that CP easily simulates resolution, so that completeness of the cutting planes refutation system follows.

Proposition 5.6.1. *CP is sound for CNF formulas.*

Proof. (Outline) Suppose that CNF formula ϕ has a CP refutation. If ϕ is satisfiable, then let σ be a satisfying truth assignment of ϕ. Instantiate each inequality in the refutation of ϕ by assigning the boolean variables their value under σ. Show by induction on the length of the refutation that each instantiated inequality in the refutation sequence holds. But then we cannot have $0 \geq 1$ as the last element of the sequence.

Theorem 5.6.1. *CP is complete for CNF formulas.*

Proof. (Outline) It is simple to simulate resolution in CP. For example, to simulate the inference

$$\frac{(x \vee y \vee z) \wedge (\overline{x} \vee y \vee u)}{y \vee z \vee u}$$

we have

1. $x + y + z \geq 1$
2. $1 - x + y + u \geq 1$
3. $1 + 2y + z + u \geq 2$ by addition of 1,2
4. $2y + z + u \geq 1$ simplification
5. $z \geq 0$
6. $u \geq 0$
7. $2y + 2z + 2u \geq 1$ by addition of 4,5,6
8. $y + z + u \geq 1$ by division by 2

Since resolution is complete for CNF formulas, so is CP.

The soundness and completeness of CP is due to [Hoo88, CCT87]. The previous proof clearly shows that if ϕ has a resolution refutation of N lines, then ϕ has a CP refutation of $O(N)$ lines, an observation due to W. Cook et al. [CCT87]. The following, deeper completeness result is due to Chvátal.

Lemma 5.6.1 ([CS88]). *For $1 \leq j \leq m$, let inequality I_j be $\sum_{i=1}^{n} a_{i,j} x_i \geq b_j$. Assume that*

$$I_1, \ldots, I_m \vdash_{CP} \sum_{i=1}^{n} c_i x_i \geq c. \tag{5.56}$$

and that for no boolean values of $x_1, \ldots, x_n$ satisfying $I_1, \ldots, I_m$ is it the case that equality holds; i.e.,

$$I_1, \ldots, I_m \models \sum_{i=1}^{n} c_i x_i \geq c + 1. \tag{5.57}$$

Then

$$I_1, \ldots, I_m \vdash_{CP} \sum_{i=1}^{n} c_i x_i \geq c + 1. \tag{5.58}$$

Proof. We prove the following claim.

CLAIM. For $1 \leq m \leq n$ and A, B disjoint subsets of $\{1, \ldots, n\}$ satisfying $|A \cup B| = m$,

$$\sum_{i=1}^{n} c_i x_i + \sum_{i \in A} x_i - \sum_{i \in B} x_i + |B| \geq c + 1$$

has a CP derivation from $I_1, \ldots, I_m$.

Proof of Claim. By reverse induction from n to 0.

BASE CASE. $m = n$

Case 1. For each $1 \leq j \leq m$, $\sum_{i \in B} a_{i,j} \geq b_j$.

It then follows that $\sum_{i \in B} c_i > c$, for otherwise, by taking $x_i = 1$ (0) if $i \in B$ ($i \notin B$), from the case assumption, we have boolean values which satisfy $I_1, \ldots, I_m$ as well as $\sum_{i=1}^{n} c_i x_i = c$. This contradicts (5.57). Letting $|x|$ here denote the absolute value of x, define $M = \max\{|c_i| : 1 \leq i \leq n\}$. Since $1 \geq x_i$, by multiplication with $(M + c_i)$, we have

$$M + c_i \geq (M + c_i) \cdot x_i$$

for each $i \in B$. By choice of M, $M \geq c_i$, so we have $M x_i \geq c_i x_i$, for $i \in A$. Adding these, and noting that $A \cup B = \{1, \ldots, n\}$, we have

$$M \cdot |A| + \sum_{i \in A} c_i \geq M \sum_{i \in A} x_i - M \sum_{i \in B} x_i + \sum_{i=1}^{n} c_i x_i.$$

Thus

$$M \cdot |B| + \sum_{i \in B} c_i + M \sum_{i \in A} x_i - M \sum_{i \in B} x_i \geq \sum_{i=1}^{n} c_i x_i.$$

By assumption (5.56), for boolean values of the x_i satisfying $I_1, \ldots, I_m$, we have $\sum_{i=1}^{n} c_i x_i \geq c$, hence $(M+1)\sum_{i=1}^{n} c_i x_i \geq (M+1)c$. By addition we obtain

$$M \cdot |B| + M \sum_{i \in A} x_i - M \sum_{i \in B} x_i + M \sum_{i=1}^{n} c_i x_i \geq Mc + c - \sum_{i \in B} c_i.$$

Dividing by M we have

$$\begin{aligned} |B| + \sum_{i \in A} x_i - \sum_{i \in B} x_i + \sum_{i=1}^{n} c_i x_i &\geq c + \lceil c - \sum_{i \in B} c_i \rceil \\ &\geq c + 1. \end{aligned}$$

Case 2. $b_{j_0} > \sum_{i=1}^{n} a_{i,j_0}$, for some $1 \leq j_0 \leq m$.

Set $M = \max\{|a_{i,j_0}| : 1 \leq i \leq n\}$. Since $1 \geq x_i$, $M + a_{i,j_0} \geq (M + a_{i,j_0}) \cdot x_i$, for $i \in B$. As well, $x_i \geq 0$, so $(M - a_{i,j_0}) \cdot x_i \geq 0$ for $i \in A$. Adding these yields

$$M \cdot |B| + M \sum_{i \in A} x_i - M \sum_{i \in B} x_i + \sum_{i \in B} a_{i,j_0} \geq \sum_{i=1}^{n} a_{i,j_0} x_i.$$

Now by the assumption I_{j_0}, we have $\sum_{i=1}^{n} a_{i,j_0} x_i \geq b_{j_0}$, so

$$M \cdot |B| + M \sum_{i \in A} x_i - M \sum_{i \in B} x_i + \sum_{i \in B} a_{i,j_0} \geq b_{j_0}.$$

Dividing by M yields

$$\begin{aligned} |B| + \sum_{i \in A} x_i - \sum_{i \in B} x_i &\geq \lceil b_{j_0} - \sum_{i \in B} a_{i,j_0} \rceil \\ &\geq 1. \end{aligned}$$

Since $\sum_{i=1}^{n} c_i x_i \geq c$, by addition

$$|B| + \sum_{i \in A} x_i - \sum_{i \in B} x_i + \sum_{i=1}^{n} c_i x_i \geq c + 1.$$

This concludes the base case of the claim.

Inductive Case. Assume the assertion of the claim holds for $m = k$. Suppose that A, B are disjoint and $|A \cup B| = k - 1$. Let $\ell \in \{1, \ldots, n\} - (A \cup B)$. Since $0 \leq x_\ell \leq 1$, by the induction hypothesis, we can deduce

$$x_\ell + \sum_{i=1}^{n} c_i x_i + \sum_{i \in A} x_i - \sum_{i \in B} x_i + |B| \geq c+1$$

and

$$-x_\ell + \sum_{i=1}^{n} c_i x_i + \sum_{i \in A} x_i - \sum_{i \in B} x_i + |B| \geq c+1-1.$$

By addition,

$$2\sum_{i=1}^{n} c_i x_i + 2\sum_{i \in A} x_i - 2\sum_{i \in B} x_i + 2|B| \geq 2(c+1) - 1$$

so by division,

$$\begin{aligned} \sum_{i=1}^{n} c_i x_i + \sum_{i \in A} x_i - \sum_{i \in B} x_i + |B| &\geq c+1+\lceil\frac{1}{2}\rceil \\ &\geq c+1. \end{aligned}$$

This concludes the proof of the claim. Taking $A = \emptyset = B$, it follows that

$$I_1, \ldots, I_m \vdash_{CP} \sum_{i=1}^{n} c_i x_i \geq c+1.$$

Theorem 5.6.2 ([CS88]). CP *is implicationally complete, i.e.,* $I_1, \ldots, I_m \models I \Rightarrow I_1, \ldots, I_m \vdash_{CP} I$.

Proof. Suppose that I_j is of the form $\sum_{i=1}^{n} a_{i,j} x_i \geq b_j$, and that I is of the form $\sum_{i=1}^{n} c_i x_i \geq c$. Let c_0 be the sum of those coefficients among $c_1, \ldots, c_n$ which are negative. Clearly $\sum_{i=1}^{n} a_i x_i \geq c_0$ is derivable. By $c - c_0$ many applications of the previous lemma, $\sum_{i=1}^{n} c_i x_i \geq c$ is derivable from $I_1, \ldots, I_m$.

Corollary 5.6.1. CP *is complete for* CNF *formulas.*

Proof. If ϕ is an unsatisfiable CNF formula represented by the linear inequalities $I_1, \ldots, I_m$, then $I_1, \ldots, I_m \models 0 \geq 1$, hence $I_1, \ldots, I_m \vdash_{CP} 0 \geq 1$.

5.6.2 Cutting Planes and the PHP

It's not hard to generalize the example proof of PHP_2^3 to show that PHP_n^{n+1} has polynomial-size CP proofs. This was first proved by W. Cook et al. in [CCT87], and in view of A. Haken's lower bound for resolution proofs of the pigeonhole principle, has the corollary that CP is strictly stronger than resolution. Known CP proofs of PHP_n^{n+1} are certainly not tree-like, so it remains an interesting open problem whether tree-like CP is stronger than tree-like resolution. In a later section, we'll give polynomial-size resolution proofs of a form of *st*-connectivity, and prove a superpolynomial lower bound for tree-like CP proofs of *st*-connectivity. In this respect, it would be interesting to resolve the question whether resolution can polynomially simulate tree-like CP.

In this section, we give polynomial size cutting plane proofs of a generalization of the pigeonhole principle, which we designate as Degen's principle, first considered in propositional logic by P. Clote and A. Setzer [CS98].

In [Deg95], W. Degen gave a natural generalization of the pigeonhole principle, which states that for positive integers m, k if f is a function mapping $\{0,\ldots,m\cdot k\}$ into $\{0,\ldots,k-1\}$ then there is $j<k$ for which $f^{-1}(j)$ has size greater than m. For ease of notation, non-negative integers will be considered as von Neumann ordinals, so that $n=\{0,\ldots,n-1\}$, and $[m]^n$ denotes the set of size n subsets of $\{0,\ldots,m-1\}$. Formulated in propositional logic, Degen's generalization is given by a family $\{D_{m,k} : m,k \in \mathbf{N}-\{0\}\}$ where $D_{m,k}$ is

$$\bigwedge_{0\leq i\leq m\cdot k}\ \bigvee_{0\leq j<k} p_{i,j} \rightarrow \bigvee_{0\leq j<k}\ \bigvee_{I\in[m\cdot k+1]^{m+1}}\ \bigwedge_{i\in I} p_{i,j}.$$

By $E_{m,k}$ we denote the CP inequalities corresponding to the CNF formula $\neg D_{m,k}$. Thus $E_{m,k}$ is

$$\sum_{j=0}^{k-1} p_{i,j} \geq 1$$

for $0\leq i\leq m\cdot k$, together with

$$-p_{i_1,j}-p_{i_2,j}-\cdots-p_{i_{m+1},j}\geq -m$$

for $0\leq j<k$ and $0\leq i_1<i_2<\cdots<i_{m+1}\leq m\cdot k$.

Theorem 5.6.3. *There are $O(k^5)$ size* CP *refutations of $E_{2,k}$.*

Proof. By assumption from $E_{2,k}$, for all $0\leq i_1<i_2<i_3\leq 2k$ and all $0\leq r<k$,

$$2\geq p_{i_1,r}+p_{i_2,r}+p_{i_3,r}.$$

Claim 1. For all $0\leq i_1<i_2<i_3<i_4\leq 2k$ and all $0\leq r<k$,

$$2\geq p_{i_1,r}+p_{i_2,r}+p_{i_3,r}+p_{i_4,r}.$$

Proof of Claim 1. Fix i_1,i_2,i_3,i_4 and r, and temporarily, set $a=p_{i_1,r}$, $b=p_{i_2,r}$, $c=p_{i_3,r}$, $d=p_{i_4,r}$. By assumption from $E_{2,k}$, we have

$2\geq a+b+c$
$2\geq b+c+d$
$2\geq a+b+d$
$2\geq a+c+d$

and so by addition

$$8 \geq 3a + 3b + 3c + 3d$$

and hence by division by 3

$$2 = \lfloor 8/3 \rfloor \geq a + b + c + d.$$

For later generalization, note that the pattern of the previous inequalities is of the following form:

$$\begin{array}{c} + + + - \\ - + + + \\ + - + + \\ + + - + \end{array}$$

where $+$ ($-$) indicates presence (absence) of the corresponding element (i.e., in the first row, there is a, b, c but no d present). In this manner, with $O(k^5)$ (i.e., order $k \cdot \binom{2k+1}{4}$) many proof lines we can show that

$$2 \geq p_{i_1,r} + \cdots + p_{i_4,r}$$

for all rows $0 \leq r < k$ and all 4-tuples $0 \leq i_1 < i_2 < i_3 < i_4 \leq 2 \cdot k$ from that row. In a similar manner, we could show by a proof of $O(k^{s+1})$ lines, that $2 \geq p_{i_1,r} + \cdots + p_{i_s,r}$, for all rows $0 \leq r < k$ and all distinct s-tuples $i_1, \ldots, i_s$. However, the overall proof would then be of $\sum_{i=5}^{2k+1} O(k^i)$ lines, hence of exponential size. For that reason, in the following claim, we consider sets $i_1, \ldots, i_s$ of a particular form. Define integers $x_1, \ldots, x_m$ to be *consecutive* if for all $1 \leq j < m$, $x_{j+1} = x_j + 1$.

Claim 2. Assume that $3 \leq s \leq 2k$ and for all $0 \leq i_1 < \cdots < i_s \leq 2k$ such that $i_2, \ldots, i_s$ are consecutive, and for all $0 \leq r < k$, it is the case that

$$2 \geq p_{i_1,r} + \cdots p_{i_s,r}.$$

Then for all $0 \leq i_1 < \cdots < i_{s+1} \leq 2k$ such that $i_2, \ldots, i_{s+1}$ are consecutive, and for all $0 \leq r < k$, it is the case that

$$2 \geq p_{i_1,r} + \cdots + p_{i_{s+1},r}.$$

Proof of Claim 2. Fix $0 \leq i_1 < \cdots < i_{s+1}$ and r. By assumption

$$\begin{aligned} 2 &\geq p_{i_1,r} + \cdots + p_{i_s,r} \\ 2 &\geq p_{i_2,r} + \cdots + p_{i_{s+1},r} \\ 2 &\geq p_{i_1,r} + p_{i_3,r} + \cdots + p_{i_{s+1},r} \\ 2 &\geq p_{i_1,r} + p_{i_2,r} + p_{i_{s+1},r} \end{aligned}$$

Note that the pattern in the previous inequalities is of the following form:

$$\begin{array}{c} + + + \cdots + - \\ - + + \cdots + + \\ + - + \cdots + + \\ + + - \cdots - + \end{array}$$

The first three inequalities hold by the assumption in the claim, and the fourth (which contains only 3 terms) holds by assumption of $E_{2,k}$. By addition, we have

$$8 \geq 3p_{i_1,r} + \cdots + 3p_{i_{s+1},r}$$

and hence by division by 3

$$2 = \lfloor 8/3 \rfloor \geq p_{i_1,r} + \cdots + p_{i_{s+1},r}.$$

By induction on s, using the base case $2 \geq p_{i_1,r} + p_{i_2,r} + p_{i_2+1,r}$ for all $0 \leq r < k$ and $0 \leq i_1 < i_2 \leq 2 \cdot k$ (given by $E_{2,k}$), and applying Claim 2 in the inductive case, it follows that for all $0 \leq r < k$,

$$2 \geq p_{0,r} + \cdots + p_{2k,r}.$$

Adding all k inequalities (one for each $0 \leq r < k$), we have

$$2k \geq \sum_{i=0}^{2k} \sum_{j=0}^{k-1} p_{i,j}.$$

However, by hypothesis $E_{2,k}$, for each fixed $0 \leq i \leq 2k$, $\sum_{j=0}^{k-1} p_{i,j} \geq 1$, and by addition of these $2k+1$ inequalities (one for each $0 \leq i \leq 2k$), we have

$$\sum_{i=0}^{2k} \sum_{j=0}^{k-1} p_{i,j} \geq 2k+1.$$

Thus we arrive at the contradiction $2k \geq 2k+1$. Rewriting the above proof in the required normal form $\sum a_{i,j} \cdot p_{i,j} \geq A$ we obtain a derivation of $0 \geq 1$ from $E_{2,k}$. Routine estimations show that the proof size is $O(k^5)$.

The previous construction easily generalizes.

Theorem 5.6.4. *Let $m \geq 2$ and $n = m \cdot k + 1$. Then there are $O(n^{m+3})$ size* CP *refutations of $E_{m,k}$, where the constant in the O-notation depends on m, and $O(n^{m+4})$ size* CP *refutations, where the constant is independent of n, m.*

Proof. We generalize the proof of the previous theorem.

Claim 3. Assume that $3 \leq s \leq mk$ and for all $0 \leq i_1 < \cdots < i_s \leq mk$ such that $i_m, \ldots, i_s$ are consecutive, and for all $0 \leq r < k$, it is the case that

$$m \geq p_{i_1,r} + \cdots + p_{i_s,r}.$$

Then for all $0 \leq i_1 < \cdots < i_{s+1} \leq mk$ such that $i_m, \ldots, i_{s+1}$ are consecutive, and for all $0 \leq r < k$, it is the case that

$$m \geq p_{i_1,r} + \cdots + p_{i_{s+1},r}.$$

Proof of Claim 3. Fix $i_1 < \cdots < i_{s+1}$ and r. We have the following $m+2$ inequalities:

$$\begin{aligned}
&m \geq p_{i_1,r} + \cdots + p_{i_s,r} \\
&m \geq p_{i_2,r} + \cdots + p_{i_{s+1},r} \\
&m \geq p_{i_1,r} + p_{i_3,r} + \cdots + p_{i_{s+1},r} \\
&m \geq p_{i_1,r} + p_{i_2,r} + p_{i_4,r} + \cdots + p_{i_{s+1},r} \\
&m \geq p_{i_1,r} + \cdots + p_{i_3,r} + p_{i_5,r} + \cdots + p_{i_{s+1},r} \\
&m \geq p_{i_1,r} + \cdots + p_{i_4,r} + p_{i_6,r} + \cdots + p_{i_{s+1},r} \\
&\vdots \\
&m \geq p_{i_1,r} + \cdots + p_{i_{m-1},r} + p_{i_{m+1},r} + \cdots + p_{i_{s+1},r} \\
&m \geq p_{i_1,r} + \cdots + p_{i_m,r} + p_{i_{s+1},r}
\end{aligned}$$

The pattern of terms in the $m+2$ inequalities above is of the form:

$$\begin{array}{ccccccccc}
+ & + & + & \cdots & + & + & + & + & - \\
- & + & + & \cdots & + & + & + & + & + \\
+ & - & + & \cdots & + & + & + & + & + \\
+ & + & - & \cdots & + & + & + & + & + \\
. & . & . & \cdots & . & . & . & . & . \\
. & . & . & \cdots & . & . & . & . & . \\
. & . & . & \cdots & . & . & . & . & . \\
+ & + & + & \cdots & + & - & + & + & + \\
+ & + & + & \cdots & + & + & - & - & +
\end{array}$$

Removal of any of the first $m-1$ summands in the term $p_{i_1,r} + \cdots + p_{i_{s+1},r}$ produces a term where $p_{i_m,r}, \ldots, p_{i_{s+1},r}$ are consecutive. This observation, with the assumption in the claim, justifies the first $m+1$ inequalities. The last inequality (which contains only $m+1$ terms) holds by assumption of $E_{m,k}$. By addition, we have

$$m \cdot (m+2) \geq (m+1) \cdot (p_{i_1,r} + \cdots + p_{i_{s+1},r})$$

and hence by division by $m+1$

$$m = \left\lfloor \frac{m(m+2)}{m+1} \right\rfloor \geq p_{i_1,r} + \cdots + p_{i_{s+1},r}.$$

Adding k inequalities $m \geq p_{0,r} + \cdots + p_{mk,r}$, we have

$$mk \geq \sum_{i=0}^{mk} \sum_{j=0}^{k-1} p_{i,j}.$$

Similarly adding the $mk+1$ inequalities $p_{i,0} + \cdots + p_{i,k-1} \geq 1$, we have

$$\sum_{i=0}^{mk} \sum_{j=0}^{k-1} p_{i,j} \geq mk + 1.$$

Finally, we have the desired contradiction $mk \geq mk + 1$. The size estimates are straightforward and left to the reader.

5.6.3 Polynomial Equivalence of CP$_2$ and CP

For an integer $q \geq 2$, the proof system CP_q is obtained from CP by restricting the division rule to division by q. The systems CP_q are quite strong, and will be shown to be p-equivalent to CP. To illustrate the idea of the proof, we present the following example of how CP_2 can simulate division by three.

Example 5.6.1. To simulate division by 3 applied to

$$9x + 12y \geq 11 \tag{5.59}$$

within CP_2, first write the coefficient of each variable with 3 as explicit factor. This gives

$$3(3x) + 3(4y) \geq 11. \tag{5.60}$$

The least power of 2 greater than 3 is 2^2. Using $x \geq 0$, $y \geq 0$ obtain $3x \geq 0$, $4y \geq 0$ which when added to (5.60) gives

$$2^2(3x) + 2^2(4y) \geq 11. \tag{5.61}$$

Two applications of division by 2 yields

$$3x + 4y \geq 2. \tag{5.62}$$

Adding (5.60) and (5.62) gives

$$2^2(3x) + 2^2(4y) \geq 13 \tag{5.63}$$

and two applications of division by 2 yields the desired inequality

$$3x + 4y \geq 3 \tag{5.64}$$

which one would obtain from (5.59) by division by 3.

Theorem 5.6.5 ([BC96]). *Let $q > 1$. Then CP_q p-simulates* CP.

Since CP trivially p-simulates CP_q, we have that CP and CP_q are p-equivalent systems, for any fixed $q > 1$.

Proof. Fix $q > 1$. We must show that an arbitrary instance of the division rule in a CP-proof can be simulated by a polynomial size CP_q proof. Without loss of generality, we assume all inequalities are of the form

$$a_1x_1 + a_2x_2 + \cdots + a_nx_n \geq b$$

where $a_1, \ldots, a_n$ and b are integers.

Suppose a cutting plane proof contains a division inference

$$\frac{c\alpha \geq M}{\alpha \geq \lceil \frac{M}{c} \rceil} \tag{5.65}$$

where $c > 1$. To prove that this can be efficiently simulated using division by q, we will describe a short CP_q proof of $\alpha \geq \lceil \frac{M}{c} \rceil$ from the assumption

$$c \cdot \alpha \geq M \tag{5.66}$$

The idea of proof is to define a sequence $s_0 \leq s_1 \leq s_2 \leq \cdots \leq \lceil \frac{M}{c} \rceil$ rapidly converging to $\lceil \frac{M}{c} \rceil$, such that from $\alpha \geq s_i$ and the assumption $c \cdot \alpha \geq M$, we obtain $\alpha \geq s_{i+1}$.

Choose p so that $q^{p-1} < c \leq q^p$. Without loss of generality, we can assume that $q^p/2 < c$, since if this does not hold, then we can find a suitable multiple mc of c such that $q^p/2 < m \cdot c \leq q^p$ and then multiply the hypothesis inequality (5.66) by m and use division by $m \cdot c$ in place of c.

The expression α is a linear combination $\sum_{i=1}^n a_ix_i$ with integer coefficients. Let s_0 equal the sum of the *negative* coefficients of α. From the axioms $x_i \geq 0$ and $-x_i \geq -1$, we can derive

$$\alpha \geq s_0 \tag{5.67}$$

without any use of the division rule. Inductively define s_i by

$$s_{i+1} = \left\lceil \frac{(q^p - c)s_i + M}{q^p} \right\rceil.$$

Assuming that $\alpha \geq s_i$ has already been derived, we show that CP_q can derive $\alpha \geq s_{i+1}$ with a short proof. First, by combining the inequality (5.66) with $\alpha \geq s_i$, CP_q can derive

$$q^p \cdot \alpha \geq (q^p - c)s_i + M$$

with no use of division. Then, with p uses of division by q, CP_q can derive $\alpha \geq s_{i+1}$.

Since the expression

$$s_{i+1} \geq \frac{q^p - c}{q^p} s_i + \frac{c}{q^p}\left(\frac{M}{c}\right)$$

is a weighted average of s_i, $\frac{M}{c}$ and $c > q^p/2$ hence $\frac{c}{q^p} > \frac{1}{2}$, it follows that $\frac{M}{c} - s_{i+1} \leq \frac{1}{2}\left(\frac{M}{c} - s_i\right)$. Since s_i is an integer, it follows that if $M/c - s_i < 1/c$, then $s_i = \lceil M/c \rceil$. Therefore, $s_i = \lceil M/c \rceil$ after $i = \log(M - c \cdot s_0)$ iterations. This completes the simulation of the inference (5.65) in CP_q; namely, the

CP_q-proof derives $\alpha \geq s_i$ for $i = 0, 1, \ldots, \log(M - c \cdot s_0)$. The fact that this CP_q-proof has length polynomially bounded by the number of symbols in inequality (5.66) is easily checked.

5.6.4 Normal Form for CP Proofs

In this section, we prove that the size of coefficients in a cutting plane refutation may be taken to be polynomial in the length of the refutation and the size of the CNF formula which is refuted. Though this result does not settle the question whether CP* and CP are p-equivalent, it is an important ingredient in Pudlák's Interpolation Theorem 5.6.7

For the purposes of this section, we make a minor modification to the syntax for cutting planes; namely, we assume that all inequalities are of the form

$$a_1 x_1 + \cdots + a_n x_n + a_{n+1} \geq 0. \tag{5.68}$$

Let $\Sigma = \{I_1, \ldots, I_p\}$ be an unsatisfiable set of linear inequalities, and suppose that the absolute value of every coefficient and constant term in each inequality of Σ is bounded by B. Let $A = p \cdot B$.

Theorem 5.6.6 ([BC96]). *Let Σ, p, A, B be as above. Let P be a CP refutation of Σ having ℓ lines. Then there is a CP refutation P' of Σ, such that P' has $O(\ell^3 \log A)$ lines and such that each coefficient and constant term appearing in P' has absolute value equal to $O(\ell 2^\ell A)$.*[18]

Proof. For the purposes of the proof, we use letters E and F, often with subscripts and superscripts to denote expressions on the left side of inequalities of the form (5.68); hence all inequalities in the proof are of the form $E \geq 0$. As well, we write $abs(b)$ to denote the absolute value of b, and $|b|$ to denote the length of the binary representation of $abs(b)$. Let $||E||$ denote the maximum of the absolute values of the coefficients and constant term appearing in E, thus $B = \max\{||E_1||, \ldots, ||E_p||\}$.

The rules of inference are unchanged, with only a slight modification to the division rule; namely, in inequality (5.68), if an integer $c > 1$ divides each of the coefficients $a_1, \ldots, a_n$, then we infer

$$\frac{a_1}{c} x_1 + \cdots + \frac{a_n}{c} x_n + \left\lfloor \frac{a_{n+1}}{c} \right\rfloor \geq 0. \tag{5.69}$$

It is easy to see that this slight modification of CP is p-equivalent with the original version. The last line of a CP refutation is now $-1 \geq 0$.

The main idea of the proof illustrated by thc following example. Suppose that

[18] Our bounds for the coefficients and line differ slightly from that given in [BC96]

$$E_1 \geq 0 \text{ is } 4x + 2y - 1 \geq 0$$
$$E_2 \geq 0 \text{ is } 5x - 8y + 2 \geq 0$$

and that F equals $500 \cdot E_1 + 32 \cdot E_2$. Suppose that E is obtained from F by applying division by 3, hence $F \geq 0$ is the inequality

$$(2000x + 1000y - 500) + (160x - 256y + 64) \geq 0$$

and so

$$F \geq 0 \text{ is } 2160x - 744y - 436 \geq 0$$
$$E \geq 0 \text{ is } 720x - 238y + 145 \geq 0.$$

In order to reduce the size of coefficients and terms, let F' be obtained from F by taking the residue modulo 3 of all coefficients and the constant term of F (modulo 3, since we are about to apply division by 3 in the next line). Since the coefficients of F are divisible by 3, so are those of F' and hence we define E' to be the result of applying division by 3 to E. This yields the following.

$$F' \geq 0 \text{ is } (500 \bmod 3) \cdot E_1 + (32 \bmod 3) \cdot E_2 \geq 0, \text{ hence}$$
$$F' \geq 0 \text{ is } 18x - 12y + 2 \geq 0$$
$$E' \geq 0 \text{ is } 6x - 4y + 0 \geq 0$$

We now turn to the formal proof. Assume that Σ consists of the inequalities $\{I_1, \ldots, I_p\}$, where each I_j is $E_j \geq 0$, and that we are given a CP refutation P. Without loss of generality, we may assume that P consists of the lines

$$E_1 \geq 0, \quad E_2 \geq 0, \quad \ldots \quad , E_p \geq 0$$

followed by

$$F_{p+1} \geq 0, \quad E_{p+1} \geq 0, \quad F_{p+2} \geq 0, \quad E_{p+2} \geq 0, \ldots, \quad F_\ell \geq 0, \quad E_\ell \geq 0, \quad F_{\ell+1} \geq 0$$

where the following conditions hold.

1. $F_{\ell+1}$ is -1.
2. Each F_{i+1} is a nonnegative linear combination of $E_1, \ldots, E_i$; i.e.,
$$F_{i+1} = b_{i,1}E_1 + b_{i,2}E_2 + \cdots + b_{i,i}E_i \tag{5.70}$$
where each $b_{i,j}$ is a nonnegative integer.
3. $E_i \geq 0$ is obtained from $F_i \geq 0$ by division by an integer $c_i > 1$.

We now describe how to form another CP derivation P' with the properties asserted in the theorem. P' will contain lines $E'_i \geq 0$ and $F'_i \geq 0$ corresponding to the lines in P. For $1 \leq i \leq p$, E'_i is equal to E_i. For $i > p$, the lines $E'_i \geq 0$ are obtained from $F'_i \geq 0$ by division c_i. It remains to describe the lines $F'_i \geq 0$.

Given that F_i is computed by (5.70), we compute F'_{p+1} as

$$F'_{p+1} = \sum_{i=1}^{p} (b_{p,i} \bmod c_{p+1}) E_i.$$

Note the coefficients of variables appearing in F'_{p+1} are divisible by c_{p+1}, since that was the case in F_{p+1}. Thus we can apply the rule of division by c_{p+1} to obtain $E'_{p+1} \geq 0$ from $F'_{p+1} \geq 0$.

CLAIM. E_{p+1} is a nonnegative linear combination of

$E'_1, \ldots, E'_{p+1}$.

This is clear, since

$$E_{p+1} = E'_{p+1} + \sum_{i=1}^{p} \lfloor b_{p,i}/c_{p+1} \rfloor E'_i.$$

We now continue inductively in this fashion in order to define F'_i and E'_i for $p+1 \leq i \leq \ell$, while maintaining the condition that E_i is a nonnegative linear combination of $E'_1, \ldots, E'_i$. Thus by induction

$$F_{i+1} = d_{i,1}E'_1 + d_{i,2}E'_2 + \cdots + d_{i,i}E'_i$$

where integers $d_{i,j}$ depend on i, and we define F'_{i+1} to equal

$$F'_{i+1} = \sum_{j=1}^{i} (d_{i,j} \bmod c_{i+1}) E'_j.$$

It follows that -1 is a nonnegative linear combination of $E'_1, \ldots, E'_\ell$. It could be that this nonnegative linear combination involves large coefficients; however, these large coefficients can be avoided in the following way. Suppose $\sum_{i=1}^{\ell} b_i E'_i = -1$ and the b_i's are nonnegative integers.

CLAIM. Let $c = \max\{||E'_1||, \ldots, ||E'_\ell||\}$. Then for $1 \leq i \leq \ell$, the absolute value of b_i is bounded by $\ell! \cdot c^\ell$.

The claim follows from [PS82b, Lemma 2.1], since the b_i's can be obtained as solutions to a linear programming problem with the constraints $b_i \geq 0$. It follows that the sizes of the b_i's are polynomially bounded by ℓ and the sizes of the coefficients and constant terms in $E'_1, \ldots, E'_\ell$.

Now let J be such that each $abs(b_i) < 2^J$, hence $|b_i| \leq J$ and $J = O(\ell \log \ell + \ell \log c)$. Instead of deriving $-1 \geq 0$ as a single nonnegative linear combination of the inequalities $E'_i \geq 0$, we use J steps, with $j = J, J-1, \ldots, 2, 1, 0$, to successively derive $G_j \geq 0$ where

$$G_j = \sum_{i=1}^{\ell} \lfloor b_i/2^j \rfloor E'_i. \tag{5.71}$$

The G_j's are not derived according to their defining equation (5.71), but instead we derive $G_{j-1} \geq 0$ from $G_j \geq 0$ and from the inequalities E'_i, by using the fact that G_{j-1} equals twice G_j plus a 0/1 linear combination of the E'_i's; namely,

$$G_{j-1} = 2 \cdot G_j + \sum_{i=1}^{\ell} \left(\lfloor b_i/2^{j-1} \rfloor \bmod 2 \right) E'_i.$$

(This is similar to the trick in repeated squaring to compute modular powers – see Exercise 1.13.11.) Since G_0 is just -1, we obtain a CP derivation P' of $-1 \geq 0$.

We now analyze the size of the coefficients which appear in P'. Note first that the absolute values of the coefficients in the G_j's must be bounded by $c \cdot \ell$, for otherwise, it would be impossible to have the final summation G_0 equal -1. To bound the size of the coefficients in E'_i, recall that $B = \max\{||E_1||, \ldots, ||E_p||\}$ and that $A = p \cdot B$.

CLAIM. For $0 \leq k < \ell - p$ we have $||E_{p+k+1}|| \leq 2^k A$.

Proof of Claim. By induction on k. Clearly

$$||F'_{p+1}|| \leq \sum_{i=1}^{p} (c_{p+1} - 1) \cdot ||E'_i||$$

and since E'_{p+1} is obtained from F'_{p+1} by applying division by c_{p+1}, we have

$$||E'_{p+1}|| \leq \left(\frac{c_{p+1} - 1}{c_{p+1}} \right) \cdot \sum_{i=1}^{p} ||E'_i|| \leq pB = A.$$

Now

$$||F'_{p+2}|| \leq \sum_{i=1}^{p+1} (c_{p+2} - 1) \cdot ||E'_i||$$

and since E'_{p+2} is obtained from F'_{p+2} by applying division by c_{p+2}, we have

$$||E'_{p+2}|| \leq \left(\frac{c_{p+2} - 1}{c_{p+2}} \right) \cdot \left[\left(\sum_{i=1}^{p} ||E'_i|| \right) + ||E'_{p+1}|| \right] \leq pB + A = 2A.$$

By induction we have

$$\begin{aligned} ||E'_{p+k+1}|| &\leq \left(\frac{c_{p+k+1} - 1}{c_{p+k+1}} \right) \cdot \left(\sum_{i=1}^{p} ||E'_i|| + \sum_{i=1}^{k} ||E'_{p+i}|| \right) \\ &\leq pB + A + 2A + \cdots + 2^{k-1} A = 2^k A. \end{aligned}$$

It follows that for $i = 0, \ldots, \ell$, we have $||E_i|| \leq 2^\ell A$, and that for $j = J, \ldots, 0$,

$$||G_j|| \leq \ell c = \ell \cdot \max\{||E_1'||, \ldots, ||E_\ell'||\} \leq \ell 2^\ell A \leq \ell 2^\ell \cdot |\Sigma|$$

and so the absolute value of every coefficient and constant term of P' is bounded by $\ell 2^\ell A$. To count the number of lines in P', note that each F_i' is obtained by at most $i+2$ additions, and hence in $O(\ell)$ lines. It follows that the derivation of the inequalities $E_1' \geq 0, \ldots, E_\ell' \geq 0$ takes at most $O(\ell^2)$ lines. In the final portion of P', each of the J inequalities $G_j \geq 0$ is derived in $O(\ell)$ lines. Since $J = O(\ell \log \ell + \ell \log c)$ and $c \leq \ell 2^\ell A$, we have $J = O(\ell(\log \ell + \ell + \log A))$ and P' has $O(\ell^2(\log \ell + \ell + \log A)) = O(\ell^3 \log A)$ many lines. This concludes the proof of Theorem 5.6.6.

The following corollary is immediate.

Corollary 5.6.2. *Let Σ be an unsatisfiable set of linear inequalities, and let n denote the size $|\Sigma|$ of Σ. If P is a CP refutation of Σ having ℓ lines, then there is a CP refutation P' of Σ, such that P' has $O(\ell^3 \log n)$ lines and such that the size of the absolute value of each coefficient and constant term appearing in P' is $O(\ell + \log n)$.*

5.6.5 Lower Bounds for CP

Recall the discussion about interpolation and its relation with boolean circuits from Section 5.4.5. In this section, following P. Pudlák [Pud97], we prove an interpolation theorem relating cutting plane refutations and *real* circuits, which then yields an exponential lower bound for CP refutations of unsatisfiable formulas concerning the Broken Mosquito Screen Problem.

Theorem 5.6.7 ([Pud97]). *Let P be a cutting plane refutation of*

$$\sum_i c_{i,m} p_i + \sum_j b_{j,m} q_j \geq A_m, \text{ for } m \in M \tag{5.72}$$

$$\sum_i c'_{i,n} p_i + \sum_k d_{k,n} r_k \geq B_n, \text{ for } n \in N \tag{5.73}$$

where $\mathbf{p}, \mathbf{q}, \mathbf{r}$ are sequences of distinct propositional variables. Then there exists a real circuit $C(\mathbf{p})$, which, for every truth assignment $\mathbf{a}$ of variables $\mathbf{p}$, satisfies

$$C(\mathbf{a}) = \begin{cases} 0 \text{ if } A(\mathbf{a}, \mathbf{q}), \text{ i.e., (5.72), is unsatisfiable} \\ 1 \text{ else (hence } B(\mathbf{a}, \mathbf{r}), \text{ i.e., (5.73), is unsatisfiable).} \end{cases}$$

Moreover, the circuit C has size bounded by a polynomial in $\sum_{m \in M} |A_m| + \sum_{n \in N} |B_n|$ and the number of lines in refutation P.

Proof. Note that addition is the only rule where variables $\mathbf{q}, \mathbf{r}$ are mixed in the conclusion. The key idea is to simulate the refutation $P(\mathbf{p}/\mathbf{a})$, of $A(\mathbf{p}/\mathbf{a}, \mathbf{q}) \wedge B(\mathbf{p}/\mathbf{a}, \mathbf{r})$, by a "refutation" $P'(\mathbf{a}) = (P_\ell(\mathbf{a}), P_r(\mathbf{a}))$, where a line of $P(\mathbf{a})$, i.e., an inequality in $\mathbf{a}, \mathbf{q}, \mathbf{r}$, is replaced by a pair of inequalities, the left inequality in variables $\mathbf{q}$ and the right inequality in $\mathbf{r}$. Let $P_\ell(\mathbf{a})$ ($P_r(\mathbf{a})$) be the sequence of left (right) inequalities. We'll show that either $P_\ell(\mathbf{a})$ refutes $A(\mathbf{a}, \mathbf{q})$, i.e., the system (5.72), or that $P_r(\mathbf{a})$ refutes $B(\mathbf{a}, \mathbf{r})$, i.e., the system (5.73).

To construct $P'(\mathbf{a}) = (P_\ell(\mathbf{a}), P_r(\mathbf{a}))$, replace the inequalities (5.72) by the pairs of inequalities

$$\sum_j b_{j,m} q_j \geq A_m - \sum_i c_{i,m} a_i, 0 \geq 0 \tag{5.74}$$

for $m \in M$ and replace (5.74) by

$$0 \geq 0, \sum_k d_{k,n} r_k \geq B_n - \sum_i c'_{i,n} a_i \tag{5.75}$$

for $n \in N$. Apply addition, multiplication and division rules pairwise, noting that there is no problem in fulfilling the criterion for application of the division rule (if the division rule is applied in refutation $P(\mathbf{a})$, then the coefficients of $\mathbf{q}$ and $\mathbf{r}$ are evenly divisible by the divisor c). There could have been a problem with divisibility of the coefficients of the $\mathbf{p}$, but variables p_i have been instantiated by boolean values a_i. Proceeding in this fashion, the refutation $P(\mathbf{a})$ is transformed into $P'(\mathbf{a})$, where inference

$$\sum_i e_i p_i + \sum_j f_j q_j + \sum_k g_k r_k \geq D \tag{5.76}$$

is transformed into an inference pair

$$\sum_j f_j q_j \geq D_0, \sum_k g_r r_k \geq D_1. \tag{5.77}$$

Moreover, by induction on the number of inferences, we show that the inference pair is as *strong* as the original inference, in that

$$D_0 + D_1 + \sum_i e_i a_i \geq D. \tag{5.78}$$

In the base case, (5.78) holds for the pairs (5.74) and (5.75); in the inductive case, (5.78) clearly is preserved when applying the addition and multiplication rules. When applying the division rule, if integer $c > 1$ evenly divides all coefficients e_i, f_j, g_k in (5.76) and by the induction hypothesis, we have

$$D_0 + D_1 + \sum_i e_i a_i \geq D$$

then

$$\left\lceil \frac{D_0}{c} \right\rceil + \left\lceil \frac{D_1}{c} \right\rceil + \frac{\sum_i e_i a_i}{c} \geq \frac{D}{c}$$

so

$$\left\lceil \frac{D_0}{c} \right\rceil + \left\lceil \frac{D_1}{c} \right\rceil + \frac{\sum_i e_i a_i}{c} \geq \left\lceil \frac{D}{c} \right\rceil.$$

Now the last inequality in $P(\mathbf{a})$ is $0 \geq 1$, so the corresponding pair of inequalities $0 \geq D_0, 0 \geq D_1$ must satisfy $D_0 + D_1 \geq D = 1$, and either $D_0 \geq 1$ or $D_1 \geq 1$. It follows that either $P_\ell(\mathbf{a})$ is a refutation of $A(\mathbf{a},\mathbf{q})$, or $P_r(\mathbf{a})$ is a refutation of $B(\mathbf{a},\mathbf{r})$, and thus we obtain a CP refutation from the instantiation of (5.72) or (5.73) by $\mathbf{a}$.

Concerning the size bound, it follows from Corollary 5.6.2 that given CP refutation P of $A(\mathbf{p},\mathbf{q}) \wedge B(\mathbf{p},\mathbf{r})$ there exists a refutation R, where all coefficients in R are polynomial in the sum of the sizes $|A_m|, |B_n|$ and number of lines of P. By applying the previous transformation to R, we obtain R' and can decide in polynomial time whether $\sum_i c_{i,m} a_i + \sum_j b_{j,m} q_j \geq A_m$, for $m \in M$, is unsatisfiable, or whether $\sum_i c'_{i,n} a_i + \sum_k d_{k,n} r_k \geq B_n$, for $n \in N$, is unsatisfiable.

Note that since the construction of circuit C follows the skeleton of given refutation P, if P is a tree-like refutation, then C has fan-out 1. This remark applies to the following corollary as well.

Corollary 5.6.3. *Under the same hypotheses, if all the coefficients $c_{i,m}$ in $A(\mathbf{a},\mathbf{q})$, i.e., system* (5.72)*, are nonnegative (all coefficients $c'_{i,m}$ in $B(\mathbf{a},\mathbf{r})$, i.e., system (5.73), are nonpositive), then there is monotonic real circuit C satisfying*

$$C(\mathbf{a}) = \begin{cases} 0 \text{ if } A(\mathbf{a},\mathbf{q}) \text{ is unsatisfiable} \\ 1 \text{ else (hence } B(\mathbf{a},\mathbf{r}) \text{ is unsatisfiable)} \end{cases}$$

whose depth is that of P, and size is bounded by a polynomial in $\sum_{m \in M} |A_m| + \sum_{n \in N} |B_n|$ and the number of lines in refutation P.

Proof. Assume first that all the $c_{i,m}$ in (5.72) are nonnegative. We need only maintain the integer values $-D_0$ in each line of the refutation, and in the last line verify whether $-D_0 \geq 0$. In the last line, if it is not the case that $-D_0 \geq 0$, then $-D_0 < 0$ and so $D_0 \geq 1$, and the circuit outputs 0, indicating that $A(\mathbf{a},\mathbf{q})$ is unsatisfiable. Otherwise, it must be that $D_1 \geq 0$ and the circuit outputs 1, indicating that $B(\mathbf{a},\mathbf{r})$ is unsatisfiable. For the initial inequalities (5.72), where

$$\sum_j b_{j,m} q_j \geq A_m - \sum_i c_{i,m} a_i$$

we have $-D_0 = \sum_i c_{i,m} a_i - A_m$, which can be computed by performing the addition $c_{i,m} \cdot a_i$ (requiring a $\log m$ depth tree of addition gates) and then applying the unary subtraction function $S_{A_m}(x) = x - A_m$ (which, for A_m fixed, is a monotonic operation). Other gates correspond to addition, multiplication by a positive constant, and division by a positive constant followed by the *floor* operation ($-D_0$ is on the left side of the inequality, so we round down). Finally, the output gate of circuit C is given by the threshold function

$$T_0(x) = \begin{cases} 1 & \text{if } x \geq 0 \\ 0 & \text{else.} \end{cases}$$

Summarizing, the required gates are

- addition $x + y$,
- multiplication by a positive constant, $m_A(x) = A \cdot x$,
- $S_A(x) = x - A$, subtraction by fixed constant A,
- $\lfloor \frac{x}{c} \rfloor$ division with floor
- T_0 threshold.

Now consider the case where all $c'_{i,n}$ are nonpositive. We maintain only the integers D_1 for each right inequality. For the initial inequalities (5.73), where

$$\sum_k d_{k,n} r_k \geq B_n - \sum_i c'_{i,n} a_i$$

we have $D_1 = B_n - \sum_i c'_{i,n} a_i$, and a similar analysis shows that the required gates are

- addition $x + y$,
- multiplication by a positive constant, $m_A(x) = A \cdot x$,
- subtraction by a fixed constant, $S_A(x) = x - A$,
- $\lceil x/c \rceil$ division with ceiling,
- T_1 threshold

where

$$T_1(x) = \begin{cases} 1 & \text{if } x \geq 1 \\ 0 & \text{else.} \end{cases}$$

All of these operations correspond to monotonic real gates, and the verification of circuit depth and size is left to the reader.

Pudlák's theorem shows that CP has the feasible interpolation property as well as the feasible monotonic interpolation property with respect to real circuits. Since it is only known that automatizability implies the feasible interpolation property, rather than the converse, it is an open question whether CP is automatizable.

In [Pud97], P. Pudlák lifted Razborov's monotonic boolean circuit lower bound for the CLIQUE problem to monotonic real circuits, and then applied his interpolation theorem to obtain an exponential size lower bound for CP proofs. About the same time, S.A. Cook modified A. Haken's monotonic boolean circuit lower bound for the *broken mosquito problem* BMS_m to the case of monotonic real circuits (see Theorem 2.4.2, with Pudlák's interpolation theorem announced an exponential lower bound for CP proofs. Here we present tautologies expressing that no graph representing a broken mosquito screen can be both good and bad, and deduce the broken mosquito screen problem, an exponential lower bound for CP proof size for BMS_m from the monotonic real circuit lower bound.

To build intuition, we'll first give inequalities, which correctly assert the existence of a graph G on $n = m^2 - 2$ vertices, such that G is both *good* and *bad* (see Definition 2.4.2). In this first formulation, there are $O\binom{m^2}{m} \sim m^{2m}$ many inequalities, an exponential number. Let $p_{i,j}$ mean that there is an edge from i to j, q_i^k mean that vertex i is in the k-th good partition class (clique), and let r_i^k mean that vertex i is in the k-th bad partition class (anticlique). The inequalities $A(\mathbf{p}, \mathbf{q})$ are given by

$$-q_i^k - q_j^k + p_{i,j} \geq -1, \text{ for all } 1 \leq i \leq n, \text{ and } 1 \leq k \leq m$$

$$m \geq q_{i_1}^k + \cdots + q_{i_{m+1}}^k,$$

$$\text{for all } 1 \leq k < m, \text{ and } 1 \leq i_1 < \cdots < i_{m+1} \leq n$$

$$m - 2 \geq q_{i_1}^m + \cdots + q_{m-1}^m, \text{ for all } 1 \leq i_1 < \cdots < i_{m-1} \leq n$$

$$\sum_{k=1}^{m} q_i^k = 1, \text{ for all } 1 \leq i \leq n.$$

The inequalities $B(\mathbf{p}, \mathbf{r})$ are given by

$$-r_i^k - r_j^k - p_{i,j} \geq -1, \text{ for all } 1 \leq i \leq n, \text{ and } 1 \leq k \leq m$$

$$m \geq r_1^k + \cdots + r_{i_{m+1}}^k,$$

$$\text{for all } 1 \leq k < m, \text{ and } 1 \leq i_1 < \cdots < i_{m+1} \leq n$$

$$m - 2 \geq r_{i_1}^m + \cdots + r_{i_{m-1}}^m, \text{ for all } 1 \leq i_1 < \cdots < i_{m-1} \leq n$$

$$\sum_{k=1}^{m} r_i^k = 1, \text{ for all } 1 \leq i \leq n.$$

Clearly $\mathbf{p}$ is positive in the $A(\mathbf{p}, \mathbf{q})$ and negative in the $B(\mathbf{p}, \mathbf{r})$, but overall size is exponential in m. A better formulation is given by letting $q_{i,j}^k$. $(r_{i,j}^k)$ mean that $k \in \{1, \ldots, n\}$ is in the j-th position of the i-th row (i.e., good partition class) forming a clique (row (i.e., bad partition class) forming an anticlique). To that end, define the relation P_m for the "pattern" or grid structure required for instances of BMS_m, i.e.,

$$P_m = \{(i,j) : 1 \leq i < m, 1 \leq j \leq m\} \cup \{(m,j) : 1 \leq j \leq m-2\}.$$

Define $A(\mathbf{p}, \mathbf{q})$ to be the inequalities

$$\sum_{(i,j) \in P_m} q_{i,j}^k = 1, \text{ for } 1 \leq k \leq n \tag{5.79}$$

$$\tag{5.80}$$

and

$$-q_{i,j}^k - q_{i,j'}^{k'} + p_{k,k'} \geq -1, \tag{5.81}$$

for $k \neq k'$, $1 \leq k, k' \leq n$, $(i,j), (i,j') \in P_m, j \neq j'$. The inequalities $B(\mathbf{p}, \mathbf{r})$ are then

$$\sum_{(i,j) \in P_m} r^k_{i,j} = 1, \text{ for } 1 \leq k \leq n \tag{5.82}$$

and

$$-r^k_{i,j} - r^{k'}_{j,j'} - p_{k,k'} \geq -2, \tag{5.83}$$

for all $1 \leq k < k' \leq n$, where $(i,j), (i,j') \in P_m, j \neq j'$. Clearly $\mathbf{p}$ is positive in the $A(\mathbf{p}, \mathbf{q})$ and negative in the $B(\mathbf{p}, \mathbf{r})$ (for our application, only one of these need be satisfied). The conjunction of the $A(\mathbf{p}, \mathbf{q})$ and $B(\mathbf{p}, \mathbf{r})$ is unsatisfiable, since this asserts the existence of a graph which is both good and bad, contradicting Lemma 2.4.1. By Pudlák's interpolation result, Corollary 5.6.3, every CP refutation of the $A(\mathbf{p}, \mathbf{q})$ and $B(\mathbf{p}, \mathbf{r})$ must contain at least

$$\frac{1.8^{\lfloor \sqrt{m/2} \rfloor}}{2}$$

many lines. There are $O(n^2) = O(m^4)$ many inequalities in the

$$A(\mathbf{p}, \mathbf{q}), B(\mathbf{p}, \mathbf{r}),$$

so relative to input size N of the formula to be refuted, we have a

$$2^{\Omega(N^{1/8})}$$

lower bound. Summarizing this, we have the

Theorem 5.6.8. *Every* CP *refutation of the propositional formulation of (the negation of) BMS_m, $n = m^2 - 2$, formalized as $A(\mathbf{p}, \mathbf{q}) \wedge B(\mathbf{p}, \mathbf{r})$ with size N by the inequalities (5.79), (5.81), (5.82), (5.83), must contain at least $\frac{1.8^{\lfloor \sqrt{m/2} \rfloor}}{2}$ many lines, and size $2^{\Omega(N^{1/8})}$.*

Another application of Pudlák's interpolation theorem was made by J. Johannsen [Joh98], who extended Theorem 5.4.15 by lifting the Karchmer-Wigderson result to monotonic real circuits and then applying Theorem 5.6.7 to give an $n^{\Omega \log n}$ size lower bound for tree-like cutting plane proofs of the weak *st*-connectivity principle, $wSTC_n$.

Theorem 5.6.9 ([Joh98]). *Every family of tree-like cutting plane refutations of $\neg wSTC_n$ has size $n^{\Omega(\log n)}$.*

Proof. Let $s(n)$ be the size of a refutation of $\neg wSTC_n$, formulated using integer linear inequalities representing $A(\mathbf{p}, \mathbf{q}) \wedge B(\mathbf{p}, \mathbf{r})$, where A asserts that if a graph on n vertices is undirected, then it has a path from s to t, and B asserts the existence of a cut between s and t. By Corollary 5.6.3, there are monotone real circuits C of size $O(s(n))$ such that $C(\mathbf{a}) = 0$ implies that

$A(\mathbf{a}, \mathbf{q})$ is refutable and $C(\mathbf{a}) = 1$ implies that $B(\mathbf{a}, \mathbf{r})$ is refutable. Restricting C to those inputs $\mathbf{a}$ which correctly encode an undirected graph G, it follows that $C(\mathbf{a}) = 1$ if and only if there is a path from s to t. By Theorem 2.4.2, it follows that $s(n)$ must be $n^{\Omega(\log n)}$.

Since Theorem 5.4.14 gives polynomial-size resolution proofs of $wSTC_n$, we have the immediate corollary.

Corollary 5.6.4. *Tree-like CP does not polynomially simulate resolution.*

Subsequent work by Bonet, Estaban, Galesi and Johannsen [BEGJ98] improved the previous result to an exponential separation. Whether there are width-related results for cutting planes, analogous to Theorems 5.4.16 and 5.4.17 is open.

The cutting plane system, like resolution, is sound and complete only for CNF formulas. By introducing new variables, which abbreviate subformulas of the formula to be refuted, we define the system *cutting planes with limited extension* (CPLE). This system is both sound and complete; i.e., every refutable formula is unsatisfiable, and every unsatisfiable propositional formula has a refutation.

Definition 5.6.1. *The constants* 0 (FALSE) *and* 1 (TRUE) *are boolean formulas of size* 1. *The propositional variables p_i are boolean formulas of size $|i|+1$, where the latter denotes the length of the binary representation of i. If B is a boolean formula of size s, then $\neg B$ is a boolean formula of size $s+1$. If B_i are boolean formulas having size s_i where $i \in I$ for finite index set I, then $(\bigvee_{i \in I} B_i)$ and $(\bigwedge_{i \in I} B_i)$ are boolean formulas of size $I + 1 + \sum_{i \in I} s_i$ (taking into account the two parentheses and $I - 1$ logical connectives).*

In the following definition, for boolean formula A, the (new) atom p_A consists of 'p' followed by the formula A and has size equal to $1 + size(A)$.

Definition 5.6.2. *By induction on depth of the (unbounded fan-in) formula A, define as follows the set $LE[A]$ of linear inequalities associated with A (the acronym LE stands for limited extension).*

- *If A is the propositional variable x_i, then*
  $LE[A] = \{p_A \geq 0, -p_A \geq -1\}$.
- *If A is $\neg B$, then*
  $LE[A] = \{p_A \geq 0, p_A + p_B \geq 1, -p_A \geq -1, -p_A - p_B \geq -1\} \cup LE[B]$.
- *If A is $\bigwedge_{i \in I} B_i$, then*
  $LE[A] = \{p_A \geq 0, -p_A \geq -1, (I-1) - \sum_{i \in I} p_{B_i} \geq -p_A,\} \cup \{p_{B_i} - p_A \geq 0 : i \in I\} \cup \bigcup_{i \in I} LE[B_i]$.
- *If A is $\bigvee_{i \in I} B_i$, then*
  $LE[A] = \{p_A \geq 0, -p_A \geq -1, \sum_{i \in I} p_{B_i} - p_A \geq 0,\} \cup \{p_A - p_{B_i} \geq 0 : i \in I\} \cup \bigcup_{i \in I} LE[B_i]$.

The system CPLE has the same rules as those of CP (i.e., addition, multiplication and division).

A CPLE refutation of the formula B (not necessarily in conjunctive normal form) is a sequence $s_0, \ldots, s_m$ of linear inequalities, such that

- s_m is $0 \geq 1$,
- for all $i \leq m$, either s_i is $p_B \geq 1$, or $s_i \in LE[B]$, or there exist $j, k < i$ such that s_i is obtained from s_j, s_k by the addition, multiplication or division rule.

We sometimes speak of $C \in LE[B]$ as an *axiom* and of $p_B \geq 1$ as the *hypothesis* in a refutation of B. The formula A is said to have a CPLE proof, if its negation $\neg A$ has a CPLE refutation. In Exercise 5.10.13, a sketch is given of polynomial-size CP refutations of a stronger version STC_n of *st*-connectivity, which asserts that every finite undirected graph G has a path from s to t, provided that s, t both have degree 1 and every other node has degree 2. This principle is much stronger than $wSTC_n$, implies the pigeonhole principle PHP_n^{n+1} and is equivalent over bounded depth Frege systems to Ajtai's equipartition principle.

5.6.6 Threshold Logic PTK

In this section, we introduce propositional threshold logic and prove a completeness theorem. It is hoped that certain lower bound results for threshold circuits may be extended to yield lower bounds for proof size of propositional threshold logic and *a fortiori* for cutting planes. Krajíček has introduced a different system FC of propositional threshold logic [Kra94b].

Definition 5.6.3. Propositional threshold logic *is given as follows. Formula depth and size are defined inductively by:*

i. a propositional variable x_i, $i \in \mathbf{N}$, is a formula of depth 0 and size 1.

ii. if F is a formula then $\neg F$ is a formula of depth $1 + dp(F)$ and size $1 + size(F)$.

iii. if $F_1, \ldots, F_n$ are formulas and $1 \leq k \leq n$ then $T_k^n(F_1, \ldots, F_n)$ is a formula of depth $1 + max\{depth(F_i) : 1 \leq i \leq n\}$ and size $(n+k) + 1 + \sum_{1 \leq i \leq n} size(F_i)$.

The interpretation of the new connective $T_k^n(F_1, \ldots, F_n)$ is that at least k of the boolean formulas $F_1, \ldots, F_n$ hold. Propositional threshold logic can be viewed as an extension of propositional logic in the connectives $\neg, \wedge, \vee$, the latter two connectives being defined by

$$\bigvee_{1 \leq i \leq n} F_i \equiv T_1^n(F_1, \ldots, F_n)$$

$$\bigwedge_{1 \leq i \leq n} F_i \equiv T_n^n(F_1, \ldots, F_n)$$

A *cedent* is any sequence $F_1, \ldots, F_n$ of formulas separated by commas. Cedents are sometimes designated by $\Gamma, \Delta, \ldots$ (capital Greek letters). A *sequent* is given by $\Gamma \vdash \Delta$, where Γ, Δ are arbitrary cedents. The size (depth) of a cedent $F_1, \ldots, F_n$ is $\sum_{1 \le i \le n} size(F_i)$ ($\max_{1 \le i \le n}(depth(F_i))$). The size (depth) of a sequent $\Gamma \vdash \Delta$ is $size(\Gamma) + size(\Delta)$ ($\max(depth(\Gamma), depth(\Delta))$). The intended interpretation of the sequent $\Gamma \vdash \Delta$ is $\wedge\Gamma \rightarrow \vee\Delta$.

An *initial sequent* is of the form $F \vdash F$ where F is any formula of propositional threshold logic. The rules of inference of PTK, the sequent calculus of propositional threshold logic, are as follows. By convention, $T^n_m(A_1, \ldots, A_n)$ is only defined if $1 \le m \le n$.

structural rules

weak left: $\dfrac{\Gamma, \Delta \mapsto \Gamma'}{\Gamma, A, \Delta \mapsto \Gamma'}$ weak right: $\dfrac{\Gamma \mapsto \Gamma', \Delta'}{\Gamma \mapsto \Gamma', A, \Delta'}$

contract left: $\dfrac{\Gamma, A, A, \Delta \mapsto \Gamma'}{\Gamma, A, \Delta \mapsto \Gamma'}$ contract right: $\dfrac{\Gamma \mapsto \Gamma', A, A, \Delta'}{\Gamma \mapsto \Gamma', A, \Delta'}$

permute left: $\dfrac{\Gamma, A, B, \Delta \mapsto \Gamma'}{\Gamma, B, A, \Delta \mapsto \Gamma'}$ permute right: $\dfrac{\Gamma \mapsto \Gamma', A, B, \Delta'}{\Gamma \mapsto \Gamma', B, A, \Delta'}$

cut rule

$$\frac{\Gamma, A \mapsto \Delta \qquad \Gamma' \mapsto A, \Delta'}{\Gamma, \Gamma' \mapsto \Delta, \Delta'}$$

logical rules

$\neg$-left: $\dfrac{\Gamma \mapsto A, \Delta}{\neg A, \Gamma \mapsto \Delta}$ $\neg$-right: $\dfrac{A, \Gamma \mapsto \Delta}{\Gamma \mapsto \neg A, \Delta}$

$\wedge$-left: $\dfrac{A_1, \ldots, A_n, \Gamma \mapsto \Delta}{T^n_n(A_1, \ldots, A_n), \Gamma \mapsto \Delta}$ for $n \ge 1$

$\wedge$-right: $\dfrac{\Gamma \mapsto A_1, \Delta \quad \cdots \quad \Gamma \mapsto A_n, \Delta}{\Gamma \mapsto T^n_n(A_1, \ldots, A_n), \Delta}$ for $n \ge 1$

$\vee$-left: $\dfrac{A_1, \Gamma \mapsto \Delta \quad \cdots \quad A_n, \Gamma \mapsto \Delta}{T^n_1(A_1, \ldots A_n), \Gamma \mapsto \Delta}$ for $n \ge 1$

$\vee$-right: $\dfrac{\Gamma \mapsto A_1, \ldots, A_n, \Delta}{\Gamma \mapsto T^n_1(A_1, \ldots, A_n), \Delta}$ for $n \ge 1$

$$T_k^n\text{-left:}\ \frac{T_k^{n-1}(A_2,\ldots,A_n),\Gamma \mapsto \Delta \qquad A_1, T_{k-1}^{n-1}(A_2,\ldots,A_n),\Gamma \mapsto \Delta}{T_k^n(A_1,\ldots,A_n),\Gamma \mapsto \Delta}$$

for $2 \leq k < n$.

$$T_k^n\text{-right:}\ \frac{\Gamma \mapsto A_1, T_{k-1}^{n-1}(A_2,\ldots,A_n),\Delta \qquad \Gamma \mapsto T_k^{n-1}(A_2,\ldots,A_n),\Delta}{\Gamma \mapsto T_k^n(A_1,\ldots,A_n),\Delta}$$

for $2 \leq k < n$.

Theorem 5.6.10. *PTK is sound.*

Proof. A *truth evaluation* is a mapping $\nu : \{x_i : i \in \mathbf{N}\} \to \{0,1\}$. By induction on formula depth, it is clear how to extend the truth evaluation ν to assign a truth value for each formula of propositional threshold logic. A formula is *valid* if it is true in every truth evaluation. Now by induction on the number of inferences in an PTK proof, it is straightforward to show that every theorem of PTK is valid. Thus PTK is sound.

Theorem 5.6.11. *PTK is complete.*

Proof. Suppose that $\wedge\Gamma \to \vee\Delta$ is valid. We construct a finite tree T, each node of which is labeled by sequents, the root of T being labeled by $\Gamma \vdash \Delta$. The tree T is constructed so that

i. if $\Gamma'' \vdash \Delta''$ is a child of $\Gamma' \vdash \Delta'$ then $size(\Gamma'' \vdash \Delta'') < size(\Gamma' \vdash \Delta')$,
ii. if $\Gamma'' \vdash \Delta''$ is a child of $\Gamma' \vdash \Delta'$ and ν is a truth evaluation such that $\nu(\Gamma'' \vdash \Delta'') = 0$, then $\nu(\Gamma' \vdash \Delta') = 0$,
iii. if $\Gamma_1 \vdash \Delta_1, \ldots, \Gamma_n \vdash \Delta_n$ are all the children of $\Gamma' \vdash \Delta'$, each of which has a proof in PTK, then there is a proof of $\Gamma' \vdash \Delta'$ in PTK,
iv. each leaf of T is of the form $\Gamma' \vdash \Delta'$ where Γ', Δ' contain only propositional variables, and moreover some propositional variable x appears both in Γ' and in Δ'.

Given an already defined node $\Gamma' \vdash \Delta'$ of T, let F be the first formula of that sequent which is not a propositional variable. If F appears in Γ', then for notational simplicity we write Γ' as F, Π rather than Π, F, Π' when F is not necessarily the first formula of cedent Γ'. Similarly for Δ'.

Case 1. F is $\neg A$, occurring in Γ'.

$$\frac{\Pi \mapsto A, \Lambda}{\neg A, \Pi \mapsto \Lambda}$$

Case 2. F is $\neg A$, occurring in Δ'.

$$\frac{A, \Pi \mapsto \Lambda}{\Pi \mapsto \neg A, \Lambda}$$

Case 3. F is $T_k^n(A_1, \ldots, A_n)$, occurring in Γ'.

$$\frac{T_k^{n-1}(A_2, \ldots, A_n), \Pi \mapsto \Lambda \qquad A, T_{k-1}^{n-1}(A_2, \ldots, A_n), \Pi \mapsto \Lambda}{T_k^n(A_1, \ldots, A_n), \Pi \mapsto \Lambda}$$

Case 4. F is $T_k^n(A_1, \ldots, A_n)$, occurring in Δ'.

$$\frac{\Pi \mapsto A_1, T_k^{n-1}(A_2, \ldots, A_n), \Lambda \qquad \Pi \mapsto T_{k-1}^{n-1}(A_2, \ldots, A_n), \Lambda}{\Pi \mapsto T_k^n(A_1, \ldots, A_n), \Lambda}$$

Conditions (i),(ii) are straightforward to check and left to the reader. Condition (iii) for cases 1 through 4 follows immediately from the relevant logical rules. If condition (iv) does not hold, then there is a leaf of tree T labeled by a sequent $\Gamma' \vdash \Delta'$ whose cedents consist only of propositional variables, but which have no variable in common. Define the truth assignment ν by

$$\nu(x) = \begin{cases} 1 \text{ if } x \text{ does not occur in } \Delta' \\ 0 \text{ otherwise} \end{cases}$$

Then $\nu(\Gamma' \vdash \Delta') = 0$, and by iterating condition (ii) along the branch consisting of all nodes of tree T between leaf $\Gamma' \vdash \Delta'$ and root $\Gamma \vdash \Delta$, it follows that $\nu(\Gamma \vdash \Delta) = 0$. But this contradicts the assumption that $\Gamma \vdash \Delta$ is valid.

Remark 5.6.1. Since the above proof does not use the cut rule, it follows that cuts may be eliminated from proofs in PTK. Also note that cut-free PTK proofs satisfy the *subformula property*; namely, every formula in a cut-free PTK proof is a subformula of a formula in the endsequent.

There is a relation between cutting plane proofs (where integers are represented in unary, i.e., CP*) and threshold logic PTK. A cutting plane inequality I of the form

$$\sum_{i=1}^{n} x_i - \sum_{i=n+1}^{m} x_i \geq k$$

can be represented by the threshold formula $\widetilde{I}$ given by

$$T_r^{n+m}(x_1, \ldots, x_n, \neg x_{n+1}, \ldots, \neg x_{n+m})$$

where $r = \max\{k + m, 0\}$.

Theorem 5.6.12 ([CJ98]). *Assume that P be a CP* derivation of inequality I from inequalities $I_1, \ldots, I_n$. Then there is a PTK proof of the sequent*

$$\widetilde{I}_1, \ldots, \widetilde{I}_n \mapsto \widetilde{I}$$

of threshold depth 1 *and size* $O(|P|^{O(1)})$.

This result immediately implies that depth 1 PTK can p-simulate CP*. J. Krajíček (personal communication) pointed out that constant-depth PTK p-simulates CP. This follows from the following three facts. By Theorem 13.1.9 of [Kra95], $I\Delta_0(\alpha)^{count}$ proves the soundness of CP. By lifting Paris-Wilkie's Theorem 9.1.3 of [Kra95] (see [Kra94b]), one can show that if $I\Delta_0(\alpha)^{count} \vdash \forall x \theta(x)$, where θ is a first order formula whose quantifiers are bounded, then FC has constant-depth polynomial-size proofs of $\{\langle\theta\rangle_n : n \in \mathbf{N}\}$. From this it follows that constant-depth FC p-simulates cutting planes. Finally, in [BC96], S.R. Buss and P. Clote show that the systems FC ([Kra94b]) and PTK p-simulate each other within a polynomial-size factor and constant-depth factor, hence constant-depth PTK p-simulates CP. It would be interesting to give a direct proof and thereby determine the exact depth required for such a polynomial simulation.

5.7 Frege Systems

Frege systems, sometimes called *Hilbert-style* systems, are propositional proof systems in the tradition of Hilbert and were introduced by Cook and Reckhow in [CR74], [CR77] in order to study the relative efficiency of proof systems from the point of view of computational complexity.

A *rule of inference* of a Frege system is of the form:

$$\text{“}\textit{from } A_1, \ldots, A_k \textit{ infer } A_0\text{”} \tag{5.84}$$

where $A_0, \ldots, A_k$ are propositional formulas with variables among $x_1, \ldots, x_m$, and $A_1, \ldots, A_k \models A_0$. (In the following, we sometimes designate variables by $p_1, \ldots, p_m$). If $k = 0$, then there are no antecedents $A_1, \ldots, A_k$ and the rule is called an *axiom*. An application of a rule of inference is of the form

$$\frac{A_1(B_1/x_1, \ldots, B_m/x_m), \ldots, A_k(B_1/x_1, \ldots, B_m/x_m)}{A_0(B_1/x_1, \ldots, B_m/x_m)} \tag{5.85}$$

where the formulas $B_1, \ldots, B_m$ have been simultaneously substituted for variables $x_1, \ldots, x_m$. To make this substitution process more explicit, rules of inference are sometimes called *rule schemas*, meaning that an application of the rule tacitly involves the simultaneous substitution of variables by propositional formulas. A *Frege proof* P is a finite sequence $F_1, \ldots, F_r$ of propositional formulas, such that for every $1 \leq i \leq r$, F_i is either an axiom, or obtained by an application of a rule of inference to earlier derived formulas F_j, $j < i$. In this case, P is said to be a proof of F_r. A *derivation* of G from $F_1, \ldots, F_k$ is a proof of G, where we additionally assume that $F_1, \ldots, F_k$ are axioms. We write $F_1, \ldots, F_k \vdash G$ to mean that there is a derivation of G from $F_1, \ldots, F_k$.

Since by definition the proof rule (5.84) is *sound*, i.e., $A_1, \ldots, A_k \models A_0$, it easily follows by induction on the number of inferences in a derivation that $F_1, \ldots, F_k \vdash G$ implies $F_1, \ldots, F_k \models G$. A proof system is *implicationally complete* if the converse holds; i.e., for any propositional formulas $F_1, \ldots, F_k, G$, it is the case that $F_1, \ldots, F_k \models G$ implies $F_1, \ldots, F_k \vdash G$. Finally, a *Frege system* $\mathcal{F}$ is given by a finite, adequate set κ of propositional connectives, together with a finite set of rules of inference, such that $\mathcal{F}$ is implicationally complete.

In the introduction to this chapter, we gave an example of a Frege system with the connectives $\neg$ and $\rightarrow$ having three axioms and the single rule of inference of modus ponens. Another example is *Shoenfield's system*, which has only the connectives $\neg, \vee$, and rules given as follows.

Excluded middle: $\dfrac{}{x \vee \neg x}$ Expansion: $\dfrac{x}{y \vee x}$

Contraction: $\dfrac{x \vee x}{x}$ Associativity: $\dfrac{x \vee (y \vee z)}{(x \vee y) \vee z}$

Cut: $\dfrac{x \vee y \qquad \neg x \vee z}{y \vee z}$

Recall from Definition 5.2.4 that a proof system T_1 *polynomially simulates* a proof system T_2 if there is a polynomial p such that for any formula F, if F has a proof P of size n in T_2, (the translation of) F has a proof Q of size at most $p(n)$ in T_1. If additionally there is a polynomial time computable function f, such that $f(P) = Q$, then T_1 is said to *p-simulate* T_2.

Theorem 5.7.1 ([CR77]). *Frege systems in the same language p-simulate each other.*

Proof. Let T_1, T_2 be two Frege systems. Let F be a formula with a proof of size n in T_1. Replace each axiom of T_1 in this proof with a proof in T_2. The resulting new proof of F in T_2 is of size $O(n)$. It is not hard to see that this association is given by a polynomial time computable function, so that T_1 additionally p-simulates T_2.

Using the technique of Theorem 5.2.2, Reckhow [Rec75] additionally showed that *any* two Frege systems p-simulate each other, not just those over the same language. The idea is to use the implicit translation of Theorem 5.2.2 to translate formulas into the De Morgan basis $\{0, 1, \neg, \vee, \wedge\}$ and then apply Theorem 5.7.1. The interested reader should consult Theorem 4.4.13 of [Kra95] for more details behind Reckhow's proof.

Theorem 5.7.2 ([Rec75]). *Gentzen systems with the cut rule and Frege systems p-simulate each other.*

Despite the fact that tree-like resolution (cutting planes) refutations can be exponentially larger than dag-like resolution (cutting planes) refutations, with Frege systems, the situation is different.

Theorem 5.7.3 ([Kra94a]). *Let $\mathcal{F}$ be a Frege system. For any tautology ϕ, let $S(\phi)$ ($S_{tree}(\phi)$) be the size of the smallest dag-like (tree-like) proof of ϕ in $\mathcal{F}$. Then there exists a polynomial p such that for every tautology ϕ, $S_{tree}(\phi) \leq p(S(\phi))$.*

The proof is sketched in the proof of Theorem 5.3.6.

5.7.1 Bounded Depth Frege Systems

In this section, we present an exponential lower bound for the size of constant-depth Frege proofs of the pigeonhole principle, a result due to Krajíček, Pudlák, Woods [KPW95] and independently Pitassi, Beame, Impagliazzo [PBI93]. This result improves A. Haken's earlier exponential size lower bound for resolution proofs of PHP_n^{n+1}, presented in Theorem 5.4.6. Our treatment follows the simplified treatment of A. Urquhart and X. Fu [UF96] very closely.

Historically, the first step was taken by M. Ajtai [Ajt94a], who used combinatorial arguments related to those for the constant-depth boolean circuit size lower bound for parity, along with forcing in non-standard models of arithmetic to prove the non-existence of polynomial-size, constant-depth Frege proofs for onto $-\, \text{PHP}_n^{n+1}$. Since PHP_n^{n+1} clearly implies onto $-\, \text{PHP}_n^{n+1}$, Ajtai's lower bound for onto $-\, \text{PHP}_n^{n+1}$ yields a lower bound for PHP_n^{n+1} as well. In [BPU92], the superpolynomial lower bound was somewhat improved, and Ajtai's use of forcing and nonstandard models was replaced by the notion of "approximate" proof. Finally, in independent work, Krajíček, Pudlák, Woods [KPW95] and Pitassi, Beame, Impagliazzo [PBI93] presented the first truly exponential lower bound for the size of constant-depth Frege proofs of the pigeonhole principle. The crucial notion of "k-evaluation" was introduced in [KPW95] and it appears that this is now the preferred approach in obtaining lower bounds for constant-depth Frege proofs. In our presentation of the exponential lower bound, we carefully follow [UF96] and [Bea], which define k-evaluations in terms of decision trees, rather than boolean algebras.

The overall idea of proof goes roughly as follows. Suppose that there exists a proof P of PHP_n^{n+1}, whose size S is subexponential in n, where every formula has depth at most d. Let *s-matching disjunction* mean a particular kind of DNF formula, to be formally defined later, whose disjuncts have size at most s and which involve only positive literals (no negated variables). By induction, we will find restrictions $\rho_1, \ldots, \rho_d$, such that for each $1 \leq i \leq d$, the union restriction $\rho_1 \ldots \rho_i$ leaves a fraction n^{ϵ_i} of the variables still unset, and for each depth i subformula A of the proof P there is an s-matching disjunction $D(A)$ satisfying $A \lceil_{\rho_1 \cdots \rho_i} \approx D(A)$. Here $\approx$ means "approximately equivalent", in the sense that $A \lceil_{\rho_1 \cdots \rho_i} \equiv D(A)$ holds over a possibly proper subset of all truth assignments, which are described by a k-evaluation. Letting $\rho = \rho_1 \cdots \rho_d$, we then have that for every subformula A of the proof P, there is an s-matching disjunction $D(A)$, which satisfies $A \lceil_{\rho} \approx D(A)$. The definition of k-evaluation is so designed that for appropriate values of k, the related notion of approximately equivalence $\approx$ is sound with respect to the rules of inference of a Frege system, in the sense that (using modus ponens as an example) if $A \upharpoonright_\rho \approx 1$ and $(A \to B) \upharpoonright_\rho \approx 1$, then $B \upharpoonright_\rho \approx 1$. Paradoxically, it will turn out that $\text{PHP}_n^{n+1} \upharpoonright_\rho \approx 0$, and so P could not have been a proof of PHP_n^{n+1}!

The size parameter S of the proof P plays a role in proving the existence of restrictions $\rho_1, \ldots, \rho_d$ as follows. Suppose that the domain D has size $n+1$ and range R has size n, and that $\text{PHP}(D, R)$ is the statement

$$\bigvee_{i \in D} \bigwedge_{j \in R} \overline{p_{i,j}} \vee \bigvee_{i' \in D, i' \neq i} \bigvee_{j \in R} (p_{i,j} \wedge p_{i',j}). \tag{5.86}$$

Note that if the variable $x_{i,j}$ is set to 1 in restriction ρ, then automatically, all variables $x_{i,j'}$ for $j' \neq j, j' \in R$ and $x_{i',j}$ for $i' \neq i, i \in D$ must be set to 0, in order for the resulting restriction $\text{PHP}(D, R) \upharpoonright_\rho \equiv \text{PHP}(D \upharpoonright_\rho, R \upharpoonright_\rho)$ to be of the proper form, and so the variables $x_{i,j}$ are not (stochastically) independent. This lack of independence of the variables, unlike the simpler boolean case, presents substantial technical difficulties in proving the switching lemma required to obtain the above restrictions.

In particular, for proof P of size S, we need to prove something roughly of the following form. Given $\rho_1, \ldots, \rho_{i-1}$, for fixed depth i subformula formula A of a formula appearing in proof P, the probability over all appropriate restrictions ρ_i that the s-matching disjunction $D(A)$ is not approximately equivalent to $A \upharpoonright_{\rho_1 \cdots \rho_{i-1} \rho_i}$ is at most $1/S$; i.e.,

$$\Pr[A \upharpoonright_{\rho_1 \cdots \rho_i} \not\approx D(A)] < \frac{1}{S}.$$

It will then follow that

$$\Pr\left[(\exists A \in P)\,(A \upharpoonright_{\rho_1 \cdots \rho_i} \not\approx D(A))\right] < S \cdot \frac{1}{S} = 1$$

and so there exists a single restriction ρ_i with the desired properties. Inductively continue.

Razborov's simplified combinatorial argument for the Håstad switching lemma (see Section 2.6.3) is a significant ingredient in the simplified lower bound argument for the pigeonhole principle in [UF96].[19]

As a matter of historical interest, we mention in passing that the original approach of [PBI93], based on the improved switching lemma and the notion of "approximate proof" from [BPU92]. The rough idea is as follows. Find an appropriate restriction ρ_1 which leaves a fraction n^{ϵ_1} of the variables still unset, and for which the depth of $A \upharpoonright_\rho$ is at most $d-1$, for each formula A of the original proof P. Argue inductively, finding restrictions $\rho_2, \ldots, \rho_{d-2}$,

[19] Independently, A. Woods discovered a similar approach which is developed in [Kra95].

such that after applying the restrictions $\rho_1 \cdots \rho_i$, the fraction n^{ϵ_i} of variables is still unset, while the depth of $A_{\rho_1 \cdots \rho_i}$ is at most $d - i$, for each formula A of proof P. A separate argument is then given for the nonexistence of an approximate proof of PHP_n^{n+1} involving formulas of depth 2. It should be mentioned as well that independently J. Krajíček [Kra95] and D. Zambella [Zam97], extended Ajtai's forcing approach, using the new switching lemma, to obtain the exponential lower bound.

Historically, the first truly exponential size lower bound for constant-depth Frege systems was given by J. Krajíček [Kra94a], who proved that depth $d+1$ Frege systems are exponentially stronger than depth d Frege systems.

Theorem 5.7.4 ([Kra94a]). *For every $d \geq 0$ and $n \geq 2$, there is an unsatisfiable set T_n^d of depth d sequents of size $O(n^{3+d})$ which have depth $d+1$ LK refutations of quasi-polynomial-size $2^{O(\log^2 n)}$, but for which every depth d LK refutation has size $2^{n^{\Omega(1)}}$.*

The sets T_n^d involve the Sipser function from Section 2.6.2 and express a weak form of the pigeonhole principle. It is an open question whether there is an exponential separation between depth d and depth $d+1$ Frege systems for sets T_n^d of clauses (or at least sequents, whose depth does not depend on d).

We now turn to the combinatorics for the exponential lower bound for PHP_n^{n+1}.

Switching Lemma for Pigeonhole Principle

Let D, R be disjoint sets of integers, with $|D| = n+1$, $|R| = n$ and $S = D \cup R$. Let $M(D, R)$, often written M_n, denote the set of *matchings* between D and R; i.e., the set of all partial injections $\pi : X \to Y$, where $X \subseteq D$, $Y \subseteq R$. We write $(i, j) \in \pi$ to mean that $i \in D, j \in R, \pi(i) = j$. At times, we will consider π extensionally as the set $\{(i,j) : i \in \text{do}(\pi), \pi(i) = j\}$. A matching π *covers* or *touches* element $k \in S = D \cup R$, if

$$(\exists j \in R)[\pi(k) = j] \vee (\exists i \in D)[\pi(i) = k].$$

The set of elements in S covered by matching π is denoted by $V(\pi)$. Depicting the matching π as an undirected bipartite graph with vertex sets D, R consisting of edges $\{i, j\}$ for $\pi(i) = j$, the set $V(\pi)$ is the set of vertices incident to an edge of the graph. Matchings $\pi \in M(D, R)$ will be identified with *restrictions* of the form $\rho : \{x_{i,j} : i \in D, j \in R\} \to \{0, 1, *\}$, where

$$\rho(x_{i,j}) = \begin{cases} 1 \text{ if } \pi(i) = j \\ 0 \text{ if } (\exists k \in R)(k \neq j \wedge \pi(i) = k) \vee (\exists k \in D)(k \neq i \wedge \pi(k) = j) \\ * \text{ otherwise.} \end{cases}$$

Context will distinguish whether π means a matching, restriction, or partial truth assignment. In particular, we sometimes write $\rho \in M(D, R)$ to mean that ρ is a restriction corresponding to a matching in $M(D, R)$. A matching π *covers* the set $X \subseteq S = D \cup R$ if every element of S is covered by π.

Matchings $\pi, \pi' \in M(D, R)$ are *compatible* if $\pi \cup \pi' \in M(D, R)$; in this case, $\pi\pi'$ is written in place of $\pi \cup \pi'$ for that matching which minimally extends both π and π'. For domain D, range R, and union $S = D \cup R$, define $D \upharpoonright_\pi = D - V(\pi)$, $R \upharpoonright_\pi = R - V(\pi)$, $S \upharpoonright_\pi = S - V(\pi)$. For the set $M(D, R)$ of matchings between D and R, $M(D, R) \upharpoonright_\pi = M(D \upharpoonright_\pi, R \upharpoonright_\pi)$. If A is a propositional formula and ρ is a restriction in $M(D, R)$, then $A \upharpoonright_\rho$ is the formula obtained by replacing $x_{i,j}$ by 1 if $\rho(x_{i,j}) = 1$, by 0 if $\rho(x_{i,j}) = 0$, and applying simplifications of the form $B \vee 0 \equiv B$, $B \vee 1 \equiv 1$, $B \wedge 0 \equiv 0$, $B \wedge 1 \equiv B$, $\neg 0 \equiv 1$, $\neg 1 \equiv 0$. If Γ is a set of propositional formulas, then $\Gamma \upharpoonright_\rho = \{A \upharpoonright_\rho : A \in \Gamma\}$. A conjunction A of the form $x_{i_1,j_1} \wedge x_{i_2,j_2} \wedge \cdots \wedge x_{i_r,j_r}$ is a *matching term* if $\pi_A = \{(i_1, j_1), \ldots, (i_r, j_r)\}$ is a matching in $M(D, R)$. The *leaf size* of A, denoted $||A||$, is the number of variables in A, i.e., $|\pi_A|$. If $\pi \in M(D, R)$, then $\wedge\pi$ is a matching term. Note that matching terms are conjunctions of variables (i.e., no negated variables appear in a matching term); this can be arranged, because $\neg x_{i,j}$ can be replaced by $\bigvee_{i' \neq i} x_{i',j} \vee \bigvee_{j' \neq j} x_{i,j'}$. An *$r$-disjunction* is a disjunction of matching terms, each term of size at most r. Formula A is a *matching disjunction over S* if A is an r-disjunction, for some r, and for each variable $x_{i,j}$ appearing in A, $i, j \in S$.

In this section, we prove an exponential lower bound for the size $|P|$ of every depth d proof P of PHP_n^{n+1} in a Frege system involving the connectives $\neg$ and $\vee$. The proof actually yields a stronger result – namely an exponential lower bound on proof size for onto – PHP_n^{n+1}, even if size of formula F were to be redefined as the number of subformulas of F. Moreover, the proof can be extended to handle other connectives for an arbitrary Frege system.

The depth of a circuit (formula) was defined in Chapter 1 as the maximum length of a path from root to leaf in the corresponding directed acyclic graph (formula tree). The lower bound result for the pigeonhole principle is proved for arbitrary fan-in disjunctions, and hence we correspondingly redefine the notion of depth.

In particular, we consider boolean formulas involving variables $x_{i,j}$, $i \in D$, $j \in R$, and only the connectives $\neg, \vee$. The collection of all such formulas is denoted by $L(D, R)$.

- 0 (FALSE) and 1 (TRUE) belong to $L(D, R)$.
- For each $i \in D$ and $j \in R$, $x_{i,j} \in L(D, R)$.
- If $A, B \in L(D, R)$, then $A \vee B \in L(D, R)$ and $\neg A \in L(D, R)$.

Conjunction $A \wedge B$ can be defined by $\neg(\neg A \vee \neg B)$. Considering a boolean formula $A \in L(D, R)$ as a tree, whose root is labeled with the primary connective of A, the *depth* of A is defined to be the maximum number of blocks of $\neg$ and $\vee$ in a path from root to leaf.

The *merged form* of a boolean formula A is a formula B using unbounded fan-in disjunctions, which is equivalent to A. Formally, the merged form of $0, 1, x_{i,j}$ is itself. If A is a disjunction in $L(D, R)$, then let $\{A_i : i \in I\}$

be the set of subformulas A_i of A, such that A_i is not a disjunction, but every subformula of A containing A_i is a disjunction. If by induction the *merged form* of A_i is B_i, then the merged form of A is $\bigvee_{i \in I} B_i$. If A is the merged form for boolean formula B, and A' is obtained by removing all double negations from A, then the depth of A (as here defined) equals the depth (as defined in Chapter 1) of A'.

Definition 5.7.1 (Matching tree). *Let D, R be disjoint sets with $|D| = n+1$, $|R| = n$ and $S = D \cup R$. A matching tree over S is a rooted tree T, whose internal (non-leaf) nodes are labeled by elements of S, and whose edges are labeled by ordered pairs (i,j), with $i \in D$, $j \in R$, such that the following conditions hold.*

1. *If T consists of a single node, then T is a matching tree.*
2. *Suppose that the root u of T is labeled by an element $i \in D$ ($j \in R$), and for each $j \in R$ ($i \in D$), there is a child node v and an edge labeled by (i,j) from root u to child v. Then T is a matching tree, provided that $T^{(i,j)}$ is a matching tree over $(D - \{i\}) \cup (R - \{j\})$, for each label (i,j) from root u to v, where $T^{(i,j)}$ denotes the subtree of T rooted at v.*

Note that though the node labels from S and leaf labels (i,j) of matching tree T are not necessarily distinct, they are so along any path from root to leaf. If p is a node of T, then the matching defined by the leaf labels from root to p is denoted by $\pi(p)$, and we have $\pi(p) \in M(D,R)$. If the leaves of T are labeled by $0,1$, then T is called a *matching decision tree.* For matching decision tree T, the complement tree T^c is obtained from T by interchanging leaf labels $0,1$. The depth of a matching tree T, denoted $d(T)$, is given by $\max\{|\pi(\ell)| : \ell \in T\}$.

Definition 5.7.2. *If $S = D \cup R$, F is a matching disjunction over S, and T is a matching decision tree over S, then T is said to represent F, denoted $T \approx F$, if for every leaf ℓ of T, $F \upharpoonright_{\pi(\ell)} \equiv 0$ if ℓ is labeled by 0, and $F \upharpoonright_{\pi(\ell)} \equiv 1$ if ℓ is labeled by 1. A matching decision tree T minimally represents F if for every leaf ℓ of T, and every node $p \neq \ell$ occurring on the path from the root of T to ℓ, $F \upharpoonright_{\pi(p)} \not\equiv 0$ and $F \upharpoonright_{\pi(p)} \not\equiv 1$.*

Clearly if T represents F, then by possibly pruning T back to nodes p, for which $F \upharpoonright_{\pi(p)}$ evaluates to 0 or 1, we can produce a tree $T' \subseteq T$ which minimally represents F.

Definition 5.7.3 (Matching disjunction $Disj(T)$). *If T is a matching tree, and p is a node of T, then $\pi(p)$ is the matching, consisting of the edge labels in the path from root of T to node p. The set of all branches of T is defined by*

$$Br(T) = \{\pi(\ell) : \ell \text{ is leaf of } T.\}.$$

If T is a matching decision tree, then

$$Br_0(T) = \{\pi(\ell) : \ell \text{ is leaf of } T \text{ with node label } 0\}$$
$$Br_1(T) = \{\pi(\ell) : \ell \text{ is leaf of } T \text{ with node label } 1\}$$

and the mapping disjunction $Disj(T)$ is defined by

$$\bigvee_{\pi \in Br_1(T)} \bigwedge_{(i,j)\in\pi} x_{i,j}.$$

Letting $Val(D, R)$ denote the collection of total truth assignments (or valuations) on the variables $x_{i,j}$ for $i \in D, j \in R$, we have

$$\{\sigma \in Val(D, R) : (\exists \rho \in Br(T))(\rho \subseteq \sigma)\} \subseteq Val(D, R)$$

and the fact that this inclusion may be proper is a key ingredient exploited in the lower bound proof of the pigeonhole principle using k-evaluations. The proof of the following lemma is not difficult, and is left to the reader.

Lemma 5.7.1. *Let T be a matching decision tree over $S = D \cup R$, ℓ a leaf of T, and $\rho \in M(D, R)$ a restriction extending $\pi(\ell)$. Then $Disj(T) \upharpoonright_\rho \equiv 0$ (1) if and only if ℓ has label 0 (1).*

Definition 5.7.4 (Restriction $T \upharpoonright_\rho$ of matching tree T). *Let $\rho \in M(D, R)$, and let T be a matching tree over $S = D \cup R$.*

1. *If T consists of a single node, then $T \upharpoonright_\rho = T$.*
2. *Suppose that T consists of more than a single node and that the root is labeled by an element of $k \in S \cap V(\pi)$ touched by ρ.*
   a) *If $k \in D$, and $j \in R$ is such that $(k, j) \in \rho$, then $T \upharpoonright_\rho = T^{(k,j)} \upharpoonright_\rho$.*
   b) *If $k \in R$, and $i \in D$ is such that $(i, k) \in \rho$, then $T \upharpoonright_\rho = T^{(i,k)} \upharpoonright_\rho$.*
3. *Suppose that T consists of more than a single node and that the root u is labeled by an element of $k \in S - V(\rho)$ not touched by ρ.*
   a) *If $k \in D$, then $T \upharpoonright_\rho$ is the tree obtained by connecting root p to every subtree $T^{(k,j)} \upharpoonright_\rho$, such that (k, j) is compatible with ρ, $j \in R$, (k, j) labels the edge from p to q, and q is a child of p in T.*
   b) *If $k \in R$, then $T \upharpoonright_\rho$ is the tree obtained by connecting root p to every subtree $T^{(i,k)} \upharpoonright_\rho$, such that (i, k) is compatible with ρ, $i \in D$, (i, k) labels the edge from p to q, and q is a child of p in T.*

The proof of the following lemma is by induction on tree depth and left to the reader.

Lemma 5.7.2. *Let D, R be disjoint sets, with $|D| = n + 1$, $|R| = n$, and $S = D \cup R$. Let T be a matching decision tree over S, and let $\rho \in M(D, R)$ be a restriction.*

1. *The tree $T \upharpoonright_\rho$ is a matching decision tree over $D \upharpoonright_\rho \cup R \upharpoonright_\rho$.*
2. *$Disj(T) \upharpoonright_\rho \equiv Disj(T \upharpoonright_\rho)$.*
3. *$(T \upharpoonright_\rho)^c = T^c \upharpoonright_\rho$.*

4. *If ℓ is a leaf of $T \upharpoonright_\rho$, then there exists a leaf ℓ' of T, such that $\pi(\ell') \subseteq \rho \cup \pi(\ell)$, where ℓ, ℓ' have the same leaf label.*
5. *If T represents matching disjunction F, then $T \upharpoonright_\rho$ represents $F \upharpoonright_\rho$.*

Let A be a formula in the language $L(D, R)$, and let T be a matching decision tree which represents A. It follows from the second part of Lemma 5.7.2 that for any leaf ℓ of T, $Disj(T) \upharpoonright_{\pi(\ell)} \equiv A \upharpoonright_\rho \equiv 0$ (1) if the label of ℓ is 0 (1), and thus that T represents $Disj(T)$, in the sense of Definition 5.7.2. It follows this observation, and from Definitions 5.7.2 and 5.7.3 that

$$\begin{aligned} T \approx 0 &\Leftrightarrow Br_0(T) = Br(T) \\ &\Leftrightarrow (\forall \pi)\,(\pi \in Br(T(A)) \rightarrow Disj(T(A)) \upharpoonright_\pi \equiv 0) \end{aligned}$$

and

$$\begin{aligned} T \approx 1 &\Leftrightarrow Br_1(T) = Br(T) \\ &\Leftrightarrow (\forall \pi)\,(\pi \in Br(T(A)) \rightarrow Disj(T(A)) \upharpoonright_\pi \equiv 1) \end{aligned}$$

Definition 5.7.5 (Full matching tree). *Let D, R be disjoint sets with*

$$|D| = n + 1, |R| = n, S = D \cup R \text{ and } X \subseteq S.$$

The full matching tree full matching tree Full(X, S) for X over S is a matching tree on S constructed as follows.

Case 1. *$X = \emptyset$. In this case, define Full($\emptyset, S$) to consist of a single node.*

Case 2. *Letting $k = \min(X)$, we have $k \in D$. By recursion, for each $j \in R$ construct Full($X - \{k, j\}, S - \{k, j\}$). Define Full($X, S$) to consist of a root p, labeled by k, and for each $j \in R$, there is an edge labeled by (k, j) connecting p to a node q which is the root of Full($X - \{k, j\}, S - \{k, j\}$).*

Case 3. *Letting $k = \min(X)$, we have $k \in R$. By recursion, for each $i \in D$ construct Full($X - \{i, k\}, S - \{i, k\}$). Define Full($X, S$) to consist of a root p, labeled by k, and for each $i \in D$, there is an edge labeled by (i, k) connecting p to a node q which is the root of Full($X - \{i, k\}, S - \{i, k\}$).*

Definition 5.7.6 (Canonical matching decision tree $TreeS(F)$). *Let D, R be disjoint sets with*

$$|D| = n + 1, |R| = n, S = D \cup R$$

and let $F = F_1 \vee \cdots \vee F_m$ be a matching disjunction over S. The canonical matching decision tree $Tree_S(F)$ for F over S is a matching decision tree on S constructed as follows.

Case 1. *If $F \equiv 0$, then $Tree_S(F)$ consists of a single node labeled 0. If $F \equiv 1$, then $Tree_S(F)$ consists of a single node labeled 1.*

Case 2. *Suppose that $F \not\equiv 0$, $F \not\equiv 1$ and that C is the first disjunct F_j for which $C \not\equiv 0$. $Tree_S(F)$ is constructed by first constructing the full matching tree $Full(V(C), S)$ of $V(C)$ over S, and replacing each leaf ℓ of that tree by the canonical matching decision tree $Tree_{S \restriction_{\pi(\ell)}}(F \restriction_{\pi(\ell)})$.*

It is difficult to motivate the definition of canonical matching decision tree. In trying to prove Lemma 5.7.3 for *any* matching decision tree, there is a technical problem which arises. Under these conditions, we cannot guarantee that $\pi_1 \cdots \pi_i$ is an initial segment of $\overline{\pi}$, or that π_i covers $V(\sigma_i)$, or that $C_i \restriction_{\rho\pi_1 \cdots \pi_i} \equiv 0$ (consult the proof of Lemma 5.7.3 for notation). It turns out that these technical problems can be resolved by proving Lemma 5.7.3 only for canonical matching decision trees.

Recall from Chapter 2 that $Code(r, s)$ is defined to be the set of sequences $\beta = (\beta_1, \ldots, \beta_k)$ where $\beta_i \in \{\downarrow, *\}^r - \{*\}^r$ for each $1 \leq i \leq k$, and where there are exactly s occurrences of $\downarrow$ in the entire sequence β, where $k \leq s$. Lemma 2.6.6 states that $|Code(r, s)| < (r/\ln 2)^s$.

For fixed n, $|D| = n + 1$, $|R| = n$, $S = D \cup R$, and $\ell < n$, let M_n^ℓ denote the set of matchings ρ, such that $n - \ell$ elements of the domain D have been set; i.e.,

$$M_n^\ell = \{\rho \in M_n : |R \restriction_\rho| = \ell\}.$$

For $F = F_1 \vee \cdots \vee F_m$ be an r-disjunction over S and $s > 0$, define

$$Bad_n^\ell(F, s) = \{\rho \in M_n^\ell : d(Tree_{S \restriction_\rho}(F \restriction_\rho)) \geq s\}.$$

With this notation, we have the following result.

Lemma 5.7.3. *There is an injection from $Bad_n^\ell(F, s)$ into*

$$\bigcup_{s/2 \leq j \leq s} M_n^{\ell - j} \times Code(r, j) \times (2\ell + 1)^s.$$

Proof. Suppose that $\rho \in Bad_n^\ell(F, s)$ and consider the leftmost path $\overline{\pi}$ of $T(F \restriction_\rho)$ of length at least s. We will define restrictions $\pi_1, \ldots, \pi_k$, where $\pi = \pi_1 \cdots \pi_k$, $\pi \subseteq \overline{\pi}$, $|\pi| = s$, and for $1 \leq i < k$, $\pi_1 \cdots \pi_i$ is an *initial segment* of the path $\overline{\pi}$. If $Bad_n^\ell(F, s)$ had been defined as $\{\rho \in M_n^\ell : d(T_{S \restriction_\rho}(F \restriction_\rho)) \geq s\}$, where $T_{S \restriction_\rho}(F \restriction_\rho)$ is a minimum depth (not necessarily canonical) matching decision tree, then we could not guarantee that $\pi_1 \cdots \pi_i$ would be an initial segment of the path stipulated by $\overline{\pi}$ in $T_{S \restriction_\rho}(F \restriction_\rho)$, for $1 \leq i < k$. The reader should attempt to carry out the proof under this assumption, to understand the importance of having considered only canonical matching decision trees.

By induction, define the following three sequences, which depend on F and $\overline{\pi}$, and satisfy the following conditions:

1. $C_1, \ldots, C_k$, where each C_i is one of the disjuncts $F_1, \ldots, F_m$ of F.
2. $\sigma_1, \ldots, \sigma_k$, where each restriction $\sigma_i \subseteq \delta_i$, $C_i = \wedge \delta_i$.

3. $\pi_1, \ldots, \pi_k$, which partition $\pi = \pi_1 \cdots \pi_k$. Additionally, for $1 \le i < k$, π_i is the minimal restriction contained in $\overline{\pi}$ which covers $V(\sigma_i)$, and $\pi_1 \cdots \pi_i$ is an initial segment of the path $\overline{\pi}$.

This is done as follows. Suppose that $C_1, \ldots, C_{i-1}, \sigma_1, \ldots, \sigma_{i-1}, \pi_1, \ldots, \pi_{i-1}$ have been defined to satisfy conditions (1)-(3) and that $|\pi_1 \cdots \pi_{i-1}| < s$.

Let C_i be the first disjunct among $F_1, \ldots, F_m$ of F, such that

$$C_i \restriction_{\rho\pi_1\cdots\pi_{i-1}} \not\equiv 0 \text{ and } C_i \restriction_{\rho\pi_1\cdots\pi_{i-1}} \not\equiv 1.$$

This must exist, since otherwise the path in $Tree_{S\restriction_\rho}(F \restriction_\rho)$ designated by $\pi_1 \cdots \pi_{i-1}$ would terminate in a leaf, violating the assumption that $|\pi_1 \cdots \pi_{i-1}| < s$

Let $\overline{\sigma_i}$ be the minimal restriction which sets $C_i \restriction_{\rho\pi_1\cdots\pi_{i-1}\overline{\sigma_i}} \equiv 1$. Let $\overline{\pi_i}$ be the minimal submatching of $\overline{\pi}$ which covers all elements touched by $\overline{\sigma_i}$.

Case 1. $|\pi_1 \cdots \pi_{i-1}\overline{\pi_i}| < s$. In this case, define $\sigma_i = \overline{\sigma_i}$ and $\pi_i = \overline{\pi_i}$.

Case 2. $|\pi_1 \cdots \pi_{i-1}\overline{\pi_i}| \ge s$. In this case, set $k = i$. Suppose that $p_1, \ldots, p_t$ is a listing of the ordered pairs of $\overline{\pi_k}$ in the order they appear in the branch of decision tree $Tree_{S\restriction_\rho}(F \restriction_\rho)$ corresponding to $\overline{\pi}$. It must be that each p_j contains an element v_j of $D \cup R$ which is the first element of $V(\overline{\sigma_k})$ not appearing in the elements of the ordered pairs of $p_1, \ldots, p_{j-1}$. If $v_j \in D$, then set q_j to be the ordered pair from $\overline{\sigma_k}$ with first coordinate v_j, while if $v_j \in R$, then set q_j to be the ordered pair from $\overline{\sigma_k}$ with second coordinate v_j. Finally, let $t_0 \le t$ be such that $|\pi_1 \cdots \pi_{k-1}| + t_0 = s$, and set $\sigma_k = \{q_1, \ldots, q_{t_0}\}$ and $\pi_k = \{p_1, \ldots, p_{t_0}\}$. Note that it may not be the case that $C_k \restriction_{\rho\pi_1\cdots\pi_{k-1}\sigma_k} \equiv 1$; nevertheless, $C_k \restriction_{\rho\pi_1\cdots\pi_{k-1}\sigma_k} \not\equiv 0$.

It follows that for $1 \le i \le k$, $\sigma_i \subseteq \delta = \wedge C_i$. For $1 \le i < k$, clearly $|\sigma_i| \le |\pi_i| \le 2 \cdot |\sigma_i|$, while $|\sigma_k| = |\pi_k|$. Define σ to be $\sigma_1 \cdots \sigma_k$ and π to be $\pi_1 \cdots \pi_k$.

Claim. For $1 \le i < k$, $\pi_1 \cdots \pi_i$ forms an initial segment of the path $\overline{\pi}$.

Proof of Claim. By induction on i. In the construction of $Tree_{S\restriction_\rho}(F \restriction_\rho)$, σ_i and π_i are paths in the full matching tree $\text{Full}(V(\sigma_i), S \restriction_{\rho\pi_1\cdots\pi_{i-1}})$, and in that construction, the leaf of the full matching tree designated by π_i is then replaced by $Tree_{S\restriction_{\rho\pi_1\cdots\pi_i}}(F \restriction_{\rho\pi_1\cdots\pi_i})$. It follows that $\pi_1 \cdots \pi_i$ forms an initial segment of the path $\overline{\pi}$.

Claim. For $1 \le i < k$, $C_i \restriction_{\rho\pi_1\cdots\pi_i} \equiv 0$. By construction of $Tree_{S\restriction_\rho}(F \restriction_\rho)$, π_i is a path in the full matching tree $\text{Full}(V(\sigma_i), S \restriction_{\rho\pi_1\cdots\pi_{i-1}})$, and so $C_i \restriction_{\rho\pi_1\cdots\pi_i} \equiv 0$ or $C_i \restriction_{\rho\pi_1\cdots\pi_i} \equiv 1$. The latter is not possible, since otherwise $F \restriction_{\rho\pi_1\cdots\pi_i} \equiv 1$, and so $\pi_1 \cdots \pi_i$ would label a path of $Tree_{S\restriction_\rho}(F \restriction_\rho)$ terminating in a leaf, contradicting the assumption that $|\pi_1 \cdots \pi_i| < s$.

Claim. For $1 \le i \le k$, $\rho\sigma_1 \cdots \sigma_i \in M_n$.

Proof of Claim. The proof is by induction on i. Suppose that for $b \ne c$, we have $(a, b) \in \rho$ and $(a, c) \in \sigma_j$, for some $1 \le j \le i$. Then the variable $x_{a,c}$

appearing in C_i would have been set to 0 in $C_i \upharpoonright_\rho$ and hence $a \notin V(\sigma_i)$, a contradiction. A similar argument holds in the case that $(a,c) \in \rho$ and $(b,c) \in \sigma_j$. It follows that $\rho\sigma_1 \in M_n$.

Suppose that by induction $\rho\sigma_1 \cdots \sigma_{i-1} \in M_n$, and that for $b \neq c$, we have $(a,b) \in \sigma_j$ and $(a,c) \in \sigma_i$, for some $1 \leq j < i$. Since π_j is the minimal submatching of $\overline{\pi}$ which covers $V(\sigma_j)$, π_j touches a, and hence $a \notin V(\sigma_i)$, a contradiction. A similar argument holds in the case that $(a,c) \in \sigma_j$ and $(b,c) \in \sigma_i$. It follows by induction that $\rho\sigma_1 \cdots \sigma_k \in M_n$.

Claim. For $1 \leq i \leq k$, $\rho\pi_1 \cdots \pi_{i-1}\sigma_i \cdots \sigma_k \ \in M_n$.

Proof of Claim. Suppose that for $b \neq c$, we have $(a,b) \in \pi_j$ and $(a,c) \in \sigma_{j'}$, for some $1 \leq j < i \leq j' \leq k$. Then $a \notin V(\sigma_{j'})$, a contradiction. A similar argument holds in the case that $(a,c) \in \pi_j$ and $(b,c) \in \sigma_{j'}$. This establishes the claim.

Since $\rho \in M_n^\ell$, there are $\ell+1$ elements of $D \upharpoonright_\rho$ and ℓ elements of $R \upharpoonright_\rho$ which are unset by ρ. Define the ordering $\prec$ on these $2\ell+1$ elements as follows, where we assume that $D \cup R$ is ordered by $\leq$. Given $a,b \in D \upharpoonright_\rho \cup R \upharpoonright_\rho$, $a \prec b$ if

- $a \in V(\sigma_i)$, $b \in V(\sigma_j)$, for $1 \leq i < j \leq k$, or
- if $a \leq b$ and $a,b \in V(\sigma_i)$, for some $1 \leq i \leq k$, or
- $a \in V(\sigma)$, $b \notin V(\sigma)$, or
- $a \leq b$ and $a,b \notin V(\sigma)$.

For $a \in D \upharpoonright_\rho \cup R \upharpoonright_\rho$, let $\text{index}_\prec(a)$ denote that number in $\{1,\ldots,2\ell+1\}$ which corresponds to the position of a in the $\prec$-ordering of $D \upharpoonright_\rho \cup R \upharpoonright_\rho$.

Define the map $G(\rho) = (G_1(\rho), G_2(\rho), G_3(\rho))$ as follows:

1. $G_1(\rho) = \rho\sigma$.
2. $G_2(\rho) = (\beta_1,\ldots,\beta_k)$, where

$$\beta_i(j) = \begin{cases} \downarrow \text{if } \sigma_i \text{ sets the } j\text{-th variable in } C_i \\ 0 \text{ else.} \end{cases}$$

3. $G_3(\rho) \in (2\ell+1)^s$ is the s-tuple $(c_1,\ldots,c_s)$ of elements in $\{1,\ldots,2\ell+1\}$ defined as follows. Suppose that π consists of the ordered pairs $p_1,\ldots,p_s$, where $p_i = (a_i,b_i)$. Extend $\prec$ to order $p_1,\ldots,p_s$ by $p_i \prec p_j$ if and only if $\min_\prec\{a_i,b_i\} \prec \min_\prec\{a_j,b_j\}$. Suppose that $\gamma : \{1,\ldots,s\} \to \{1,\ldots,s\}$ is that permutation for which $p_{\gamma(i)} \prec p_{\gamma(i+1)}$, for $1 \leq i < s$. Finally, for $1 \leq i \leq s$, set $c_i = \text{index}_\prec(\max_\prec\{a_{\gamma(i)}, b_{\gamma(i)}\})$. This defines $G_3(\rho)$.

Now let $j = |\rho|$.

Claim. $G(\rho) \in M_n^{\ell-j} \times Code(r,j) \times (2\ell+1)^s$, where $\frac{s}{2} \leq j \leq s$.

Proof of Claim. By a previous claim, $\rho\sigma \in M_n$, so $G_1(\rho) = \rho\sigma \in M_n^{\ell-j}$. Since $\sigma_i \subseteq \delta_i = \wedge C_i$, $G_2(\rho) \in Code(r,j)$. By definition, $G_3(\rho) \in (2\ell+1)^s$. As earlier mentioned, for $1 \leq i < k$, $|\sigma_i| \leq |\pi_i| \leq 2 \cdot |\sigma_i|$, and $|\sigma_k| = |\pi_k|$.

Thus $|\sigma| \leq |\pi| \leq 2 \cdot |\sigma|$, hence $\frac{|\pi|}{2} \leq |\sigma| \leq |\pi|$, and recalling that $|\sigma| = j$ and $|\pi| = s$, we have $\frac{s}{2} \leq j \leq s$.

CLAIM. G is an injection from $Bad_n^\ell(F, s)$ into

$$\bigcup_{s/2 \leq j \leq s} M_n^{\ell-j} \times Code(r, j) \times (2\ell + 1)^s.$$

Proof of Claim. D,R, $F = F_1 \vee \cdots \vee F_m$, and s are known, as they were fixed at the beginning of the proof of the lemma. Given $G(\rho)$, we explain how to unambiguously reconstruct ρ, thus proving injectivity of G. From $G_1(\rho) = \rho\sigma$, we determine the sets of $S = D \cup R$ unset by $\rho\sigma$.

Suppose that in the reconstruction process, $C_1, \ldots, C_{i-1}$, $\sigma_1, \ldots, \sigma_{i-1}$, $\pi_1, \ldots, \pi_{i-1}$, $\rho\pi_1 \cdots \pi_{i-1}\sigma_i \cdots \sigma_k$ and the elements $V(\sigma_1), \ldots, V(\sigma_{i-1})$ have been found. From this, the position of the elements of $V(\sigma_1) \cup \cdots \cup V(\sigma_{i-1})$ in the $\prec$-ordering is uniquely determined.

SUBCLAIM. C_i is the first disjunct of F which satisfies $C_i \upharpoonright_{\rho\pi_1\cdots\pi_{i-1}\sigma_i\cdots\sigma_k} \not\equiv 0$.
Proof of Subclaim. By construction, C_i is the first disjunct of F for which $C_i \upharpoonright_{\rho\pi_1\cdots\pi_{i-1}} \not\equiv 0$ and $C_i \upharpoonright_{\rho\pi_1\cdots\pi_{i-1}} \not\equiv 1$. If F_j is a disjunct of F occurring before C_i, then $F_j \upharpoonright_{\rho\pi_1\cdots\pi_{i-1}} \equiv 0$ or 1. The latter possibility cannot be realized, for otherwise $\pi_1 \cdots \pi_{i-1}$ would label a path in $Trees_{S \upharpoonright \rho}(F \upharpoonright_\rho)$ terminating in a leaf, contradicting the fact that $|\pi_1 \cdots \pi_{i-1}| < s$. Hence $F_j \upharpoonright_{\rho\pi_1\cdots\pi_{i-1}} \equiv 0$, and so $F_j \upharpoonright_{\rho\pi_1\cdots\pi_{i-1}\sigma_i\cdots\sigma_k} \equiv 0$. By construction, for $1 \leq i < k$, $C_i \upharpoonright_{\rho\pi_1\cdots\pi_{i-1}\sigma_i} \equiv 1$ and $C_k \upharpoonright_{\rho\pi_1\cdots\pi_{k-1}\sigma_k} \not\equiv 0$, hence for all $1 \leq i \leq k$, $C_i \upharpoonright_{\rho\pi_1\cdots\pi_{i-1}\sigma_i} \not\equiv 0$, so $C_i \upharpoonright_{\rho\pi_1\cdots\pi_{i-1}\sigma_i\cdots\sigma_k} \not\equiv 0$. This concludes the proof of the subclaim.

After determining C_i, the variables set by σ_i can be found using the term β_i from $G_2(\rho) = (\beta_1, \ldots, \beta_k)$. Consulting $G_1(\rho) = \rho\sigma$ on these variables determines the restriction σ_i. From $V(\sigma_i)$ and $V(\sigma_1), \ldots, V(\sigma_{i-1})$, we can determine the position of elements of $V(\sigma_1) \cup \cdots \cup V(\sigma_i)$ in the ordering $\prec$. Now π_i touches every element of $V(\sigma_i)$, hence every ordered pair in π_i contains at least one element from $V(\sigma_i)$. Using $\text{index}_\prec(a)$, for $a \in V(\sigma_i)$, we can determine the lower index element from each ordered pair of π_i – indeed, begin with the $\min_\prec$ element a of $V(\sigma_i)$, which must be touched by π_i, apply $G_3(\rho)$ to find the corresponding element b with high index, and depending on whether $a \in D, b \in R$ or $b \in D, a \in R$, we have determined that (a, b) resp. (b, a) belongs to π, and continue inductively, taking the next element of $V(\sigma_i)$ not treated, etc. This determines π_i, whence by replacement of σ_i by π_i, we find $\rho\pi_1 \cdots \pi_i\sigma_{i+1} \cdots \sigma_k$. This completes the proof of claim that G is an injection, thus completing the proof of Lemma 5.7.3.

Lemma 5.7.4 (Matching switching lemma). *Let $F = F_1 \vee \cdots \vee F_m$ be an r-disjunction over $D \cup R$, where $|D| = n+1$ and $|R| = n$. Let $\ell \geq 10$, and set $p = \frac{\ell}{n}$. If $r \leq \ell$ and $p^4 n^3 \leq \frac{1}{10}$, then*

$$\frac{|Bad_n^\ell(F, 2s)|}{|M_n^\ell|} \leq (9p^4n^3r)^s.$$

Proof. By Lemma 5.7.3, $|Bad_n^\ell(F, 2s)| \leq |\cup_{s \leq j \leq 2s} M_n^{\ell-j} \times Code(r, j) \times (2\ell + 1)^{2s}|$, and we derive an upper bound for the latter as follows.

Each matching π in M_n^ℓ is determined by first choosing ℓ out of n elements in R which will not be the image of any element of D, and then for each element y among the $n-\ell$ remaining elements of R choosing a distinct element x of D and placing the ordered pair (x, y) in π. Recall Definition 2.6.2 where the falling factorial

$$a^{\underline{m}} = a(a-1)\cdots(a-(m-1)) = m!\binom{a}{m}$$

was defined, which satisfies $a^{\underline{m+n}} = a^{\underline{m}}(a-m)^{\underline{n}}$. With this notation, it follows that

$$\begin{aligned} |M_n^\ell| &= \binom{n}{\ell}(n+1)^{\underline{n-\ell}} \\ &= \frac{n^{\underline{\ell}}(n+1)^{\underline{n-\ell}}}{\ell!} \end{aligned}$$

and so

$$\begin{aligned} \frac{|M_n^{\ell-j}|}{|M_n^\ell|} &= \frac{n^{\underline{\ell-j}}(n+1)^{\underline{n-\ell+j}}\ell!}{(\ell-j)!n^{\underline{\ell}}(n+1)^{\underline{n-\ell}}} \\ &= \frac{(\ell+1)^{\underline{j}}\ell!}{(\ell-j)!(n-\ell+j)^{\underline{j}}} \\ &= \frac{(\ell+1)^{\underline{j}}\ell^{\underline{j}}}{(n-\ell+j)^{\underline{j}}} \\ &= \left(\frac{\ell(\ell+1)}{n-\ell}\right)^j. \end{aligned}$$

By Lemma 2.6.6, $|Code(r, j)| < (r/\ln 2)^j < (1.5r)^j$, and so the quantity

$$|\bigcup_{s \leq j \leq 2s} M_n^{\ell-j} \times Code(r, j) \times (2\ell+1)^{2s}| \tag{5.87}$$

is bounded above by

$$\sum_{s \leq j \leq 2s} \left(\frac{\ell(\ell+1)}{n-\ell}\right)^j (r/\ln 2)^j (2\ell+1)^{2s}$$

hence by

$$(2\ell+1)^{2s} \sum_{s \leq j \leq 2s} \left(\frac{1.5r\ell(\ell+1)}{n-\ell}\right)^j. \tag{5.88}$$

Since $r \leq \ell$, $p = \frac{\ell}{n}$, $p^4 n^3 \leq \frac{1}{10}$ and $10 \leq \ell$, we have

$$\begin{aligned}
\frac{1.5r\ell(\ell+1)}{n-\ell} &\le \frac{1.5\ell^2(\ell+1)}{n-\ell} \\
&= \frac{\ell^3}{n} \cdot \frac{1.5(1+1/\ell)}{1-p} \\
&= \left(\frac{\ell}{n}\right)^3 \cdot n^2 \cdot \frac{1.5(1+1/\ell)}{1-p} \\
&\le p^3 n^2 \cdot \left(\frac{\ell}{n} \cdot \frac{n}{\ell}\right) \cdot \frac{1.65}{0.9999} \\
&\le \frac{p^4 n^3}{\ell} \cdot 1.6502 < 0.016502
\end{aligned}$$

hence $\frac{1.5r\ell(\ell+1)}{n-\ell} < 0.016502$, and so

$$\begin{aligned}
\sum_{s \le j \le 2s} \left(\frac{1.5r\ell(\ell+1)}{n-\ell}\right)^j &= \left(\frac{1.5r\ell(\ell+1)}{n-\ell}\right)^s \sum_{0 \le j \le s} \left(\frac{1.5r\ell(\ell+1)}{n-\ell}\right)^j \\
&\le \left(\frac{1.5r\ell(\ell+1)}{n-\ell}\right)^s \sum_{0 \le j \le s} (0.016502)^j \\
&\le \left(\frac{1.5r\ell(\ell+1)}{n-\ell}\right)^s \cdot \frac{1}{1-0.016502} \\
&\le 1.01678 \cdot \left(\frac{1.5r\ell(\ell+1)}{n-\ell}\right)^s . \qquad (5.89)
\end{aligned}$$

It follows that

$$\begin{aligned}
\frac{|Bad_n^\ell(F,2s)|}{|M_n^\ell|} &\le 1.01678(2\ell+1)^{2s} \left(\frac{1.5r\ell(\ell+1)}{n-\ell}\right)^s \\
&\le 1.01678 \left(\frac{1.5r\ell(\ell+1)(2\ell+1)^2}{n-\ell}\right)^s . \qquad (5.90)
\end{aligned}$$

Recalling that $10 \le \ell$, and $\frac{1}{1-p} \le 0.9999$, we have

$$\begin{aligned}
\frac{1.5r\ell(\ell+1)(2\ell+1)^2}{n-\ell} &\le \frac{6(\ell+1)^3 r\ell}{n-\ell} \\
&\le \frac{\ell^4 r}{n} \cdot \frac{6(1+1/\ell)^3}{1-p} \\
&\le \frac{\ell^4 r}{n} \cdot \frac{6(1.1)^3}{0.9999} \\
&\le \frac{8\ell^4 r}{n} = 8p^4 n^3 r. \qquad (5.91)
\end{aligned}$$

From equations (5.90) and (5.91) we obtain that

$$\frac{|Bad_n^\ell(F,2s)|}{|M_n^\ell|} \le 1.01678(8p^4n^3r)^s \le (9p^4n^3r)^s.$$

This completes the proof of Lemma 5.7.4.

k-Evaluations and the Lower Bound

The notion of k-evaluation is crucial to the lower bound proof for the pigeonhole principle in this section. To gain intuition for this notion, it is important to better understand what it means for a matching decision tree T to *represent* a matching disjunction, as defined earlier in Definition 5.7.2.

In Section 1.11.5 of Chapter 1, we defined how a decision tree T, and more generally a decision diagram, *represents* a boolean formula or function F, in that for all total truth assignments σ, there exists a leaf ℓ of T such that $\pi(\ell) \subseteq \sigma$ and $F \restriction_\sigma \equiv 0$ (1) if ℓ is labeled by 0 (1). This is equivalent to the condition that $F \equiv Disj(T)$. In Chapter 2, for instance in Definition 5.7.2, we defined how a decision tree T *represents* a boolean formula *over* a possibly proper set of $\mathcal{C}$ of total truth assignments, in that for all total truth assignments $\sigma \in \mathcal{C}$, there exists a leaf ℓ of T such that $\pi(\ell) \subseteq \sigma$ and $F \restriction_\sigma \equiv 0$ (1) if ℓ is labeled by 0 (1).[20] This is equivalent to the condition that $F \equiv_\mathcal{C} Disj(T)$.

In the notation of Definition 5.7.2, the notion that T represents F, as used in Chapters 1 and 2, would require the following conditions:

1. For every total truth assignment σ (or if relevant, $\sigma \in \mathcal{C}$), there is a path in T from root to leaf ℓ determined by σ, such that σ extends $\pi(\ell)$, i.e., $\pi(\ell) \subseteq \sigma$.
2. For every leaf ℓ of T, $F \restriction_{\pi(\ell)} \equiv 0$ if ℓ is labeled by 0, and $F \restriction_{\pi(\ell)} \equiv 1$ if ℓ is labeled by 1.

In contrast, Definition 5.7.2 only requires condition (2), but not (1), so that $F \approx Disj(T)$, but not necessarily $F \equiv Disj(T)$ or $F \equiv_\mathcal{C} Disj(T)$. The point is that there may exist total truth assignments which do not extend any restriction $\pi(\ell)$ for any leaf ℓ of T. This is a key property in the proof of the exponential size lower bound for constant-depth Frege proofs of the pigeonhole principle, where one might be tempted to consider the class $\mathcal{C}$ of *critical* truth assignments from the proof of Theorem 5.4.7. If T is the matching decision tree for $x_{i,j}$, as given in Definition 5.7.7, and σ is critical at i, so that $\sigma(x_{i,k}) = 0$ for all $1 \leq k \leq n$, then for no leaf ℓ of T is it the case that σ is compatible with $\pi(\ell)$. Note that the Compatibility Lemma 5.7.5 stipulates a condition when a partial truth assignment will be compatible with a matching decision tree.

We now introduce the key notion of k-evaluation, due to [KPW95], and as modified in [UF96].

Definition 5.7.7 (k-evaluation). *Let D, R be disjoint sets, $|D| = n+1$, $|R| = n$, $S = D \cup R$, and fix $k > 0$. Let Γ be a set of boolean formulas in $L(D,R)$ closed under subformulas. A k-evaluation on Γ is an assignment T of matching decision trees $T(A)$ for each formula $A \in \Gamma$, which satisfies the following:*

[20] In the terminology of Chapter 2, T represents F provided that for all $\sigma \in \mathcal{C}$ (e.g. all σ corresponding to a k-layered graph), the path in T from root to leaf determined by σ has leaf labeled by 1 (0) iff $\sigma \models F$ ($\sigma \not\models F$).

1. *$T(A)$ has depth at most k, for each $A \in \Gamma$.*
2. *$T(1)$ resp. $T(0)$ is the tree consisting of a single node, which is labeled* 1 *resp.* 0.
3. *$T(x_{i,j})$ is the tree of depth* 1*, whose root is labeled by i, and which contains an edge from the root to a leaf labeled by (i, k), for each $k \in R$. The leaf connected to root by edge labeled (i, j) has label* 1*, while all other leaves have label* 0.
4. *$T(\neg A) = T(A)^c$.*
5. *If A is a disjunction, and $\bigvee_{i=1}^{m} A_i$ is the merged form of A, then $T(A)$ represents $\bigvee_{i=1}^{m} Disj(T(A_i))$.*

Since a matching decision tree T represents $Disj(T)$, in the sense of Definition 5.7.2, it follows that if T has depth d, then $Disj(T)$ is an OR of AND's of size d. Thus a d-evaluation of a set Γ of formulas represents every formula of Γ as a DNF formula, where every conjunction is over at most d variables.

Note that $Disj(T(x_{i,j}))$ is $x_{i,j}$, and that $Disj(T(\neg x_{i,j}))$ is $\bigvee_{k \neq j, k \in R} x_{i,k}$, hence we have that $x_{i,j} \equiv Disj(T(x_{i,j}))$, but $\neg x_{i,j} \not\equiv Disj(T(\neg x_{i,j}))$, even when equivalence is restricted to range over the set of all critical truth assignments.

Recall the intuition behind the lower bound proof for the pigeonhole principle. If P is a proof of subexponential size S of PHP_n^{n+1}, where each formula has depth at most d, then inductively we find restrictions $\rho_1, \rho_2, \ldots, \rho_d \in M_n$, with $\rho = \rho_1 \cdots \rho_d$, for which there is a k-evaluation (k appropriately chosen) for the set $\Gamma \restriction_\rho$ of all subformulas of formulas $A \restriction_\rho$, for A appearing in proof P. The k-evaluation $T(A \restriction_\rho)$ plays the role of the space of all truth assignments for formula $A \restriction_\rho$, where we note that by earlier remarks, there may be truth assignments which are incompatible with every restriction $\pi(\ell)$, where ℓ is a leaf of $T(A \restriction_\rho)$. It will turn out that provided $kc \leq n$, where c is the maximum number of subformulas occurring in any rule of inference, the k-evaluation is nevertheless sound with respect to the axioms and proof rules of the Frege system. This means that if A_0 is implied by $A_1, \ldots, A_r$, then $A_0 \restriction_\rho$ is implied by $A_1 \restriction_\rho, \ldots, A_r \restriction_\rho$, and so if $T(A_1) \approx 1, \ldots, T(A_r) \approx 1$, then $T(A_0) \approx 1$. On the other hand, since the depth of the matching decision tree $T(\text{PHP}_n^{n+1} \restriction_\rho)$ is less than $n - |\rho|$, not everything about matchings in $M(D \restriction_\rho, R \restriction_\rho)$ can be determined, and so paradoxically we have $T(\text{PHP}_n^{n+1} \restriction_\rho) \approx 0$. Thus P can not have been a proof of PHP_n^{n+1}.

The following lemma, although simple, it is important in proving that k-evaluations are sound with respect to Frege rules in the language $L(D, R)$, provided that $ck \leq n$, where c is an upper bound for the number of subformulas in any rule R of the Frege system.

Lemma 5.7.5 (Compatibility). *Let D, R be disjoint sets, $|D| = n + 1$, $|R| = n$, $S = D \cup R$, $\rho \in M_n$, and let T be a matching decision tree over S, such that $|\rho| + d(T) \leq n$. Then there exists $\sigma \in Br(T)$ which is compatible with ρ, i.e., $\rho\sigma \in M_n$.*

Proof. If $\rho = \emptyset$, then the result is trivial. Assume that $\rho \neq \emptyset$, so that $d(T) < n$. Beginning at the root p_1 of T, successively choose nodes $p_1, p_2, \ldots$, whose edge labels determine a path π from root to leaf as follows. Suppose that nodes $p_1, \ldots, p_t$ have been chosen thus far in the construction, so that $\rho\pi(p_t) \in M_n$, where $\pi(p_t)$ denotes the path so far determined (note that $\pi(p_1) = \emptyset$). If p_t is a leaf of T, then set $\pi = \pi(p_t)$ and we are finished. Otherwise, p_t is an internal node of T, so $|\pi(p_t)| < n$ and by definition of matching tree, p_t must be labeled by an element $k \in S$. If $k \in D$ ($k \in R$), then there are edges in T from p_t to child nodes q, which are labeled by (k, j) ((i, k)) for each $j \in R$ ($i \in D$) which satisfies $\pi(p_t) \cup \{(k, j)\} \in M_n$ ($\pi(p_t) \cup \{(i, k)\} \in M_n$). At least one of these edge labels must be compatible with $\pi(p_t)$, so for such an edge, say from p_t to node q, set $p_{t+1} = q$, and continue.

Lemma 5.7.6 (Soundness of k-evaluation). *Let $\mathcal{F}$ be a Frege system, and let c be an upper bound for the number of subformulas occurring in any rule R of $\mathcal{F}$. Let P be a proof in $\mathcal{F}$ in the language $L(D, R)$, where D, R are disjoint sets, $|D| = n + 1$, $|R| = n$, $S = D \cup R$, and let Γ be the set of all subformulas of all formulas in P. If T is a k-evaluation for Γ, and $k \cdot c \leq n$, then for any line A in P, $T(A) \approx 1$.*

Proof. Equivalently stated, the conclusion of the lemma is that

$$(\forall \pi)\,(\pi \in Br(T(A)) \to Disj(T(A)) \restriction_\pi \equiv 1)\,.$$

The proof is by induction on the number of inferences in P. Suppose that

$$\frac{A_1(B_1/p_1, \ldots, B_m/p_m), \ldots, A_r(B_1/p_1, \ldots, B_m/p_m)}{A_0(B_1/p_1, \ldots, B_m/p_m)}$$

is an instance of an application of rule of inference R of $\mathcal{F}$,

$$\frac{A_1(p_1, \ldots, p_m), \ldots, A_r(p_1, \ldots, p_m)}{A_0(p_1, \ldots, p_m)}$$

where by assumption the number of subformulas appearing in rule R is at most c. Assume by induction that the assertion of the lemma holds for $A_1(B_1/p_1, \ldots, B_m/p_m), \ldots, A_r(B_1/p_1, \ldots, B_m/p_m)$; i.e.,

$$T(A_i(B_1/p_1, \ldots, B_m/p_m)) \approx 1$$

for $1 \leq i \leq r$.

Let $\pi \in Br(T(A_0(B_1/p_1, \ldots, B_m/p_m)))$ be arbitrary. We must show that $Disj(T(A_0(B_1/p_1, \ldots, B_m/p_m))) \restriction_\pi \equiv 1$. To that end, let $\Gamma = \{C_1, \ldots, C_s\}$, consist of all subformulas of the rule R, with the exception of A_0. Since the number of subformulas appearing in rule R is at most c, we have $s < c$. Let D_i abbreviate $C_i(B_1/p_1, \ldots, B_m/p_m)$, for $1 \leq i \leq s$. Define ρ_0 to be π. Since $ck \leq n$ and $d(T(D_i)) \leq k$ for $1 \leq i \leq s$, we can apply Lemma 5.7.5 to inductively find $\rho_1 \in Br(T(D_1)), \ldots, \rho_s \in Br(T(D_s))$, such that $\rho_0 \ldots \rho_i \in M_n$ for $1 \leq i \leq s$. Define $\rho = \rho_0 \cdots \rho_s$.

Since each ρ_i consists of the edge labels of a branch in $Br(T(D_i))$ from root to leaf $\ell_i \in T(D_i)$, for $1 \leq i \leq s$, $D_i \restriction_{\rho_i} \equiv Disj(T(D_i)) \restriction_\rho \equiv 0$ (1) if the label of ℓ is 0 (1). The restriction ρ extends ρ_i, so $D_i \restriction_\rho \equiv Disj(T(D_i)) \restriction_\rho \equiv 0$ or 1. By the definition of k-evaluation $Disj(T(0)) \restriction_\rho \equiv 0$ and $Disj(T(1)) \restriction_\rho \equiv 1$. If $\neg A \in \Gamma$, then $Disj(\neg A) \restriction_\rho \equiv 0 \Leftrightarrow Disj(A) \restriction_\rho \equiv 1$. If $A \vee B \in \Gamma$, then $Disj(A \vee B) \restriction_\rho \equiv 1$ if and only if $Disj(A) \restriction_\rho \equiv 1$ or $Disj(B) \restriction_\rho \equiv 1$. It follows that a consistent truth valuation V can be defined for all formulas of Γ, where $V(A) = 0$ if $Disj(T(A)) \restriction_\rho \equiv 0$ and $V(A) = 1$ if $Disj(T(A)) \restriction_\rho \equiv 1$. By assumption, we have

$$V(A_1) = \cdots = V(A_s) = 1$$

and since the rule R is sound, it follows that $V(A_0) = 1$. This means that $Disj(T(A_0)) \restriction_\rho \equiv 1$. We began the proof with the assumption that $\pi \in Br(T(A_0))$, hence $Disj(T(A_0)) \restriction_\pi \equiv 0$ or 1. Since $\pi \subseteq \rho$, it must be that $Disj(T(A_0)) \restriction_\pi \equiv 1$. It follows that $T(A_0(B_1/p_1, \ldots, B_m/p_m)) \approx 1$. This concludes the proof of the lemma.

Lemma 5.7.7 **($T(\text{PHP}_n^{n+1}) \approx 0$).** *Let $k \leq n-2$, and T be a k-evaluation for the set Γ consisting of all subformulas of* PHP_n^{n+1}. *Then* $T(\text{PHP}_n^{n+1}) \approx 0$.

Proof. The assertion of the lemma is

$$(\forall \pi) \left(\pi \in Br(T(\text{PHP}_n^{n+1})) \to Disj(T(\text{PHP}_n^{n+1})) \restriction_\pi \equiv 0\right).$$

We prove the assertion of the lemma for onto $-$ PHP_n^{n+1}, an even stronger assertion. The onto version of PHP_n^{n+1} is a disjunction of the following.

1. $(x_{i_1,j} \wedge x_{i_2,j})$, where $i_1, i_2 \in D$ are distinct, and $j \in R$.
2. $(x_{i,j_1} \wedge x_{i,j_2})$, where $i \in D$, and and $j_1, j_2 \in R$ are distinct.
3. $\bigwedge_{j \in R} \neg x_{i,j}$, for $i \in D$.
4. $\bigwedge_{i \in D} \neg x_{i,j}$, for $j \in R$.

Since $\wedge$ is an abbreviation for $\neg \vee \neg$, as formulated in the language $L(D, R)$, onto $-$ PHP_n^{n+1} is the disjunction of the following.

1. $\neg(\neg x_{i_1,j} \vee \neg x_{i_2,j})$, where $i_1, i_2 \in D$ are distinct, and $j \in R$.
2. $\neg(\neg x_{i,j_1} \vee \neg x_{i,j_2})$, where $i \in D$, and and $j_1, j_2 \in R$ are distinct.
3. $\neg \bigvee_{j \in R} x_{i,j}$, for $i \in D$.
4. $\neg \bigvee_{i \in D} x_{i,j}$, for $j \in R$.

Since $T(\text{onto} - \text{PHP}_n^{n+1})$ represents the disjunction of

$$Disj(T(\neg(\neg x_{i_1,j} \vee \neg x_{i_2,j})))$$
$$Disj(T(\neg(\neg x_{i,j_1} \vee \neg x_{i,j_2})))$$
$$Disj(T(\neg \bigvee_{j \in R} x_{i,j}))$$

$$Disj(T(\neg \bigvee_{i \in D} x_{i,j}))$$

over appropriate i, j, i_1, i_2, j_1, j_2, it suffices to show that for each formula A in (1)-(4), the leaves of $T(A)$ are labeled by 0, and hence $Disj(T(A)) \equiv 0$.

Case 1. $T(\neg(\neg x_{i_1,j} \vee \neg x_{i_2,j}))$ has all its branches labeled by 0 iff $T(\neg x_{i_1,j} \vee \neg x_{i_2,j})$ has all its branches labeled by 1. By definition $T(\neg x_{i_1,j} \vee \neg x_{i_2,j})$ represents $Disj(T(\neg x_{i_1,j})) \vee Disj(T(\neg x_{i_2,j}))$, which since $T(\neg x_{i_1,j}) = T^c(x_{i_1,j})$ and $T(\neg x_{i_2,j}) = T^c(x_{i_2,j})$, is just

$$\bigvee_{j' \neq j, j' \in R} x_{i_1,j'} \vee \bigvee_{j' \neq j, j' \in R} x_{i_2,j'}. \tag{5.92}$$

Let ℓ be a leaf of $T(\neg x_{i_1,j} \vee \neg x_{i_2,j})$. We will show that $\pi(\ell)$ satisfies (5.92).

Suppose that $\pi(\ell)$ does not contain (i_1, j). Since $|\pi(\ell)| \leq k \leq n-2$, there is $j' \in R$, $j' \neq j$, for which $\pi(\ell) \cup \{(i_1, j')\} \in M_n$. If $\pi(\ell)$ contains (i_1, j), then it cannot contain (i_2, j) and so a similar argument shows that for some $j' \neq j$, $\pi(\ell) \cup \{(i_2, j')\} \in M_n$. Thus an extension of $\pi(\ell)$ satisfies the disjunction (5.92), so the label of ℓ must be 1. This concludes the treatment of Case 1.

Case 2. $T(\neg(\neg x_{i,j_1} \vee \neg x_{i,j_2}))$ has all its branches labeled by 0 by an argument analogous to that of Case 1.

Case 3. $T(\neg \bigvee_{j \in R} x_{i,j})$ has all its branches labeled by 0 iff $T(\bigvee_{j \in R} x_{i,j})$ has all its branches labeled by 1. Now

$$T(\bigvee_{j \in R} x_{i,j}) \text{ represents } \bigvee_{j \in R} Disj(T(x_{i,j})).$$

Since $Disj(T(x_{i,j}))$ is just $x_{i,j}$, we must show every branch of $T(\bigvee_{j \in R} x_{i,j})$ satisfies

$$\bigvee_{j \in R} x_{i,j}. \tag{5.93}$$

Let ℓ be a leaf of $T(\bigvee_{j \in R} x_{i,j})$. If $(i, j) \in \pi(\ell)$ for some $j \in R$, then clearly $\pi(\ell)$ satisfies (5.93). If $(i, j) \notin \pi(\ell)$ for all $j \in R$, then since $|\pi(\ell)| \leq k \leq n-2$, there is some $j' \in R$ for which $\pi(\ell) \cup \{(i, j')\} \in M_n$. Thus there is an extension of $\pi(\ell)$ which satisfies (5.93), and so in all cases the label of ℓ must be 1. Note that here we only need the hypothesis that $k \leq n-1$. This concludes the treatment of Case 1.

Case 4. $T(\neg \bigvee_{j \in D} x_{i,j})$ has all its branches labeled by 0. This case is handled in an analogous manner to that of Case 3.

Theorem 5.7.5 (Constructing k-evaluations). *Let $d \geq 0$, $0 < \epsilon < \frac{1}{5}$, $0 < \delta < \epsilon^d$, and let Γ be a set of propositional formulas in $L(D, R)$ of depth at most d closed under subformulas. If $|\Gamma| < 2^{n^\delta}$, $\gamma = \lceil n^{\epsilon^d} \rceil$ and n is* *sufficiently large, then there exists a matching* $\rho \in M_n^\gamma$ *for which there exists a* $2n^\delta$*-evaluation of* $\Gamma \upharpoonright_\rho$.

Proof. The proof is by induction on d. For $d = 0$, since depth 0 formulas are either constants 0, 1 or variables $x_{i,j}$, the depth of the canonical matching decision tree $Trees_S(A)$ is at most 1. Thus it suffices to take $\rho = \emptyset$.

Assume that assertion of the lemma holds for depth d. Let Γ be a set of formulas of depth at most $d+1$, closed under subformulas, and assume that $|\Gamma| \leq 2^{n^\delta}$, where $0 < \delta < \epsilon^{d+1}$. Let Δ be the set of formulas in Γ whose depth is at most d. Let $\gamma = \lceil n^{\epsilon^d} \rceil$. By the inductive hypothesis, there exists a restriction $\rho \in M_n^\gamma$, for which there exists a $2n^\delta$-evaluation T of $\Delta \upharpoonright_\rho$. We will define a restriction $\pi \in M_n^\gamma$ extending ρ and a $2n^\delta$-evaluation $\overline{T}$ of $\Gamma \upharpoonright_\pi$ which satisfy the requirements of the Lemma.

Suppose that A is a formula of depth $d+1$, whose merged form is $\bigvee_{i \in I} A_i$. Let $\gamma' = \lceil n^{\epsilon^{d+1}} \rceil$, and in the Matching Switching Lemma 5.7.4 replace D by $D \upharpoonright_\rho$, R by $R \upharpoonright_\rho$, n by $\gamma = \lceil n^{\epsilon^d} \rceil$, ℓ by $\gamma' = \lceil n^{\epsilon^{d+1}} \rceil$, r by $\lfloor 2n^\delta \rfloor$, and s by n^δ. Letting $p = \frac{\ell}{n}$, note that for sufficiently large n, $p^4 \lceil n^{\epsilon^d} \rceil^3 < n^{-\epsilon^d/5} \leq 1/10$, where $\lfloor 2n^\delta \rfloor \leq \lceil n^{\epsilon^{d+1}} \rceil$ since $\delta < \epsilon^{d+1}$, so we can apply Lemma 5.7.4. It follows that

$$\frac{|Bad_\gamma^{\gamma'}(\bigvee_{i \in I} Disj(T(A_i \upharpoonright_\rho)), 2n^\delta)|}{|M_\gamma^{\gamma'}|} \tag{5.94}$$

is bounded above by $9n^{-\epsilon^d/5} \cdot \lfloor 2n^\delta \rfloor^{n^\delta}$. Since $\delta < \epsilon^{d+1} < \epsilon^d/5$, for sufficiently large n it is the case that $9n^{-\epsilon^d/5} \cdot \lfloor 2n^\delta \rfloor < \frac{1}{2}$, and so (5.94) is bounded above by 2^{-n^δ}. Thus that there exists a restriction $\rho' \in M_\gamma^{\gamma'}$ such that for every disjunction $A \in \Gamma$ of depth $d+1$,

$$|Trees_{\upharpoonright_{\rho\rho'}}(\bigvee_{i \in I} Disj(T(A_i \upharpoonright_\rho)) \upharpoonright_{\rho'})| < 2n^\delta.$$

Define $\pi = \rho\rho'$. By construction, $\pi \in M_n^{\gamma'}$.

Define the $2n^\delta$-evaluation $\overline{T}$ as follows. Since T is a $2n^\delta$-evaluation on $\Delta \upharpoonright_\rho$, $T \upharpoonright_{\rho'}$ is a $2n^\delta$-evaluation on $\Delta \upharpoonright_{\rho\rho'} = \Delta \upharpoonright_\pi$. For formulas $A \in \Gamma$ of depth at most d, define $\overline{T}(A \upharpoonright_\pi) = T \upharpoonright_{\rho'}(A \upharpoonright_\pi)$.

Suppose now that A is a formula in Γ of depth $d+1$, which is of the form $\neg\neg\cdots\neg B$, where the primary connective of B is $\vee$. Then define $\overline{T}(A \upharpoonright_\pi)$ as either $T'(B \upharpoonright_\pi)$ or $(T'(B \upharpoonright_\pi))^c$, according to whether there are an even or odd number of negations before B. For disjunctive formulas A in Γ of depth $d+1$, whose merged form is $\bigvee_{i \in I} A_i$ set

$$\overline{T}(A \upharpoonright_\pi) = Trees_{\upharpoonright_\pi}(\bigvee_{i \in I} Disj(T(A_i \upharpoonright_\rho)) \upharpoonright_{\rho'})$$

It can be verified that $\overline{T}$ is a $2n^\delta$-evaluation $\overline{T}$ satisfying the assertion of the Theorem.

A lower bound for the pigeonhole principle is given in the next theorem.

Theorem 5.7.6 ([PBI93, KPW95]). *Let $\mathcal{F}$ be a Frege system, $d > 4$ and $\delta < (1/5)^d$. Then for sufficiently large n, every depth d proof in $\mathcal{F}$ of* PHP_n^{n+1} *has circuit size at least 2^{n^δ}.*

Proof. We follow the exposition in [UF96]. Assume that $0 < \delta < (1/5)^d$, and suppose, in order to obtain a contradiction, that $P = (F_1, \ldots, F_s)$ is a proof of PHP_n^{n+1} in $\mathcal{F}$ of depth d and size $s \leq 2^{n^\delta}$.

Let Γ be the set of all subformulas in proof P and choose $\epsilon < 1/5$ and $\delta < \epsilon^d$. By Theorem 5.7.5 there exists $\rho \in M_n^\gamma$ with $\gamma = \lceil n^{\epsilon^d} \rceil$, and a $2n^\delta$-evaluation T of $\Gamma \restriction_\rho$. Then $P \restriction_\rho = (F_1 \restriction_\rho, F_2 \restriction_\rho, \ldots, F_s \restriction_\rho)$ is a proof in $\mathcal{F}$ in the language $L(D \restriction_\rho, R \restriction_\rho)$.

Assume that the number of subformulas in every rule R of Frege system $\mathcal{F}$ is bounded above by c. Now $\delta < \epsilon^d$, so for n sufficiently large we have that $(2n^\delta) \cdot c \leq n^{\epsilon^d}$. By Lemma 5.7.6, it follows that for $1 \leq i \leq s$, $T(F_i \restriction_\rho) \approx 1$. However, $\text{PHP}_n^{n+1} \restriction_\rho \equiv \text{PHP}(D \restriction_\rho, R \restriction_\rho)$, and so by Lemma 5.7.7 $T(\text{PHP}_n^{n+1} \restriction_\rho) \approx 0$, a contradiction if $F_1, \ldots, F_s$ were a proof of PHP_n^{n+1}. This concludes the proof of the theorem.

Ramsey's theorem is a generalization of the pigeonhole principle, which states that for n, m, k arbitrary integers, there exists a sufficiently large integer N such that for any m-coloring of the n-size subsets of $\{0, \ldots, N\}$, there exists a subset of $\{0, \ldots, N\}$ of size k, all of whose n-size subsets have the same color. The size k subset with this property is called monochromatic (or homogeneous). It is well-known that the least value of N satisfying the previous assertion, as a function of n, m, k is exponential in these parameters. By expressing the statement "for any m-coloring of the n-size subsets of N there is a k-size monochromatic subset", where k is an appropriate function involving the logarithm of N, one can express Ramsey's theorem in propositional logic. In [Pud91], P. Pudlák gives polynomial-size constant-depth Frege proofs for the following formalization of Ramsey's theorem. Write $X \subseteq m$ to abbreviate $X \subseteq \{1, \ldots, m\}$, and $|X|$ to denote the size of X. For m, k integers, $R(m, k)$ is the statement:

$$\bigvee_{X \subseteq m, |X| = k} \Big(\bigwedge_{i,j \in X, i<j} p_{i,j} \vee \bigwedge_{i,j \in X, i<j} \neg p_{i,j} \Big).$$

This may seem contradictory, since Ramsey's theorem generalizes the pigeonhole principle. However, it is known that homogeneous sets may be logarithmic in the size of n, and so in stating Ramsey's theorem in propositional logic, the existence of only logarithmic size homogeneous sets is posited, hence the usual argument can be formalized. This is akin to the observation that $\text{PHP}_{\log n}^n$ has polynomial-size resolution proofs. Finally, it should be mentioned that in [PWW88], Paris, Wilkie and Woods proved that the weak pigeonhole principle $\text{PHP}_n^{n^2}$ has constant-depth quasipolynomial size Frege proofs – even stronger, there exists constant-depth Frege proofs of size $n^{\log^{(k)} n}$ for each $k \geq 1$, where the iterated logarithm is defined by $\log^{(k+1)} n = \log(\log^{(k)} n)$.

In [Ajt94b], M. Ajtai proved that in the context of constant-depth Frege systems, the pigeonhole principle is strictly weaker than the *parity principle* and more generally than a modular counting principle defined below. The parity principle $Count_2^n$ states that, assuming that n is odd, there is no partition of $\{1, \ldots, n\}$ into disjoint equivalence classes of size 2. The modular counting principle $Count_r^n$ similarly states that, assuming $n \not\equiv 0 \bmod r$, there is no partition of $\{1, \ldots, n\}$ into size r disjoint subsets. This is formalized as follows.

Definition 5.7.8 ([BIK+96]). *For $2 \leq r \leq n$, let $[n]^r$ denote the set of subsets of $\{1, \ldots, n\}$ of size r. The formula $Count_r^n$ is*

$$\bigvee_{i=1}^{n} \bigwedge_{i \in e} \neg x_e \vee \bigvee_{e \perp f} (x_e \wedge x_f)$$

where $e, f \in [n]^r$ vary over r-element subsets of $\{1, \ldots, n\}$, and $e \perp f$ means that $e, f \in [n]^r$ are distinct sets with non-empty intersection. $Count_{r,i}$ is the set

$$\{Count_{r,i}^n : n \equiv i \bmod r, n \in \mathbf{N}\}$$

and $Count_r = \cup_{i=1}^{r-1} Count_{r,i}^n$. If $\Phi = \{\phi_e : e \in [n]^r\}$ is a set of propositional formulas, then $Count_r^n(\Phi)$ is the set of instances $Count_r^n[\langle \phi_e / x_e : e \in [n]^r \rangle]$ for $n \not\equiv 0 \bmod r$, $n \in \mathbf{N}$, obtained by the simultaneous substitution of variables x_e by ϕ_e for $e \in [n]^r$.

It is not hard to prove that if r evenly divides s, then there are polynomial-size constant-depth Frege proofs of $Count_r$ from $Count_s$ (see Exercise 5.10.5). The following theorem, whose proof is omitted for reasons of space, entirely settles the issue of modular counting in the context of constant-depth Frege systems. (See [BIK+96] for stronger results than what is here stated.)

Theorem 5.7.7 ([BIK+96]). *Let $2 \leq p, q$. Then there are constant-depth polynomial-size Frege proofs of $Count_q^n$, for $n \not\equiv 0 \bmod q$, from instances of $Count_p$ if and only if every prime factor of q also divides p.*

The separation is now known to be exponential.

Theorem 5.7.8 ([BIK+97]). *Let $2 \leq p, q$. If there is a prime factor of p which does not divide q, then the size of any constant-depth Frege proof of $Count_p^n$ from instances of $Count_q$ is at least $2^{n^{\Omega(1)}}$.*

We close this section with a theorem of Bonet, Domingo, Gavalda, Maciel, and Pitassi [BDG+99] which partially answers the question whether constant-depth Frege systems are automatizable.

Theorem 5.7.9 ([BDG$^+$99]). *Assume that for every $\epsilon > 0$, the Diffie-Hellman cryptographic function cannot be computed by circuits of size 2^{n^ϵ}. Then constant-depth Frege systems do not have the feasible interpolation property and hence are not automatizable.*

For details, see [BDG$^+$99], which extended a first result in this direction by Krajíček and Pudlák [KP95], who show, assuming that the cryptosystem RSA is secure, that extended Frege systems of Section 5.7.2 do not have the feasible interpolation property, and hence are not automatizable.

5.7.2 Extended Frege Systems

An *extended*[21] *Frege* system [CR77] is a Frege system with the additional *extension rule*, which allows the introduction of a new propositional variable p and the inference $p \equiv G$, where G is an arbitrary formula not containing p. A *proof* in an extended Frege system is a sequence $F_1, \ldots, F_n$ where each F_i is either inferred from previous formulas F_j, $j < i$, by a Frege rule, or where F_i is of the form $p \equiv G$, where p does not appear in $F_1, \ldots, F_{i-1}, F_n, G$. Thus extended Frege systems allow the introduction of abbreviations.

A *substitution Frege* system is a Frege system with the additional substitution rule

$$\frac{F}{F(G/p)}$$

where $F(G/p)$ is obtained from G by substituting each occurrence of the propositional variable p by formula G. Substitution was introduced by G. Frege in his *Begriffschrift* [Fre67], although it was not explicitly stated until his "Grundgesetze der Arithmetik" [Fre93]. Clearly, substitution would not be necessary, if we were to use unrestricted axiom schemata. See also p. 157 of [Chu56].

For extended Frege systems, the extension of Theorem 5.7.1 to arbitrary languages can be proved in a different manner, without relying on Theorem 5.2.2.

Theorem 5.7.10 ([Rec75]). *Extended Frege systems (not necessarily in the same language) polynomially simulate each other.*

Proof. Let $\mathcal{F}, \mathcal{F}'$ be two arbitrary extended Frege systems the first in the language $L = \{\neg, \vee, \wedge\}$ and the second in a language L'. We give a transformation τ of formulas of $\mathcal{F}$ into formulas of $\mathcal{F}'$ such that for any formula ϕ in L, $\phi \equiv \tau(\phi)$ is a tautology, $|\tau(\phi)| \leq c|\phi|$, and if n is the minimal length of a proof of ϕ in $\mathcal{F}$ then $\tau(\phi)$ has a proof in $\mathcal{F}'$ of length $\leq cn$, where c is some constant. This clearly implies the theorem.

[21] In [Pud98], this is called *extension Frege system*, which is more correct terminology. We defer rather to the widespread usage and historical precedence of *extended Frege system*.

Since the language L' is complete there exists formulas

$$\phi_\neg(p), \phi_\wedge(p,q), \phi_\vee(p,q)$$

in the language L' which are tautologically equivalent to $\neg p, p \wedge q, p \vee q$, respectively. We show how to construct new formulas

$$\psi_\neg(p), \psi_\wedge(p,q), \psi_\vee(p,q)$$

in the language L' which are also tautologically equivalent to $\neg p, p \wedge q, p \vee q$, respectively, but with the additional requirement that the variables p, q occur exactly once in these formulas. Assuming this the required translation is easy (see exercise 1). So next we concentrate on the construction of the formulas $\phi_\neg, \psi_\wedge, \psi_\vee$.

First consider the formula $\phi_\neg(p)$. By replacing (if necessary) every occurrence of any variable other than p by TRUE or 1, we may assume without loss of generality that p is the only propositional variable occurring in $\phi_\neg(p)$. Let n be the number of occurrences of p in $\phi_\neg(p)$ and assume (without loss of generality) that these occurrences are ordered. Let ϕ^i be obtained from $\phi_\neg(p)$ by replacing the first i occurrences of p by 1 and the last $n-i$ occurrences by FALSE or 0. Clearly, each ϕ^i is either a tautology or unsatisfiable and ϕ^0 is a tautology and ϕ^{n+1} is unsatisfiable. Hence there exists an $0 \leq i < n$ such that ϕ^i is a tautology and ϕ^{i+1} is unsatisfiable. Now let $\psi_\neg$ be obtained from $\phi_\neg(p)$ by replacing the first i occurrences of p by 1 and the last $n-i-1$ occurrences of p by 0 while leaving the $i+1$st occurrence unchanged.

Next we consider the formula $\phi_\wedge(p,q)$. As before we may assume that p, q are the only propositional variables occurring in $\phi_\wedge(p,q)$. Let m, n be the number of occurrences of p, q in $\phi_\neg(p)$, respectively, and assume (without loss of generality) that these occurrences are ordered. Let $\phi^{i,j}$ be obtained from $\phi_\wedge(p,q)$ by replacing the first i occurrences of p by 1, the last $m-i$ occurrences by 0 the first j occurrences of q by 1, the last $n-j$ occurrences by 0. Let $P(i,j,i',j')$ abbreviate the assertion: among the four formulas $\phi^{i,j}, \phi^{i,j'}, \phi^{i',j}, \phi^{i',j'}$ either exactly three are tautologies and the other unsatisfiable or exactly three are unsatisfiable and the other tautology. We claim there exist i, j such that $P(i,j,i+1,j+1)$. Indeed to find i notice that

- $P(0,0,m,n)$, and
- for all $i_1 < i_2 < i_3, j_1 < j_2$, if $P(i_1,j_1,i_3,j_2)$ then

$$\text{either } P(i_1,j_1,i_2,j_2) \text{ or } P(i_2,j_1,i_3,j_2).$$

A similar argument gives j. Now construct $\theta_\wedge(p,q)$ from $\phi_\wedge(p,q)$ by replacing the first i occurrences of p by 1, the last $m-i-1$ occurrences of p by 0 and the first j occurrences of q by 1, the last $n-j-1$ occurrences of q by 0. The required formula $\psi_\wedge$ can be constructed from $\theta_\wedge$ by adding extra negation symbols to the variable p, q as necessary in order to make the resulting formula tautologically equivalent to $p \wedge q$. This together with Exercise 1 completes the proof of the theorem.

We now study whether extended and substitution Frege systems can polynomially simulate each other.

Theorem 5.7.11 ([CR77]). *Every substitution Frege system SF can polynomially simulate every extended Frege system EF.*

Proof. We prove the assertion, assuming that the language of the extended and substitution Frege systems is the same. For different languages, Reckhow's implicit translation can be used (see remarks after Theorem 5.7.1). Consider an extended Frege proof P of ϕ which is using the abbreviations $p_i \equiv \phi_i$, for $i = 1, \ldots, n$. We remove the abbreviations by showing that there is a Frege proof, say P', of size polynomial in the size of P (in fact of size $O(|P|^2)$) of the formula

$$\bigwedge_{i \leq n} (p_i \equiv \phi_i) \rightarrow \phi. \tag{5.95}$$

This is proved by induction on the proof P. Let $P = \Phi_1, \ldots, \Phi_k$, with $\Phi_k = \phi$. The derivation P' consists of sequences $P'_1, \ldots, P'_k$ as follows.

If Φ_j is an axiom then P'_j consists of the axiom Φ_j, a derivation $\Phi_j \rightarrow (\bigwedge_{i=1}^n p_i \equiv \phi_i) \rightarrow \Phi_j$ plus the formula $(\bigwedge_{i=1}^n p_i \equiv \phi_i) \rightarrow \Phi_j$.

If Φ_j is inferred from Φ_r and $\Phi_s = \Phi_r \rightarrow \Phi_j$ by modus ponens then let P'_j be a derivation of the tautology

$$\left[(\bigwedge_{i=1}^n p_i \equiv \phi_i) \rightarrow \Phi_s\right] \rightarrow \left[(\bigwedge_{i=1}^n p_i \equiv \phi_i) \rightarrow \Phi_r\right] \rightarrow \left[(\bigwedge_{i=1}^n p_i \equiv \phi_i) \rightarrow \Phi_j\right]$$

If Φ_j is the substitution $p_m \equiv \phi_m$ then P'_j will be the concatenation of derivations of the following formulas

1. $(\bigwedge_{i=m}^n p_i \equiv \phi_i) \rightarrow p_m \equiv \phi_m$
2. $[(\bigwedge_{i=m}^n p_i \equiv \phi_i) \rightarrow p_m \equiv \phi_m] \rightarrow \left[(\bigwedge_{i=m-1}^n p_i \equiv \phi_i) \rightarrow p_m \equiv \phi_m\right]$
3. $\left(\bigwedge_{i=m-1}^n p_i \equiv \phi_i\right) \rightarrow p_m \equiv \phi_m$ (by modus ponens).

Now repeat (2) and (3) until we obtain a derivation for $(\bigwedge_{i=1}^n p_i \equiv \phi_i) \rightarrow p_m \equiv \phi_m$.

Now the simulation result is easy. Beginning with $i = 1$, append, for each $1 \leq i \leq n$, the following formulas to the Frege proof of (5.95):

$$(p_i \equiv p_i) \wedge \bigwedge_{i<j\leq n} (p_j \equiv \phi_j) \rightarrow \phi$$

$$\left((p_i \equiv p_i) \wedge \bigwedge_{i<j\leq n} (p_j \equiv \phi_j) \rightarrow \phi\right) \rightarrow \left(\bigwedge_{i<j\leq n} (p_j \equiv \phi_j) \rightarrow \phi\right)$$

$$\bigwedge_{i<j\leq n} (p_j \equiv \phi_j) \rightarrow \phi$$

where the first formula is a substitution inference, the second is a tautology and the third is obtained from the first two by modus ponens. Clearly, the resulting substitution proof is of size $O(|P|^3)$.

Originally, Cook and Reckhow conjectured in [CR77] that the converse of Theorem 5.7.11 is false. However, Dowd and later independently, Krajíček and Pudlák [KP89] proved that this is not the case.

Theorem 5.7.12 ([Dow85],[KP89]). *Every extended Frege system EF polynomially simulates every substitution Frege system SF.*

Proof. As in the last theorem, we assume that the substitution and extended Frege system have the same language. Indeed, if EF and SF have different languages, then by the method of Reckhow's implicit translation (see remarks after Theorem 5.7.1), EF is polynomially equivalent with EF', the latter which can be chosen to have the same language as SF.

Below, we follow closely the proof of Krajíček and Pudlák given in [KP89]. Let $P = \phi_1, \ldots, \phi_n$ be a substitution Frege proof of ϕ with substitutions using only the variables $\mathbf{p} = p_1, \ldots, p_m$. For $i < n$ consider new propositional variables $\mathbf{q}_i = q_{i,1}, \ldots, q_{i,m}$ and for $i = n$ put $\mathbf{q}_n = q_{n,1}, \ldots, q_{n,m}$. Put $\psi_i = \phi_i(\mathbf{p}/\mathbf{q}_i)$ and define the following vector $\boldsymbol{\beta}_j$ of propositional formulas:

- if ϕ_j is either an axiom or has been obtained from previous formulas by modus ponens then let $\boldsymbol{\beta}_j = \mathbf{q}_j$,
- if ϕ_j has been obtained from ϕ_i by the substitution

$$\frac{\phi_i(\mathbf{p})}{\phi_i(\mathbf{p}/\boldsymbol{\alpha})}$$

  then let $\boldsymbol{\beta}_j = \boldsymbol{\alpha}(\mathbf{p}/\mathbf{q}_j)$.

Define

$$\Psi_{i,j} = \psi_i \wedge \psi_{i+1} \wedge \cdots \wedge \psi_j, \text{ for } 0 \le i \le j+1 \le n,$$

where $\Psi_{i,i-1}$ is a true sentence. We now show how to simulate the above substitution Frege proof by an extended Frege proof. First introduce the propositional variables $\mathbf{q}_{n-1}, \ldots, \mathbf{q}_1$ via the abbreviations

$$q_{i,k} \equiv \bigvee_{j=i}^{n-1} \Psi_{i+1,j} \wedge \neg\psi_{i+1} \wedge \beta_{i+1,k}.$$

It is now obvious that $i < j$,

$$\Psi_{i+1,j-1} \wedge \neg\psi_j \to (q_{i,k} \equiv \beta_{j,k})$$

has a polynomial-size Frege proof. Hence the same is true for the formulas

$$\Psi_{i+1,j-1} \wedge \neg\psi_j \to (\phi_i(\mathbf{p}/\mathbf{q}_i) \equiv \phi_i(\mathbf{p}/\boldsymbol{\beta}_j))$$

which can also be rewritten in the form

$$\Psi_{i+1,j-1} \wedge \neg\psi_j \rightarrow (\psi_i \equiv \psi_i(\mathbf{p}/\boldsymbol{\beta}_j)). \tag{5.96}$$

Now we can show that each of the formulas

$$\psi_1, \psi_2, \ldots, \psi_n$$

have extended Frege proofs. This suffices since $\psi_n = \phi_n$. By induction, assume that each of the formulas $\psi_1, \psi_2, \ldots, \psi_{j-1}$ have extended Frege proofs. We prove that the same is true of ψ_j. If ϕ_j is an axiom so is ψ_j.

If ϕ_j is obtained by application of modus ponens from two previous formulas, i.e.,

$$\frac{\phi_u \;\; \phi_v}{\phi_j}, \quad u, v < j,$$

then first derive $\Psi_{u+1,j-1}$ and $\Psi_{v+1,j-1}$ and using (5.96) with $i = u, i = v$ obtain

$$\neg\psi_j \rightarrow (\phi_u(\boldsymbol{\beta}_j) \wedge \phi_v(\boldsymbol{\beta}_j))$$

Applying modus ponens to $\phi_u(\boldsymbol{\beta}_j), \phi_v(\boldsymbol{\beta}_j)$ we obtain $\neg\psi_j \rightarrow \phi_j(\boldsymbol{\beta}_j)$ which is the same as $\neg\psi_j \rightarrow \psi_j$, which proves ψ_j.

If ϕ_j is obtained by application of substitution to a previous formula, i.e.,

$$\frac{\phi_i(\mathbf{p})}{\phi_i(\mathbf{p}/\boldsymbol{\alpha})}, i < j,$$

then using (5.96) derive $\neg\psi_j \rightarrow \phi_i(\boldsymbol{\beta}_j)$, but by definition

$$\phi_i(\boldsymbol{\beta}_j) = \phi_i(\boldsymbol{\alpha}(\mathbf{q}_j)) = \phi_i(\mathbf{q}_j) = \psi_j$$

This completes the proof of the theorem.

For further results see also [KP90]. Recall the system ER of extended resolution from Definition 5.4.2.

Theorem 5.7.13 ([Coo76]). *The system ER of extended resolution polynomially simulates every extended Frege system.*

Proof. (Outline) In [Coo76], S.A. Cook used Cobham's function algebraic characterization of polynomial time P, which we later give in Theorem 6.3.6, in order to define a free variable equational calculus PV and to prove the so-called ER-Simulation Theorem. This theorem states that if '$f = g$' is provable in PV, where f, g are formal symbols for polynomial time computable functions, then the family $\{|f = g|_m^n : n, m \in \mathbf{N}\}$ has polynomial size extended resolution proofs. Here, $|f = g|_n$ is a propositional formula, which asserts that f equals g for inputs whose binary representation have length at most

n. Moreover, Cook showed that PV proves the soundness of ER in the following sense. If $\mathrm{Pr}_{ER}(x,y)$ is a formula[22] in PV which encodes the statement "x is an extended resolution proof of y", then $PV \vdash \mathrm{Pr}_{ER}(x,y) \rightarrow \mathrm{TAUT}(y)$. The method of proof easily yields the same for every extended Frege system EF; i.e., $PV \vdash \mathrm{Pr}_{EF}(x,y) \rightarrow \mathrm{TAUT}(y)$. If $\{\phi_n : n \in \mathbf{N}\}$ is a family of tautologies, such that ϕ_n has a proof P_n in the system EF of size polynomial in n, then $PV \vdash PR_{EF}(\ulcorner P_n \urcorner, \ulcorner \phi_n \urcorner)$ for each n, hence there are polynomial-size ER proofs of ϕ_n.

In his original proof of Theorem 5.7.12, M. Dowd proved that $PV \vdash \mathrm{Pr}_{SF}(x,y) \rightarrow \mathrm{TAUT}(y)$, and using the ER-Simulation Theorem mentioned in the previous proof, inferred that ER polynomially simulates SF.

5.7.3 Frege Systems and the PHP

Our previous investigations have left one important point unclear. What is the status of Frege systems with respect to extended Frege systems? Is there a sentence provable in the latter but not in the former type of systems? To answer this question several authors initiated the study of the pigeonhole principle as a tool for separating proof systems. The first theorem in this respect is the following.

Theorem 5.7.14 ([CR77]). PHP *has polynomial-size extended Frege proofs.*

Proof. Let $f : \{1,2,\ldots,n\} \rightarrow \{1,2,\ldots,n-1\}$. We want to show that for some $i \neq j$, $f(i) = f(j)$. Define by induction a sequence of functions

$$f_n := f, f_{n-1}, \ldots, f_1,$$

where $f_i : \{1,2,\ldots,i\} \rightarrow \{1,2,\ldots,i-1\}$ and f_i is defined from f_{i+1} by

$$f_i(x) = \begin{cases} f_{i+1}(x) & \text{if } f_{i+1}(x) \neq i \\ f_{i+1}(i+1) & \text{otherwise} \end{cases} \tag{5.97}$$

It is easy to show that if f_{i+1} is one-to-one so is f_i. This implies that for all i, $\neg\mathrm{PHP}^{i+2}_{i+1} \rightarrow \neg\mathrm{PHP}^{i+1}_{i}$ and hence also $\neg\mathrm{PHP}^{n+1}_{n} \rightarrow \neg\mathrm{PHP}^{2}_{1}$. Since PHP^{2}_{1} is valid so must be PHP^{n+1}_{n}.

All that is left is to formalize this proof in an extended Frege system. Introduce a new propositional variable $p^i_{x,y}$ representing the function $f_i(x) = y$. This function is defined from f_{i+1} via equation (5.97), which leads to the following abbreviations

$$\begin{aligned} p^n_{x,y} &\equiv p_{x,y} \\ p^i_{x,y} &\equiv p^{i+1}_{x,y} \vee \left(p^{i+1}_{x,i} \wedge p^{i+1}_{i+1,y}\right). \end{aligned}$$

[22] This is formalized as a function equality $f = g$, where f, g are symbols for polynomial time computable functions.

Substituting $p^i_{x,y}$ in place of $p_{x,y}$ in formula (5.4) we obtain the propositional formula ϕ_i

$$\bigwedge_{1\le x\le i}\ \bigvee_{1\le y\le i-1} p^i_{x,y} \rightarrow \bigvee_{1\le x<x'\le i}\ \bigvee_{1\le y\le i-1} (p_{x,y}\wedge p_{x',y}).$$

It is obvious that there are extended Frege proofs of the formulas $\neg\mathrm{PHP}^{n+1}_n \rightarrow \neg\phi_n$ and $\neg\phi_{i+1} \rightarrow \neg\phi_i$. Each formula is obtained from previous formulas using one of the above $O(n^3)$ substitutions; moreover each formula in the proof has size $O(n^3)$. The resulting proof of $\neg\phi_{i+1} \rightarrow \neg\phi_i$ has size $O(n^6)$. Combining

$$\neg\mathrm{PHP}^{n+1}_n \rightarrow \neg\phi_n, \neg\phi_n \rightarrow \neg\phi_{n-1}, \dots, \neg\phi_2 \rightarrow \neg\phi_1,$$

using modus ponens and the obvious fact that there is an extended Frege proof of ϕ_1 we obtain an extended Frege proof of PHP^{n+1}_n of size $O(n^7)$.

Historically, one of the reasons for studying the pigeonhole principle was in order to "separate" extended Frege systems from Frege systems. In fact since it had been pointed out by several people that no polynomial-size Frege proof of this principle was known, it was conjectured that PHP^{n+1}_n might separate Frege from extended Frege systems. This conjecture was disproved when S.R. Buss proved that the pigeonhole principle has polynomial-size Frege proofs. Since then, despite intensive work, no suitable candidates have been found of natural combinatorial principles which might separate Frege from extended Frege systems. Whether Frege systems can polynomially simulate extended Frege systems is closely akin to whether polynomial-size boolean formulas (i.e., NC^1) have equal computation power to polynomial-size boolean circuits (i.e., P/poly), though no formal equivalence has been proved. As a consequence the following question still remains open.

Question 5.7.1. Do Frege systems polynomially simulate extended Frege systems?

We give Buss' solution in Theorem 5.7.15. As in the proof of Theorem 5.7.14 the proof proceeds by contradiction. Let $f : \{0,1,\dots,n\} \rightarrow \{0,1,\dots,n-1\}$ be a function which is one-to-one. For each i let f_i denote the restriction of f onto the set $\{0,1,\dots,i\}$. A contradiction is then obtained by formalizing with polynomial-size Frege proofs the assertions

- $|\mathrm{range}(f_{i-1})| < |\mathrm{range}(f_i)|$, and
- $|\mathrm{range}(f_i)| \ge i$,

for $i = n, n-1, \dots, 2$, since these two assertions imply the absurd fact

$$|\mathrm{range}(f_n)| = n-1 \ge n.$$

To obtain polynomial-size proofs of the above assertions it will be necessary to formalize and determine the complexity of several elementary functions and relations (among natural numbers) such as addition, $<$, $\leq$, as well as counting predicates like $E_{k,b}(x_1, \ldots, x_r)$ (meaning there are exactly k occurrences of the bit b in the sequence of bits $x_1, \ldots, x_r$), etc.

Theorem 5.7.15 ([Bus87b]). *The pigeonhole principle has polynomial-size Frege proofs.*

Proof. We follow closely the proof of S.R. Buss [Bus87b]. We represent natural numbers by sequences of propositional formulas in the following way: if the binary representation of n is $\sum_i 2^i \cdot n_i$ then we represent n by the sequence $\bar{n} := \phi_0, \phi_1, \ldots$ of propositional formulas, where

$$\phi_i = \begin{cases} \text{TRUE or } 1 \text{ if } n_i = 1 \\ \text{FALSE or } 0 \text{ otherwise} \end{cases}$$

Let $\phi \oplus \psi$ abbreviate $(\phi \wedge \neg\psi) \vee (\psi \wedge \neg\psi)$. Keeping the above representation of natural numbers in mind it is easy to define addition modulo 2^{r+1}. For $\phi_0^i, \phi_1^i, \ldots, \phi_r^i$ propositional formulas, $i = 0, 1, 2$, define the propositional formula $Add_r(\phi^0, \phi^1, \phi^2)$ by the conjuction of the following formulas:

- $\phi_0^0 \equiv \phi_0^1 \oplus \phi_0^2$
- $\phi_i^0 \equiv \phi_i^1 \oplus \phi_i^2 \oplus \bigvee_{0 \leq j \leq i-1} [\phi_j^1 \wedge \phi_j^2 \wedge \bigwedge_{j<k<i} (\phi_k^1 \wedge \phi_k^2)]$.

Here the idea being that if ϕ^i is the representation of the natural number $n^i, i = 0, 1, 2$, then $Add_r(\phi^0, \phi^1, \phi^2)$ is true exactly when $n^0 = n^1 + n^2 \bmod 2^{r+1}$. It is also clear from the above definitions that if the length of ϕ^1 and ϕ^2 is $\leq r$ then the formula ϕ^0 defined above must satisfy

$$\| \phi^0 \| = O(r^3 \cdot \max\{\| \phi^1 \|, \| \phi^2 \|\}),$$

where $\| \phi^i \| = \max\{|\phi_j^i| : j = 1, \ldots, r\}$. We can also define propositional formulas for equality, and inequality

- $\phi^0 =_r \phi^1 \equiv \bigwedge_{0 \leq i \leq r} (\phi_i^0 \equiv \phi_i^1)$
- $\phi^0 <_r \phi^1 \equiv \bigvee_{0 \leq i \leq r} \left[\neg\phi_i^0 \wedge \phi_i^1 \wedge \bigwedge_{i<j \leq r} (\phi_j^0 \equiv \phi_j^1) \right]$
- $\phi^0 \leq_r \phi^1 \equiv \phi^0 =_r \phi^1 \vee \phi^0 <_r \phi^1$

Now it can be shown that the predicates and functions defined above satisfy the intended properties of addition, $<$ and $\leq$ and moreover these properties have polynomial-size Frege proofs. More formally we have the following lemma.

Lemma 5.7.8. *There are Frege proofs of size $O(r^5), O(r^8), O(r^8)$ for the following formulas respectively*

1. $Add_r(p^0, p^1, p^2) \wedge Add_r(q^0, p^1, p^2) \rightarrow p^0 =_r q^0$.
2. $p^1 \leq_r q^1 \wedge p^2 \leq_r q^2 \wedge \neg q_r^1 \wedge \neg q_r^2 \wedge Add_r(p^0, p^1, p^2) \wedge Add_r(q^0, q^1, q^2) \rightarrow p^0 \leq_r q^0$.

3. $p^1 \leq_r q^1 \wedge p^2 <_r q^2 \wedge \neg q_r^1 \wedge \neg q_r^2 \wedge Add_r(p^0, p^1, p^2) \wedge Add_r(q^0, q^1, q^2) \rightarrow p^0 <_r q^0$.

Proof. We only give the proof of (3). Assuming $p^1 \leq_r q^1$ we have that either $p^1 <_r q^1$ or $p^1 =_r q^1$. We consider only the case $p^1 =_r q^1$. Since $p^2 <_r q^2$ there exists a k such that $p_k^2 \wedge q_k^2 \wedge \bigwedge_{k<j\leq r}(p_j^2 \equiv q_j^2)$. Since by assumption q_r^2 we must have $k < r$ and the Frege proof splits into r parts depending on the value of k. Consider the formula $Carry_i(x, y)$ defined by $\bigvee_{0\leq j<i}\left[x_j \wedge y_j \wedge \bigwedge_{j<e<i}(x_e \oplus y_e)\right]$ expressing that there is a carry in the 2^i column when adding x and y. Depending on the values of $Carry_k(p^1, p^2)$ and $Carry_k(q^1, q^2)$ the Frege proof splits into four cases. In the first three cases we have that $Carry_k(p^1, p^2) \rightarrow Carry_k(q^1, q^2)$ in which case it is not hard to give a Frege proof of $\bigvee_{m\geq k}\left[\neg p_m^0 \wedge q_m^0 \wedge \bigwedge_{m<j\leq r}(p_j^0 \equiv q_j^0)\right]$ of size $O(r^3)$ so $p^0 <_r q^0$ holds. In the fourth case $Carry_k(p^1, p^2) \wedge \neg Carry_k(q^1, q^2)$ holds and there is a Frege proof of $\bigwedge_{k\leq j\leq r}((p_j^0 \equiv q_j^0)$ of size $O(r^3)$. Then we can prove for $m = k, k-1, \ldots, 0$ that

$$\bigwedge_{m\leq n<k}\left[q_n^0 \wedge \neg p_n^0 \wedge \bigwedge_{n<\leq r}(p_j^0 \equiv q_j^0)\right] \vee \left[\bigwedge_{m<\leq r}(p_j^0 \equiv q_j^0) \wedge Carry_m(p^1, p^2) \wedge \neg Carry_m(q^1, q^2)\right]$$

has a Frege proof of size $O(r^4)$. Since $Carry_m(p^1, p^2)$ we obtain

$$\bigvee_{0\leq n\leq r}\left[q_n^0 \wedge \neg p_n^0 \wedge \bigwedge_{n<j\leq r}(p_j^0 \equiv q_j^0)\right],$$

i.e., $p^0 <_r q^0$, as desired. This completes the proof of Lemma 5.7.8.

The next step in the argument is to show that carry-save addition has polynomial-size Frege proofs. Here we use the well-known "3 for 2" trick employed in Theorem 1.7.2. Intuitively, $CSum(n_0, n_1, n_2, n_3, n_4)$ is satisfied if $n_2 + n_3 + n_4 = n_0 + n_1$, n_0 is the bitwise sum modulo 2 of n_2, n_3, n_4 and n_1 represents the carries which are saved. More formally the predicate $CSum_r(\phi^0, \phi^1, \phi^2, \phi^3, \phi^4)$ is defined by the conjuction of the following formulas:

- $\phi_i^0 \equiv \phi_i^2 \oplus \phi_i^3 \oplus \phi_i^4$
- $\phi_0^1 \equiv 0$
- $\phi_i^1 \equiv (\phi_{i-1}^2 \wedge \phi_{i-1}^3) \vee (\phi_{i-1}^2 \wedge \phi_{i-1}^4) \vee (\phi_{i-1}^3 \wedge \phi_{i-1}^4))$

where $i = 1, \ldots, r$. Then we can define $CSAdd_r(\phi^0, \phi^1, \phi^2, \phi^3, \phi^4, \phi^5)$ (carry-save addition) by $CSum_r(\phi^0, \phi^1, \psi^0, \psi^1, \phi^5)$, with

- ψ_i^0 is $\phi_i^2 \oplus \phi_i^3 \oplus \phi_i^4$
- ψ_0^1 is 0
- ψ_i^1 is $(\phi_{i-1}^2 \wedge \phi_{i-1}^3) \vee (\phi_{i-1}^2 \wedge \phi_{i-1}^4) \vee (\phi_{i-1}^3 \wedge \phi_{i-1}^4)$,

where $i = 1, \ldots, r$. It can be shown that there is a polynomial-size Frege proof of the equivalence of addition and carry-save addition.

Lemma 5.7.9. *The formula*

$$\begin{aligned}&CSAdd_r(p^0, \ldots, p^5) \wedge \\ &Add_r(p^6, p^0, p^1) \wedge Add_r(p^7, p^2, p^3) \wedge Add_r(p^8, p^4, p^5) \wedge Add_r(p^9, p^7, p^8) \\ &\to p^6 =_r p^8.\end{aligned}$$

has a Frege proof of size $O(r^6)$.

The proof of Lemma 5.7.9 is easy. It is now possible to define the counting predicates $E_{k,b}(x_1, \ldots, x_r)$. We want to be able to count how many ϕ_is are true among a set $\phi_0, \phi_1, \ldots, \phi_{n-1}$ of propositional formulas by a polynomial-size Frege proof. Throughout we assume that $n = 2^{r-1}$, for some $r \geq 1$. If $s^{i,j} = s_0^{i,j}, \ldots, s_r^{i,j}, c^{i,j} = c_0^{i,j}, \ldots, c_r^{i,j}$ are propositional formulas, $0 \leq i < r$ and $0 \leq j < n2^{-i}$, then the formula $VSum_r(s, c)$ is defined by the formula $VSum_{r,r-1}(s, c)$, where $VSum_{r,k}(s, c)$ is defined as follows:

$$\bigwedge_{i=1}^{k} \bigwedge_{j=0}^{n2^{-i}-1} CSAdd_r(s^{i,j}, c^{i,j}, s^{i-1,2j}, c^{i-1,2j}, s^{i-1,2j+1}, c^{i-1,2j+1})$$

To define counting we make the following observations. Suppose $\phi_i, c_k^{i,j}, s_k^{i,j}$ have been assigned truth values so that $\phi_i \equiv s_0^{0,i}$ and each $s_{k+1}^{0,i}, c_{k+1}^{0,i}$ are assigned false so that $VSum(s, c)$ is valid. If $S^{i,j}, C^{i,j}$ are the numbers represented by $s^{i,j}, c^{i,j}$, respectively, then it can be shown by induction on i that $S^{i,j} + C^{i,j}$ is equal to the number of true ϕ_ks with $2^i j \leq k < 2^i(j + 1)$. Consequently, $S^{r-1,0} + C^{r-1,0}$ is equal to the total number of true ϕ_ks. This leads to the definition of the counting predicate $Count_r(a, s, c, \phi)$ as $Count_{r,r-1}(a, s, c, \phi)$, where $Count_{r,t}(a, s, c, \phi)$ is defined to be the conjunction of $VSum_{r,t}(s, c)$ and the following formula

$$\bigwedge_{i=0}^{t} \bigwedge_{j=0}^{n2^{-i}-1} Add_r(a^{i,j}, s^{i,j}, c^{i,j}) \wedge \bigwedge_{j=0}^{n-1} \left[(s_0^{0,j} \equiv \phi_j) \wedge \bigwedge_{k=1}^{r} (\neg s_k^{0,j} \wedge \neg c_k^{0,j}) \right].$$

Now it is easy to prove the following lemma.

Lemma 5.7.10. *Assume* $r \geq 1, n = 2^{r-1}$.

1. *Let* $s_k^{0,j}, c_k^{0,j}$ *be propositional formulas of size* $\leq m$ *and let* $s_k^{i,j}, s_k^{i,j}$ *be the natural formulas for which* $VSum_r(s, c)$ *holds. Then* $|s_k^{i,j}|, |c_k^{i,j}| = m \cdot n^{O(1)}$.
2. *Suppose that* $\phi_0, \ldots, \phi_{n-1}$ *are propositional formulas of size* $\leq m$ *and let* $s_k^{i,j}, c_k^{i,j}, a_k^{i,j}$ *be the natural formulas for which* $Count_r(a, s, c, \phi)$ *holds. Then* $|a_k^{i,j}|, |s_k^{i,j}|, |c_k^{i,j}| = m \cdot n^{O(1)}$.

Now we can prove the following result.

Lemma 5.7.11. *Let*

$$\phi_r = \bigwedge_{j=0}^{n-1} (q'_j \to q_j) \wedge Count_r(a,s,c,q) \wedge Count_r(b,t,d,q).$$

Then there are polynomial-size Frege proofs of the formulas

1. $\phi \to a^{r-1,0} \leq_r b^{r-1,0}$
2. $\bigvee_{j=0}^{n-1}(q_j \wedge \neg q'_j) \wedge \phi \to a^{r-1,0} <_r b^{r-1,0}$.

Proof. Let $\phi_{m,r} = \phi_r \wedge q_m \wedge \neg q'_m$. By Lemmas 5.7.8, 5.7.9 we have that for all i,j such that $j2^i \leq m < (j+1)2^i$ there is a polynomial-size Frege proof of $\phi_{m,r} \to a^{r-1,0} <_r b^{r-1,0}$. These n proofs, for all values of m, can be combined to give a polynomial-size proof of (2). For (1), use Lemmas 5.7.8, 5.7.9 to give polynomial size proofs of $\phi_r \to a^{i,j} \leq_r b^{i,j} \wedge b^{i,j} \leq_r 2^i$ (here by 2^i we understand the sequence of boolean formulas coded by 2^i). Now use these $O(n^2)$ intermediate steps in order to get a proof of $\phi_r \to a^{r-1,0} \leq_r b^{r-1,0}$. This completes the proof of Lemma 5.7.11.

Now we can complete the proof of Theorem 5.7.15 as indicated in the remarks preceding the main theorem. Assume that $n = 2^{r-1}$, for some $r \geq 1$. Introduce new propositional variables defined by $q_j^m \equiv \bigvee_{0 \leq k \leq m} p_{k,j}$, where PHP_n^{n+1} is given by formula (5.4). Introduce variables $a_k^{m,i,j}, s_k^{m,i,j}, c_k^{m,i,j}$ so that $Count_r(a^m, s^m, c^m, q^m)$ holds. If we admit the interpretation of $p_{i,j}$ as the graph of the function f then $a^{m,r-1,0}$ represents the number of holes j mapped onto by the first $m+1$ pigeons. Now there is a simple proof of $\neg\text{PHP}_n^{n+1} \to \bigvee_{0 \leq j < n} q_j^0$. Hence by Lemma 5.7.11 there is a polynomial size proof of $\neg\text{PHP}_n^{n+1} \to 0 <_r a^{0,r-1,0}$. Similarly by Lemma 5.7.11 there exist n polynomial-size proofs of $\neg\text{PHP}_n^{n+1} \to a^{m,r-1,0} <_r a^{m+1,r-1,0}$, where $m < n$. Now there are polynomial-size proofs of $\neg\text{PHP}_n^{n+1} \to m <_r a^{m,r-1,0}$ and in particular of $\neg\text{PHP}_n^{n+1} \to n <_r a^{n,r-1,0}$. As in the proof of Lemma 5.7.11 we can show that $a^{m,r-1,0} <_r n$. Hence $\neg\text{PHP}_n^{n+1} \to n <_r n$. This gives an extended Frege proof of PHP_n^{n+1}. Since this extended Frege proof has size bounded by a polynomial in n and by Lemma 5.7.10 the propositional variables introduced by extension have size polynomial in n the proof of the theorem is complete.

5.8 Open Problems

In this section, we list some open problems in the area of propositional proof systems. Problems not ascribed to any person are common knowledge, or obvious to anyone working in the field. The interested reader should consult [CK93] and [BP98] for other lists of open problems.

1. (S. Buss and G. Turán) What is the resolution refutation size of $\mathrm{PHP}_n^{n^2}$?
   (J. Paris and A. Wilkie) Does the weak pigeonhole principle $\mathrm{PHP}_n^{n^2}$ have polynomial-size constant-depth Frege proofs?
   (J. Krajíček) Does Ramsey's theorem have polynomial-size resolution proofs?
2. (J. Krajíček) Are there sets of clauses (or at least formulas, whose depth does not depend on d) which exponentially separate depth d Frege systems from depth $d+1$ Frege systems?
3. (P. Pudlák) Give optimal lower bounds for the intuitionistic propositional logic.
4. (A. Carbone) Are there polynomial-size proofs of PHP_n^{n+1} in the monotone sequent calculus MLK?
5. (S.A. Cook) Do Frege systems polynomially simulate extended Frege systems? This question is well known to be equivalent to asking whether Gentzen propositional calculus LK polynomially simulates Tseitin's system ER of extended resolution (this follows from Theorem 5.7.2 and Theorem 5.7.13). Cook has proposed that a candidate for a problem separating Frege and extended Frege proof systems is the set of tautologies formalizing the theorem $AB = I \Rightarrow BA = I$, where A and B are $n \times n$ matrices over a fixed field F and I is the $n \times n$ identity matrix (for instance, one might take F to be $GF(2)$). S.A. Cook has observed that polynomial-size EF proofs of this result can be obtained by formalizing the standard proof involving Gaussian elimination (e.g. computing the inverse, we have $AB = I \Rightarrow B = A^{-1} \Rightarrow BA = I$). A related open question is whether these tautologies have quasi-polynomial-size Frege proofs. The existence of such is possibly suggested by the fact that computing the inverse of a matrix is well known to be reducible to matrix powering, and so matrix inverse can be expressed with quasi-polynomial-size formulas.
6. Determine optimal depth for constant-depth PTK p-simulations of CP. Does CP* polynomially simulate CP? Are there polynomial-size CP proofs of the Tseitin graph theoretic formulas? Give lower bounds for the size of CP refutations of random $3-\mathrm{CNF}$ and more generally $k-\mathrm{CNF}$ formulas, as a function of clause density Δ. Are there width-related results for CP, analogous to the results of Ben-Sasson and Wigderson concerning resolution and width?
7. Which of the following systems are automatizable: R (resolution), PCR (a hybrid system of polynomials calculus with resolution), CP (cutting planes). Current techniques might suffice for the latter, since it is known that PTK polynomially simulates CP, and that (under a suitable cryptographic assumption) constant-depth PTK (sometimes called TC^0-Frege) is not automatizable.
8. Prove superpolynomial (exponential) lower bounds for the following systems.

a) (A.A. Razborov) Tree-like monotone calculus MLK.
b) Constant depth Frege systems in the language $\neg, \wedge, \vee, \oplus$, where the latter three connectives have arbitrary arity – this system is also called $\text{AC}^0[2]$-Frege.
c) As well, more generally, constant-depth Frege systems within the language $\neg, \wedge, \vee, \text{MOD}_p$, where the latter three connectives have arbitrary arity, and

$$\text{MOD}_p^n(x_1, \ldots, x_n) = \begin{cases} 0 \text{ if } \sum_{i=1}^n x_i \equiv 0 \bmod p \\ 1 \text{ otherwise.} \end{cases}$$

This system is usually called $\text{AC}^0[p]$-Frege.
d) Constant depth propositional threshold calculus PTK proofs – this system is polynomially equivalent (additionally constant-depth corresponds to constant-depth) with Krajíček's FC, and with TC^0-Frege.

9. (E. Ben-Sasson, R. Impagliazzo) Give lower bounds for polynomial calculus refutation size for random $3 - \text{CNF}$, and more generally random $k - \text{CNF}$ formulas, over fields F of characteristic 2.
10. (R. Impagliazzo) A proof system T is said to be *closed under renaming*, if whenever a formula ϕ has a proof in T of size s, then the formula ϕ' has a proof in T of size s, where ϕ' is obtained by replacing variables of ϕ by other variables or the constants $0, 1$. For instance, if $\phi(x, y, z)$ has a proof of size s, then $\phi(z, x, 1)$ should also have a proof of size s.
Find a proof system that is automatizable, closed under renaming, and which has polynomial-size proofs of the pigeonhole principle. R. Impagliazzo has pointed out, that one consequence of the existence of such a proof system T is that T automatically yields a polynomial time algorithm for perfect matching in graphs, as well as a polynomial time algorithm for separating k colorable graphs from graphs having a $k+1$ clique. The existence of such polynomial time algorithms is known, but depends on Lovasz's Theta function. Thus, such a proof system T would provide a unified approach in design of such algorithms.
11. (P. Beame and T. Pitassi) In [LS95], L. Lovasz and A. Schrijver introduced a proof system related to combinatorial optimization. Are there polynomial size proofs in the Lovasz-Schrijver system of Ajtai's parity principle $Count_2^n$ (see Exercise 5.10.5 for the definition of this principle). Are there polynomial-size refutations of Tseitin's unsatisfiable graph formulas in the Lovasz-Schrijver system?

5.9 Historical and Bibliographical Remarks

Research in propositional proof systems is vigorously pursued at this time, so we can only give a 'Leitfaden' to some of the most significant results and directions. Omission of mention of even very significant results is due to space considerations. For more, the reader should consult J. Krajíček's monograph

[Kra95], A. Urquhart's survey article [Urq95], P. Pudlák's Handbook article [Pud98], and the volume edited by P. Beame and S. Buss [BB98].

For general background on proof theory, the reader should consult G. Takeuti's classic text [Tak75, Tak75], or that of J.-Y. Girard [Gir87], as well as the newer collection of articles from the *Handbook of Proof Theory*, such as [Bus98a]. See E. Eder's text [Ede92] for a proof of the polynomial equivalence of Gentzen sequent calculus with natural deduction systems and with Frege systems, and related material.

In this chapter, with the exception of the proof of Theorem 5.7.13, we have tried to avoid treatment of bounded arithmetic, free variable equational theories, etc. The general theory of bounded arithmetic provides a very powerful vantage point for understanding propositional logic – for instance, Ajtai's original superpolynomial lower bound for the size of constant-depth Frege proofs of PHP_n^{n+1} used a battery of techniques from bounded arithmetic. The interested reader should consult J. Krajíček's excellent text [Kra95]. It should be pointed out that function algebras, treated in the next chapter, provide much of the underpinnings of bounded arithmetic systems and related free-variable equational calculus. Interesting collections of papers on feasible mathematics and complexity of proof systems can be found in [CR94], [CK93], and [BB98]. Finally, we would like to thank P. Beame, S. Buss, S. Cook, R. Impagliazzo, P. Pudlák, and A. Razborov for suggesting certain open problems in Section 5.8 and especially S. Buss for suggesting that $\Gamma_n \mapsto \Delta_n$ in Theorem 5.3.3 might have polynomial-size tree-like proofs with cut.

5.10 Exercises

Exercise 5.10.1. By providing derivations of appropriate formulas which express the rules of Gentzen sequent calculus LK, prove that Lukasiewicz's system L given in Section 5.1 is sound and complete, i.e., for any propositional formula ϕ, $\models \phi \Leftrightarrow \vdash_L \phi$.

Exercise 5.10.2. This exercise concerns whether a set of propositional logic connectives is adequate.

1. Show that the set $\{|\}$ of connectives is adequate, where $x|y$ is defined by $\neg(x \wedge y)$. The symbol $|$ is called the *Sheffer stroke*, and more commonly the NAND operator.
2. Prove that the set $\{T, F, \wedge, \vee\}$ of connectives is not an adequate set. HINT. Define $x_1, \ldots, x_n \leq y_1, \ldots, y_n$ if and only if for all $1 \leq i \leq n$, $x_i \leq y_i$. A boolean function is said to be *monotone*, if for all $x_1, \ldots, x_n \leq y_1, \ldots, y_n$ it is the case that $f(x_1, \ldots, x_n) \leq f(y_1, \ldots, y_n)$. By induction on formula length, show that if a boolean function f is represented by a formula ϕ in connective set $\text{T}, \text{F}, \wedge, \vee$, then f is monotone.

3. Prove that the set $\{T, F, \equiv, \neg\}$ of connectives is not an adequate set.
4. ($\star$) Define a Frege system based solely on connective $|$, and prove soundness and implicational completeness.
5. ($\star$) Define a schematic system which is sound and complete for all monotonic propositional tautologies in the connectives $\text{T}, \text{F}, \wedge, \vee$.

Exercise 5.10.3. Using the translation

$$x \oplus y \equiv (x \wedge \neg y) \vee (\neg x \wedge y)$$

represent $x \oplus (y \oplus z)$ using the boolean connectives. Estimate the number of nodes in the boolean formula tree representing $x_1 \oplus \cdots \oplus x_n$, where $\oplus$ is associated to the right. Now, using the 2-3 trick, represent $x \oplus (y \oplus z)$ in a more efficient manner using the boolean connectives.

Exercise 5.10.4. For any sets of formulas Γ, Δ let $shc(\Gamma \mapsto \Delta)$ denote the size of the shortest tree-like cut-free Gentzen proof of the sequent $\Gamma \mapsto \Delta$. Show that

1. For each propositional formula ϕ, there exists a tree-like, cut-free proof of $\phi \mapsto \phi$ of length $O(|\phi_i|)$ and size $O(|\phi_i|)$.
2. $shc(\phi \vee \psi, \Gamma \mapsto \Delta) \geq shc(\phi, \Gamma \mapsto \Delta)$
3. $shc(\Gamma \mapsto \phi \wedge \psi, \Delta) \geq shc(\Gamma \mapsto \phi, \Delta)$
4. $shc(\phi \supset \psi, \Gamma \mapsto \Delta) \geq shc(\psi, \Gamma \mapsto \Delta), shc(\Gamma \mapsto \phi, \Delta)$

Exercise 5.10.5. Let k, n be positive integers, such that n is not divisible by k. The *k-modular counting principle* for n, also called the *k-equipartition principle* for n, denoted $Count_k^n$, states that (assuming $n \not\equiv 0 \bmod k$) there is no partition of $\{1, \ldots, n\}$ into disjoint sets of size k. In propositional logic $\neg Count_2^3$ can be expressed by

$$\bigwedge_{1 \leq i \leq 3} \bigvee_{j \in \binom{3}{2}} x_{i,j} \wedge \bigwedge_{1 \leq i \leq 3} \bigwedge_{j,j' \in \binom{3}{2}} (\overline{x}_{i,j} \vee \overline{x}_{i',j}).$$

1. Give a resolution refutation of $\neg Count_2^3$.
2. Give constant-depth Frege proofs of onto $- \text{PHP}_n^{n+1}$, assuming $\{Count_2^n : n \not\equiv 0 \bmod 2, n \in \mathbf{N}\}$.
3. ([BR98])
   a) Prove that over constant-depth Frege systems, the relational and the functional version of PHP_n^{n+1} are equivalent.
   b) Generalizing the proof of (2) of this exercise, prove that onto$-\text{PHP}_n^{n+1}$ has polynomial-size constant-depth Frege proofs, assuming $Count_p$.
   c) Prove an exponential lower bound for the size of PHP_n^{n+1} in constant-depth Frege systems, assuming instances of $Count_p$.
   d) Deduce an exponential separation between the general version and the onto version of the pigeonhole principle. (See the paper by P. Beame and S. Riis [BR98] for details.)

4. (J. Krajíček [Kra]) Let p be a prime and r a positive integer, not divisible by p. Give non-constant degree lower bounds for $Count_r$ over the field $GF(p)$.

Exercise 5.10.6. Show that Gentzen sequent calculus LK (with cuts) polynomially simulates resolution.

Exercise 5.10.7 ([BP96]). Using the technique of repeatedly killing off large clauses, as explained in the proof of Theorem 5.4.8, give a simpler proof of Theorem 5.4.7.

Exercise 5.10.8. Using a graph-transformation due to Kirkpatrick [Kir74], the proof of Theorem 5.4.9 can be adapted to produce examples of formulas with clauses of size exactly 3. This can be done by replacing each vertex of G_n of degree k by a ring of k vertices and joining the resulting rings by edges corresponding to the old vertices of G_n. The resulting graphs G'_n give rise to clauses satisfying the required properties (see [Urq87], [Gal77b], [Gal77a]).

Exercise 5.10.9. ([Tse68]) Resolution as a proof method is applicable only to propositional formulas in conjunctive normal form. To have a resolution method applicable to any formula, not just those in conjunctive normal form, we argue as follows. For each formula ϕ introduce a propositional variable p_ϕ with the assumption that if ϕ is atomic then $p_\phi = \phi$. The limited extension of a formula ϕ, denoted by $LE(\phi)$, is defined as follows on subformulas ψ of ϕ:

- if $\psi = \neg\chi$ then $\{p_\psi, p_\chi\}, \{\neg p_\psi, \neg p_\chi\} \in LE(\phi)$.
- if $\psi = \chi_1 \vee \chi_2$ then $\{p_\psi, \neg p_{\chi_1}\}, \{p_\psi, \neg p_{\chi_2}\}, \{\neg p_\psi, p_{\chi_1}, p_{\chi_2}\} \in LE(\phi)$.
- if $\psi = \chi_1 \wedge \chi_2$ then $\{\neg p_\psi, p_{\chi_1}\}, \{\neg p_\psi, p_{\chi_2}\}, \{p_\psi, \neg p_{\chi_1}, \neg p_{\chi_2}\} \in LE(\phi)$.

Show that

1. ϕ is satisfiable $\Leftrightarrow LE(\phi) \cup \{\{p_\phi\}\}$ is satisfiable.
2. ϕ is a tautology $\Leftrightarrow LE(\neg\phi) \cup \{\{\neg p_\phi\}\}$ is not satisfiable.
3. ϕ is a tautology $\Leftrightarrow$ there is a resolution derivation of $\Box$ from $LE(\neg\phi) \cup \{\{\neg p_\phi\}\}$.

Exercise 5.10.10 ([Kra97a]). Prove Krajíček's interpolation Theorem 5.4.13, which is stated as follows. Suppose that propositional variables $\mathbf{p}$ are the common variables in clauses

$$A_1(\mathbf{p},\mathbf{q}), \ldots, A_m(\mathbf{p},\mathbf{q}), B_1(\mathbf{p},\mathbf{r}), \ldots, B_\ell(\mathbf{p},\mathbf{r}),$$

and that there is a resolution refutation P of $A(\mathbf{p},\mathbf{q}) \wedge B(\mathbf{p},\mathbf{r})$ of depth d and size s. Then there is a boolean circuit C of depth d and size $O(s)$ for which

$$C(\mathbf{p}) = \begin{cases} 0 \text{ if } A(\mathbf{p},\mathbf{q}) \text{ is refutable} \\ 1 \text{ else if } B(\mathbf{p},\mathbf{r}) \text{ is refutable} \end{cases}$$

Moreover, if P is a proof tree, then the circuit C has fan-out 1. Additionally, if the propositional variables $\mathbf{p}$ are all positive in $A(\mathbf{p}, \mathbf{q})$, or all negative in $B(\mathbf{p}, \mathbf{r})$, then the interpolating circuit is a monotone boolean circuit.

Exercise 5.10.11. In this exercise we study the complexity of Craig's interpolation theorem in propositional logic with logical symbols $\vee, \wedge, \neg$ (note that the logical symbol $\rightarrow$ is not in the language). A Craig interpolant for a valid implication $\phi \rightarrow \psi$ is a formula θ such that both $\phi \rightarrow \theta$ and $\theta \rightarrow \psi$ are valid and θ uses only propositional variables which are in both ϕ and ψ.

1. ($\star$) (**[Mun82]**) Show that if $\phi \rightarrow \psi$ is a valid implication then there is a Craig interpolant θ such that $|\theta| \leq 2^{O(|\phi|+|\psi|)}$.
2. ($\star$) (**[Mun83]**) For any sentence ϕ let d_ϕ denote the delay complexity (i.e., the depth of the smallest depth circuit with bounded fan-in and fan-out 1 in the basis $\vee, \wedge, \neg$ computing a boolean function f) of the boolean function f_ϕ represented by the sentence ϕ. Show that for infinitely many d there exist valid implications $\phi \rightarrow \psi$ with $d_\phi, d_\psi \leq d$ such that any Craig interpolant θ of $\phi \rightarrow \psi$ satisfies
$$d_\theta \geq d + \frac{\log(d/2)}{3}.$$
3. It is an open problem to determine whether or not the Craig interpolant of two formulas can have polynomial growth in the size of the two formulas.
4. ($\star$) (**[Mun83]**) In predicate calculus it is possible to construct valid implications $\phi \rightarrow \psi$ of size $O(n)$ all of whose interpolants have size 2_n, where 2_n is defined as follows by induction on n: $2_1 = 2$ and $2_{k+1} = 2^{2_k}$.

Exercise 5.10.12 ([Pud97]). Consider the graph theoretic principle:

> *Either a graph on n vertices does not contain a size m clique, or it does not have an $m-1$ coloring (i.e., a coloring of the vertex set by $m-1$ colors, such that no vertices joined by an edge have the same color.*

1. Express the negation of this statement by a family $\mathcal{F}$ of linear inequalities of size $O(n^2m^2)$.
HINT. Let $p_{i,j}$ express the edge relation, let variables $q_{k,i}$ code an injective mapping from $\{1, \cdots, m\}$ into the vertex set, and let variables $r_{i,\ell}$ code the $m-1$ coloring of the vertex set.
2. Prove an exponential lower bound for the number of inferences in a cutting plane refutation of $\mathcal{F}$.
HINT. Lift the exponential lower bound [Raz87b] for CLIQUE recognition from monotonic boolean circuits to monotonic real circuits, and apply the Interpolation Theorem 5.6.7.

Exercise 5.10.13 ([BC96]). This exercise concerns cutting plane proofs of a strong form of the *st*-connectivity principle.

1. The *unique endnode principle* states that there exists a finite, simple undirected graph of valence 2 having a unique endnode. Let UEP_n, the unique endnode principle on graphs with n vertices, denote the family of linear inequalities (5.98) through (5.101). Give polynomial-size CP refutations of UEP_n.

$$r_{i,i} = 0, \text{for all } 0 \le i < n \tag{5.98}$$

$$r_{i,j} = r_{j,i}, \text{for all } 0 \le i, j < n \tag{5.99}$$

$$\sum_{0 \le j < n} r_{0,j} = 1 \tag{5.100}$$

$$\sum_{0 \le j < n} r_{i,j} = 2, \text{for all } 0 < i < n. \tag{5.101}$$

Here, an equality $A = B$ abbreviates the two inequalities $A \geq B$, $B \geq A$.
HINT. For $0 \le k < n$, let

$$B_k = \sum_{i<j\le k<n} r_{i,j} \tag{5.102}$$

$$S_k = \sum_{i\le k<j<n} r_{i,j}. \tag{5.103}$$

Thus B_k is the number of edges $r_{i,j}$ for $i < j$ *both* of whose endpoints are bounded by k, while S_k is the number of edges $r_{i,j}$ for $i < j$ whose endpoints *straddle* k. Prove by induction on $0 \le k < n$ that $2 \cdot B_k + S_k = 2k + 1$, and deduce the desired statement when $k = n - 1$. For details, see [BC96].
2. The strong *st*-connectivity principle states that if G is a finite undirected graph, with two designated vertices s, t of degree 1 and other vertices of degree 2, then there is a path from s to t. Give a propositional formalization $\neg STC_n$ of the negation of this principle for graphs, whose vertex size is n, and give polynomial-size CPLE refutations of the strong *st*-connectivity principle, by relating UEP_n to STC_n.
3. (**[BCE+95]**) Show that

$$\{\neg UEP_n : n \in \mathbf{N}\}$$

is equivalent to the parity principle $\{2-EQ_n : n \in \mathbf{N}\}$ modulo constant-depth, polynomial-size Frege proofs.

Exercise 5.10.14. This exercise concerns applications of the carry-save addition technique from Theorem 5.7.15 within Frege systems.

1. (**[Goe92]**) Prove that any Frege system $\mathcal{F}$ p-simulates the cutting planes system CP.
HINT. Use carry-save boolean circuits of logarithmic depth to define multiplication (division by repeated addition, and thus represent inequality

I from CP by a propositional formula $Rep(I)$. If there is a CP refutation $I_1, \ldots, I_m$ of the system of linear inequalities representing formula A, then there are Frege proofs, of size polynomial in the refutation size of A, that

$$\vdash_{\mathcal{F}} A \supset Rep(I_1)$$
$$\vdots$$
$$\vdash_{\mathcal{F}} A \supset Rep(I_m)$$
$$\vdash_{\mathcal{F}} A \supset 0$$

2. (**[Clo95], [BC96]**) The extension CP^+ of CP is obtained by allowing fractional expressions (with the ceiling operator) and then removing the requirement of divisibility of all variable coefficients in the division rule. Give a formal definition of CP^+ and prove that CP^+ and any Frege system $\mathcal{F}$ p-simulate each other.
HINT. For the direction that CP^+ p-simulates $\mathcal{F}$, represent $A \vee B$ by $\lceil \frac{A+B}{2} \rceil \geq 1$. Define division by a fixed constant using carry-save circuits for multiplication, and then proceed as in the previous part.

Exercise 5.10.15 ([CEI96]). Prove that the polynomial calculus

- polynomially $n^{O(1)}$ simulates Horn clause resolution,
- quasipolynomially $2^{\log^{O(1)} n}$ simulates tree-like resolution, and
- weakly exponentially $2^{o(n)}$ simulates resolution.

HINT. Theorem 5.4.11 and Theorem 5.4.12 give width upper bounds for tree-like (dag-like) resolution proofs, given size upper bounds, and hence provide an upper bound for the runtime of Algorithm A 5.4.2. From width w resolution refutations, by using Theorem 5.5.8, we obtain degree $2w$ polynomial calculus refutations. Historically, Ben-Sasson and Wigderson's width bounds from Theorems 5.4.11 and 5.4.12 were in fact modeled on similar proofs from [CEI96].

Exercise 5.10.16. State and prove the analogue of Theorem 5.5.22 for arbitrary $k \geq 3$ – i.e., using the techniques of Section 5.5.6, prove lower bounds for polynomial calculus refutations of random k-CNF formulas.

Exercise 5.10.17. Prove Theorem 5.5.23, by stating and proving the analogues of the appropriate lemmas, in particular of Lemma 5.5.4.

Exercise 5.10.18. This exercise concerns extended Frege systems from Section 5.7.2.

1. Explicitly construct the translation τ in Theorem 5.7.10 and prove that it provides the required simulation of $\mathcal{F}$ by $\mathcal{F}'$.
2. Use induction on the length of proofs in order to convert the extended Frege proof P of ϕ (using only the abbreviations $p_i \equiv \phi_i$, for $i = 1, \ldots, n$) into a Frege proof of size $O(|P|^2)$ of the formula $\bigwedge_{i \leq n}(p_i \equiv \phi_i) \rightarrow \phi$.

3. Show that although replacing each p_i with ϕ_i will convert the extended Frege proof into a Frege proof, the resulting simulation can yield a Frege of exponential size.

6. Machine Models and Function Algebras

> Like so many other mathematical ideas, especially the more profoundly beautiful and fundamental ones, the idea of computability seems to have a kind of *Platonic reality* of its own.
>
> *R. Penrose [Pen89]*

6.1 Introduction

A recurring theme in theory of computation is that of a *function algebra*[1] i.e., a smallest class of functions containing certain initial functions and closed under certain operations (especially substitution and primitive recursion).[2] In 1904, G.H. Hardy [Har04] used related concepts to define sets of real numbers of cardinality $\aleph_1$. In 1923, Th. Skolem [Sko23] introduced the primitive recursive functions, and in 1925, as a technical tool in his claimed sketch proof of the continuum hypothesis, D. Hilbert [Hil25] defined classes of higher type functionals by recursion. In 1928, W. Ackermann [Ack28] furnished a proof that the diagonal function $\varphi_a(a,a)$ of Hilbert [Hil25], a variant of the Ackermann function, is not primitive recursive. In 1931, K. Gödel [Göd31] defined the primitive recursive functions, there calling them "rekursive Funktionen", and used them to arithmetize logical syntax via Gödel numbers for his incompleteness theorem. Generalizing Ackermann's work, in 1936 R. Péter [Pét36] defined and studied the k-fold recursive functions. The same year saw the introduction of the fundamental concepts of Turing machine (A.M. Turing [Tur37]), λ-calculus (A. Church [Chu36]) and μ-recursive functions (S.C. Kleene [Kle36a]). By restricting the scheme of primitive recursion to allow only limited summations and limited products, the *elementary functions* were introduced in 1943 by L. Kalmár [Kal43]. In 1953, A. Grzegorczyk [Grz53] studied the classes $\mathcal{E}^k$ obtained by closing certain fast growing

[1] With some additions (such as parallel programming examples) and deletions, the material for this chapter is reprinted from P. Clote [Clo98], with the kind permission of Elsevier Science.

[2] In [Hil25], Hilbert stated that "*substitution* (i.e., replacement of an argument by a new variable or function) and *recursion* (the scheme of deriving the function value for $n+1$ from that of n)" are "the elementary operations for the construction of functions".

"diagonal" functions under composition and *bounded primitive recursion* or *bounded minimization.*

H. Scholz's 1952 [Sch52] question concerning the characterization of *spectra* $\{n \in \mathbf{N} : (\exists \text{ model } M \text{ of } n \text{ elements})(M \models \phi)\}$ of first order sentences ϕ, which was shown in 1974 by N. Jones and A. Selman [JS74] to equal $\text{NTIME}(2^{O(n)})$, was the starting point for J.H. Bennett's work [Ben62] in 1962. Among other results, Bennett introduced the key notions of *positive extended rudimentary* and *extended rudimentary* (equivalent to the notions of nondeterministic polynomial time NP and the polynomial time hierarchy PH), characterized the spectra of sentences of higher type logic as exactly the Kalmár elementary sets, and proved that *rudimentary* coincides with Smullyan's notion of *constructive arithmetic* (those sets definable in the language $\{0, 1, +, \cdot, \leq\}$ of arithmetic by first order bounded quantifier formulas). Only much later in 1976 did C. Wrathall [Wra76] connect these concepts to computer science by proving that the linear time hierarchy LTH coincides with rudimentary, hence constructive arithmetic, sets. In 1963 R. W. Ritchie [Rit63] proved that Grzegorczyk's class $\mathcal{E}^2$ is the collection of functions computable in linear space on a Turing machine. In 1965, A. Cobham [Cob65] characterized the polynomial time computable functions as the smallest function algebra closed under Bennett's scheme of *bounded recursion on notation.*[3] These arithmetization techniques led to a host of characterizations of sequential computational complexity classes by machine-independent function algebras in the work of D. B. Thompson [Tho72] in 1972 on polynomial space, of K. Wagner [Wag79] in 1979 on general time complexity classes. Function algebra characterizations of *parallel* complexity classes were given more recently by P. Clote [Clo90] in 1990 and B. Allen [All91] in 1991, while certain small *boolean circuit* complexity classes were treated by P. Clote and G. Takeuti [CT94] in 1995. Higher type analogues of certain characterizations were given in 1976 by K. Mehlhorn [Meh76], in 1991 by S. Cook and B. Kapron [KC96, CK90] for sequential computation, and in 1993 by the P. Clote, A. Ignjatovic, B. Kapron [CIK93] for parallel computation. (See the next chapter for a strong extension of the latter result.) In 1995 H. Vollmer and K. Wagner [VW96] characterized Valiant's class $\#P$ by a function algebra. Though distinct, the arithmetization techniques of function algebras are related to those used in proving numerous results like *(i)* NP equals generalized first order spectra (R. Fagin [Fag74]), *(ii)* the characterization of complexity classes via finite models (the program of *descriptive complexity theory* investigated by R. Fagin [Fag90], N. Immerman [Imm87, Imm89], Y. Gurevich and S. Shelah [GS86], and others).

From this historical overview, it clearly emerges that *function algebras* and *computation models* are intimately related as the software (class of programs) and hardware (machine model) counterparts of each other. Histor-

[3] According to [Meh76], K. Weihrauch independently proved a similar characterization in 1972.

ically, Cobham's machine independent characterization of the polynomial time computable functions was the start of modern complexity theory, indicating a robust and mathematically interesting field. As outlined in the next chapter, current work on type 2 and higher type function algebras suggests directions for the extension of complexity theory to higher type computation. The development of function algebras is potentially important in computer science for programming language design. New kinds of operations used in defining function algebras could possibly be incorporated in *small*, non-universal programming languages for dedicated purposes. All the function algebras defined in this chapter could be used to define free variable equational calculi. For instance, S. Cook's system PV [Coo71] comes from Theorem 6.3.6, P. Clote's systems AV, ALV, ALV' [Clo92a, Clo93] come from Theorems 6.3.10 and 6.3.11, and J. Johannsen's [Joh96] systems $TV, A2V$ come from Theorem 6.3.5, while M. O'Donnell [O'D85] has proposed equational calculus as a programming language.

In this chapter, we will survey a selection of results which illustrate the arithmetization techniques used in characterizing certain computation models by function algebras.

6.2 Machine Models

Despite the immense diversity of abstract machine models and complexity classes (see for instance [vEB90] or [WW86]), only the most natural and robust models and classes will be treated in this chapter. Many of the following machine models are familiar. Nevertheless, definitions are given in sufficient detail to provide an idea of the required initial functions and closure operations which permit function algebra characterizations of complexity classes.

6.2.1 Turing Machines

In proving the recursive unsolvability of Hilbert's *Entscheidungsproblem* (independently established as well by A. Church [Chu36] using the λ-calculus), A.M. Turing [Tur37] introduced the Turing machine, largely motivated by his attempt to make precise the notion of computable (real) number, i.e., "those whose decimals which are calculable by finite means". Considering the "computer" as an idealized human clerk, Turing argued that the "behavior of the computer at any moment is determined by the symbols which he is observing, and his 'state of mind' at that moment", and specified that the number of "states of mind" should be finite, since "human memory is necessarily limited". Formally, we have the following:

Definition 6.2.1. *A multitape Turing machine* (TM) *M is specified by a six-tuple* $(Q, \Sigma, \Gamma, \delta, q_0, k)$ *where* $k \in \mathbf{N}$,

- *Q is a finite set of* states *containing the accept and reject states q_A, q_R, as well as the start state q_0,*
- *Σ [resp. Γ] is a finite read-only input [resp. read-write work] tape alphabet not containing the blank symbol B,*
- *δ is the transition function and maps*

$$(Q - \{q_A, q_R\}) \times (\Sigma \cup \{B\}) \times (\Gamma \cup \{B\})^k$$

into

$$Q \times (\Gamma \cup \{B\})^k \times \{-1, 0, 1\}^{k+1}.$$

A Turing machine is assumed to have a one-way infinite input tape and k one-way infinite work tapes. The work tapes are initially blank, while on input $w = w_1 \cdots w_n$ with $w_i \in \Sigma$, the initial input tape is of the form below.

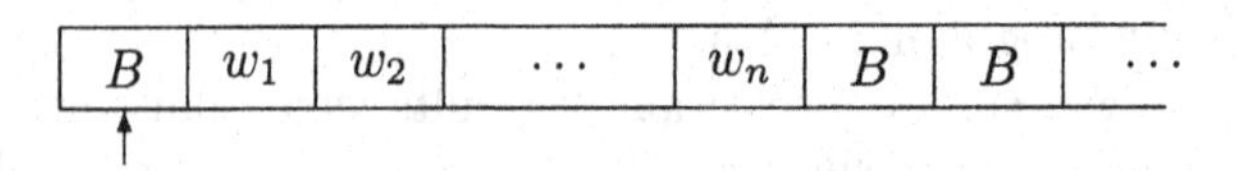

Each work tape has a tape head (above indicated by an arrow) capable of reading the symbol in the currently scanned square, writing a symbol in that square and remaining stationary or moving one square left or right. The leftmost cell is the 0-th cell. Since the input tape is read-only, the input tape head can scan a tape cell and remain stationary or move one square left or right. A *configuration* is a member of $Q \times (\Sigma \cup \{B\})^* \times (\Gamma \cup \{B\})^{*k} \times \mathbf{N}^{k+1}$, and indicates the current state, tape contents, and head positions. Alternately, a configuration can be abbreviated by underscoring the symbols currently scanned by a tape head, in order to indicate the current tape head position. For instance, $(q, Bab\underline{a}B, Bb\underline{b}B)$ abbreviates the configuration of a TM in state q, with an input tape, whose head currently scans an a, and one work tape, whose head currently scans a b. A halted configuration is one whose state is q_A or q_R.

Let

$$\alpha = (q, BxB, \alpha_1, \ldots, \alpha_k, n_0, n_1, \ldots, n_k)$$
$$\beta = (r, BxB, \beta_1, \ldots, \beta_k, m_0, m_1, \ldots, m_k)$$

be configurations for M on input x. Then β is the *next configuration* after α in M's computation on $x \in \Sigma^*$, denoted $\alpha \vdash_M \beta$, if the following conditions are satisfied:

1. the n_0-th cell of the input tape BxB contains symbol a,
2. for $1 \leq i \leq k$ the following hold:
   a) $\sigma_i, \tau_i \in \Gamma \cup \{B\}$ and $u_i, v_i, w_i \in (\Gamma \cup \{B\})^*$
   b) $\alpha_i = u_i \sigma_i v_i$ and $\beta_i = u_i \tau_i w_i$
   c) $|u_i| = n_i$ (Recall that the leftmost cell is the 0-th cell, so the n-th cell has n cells to its left. This implies that σ_i [resp. τ_i] is the contents of the n_i-th cell of the i-th tape in configuration α [resp. β].)

3. $\delta(q, a, \sigma_1, \ldots, \sigma_k) = (r, \tau_1, \ldots, \tau_k, m_0 - n_0, m_1 - n_1, \ldots, m_k - n_k)$, where for $1 \leq i \leq k$:
   a) $m_i < |\beta_i|$
   b) either $v_i = w_i$ or $v_i = \lambda$ (the empty word), $w_i = B$, and $m_i = n_i + 1$.

The reflexive, transitive closure of $\vdash_M$ is denoted by $\vdash_M^*$, and a configuration C is said to *yield* a configuration D *in n-steps*, denoted $C \vdash_M^n D$, if there are $C_1, \ldots, C_n$ such that $C = C_1 \vdash_M C_2 \vdash_M \cdots \vdash_M C_n = D$, while C *yields* D if $C \vdash_M^* D$. A Turing machine M *accepts* a language $L \subseteq \Sigma^*$, denoted by $L = L(M)$, if L is the collection of words w such that the *initial configuration* $(q_0, \underline{B}wB, \underline{B}, \ldots, \underline{B})$ yields $(q_A, \underline{B}wB, \underline{B}, \ldots, \underline{B})$; a word w is *accepted* in n steps if $(q_0, \underline{B}wB, \underline{B}, \ldots, \underline{B}) \vdash_M^n (q_A, \underline{B}wB, \underline{B}, \ldots, \underline{B})$. The machine M accepts $L \subseteq \Sigma^*$ in *time* $T(n)$ (resp. *space* $S(n)$) if $L = L(M)$ and for each word $w \in L(M)$ of length n, w is accepted in at most $T(n)$ steps (resp. the maximum number of cells visited on each of M's work tapes is $S(n)$). A language $L \subseteq \Sigma^*$ is *decided* by M in *time* $T(n)$ (resp. *space* $S(n)$) if L [resp. $\Sigma^* - L$] is the collection of words for which M halts in state q_A [resp. q_R], and for each word $w \in \Sigma^*$ of length n, M halts in at most $T(n)$ steps (resp. the maximum number of cells visited on each of M's work tapes is $S(n)$). This article concerns complexity classes, so for the most part we identify the notions of acceptance and decision (for most of the complexity classes here considered, machines of a certain complexity class can be clocked so as to reject a word which is not accepted).

Recall that

$$O(f) = \{g : (\exists c > 0)(\exists n_0)(\forall n \geq n_0)[g(n) \leq c \cdot f(n)]\},$$
$$\Omega(f) = \{g : (\exists c > 0)(\exists n_0)(\forall n \geq n_0)[f(n) \leq c \cdot g(n)]\}$$
$$\Theta(f) = O(f) \cap \Omega(f)$$

so that $n^{O(1)}$ denotes the set of all polynomially bounded functions. If T, S are one-place functions, then

$$\text{DTIME}(T(n)) = \{L \subseteq \Sigma^* : L \text{ accepted by a TM in time } O(T(n))\}$$
$$\text{DSPACE}(S(n)) = \{L \subseteq \Sigma^* : L \text{ accepted by a TM in space } O(S(n))\}$$
$$\text{PTIME} = \text{P} = \text{DTIME}(n^{O(1)})$$
$$\text{PSPACE} = \text{DSPACE}(n^{O(1)})$$
$$\text{ETIME} = \cup_{c \geq 1}\text{DTIME}(2^{c \cdot n}) = \text{DTIME}(2^{O(n)})$$
$$\text{EXPTIME} = \cup_{c \geq 1}\text{DTIME}(2^{n^c}) = \text{DTIME}(2^{n^{O(1)}}).$$

Finally, DTimeSpace$(T(n), S(n))$ is defined as

$$\{L \subseteq \Sigma^* : L \text{ accepted by a TM in time } O(T(n)) \text{ and space } O(S(n))\}.$$

Definition 6.2.2. *A nondeterministic multitape Turing machine (*NTM*) M is specified by $(Q, \Sigma, \Gamma, \Delta, q_0, k)$ where $Q, \Sigma, \Gamma, q_0, k$ are as in Definition 6.2.1 and the* transition relation *Δ is contained in*

$$((Q - \{q_A, q_R\}) \times (\Sigma \cup \{B\}) \times (\Gamma \cup \{B\})^k) \\ \times (Q \times (\Gamma \cup \{B\})^k \times \{-1, 0, 1\}^{k+1}).$$

If α, β are configurations in the computation of the nondeterministic Turing machine (NTM) M on input x, then write $\alpha \vdash_M \beta$ if

$$(q, a, \sigma_1, \ldots, \sigma_k, r, \tau_1, \ldots, \tau_k, m_0 - n_0, m_1 - n_1, \ldots, m_k - n_k) \in \Delta,$$

where $\sigma_i, \tau_i, a, n_i, m_i$ are as in the deterministic case.

With this change, the notions of configuration, yield and acceptance are analogous to the previously defined notions. A nondeterministic computation corresponds to a *computational tree* whose root is the initial configuration, whose leaves are halted computations, and whose internal nodes α have as children those configurations β obtained in one step from α, $\alpha \vdash_M \beta$. A word $w \in \Sigma^*$ is *accepted* if there is an accepting path in the computation tree, though many non-accepting paths may exist. A NTM M accepts a word of length n in *time* $T(n)$ [resp. *space* $S(n)$] if the depth of the associated computation tree is at most $T(n)$ [resp. for each configuration α in the computation tree the number of cells used on each work tape is at most $S(n)$]. NTIME$(T(n))$ [resp. NSPACE$(S(n))$] is the collection of languages $L \subseteq \Sigma^*$ accepted by a NTM in time $O(T(n))$ [resp. space $O(S(n))$]; NP $=$ NTIME$(n^{O(1)})$.

Similarly, NTIMESPACE$(T(n), S(n))$ is the set of languages $L \subseteq \Sigma^*$ accepted by a NTM in time $O(T(n))$ and space $O(S(n))$.

With the previous definitions, any computation depending on all bits of the input requires at least linear time, the minimum amount of time taken to scan the input. However, by allowing a TM to access its input bitwise via pointers or random access, sublinear runtimes can be achieved, as shown by Chandra et al. [CKS81].

Definition 6.2.3. *A Turing machine M with* random access *(*RATM*) is given by a finite set Q of states, an input tape having no tape head, k work tapes, an* index query *tape and an* index answer *tape . To permit random access, the alphabet Γ is always assumed to contain the symbols* 0, 1. *Except for the input tape, all other tapes have a tape head. M contains a distinguished input query state q_I, in which state M writes into the leftmost cell of the index answer tape that symbol which appears in the k-th input tape cell, where $k = \sum_{i<m} k_i \cdot 2^i$ is the integer whose binary representation is given by the contents*

B	k_{m-1}	k_{m-2}	$\cdots$	k_0	B	$\cdots$

of the query index tape. Unlike the oracle Turing machine in Definition 6.2.4, the query index tape is not automatically erased after making an input bit query. A logtime RATM *runs in time $O(\log n)$, where n is the length of the input. (All logarithms are with respect to base 2.)*

Logtime on a RATM is not so weak, and can compute certain simple functions, as shown by the next result. In the following, the function value $f(u) = v$ is computed by a logtime Turing machine in the sense that on input (k, u), the machine outputs the k-th bit of v in time logarithmic in the length of the input.

Fact 6.2.1 ([BIS90]). Given an input of length n, a deterministic logtime RATM can

(i) compute the length of its input,
(ii) add and subtract integers of $O(\log n)$ bits,
(iii) decode a simple pairing function on strings of length $O(n)$.

Proof. Since a RATM has no output tape, we adopt the convention that M computes the function $f : \Sigma^* \to \Sigma^*$ if $|f(x)|$ is bounded by a polynomial in $|x|$, and for all bits, the i-th bit of $f(x)$ is a iff M accepts (x, i, a). The proof of *(i)* uses binary search, and according to [Bus87a], appears to have been first noticed by M. Dowd. The proof of *(ii)* is clear, since addition and subtraction take time linear in the input length. In *(iii)*, for $u, v \in \Sigma^*$ the pair (u, v) can be encoded by $\tau(|u|)11\tau(|v|)11uv$, where τ replaces each 0 [resp. 1] by 00 [resp. 01]. Decoding can then be done by using addition and random access.

A.M. Turing [Tur37] introduced the notion of *relative computation* using an *oracle Turing machine.*

Definition 6.2.4. *Let $B \subseteq \Gamma^*$. An oracle Turing machine (*OTM*) with oracle B is a Turing machine M which in addition to a read-only input tape, a distinguished output tape and finitely many work tapes, has a one-way infinite* oracle query tape . *The machine M has oracle answer states q_{yes} , q_{no} as well as a special oracle query state $q_?$ in which it queries whether the current contents of the oracle query tape belongs to oracle B. The transition function δ of M is a mapping from*

$$(Q - \{q_A, q_R, q_?\}) \times (\Sigma \cup \{B\}) \times (\Gamma \cup \{B\})^{k+1}$$

into

$$Q \times (\Gamma \cup \{B\})^{k+1} \times \{-1, 0, 1\}^{k+2}.$$

A computation is defined as previously, except that if M is in state $q_?$ then the machine queries whether the word given by the current contents of the oracle query tape belongs to B. Dependent on the outcome of the oracle query, M goes into state q_{yes} or q_{no}, and simultaneously erases the query tape and *places the oracle tape head at the leftmost square. This entire sequence of events takes place in one step. Finally, nondeterministic oracle Turing machines are analogously defined by adding the oracle apparatus to the* NTM *model.*

For $A \subseteq \Sigma^*$ and $B \subseteq \Gamma^*$, write $A \leq_T B$ if A can be decided by an oracle Turing machine with oracle B. Similarly write $A \leq_T^P B$ [resp. $A \leq_T^{NP} B$] if A can be computed by a deterministic [resp. nondeterministic] oracle Turing machine with oracle B in polynomial time. Let $\Sigma_0^P = P$ and Σ_{n+1}^P be

$$\{A : (\exists B \in \Sigma_n^P)(A \leq_T^{NP} B)\}.$$

In [CKS81], A. Chandra, D. Kozen and L. Stockmeyer introduced the *alternating Turing machine* (ATM) , a model suitable for formalizing divide and conquer algorithms.[4] When used with random access, this model allows sublinear runtimes and can be viewed as a kind of parallel computational device; in particular, uniform boolean circuit families, another parallel computation model, can be related to ATM's.

Definition 6.2.5. *An alternating multitape Turing machine* (ATM) *M is specified by $(Q, \Sigma, \Gamma, \Delta, q_0, k, \ell)$ where $\ell : (Q - \{q_A, q_R\}) \to \{\wedge, \vee\}$ and $Q, \Sigma, \Gamma, \Delta, q_0, k$ are as in Definition 6.2.2 of a nondeterministic machine.*

The function ℓ labels non-halting states as *universal* ($\wedge$) and *existential* ($\vee$). An *accepting computation tree* T is a subtree of the computation tree of M on x such that for any configuration $\alpha \in T$,

- the root of T is the initial configuration of M on x,
- if α is a leaf of T, then α is an *accepting* configuration,
- if α is universal, then for all β, $\alpha \vdash_M \beta \Rightarrow \beta \in T$, and
- if α is existential, then there exists $\beta \in T$ for which $\alpha \vdash_M \beta$.

The ATM M *accepts* input x if there is a non-empty accepting computation tree of M on x; otherwise x is rejected . $L(M)$ denotes the set of $x \in \Sigma^*$ accepted by M. The language $L(M)$ is accepted by M in *time* $T(n)$ [resp. *space* $S(n)$] if for each $w \in L(M)$ of length n, there is an accepting computation tree T of depth at most $T(n)$ [resp. in which at most $S(n)$ cells are used for each of the work tapes and index tapes at any node in the tree T]. The number of *alternations* M makes in an accepting computation tree T is defined to be the maximum number of alternations between existential and universal nodes in a path from the root to a leaf.

Convention 6.2.1. From now on, unless otherwise indicated, for any sublinear runtime $T(n) = o(n)$, the intended Turing machine model is RATM, while for runtimes $T(n) = \Omega(n)$, the intended Turing machine model is the

[4] Divide and conquer algorithms are generally space efficient. The *parallel computation thesis* states that sequential space equals parallel time (see [Bor73]). In this sense, ATM's provide a parallel computation model.

conventional TM. This convention applies to deterministic, nondeterministic, and alternating Turing machines. While it is a simple exercise to show that PTIME is the same class, regardless of model, it appears to be an open problem to determine the relationship between DTIME($T(n)$) on TM and RATM, for $T(n) = \Omega(n)$.

Definition 6.2.6.

$$\begin{aligned}
&\text{ATIME}(T(n)) = \{L \subseteq \Sigma^* : L \textit{ accepted by an } \text{ATM} \textit{ in time } O(T(n))\} \\
&\text{ASPACE}(S(n)) = \{L \subseteq \Sigma^* : L \textit{ accepted by an } \text{ATM} \textit{ in space } O(S(n))\} \\
&\text{ALOGTIME} = \text{ATIME}(O(\log n)) \\
&\text{APOLYLOGTIME} = \cup_{k\geq 1}\text{ATIME}(O(\log^k n)) \\
&\text{ALINTIME} = \text{ATIME}(O(n)).
\end{aligned}$$

The logtime hierarchy LH *[resp. the* linear time hierarchy LTH, *resp. the* polynomial time hierarchy PH *] is the collection of languages $L \subseteq \Sigma^*$, for which L is accepted by an* ATM *in time $O(\log n)$ [resp. $O(n)$, resp. $n^{O(1)}$] with at most $O(1)$ alternations.*[5] *Σ_k-TIME($T(n)$) is the collection of languages accepted by an* ATM *in time $O(T(n))$ with at most k alternations, beginning with an existential state.*

The class ALOGTIME is surprisingly powerful. It is not difficult to see that it contains all of the regular languages. This follows, by noting that the construction in the proof of Proposition 1.6.1 can be formalized to yield an ALOGTIME algorithm. Much more striking are the results of D. Barrington[6] and S. Buss. First, define a language L to be *complete* for ALOGTIME under DLOGTIME reductions, if $L \in$ ALOGTIME and for any $L' \in$ ALOGTIME, there is a logtime many-one function f with the property that $|f(u)|$ is polynomial in $|u|$, and $u \in L'$ iff $f(u) \in L$. Here, the function value $f(u) = v$ is computed by a logtime Turing machine in the sense that on input (k, u), the machine outputs the k-th bit of v in time logarithmic in the length of the input.

Theorem 6.2.1 ([Bar89]). *Let G be any finite non-solvable permutation group (for example S_5). Then the word problem*

$$\{(\sigma_1, \ldots, \sigma_n) : \sigma_i \in G, \sigma_1 \circ \sigma_2 \circ \cdots \circ \sigma_n = id\}$$

for G is complete for ALOGTIME *under* DLOGTIME *reductions.*

Proof. (Outline) If G is a group, then the *commutator* of elements $a, b \in G$ is the element $aba^{-1}b^{-1}$. The *commutator subgroup* of G, denoted by $[G, G]$, is the subgroup of G generated by all the commutators of G. For any group

[5] It follows from [FSS84, Ajt83] that LH is a hierarchy, where the collection of languages accepted by k-alternations is properly contained in the collection of languages accepted by $k+1$-alternations. The question of whether LTH or PH is a proper hierarchy is still open.

[6] D. Barrington changed his name to D. Mix Barrington, so that some articles appear under the former name and some under the latter name.

G, define $G^{(0)} = G$, and $G^{(n+1)} = [G^n, G^n]$. By definition, a group G is solvable if there is a finite series $G = G^{(0)} \geq G^{(1)} \geq \cdots \geq G^{(n)} = \{e\}$. If G is finite, then G is non-solvable if and only if $G = G^{(0)} \geq G^{(1)} \geq \cdots \geq G^{(n)} = G^{(n+1)} \neq \{e\}$, i.e., $G^{(n)}$ is non-trivial and equal to its commutator subgroup. For example, the groups A_k, S_k for $k \geq 5$ are non-solvable.

Assume now that G is a non-solvable group with series $G = G^{(0)} \geq \cdots \geq G^{(n)} = H$, and that H is non-trivial and equal to its commutator subgroup $[H, H]$; i.e., there exists m such that every element of H can be expressed as a product $\Pi_{i=1}^{m} a_i b_i a_i^{-1} b_i^{-1}$ of m commutators of H. Using this observation, Barrington showed how to represent conjunctions and disjunctions as a word problem over H.

Namely, given a non-identity element $g \in H$ and an alternating AND/OR computation tree $T(x)$ for the computation of M on x, describe a word $w_{T(x)}$ in the elements of H such that

$$w_{T(x)} = \begin{cases} e \text{ if } M \text{ accepts } x \\ g \text{ else.} \end{cases}$$

This is done by induction on depth of node A in $T(x)$. Recall that $H = [H, H]$ and every element of H can be written as the product of m commutators of H. Then Barrington observed that if $A = (B \vee C)$ then

$$\begin{aligned} w_A(g) &= w_{B \vee C}(g) \\ &= \prod_{i=1}^{m} w_B(b_i) w_C(c_i) w_B(b_i^{-1}) w_C(c_i^{-1}). \end{aligned}$$

Similarly, $B \wedge C$ and $\neg B$ can be expressed. Inductively one forms the word $w_{T(x)}$ whose product equals e exactly when M accepts x.

From the above discussion, with a close look at uniformity issues, it follows that the word problem for a finite non-solvable permutation group is hard for ALOGTIME. On the other hand, the word problem is clearly in ALOGTIME, since one can compose n permutations by associating them with the leaves of a binary tree, whose internal nodes compute the composition of their children.

Theorem 6.2.2 ([Bus87a, Bus93]). *The boolean formula valuation problem*

$$\{\Theta : \Theta \text{ is a true variable-free propositional logic formula}\}$$

is complete for ALOGTIME *under* DLOGTIME *reductions.*

The proof of Theorem 6.2.2 is long and difficult, so will not be sketched here. The results of Barrington and Buss are complementary in the sense that the word problem for S_5 is clearly in ALOGTIME, but not obviously complete, while the boolean formula evaluation problem is clearly complete but not obviously in ALOGTIME.

In [Sav70], W. Savitch proved that $\text{NSPACE}(S(n)) \subseteq \text{DSPACE}(S^2(n))$, for any space constructible $S(n) \geq \log n$. The following theorem, due to Chandra, Kozen and Stockmeyer [CKS81], is in part a generalization of Savitch's result that PSPACE = NPSPACE, and relates alternating time and space to deterministic time and space.

Theorem 6.2.3 ([CKS81]). *If $f(n) \geq n$, then*

$$\text{ATIME}(f(n)) \subseteq \text{DSPACE}(f(n)) \subseteq \text{NSPACE}(f(n)) \subseteq \cup_{c>0}\text{ATIME}(c \cdot f(n)^2).$$

If $f(n) \geq \log n$, then

$$\text{ASPACE}(f(n)) \subseteq \cup_{c>0}\text{DTIME}(c^{f(n)}).$$

From definitions, it is clear that

$$\text{LH} \subseteq \text{ALOGTIME} \subseteq \text{LOGSPACE} \subseteq \text{PTIME} \subseteq \text{PH} \subseteq \text{PSPACE}$$

and

$$\text{LH} \subseteq \text{LTH} \subseteq \text{ALINTIME} \subseteq \text{DLINSPACE} \subseteq \text{PSPACE}.$$

Theorem 2.5.2 reduces integer multiplication to parity, and Theorem 2.6.2 and Corollary 2.6.1 provide an exponential lower bound for constant depth circuits computing parity. *A fortiori* integer multiplication does not belong to LH, a uniform version of AC^0. (Since integer multiplication, $\times$, is a function, what is meant is that $\times \notin \text{FLH}$, where the latter is the class of functions of polynomial growth rate, whose bitgraph belongs to LH; this is defined later.) Note that Buss [Bus92] has even shown that the graph of multiplication does not belong to LH. Since the graph of integer multiplication belongs to ALOGTIME, the first containment above is proper. With this exception, nothing else is known about whether the other containments are proper.

All the previous machine models concern language recognition problems. Predicates $R \subseteq (\Sigma^*)^k$ can be recognized by allowing input of the form

$$Bx_1Bx_2B\cdots Bx_nB$$

consisting of n inputs $x_i \in \Sigma^*$, each separated by the blank $B \notin \Sigma$. By adding a write-only output tape with a tape head capable only of writing and moving to the right, and by allowing input of the form $Bx_1Bx_2B\cdots Bx_nB$, a TM or RATM can compute an n-place function. In the literature, function classes such as the polynomial time computable functions were so introduced. To provide uniform notation for such function classes, along with newer classes of sublinear time computable functions, we proceed differently.

Definition 6.2.7. *A function $f(x_1,\ldots,x_n)$ has* polynomial growth *resp.* linear growth *resp.* logarithmic growth *if*

$$|f(x_1,\ldots,x_n)| = O(\max_{1\leq i\leq n} |x_i|^k), \textit{for some } k$$

resp.

$$|f(x_1, \ldots, x_n)| = O(\max_{1 \leq i \leq n} |x_i|)$$

resp.

$$|f(x_1, \ldots, x_n)| = O(\log(\max_{1 \leq i \leq n} |x_i|).$$

The graph G_f *satisfies* $G_f(\mathbf{x}, y)$ *iff* $f(\mathbf{x}) = y$. *The* bitgraph B_f *satisfies* $B_f(\mathbf{x}, i)$ *iff the i-th bit of* $f(\mathbf{x})$ *is* 1. *If* $\mathcal{C}$ *is a complexity class, then* $\mathcal{FC}$ *[resp. Lin*$\mathcal{FC}$ *resp. Log*$\mathcal{FC}$ *] is the class of functions of polynomial [resp. linear resp. logarithmic] growth whose bitgraph belongs to* $\mathcal{C}$. *In this chapter,* $\mathcal{GC}$ *will abbreviate Lin*$\mathcal{FC}$. *The iteration* $f^{(n)}(x)$ *is defined by induction on* n: $f^{(0)}(x) = x$, $f^{(n+1)}(x) = f(f^{(n)}(x))$. *With this notation, the iteration* $\log^{(n)} x$ *should not be confused with the power* $\log^n x = (\log x)^n$.

There are other extensions of the Turing machine model not covered here, such as the *probabilistic* Turing machine (yielding classes such as R and BPP, see [vEB90]), the *genetic* Turing machine (defined by P. Pudlák [Pud94], who showed that polynomial time bounded genetic TM's compute exactly PSPACE), and the *quantum* Turing machine (first introduced by D. Deutsch [Deu85], and for which P. Shor [Sho97] proved that integer factorization is computable in bounded error probabilistic quantum polynomial time BQP).

6.2.2 Parallel Machine Model

"Having one processor per data element changes the way one thinks."

W.D. Hillis and G.L. Steele, Jr. [HS86]

Emerging around 1976/77 from the work of Goldschlager [Gol77, Gol82], Fortune–Wyllie [FW78], and Shiloach–Vishkin [SV82], the *parallel random access machine* (PRAM) provides an abstract model of parallel computation for algorithm development. While existent "massively parallel" computers generally require a specific communication network (e.g. hypercube, mesh, etc.) for message passing between processors (and such details are of immense practical importance), the PRAM abstracts out all such processor communication details and postulates a global shared memory. Individual processors of a PRAM additionally have local memory, and while operating synchronously on the same program, are capable of performing arithmetic and logical operations as well as local and global read/write in both direct and indirect addressing mode. Processors may have different data stored in their local memories and have access to their unique processor identity number PID. Thus the effect of an instruction like "add the contents of the PID-th global memory register to local memory register 2 and store in local memory register 7" may be quite different in different processors. Different models of PRAM have been studied, depending on the strength of local arithmetic operations allowed, and whether simultaneous read/write in the same global memory register is allowed by several processors. This yields EREW, CREW, and CRCW models, according to whether exclusive read, exclusive write, concurrent read or concurrent write are allowed. An excellent survey of parallel algorithms and models is R.M. Karp and V. Ramachandran [KR90]. The formal development follows.

A *concurrent random access machine* CRAM has a sequence $R_0, R_1, \ldots$ of random access machines which operate in a synchronous fashion in parallel. Each R_i has its own local memory, an infinite collection of registers, each of which can hold an arbitrary non-negative integer. Global memory consists of an infinite collection of registers accessible to all processors, which are used for reading the input, processor message passing, and output. Global registers are designated $M_0^g, M_1^g, M_2^g, \ldots$, and local registers by $M_0, M_1, M_2, \ldots$ – local registers of processor P_i might be denoted $M_{i,0}, M_{i,1}, \ldots$. A global memory register can be read simultaneously by several processors (*concurrent read*, rather than *exclusive read*). In the case where more than one processor may attempt to write to the same global memory register, the lowest numbered processor succeeds (*priority resolution* of write conflict in this *concurrent write* rather than *exclusive write* model). An input x is initially given bitwise in the global registers, the register M_i^g holding the i-th bit of x. All other registers initially contain the blank symbol B (different from $0, 1$) which designates that the register is empty. Similarly at termination, the output y is specified in the global memory, the register M_i^g holding the i-th bit of y. At termination of a computation all other global registers contain the blank symbol. [The input/output convention of one integer per global memory register yields an equivalent model for the complexity classes here considered.] Let *res* (result), *op*0, *op*1, *op*2 (operands 0,1,2) be non-negative integers. If any register occurring on the right side of an instruction contains 'B', then the register on the left side of the instruction will be assigned the value 'B' (undefined).

Instructions are as follows.

M_{res} = `constant`
M_{res} = `processor number`
M_{res} = M_{op1}
M_{res} = $M_{op1} + M_{op2}$
M_{res} = $M_{op1} \dot{-} M_{op2}$
M_{res} = `MSP`(M_{op1}, M_{op2})
M_{res} = `LSP`(M_{op1}, M_{op2})
M_{res} = $*M_{op1}$
M_{res} = $*M_{op1}^g$
$*M_{res}$ = M_{op1}
$*M_{res}^g$ = M_{op1}
`GOTO label`
`GOTO label IF` M_{op1} = M_{op2}
`GOTO label IF` $M_{op1} \leq M_{op2}$
`HALT`

Cutoff subtraction is defined by $x \dot{-} y = x - y$, provided that $x \geq y$, else 0. The shift operators `MSP` and `LSP` are defined by

- $\mathtt{MSP}(x, y) = \lfloor x/2^y \rfloor$, provided that $y < |x|$, otherwise 'B',
- $\mathtt{LSP}(x, y) = x - 2^y \cdot (\lfloor x/2^y \rfloor)$, provided that $y \leq |x|$, otherwise 'B'.

The CRAM model is due to N. Immerman [Imm89], though there slightly different conventions are made.

Instructions with '$*$' concern indirect addressing. The instruction $M_{res} = *M_{op1}$ assigns to local register M_{res} the contents of local register with address given by the value M_{op1}. Similarly, $M_{res} = *M^g_{op1}$ performs an indirect read from global memory into local memory. The instruction $*M_{res} = M_{op1}$ assigns the value of local register M_{op1} to the local register whose address is given by the current contents of the local register M_{res}. Similarly, $*M^g_{res} = M_{op1}$ performs an indirect write into global memory.

In summary, the CRAM has instructions for *(i)* local operations — addition, cutoff subtraction, shift, *(ii)* global and local indirect reading and writing, *(iii)* control instructions — GOTO, conditional GOTO and HALT. A program is a finite sequence of instructions, where each individual processor of a CRAM has the same program. Each instruction has unit-cost (*uniform time cost*). During the course of a computation, only finitely many *active* processors perform computations. An input x of length n is *accepted* by a CRAM M in time $T(n)$ with $P(n)$ many active processors, if M halts after $T(n)$ time where processors $P_0, \ldots, P_{n-1}$ synchronously execute the program. The class TIMEPROC$(T(n), P(n))$ consists of those languages accepted by a CRAM in time $T(n)$ with $P(n)$ many processors.

Example 6.2.1. The following is a CRAM program for computing $|x| = \lceil \log_2(x + 1) \rceil$, where comments begin by '%'.

Let $M_{res} = \mathrm{BIT}(M_{op1}, M_{op2})$ be the instruction which, for $i = M_{op2}$ computes the coefficient of 2^i in the binary representation of the integer stored in M_{op1}, provided that $i < |M_{op1}|$, and otherwise returns the value 'B'.

```
1  M1 = processor number
2  M2 = *M1^g     % in Pi, Mi = Mi^g
3  if (M2 = B) then M0^g = M1
4  M3 = M0^g      % in Pi, M3 = least i [ Mi^g = B ] = |x|
5  *M1^g = B        % erase global memory
6  M4 = 1
7  M4 = M1 + 1
8  M4 = M3 ∸ M4
9  M5 = MSP(M3,M4)
10 M6 = MSP(M5,1)
11 M6 = M6 + M6
12 M4 = M5 ∸ M6
13  *M1^g = M4  % output placed in global memory
14  HALT
```

Processor bound: $P(|x|) = |x|$.

Strictly speaking, line 3 is not syntactically allowed, but can easily be implemented with a few extra lines of code, and will not affect the time or processor bound. Lines 6–12 ensure that $M_4 = \text{BIT}(M_3, M_3 \dot{-} (M_1 + 1))$, so that in processor P_i, $M_4 = \text{BIT}(|x|, |x| \dot{-} (i + 1))$.

6.2.3 Example Parallel Algorithms

In this section, we present some sample parallel algorithms, especially concerning string matching. Our treatment in this section, especially Algorithms 6.2.4 and 6.2.5, derive from unpublished course notes of E. Kaltofen at Rensellaer University.

When the type of local arithmetic and logical operations allowed in the instruction set are not important, algorithms are generally written in a pseudocode reminiscent of C^*, a version of parallel C implemented on the Connection Machine. For instance, Algorithm 6.2.1 computes $\max(x_1, \ldots, x_n)$ of n integers in constant time with $O(n^2)$ processors.

Algorithm 6.2.1 (Algorithm for maximum, [Kuc82]).

(1) for all $\binom{n}{2}$ pairs $1 \leq i < j \leq n$ in parallel

$$a_{i,j} = \begin{cases} 1 \text{ if } x_i < x_j \\ 0 \text{ else} \end{cases}$$

(2) for $i = 1$ to n in parallel
 $m_i = 0$

(3) for $1 \leq i < j \leq n$ in parallel
 if $a_{i,j} = 1$ then $m_i = 1$

(4) for $i = 1$ to n in parallel
 if $m_i = 0$ then $m = i$

(5) $\max = x_m$

Time = $O(1)$, Processors = $O(n^2)$

If T and P are functions then CRCW(T, P) is the class of languages and functions computable on a concurrent read, concurrent write parallel random access machine with time bound $T(n)$ and processor bound $P(n)$, where n is the length of the input. Similarly, we define EREW(T, P), and CREW(T, P) for the exclusive read, exclusive write and concurrent read, exclusive write models of parallel random access machines. The *cost* of an algorithm running on a PRAM running in time $T(n)$ with $P(n)$ processors is $T(n) \cdot P(n)$. Let $T(n)$ be the runtime for the fastest known sequential algorithm solving a problem P. Since the execution of a parallel algorithm can always be simulated by a sequential machine, a parallel algorithm for the problem P is called *optimal* (resp. *efficient*) if its cost $C(n)$ is $O(T(n))$ (resp. $O(T(n) \cdot \log^{O(1)}(n))$).

The following is an EREW algorithm to find the sum of n integers, using an obvious idea of "binary tree". The time is $O(\log(n))$ and the number of processors used is $O(n)$.

Algorithm 6.2.2 (Iterated sum).
INPUT. Array A with entries $A[0], \ldots, A[n-1]$, where $n = 2^k$.
OUTPUT. $\sum_{i<n} A[i]$.

```
for i = 1 to |n|
    for j = 0 to n − 1 in parallel
        if j mod 2^i = 0 then
            A[j] = A[j + 2^(i−1)] + A[j];
write(A[0]);
```

In the above algorithm, at the i-th step, there are $n/2^i$ active processors. In order to optimize processor use, one can similarly calculate all *prefix sums.*[7]

Algorithm 6.2.3 (Prefix sums, [LF80]).
INPUT. Array A with entries $x_0, \ldots, x_{n-1}$, where $n = 2^k$.
OUTPUT. Array A with entries $A[i] = x_0 \circ \cdots \circ x_{i-1}$, where $\circ$ is a given binary associative operation.

```
for i = 1 to |n|
    for j = 0 to n − 1 in parallel
        if j > 2^(i−1) then
            A[j] = A[j] ∘ A[j + 2^(i−1)];
write(A[0])
```

There are many immediate applications of the prefix sums technique, where $\circ$ may be a logical operation like conjuction, disjunction, exclusive-or, or an arithmetic operation like addition, multiplication of integers and matrices. As we saw in Proposition 1.6.1, an application of this technique shows that every regular language is in NC^1.

Another application of prefix sums yields a logtime string matching algorithm. Let $\Sigma = \{a_1, \ldots, a_k\}$ be a finite alphabet. For given words $u = u_1 \cdots u_m$ and $v = v_1 \cdots v_n$ of Σ^*, we would like to test in parallel if u is a subword of v, i.e., if $m \leq n$ and there exists $0 \leq i \leq n - m$ such that $u_1 \cdots u_m = v_{i+1} \cdots v_{i+m}$. This leads to the following obvious parallel algorithm.

```
for i = 0 to n − m in parallel {
      FLAG_i = true
      for j = 1 to m in parallel
          if u_j ≠ v_{i+j} then FLAG_i = FALSE
      }
 FLAG = ⋁_{i=0}^{n−m} FLAG_i
return FLAG
```

Since addition, subtraction and BIT belong to AC^0, the above algorithm places the problem of string matching in AC^0. However, when implemented on a CRCW parallel random access machine, we require mn, hence, for $m \leq n$, $O(n^2)$ processors. This is similar to the runtime of $O(mn)$ on a sequential machine for the brute force algorithm below.

[7] This operation corresponds to an atomic operation called SCAN, implemented on the Connection Machine.

```
continue  = (m ≤ n)
 FLAG = FALSE
i = 0
while continue {
    FLAG = TRUE
   for j = 1 to m
       if u_j ≠ v_{i+j} then FLAG = FALSE
   }
return FLAG
```

To obtain a linear number of processors, we identify words in Σ^* with $k \times k$ matrices, compute inverses of matrices and apply the prefix sums technique. Let $\mathcal{M}_k$ be the collection of all $k \times k$ matrices with integer entries. Temporarily, let $\circ$ denote the concatenation operation between words in Σ^*, and let $\times$ denote the operation of matrix multiplication. (Often the symbol $\circ$ will be omitted, so that we may write uv, in place of $u \circ v$, for $u, v \in \Sigma^*$.) We will define a language homomorphism Φ from $(\Sigma^*, \circ)$ into $(\mathcal{M}_k, \times)$ by defining Φ on the empty word λ and on all words a_i of length 1, and then extending Φ by homomorphism. Let $\Phi(\lambda)$ be the $k \times k$ identity matrix I_k, and for $1 \le r \le k$, let $\Phi(a_r)$ be the matrix obtained from I_k by setting the r-th row to be the sum of all rows. Formally, $\Phi(a_r) = (m_{i,j})$, where

$$m_{i,j} = \begin{cases} 1 & \text{if } (i = j) \text{ or } (i = r) \\ 0 & \text{else.} \end{cases}$$

For $M \in \mathcal{M}_k$, it is easy to check that the product matrix $\Phi(a_r) \times M$ is obtained from M by setting the r-th row to be the sum of all rows of M. Now extend Φ by homomorphism: $\Phi(u \circ v) = \Phi(u) \times \Phi(v)$. Since both concatenation and matrix multiplication are associative operations, Φ is well-defined. The following claim is easily established by induction, and so is left to the reader.

Claim 1. If $M \in \Phi(\Sigma^*)$, then the determinant of M is 1, all entries of M are non-negative, and for each row, there is at least one entry which is strictly positive.

Claim 2. For $n \in \mathbf{N}$, the restriction of Φ to Σ^n is injective.

Proof of Claim. By induction on n, we show that

$$(\forall u, v \in \Sigma^n)\,(\Phi(u) = \Phi(v) \Rightarrow u = v)\,. \tag{6.1}$$

Since only the empty word λ has length 0, trivially (6.1) holds for $n = 0$. Assume now that (6.1) holds for $n = m$. Suppose, in order to obtain a contradiction, that $u = u_1 \cdots u_{m+1}$ and $v = v_1 \cdots v_{m+1}$ are distinct words over Σ and $\Phi(u) = \Phi(v)$.

Case 1. $u_1 = v_1$. Then since

$$\Phi(u_2 \cdots u_{m+1}) = \Phi(u_1)^{-1} \times \Phi(u) = \Phi(v_1)^{-1}\Phi(v) = \Phi(v_2 \cdots v_{m+1})$$

it follows that $u_2 \cdots u_{m+1}$ and $v_2 \cdots v_{m+1}$ are distinct words of length m yet $\Phi(u_2 \cdots u_{m+1}) = \Phi(v_2 \cdots v_{m+1})$. This contradicts the induction hypothesis (6.1) when $n = m$.

Case 2. $u_1 \neq v_1$. Without loss of generality, assume that $u_1 = a_1$ and $v_1 = a_2$. Define

$$\begin{aligned}
\bar{B} &= (\bar{b}_{i,j}) = \Phi(u) \\
\bar{C} &= (\bar{c}_{i,j}) = \Phi(v) \\
B &= (b_{i,j}) = \Phi(u_2 \cdots u_{m+1}) \\
C &= (c_{i,j}) = \Phi(v_2 \cdots v_{m+1}).
\end{aligned}$$

Since the first row of $\bar{B}$ is formed by summing all rows of B, and the second row of $\bar{C}$ is formed by summing all rows of C, we have $\bar{b}_{1,j} = \sum_{i=1}^{k} b_{i,j}$, $\bar{b}_{2,j} = b_{2,j}$, $\bar{c}_{1,j} = c_{1,j}$ and $\bar{c}_{2,j} = \sum_{i=1}^{k} c_{i,j}$. If $\bar{B}$ and $\bar{C}$ are equal, then

$$\begin{aligned}
c_{1,j} = \bar{c}_{1,j} = \bar{b}_{1,j} &= b_{1,j} + b_{2,j} + \textstyle\sum_{i=3}^{k} c_{i,j} \\
b_{2,j} = \bar{b}_{2,j} &= \bar{c}_{2,j} = \textstyle\sum_{i=1}^{k} c_{i,j}.
\end{aligned}$$

It follows that

$$b_{1,j} = -1(c_{2,j} + 2 \cdot \sum_{i=3}^{k} c_{i,j}).$$

By Claim 1, all entries of B are non-negative, so it must be that $b_{1,j} = 0$ for $1 \leq j \leq k$. But this implies that the first row of B contains no strictly positive entry, contradicting Claim 1. This contradiction establishes Claim 2. □

For every word $v \in \Sigma^*$, $\Phi(v)$ has determinant 1 and hence is invertible. It follows that for $i \leq j$,

$$\begin{aligned}
\Phi(v_i \cdots v_j) &= \Phi(v_1 \cdots v_{i-1})^{-1} \times \Phi(v_1 \cdots v_{i-1}) \times \Phi(v_i \cdots v_j) \\
&= \Phi(v_1 \cdots v_{i-1})^{-1} \times \Phi(v_1 \cdots v_j)
\end{aligned}$$

which yields the following more efficient parallel string matching algorithm.

Algorithm 6.2.4 (Parallel string matching, [KR87]).
INPUT. Words $u = u_1 \cdots u_m$ and $v = v_1 \cdots v_n$ in Σ^*, with $m \leq n$.
OUTPUT. TRUE if u is a subword of v, otherwise FALSE.

```
for i = 1 to m in parallel
        compute Φ(u_i)
for i = 1 to n in parallel
        compute Φ(v_i)
compute Φ(u) by prefix sums
for i = 1 to n in parallel
        compute Φ(v_1 ··· v_i) by prefix sums
 FLAG =  FALSE
for i = 0 to n − m in parallel {
```

```
        if Φ(u_1 ··· u_m) = Φ(v_1 ··· v_i)^{-1} × Φ(v_1 ··· v_{i+m})
            FLAG = TRUE
    }
return FLAG
```

In the last `for`-loop, one can use a broadcast to determine whether to set FLAG to TRUE, to avoid concurrent writes. Since the broadcast and prefix sums can be executed in time $O(log(n))$, the above is an EREW$(O(log(n)), O(n))$ and NC^1 algorithm for testing string matching.

In practice, many efficient string matching algorithms are known. In particular the Knuth-Morris-Pratt algorithm (see [Sed83]) is a sequential algorithm, which because it does not require any re-reading of any part of the word v, is particularly useful when v is an external file. The parallel algorithm given here does not have this useful property.

CLAIM 3. For $w \in \Sigma^m$, every entry of $\Phi(w)$ is less than $k^m = 2^{O(m)}$.

The claim is immediate by induction, noting that since multiplication of any matrix M by $\Phi(a_i)$, $1 \leq i \leq k$, at most multiplies the entries by the factor k.

For practical applications, $k = 128$ (number of ASCII characters), and so when searching for a word u of length 10 in a string v of length 10^3, we must store matrices of size $2^7 \times 2^7$ having entries bounded by $2^{70} \approx 10^{21}$. Algorithm 6.2.4 is thus grossly impractical. By working with modular arithmetic, modulo a small prime p, where p is chosen randomly from the first

$$\left\lceil \frac{nm \log_2 k}{\epsilon} \right\rceil$$

primes, we obtain a randomized algorithm which is guaranteed correct with probability $1 - \epsilon$. To that end, for $u \in \Sigma^*$, if $\Phi(u) = (b_{i,j})$ and p is a prime number, then let $\Phi_p(u) = (b_{i,j} \bmod p)$; i.e., all entries of $\Phi_p(u)$ are the modular remainders of the corresponding entries of $\Phi(u)$.

Lemma 6.2.1 ([KR87]). *Let Σ be a finite alphabet. Let $n, m \in \mathbf{N}$ satisfy $1 \leq m \leq n$. Then there exists N, such that for all words $u_1 \ldots u_m \in \Sigma^m$, $v_1 \cdots v_n \in \Sigma^n$ and for all prime numbers p, if $p \not| N$ then for all $0 \leq i \leq n-m$,*

$$\Phi_p(u_1 \cdots u_m) = \Phi_p(v_{i+1} \cdots v_{i+m}) \to u_1 \cdots u_m = v_{i+1} \cdots v_{i+m}.$$

Proof. For each $0 \leq i \leq n - m$, let M_i be the $k \times k$ matrix given by $\Phi(v_{i+1} \cdots v_{i+m}) - \Phi(u_1 \cdots u_m)$, and N_i be the smallest absolute value of a non-zero entry of M_i, if M_i is not the zero matrix, otherwise $N_i = 0$. Let $I = \{i : 0 \leq i \leq n - m, N_i \neq 0\}$ and define $N = \prod_{i \in I} N_i$. By Claim 3, each entry of M_i is less than $k^m = 2^{O(m)}$ and so $N < (k^m)^n = k^{mn}$. By construction, if p is a prime, $p \not| N$ and $u_1 \cdots u_m \neq v_{i+1} \cdots v_{i+m}$, then

$$\Phi_p(v_{i+1} \cdots v_{i+m}) \neq \phi_p(u_1 \cdots u_m).$$

This concludes the proof of the lemma.

The next result is reminiscent of Theorem 1.7.8 and of Theorem 6.3.22, where in both instances, a result was checked modulo sufficiently many primes. In contrast with those results, which rely on the prime number theorem, the following result uses a simple estimate, to show that the probability of randomly choosing a prime p, such that Φ_p is not injective is small. In the remainder of this section, let p_r denote the r-th prime, and $P_r = \{p_1, \ldots, p_r\}$.

Theorem 6.2.4 ([KR87]). *Let Σ be a finite alphabet of size k. Let $m, n \in \mathbf{N}$ satisfy $1 \leq m \leq n$, and let $u_1 \cdots u_m \in \Sigma^m$, $v_1 \cdots v_n \in \Sigma^n$. Given $\epsilon > 0$, let*

$$r = \lceil nm \log_2 k \cdot \epsilon^{-1} \rceil.$$

Then the probability that for $p \in P_r$ there exists $0 \leq i \leq n - m$ such that

$$\Phi_p(u_1 \cdots u_m) = \Phi_p(v_{i+1} \cdots v_{i+m}) \wedge u_1 \cdots u_m) \neq v_{i+1} \cdots v_{i+m}$$

is bounded by ϵ.

Proof. Given $\Sigma, u_1 \cdots u_m, v_1 \cdots v_n, r, \epsilon$, let N be given by the previous lemma. It follows that

$$\{p \in P_r : \text{(6.2.4) holds for some } 0 \leq i \leq n - m\} \subseteq \{p \in P_r : p|N\}.$$

CLAIM. $|\{p : p \text{ is prime}, p|N\}| \leq \log_2(N)$.[8]

Proof of Claim. If the claim does not hold, then

$$N \geq \prod\{p : p|N\} > 2^{\log_2 N} = N$$

which is a contradiction.

The probability that for $p \in P_r$ there exists $0 \leq i \leq n - m$ such that

$$\Phi_p(u_1 \cdots u_m) = \Phi_p(v_{i+1} \cdots v_{i+m}) \wedge u_1 \cdots u_m) \neq v_{i+1} \cdots v_{i+m}$$

is bounded above by

$$\frac{|\{p \in P_r : p|N\}|}{r} \leq \frac{\log_2 N}{r}.$$

Now $N < k^{mn}$ so $\log_2 N < mn \log_2 k$, and recalling that $r = mn \log_2 k \cdot \epsilon^{-1}$, it follows that $\frac{\log_2 N}{r} < \epsilon$. This concludes the proof of the theorem.

Using Theorem 6.2.4 we have the following randomized parallel algorithm for string matching. First, some notation. Let P_r denote the set of the first r prime numbers. For prime p, let $\times_p$ denote the operation of matrix multiplication over the field $GF(p)$. For $1 \leq i \leq k$, define $\Phi_p(a_i) = \Phi(a_i)$, and extend $\Phi_p : (\Sigma^*, \circ) \to (\mathcal{M}_k, \times_p)$ to be the homomorphism defined by $\Phi_p(u \circ v) = \Phi_p(u) \times_p \Phi_p(v)$.

[8] A stronger result follows from the prime number theorem. Let $\omega(N)$ be the number of distinct prime divisors of N. Then for N equal to the product of the first r primes, $\omega(N) \sim \ln / \ln \ln N$, while for arbitrary N, on average, $\omega(N) \sim \ln \ln N$. See Theorem 430 of [HW79].

Algorithm 6.2.5 (Randomized string matching, [KR87]).
INPUT. Finite alphabet $\Sigma = \{a_1, \ldots, a_k\}$, integers $1 \leq m \leq n$, words $u = u_1 \cdots u_m \in \Sigma^m$, $v = v_1 \cdots v_n \in \Sigma^n$, and error tolerance $\epsilon > 0$.
OUTPUT. Whether u is a subword of v.

Note that if the algorithm returns an affirmative answer, then u is a subword of v with probability at least $1 - \epsilon$, while if the algorithm returns a negative answer, then u is not a subword of v with probability 1.

```
r = ⌈mn · log₂ k · ε⁻¹⌉
choose p ∈ P_r at random
for i = 1 to m in parallel
        compute φ_p(u_i)
for i = 1 to n in parallel
        compute φ_p(v_i)
compute φ_p(u) using prefix sums
for i = 1 to n in parallel
        compute φ_p(v_1 ··· v_i)  using prefix sums
FLAG = FALSE
for i = 0 to n − m in parallel {
        compute φ_p(v_1 ··· v_i)⁻¹
        φ(v_{i+1} ··· v_{i+m}) = φ(v_1 ··· v_i)⁻¹  × φ(v_1 ··· v_{i+m})
        if φ_p(u_1 ··· u_m) = φ(v_{i+1} ··· v_{i+m})
         FLAG = TRUE
        }
return FLAG
```

For example, in order to ensure an error of at most 0.01 when $k = 2^7$ and u, v are ASCII words with respective lengths 10 and 10^3, we can take $r = 10 \cdot 10^3 \cdot 7/10^{-2} = 7 \cdot 10^6$. Algorithm 6.2.5 thus requires an initial computation of the first 7 million prime numbers (unless, we additionally use a randomized primality testing algorithm)! Surely a more efficient distributed programming strategy is to run a fast sequential string matching algorithm simultaneously on different processors which are assigned to different "chunks" of the file v. However, the parallel string matching algorithms presented here illustrate an often recurring point in the design of parallel algorithms – namely, the design of parallel NC algorithms often require quite different ideas and substantially deeper mathematics than that used in the design of sequential algorithms. This point is very clearly made in the work of [BLS87], where an NC algorithm for testing permutation group membership uses results such as the Jordan conjecture and O'Nan-Scott Theorem 3.7.2 from the classification of finite simple groups!

There is a parallel matching algorithm due to Kedem, Landau, and Palem [KLP89].

6.2.4 *LogP* Model

The first models of Connection Machine CM1, CM2, built in the early 1980's by Thinking Machines Inc., arguably had as underlying virtual machine the

PRAM, where weak, off-the-shelf processors with little memory,[9] were assumed to communicate (via a fast router) with any other processor. Even the priority write conflict resolution model of CRCW-PRAM was supported by the Connection Machine with its SEND instruction in PARIS (parallel instruction set) assembly language. Thus the PRAM provided a good theoretical model, getting to the core of parallel computation, without worrying about communication costs between processors, and initially many abstract PRAM algorithms were directly implemented on the Connection Machine.

Over the years, one of the trends in the high-end computer industry has been to develop massively parallel computers, built from powerful individual processors having substantial memory, interconnected via a network of limited bandwidth (as exemplified in the CM5 and other machines). Thus communication costs could no longer be neglected in parallel algorithm design for current parallel computers. For such reasons, the LogP model, developed in [CKP+96], attempts to define an abstract parallel model of computation, which more realistically accounts for communication and processing costs. The parameters L, o, g, P are defined as follows.

L: Latency, or Upper bound on delay, in transmitting a word. In practice, $L = Hr + \lceil \frac{M}{w} \rceil$, where H is the maximum distance of a route (number of hops) in the interconnection network between 2 processors, r is the delay through each intermediate processing route, M is fixed message size, and w is the channel width.

o: Overhead, or bound on the time a processor uses to transmit [receive] a message to [from] the network.

g: Gap, or initiation rate; i.e., the minimum time interval between consecutive message transmissions or receptions at a processor. Large gap machines are effective on algorithms, for which the ratio of computation to communication is large.

P: Number of processors.

With this model, the time to transmit a message is thus $2o + L$, and the available bandwidth per processor is $\lceil \frac{L}{g} \rceil$. In [CKP+96], it is argued that the LogP model encourages the algorithm designer to consider data layout as part of the design problem, since this influences the communication.

6.2.5 Circuit Families

We begin by recalling some definitions from Chapter 1. Let $G = (V, E)$ be a finite directed graph, with $E \subseteq V \times V$. The *in-degree* or *fan-in* [resp. *out-degree* or *fan-out*] of node x is the size of $\{i \in V : (i, x) \in E\}$ [resp. $\{i \in V : (x, i) \in E\}$]. A *circuit* C_n is a labeled, directed acyclic graph whose nodes of in-degree 0 are called *input* nodes and are labeled by one of

[9] Early models of the CM1 had $2^{16} \approx 64,000$ processors, each with 8 Kbytes of primary memory.

$0, 1, x_1, \ldots, x_n$, and whose nodes v of in-degree $k > 0$ are called *gates* and are labeled by a k-place function from a *basis* set of boolean functions. A circuit has a unique *output* node of out-degree 0.[10] A family $\mathcal{C} = \{C_n : n \in \mathbf{N}\}$ of circuits has *bounded fan-in* if there exists k, for which all gates of all C_n have in-degree at most k; otherwise $\mathcal{C}$ has *unbounded* or *arbitrary* fan-in.

Boolean circuits have basis $\wedge$, $\vee$, $\neg$, where $\wedge$, $\vee$ may have fan-in larger than 2 (as described below, the AC^k [resp. NC^k] model concerns unbounded fan-in [resp. fan-in 2] boolean circuits). A *threshold* gate $\text{TH}_{k,n}$ outputs 1 if at least k of its n inputs is 1. A *modular counting* gate $\text{MOD}_{k,n}$ outputs 1 if the sum of its n inputs is evenly divisible by k. A *parity* gate $\oplus$ outputs 1 if the number of input bits equal to 1 is even, where as for $\wedge$, $\vee$ the fan-in may be restricted to 2, or arbitrary, depending on context.

An input node v labeled by x_i computes the boolean function

$$f_v(x_1, \ldots, x_n) = x_i.$$

A node v having in-edges from $v_1, \ldots, v_m$, and labeled by the m-place function g from the basis set, computes the boolean function

$$f_v(x_1, \ldots, x_n) = g(f_{v_1}(x_1, \ldots, x_n), \ldots, f_{v_m}(x_1, \ldots, x_n)).$$

The circuit C_n *accepts* the word $x_1 \cdots x_n \in \{0,1\}^n$ if $f_v(x_1, \ldots, x_n) = 1$, where f_v is the function computed by the unique output node v of C_n. A family $(C_n : n \in \mathbf{N})$ of circuits *accepts* a language $L \subseteq \{0,1\}^*$ if for each n, $L^n = L \cap \{0,1\}^n$ consists of the words accepted by C_n.

The *depth* of a circuit is the length of the longest path from an input to an output node, while the *size* is the number of gates. A language $L \subseteq \{0,1\}^*$ belongs to $\text{SizeDepth}(S(n), D(n))$ over basis B if L consists of those words accepted by a family $(C_n : n \in \mathbf{N})$ of circuits over basis B, where $size(C_n) = O(S(n))$ and $depth(C_n) = O(D(n))$.

A boolean circuit which computes the function $f(x_1, x_2) = x_1 \oplus x_2$ is as in Figure 6.1.

Example 6.2.2. The function $\max(a_0, \ldots, a_{n-1})$ of n integers, each of size at most m, can be computed by a boolean circuit as follows. Assume the integers a_i are distinct (a small modification is required for non-distinct integers). Then the k-th bit of $\max(a_0, \ldots, a_{n-1})$ is 1 exactly when

$$(\exists i < n)(\forall j < n)(j \neq i \rightarrow a_j \leq a_i \wedge \text{BIT}(k, a_i) = 1).$$

This bounded quantifier formula is translated into a boolean circuit by

[10] The usual convention is that a circuit may have any number of output nodes, and hence compute a function $f : \{0,1\}^n \to \{0,1\}^m$. In this chapter, we adopt the convention that a circuit computes a boolean function $f : \{0,1\}^n \to \{0,1\}$. An m-output circuit C computing function $g : \{0,1\}^n \to \{0,1\}^m$ can then be simulated by a circuit computing the boolean function $f : \{0,1\}^{n+m} \to \{0,1\}$ where $f(x_1, \ldots, x_n, 0^{m-i}1^i) = 1$ iff the i-th bit of $g(x_1, \ldots, x_n)$ is 1.

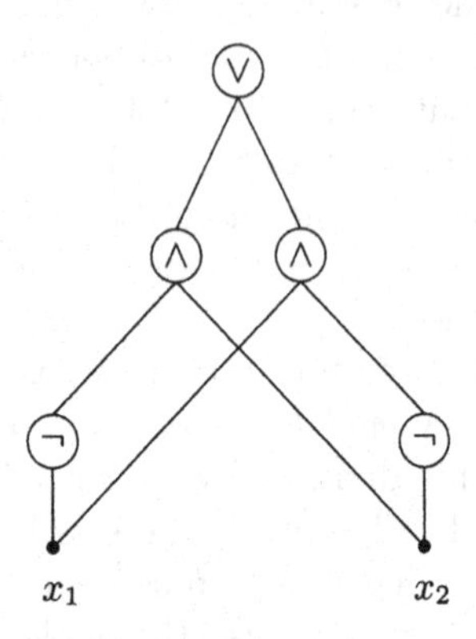

Fig. 6.1. Exclusive-or

$$\bigvee_{i<n} \bigwedge_{j<n, j \neq i} \bigvee_{\ell<n} \bigwedge_{\ell<p<m} (\text{BIT}(p, a_j) = \text{BIT}(p, a_i) \\ \wedge \text{BIT}(\ell, a_j) = 0 \wedge \text{BIT}(\ell, a_i) = 1.$$

Note that by Algorithm 6.2.1, $\max(a_0, \ldots, a_{n-1})$ is computed by a CRAM in constant time with a polynomial number of processors and by Example 6.2.2 $\max(a_0, \ldots, a_{n-1})$ is computed by a constant depth polynomial size family of boolean circuits. As this suggests, there is a relation between time/processors for a CRAM and depth/size for a boolean circuit family. The exact relation between the two models is given in Theorem 6.2.5.

Without a uniformity condition, circuit families of depth 2 and size 1 can accept non-recursive languages (e.g. all inputs are accepted [resp. rejected] if the n-th circuit is of the form $x_1 \vee \neg x_1$ [resp. $x_1 \wedge \neg x_1$]). Various notions of uniformity have been suggested (PTIME-uniformity [BCH86], LOGSPACE-uniformity [Bor73], U_{E^*}-uniformity [Ruz81], etc.), but the most robust (and strictest) appears to be that of LOGTIME-uniformity [Bus86a, BIS90], which is adopted in this chapter.

Definition 6.2.8 ([Ruz81], [BIS90]). *The* direct connection language *(abbreviated* DCL*) of a circuit family $(C_n : n \in \mathbf{N})$ is the set of $(a, b, \ell, 0^n)$, where a is the parent of b in the circuit C_n, and the label of gate a is ℓ. A circuit family is* LOGTIME*-uniform if its associated* DCL *belongs to* DLOGTIME. *For $k \geq 0$,* AC^k *[resp.* NC^k *] is the class of languages accepted by* LOGTIME*-uniform* SIZEDEPTH$(n^{O(1)}, O(\log^k n))$ *over the boolean basis, where* $\wedge,\vee$ *have arbitrary fan-in [resp. fan-in 2], and* $\text{NC} = \cup_k \text{AC}^k = \cup_k \text{NC}^k$. ACC$(k)$ *is the class of languages accepted by* LOGTIME*-uniform* SIZEDEPTH$(n^{O(1)}, O(1))$ *over the basis* $\wedge$, $\vee$, $\neg$, $\text{MOD}_{k,n}$, *where* $\wedge,\vee$ *have unbounded fan-in, and*

$\text{ACC} = \cup_{k\geq 2}\text{ACC}(k)$. TC^0 *is the class of languages in* LOGTIME*-uniform* $\text{SIZEDEPTH}(n^{O(1)}, O(1))$ *over the basis* $\text{TH}_{k,n}$.[11]

In [SV84], L. Stockmeyer and U. Vishkin related PRAM time and processors to boolean circuit depth and size. The LOGTIME-uniform version of that result was proved by N. Immerman [Imm89] and follows.

Theorem 6.2.5. *For* $k \geq 0$, AC^k *equals* $\text{TIMEPROC}(O(\log^k n), n^{O(1)})$ *on a* CRAM.

The following containments are known:

$$\text{NC}^k \subseteq \text{AC}^k \subseteq \text{NC}^{k+1}$$

and

$$\text{NC}^1 = \text{ALOGTIME} \subseteq \text{LOGSPACE} \subseteq \text{NLOGSPACE} \subseteq \text{AC}^1.$$

None of the inclusions are known to be strict or not. For more information on circuits, see the excellent survey article by R. Boppana and M. Sipser [BS90].

From this point on, we will assume that all language and function complexity classes are over the alphabet $\{0, 1\}$.

6.3 Some Recursion Schemes

Kleene's normal form theorem [Kle36a] states that for each recursive (partial) function f there is an index e for which $f(\mathbf{x}) = U(\mu y[T(e, \mathbf{x}, y) = 0])$, where T, U are primitive recursive. The proof relies on arithmetizing computations via Gödel numbers, a technique introduced in [Göd31] by Gödel, and with which Turing computable functions can be shown equivalent to μ-recursive functions. Since then, there have been a number of *arithmetizations* of machine models [Kle36a, Kle36b, Ben62, Cob65, Rit63, Tho72, Wag79, Clo90, CT94], etc. Key to all of these results is the availability in a function algebra $\mathcal{F}$ of a conditional function, a pairing function, and some string manipulating functions, in order to show that the function $\text{NEXT}_M(x, c) = d$ belongs to $\mathcal{F}$. Here, c, d encode configurations of machine M on input x and d is the configuration obtained in one step from configuration c.

Definition 6.3.1. *An* operator *(here also called* operation*) is a mapping from functions to functions. If* $\mathcal{X}$ *is a set of functions and* OP *is a collection of operators, then* $[\mathcal{X}; \text{OP}]$ *denotes the smallest set of functions containing*

---

[11] In this chapter, circuit classes such as AC^k, NC^k, NC and TC^0 sometimes denote both language classes, though more often function classes, where the intended meaning is clear from context. That is, we write NC in place of $\mathcal{F}\text{NC}$, etc. NC is an acronym for "Nick's Class", as this class was first studied by N. Pippenger. AC^k was studied by W.L. Ruzzo, using the alternating Turing machine model.

$\mathcal{X}$ and closed under the operations of OP. *The set* $[\mathcal{X}; \text{OP}]$ *is called a* function algebra. *In a straightforward inductive manner, define* representations *or* names *for functions in* $[\mathcal{X}; \text{OP}]$. *The* characteristic function $c_P(\mathbf{x})$ *of a predicate P satisfies*

$$c_p(\mathbf{x}) = \begin{cases} 1 & \text{if } P(\mathbf{x}) \\ 0 & \text{else,} \end{cases} \tag{6.2}$$

where P is often written in place of c_P. If $\mathcal{F}$ is a class of functions, then $\mathcal{F}_*$ is the class of predicates whose characteristic function belongs to $\mathcal{F}$.[12]

Definition 6.3.2. *Let $\mathcal{F} = [f_1, f_2, \ldots; O_1, O_2, \ldots]$ be a function algebra. Let O denote operator O_{i_0}, and fix a representation R of $f \in \mathcal{F}$. The rank $rk_{O,R}(f)$ of applications of O in the representation R of $f \in \mathcal{F}$ is defined by induction. If f is an initial function $f_1, f_2, \ldots$ then $rk_{O,R}(f) = 0$. Suppose that f is defined by application of operator O_i to functions $g_1, \ldots, g_m$ where $rk_{O,R}(g_j) = r_j$ for $1 \leq j \leq m$. If $i = i_0$ then $rk_{O,R}(f) = 1 + \max\{r_1, \ldots, r_m\}$; otherwise $rk_{O,R}(f) = \max\{r_1, \ldots, r_m\}$. The O-*rank *$rk_O(f)$ of a function $f \in \mathcal{F}$ is the minimum of $rk_{O,R}(f)$ over all representations R of f in $\mathcal{F}$.*

Operations which have been studied in the literature include composition, primitive recursion, minimization, and their variants including bounded composition, bounded recursion, bounded recursion on notation, bounded minimization, simultaneous recursion, multiple recursion, course-of-values recursion, divide and conquer recursion, safe and tiered recursion, etc.[13]

Since newer results concerning smaller complexity classes yield older results concerning larger classes as corollaries, we begin with a function algebra introduced by P. Clote for the class $\mathcal{F}$LH of functions in the logtime hierarchy.

6.3.1 An Algebra for the Logtime Hierarchy LH

Definition 6.3.3. *The* successor *function satisfies $s(x) = x + 1$; the* binary successor *functions s_0, s_1 satisfy $s_0(x) = 2 \cdot x$, $s_1(x) = 2 \cdot x + 1$; the n-place projection functions $I_k^n(x_1, \ldots, x_n) = x_k$; I denotes the collection of all projection functions.*

Definition 6.3.4. *The function f is defined by* composition *(*COMP*) from the functions $h, g_1, \ldots, g_m$ if*

$$f(x_1, \ldots, x_n) = h(g_1(x_1, \ldots, x_n), \ldots, g_m(x_1, \ldots, x_n)).$$

[12] In [Grz53], Grzegorczyk defined $\mathcal{F}_*$ as the collection of predicates P for which there is a function $f \in \mathcal{F}$ satisfying $P(\mathbf{x}) \iff f(\mathbf{x}) = 0$. For the function classes here considered, these are equivalent definitions.

[13] In this chapter, for uniformity of notation, a number of operations are introduced as *bounded* instead of *limited* operations. For example, Grzegorczyk's schemes of *limited recursion* and *limited minimization* are here called *bounded recursion* and *bounded minimization*.

The function f *is defined by* primitive recursion (PR) *from functions* g, h *if*

$$f(0,\mathbf{y}) = g(\mathbf{y}),$$
$$f(x+1,\mathbf{y}) = h(x,\mathbf{y},f(x,\mathbf{y})).$$

The collection $\mathcal{PR}$ *of primitive recursive functions is* $[0, I, s; \text{COMP}, \text{PR}]$. *The function* f *is defined by* iteration (ITER) *from functions* g *if*

$$f(0) = 0$$
$$f(x+1) = g(f(x)).$$

Theorem 6.3.1 ([Rob47]). *Define the operation* ADD *by*

$$\text{ADD}(f,g)(x) = f(x) + g(x), \text{ and let } q(x) = x - \lfloor\sqrt{x}\rfloor^2.$$

Let $\mathcal{PR}_1$ *denote the collection of one-place primitive recursive functions. Then* $\mathcal{PR}_1$ *equals* $[0, s, q; \text{COMP}, \text{ITER}, \text{ADD}]$.

In [Ass87] G. Asser presented a version of the previous theorem for primitive recursive word functions of one variable.

Primitive recursion defines $f(x+1)$ in terms of $f(x)$, so that the computation of $f(x)$ requires approximately $2^{|x|}$ many steps, an exponential number in the length of x. To define smaller complexity classes of functions, Bennett [Ben62] introduced the scheme of *recursion on notation*, which Cobham [Cob65] later used to characterize the polynomial time computable functions.

Definition 6.3.5. *Assume that* $h_0(x,\mathbf{y}), h_1(x,\mathbf{y}) \leq 1$. *The function* f *is defined by* concatenation recursion on notation (CRN) *from* g, h_0, h_1 *if*

$$f(0,\mathbf{y}) = g(\mathbf{y})$$
$$f(s_0(x),\mathbf{y}) = s_{h_0(x,\mathbf{y})}(f(x,\mathbf{y})), \quad \text{if } x \neq 0$$
$$f(s_1(x),\mathbf{y}) = s_{h_1(x,\mathbf{y})}(f(x,\mathbf{y})).$$

This scheme can be written in the abbreviated form

$$f(0,\mathbf{y}) = g(\mathbf{y})$$
$$f(s_i(x),\mathbf{y}) = s_{h_i(x,\mathbf{y})}(f(x,\mathbf{y})).$$

The scheme CRN was introduced by P. Clote in [Clo90], though motivated by a similar scheme due to J. Lind [Lin74]. If concatenation of the empty string is allowed, or if the condition $h_i(x,\mathbf{y}) \leq 1$ is dropped (i.e., if concatenation of $f(x,\mathbf{y})$ with an arbitrary string $h_i(x,\mathbf{y})$ is allowed as in Lind's scheme), then the resulting scheme is provably stronger (i.e., parity $\oplus_{i=1}^{n} x_i$ is easily defined using Lind's version, although parity is known not to belong to LH – see Exercise 6.6.19).

Definition 6.3.6. *The length of* x *in binary satisfies* $|x| = \lceil\log(x+1)\rceil$; $||x||$ *is defined as* $|(|x|)|$, *etc.;* $\text{MOD2}(x) = x - 2 \cdot \lfloor\frac{x}{2}\rfloor$; *the function* $\text{BIT}(i,x) =$

$\text{MOD2}(\lfloor \frac{x}{2^i} \rfloor)$ *yields the coefficient of* 2^i *in the binary representation of* x; *the* smash *function satisfies* $x\#y = 2^{|x|\cdot|y|}$. *The algebra* A_0 *is defined to be*

$$[0, I, s_0, s_1, \text{BIT}, |x|, \#; \text{COMP}, \text{CRN}].$$

Arbitrary constants belong to A_0. For instance the integer 6 has binary representation 110 and is represented by $s_0(s_1(s_1(0)))$. The auxiliary reverse function $rev0(x, y)$ gives the $|y|$ many least significant bits of x written in reverse. Let

$$\begin{aligned} rev0(x, 0) &= 0 \\ rev0(x, s_i(y)) &= s_{\text{BIT}(|y|,x)}(rev0(x, y)). \end{aligned} \tag{6.3}$$

The reverse of the binary notation for x is given by $rev(x) = rev0(x, x)$. For instance the integer 10 has binary notation 1010 whose reverse is 101, corresponding to the integer 5, so $rev(10) = 5$. The following computation may be helpful, where $\overline{w}$ temporarily denotes the integer having binary representation w.

$$\begin{aligned} rev(10) &= rev(\overline{1010}) \\ &= rev0(\overline{1010}, \overline{1010}) \\ &= s_{\text{BIT}(|\overline{101}|,\overline{1010})}(rev0(\overline{1010}, \overline{101})) \\ &= s_1(s_{\text{BIT}(|\overline{10}|,\overline{1010})}(rev0(\overline{1010}, \overline{10}))) \\ &= s_1 s_0(s_{\text{BIT}(|\overline{1}|,\overline{1010})}(rev0(\overline{1010}, \overline{1}))) \\ &= s_1 s_0 s_1(s_{\text{BIT}(\overline{0},\overline{1010})}(rev0(\overline{1010}, \overline{0}))) \\ &= s_1 s_0 s_1 s_0(0) \\ &= 5 \end{aligned}$$

Let

$$\begin{aligned} ones(0) &= 0 \\ ones(s_i(x)) &= s_1(ones(x)) \end{aligned} \tag{6.4}$$

so that $ones(x) = 2^{|x|} - 1$ whose binary representation consists of $|x|$ many 1's. Let

$$\begin{aligned} pad(x, 0) &= x \\ pad(x, s_i(y)) &= s_0(pad(x, y)) \end{aligned} \tag{6.5}$$

so that $pad(x, y) = 2^{|y|} \cdot x$ whose binary representation is that of x with $|y|$ many 0's appended to the right. Kleene's signum functions sg, $\overline{sg}$, which satisfy $sg(x) = \min(x, 1)$ and $\overline{sg}(x) = 1 - sg(x)$, are defined by

$$\begin{aligned} sg(x) &= \text{BIT}(0, ones(x)) \\ \overline{sg}(x) &= \text{BIT}(0, pad(1, x)). \end{aligned} \tag{6.6}$$

The conditional function, easily defined using stronger recursion schemes,

$$cond(x, y, z) = \begin{cases} y & \text{if } x = 0 \\ z & \text{else} \end{cases} \tag{6.7}$$

is here defined using the auxiliary functions *cond*0, *cond*1, *cond*2. Define

$$\begin{aligned} &cond0(0, y) = 0 \\ &cond0(s_i(x), y) = s_{\mathrm{BIT}(0,y)}(cond0(x, y)) \end{aligned} \tag{6.8}$$

$$cond1(x, y) = sg(cond0(x, y)) \tag{6.9}$$

$$\begin{aligned} &cond2(x, 0) = 0 \\ &cond2(x, s_i(y)) = s_{cond1(x, s_i(y))}(cond2(x, y)) \end{aligned} \tag{6.10}$$

so that

$$cond0(x, y) = \begin{cases} 0 & \text{if } \mathrm{BIT}(0, y) = 0 \\ 2^{|x|} - 1 & \text{else} \end{cases} \tag{6.11}$$

$$cond1(x, y) = \begin{cases} 0 & \text{if } x = 0 \\ \mathrm{BIT}(0, y) & \text{else} \end{cases} \tag{6.12}$$

$$cond2(x, y) = \begin{cases} 0 & \text{if } x = 0 \\ y & \text{else.} \end{cases} \tag{6.13}$$

The concatenation function $x * y = 2^{|y|} \cdot x + y$ is defined by

$$\begin{aligned} &x * 0 = x \\ &x * s_i(y) = s_i(x * y). \end{aligned} \tag{6.14}$$

Then the conditional function *cond* is defined by

$$cond(x, y, z) = cond2(\overline{sg}(x), y) * cond2(x, z).$$

With *cond* one can form (characteristic functions of) predicates by applying boolean operations AND, OR, NOT to other predicates. Additionally, using *cond*, one can introduce functions using *definition by cases*

$$f(\mathbf{x}) = \begin{cases} g_1(\mathbf{x}) & \text{if } P_1(\mathbf{x}) \\ g_2(\mathbf{x}) & \text{if } P_2(\mathbf{x}) \\ \vdots & \\ g_n(\mathbf{x}) & \text{if } P_n(\mathbf{x}) \end{cases} \tag{6.15}$$

where predicates $P_1, \ldots, P_n$ are disjoint and exhaustive. A *sharply bounded quantifier* is of the form $(\exists x \leq |y|)$ or $(\forall x \leq |y|)$.

Lemma 6.3.1. *$(A_0)_*$ is closed under sharply bounded quantifiers.*

Proof. Suppose that the predicate $R(x, \mathbf{z})$ belongs to A_0 and that $P(y, \mathbf{z})$ is defined by $(\exists x \leq |y|)R(x, \mathbf{z})$. Define

$$f(0, \mathbf{z}) = 0$$
$$f(s_i(x), \mathbf{z}) = s_{R(|x|, \mathbf{z})}(f(x, \mathbf{z})).$$

Then $P(y, \mathbf{z}) = sg(f(s_1(y), \mathbf{z}))$ belongs to A_0. Bounded universal quantification can be derived from bounded existential quantification using $\overline{sg}$.

Definition 6.3.7. *The function f is defined by* sharply bounded minimization *(*SBMIN*) from the function g, denoted by $f(x, \mathbf{y}) = \mu i \leq |x|[g(i, \mathbf{y}) = 0]$, if*

$$f(x, \mathbf{y}) = \begin{cases} \min\{i \leq |x| : g(i, \mathbf{y}) = 0\} & \textit{if such exists} \\ 0 & \textit{else.} \end{cases}$$

Sharply bounded maximization *(*SBMAX*) is analogously defined.*

From Lemma 6.3.1, it follows that A_0 is closed under SBMIN and SBMAX . Namely, define

$$k(0, \mathbf{y}) = 0$$
$$k(s_i(z), \mathbf{y}) = s_{h(z, \mathbf{y})}(k(z, \mathbf{y}))$$

where

$$h(z, \mathbf{y}) = \begin{cases} 0 \text{ if } (\exists x \leq |z|)[g(x, \mathbf{y}) = 0] \\ 1 \text{ else.} \end{cases}$$

Then $f(x, \mathbf{y}) = \mu i \leq |x|[g(i, \mathbf{y}) = 0]$ is defined by

$$f(x, \mathbf{y}) = \begin{cases} 0 & \text{if } g(0, \mathbf{y}) = 0 \text{ or } \neg(\exists i \leq |x|)[g(i, \mathbf{y}) = 0] \\ |rev(k(s_1(x), \mathbf{y}))| & \text{else.} \end{cases}$$

The integer x is a beginning of y, denoted xBy, if the binary representation of x is an initial segment (from left to right) of the binary representation of y; formally xBy iff $x = 0$ or $x, y > 0$ and

$$(\forall i \leq |x|)[\text{BIT}(i, rev(s_1(x))) = \text{BIT}(i, rev(s_1(y)))].$$

Thus the predicate $B \in A_0$. Similarly, predicates xPy (x is part of y, i.e., a subword of y) and xEy (x is an end of y) can be shown to belong to A_0.

To show the closure of A_0 under part-of quantifiers $(\exists xBy)$, $(\exists xPy)$, $(\exists xEy)$, etc. define the *most significant part* function MSP by

$$\begin{aligned} &\text{MSP}(0, y) = 0 \\ &\text{MSP}(s_i(x), y) = s_{\text{BIT}(y, s_i(x))}(\text{MSP}(x, y)) \end{aligned} \tag{6.16}$$

and the *least significant part* function LSP by

$$\text{LSP}(x, y) = \text{MSP}(rev(\text{MSP}(rev(s_1(x)), |\text{MSP}(x, y)|)), 1). \tag{6.17}$$

These functions satisfy $\text{MSP}(x,y) = \lfloor \frac{x}{2^y} \rfloor$ and $\text{LSP}(x,y) = x \bmod 2^y$, where $x \bmod 1$ is defined to be 0. For later reference, define the unary analogues *msp lsp*

$$msp(x,y) = \lfloor x/2^{|y|} \rfloor = \text{MSP}(x,|y|) \tag{6.18}$$
$$lsp(x,y) = x \bmod 2^{|y|} = \text{LSP}(x,|y|), \tag{6.19}$$

and note that *lsp* is definable from *msp*, *rev* as follows

$$lsp(x,y) = msp(rev(msp(rev(s_1(x)), msp(x,y))), 1). \tag{6.20}$$

Using MSP, LSP together with ideas of the proof of the previous lemma, the following is easily shown.

Lemma 6.3.2. $(A_0)_*$ *is closed under part-of quantifiers.*

Using part-of quantification, the inequality predicate $x \le y$ can be defined by

$$|x| < |y|$$
$$\text{OR}$$
$$|x| = |y| \text{AND}$$
$$(\exists u B x)[uBy \wedge \text{BIT}(|x| \dot{-} |u| \dot{-} 1, y) = 1 \wedge \text{BIT}(|x| \dot{-} |u| \dot{-} 1, x) = 0]$$

where $|x| < |y|$ has characteristic function $sg(\text{MSP}(y,|x|))$. Note that $|x| \dot{-} |u| \dot{-} 1$ can be expressed by $|msp(msp(x,u),1)| = ||\lfloor \frac{msp(x,u)}{2} \rfloor||$.

Addition $x + y$ can be defined in A_0 by applying CRN to $sum(x,y,z)$, whose value is the $|z|$-th bit of $x + y$. In adding x and y, the $|z|$-th bit of the sum depends whether a *carry* is *generated* or *propagated.* Define the predicates GEN, PROP by having $\text{GEN}(x,y,z)$ hold iff the $|z|$-th bit of both x and y is 1 and $\text{PROP}(x,y,z)$ hold iff the $|z|$-th bit of either x or y is 1. Define $carry(x,y,0) = 0$ and $carry(x,y,s_i(z))$ to be 1 iff

$$(\exists u B z)[\text{GEN}(x,y,u) \wedge (\forall v B z)[|v| > |u| \rightarrow \text{PROP}(x,y,v)]].$$

Then $sum(x,y,z) = x \oplus y \oplus carry(x,y,z)$ where the EXCLUSIVE-OR $x \oplus y$ is defined by $cond(x, cond(y,0,1), cond(y,1,0))$. Using the 2's complement trick, *modified subtraction* $x \dot{-} y = \max(x - y, 0)$ can be shown to belong to A_0. In order to arithmetize machine computations, pairing and sequence encoding functions are needed. To that end, define the *pairing* function $\tau(x,y)$ by

$$\tau(x,y) = (2^{\max(|x|,|y|)} + x) * (2^{\max(|x|,|y|)} + y). \tag{6.21}$$

Noting that $2^{\max(|x|,|y|)} = cond(msp(x,y), pad(1,y), pad(1,x))$, this function is easily definable from *msp*, *cond*, *pad*, $*$, $+$ hence belongs to A_0. As an example, to compute $\tau(4,3)$, note that $\max(|4|,|3|) = 3$ and so one concatenates $1\underline{100}$ with $1\underline{011}$, where the underlined portions represent 4 resp. 3 in binary. Define the functions TR [resp. TL] which truncate the rightmost [resp. leftmost] bit:

$$\text{TR}(x) = \text{MSP}(x, 1) = \left\lfloor \frac{x}{2} \right\rfloor \tag{6.22}$$

$$\text{TL}(x) = \text{LSP}(x, |\text{TR}(x)|) = \text{TR}(rev(\text{TR}(rev(s_1(x))))) \tag{6.23}$$

where the latter definition is used later to show that TL belongs to a certain subclass of A_0. The left π_1 and right π_2 projections are defined by

$$\pi_1(z) = \text{TL}\left(\text{MSP}\left(z, \left\lfloor \frac{|z|}{2} \right\rfloor\right)\right) \tag{6.24}$$

$$\pi_2(z) = \text{TL}\left(\text{LSP}\left(z, \left\lfloor \frac{|z|}{2} \right\rfloor\right)\right) \tag{6.25}$$

and satisfy $\tau(\pi_1(z), \pi_2(z)) = z$, $\pi_1(\tau(x,y)) = x$ and $\pi_2(\tau(x,y)) = y$. An n-tuple $(x_1, \ldots, x_n)$ can be encoded by $\tau_n(x_1, \ldots, x_n)$, where $\tau_2 = \tau$ and

$$\tau_{k+1}(x_1, \ldots, x_{k+1}) = \tau(x_1, \tau_k(x_2, \ldots, x_{k+1})).$$

At this point, it should be mentioned that by using the functions so far defined, Turing machine configurations (TM and RATM) are easily expressed in A_0, and even in subalgebras of A_0. A *configuration* of RATM is of the form $(q, u_1, \ldots, u_{k+2}, n_1, \ldots, n_{k+2})$ where $q \in Q$, $u_i \in (\Gamma \cup \{B\})^*$ and $n_i \in \mathbf{N}$. The u_i represent the contents of the k work tapes and of the index query and the index answer tapes, and the n_i represent the head positions on the tapes (the input tape has no head). Since the input is accessed through random access, the input does not form part of the configuration of the RATM. Let ℓ_i [resp. r_i] represent the contents of the left portion [resp. the reverse of the right portion] of the i-th tape (i.e., tape cells of index $\leq n_i$ [resp. $> n_i$]). Assuming some simple binary encoding of $\Gamma \cup \{B\}$, a RATM configuration can be represented using the tupling function by

$$\tau_{2k+5}(q, \ell_1, r_1, \ldots, \ell_{k+2}, r_{k+2}).$$

Let $\text{INITIAL}_M(x)$ be the function mapping x to the initial configuration of RATM M on input x. For configurations α, β in the computation of RATM M on x, let predicate $\text{NEXT}_M(x, \alpha, \beta)$ hold iff $(x, \alpha) \vdash_M (x, \beta)$.

If M is a TM with input x, then a configuration can be similarly represented by $\tau_{2k+3}(q, \ell_0, r_0, \ldots, \ell_k, r_k)$ where $initial_M(x)$, $next_M(x, \alpha, \beta)$ are the counterparts for Turing machine computations without random access.

Lemma 6.3.3. *INITIAL_M, NEXT_M belong to $[0, I, s_0, s_1, \text{BIT}, |x|; \text{COMP}, \text{CRN}]$. Moreover, τ, π_1, π_2, $initial_M$, $next_M$ belong to*

$$[0, I, s_0, s_1, \text{MOD2}, msp; \text{COMP}, \text{CRN}].$$

Proof. Using s_0, s_1, *pad*, $*$, $\lfloor x/2 \rfloor$, *cond*, BIT, MSP, LSP, the pairing and tupling functions, etc. it is routine to show that INITIAL_M, NEXT_M are definable in A_0 without use of the smash function. For instance, a move of the first tape head to the right would mean that in the next configuration $\ell_1' = 2 \cdot \ell_1 + \text{MOD2}(r_1)$ and $r_1' = \lfloor r_1/2 \rfloor$.

Temporarily, let $\mathcal{F}$ designate the algebra

$$[0, I, s_0, s_1, \text{MOD2}, msp; \text{COMP}, \text{CRN}].$$

Using MOD2 and msp appropriately, functions from (6.3) through (6.15) can be introduced in $\mathcal{F}$. For instance, in (6.3)

$$rev0(x, s_i(y)) = s_{\text{MOD2}(msp(x,y))}(rev0(x, y)).$$

Part-of quantifiers, the pairing function (6.21), its left, right projections (6.24) can be defined in $\mathcal{F}$, by using msp, lsp appropriately in place of MSP, LSP. For instance, to define the projections of the pairing function, define auxiliary functions g, h as follows:

$$\begin{aligned} g(0, x) &= 0 \\ g(s_i(z), x) &= s_{\text{BIT}(z*z, ones(x))}(g(z, x)) \\ h(x) &= rev(g(x, x)). \end{aligned}$$

Then $|h(x)| = \lfloor \frac{|x|}{2} \rfloor$ and for x of even length (i.e., $ones(h(x)) * ones(h(x)) = ones(x)$), the left and right projections of the pairing function are defined by

$$\begin{aligned} \pi_1(x) &= msp(x, h(x)) \\ \pi_2(x) &= lsp(x, h(x)). \end{aligned}$$

From this, the function $initial_M$ and predicate $next_M$ are now routine to define.

We can now describe how short sequences of small numbers are encoded in A_0. To illustrate the idea, what follows is a first approximation to the sequence encoding technique. Generalizing the pairing function, to encode the sequence (3,9,0,4) first compute $\max\{|3|, |9|, |0|, |4|\}$. Temporarily let t denote the integer having binary representation

$$1\underline{0011}1\underline{1001}1\underline{0000}1\underline{0100}$$

where the underlined portions correspond to the binary representations of 3,9,0,4. Now the length ℓ of sequence $(3, 9, 0, 4)$ is 4, the *block size* BS is 5, and $|t| = \ell \cdot \text{BS}$. Define, as a first approximation, the *sequence number* $\langle 3, 9, 0, 4 \rangle$ by $\tau(t, \ell)$.

Given the sequence number $z = \langle 3, 9, 0, 4 \rangle$, the Gödel β function decoding the sequence is given by

$$\beta(0, z) = \pi_2(z) = \ell = 4.$$

The blocksize $\text{BS} = \lfloor |\pi_1(z)| / \pi_2(z) \rfloor = \lfloor 20/4 \rfloor = 5$, and for $i = 1, \ldots, 4$

$$\beta(i, z) = \text{LSP}(\text{MSP}(\pi_1(z), (\ell - i) \cdot \text{BS}), \text{BS} - 1).$$

Thus $\beta(1, z) = \text{LSP}(\text{MSP}(\pi_1(z), 3 \cdot 5), 4) = 3$, etc. All the above operations belong to A_0, with the exception of multiplication and division (which provably do not belong to A_0). However, multiplication and division by powers of 2 is possible in A_0, so the previously described sequence encoding technique is slightly modified. The sequence $(a_1, \ldots, a_n)$ is encoded by $z = \langle a_1, \ldots, a_n \rangle$ where

$$\begin{aligned} z &= \tau(t, n) \\ \text{BS} &= \max\{2^{||a_i||} : 1 \leq i \leq n\} \\ t &= h(N) \end{aligned}$$

where

$$\begin{aligned} |N| &= n \cdot \text{BS} \\ h(0) &= 0 \\ h(s_i(x)) &= s_{g(x)}(h(x)) \end{aligned}$$

and

$$g(x) = \begin{cases} 1 & \text{if } |x| \bmod \text{BS} = 0 \\ \text{BIT}((\text{BS} \dot{-} 1) \dot{-} (|x| \bmod \text{BS}), a_{\lfloor |x|/\text{BS} \rfloor + 1}) & \text{else.} \end{cases}$$

Finally define

$$\ell h(z) = \beta(0, z) = \begin{cases} \pi_2(z) & \text{if } z \text{ encodes a pair} \\ 0 & \text{else} \end{cases} \tag{6.26}$$

and for $1 \leq i \leq \beta(0, z)$

$$\beta(i, z) = \text{LSP}\left(\text{MSP}\left(\pi_1(z), (\ell h(z) \dot{-} i) \cdot \left\lfloor \frac{|\pi_1(z)|}{\ell h(z)} \right\rfloor\right), \left\lfloor \frac{|\pi_1(z)|}{\ell h(z)} \right\rfloor \dot{-} 1\right). \tag{6.27}$$

Suppose that $z = \tau(t, n)$ codes a sequence of length n, where $|t| = \text{BS} \cdot n$ and the block size $\text{BS} = 2^m$ for some m. The exponent m can be computed, since $m = \mu x \leq ||a|| \; [\text{MSP}(|t|, x) = n]$, and A_0 is closed under sharply bounded minimization. Using this observation, it is clear that the β function belongs to A_0. Using the techniques introduced, the following can be proved.

Theorem 6.3.2 ([Clo93]). *If $f \in A_0$ then there exists $g \in A_0$ such that for all x,*

$$g(x, \mathbf{y}) = \langle f(0, \mathbf{y}), \ldots, f(|x| - 1, \mathbf{y}) \rangle.$$

The following two lemmas, together with the sequence encoding machinery of A_0, will allow us soon to establish that $A_0 = \mathcal{F}\text{LH}$.

Lemma 6.3.4. *For every $k, m > 1$,*

$$\text{DTIMESPACE}(\log^k(n), \log^{1-1/m}(n)) \subseteq A_0.$$

Proof. Let M be a RATM running in time $\log^k(n)$ and space $\log^{1-1/m}(n)$. For each $i \leq m \cdot k$, define a predicate $\text{NEXT}_{M,i}$ belonging to A_0 such that

$$\text{NEXT}_{M,i}(x,c,d) \iff d \text{ follows } c \text{ in at most } \log^{i/m}(n) \text{ steps} \tag{6.28}$$

where c, d are encodings of configurations in the computation of M on input x, and $n = |x|$. By Lemma 6.3.3, the predicate $\text{NEXT}_{M,0}$ belongs to A_0 and satisfies (6.28). Assume that $\text{NEXT}_{M,i} \in A_0$ has been defined and satisfies (6.28). Define the formula $\text{NEXT}_{M,i+1}(x,c,d)$ by

$$\begin{aligned}(\exists s \leq |x|^3)(\forall j < ||x||^{1/m} - 1)\ [s = \langle s_0, \ldots, s_{||x||^{1/m}-1}\rangle \wedge \\ c = s_0 \wedge d = s_{||x||^{1/m}-1} \wedge \\ \text{NEXT}_{M,i}(s_j, s_{j+1})].\end{aligned} \tag{6.29}$$

Since for all $j < ||x||^{1/m}$, $|s_j| \leq ||x||^{1-1/m}$, $|s| \leq (||x||^{1-1/m} + 1) \cdot ||x||^{1/m} \leq 2 \cdot ||x||$. This establishes the validity of the bound $s \leq |x|^3$ in the definition of $\text{NEXT}_{M,i+1}$. It follows that M accepts input x iff $\text{NEXT}_{M,m \cdot k}(x,c,d)$ holds, where c and d respectively are the initial configuration and the terminal accepting configuration in the computation of M on x.

The following result for LH is similar.

Lemma 6.3.5. DSPACE$(\log\log(n))$ *on a* RATM *is contained in* LH.

Proof. Using the logtime computable pairing function from Fact 6.2.1, one can define a logtime predicate $\text{NEXT}_{M,0}$ which identifies consecutive configurations in the computation of the RATM M running in polylogarithmic time $(\log^{O(1)}(n))$ and simultaneous sublogarithmic space $(\log^{1-\epsilon}(n))$. As in the preceding lemma, by using ATM existential and universal branching, the predicate $\text{NEXT}_{M,i}$ can be shown to belong to LH. Thus

$$\text{DTIMESPACE}(\log^k(n), \log^{1-1/m}(n)) \subseteq \text{LH}.$$

Since there are only $2^{c \cdot \log\log n} = \log^c(n)$ many possible configurations for some constant c, it follows that DSPACE$(O(\log\log(n)))$ on RATM is contained in

$$\text{DTIMESPACE}(\log^k(n), \log^{1-1/m}(n))$$

on RATM, hence in LH.

Theorem 6.3.3 (Clote). $A_0 = \mathcal{F}\text{LH}$.

Proof. Consider the direction $A_0 \subseteq \mathcal{F}\text{LH}$. It follows from Fact 6.2.1 that 0, s_0, s_1, $|x|$ are computable in logtime. To compute $\text{BIT}(i,x)$, the machine M_1 on input $BiBxB$ writes the bits of i onto a work tape, computes $|i|$ and writes $i + |i| + 1$ onto its input query tape and reads the input answer tape. To compute the i-th bit of $I^n_k(x_1, \ldots, x_n)$, the machine M_2 on input $BiBx_1Bx_2B \cdots Bx_nB$ uses existential and universal states find the locations of the input separators B, computes $m = i + k + \sum_{j<i} |x_j|$ and returns the m-th bit of the input. To compute the the i-th bit of $x\#y = 2^{|x|\cdot|y|}$, the machine M_3 outputs 1 if $i = |x| \cdot |y|$, else 0. Since the product $|x| \cdot |y|$ can be computed in DSPACE($\log\log n$) on a RATM, hence in LH by Lemma 6.3.5, it follows that $\# \in \mathcal{F}$LH.

To see that $\mathcal{F}$LH is closed under composition, suppose that $f(x)$ equals $g(h_1(x), h_2(x))$, where the bitgraphs of g, h_1, h_2 are computed by the ATM M_g, M_{h_1}, M_{h_2} running in logtime with constantly many alternations. The bitgraph of f is then computed by the ATM M_f obtained from M_g as follows. Recall that M_g expects input of the form By_1By_2B, where $y_1, y_2 \in \{0,1\}^*$. Whenever M_g requests the i-th bit of its input, M_f computes $|h_1(x)|$, $|h_2(x)|$, and then executes the following code.

```
if       i = 0 then
             return B
else if  i ≤ ||h_1(x)|| then
             return M_{h_1}(i − 1, x)
else if  i = ||h_1(x)|| + 1 then
             return B
else if  i ≤ ||h_1(x)|| + 1 + ||h_2(x)|| then
             return M_{h_2}(i − ||h_1(x)|| − 2, x)
else
             return B
```

Inequalities like $i \leq |h_1(x)|$ can be decided by checking whether $|i| < ||h_1(x)||$ or $|i| = ||h_1(x)||$ and i precedes $|h_1(x)|$ in lexicographic order (i.e., $iB|h_1(x)|$). Values $M_{h_1}(i-1, x)$ can be computed by simulating M_{h_1}, providing bits of input $i - 1$ when required, etc. It is similarly easy to see that $\mathcal{F}$LH is closed under CRN. It follows that $A_0 \subseteq \mathcal{F}$LH.

Consider the direction that $\mathcal{F}\text{LH} \subseteq A_0$. A first attempt to arithmetize the computation of the logtime bounded RATM M on input x might be to use Lemma 6.3.3 together with sequence numbers. However, this encoding of M's computation cannot be done in A_0 because there are $O(\log n)$ many configurations in M's computation, with each configuration of size $O(\log n)$, thus requiring sequence numbers of size $O(\log^2 n)$, and quantification over such values is not sharply bounded. However, the integer s, which encodes the sequence of *instructions* executed (rather than configurations), *is* bounded by a polynomial in n and so can be expressed within the scope of a sharply bounded quantifier. What then remains to be shown is the existence of functions in A_0 which recognize whether a sequence of instructions corresponds to a correct computation.

Suppose that $M = (Q, \Sigma, \Gamma, \Delta, q_0, k+2, \ell)$, is a Σ_m-RATM, running in time $c \cdot |n|$, where $n = |x|$. For notational simplicity, assume $c = 1$ and that $\Sigma = \{0, 1\}$. An instruction of M is of the form

$$(q, a_1, \ldots, a_{k+2}, q', b_1, \ldots, b_{k+2}, d_1, \ldots, d_{k+2})$$

belonging to the transition relation Δ, where q is the current state, and $a_1, \ldots, a_{k+2} \in (\Gamma \cup \{B\})$ are the symbols currently read on the k work tapes and input query and answer tapes, q' is the next state, $b_1, \ldots, b_{k+2} \in (\Gamma \cup \{B\})$ are the symbols printed on the work tapes and input query and answer tapes, and $d_i \in \{-1, 0, 1\}$ is the direction of head movement on tape $1 \leq i \leq k+2$.

Using the earlier sequence encoding, one can code the sequence of $\log n$ instructions by an integer bounded by $|x|^{O(1)}$. Thus the Σ_m machine M accepts input x iff

$$(\exists y_1 \leq |x|^d)(\forall y_2 \leq |x|^d)(\exists y_3 \leq |x|^d) \cdots (Q y_m \leq |x|^d) \Theta(x, y_1, \ldots, y_m),$$

where Q is $\forall$ (resp. $\exists$) if m is even (resp. odd) and $\Theta(x, \mathbf{y})$ says that

if

(i) for $i = 1, 2, \ldots, m$, each y_i encodes a sequence of instructions from M's program, where the states occurring in y_i are existential (resp. universal) if i is odd (resp. even),

and

(ii) the sequence of instructions coded by $y_1, \ldots, y_m$ determines a correct computation of M,

then

(iii) this computation is accepting.

Using BIT, MSP, LSP, etc. it is not difficult to express *(i)* and *(iii)* in A_0. It remains to see how to formulate *(ii)* in A_0. If $y_1, \ldots, y_m$ encodes a sequence s of $m \cdot \log n$ instructions from M's program, then s corresponds to a correct computation of M, provided that

- The state in the first instruction is q_0, the state in the last instruction is q_A, and for all $0 \leq r < m \cdot \log n - 1$, the new state in the r-th instruction is the old state in the $r+1$-st instruction.
- For all tape cells, and all $r < m \cdot \log n$, if the r-th instruction is

  $$(q, a_1, \ldots, a_{k+2}, q', b_1, \ldots, b_{k+2}, d_1 \ldots, d_{k+2})$$

  then for $1 \leq j \leq k+2$, a_j is the symbol written on the j-th work tape at the last visit of the position p_j, where p_j is the current head position of the j-th work tape, provided that this position has previously been visited, and $a_j = B$ otherwise. Moreover, if q is the input query state q_I, then $b_{k+2} = \text{BIT}(i, x)$, where i is the content of the input query tape.

Note that the function SBBITSUM (sharply bounded bitsum)

$$\text{SBBITSUM}(x, y) = \begin{cases} \sum_{i<|y|} \text{BIT}(i, y) & \text{if } y \leq |x| \\ |x| + 1 & \text{else} \end{cases}$$

is computable in $\log^2 n$ time and log log space, hence by Lemma 6.3.4, belongs to the algebra A_0. Using SBBITSUM, one can determine whether, given i_0, i_1, j, at instruction i_0 the head of tape j is in the same position as at instruction i_1 in the execution of M on input x. It follows that $\mathcal{F}\text{LH} \subseteq A_0$.

Theorem 6.3.4 ([BIS90]). *Logtime uniform* AC^0 *equals the logtime hierarchy* LH.

Theorem 2.5.2 gives an AC^0 reduction of integer multiplication to parity, and Theorem 1.4.1 (7) gives an AC^0 reduction of parity to majority. By transitivity, it follows that integer multiplication is AC^0 reducible to MAJ, where we recall that $\text{MAJ}(x)$ is 1 if $\sum_{i<|x|} \text{BIT}(i,x) \geq \lceil |x|/2 \rceil$, else 0. In [BIS90], Barrington et al. observed that these reductions hold for logtime uniform AC^0, i.e., the logtime hierarchy. The following characterization of polysize, constant depth threshold circuits TC^0 is proved by formalizing these reductions, using the previous techniques.

Theorem 6.3.5 ([CT94]).

$$\text{TC}^0 = [0, I, s_0, s_1, |x|, \text{BIT}, \times, \#; \text{COMP}, \text{CRN}].$$

6.3.2 Bounded Recursion on Notation

Cobham's original characterization of $\mathcal{F}$PTIME was in terms of functions on words in a finite alphabet, rather than integers. To our knowledge, the first published proof of Cobham's result, additionally formulated for functions on the integers, appeared in [Ros84]. The following proof is new and simple, since we have arithmetized Turing machine computations in the algebra A_0.

Definition 6.3.8. *The function f is defined by* bounded recursion on notation *(abbreviated* BRN*) from g, h_0, h_1, k if*

$$\begin{aligned} f(0,\mathbf{y}) &= g(\mathbf{y}), \\ f(s_0(x),\mathbf{y}) &= h_0(x,\mathbf{y},f(x,\mathbf{y})), \text{ if } x \neq 0 \\ f(s_1(x),\mathbf{y}) &= h_1(x,\mathbf{y},f(x,\mathbf{y})) \end{aligned}$$

provided that $f(x,\mathbf{y}) \leq k(x,\mathbf{y})$ for all $x, \mathbf{y}$.

Theorem 6.3.6 ([Cob65], [Ros84]).

$$\mathcal{F}\text{PTIME} = [0, I, s_0, s_1, \#; \text{COMP}, \text{BRN}].$$

Proof. Temporarily denote the algebra $[0, I, s_0, s_1, \#; \text{COMP}, \text{BRN}]$ by $\mathcal{F}$. Consider first the inclusion from left to right. Let M be a TM with input x, running in polynomial time $p(|x|)$. Using BRN the functions MOD2,TR,msp can be defined in $\mathcal{F}$ as follows: $\text{MOD2}(0) = 0$, $\text{MOD2}(s_0(x)) = 0$, $\text{MOD2}(s_1(x)) = 1$; $\text{TR}(0) = 0$, $\text{TR}(s_i(x)) = x$; $msp(x,0) = x$, $msp(x,s_i(y)) = \text{TR}(msp(x,y))$, where $\text{MOD2}(x)$, $\text{TR}(x)$, $msp(x,y)$ are bounded by x. It follows that

$$[0, I, s_0, s_1, \text{MOD2}, msp; \text{COMP}, \text{CRN}] \subseteq \mathcal{F}$$

so by Lemma 6.3.3, $initial_M$, $next_M$ belong to $\mathcal{F}$. By suitably composing $0, s_0, s_1, \#$, there is a function $k \in \mathcal{F}$ satisfying $p(|x|) \leq |k(x)|$ for all inputs x. Using BRN, define

$$\begin{aligned} Run_M(x, 0) &= initial_M(x) \\ Run_M(x, s_i(y)) &= next_M(x, Run_M(x, y)). \end{aligned}$$

The value computed by M on input x can be obtained from $Run_M(x, k(x))$ by π_1, π_2.

The inclusion from right to left is proved by an easy induction on term formation in the Cobham algebra.

Using the same techniques, one can characterize the class

$$\mathcal{G}\text{TIMESPACE}\left(n^{O(1)}, O(n)\right)$$

of polynomial time linear space computable functions of linear growth as follows. The first assertion is due to D.B. Thompson [Tho72] (recall that $*$ is concatenation), and the other assertion follows by an alternate function in bounding the recursion on notation.

Theorem 6.3.7.

$$\begin{aligned} \mathcal{G}\text{TIMESPACE}(n^{O(1)}, O(n)) &= [0, I, s_0, s_1, *; \text{COMP}, \text{BRN}] \\ &= [0, I, s_0, s_1, \times; \text{COMP}, \text{BRN}]. \end{aligned}$$

Definition 6.3.9. *The function f is defined from functions g, h_0, h_1, k by* sharply bounded recursion on notation[14] *(abbreviated* SBRN*) if*

$$\begin{aligned} f(0, \mathbf{y}) &= g(\mathbf{y}) \\ f(s_0(x), \mathbf{y}) &= h_0(x, \mathbf{y}, f(x, \mathbf{y})), \ \text{if } x \neq 0 \\ f(s_1(x), \mathbf{y}) &= h_1(x, \mathbf{y}, f(x, \mathbf{y})), \end{aligned}$$

provided that $f(x, \mathbf{y}) \leq |k(x, \mathbf{y})|$ for all $x, \mathbf{y}$.

In [Lin74], J. Lind characterized $\mathcal{F}$LOGSPACE functions on words $w \in \Sigma^*$ as the smallest class of functions containing the initial functions $c_=$ (characteristic function of equality), $*$ (string concatenation) and closed under the operations of explicit transformation, log bounded recursion on notation, and a (provably stronger) version of concatenation on notation. An arithmetic version of Lind's characterization is the following.

[14] In [CT94], this scheme was denoted B_2RN.

Theorem 6.3.8.

$$\begin{aligned}\mathcal{F}\text{LOGSPACE} &= [0, I, s_0, s_1, |x|, \text{BIT}, \#; \text{COMP}, \text{CRN}, \text{SBRN}] \\ &= [0, I, s_0, s_1, \text{MOD2}, msp, \#; \text{COMP}, \text{CRN}, \text{SBRN}].\end{aligned}$$

The first statement appeared in [Clo88, CT94] and the second can be proved using similar techniques.

Recently, function algebras have been found for small parallel complexity classes. Consider the following variants of recursion on notation.

Definition 6.3.10. *The function f is defined by* k-bounded recursion on notation *(k-BRN) from g, h_0, h_1 if*

$$\begin{aligned} f(0, \mathbf{y}) &= g(\mathbf{y}) \\ f(s_0(x), \mathbf{y}) &= h_0(x, \mathbf{y}, f(x, \mathbf{y})), \text{ if } x \neq 0 \\ f(s_1(x), \mathbf{y}) &= h_1(x, \mathbf{y}, f(x, \mathbf{y})) \end{aligned}$$

provided that $f(x, \mathbf{y}) \leq k$ holds for all $x, \mathbf{y}$, where k is a constant.

Definition 6.3.11. *The function f is defined by* weak bounded recursion on notation *(WBRN) from g, h_0, h_1, k if $F(x, \mathbf{y})$ is defined from g, h_0, h_1, k by* BRN *and $f(x, \mathbf{y}) = F(|x|, \mathbf{y})$; i.e.,*

$$\begin{aligned} F(0, \mathbf{y}) &= g(\mathbf{y}) \\ F(s_0(x), \mathbf{y}) &= h_0(x, \mathbf{y}, F(x, \mathbf{y})), \text{ if } x \neq 0 \\ F(s_1(x), \mathbf{y}) &= h_1(x, \mathbf{y}, F(x, \mathbf{y})) \\ f(x, \mathbf{y}) &= F(|x|, \mathbf{y}) \end{aligned}$$

provided that $F(x, \mathbf{y}) \leq k(x, \mathbf{y})$ holds for all $x, \mathbf{y}$.

The characterization of polynomial size, constant depth boolean circuits with parity gates (resp. MOD6 gates) uses sequence encoding techniques of A_0 together with logtime hierarchy analogues of work of Handley, Paris, Wilkie [HPW84].

Theorem 6.3.9 ([CT94]).

$$\text{ACC}(2) = [0, I, s_0, s_1, |x|, \text{BIT}, \#; \text{COMP}, \text{CRN}, 1 - \text{BRN}] \tag{6.30}$$

$$\text{ACC}(6) = [0, I, s_0, s_1, |x|, \text{BIT}, \#; \text{COMP}, \text{CRN}, 2 - \text{BRN}] \tag{6.31}$$

$$\text{ACC}(6) = [0, I, s_0, s_1, |x|, \text{BIT}, \#; \text{COMP}, \text{CRN}, 3 - \text{BRN}]. \tag{6.32}$$

Proof. We first prove (6.30). Temporarily define C to be the smallest class of functions containing the initial functions and closed under composition, CRN and $1 - $ BRN. Define $summod2(x) = (\sum_{i<|x|} \text{BIT}(i, x)) \bmod 2$. Techniques of this chapter yield that $\text{ACC}(2) = D$, where D is the smallest class of functions containing the initial functions, with *summod*2 and closed under the operations of composition and CRN.

We first show that $D \subseteq C$. The function *summod2* is immediately definable using $1 - \text{BRN}$ as follows

$$\begin{aligned} \mathit{summod2}(0) &= 0 \\ \mathit{summod2}(s_i(x)) &= \mathit{cond}(\mathit{summod2}(x), 1, 0) \end{aligned}$$

with $\mathit{summod2}(x) \leq 1$. Hence $\text{ACC}(2) \subseteq C$.

We now show that $C \subseteq \text{ACC}(2)$. The initial functions are in A_0, and $\text{ACC}(2)$ is closed under composition and CRN. Modifying a proof of Handley, Paris and Wilkie [HPW84], we show how to define $1-\text{BRN}$ in $\text{ACC}(2)$. Suppose that

$$\begin{aligned} f(0, x) &= g(x) \\ f(s_0(y), x) &= h_0(y, x, f(y, x)), \text{provided } y \neq 0, \\ f(s_1(y), x) &= h_1(y, x, f(y, x)). \end{aligned}$$

and $f(y, x) \leq 1$ for all y, x. This is equivalent to the definition

$$\begin{aligned} &f(0, x) = g(x) \\ &f(y, x) = h(y, x, f(\lfloor y/2 \rfloor, x)), \text{provided } y \neq 0. \end{aligned}$$

where $f(y, x) \leq 1$ for all y, x. Then for x fixed, let $H_y^x : \{0,1\} \to \{0,1\}$ be defined by

$$\begin{aligned} &H_0^x(i) = g(x), \text{ for } i \leq 1, \\ &H_y^x(i) = h(y, x, i), \text{ for } i \leq 1, y > 0. \end{aligned}$$

Note that $H_0^x(0) = H_0^x(1)$.

We would like to show $\mathcal{H} \in \text{ACC}(2)$, where $\mathcal{H}$ is defined by

$$\mathcal{H}(y, x) = H_y^x \circ H_{\text{MSP}(y,1)}^x \circ H_{\text{MSP}(y,2)}^x \circ \cdots \circ H_{\text{MSP}(y,|y|)}^x,$$

for in this case, $f(y, x) = \mathcal{H}(y, x)$. Now $\mathcal{H}(y, x) = 0$ if and only if

EITHER

$$\begin{aligned} &(\exists i \leq |y|)[H_{\text{MSP}(y,i)}^x(0) = 0 = H_{\text{MSP}(y,i)}^x(1) \wedge \\ &(\forall j < |y|)[j < i \supset H_{\text{MSP}(y,j)}^x(0) \neq H_{\text{MSP}(y,j)}^x(1)] \wedge \\ &\{j < |y| : j < i \wedge H_{\text{MSP}(y,j)}^x(0) \neq 0\} \text{has even cardinality}] \end{aligned}$$

OR

$$\begin{aligned} &(\exists i \leq |y|)[H_{\text{MSP}(y,i)}^x(0) = 1 = H_{\text{MSP}(y,i)}^x(1) \wedge \\ &(\forall j < |y|)[j < i \to H_{\text{MSP}(y,j)}^x(0) \neq H_{\text{MSP}(y,j)}^x(1)] \wedge \\ &\{j < |y| : j < i \wedge H_{\text{MSP}(y,j)}^x(0) \neq 0\} \text{has odd cardinality}] \end{aligned}$$

It is not difficult to show that the above can be expressed in $\text{ACC}(2)$.

Next, we prove (6.31). We treat three cases.

Case 1. $(\forall i \leq |x|)[f(i, \mathbf{y}) \notin S_3]$

Let $r(i, \mathbf{y}) = \mu z[z \notin rng(f(i, \mathbf{y}))]$. Define $g_1(i, \mathbf{y}) : \{0,1,2\} \to \{0,1\}$, and $g_2(i, \mathbf{y}) : \{0,1\} \to \{0,1,2\}$ by

$$g_1(i, \mathbf{y}) = \begin{cases} z & \text{if } z \leq r(i, \mathbf{y}) \\ z+1 & \text{else} \end{cases}$$

and

$$g_2(i, \mathbf{y}) = \begin{cases} z & \text{if } z < r(i, \mathbf{y}) \\ z-1 & \text{else.} \end{cases}$$

Note that for all $i \leq |x|$, $(g_2(i, \mathbf{y}) \circ g_1(i, \mathbf{y})$ is the identity permutation in S_3, so that $g_2(i, \mathbf{y}) \circ g_1(i, \mathbf{y}) \circ f(i, \mathbf{y}) = f(i, \mathbf{y})$. Now

$$\begin{aligned} g(x, \mathbf{y}) &= \Pi_{i \leq |x|} f(i, \mathbf{y}) \\ &= f(|x|, \mathbf{y}) \circ \cdots \circ f(0, \mathbf{y}) \\ &= g_2(|x|, \mathbf{y}) \circ g_1(|x|, \mathbf{y}) \circ f(|x|, \mathbf{y}) \circ g_2(|x|-1, \mathbf{y}) \circ \cdots \circ \\ &\quad \circ \cdots \circ g_1(1, \mathbf{y}) \circ f(1, \mathbf{y}) \circ g_2(0, \mathbf{y}) \circ g_1(0, \mathbf{y}) \circ f(0, \mathbf{y}) \\ &= g_2(|x|, \mathbf{y}) \circ h(|x|-1, \mathbf{y}) \circ h(|x|-2, \mathbf{y}) \circ \cdots \circ \\ &\quad \circ \cdots \circ h(0, \mathbf{y}) \circ g_1(1, \mathbf{y}) \circ f(0, \mathbf{y}) \end{aligned}$$

where $h(i, \mathbf{y}) \in {}^2 2$ is defined by

$$h(i, \mathbf{y}) = g_1(i+1, \mathbf{y}) \circ f(i+1, \mathbf{y}) \circ g_2(i, \mathbf{y}).$$

Now by the first part of this theorem, $H(x, \mathbf{y}) = \Pi_{i \leq |x|} h(i, \mathbf{y}) \in \text{ACC}(2)$, hence it is easily seen that $H(x, \mathbf{y}) \in \text{ACC}(6)$. This completes the discussion of Case 1.

Case 2. $(\forall i \leq |x|)(f(i, \mathbf{y}) \in S_3$.

Let A_3 temporarily denote the even permutations of S_3, so that A_3 consists of the identity permutation $e : \{0,1,2\} \to \{0,1,2\}$, together with the two cycles

$$(0,1,2) = \begin{pmatrix} 0 & 1 & 2 \\ 2 & 3 & 1 \end{pmatrix}$$

and

$$(0,2,1) = \begin{pmatrix} 0 & 1 & 2 \\ 2 & 0 & 1 \end{pmatrix}$$

Clearly every permutation in S_3 belongs to A_3 or to $(0,1)A_3$, hence can be written as a product $\sigma \circ \tau$, where $\sigma \in \{e, (0,1)\}$ and $\tau \in A_3$. Since $A_3 \lhd S_3$ is a normal subgroup, it is clear that for $\sigma, \sigma' \in \{e, (0,1)\}$ and $\tau \in A_3$ there exists $\rho \in A_3$ such that $\sigma\tau\sigma' = \sigma\sigma'\rho$. Figure 1 presents a table of such values.

For $i \leq |x|$, define $\sigma(i, \mathbf{y}) \in \{e, (0,1)\}$, $\tau(i, \mathbf{y}) \in A_3$ such that $f(i, \mathbf{y}) = \sigma(i, \mathbf{y}) \circ \tau(i, \mathbf{y})$. Abbreviate $\sigma(i, \mathbf{y})$ by σ_i and $\tau(i, \mathbf{y})$ by τ_i. For $n \leq |x|$,

σ	τ	σ'	$\sigma \circ \sigma'$	ρ
e	e	e	e	e
e	e	(0,1)	(0,1)	e
(0,1)	e	e	(0,1)	e
(0,1)	e	(0,1)	e	e
e	(0,1,2)	e	e	(0,1,2)
e	(0,1,2)	(0,1)	(0,1)	(0,2,1)
(0,1)	(0,1,2)	e	(0,1)	(0,1,2)
(0,1)	(0,1,2)	(0,1)	e	(0,2,1)
e	(0,2,1)	e	e	(0,2,1)
e	(0,2,1)	(0,1)	(0,1)	(0,1,2)
(0,1)	(0,2,1)	e	(0,1)	(0,2,1)
(0,1)	(0,2,1)	(0,1)	e	(0,1,2)

Fig. 6.2. Values satisfying $\sigma\tau\sigma' = \sigma\sigma'\rho$

define $h(n, \mathbf{y})$ to be ρ_n, where the entry $\sigma_n \circ \tau_n \circ \Pi_{i<n}\sigma_i = (\Pi_{i\leq n}\sigma_i) \circ \rho_n$ appears in Figure 1. Since $\sigma_n(3) = 3$, σ_n may be considered an element of S_2, and $\Pi_{i<n\leq|x|}\sigma_i$ can be defined by $1 - \text{BRN}$, hence by Theorem 6.3.9, is in $\text{ACC}(2)$. Using conditionals and table lookup, it follows that $h \in \text{ACC}(2)$, hence $h \in \text{ACC}(6)$. We now have

$$\Pi_{i\leq|x|}f(i, \mathbf{y}) = \Pi_{i\leq|x|}\sigma_i \circ \Pi_{i\leq|x|}\rho_i.$$

Now $A_3 \simeq \mathbf{Z}_3$ is cyclic and it is straightforward to compute $\Pi_{i\leq|x|}\rho_i$ using counting modulo 3 (hence counting modulo 6). Thus $\Pi_{i\leq|x|}f(i, \mathbf{y}) \in \text{ACC}(6)$. This completes discussion of Case 2.

Case 3. Otherwise. In this case, we argue how the product of (an effectively given) sequence of elements in ^{33} can be written as the product of (an effectively given) sequence of elements in S_3 together with (an effectively given) sequence of elements of ${}^33 \setminus S_3$, thus reducing to the two previous cases. It is not at all obvious why a sequence of interspersed elements from S_3 and ${}^33 \setminus S_3$ can be so written. For more description of this technique, see [HPW84] where it was first introduced.

We claim that there exists a function $h(i, \mathbf{y})$ in $\text{ACC}(6)$ satisfying $(\forall n \leq |x|)[\Pi_{i\leq n}h(i, \mathbf{y}) = \Pi_{i\leq n}f(i, \mathbf{y})]$, such that for some n_0

$$\forall i < n_0(h(i, \mathbf{y}) \in S_3 \wedge \forall i \geq n_0(h(i, \mathbf{y}) \in {}^33 \setminus S_3).$$

Let $n_0 = \mu i \leq |x|[f(i, \mathbf{y}) \notin S_3]$. For $n < n_0$, define $h(n, \mathbf{y}) = f(n, \mathbf{y})$.

To compute $h(n, \mathbf{y})$ for $n \geq n_0$, proceed as follows. For $i < |x|$, define $g(i, \mathbf{y}) \in S_3$ by

$$g(i, \mathbf{y}) = \begin{cases} f(i, \mathbf{y}) \text{ if } f(i, \mathbf{y}) \in S_3 \\ e \quad\quad \text{else .} \end{cases}$$

By Case 2, for $n \leq |x|$, $\Pi_{i \leq n} g(i, \mathbf{y}) \in \text{ACC}(6)$. Let $n_1 = \max\{i \leq |x| : n_0 \leq i < n, f(i, \mathbf{y}) \notin S_3\}$. Thus for $n_1 < i \leq n$, it is the case that $f(i, \mathbf{y}) \in S_3$. We will now define integers $r, r', k_0 \in \{0, 1, 2\}$ as follows. Since $f(n_1, \mathbf{y}) \notin S_3$, let $r = \mu z[z \notin rng(f(n_1, \mathbf{y}))]$. Now set

$$\begin{aligned} r' &= f(n-1, \mathbf{y}) \circ \cdots \circ f(n_1 + 1, \mathbf{y})(r) \\ &= \Pi_{i<n} g(i, \mathbf{y}) \circ (\Pi_{i \leq n_1} g(i, \mathbf{y}))^{-1}(r). \end{aligned}$$

Then it follows that r' can be computed in $\text{ACC}(6)$. Let $k_0 = \mu z[z \neq r']$ and define h by

$$h(n, \mathbf{y})(k) = \begin{cases} f(n, \mathbf{y})(k) & \text{if } k \neq r' \\ f(n, \mathbf{y})(k_0) & \text{else.} \end{cases}$$

Define h_0 by

$$h_0(n, \mathbf{y}) = \begin{cases} f(n, \mathbf{y}) & \text{if } n < n_0 \\ e & \text{else .} \end{cases}$$

Define h_1 by $h_1(n) = h(n_0 + n)$. Then Case 1 applies to h_1, for $n \leq |x|$, we have $\Pi_{i \leq n} h_1(i, \mathbf{y}) \in \text{ACC}(6)$.

Now for $n \leq |x|$,

$$\Pi_{i \leq n} h(i, \mathbf{y}) = \begin{cases} \Pi_{i \leq n} h_0(i, \mathbf{y}) & \text{if } n \leq n_0 \\ \Pi_{i \leq n - n_0} h_1(i, \mathbf{y}) \circ \Pi_{i < n_0} h_0(i, \mathbf{y}) & \text{else.} \end{cases}$$

It follows that for $n \leq |x|$,

$$\Pi_{i \leq n} h(i, \mathbf{y}) = \Pi_{i \leq n} f(i, \mathbf{y})$$

and that $F(x, \mathbf{y}) = \Pi_{i \leq |x|} f(i, \mathbf{y})$ belongs to $\text{ACC}(6)$. This completes the discussion of Case 3.

A word should be said about the formalization of this argument. As previously mentioned, sharply bounded quantifiers $(\exists x \leq |y|)$, $(\forall x \leq |y|)$, and applications of the sharply bounded minimization operator $(\mu i \leq |x|) R(i, x, \mathbf{y})$ can easily be simulated using CRN. In this proof, functions $\sigma : \{0, 1, 2\} \to \{0, 1, 2\}$ are represented by their images $\sigma(0), \sigma(1), \sigma(2)$ preceded by 11. Thus

$$\begin{pmatrix} 0 & 1 & 2 \\ 1 & 2 & 0 \end{pmatrix}$$

is represented by 11011000, consisting of 8 bits. Using MSP, LSP, *cond* and other functions in A_0, along with the function $summod6(x)$ is equal to $(\sum_{i<|x|} \text{BIT}(i, x)) \bmod 6$, it is straightforward though tedious to formalize the previous argument.

The following characterization of $\mathcal{F}$ALOGTIME uses earlier techniques with a formalization of Barrington's trick [Bar89] in Theorem 6.2.1 of expressing boolean connectives AND, OR by permutation group word problems.

Theorem 6.3.10 ([Clo93]).

$$\mathcal{F}\text{ALOGTIME} = [0, I, s_0, s_1, |x|, \text{BIT}, \#; \text{COMP}, \text{CRN}, 4 - \text{BRN}].$$

Proof. Consider first the inclusion from right to left. $\mathcal{F}$ALOGTIME is easily seen to be closed under the given initial functions and the operations of composition and CRN. Suppose that g, h_0, h_1 are ALOGTIME computable and that f is defined by k-BRN from g, h_0, h_1. Given $x, \mathbf{u}$, define $\alpha(x, \mathbf{u}) : \{0, \ldots, k\} \to \{0, \ldots, k\}$ by

$$\alpha(x, \mathbf{u})(i) = h_{\text{MOD2}(x)}(\lfloor x/2 \rfloor, \mathbf{u}, i).$$

Given $x, \mathbf{u}$, let m be the least power of 2 greater than or equal to $|x|$. For $1 \leq i \leq |x|$, let the i-th leaf of a perfect binary tree of depth $\log_2(m)$ be $\alpha(\lfloor x/2^i \rfloor, \mathbf{u})$ and for $|x| < i \leq m$, let the i-th leaf be the identity function $id_k : \{0, \ldots, k\} \to \{0, \ldots, k\}$. Let each interior node of the binary tree have as value the function obtained by composing the functions associated with that node's children. Specifically, if H_ℓ, H_r are $n+2$-place functions associated with the left, right children of a node, then associate the function $H(x, \mathbf{u}, i) =_{\text{df}} H_r(x, \mathbf{u}, H_\ell(x, \mathbf{u}, i))$ to the node. Composition of functions whose domain and range are contained in $\{0, \ldots, k\}$ can be implemented by finite table look-up. Evaluating the binary tree yields a finite function, which when composed with the initial value $g(\mathbf{u})$, yields the value $f(x, \mathbf{u})$. From this description, standard techniques allow the construction of an alternating Turing machine which computes the bitgraph of f in logarithmic time.

Consider now the direction from left to right, and temporarily denote the function algebra by B. By work of [BIS90, Bar89], the word problem for the group S_5 is complete for ALOGTIME under LH reductions. We know that LH functions are exactly those in A_0. Define the five functions $f_0, \ldots, f_4$ mapping $\mathbf{N}$ into $\{0, 1, 2, 3, 4\}$ by

$$f_i(x) = \sigma_m \circ \cdots \circ \sigma_1(i)$$

for $0 \leq i \leq 4$. Here σ_i is a permutation on the set $\{0, 1, 2, 3, 4\}$, and the input x codes in a natural manner a sequence $(\sigma_1, \ldots, \sigma_m)$ of m permutations in S_5. Now if $A \subseteq \mathbf{N}$ is in ALOGTIME, then let $g \in \mathcal{F}$LH be such that $x \in A$ holds iff $g(x)$ is a sequence of permutations in S_5 evaluating to the identity permutation in S_5. Thus

$$x \in A \iff \bigwedge_{0 \leq i \leq 4} f_i(g(x)) = i.$$

By hypothesis, $g \in \mathcal{F}$LH, so $g \in A_0$ and the f_i are defined by $4 - \text{BRN}$. It follows that the characteristic function of A belongs to B. An arbitrary function f of $\mathcal{F}$ALOGTIME is one whose bitgraph $B_f(x, i) \equiv \text{BIT}(i, f(x)) = 1$ belongs to ALOGTIME. Since B is closed under concatenation recursion on notation, it now follows that $f \in B$.

A natural question arising from work in vectorizing compilers is whether there is an effective procedure to efficiently *parallelize* sequential code (i.e., from sequential code, generate optimal code for a parallel machine). Though we are not aware of details having been worked out, it seems clear that the non-existence of such a procedure must follow from the unsolvability of the halting problem. More importantly, it is not known whether NC is properly contained in PTIME, with modular powering $a^b \bmod m$ being a candidate to separate the classes. Though effective optimal parallelization of sequential code is hopeless, it may seem surprising that certain well-known parallel complexity classes can be characterized in a sequential manner. From the following theorem it follows that NC is characterized by a fragment of the PASCAL language allowing only `for`-loops of the form

```
for i = 1 to |x| if P then y = 2*y else y = 2*y+1;
for i = 1 to ||x|| if <statement>;
```

Using repeated squaring (see proof of Theorem 6.3.22), modular powering is evidently a polynomial time algorithm, yet cannot obviously be written using the above two `for`-loops.

Theorem 6.3.11 ([Clo90]).

$$\text{NC} = [0, I, s_0, s_1, |x|, \text{BIT}, \#; \text{COMP}, \text{CRN}, \text{WBRN}]$$
$$\text{AC}^k = \{f \in \text{NC} : rk_{\text{WBRN}}(f) \leq k\}.$$

It should be mentioned that independently and at about the same time, B. Allen [All91] characterized NC by a function algebra using a form of *divide and conquer recursion*, and noticed without giving details that over a basis of appropriate initial functions, NC could also be characterized by the scheme of WBRN.[15] A precise statement of Allen's characterization is given later in Theorem 6.3.35.

Using such techniques, two characterizations of NC^k were given in [Clo90, CT94]. Levels of a natural time-space hierarchy between $\mathcal{F}$PTIME and $\mathcal{F}$PSPACE were characterized in [Clo92b].

6.3.3 Bounded Recursion

In 1953, A. Grzegorczyk [Grz53] investigated a hierarchy of subclasses $\mathcal{E}^n$ of primitive recursive functions, defined as the closure of certain initial functions under composition and bounded recursion.

Definition 6.3.12. *The function f is defined by* bounded recursion *(BR) from functions g, h, k if*

$$f(0, \mathbf{y}) = g(\mathbf{y})$$
$$f(x+1, \mathbf{y}) = h(x, \mathbf{y}, f(x, \mathbf{y}))$$

provided that $f(x, \mathbf{y}) \leq k(x, \mathbf{y})$ holds for all $x, \mathbf{y}$.

---
[15] See remark at bottom of p. 13 of [All91].

Definition 6.3.13. *Define the following* principal *functions*

$$\begin{aligned} f_0(x) &= s(x) = x + 1 \\ f_1(x, y) &= x + y \\ f_2(x, y) &= (x + 1) \cdot (y + 1) \\ f_3(x) &= 2^x \\ f_{n+1}(x) &= f_n^{(x)}(1), \ \textit{for } n \geq 3 \end{aligned}$$

Let $\mathcal{E}f$ *denote* $[0, I, s, \max, f; \text{COMP}, \text{BR}]$*; define* $\mathcal{E}^0$ *to be* $[0, I, s; \text{COMP}, \text{BR}]$ *and for* $n > 0$ *define* $\mathcal{E}^n$ *to be* $\mathcal{E}f_n$*.*

Remark 6.3.1. Grzegorczyk's [Grz53] original functions were defined by

$$\begin{aligned} g_0(x, y) &= x + 1 \\ g_1(x, y) &= x + y \\ g_2(x, y) &= (x + 1) \cdot (y + 1) \\ g_{n+1}(0, y) &= g_n(y + 1, y + 1), \text{for } n \geq 3, \\ g_{n+1}(x + 1, y) &= g_{n+1}(x, g_{n+1}(x, y)). \end{aligned}$$

The unary functions f_n, for $n \geq 3$, were taken from [Bel82]. The definition $\mathcal{E}f$ of the elementary closure of f is derived from that of [Bel82] $\mathcal{E}f = [0, I, s, f; \text{COMP}, \text{BR}]$ by the addition of max. If one were to *incorrectly* define $\mathcal{E}^3$ to be $[0, I, s, f_3; \text{COMP}, \text{BR}]$ (i.e., without the function max), then it would follow that for every n-ary function $g \in \mathcal{E}^3$, there would exist a unary function $b(x)$ and an index $1 \leq i \leq n$ such that for all $x_1, \ldots, x_n$ it is the case that $f(x_1, \ldots, f_n) \leq b(x)$, where b is obtained from f_3, s by composition. This was pointed out to us by F. Felix Lara and A. Fernandez Margarit (personal communication). Of course, this problem would not arise if we were to take Grzegorczyk's original binary functions g_n.

A number of characterizations of Grzegorczyk's classes $\mathcal{E}^n$ $\mathcal{E}^n$, for $n \geq 3$, have been given. A. Meyer and D. Ritchie [MR67] characterized $\mathcal{E}^n$ in terms of certain *loop* programming languages, H. Schwichtenberg [Sch69] investigated the number of nested bounded recursions used in function definitions (the so-called *Heinermann hierarchy*), S.S. Muchnick [Muc76] investigated *vectorized* Grzegorczyk classes (essentially related to simultaneous bounded recursion schemes), etc. The following theorem is due to H. Schwichtenberg [Sch69] for $n \geq 3$ and to H. Müller [Mül74] for $n = 2$.[16]

Theorem 6.3.12 ([Sch69], [Mül74]). *Let* $\mathcal{H}_n$ *be the set*

$$\{f : f \textit{ primitive recursive}, rk_{\text{PR}}(f) \leq n\}.$$

Then for $n \geq 2$, $\mathcal{H}_n = \mathcal{E}^{n+1}$.

[16] It should be mentioned that [Sch69] used slightly different functions f_i; there f_i is the i-th Ackermann branch A_i.

In [Grz53] Grzegorczyk proved that for all $n \geq 0$, $\mathcal{E}^n$ is properly contained in $\mathcal{E}^{n+1}$ by demonstrating that $f_{n+1} \notin \mathcal{E}^n$. Concerning the relational classes, he showed that for $n \geq 2$, $\mathcal{E}^n_*$ is properly contained in $\mathcal{E}^{n+1}_*$, and asked whether $\mathcal{E}^0_* \subset \mathcal{E}^1_* \subset \mathcal{E}^2_*$. This question remains open. In fact, LTH $\subseteq \mathcal{E}^0_*$ and $\mathcal{E}^2_* =$ LINSPACE,[17] so Grzegorczyk's question is related to the yet open problem whether the linear time hierarchy is properly contained in linear space. An interesting partial result concerning the containment of the first two relational classes is the following.

Theorem 6.3.13 ([Bel82]). *For $s \geq 1$, define $\beta_s(x) = \max(1, x + \lceil x^{1-1/s} \rceil)$. Then for $s \geq 1$, $\mathcal{E}^0_* = (\mathcal{E}\beta_s)_*$. Additionally, $\mathcal{E}^2_* = \mathcal{E}^1_*$ implies $\mathcal{E}^2_* = \mathcal{E}^0_*$*

To obtain this result, Bel'tyukov introduced the *stack register machine*, a machine model capable of describing $(\mathcal{E}f)_*$. The stack register machine, a variant of the successor register machine, has a finite number of *input registers* and *stack* registers $S_0, \ldots, S_k$ together with a *work* register W. Branching instructions

if $p(x_1, \ldots, x_m) = q(x_1, \ldots, x_m)$ **then** I_i **else** I_j

allow to jump to different instructions I_i, I_j depending on the comparison of two polynomials whose variables are current register values. Storage instructions

W = S_i

allow a value to be saved from a stack register to the work register. Incremental instructions

S_i = S_i + 1

perform the only computation, and have a side effect of setting to 0 all S_j for $j < i$. A program is a finite list of instructions, where for each i there is at most one incremental instruction for S_i.

Apart from characterizing $\mathcal{E}^2_*$ or LINSPACE, Bel'tyukov characterized the linear time hierarchy LTH. The papers of Paris, Wilkie [PW85] and Handley, Paris, Wilkie [HPW84] study counting classes between LTH and LINSPACE defined by stack register machines. Work in [Clo97a] and [Han94b, Han94a] further studies the effect of nondeterminism for this model.

Lemma 6.3.6 ([Grz53]). *The functions $x \dot{-} y$, $sg(x)$, $\overline{sg}(x)$, $sg(x) \cdot y$, $\overline{sg}(x) \cdot y$ belong to $\mathcal{E}^0$. If $f \in \mathcal{E}^0$ then $\sum_{i \leq x} sg(f(i))$, $\sum_{i \leq x} \overline{sg}(f(i))$, $\Pi_{i \leq x} sg(f(i))$ and $\Pi_{i \leq x} \overline{sg}(f(i))$ belong to $\mathcal{E}^0$.*

Definition 6.3.14. *The function f is defined by* bounded minimization *(BMIN) from the function g, denoted by $f(x, \mathbf{y}) = \mu i \leq x[g(i, \mathbf{y}) = 0]$, if*

$$f(x, \mathbf{y}) = \begin{cases} \min\{i \leq x : g(i, \mathbf{y}) = 0\} & \text{if such exists} \\ 0 & \text{else.} \end{cases}$$

[17] See Corollary 6.3.2.

Corollary 6.3.1 ([Grz53]). *For $n \geq 0$, $\mathcal{E}_*^n$ is closed under boolean connectives and bounded quantification, and $\mathcal{E}^n$ is closed under bounded minimization.*

Proof. The predicate $\neg P(\mathbf{x})$ has characteristic function

$$\overline{sg}(c_P(\mathbf{x})),$$

the predicate $P(\mathbf{x}) \vee Q(\mathbf{x})$ has characteristic function

$$\overline{sg}(\overline{sg}(c_P(\mathbf{x})) \cdot \overline{sg}(c_Q(\mathbf{x}))),$$

while $(\exists i \leq x)R(i, \mathbf{y})$ has characteristic function

$$sg(s(x) \dot{-} \sum_{i \leq x} \overline{sg}(c_R(i, \mathbf{y}))),$$

and $(\forall i \leq x)R(i, \mathbf{y})$ has characteristic function

$$\overline{sg}(s(x) \dot{-} \sum_{i \leq x} c_R(i, \mathbf{y})).$$

To define $f(x, \mathbf{y}) = \mu i \leq x[g(i, \mathbf{y}) = 0]$, define the auxiliary function h by

$$\begin{aligned} h(i, \mathbf{y}) &= \sum_{j \leq i} \overline{sg}(g(j, \mathbf{y})) \\ &= \text{cardinality of } \{j \leq i : g(j, \mathbf{y}) = 0\} \end{aligned}$$

so $\overline{sg}(h(i, \mathbf{y})) = 1$ iff $(\forall j \leq i)(g(j, \mathbf{y}) \neq 0)$, and

$$\sum_{i \leq x} \overline{sg}(h(i, \mathbf{y})) = \begin{cases} \mu i \leq x[g(i, \mathbf{y}) = 0] & \text{if } (\exists i \leq x)(g(i, \mathbf{y}) = 0) \\ x + 1 & \text{else.} \end{cases}$$

Then

$$f(x, \mathbf{y}) = \overline{sg}\left(\left(\sum_{i \leq x} \overline{sg}(h(i, \mathbf{y}))) \dot{-} x\right) \cdot \sum_{i \leq x} \overline{sg}(h(i, \mathbf{y})\right).$$

The following characterization of LINSPACE in terms of the Grzegorczyk hierarchy was proved by R.W. Ritchie [Rit63].

Theorem 6.3.14. $\mathcal{F}$LINSPACE $= \mathcal{E}^2$.

Proof. Consider first the direction from right to left. The initial functions of $\mathcal{E}^2$ are computable in LINSPACE, and $\mathcal{F}$LINSPACE is closed under composition and bounded recursion.

Now consider the direction from left to right. We first claim that

$$[0, I, s_0, s_1, \text{MOD2}, msp; \text{COMP}, \text{CRN}] \subseteq \mathcal{E}^2.$$

Clearly $s_0, s_1, cond \in \mathcal{E}^2$ and $\mathcal{E}^2$ is closed under bounded quantification. Now $\overline{sg}(0) = 1$, $\overline{sg}(s(x)) = 0$, $\text{MOD2}(0) = 0$, $\text{MOD2}(s(x)) = \overline{sg}(\text{MOD2}(x))$, so that $\text{MOD2} \in \mathcal{E}^2$. Using BR define the following functions in $\mathcal{E}^2$:

$$\begin{aligned}
\lfloor x/2 \rfloor &= \mu y \leq x[y + y = x \vee y + y + 1 = x] \\
\text{MSP}(x, 0) &= x \\
\text{MSP}(x, i+1) &= \lfloor \text{MSP}(x, i)/2 \rfloor \\
\text{BIT}(i, x) &= \text{MOD2}(\text{MSP}(x, i)).
\end{aligned}$$

Temporarily define the auxiliary function h by

$$\begin{aligned}
h(x, 0) &= 0 \\
h(x, i+1) &= \begin{cases} h(x, i) + 1 & \text{if } \text{BIT}(i, x) = 1 \\ h(x, i) & \text{else.} \end{cases}
\end{aligned}$$

Note that $ones(x) = 2^{|x|} - 1$ is defined by

$$\mu y \leq s_0(x)[(\forall i \leq x)(\text{BIT}(i, y) = 1 \leftrightarrow (\exists j \leq x)(i \leq j \wedge \text{BIT}(j, x) = 1))].$$

Then $|x| = h(ones(x), x)$ and $msp(x, y) = \text{MSP}(x, |y|)$ belong to $\mathcal{E}^2$.

Suppose that f is defined from g, h_0, h_1 by CRN, where $g, h_0, h_1 \in \mathcal{E}^2$. Then $f(x, \mathbf{y})$ is $\mu z \leq g(\mathbf{y}) \cdot (2x + 1) + 2x[(6.33) \vee (6.34) \vee (6.35) \vee (6.36)]$ where

$$|z| = |g(\mathbf{y})| + |x| \tag{6.33}$$

$$\text{MSP}(z, |x|) = g(\mathbf{y}) \tag{6.34}$$

$$\begin{aligned}
(\forall i, j < |x|)(j = |x| - i - 1 \wedge \text{BIT}(j, x) = 0 \\
\rightarrow \\
\text{BIT}(j, z) = h_0(\text{MSP}(x, j+1), \mathbf{y}))
\end{aligned} \tag{6.35}$$

$$\begin{aligned}
(\forall i, j < |x|)(j = |x| - i - 1 \wedge \text{BIT}(j, x) = 1 \\
\rightarrow \\
\text{BIT}(j, z) = h_1(\text{MSP}(x, j+1), \mathbf{y})).
\end{aligned} \tag{6.36}$$

The above bound on f suffices, since $f(x, \mathbf{y}) \leq g(\mathbf{y}) \cdot 2^{|x|} + 2^{|x|} - 1$, and the latter is at most $g(\mathbf{y}) \cdot (2x + 1) + 2x$. By Corollary 6.3.1, $\mathcal{E}^2$ is closed under BMIN, so $f \in \mathcal{E}^2$. It follows that

$$[0, I, s_0, s_1, \text{MOD2}, msp; \text{COMP}, \text{CRN}] \subseteq \mathcal{E}^2.$$

Now let M be a linear space bounded multitape Turing machine computing a function f. By Lemma 6.3.3, $initial_M$ and $next_M$ belong to $\mathcal{E}^2$. Define the function T by

$$\begin{aligned}
T(x, 0) &= initial_M(x) \\
T(x, y+1) &= next_M(T(x, y)).
\end{aligned}$$

From the linear space bound, there exists a constant c such that $|T(x,y)| \leq c \cdot |x|$, and so $T(x,y) \leq (x+1)^c + 1$. Thus T is definable using bounded recursion from functions belonging to $\mathcal{E}^2$. It follows that $\mathcal{F}$LINSPACE $\subseteq \mathcal{E}^2$.

Corollary 6.3.2 ([Rit63]). LINSPACE $= \mathcal{E}^2_*$.

The following well-known fact follows from Ritchie's arithmetization techniques together with the observations that a TM with space bound $S(n)$ has time bound $2^{O(S(n))}$ and that $2^x \in \mathcal{E}^k$ for $k \geq 3$.

Theorem 6.3.15. *For $k \geq 3$,*

$$\mathcal{E}^k = \text{DTIME}(\mathcal{E}^k) = \text{DSPACE}(\mathcal{E}^k).$$

Similar techniques yield a characterization of PSPACE, where the first assertion below appears in [Tho72], and the second is easily obtained from the techniques of this chapter.

Theorem 6.3.16 ([Tho72]).

$$\begin{aligned} \mathcal{F}\text{PSPACE} &= [0, I, s, \#; \text{COMP}, \text{BR}] \\ &= [0, I, s, \max, x^{|x|}; \text{COMP}, \text{BR}]. \end{aligned}$$

Definition 6.3.15. *Let k be an integer. The function f is defined by k-*bounded recursion *(k-BR) from functions g, h, k if*

$$\begin{aligned} f(0, \mathbf{y}) &= g(\mathbf{y}) \\ f(x+1, \mathbf{y}) &= h(x, \mathbf{y}, f(x, \mathbf{y})) \end{aligned}$$

provided that $f(x, \mathbf{y}) \leq k$ holds for all $x, \mathbf{y}$, where k is a constant.

The following characterization results from the method of the proof of Barrington's Theorem 6.2.1, arithmetization techniques of this chapter, and Theorem 6.2.3 implying that ATIME$(n^{O(1)})$ = PSPACE. In [CF88], J.-Y. Cai and M. Furst give a related characterization of PSPACE using *safe-storage* Turing machines, a model related to Bel'tyukov's earlier stack register machines. The next result follows from a characterization of PSPACE by a variant of the stack register machine model.

Theorem 6.3.17 ([Clo97a]). *For $k \geq 4$,*

$$\text{PSPACE} = [0, I, s_0, s_1, |x|, \text{BIT}, \#; \text{COMP}, \text{CRN}, k\text{-BR}]_*.$$

In [Wag79] K. Wagner extended Ritchie's characterization to more general complexity classes.

Theorem 6.3.18 ([Wag79]). *Let f be an increasing function such that for some $r > 1$ and for all but finitely many x, it is the case that $f(x) \geq x^r$. Let $\mathcal{F}$ temporarily denote the algebra $[|f(2^n)|; \text{COMP}]$. Then recalling that $f_2(x, y) = (x+1) \cdot (y+1)$,*

$$\text{DSPACE}(\mathcal{F}) = [0, I, s, \max, f; \text{COMP}, \text{BR}]_* = [0, I, s, f_2, f; \text{COMP}, \text{BR}]_*.$$

Let $|x|_0 = x$, $|x|_{k+1} = ||x|_k|$.

Corollary 6.3.3 ([Wag79]). *For $k \geq 1$,*

$$\text{DSPACE}(n \cdot (\log^{(k)} n)^{O(1)}) = [0, I, s, \max, x^{|x|_{k+1}}; \text{COMP}, \text{BR}]_*.$$

Definition 6.3.16 ([Wag79]). *The function f is defined by* weak bounded primitive recursion *(WBPR) from g, h, k if $f(x, \mathbf{y}) = F(|x|, \mathbf{y})$, where F is defined by bounded primitive recursion, i.e.,*

$$\begin{aligned} F(0, \mathbf{y}) &= g(\mathbf{y}) \\ F(x+1, \mathbf{y}) &= h(x, \mathbf{y}, F(x, \mathbf{y})) \\ f(x, \mathbf{y}) &= F(|x|, \mathbf{y}) \end{aligned}$$

provided that $F(x, \mathbf{y}) \leq k(x, \mathbf{y})$ holds for all $x, \mathbf{y}$.

Provided the proper initial functions are chosen, WBPR is equivalent with BRN. Using this observation, Wagner characterized certain general complexity classes as follows.

Theorem 6.3.19 ([Wag79]). *Let f be an increasing function such that $f(x) \geq x^r$ for some $r > 1$, and temporarily let $\mathcal{F}$ denote the algebra $[|f(2^n)|; \text{COMP}]$ and $\mathcal{G}$ denote $\{g^k : g \in \mathcal{F}, k \in \mathbf{N}\}$. Then*

$$\begin{aligned} \text{DTIMESPACE}(\mathcal{G}, \mathcal{F}) &= [0, I, s_0, s_1, \max, f; \text{COMP}, \text{BRN}]_* \\ &= [0, I, s, \max, 2 \cdot x, f; \text{COMP}, \text{BRN}]_* \\ &= [0, I, s, +, f; \text{COMP}, \text{BRN}]_* \\ &= [0, I, s, \max, 2 \cdot x, \lfloor x/2 \rfloor, \dot{-}, f; \text{COMP}, \text{WBPR}]_* \\ &= [0, I, s, \lfloor x/2 \rfloor, +, \dot{-}, f; \text{COMP}, \text{WBPR}]_*. \end{aligned}$$

The class DTIMESPACE$(n^{O(1)}, O(n))$ of simultaneous polynomial time and linear space can be characterized from the previous theorem by taking $f(x) = x^2$. As referenced in [WW86], S.V. Pakhomov [Pak79] has characterized general complexity classes DTIMESPACE(T, S), DTIME(T), and DSPACE(S) for suitable classes S, T of unary functions. The class QL $=$ DTIME$(n \cdot (\log n)^{O(1)})$ of *quasilinear time* was studied by C.P. Schnorr in [Sch78]. In analogy, let *quasilinear space* be the class DSPACE$(n \cdot (\log n)^{O(1)})$. Though Corollary 6.3.3 characterizes quasilinear space via a function algebra, there appears to be no known function algebra for *quasilinear time.* In [GS89], Y. Gurevich and S. Shelah studied the class NLT (*nearly linear time*) of functions computable in time $O(n \cdot (\log n)^{O(1)})$ on a random access Turing machine RTM, which is allowed to change its input tape.

Definition 6.3.17 ([GS89]). *A* RTM *is a Turing machine with one-way infinite main tape, address tape and auxiliary tape, such that the head of the main tape is at all times in the cell whose position is given by the contents of the address tape. Instructions of a* RTM *are of the form*

$$(q, a_0, a_1, a_2) \to (q', b_0, b_1, b_2, d_1, d_2)$$

where q, q' are states, $a_i, b_i \in (\Sigma \cup \{B\})$, and $d_i \in \{-1, +1\}$. For such an instruction, if the machine is in state q reading a_0, a_1, a_2 on the main, address and auxiliary tapes, then the machine goes to state q', writes b_0, b_1, b_2 on the respective tapes, and the head of the address [resp. auxiliary] tape goes one square to the right if $d_1 = 1$ [resp. $d_2 = 1$] otherwise one square to the left.

In [GS89], Gurevich and Shelah show the robustness of NLT by proving the equivalence of this class with respect to different machine models, and give a function algebra for NLT. Their algebra, defined over words from a finite alphabet, is the closure under composition of certain initial functions and weak iterates $f^{(|x|)}(x)$ of certain string manipulating initial functions.

6.3.4 Bounded Minimization

In [Grz53], Grzegorczyk considered function classes defined by bounded minimization, defined in Definition 6.3.14. Recall Grzegorczyk's original principal functions from Remark 6.3.1.

Definition 6.3.18 ([Har75, Har79]). *For $n = 0, 1, 2$, define*

$$\mathcal{M}^n = [0, I, s, g_n; \text{COMP}, \text{BMIN}]$$

and for $n \geq 3$, define

$$\mathcal{M}^n = [0, I, \max, s, x^y, g_n; \text{COMP}, \text{BMIN}].$$

Though implicitly asserted in [Grz53], the details for the proof of the following result, which follow those in the proof of Theorem 6.3.28, are given by K. Harrow [Har79]. The idea of the proof is simply to encode via sequence numbers a definition by bounded primitive recursion and apply the bounded minimization operator.

Theorem 6.3.20 ([Grz53], [Har79]). *For $n \geq 3$, $\mathcal{E}^n = \mathcal{M}^n$.*

In the literature, the algebra RF of *rudimentary functions* is sometimes defined by

$$\text{RF} = [0, I, s, +, \times; \text{COMP}, \text{BMIN}].$$

As noticed in [Har79], it follows from J. Robinson's [Rob49] bounded quantifier definition of addition from successor and multiplication that $\mathcal{M}^2 = \text{RF}$. As is well-known, there is a close relationship between (bounded) minimization and (bounded) quantification. This is formalized as follows.

Terms in the first order language of $0, s, +, \cdot, \leq$ of arithmetic are defined inductively by: 0 is a term; $x_0, x_1, \ldots$ are terms; if t, t' are terms, then $s(t)$, $t + t'$ and $t \cdot t'$ are terms. Atomic formulas are of the form $t = t'$ and $t \leq t'$, where t, t' are terms. The set Δ_0 of bounded quantifier formulas is defined inductively by: if ϕ is an atomic formula, then $\phi \in \Delta_0$; if $\phi, \theta \in \Delta_0$ then $\neg\phi$, $\phi \wedge \theta$, and $\phi \vee \theta$ belong to Δ_0; if $\phi \in \Delta_0$ and t is a term, then $(\exists x \leq t)\phi(x, t)$ and $(\forall x \leq t)\phi(x, t)$ belong to Δ_0. A k-ary relation $R \subseteq \mathbf{N}^k$ belongs to Δ_0^N if there is a Δ_0 formula ϕ for which $R = \{(a_1, \ldots, a_n) : \mathbf{N} \models \phi(a_1, \ldots, a_n)\}$.

Definition 6.3.19. *A predicate $R \subseteq \mathbf{N}^k$ belongs to* CA (constructive arithmetic*), a notion due to R. Smullyan, if there is $\phi(\mathbf{x}) \in \Delta_0$ such that $R(a_1, \ldots, a_k)$ holds iff $\mathbf{N} \models \phi(a_1, \ldots, a_k)$. Following Definition 6.2.7, a function $f(\mathbf{x}) \in \mathcal{G}$CA if the bitgraph $B_f \in$ CA and f is of linear growth.*[18]

Definition 6.3.20. Presburger arithmetic *(*PRES*) is the collection of all predicates $R \subseteq \mathbf{N}^k$ for which there exists a first order formula $\phi(\mathbf{x})$ in the language $0, s, +$ of arithmetic such that $R(a_1, \ldots, a_k)$ holds iff $\mathbf{N} \models \phi(a_1, \ldots, a_k)$.*

The following theorem is proved by using quantifier elimination for Presburger arithmetic to show the equivalence between first order formulas and bounded formulas in a richer language allowing congruences, and then exploiting the correspondence between bounded quantification and bounded minimization.

Theorem 6.3.21 ([Har75]). *$\mathcal{M}^1_*$ equals the collection of Presburger definable sets.*

From J. Robinson's definition of addition from successor and multiplication, the following easily follows.

Proposition 6.3.1 ([Har75]). *$\mathcal{M}^2_* =$ CA and $\mathcal{M}^2 = \mathcal{G}$CA.*

While it is obvious that $\text{RF}_* =$ CA, it is non-trivial and surprising that CA equals the linear time hierarchy. In [Ben62], J.H. Bennett showed that the collection of constructive arithmetic sets (Δ_0 definable) is equal to RUD, the class of *rudimentary* sets in the sense of [Smu61]. Later, C. Wrathall [Wra76] proved that the rudimentary sets are exactly those in the linear time hierarchy LTH.

Theorem 6.3.22 ([Ben62]). *The ternary relation $G(x, y, z)$ for the graph $x^y = z$ of exponentiation is in constructive arithmetic.*

Proof. Using the technique of *repeated squaring* to compute the exponential $x^y = z$, the idea is to encode the computation in a Δ_0 manner where all

[18] In the literature, especially in [PW85], a function f is defined to be Δ_0^N if its graph G_f belongs to Δ_0^N and f is of linear growth. It easily follows from Corollary 6.3.4 that $f \in \mathcal{G}$CA $\iff f \in \Delta_0^N$.

quantifiers are bounded by a polynomial in z. Suppose that $x, y > 1$ and that $y = \sum_{i<n} y_i \cdot 2^i$ is the binary representation of y. The following algorithm computes $z = x^y$ by repeatedly applying the fact that $x^{2y} = (x^y)^2$ and $x^{2y+1} = (x^y)^2 \cdot x$. Throughout the rest of the proof, let n denote $|y|$.

```
z = 1
for i = 1 to n
    if y_{n-i} = 0 then z = z^2 else z = z^2· x
```

To encode the computation, for $0 \leq i \leq n$, let $a_i = \text{MSP}(y, |y| - i)$ and $b_i = x^{a_i}$. The binary representation of a_i consists of the i leftmost bits of y, while b_i equals the value of z at the end of the i-th pass through the above **for**-loop. Except for trivial cases where x, y take values among $0, 1$ it follows that $x^y = z$ if and only if there exist sequences $(a_0, \ldots, a_n)$ and $(b_0, \ldots, b_n)$ for which

$$a_0 = 0, b_0 = 1, a_n = y, b_n = z, (\forall i < n)(a_{i+1} \in \{2a_i, 2a_i + 1\})$$

and

$$(\forall i < n)((a_{i+1} = 0 \rightarrow b_{i+1} = b_i^2) \wedge (a_{i+1} = 1 \rightarrow b_{i+1} = b_i^2 \cdot x)).$$

Thus the graph of exponentiation is Δ_0 provided that sequences $(a_0, \ldots, a_n)$, $(b_0, \ldots, b_n)$ can be encoded in a manner where all quantifiers are bounded by a polynomial in z.

To do this, we will find relatively prime $m_0 < m_1 < \cdots < m_n$ satisfying $a_i, b_i < m_i$ and apply the Chinese remainder theorem to obtain $M = \Pi_{i \leq n} m_i$ and unique $A, B < M$ for which

$$A \equiv a_i \pmod{m_i}$$
$$B \equiv b_i \pmod{m_i}$$

for $i \leq n$. By choosing the m_i to be prime powers of distinct primes, and m_{i+1} to be the smallest prime power divisor of M greater than m_i, one can determine the m_i from M in a Δ_0 manner. Formally, define the predicates $\text{PRIME}(p)$, $\text{MPP}(m, M)$, $\text{I}(m, M)$, $\text{N}(m, m', M)$, and $\text{F}(m, M)$ as follows.

1. $\text{PRIME}(p)$ means that p is prime:

$$2 \leq p \wedge (\forall q < p)(q | p \rightarrow q = 1).$$

2. $\text{MPP}(m, M)$ means that m is a maximal prime power divisor of M:

$$m | M \wedge (\exists p \leq m)(\text{PRIME}(p) \wedge p | m$$
$$\wedge (\forall q < m)(q | m \rightarrow q \in \{1, p\}) \wedge p \cdot m \not| M).$$

3. $\text{I}(m, M)$ means that $m = m_0$ is the *initial* (least) maximal prime power divisor of M:

$$(m = 1 \wedge M \in \{0, 1\}) \vee (\text{MPP}(m, M)$$
$$\wedge (\forall m' \leq M)(\text{MPP}(m', M) \rightarrow m \leq m').$$

4. $\mathrm{N}(m, m', M)$ means that $m' = m_{i+1}$ is the *next* maximal prime power divisor of M after $m = m_i$:

$$\mathrm{MPP}(m, M) \wedge \mathrm{MPP}(m', M) \wedge m < m' \wedge$$
$$(\forall m'' < m')(\mathrm{MPP}(m'', M) \rightarrow m'' \leq m).$$

5. $\mathrm{F}(m, M)$ means that $m = m_n$ is the *final* (greatest) maximal prime power divisor of M:

$$(m = 1 \wedge M \in \{0, 1\}) \vee (\mathrm{MPP}(m, M) \wedge$$
$$(\forall m' \leq M)(\mathrm{MPP}(m', M) \rightarrow m' \leq m).$$

Assume that neither x nor y take values among $0, 1$. Then $a_i < 2^i$ and $b_i \leq x^{2^i}$ for $i \leq n$. Define $m_0, \ldots, m_n$ to be an increasing sequence of prime powers of distinct primes as follows. Let $m_0 = 2$. Given $m_0, \ldots, m_{k-1}$ let $m_k = p^\alpha$, where p is the least prime not dividing $\Pi_{i<k} m_i$ and α is the least integer for which $p^\alpha > x^{2^k}$. By the prime number theorem, $p = O(k \cdot \ln k) < k^2 \leq x^{2^k}$, and by choice of α, it is the case that $p^{\alpha-1} < x^{2^k}$, and so

$$p^\alpha = p \cdot p^{\alpha-1} < x^{2^k} \cdot x^{2^k} \leq x^{2^{k+1}}.$$

An inductive proof yields that $\Pi_{i<k} m_i \leq x^{2^{k+1}}$ for all $k > 0$, hence

$$M = \Pi_{i \leq n} m_i \leq x^{2^{n+2}} = (x^{2^{n-1}})^8 \leq z^8.$$

It now follows that the relation $x^y = z$ is Δ_0 definable.

The main lines of this proof were influenced by Wilkie's presentation in [Wil83]. See [HP93] for other proofs.

Corollary 6.3.4. *The function algebra*

$$[0, I, s_0, s_1, |x|, \mathrm{BIT}; \mathrm{COMP}, \mathrm{CRN}]$$

is contained in $\mathcal{M}^2$.

Proof. Note that $s_0(x) = x + x$, $s_1(x) = x + x + 1$,

$$|x| = \mu y \leq x[(\exists z \leq 2 \cdot x)(2^y = z \wedge x < z \wedge \lfloor z/2 \rfloor \leq x)]$$

and that

$$\mathrm{BIT}(i, x) = \mu y \leq 1[(\exists u, v \leq x)(|u| = i + 1 \wedge v = 2^{i+1} \wedge v|(x - u))]$$

so that $s_0, s_1, |x|, \mathrm{BIT}$ belong to $\mathcal{M}^2$. Using these functions and bounded minimization, it is easy to show that $\mathcal{M}^2$ is closed under CRN.

The following is proved in a manner similar to that of Lemma 6.3.4 and Lemma 6.3.5.

Lemma 6.3.7 ([Nep70]). *For every $k, m > 1$,*

$$\text{NTIMESPACE}(n^k, n^{1-1/m})$$

on a TM *is contained in* CA. *Moreover,* NSPACE$(O(\log(n))) \subseteq$ LTH.

Theorem 6.3.23. LTH = CA.

Proof. Consider first the direction from left to right. It follows from Lemmas 6.3.4 and 6.3.3 that $initial_M$ and $next_M$ are CA. Now proceed in a similar fashion as in the proof of Theorem 6.3.3.

The direction from right to left is proved by induction on the length of Δ_0 formulas. Addition, inequality $\leq$, and multiplication are computable in LOGSPACE, and $\mathcal{F}$LOGSPACE is closed under composition. By Lemma 6.3.7 it follows that atomic Δ_0 formulas define relations in LTH. Bounded quantifiers can be handled by existential and universal branching of an alternating Turing machine.

Corollary 6.3.5. $\mathcal{M}^2_* =$ LTH, *and* $\mathcal{M}^2 = \mathcal{F}$LTH.

Though the linear time hierarchy equals the bounded arithmetic hierarchy, there is no known exact level-by-level result. The sharpest result we know is due to A. Woods [Woo86].

If Γ is a class of first order formulas, then Γ^N denotes the collection of predicates definable by a formula in Γ. Let $\Sigma_{0,m}$ denote the collection of bounded quantifier formulas of the form $(\exists \mathbf{x}_1 \leq y)(\forall \mathbf{x}_2 \leq y) \ldots (Q\mathbf{x}_m \leq y)\phi$ where ϕ is a quantifier free formula in the first order language $0, 1, +, \cdot, \leq$. Thus $\Sigma_{0,0}$ is the collection of quantifier free formulas. In the following theorem, recall the definition of $\Sigma_n - \text{TIME}(T(n))$ from Definition 6.2.6.

Theorem 6.3.24 ([Woo86]). *For $m \geq 1$, $\Sigma^N_{0,m} \subseteq \Sigma_{m+2} - \text{TIME}(n)$.*

Proof. (Outline) The inclusion $\Sigma^N_{0,0} \subseteq \Sigma_2 - \text{TIME}(O(n))$ is shown as follows. Given an atomic formula $\phi(n)$, suppose that all terms appearing in $\phi(n)$ are bounded by a polynomial in n. By the prime number theorem, there exists a constant c such that the product of the first $c \cdot \ln(n)$ primes is larger than the values of all terms occurring in $\phi(n)$. Using non-determinism guess all terms and subterms appearing in the given quantifier free formula, guess the first $c \cdot \ln(n)$ many prime numbers p and the residues modulo p of the products of subterms occurring in a term, and branching universally, verify that the computations are correct. Now, by the Chinese remainder theorem, the computations are correct iff they are correct modulo the primes.

Since the negation of a quantifier free formula is quantifier free, it follows that

$$\Sigma^N_{0,0} \subseteq \Sigma_2 - \text{TIME}(O(n)) \cap \Pi_2 - \text{TIME}(O(n)).$$

Now induct on the number of quantifier blocks.

By Corollary 6.3.5 and Theorem 6.3.14, $\mathcal{M}^2_* = \text{LTH} \subseteq \text{LINSPACE} = \mathcal{E}^2_*$. While LINSPACE is clearly closed under *counting*, this may not be the case for LTH. A typical open question is whether $\pi(x) \in \mathcal{M}^2$, where $\pi(x)$ is the number of primes less than x. In [PW85, HPW84], J. Paris, A. Wilkie and later W. Handley studied the effect of adding k-bounded recursion to LTH. Using the techniques of Barrington, Paris, Wilkie and Handley, together with those of this chapter, the following result can be proved.

Theorem 6.3.25 ([Clo97a]). *For any $k \geq 4$,*

$$\text{ALINTIME} = [0, I, s, +, \dot{-}, \times; \text{COMP}, \text{BMIN}, k\text{-BR}]_*.$$

As in Corollary 6.3.5, $\mathcal{F}$PH can similarly be characterized.

Theorem 6.3.26 (Folklore).

$$\begin{aligned}\mathcal{F}\text{PH} &= [0, I, s, +, \dot{-}, \times, \#; \text{COMP}, \text{BMIN}] \\ &= [0, I, s, +, \dot{-}, \times, \#; \text{COMP}, \text{BRN}, \text{BMIN}] \\ &= [0, I, s_0, s_1, \#; \text{COMP}, \text{BRN}, \text{BMIN}].\end{aligned}$$

The last assertion of this theorem was sharpened by S. Bellantoni as follows. Following S. Buss [Bus86a] let $\square^P_i$ denote the class of functions computed in polynomial time on a Turing machine with oracle A, for some set $A \in \Sigma^P_i$. With this notation, $\mathcal{F}\text{PH} = \cup_i \square^P_i$.

Definition 6.3.21 ([Bel94]). *Let μFP_i denote the algebra*

$$\{f \in [0, I, s_0, s_1, \#; \text{COMP}, \text{BRN}, \text{BMIN}] : rk_{\text{BMIN}}(f) \leq i\}.$$

Theorem 6.3.27 ([Bel94]). *For $i \geq 0$, $\square^P_i = \mu FP_i$.*

6.3.5 Miscellaneous

Definition 6.3.22 ([Grz53], [Con73]). *The function f is defined by* bounded summation *(*BSUM*) [resp.* bounded product *(*BPROD*)] from g, k if $f(x, \mathbf{y})$ equals*

$$\sum_{i=0}^{x} g(i, \mathbf{y}) \qquad [\text{resp. } \textstyle\prod_{i=0}^{x} g(i, \mathbf{y})]$$

provided that $f(x, \mathbf{y}) \leq k(x, \mathbf{y})$ for all $x, \mathbf{y}$.

The function f is defined by sharply bounded summation *(*SBSUM*) [resp.* sharply bounded product *(*SBPROD*)] from g, k if $f(x, \mathbf{y})$ equals*

$$\sum_{i=0}^{|x|} g(i, \mathbf{y}) \qquad [\text{resp. } \textstyle\prod_{i=0}^{|x|} g(i, \mathbf{y})]$$

provided that $f(x, \mathbf{y}) \leq k(x, \mathbf{y})$ *for all* $x, \mathbf{y}$.[19]

The elementary functions were first introduced by Kalmár [Kal43] and Csillag [Csi47].

Definition 6.3.23. *The class* $\mathcal{E}$ *of* elementary *functions is the algebra*

$$[0, I, s, +, \dot{-}; \text{COMP}, \text{BSUM}, \text{BPROD}].$$

The elementary functions have many alternate characterizations, among them that $\mathcal{E} = \mathcal{E}^3$.

Theorem 6.3.28 ([Grz53]).

$$\begin{aligned} \mathcal{E} &= [0, I, s, f_3; \text{COMP}, \text{BR}] \\ &= [0, I, s, \dot{-}, x^y; \text{COMP}, \text{BMIN}] \\ &= [0, I, s, \dot{-}, \times, x^y; \text{COMP}, \text{BSUM}]. \end{aligned}$$

Grzegorczyk asked whether $\mathcal{E}$ had a *finite basis*, i.e., a finite number of functions, whose closure with I under composition equals $\mathcal{E}$. As surveyed in [WW86], D. Rödding first gave a positive answer, which was refined by C. Parsons. In [Mar80], S.S. Marčenkov gave a particularly elegant characterization of $\mathcal{E}$ as $[0, I, s, \lfloor x/y \rfloor, x^y, \phi(x,y); \text{COMP}]$, where $\phi(x, y)$ is 0 for $x \leq 1$, and otherwise is the least index i for which $a_i = 0$ in the radix x representation of y, i.e., $y = \sum_{i=0}^{\infty} a_i \cdot x^i$, where $0 \leq a_i < x$ for all i. In the following theorem, the first statement is due to S.S. Marčenkov [Mar80], while the second statement to J.P. Jones and Y. Matijasevič [JM82].

Theorem 6.3.29.

$$\begin{aligned} \mathcal{E}^3_* &= ([0, I, s, \dot{-}, \lfloor x/y \rfloor, x^y; \text{COMP}])_* \\ &= ([0, I, +, \dot{-}, \lfloor x/y \rfloor, x!, 2^x; \text{COMP}])_*. \end{aligned}$$

In [Con73], R. Constable defined the class $\mathcal{K}$ by

$$[0, I, s, +, \dot{-}, \times, \lfloor x/y \rfloor; \text{COMP}, \text{SBSUM}, \text{SBPROD}],$$

a polynomial analogue of the definition of Kalmár elementary functions. The class $\mathcal{K}(f)$ is defined as above, but with f as an additional initial function. Let $FP(f)$ denote the collection of functions polynomial time computable in f ($FP(f)$ can equivalently be defined as the set of type 1 functions in $BFF(f)$; see Definition 7.2.6).

On p. 118 of [Con73], the following claim is stated as a theorem without proof.

[19] Sharply bounded summation [resp. product] is called *weak sum* [resp. *product* in [WW86], and bounded summation [resp. product] is called *limited sum* [resp. *product* in [Grz53].

Claim 4. For all non-decreasing f, $K(f) = FP(f)$.

The statement $\mathcal{F}\text{PTIME} = \mathcal{K}$ was then claimed as a corollary in [Con73]. This statement was again cited as a theorem (without proof) in [WW86].

It now appears that this assertion is doubtful, since $\mathcal{K} \subseteq \text{NC}$ and it is currently conjectured that NC is properly contained in $\mathcal{F}$PTIME. Moreover, using an oracle separation of NC^A from P^A, P. Clote [Clo96] provided a counterexample to the previous claim.

Theorem 6.3.30 ([Clo96]). *There exists a non-decreasing function f for which $K(f) \neq FP(f)$.*

Nevertheless, Constable's class $\mathcal{K}$ is very natural, suggesting the following question.

Question 6.3.1. What complexity class corresponds to

$$[0, I, s, +, \dot{-}, \times, \lfloor x/y \rfloor; \text{COMP}, \text{SBSUM}, \text{SBPROD}]?$$

H.-J. Burtschick (personal correspondence) suggested that polynomial size uniform arithmetic circuits could be related to the class $\mathcal{K}$.

Somewhat related is the recent work on counting classes. The class $\#P$, introduced by Valiant [Val79], is the set of functions f, for which there exists a nondeterministic polynomial time bounded Turing machine M, such that $f(x)$ is the number of accepting paths in the computation tree of M on input x. Unless $P = NP$, it is unlikely that $\#P$ is closed under composition. Using the arithmetization of boolean formulas from A. Shamir (see [BF91]), H. Vollmer and K. Wagner gave the following characterization of $\#P$.[20]

Theorem 6.3.31 ([VW96]).

$$\begin{aligned}\#P &= [[0, I, S, +, \dot{-}, \times, \lfloor x/y \rfloor, \#; \text{COMP}, \text{SBPROD}]; \text{BSUM}] \\ &= [[0, I, S, +, \dot{-}, \times, \#; \text{COMP}]; \text{SBPROD}, \text{BSUM}].\end{aligned}$$

Definition 6.3.24. *Let $\mathcal{R}_k$ be the smallest class of functions definable from the constant functions $0, \ldots, k$, the projections I, the characteristic functions of the* graphs *of $+, \times, =$, and closed under composition and bounded recursion.*

The following result was proved using the Paris-Wilkie modification of Bel'tyukov's stack register machines.

Theorem 6.3.32 ([PW85]). $(\mathcal{R}_2)_* = (\mathcal{R}_3)_*$.

The next theorem follows from work of P. Clote in [Clo97a] and is based on Barrington's trick.

[20] Using the notation of [VW96], the characterization of the class $\#P$ reads $\#P = \left[[+, \dot{-}, \times, :]_{\text{Sub}}, \text{WProd}\right]_{\text{Sum}}$, and $\#P = [[+, \dot{-}, \times]_{\text{Sub}}]_{\text{WProd,Sum}}$. This formulation is equivalent to that given in Theorem 6.3.31.

Theorem 6.3.33 ([Clo97a]). *For* $n \geq 4$,

$$(\mathcal{R}_n)_* = (\mathcal{R}_{n+1})_* = \text{ALINTIME}.$$

In [Kut88], Kutyłowski considered oracle versions of the Paris-Wilkie work.

Definition 6.3.25 ([Kut88]). *f is a k-function*[21] *if for all* $x_1, \ldots, x_n$

$$f(x_1, \cdots, x_n) = f(\min(x_1, k), \cdots, \min(x_n, k)) \leq k.$$

*For a family $\mathcal{F}$ of functions, $\mathcal{W}_k(\mathcal{F})$ is the smallest class of functions containing I, $\mathcal{F}$, all k-functions and closed under composition and k-bounded recursion. The function f is defined from g, h by m-*counting *if*

$$f(0, \mathbf{x}) = g(\mathbf{x})$$
$$f(n+1, \mathbf{x}) = (f(n, \mathbf{x}) + h(n, \mathbf{x})) \ (\mathit{mod}\ m)$$

The class $\mathcal{CW}_k(\mathcal{F})$ is the smallest class of functions containing I, $\mathcal{F}$, all m-functions for $m \in \mathbf{N}$ and closed under composition, k-bounded recursion and arbitrary counting.

Theorem 6.3.34 ([Kut88]). *For every class $\mathcal{F}$ of functions, $\mathcal{W}_2(\mathcal{F})_* = \mathcal{W}_3(\mathcal{F})_*$. For every $k \geq 3$, there exists a family $\mathcal{F}$ of functions, such that $\mathcal{W}_k(\mathcal{F})_* \subset \mathcal{W}_{k+1}(\mathcal{F})_*$. For every $k \geq 3$, there is a family $\mathcal{F}$ of functions such that $\mathcal{CW}_k(\mathcal{F})_* \subset \mathcal{CW}_{k+1}(\mathcal{F}_*)$.*

Parallel algorithms often employ a divide and conquer strategy. B. Allen [All91] formalized this approach to characterize NC.

Definition 6.3.26. *The* front half $\text{FH}(x)$ *is defined by* $\text{MSP}(x, \lfloor |x|/2 \rfloor)$ *and the* back half $\text{BH}(x)$ *by* $\text{LSP}(x, \lfloor |x|/2 \rfloor)$. *The function f is defined by* polynomially bounded branching recursion *(*PBBR*) from functions g, h if there exists a polynomial p such that*

$$f(0, \mathbf{y}) = g_0(\mathbf{y})$$
$$f(1, \mathbf{y}) = g_1(vecy)$$
$$f(x, \mathbf{y}) = h(x, \mathbf{y}, f(\text{FH}(x), \mathbf{y}), f(\text{BH}(x), \mathbf{y})), \text{ if } x > 1$$

provided that $|f(x, \mathbf{y})| \leq p(\max(|x|, |y_i|))$ for all $x, \mathbf{y}$. Let $Seq(x) = 0$ if x encodes a sequence[22] *else 0. If x encodes a sequence $(x_1, \ldots, x_n)$ and f is a one-place function, then the operation* MAP *is defined by* $\text{MAP}(f, x) = \langle f(x_1), \ldots, f(x_n) \rangle$. *Define the* bounded shift left function *by* $\text{SHL}(x, i, y) = x \cdot 2^{\min(i, |y|)}$.

[21] What is here called a k-function is called a $k+1$-function in [Kut88]. As our definition of k-bounded recursion corresponds to Kutyłowski's definition of $k+1$-bounded recursion, the indices of $\mathcal{W}_k(\mathcal{F})$ and $\mathcal{CW}_k(\mathcal{F})$ differ by 1 from [Kut88].

[22] Here, we use the earlier defined sequence numbers, though Allen [All91] uses a different sequence encoding technique.

Theorem 6.3.35 ([All91]). NC *is characterized by the function algebra*

$$[0, I, s, +, \dot{-}, |x|, \text{BIT}, cond, c_{\leq}, Seq, \beta, \text{MSP}, \text{SHL}; \text{COMP}, \text{MAP}, \text{PBBR}].$$

Allen explicitly did not attempt to find the smallest set of initial functions, but went on to develop a proof theory for NC functions, similar in spirit to that of S. Buss [Bus86a]. Independently and at the same time, an equivalent theory of arithmetic for NC functions was given by P. Clote [Clo89], appearing in Clote–Takeuti [CT92].

In [Pit98] F. Pitt considered a variant of B. Allen's polynomial bounded branching recursion, where the function value is bounded by a constant.

Definition 6.3.27 (F. Pitt [Pit98]).
*The function f is defined by k-*bounded tree recursion on notation *(k-BTRN) from functions g, h if*

$$f(0, \mathbf{y}) = g_0(\mathbf{y})$$
$$f(1, \mathbf{y}) = g_1(\mathbf{y})$$
$$f(x, \mathbf{y}) = h(x, \mathbf{y}, f(\text{FH}(x), \mathbf{y}), f(\text{BH}(x), \mathbf{y})), \text{ if } x > 1$$

provided that $f(x, \mathbf{y}) \leq k$, for all $x, \mathbf{y}$. When k is unspecified, the scheme k-BTRN is meant to allow all constants $k \in \mathbf{N}$.

Theorem 6.3.36 ([Pit98]).

$$\mathcal{F}\text{ALOGTIME} = [0, I, s_0, s_1, |x|, \#, \text{MSP}, \text{LSP}; \text{COMP}, \text{CRN}, k - \text{BTRN}].$$

The theorem is proved by showing that FH, BH can be defined from the initial functions, and then by defining the function TREE in the above function algebra, where TREE is a function evaluating a full binary tree with alternating levels of AND's and OR's, and whose leaves are the bits of a given input x (see [Clo90, Clo92a] for details of definition). In [Clo90], P. Clote characterized ALOGTIME as $[0, I, s_0, s_1, |x|, \text{BIT}, \#, \text{TREE}; \text{COMP}, \text{CRN}]$, so the proof sketch is complete.

It is often useful to define two or more functions simultaneously. Simultaneous versions of recursion, recursion on notation, k-bounded recursion on notation, etc. are defined in the obvious manner. For example, simultaneous recursion and simultaneous recursion on notation are defined as follows.

Definition 6.3.28. *The functions $f_1, \ldots, f_n$ are defined from the functions $g_1, \ldots, g_n$, $h_1, \ldots, h_n$ by* simultaneous recursion *if*

$$f_i(0, \mathbf{y}) = g_i(\mathbf{y}), \text{ for } 1 \leq i \leq n$$
$$f_i(x+1, \mathbf{y}) = h_i(x, \mathbf{y}, f_1(x, \mathbf{y}), \ldots, f_n(x, \mathbf{y})), \text{ for } 1 \leq i \leq n.$$

If additionally $f_i(x, \mathbf{y}) \leq k_i(x, \mathbf{y})$ for $1 \leq i \leq n$, then the f_i are defined by simultaneous bounded recursion *from* $\mathbf{g}$, $\mathbf{h}$, $\mathbf{k}$.

The functions $f_1, \ldots, f_n$ are defined from functions $g_1, \ldots, g_n$, $h_1^0, \ldots, h_n^0$ and $h_1^1, \ldots, h_n^1$ by simultaneous recursion on notation *if for $1 \leq i \leq n$*

$$f_i(0, \mathbf{y}) = g_i(\mathbf{y})$$
$$f_i(s_0(x), \mathbf{y}) = h_i^0(x, \mathbf{y}, f_1(x, \mathbf{y}), \ldots, f_n(x, \mathbf{y})), \textit{provided } x \neq 0$$
$$f_i(s_1(x), \mathbf{y}) = h_i^1(x, \mathbf{y}, f_1(x, \mathbf{y}), \ldots, f_n(x, \mathbf{y})).$$

If additionally $f_i(x, \mathbf{y}) \leq k_i(x, \mathbf{y})$ *for* $1 \leq i \leq n$, *then the* f_i *are defined by* simultaneous bounded recursion on notation *from* $\mathbf{g}$, $\mathbf{h^0}$, $\mathbf{h^1}$, $\mathbf{k}$.

A function algebra $\mathcal{F}$, whose primary closure operation is a certain form of recursion, can often be proved to be closed under the simultaneous version of that form of recursion, by using the pairing function τ and its projections π_1, π_2. For instance, the following is straightforward to establish.

Proposition 6.3.2 ([KC96]). *The Cobham algebra* $[0, I, s_0, s_1, \#; \text{COMP}, \text{BRN}]$ *is closed under simultaneous bounded recursion on notation.*

Proof. For notational simplicity, suppose that $n = 2$. Define

$$\begin{aligned} f(0, \mathbf{y}) &= \quad \tau(g_1(\mathbf{y}), g_2(\mathbf{y})) \\ f(s_i(x), \mathbf{y}) &= \tau(h_1^i(x, \mathbf{y}, \pi_1(f(x, \mathbf{y})), \pi_2(f(x, \mathbf{y}))), \\ &\qquad h_2^i(x, \mathbf{y}, \pi_1(f(x, \mathbf{y})), \pi_2(f(x, \mathbf{y})))) \end{aligned}$$

where $f(x, \mathbf{y}) \leq \tau(k_1(x, \mathbf{y}), k_2(x, \mathbf{y}))$. Then $f_1(x, \mathbf{y})$ is $\pi_1(f(x, \mathbf{y}))$ and $f_2(x, \mathbf{y})$ is $\pi_2(f(x, \mathbf{y}))$.

The Fibonacci sequence $1, 1, 2, 3, 5, 8, \ldots$ is defined by $Fib(0) = Fib(1) = 1$, and $Fib(n + 2) = Fib(n) + Fib(n + 1)$. This is a special case of course-of-values recursion.

Definition 6.3.29. *The function* f *is defined from functions* g, h *by* course-of-values recursion *(*VR*) if*

$$f(0, \mathbf{y}) = g(\mathbf{y})$$
$$f(x + 1, \mathbf{y}) = h(x, \mathbf{y}, \langle f(0, \mathbf{y}), \ldots, f(x, \mathbf{y})\rangle).$$

The class $\mathcal{PR}$ of primitive recursive functions is easily seen to be closed under VR.

Definition 6.3.30. *The function* f *is defined from functions* g, h, r, k *by* bounded 2-value recursion *(*BVR*) if*

$$f(0, \mathbf{y}) = g(\mathbf{y})$$
$$f(x + 1, \mathbf{y}) = h(x, \mathbf{y}, f(x, \mathbf{y}), f(r(x, \mathbf{y}), \mathbf{y}))$$

provided that $f(x, \mathbf{y}) \leq k(x, \mathbf{y})$ *and* $r(x, \mathbf{y}) < x$ *for all* $x, \mathbf{y}$.

Theorem 6.3.37 ([Mon77]). *Let* $f_2(x, y) = (x + 1) \cdot (y + 1)$. *Then*

$$\{f \in \text{ETIME} : f \text{ has linear growth rate}\} = [0, I, s, f_2; \text{COMP}, \text{BVR}].$$

Proof. Our exposition follows [WW86]. Temporarily, let $\mathcal{F}$ denote $\{f \in$ ETIME : f has linear growth rate$\}$ and $\mathcal{G}$ denote $[0, I, s, f_2; \text{COMP}, \text{BVR}]$. Consider first the inclusion $\mathcal{F} \subseteq \mathcal{G}$. Suppose that M is a TM which computes the bitgraph B_f of $f : \mathbf{N}^k \to \mathbf{N}$ in time $2^{c \cdot n}$. For notational simplicity, suppose $k = 1$ and $|f(x)| \leq d \cdot |x|$.
CLAIM. $B_f \in \mathcal{G}$.
Since $\sum_{i<2^{c \cdot n}} i \leq 2^{2c \cdot n}$, without loss of generality, M's head movements before halting may be assumed to be of the form (see Figure 6.3). For notational

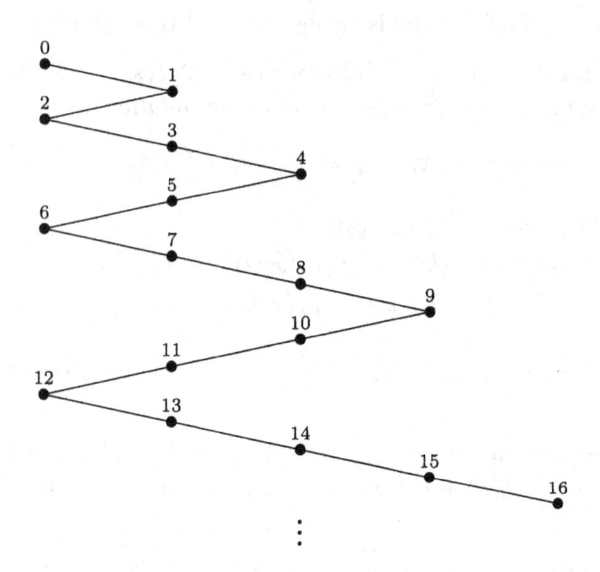

Fig. 6.3. M's head movements before halting

simplicity, assume that M has only one tape, and that the transition function

$$\delta : (Q - \{q_A, q_R\}) \times (\Sigma \cup \{B\}) \to Q \times (\Sigma \cup \{B\}) \times \{-1, 0, 1\}$$

satisfies $\delta(q, \sigma) = (state(q, \sigma), symbol(q, \sigma), direction(q, \sigma)$ for suitable functions *state*, *symbol*, *direction*. Let $h(t)$ be M's head position at the *beginning* of step t; let $s(t, x)$ be the state of M at the *completion* of step t on input x; let $a(t, x)$ be the symbol written by M on cell $h(t)$ during step t. Let

$$p_0(t, t') = \begin{cases} \max\{t'' : t'' \leq t' \wedge h(t'') = h(t)\} & \text{if such exists} \\ t' + 1 & \text{else} \end{cases}$$

and $p(t) = p_0(t, t \dot{-} 1)$. Let $sqrt(x) = \lfloor \sqrt{x} \rfloor$. Note that

$$h(t) = \begin{cases} sqrt(x) + sqrt(x)^2 \dot{-} x \text{ if } x \leq sqrt(x)^2 + sqrt(x) \\ x \dot{-} sqrt(x) \dot{-} sqrt(x)^2 \text{ else.} \end{cases}$$

Using BMIN, $sqrt$ is definable by $sqrt(x) = \mu y \leq x[x < (y+1)^2]$ and so $p, h \in \mathcal{M}^2 \subseteq \mathcal{G}$. Define the functions $s(t,x)$ and $a(t,x)$ by

$$s(0,x) = state(q_0, B)$$

$$s(t+1,x) = \begin{cases} state(s(t,x), a(p(t+1),x)) & \text{if } p(t+1) \leq t \\ state(s(t,x), \text{BIT}(|x| \dot{-} h(t+1), x)) & \text{else and} \\ & 1 \leq h(t+1) \leq |x| \\ state(s(t,x), B) & \text{else} \end{cases}$$

$$a(0,x) = symbol(q_0, B)$$

$$a(t+1,x) = \begin{cases} symbol(s(t,x), a(p(t+1),x)) & \text{if } p(t+1) \leq t \\ symbol(s(t,x), \text{BIT}(|x| \dot{-} h(t+1), x)) & \text{else and } 1 \leq \\ & h(t+1) \leq |x| \\ symbol(s(t,x), B) & \text{else.} \end{cases}$$

Instead of defining the functions $s(t,x)$, $a(t,x)$ by simultaneous bounded recursion, define $F(t,x) = \tau(s(t,x), a(t,x))$ by bounded 2-value recursion in the obvious manner. Since $\tau, \pi_1, \pi_2 \in \mathcal{M}^2 \subseteq \mathcal{G}$, it is now routine to complete the proof of the claim that $B_f \in \mathcal{G}$.

Define

$$f(x) = \mu y \leq (x+1)^{d+1} \cdot 2^d [(\forall i \leq d \cdot |x|)((x,i) \in B_f \leftrightarrow \text{BIT}(i,y) = 1)].$$

Since BR is included in BVR, by the proof of Theorem 6.3.14, BIT $\in \mathcal{G}$. By Corollary 6.3.1, $\mathcal{G}$ is closed under bounded quantification and bounded minimization, so it follows that $f \in \mathcal{G}$.

Consider now the inclusion $\mathcal{G} \subseteq \mathcal{F}$. By induction, all functions of $\mathcal{G}$ are of linear growth rate. The functions $0, I^n_k, s, f_2$ are computable in exponential time, and because functions of $\mathcal{F}$ are of linear growth rate, $\mathcal{F}$ is closed under composition. If f is defined by BVR from g, h, r, k, then when computing $f(x+1, \mathbf{y})$, an exponential time bounded machine M has sufficient space to store the entire sequence $f(0,\mathbf{y}), \ldots, f(x,\mathbf{y})$ of previous values on a work tape. It follows that any function of the algebra $\mathcal{G}$ is computable in exponential time.

A more powerful version of simultaneous recursion was introduced in [KC96].

Definition 6.3.31. *The functions $f_1, \ldots, f_n$ are defined from the functions $g_1, \ldots, g_n$, $h^0_1, \ldots, h^0_n$, $h^1_1, \ldots, h^1_n$ and $k_1, \ldots, k_n$ by* multiple bounded recursion on notation *if the f_i are defined by simultaneous recursion on notation from $\mathbf{g_i}$, $\mathbf{h^0_i}$, $\mathbf{h^1_i}$ and moreover*

$$f_1(x,\mathbf{y}) \leq k_1(x,\mathbf{y})$$
$$f_i(x,\mathbf{y}) \leq k_i(x,\mathbf{y}, f_1(x,\mathbf{y}), \ldots, f_{i-1}(x,\mathbf{y})), \text{ for } 2 \leq i \leq n.$$

The following non-trivial closure property has an important application in the Kapron–Cook characterization of type 2 polynomial time computations described in the next section.

Theorem 6.3.38 ([KC96]). *The Cobham algebra*

$$[0, I, s_0, s_1, \#; \text{COMP}, \text{BRN}]$$

is closed under multiple bounded recursion on notation.

6.3.6 Safe Recursion

All the function algebras from the previous subsection are defined from specific initial functions, using some version of bounded recursion. Without any bound, even schemes such as WBRN can generate all the primitive recursive functions. Recently, certain *unbounded* recursion schemes have been introduced which distinguish between variables as to their position in a function $f(x_1, \ldots, x_n; y_1, \ldots, y_m)$. Variables x_i occurring to the left of the semi-colon are called *normal*, while variables y_j to the right are called *safe*. By allowing only recursions of a certain form, which distinguish between normal and safe variables, particular complexity classes can be characterized. *Normal* values are considered as known in totality, while *safe* values are those obtained by impredicative means (i.e., via recursion). Sometimes, to help distinguish normal from safe positions, the letters $u, v, w, x, y, z, \ldots$ denote normal variables, while $a, b, c, \ldots$ denote safe variables. This terminology, due to Bellantoni–Cook [BC92], was chosen to indicate that a *safe* position is one where it is safe to substitute an impredicative value. Related *tiering* notions, though technically different, have occurred in the literature, as in Clote–Takeuti [CT86] (k sorted variables used in defining k-fold multiple exponential time), but most especially in H. Simmons [Sim88] (*control* variables, i.e., those used for recursion, are distinguished from usual variables; by separating their function, one prevents diagonalization as in the Ackermann function) and in D. Leivant [Lei89, Lei91, Lei93, Lei94] (stratified polymorphism, second order system $L_2(QF^+)$ corresponding to polynomial time computable functions, stratified functional programs, ramified recurrence over 2 tiered word algebras corresponding to polynomial time). Of these, [Sim88] and [Lei91] are the most related to the Bellantoni–Cook work described below.

If $\mathcal{F}$ and $\mathcal{O}$ are collections of initial functions and operations which distinguish normal and safe variables, then $\text{NORMAL} \cap [\mathcal{F}; \mathcal{O}]$ denotes the collection of all functions $f(\mathbf{x};) \in [\mathcal{F}; \mathcal{O}]$ which have only normal variables. Similarly, $(\text{NORMAL} \cap [\mathcal{F}; \mathcal{O}])_*$ denotes the collection of predicates whose characteristic function $f(\mathbf{x};)$ has only normal variables and belongs to $[\mathcal{F}; \mathcal{O}]$.

Define the following initial functions by

0-ary constant	0
projections	$I_j^{n,m}(x_1,\ldots,x_n;a_1,\ldots,a_m) = \begin{cases} x_j & \text{if } 1 \le j \le n \\ a_{j-n} & \text{if } n < j \le n+m \end{cases}$
successors	$S_0(;a) = 2 \cdot a,\ S_1(;a) = 2 \cdot a + 1$
binary predecessor	$P(;a) = \lfloor a/2 \rfloor$
conditiona)	$C(;a,b,c) = \begin{cases} b & \text{if } a \bmod 2 = 0 \\ c & \text{else.} \end{cases}$

Definition 6.3.32 ([BC92]). *The function f is defined by* safe composition *(SCOMP) from $g, u_1, \ldots, u_n, v_1, \ldots, v_m$ if*

$$f(\mathbf{x};\mathbf{a}) = g(u_1(\mathbf{x};),\ldots,u_n(\mathbf{x};);v_1(\mathbf{x};\mathbf{a}),\ldots,v_m(\mathbf{x};\mathbf{a})).$$

If $h(x;y)$ is defined, then SCOMP allows one to define

$$f(x,y;) = h(I_1^{2,0}(x,y;);I_2^{2,0}(x,y;)) = h(x;y).$$

However, one *cannot* similarly define $g(;x,y) = h(x;y)$.

Definition 6.3.33. *The function f is defined by* safe recursion on notation[23] *(SRN) from the functions g, h_0, h_1 if*

$$f(0,\mathbf{y};\mathbf{a}) = g(\mathbf{y};\mathbf{a})$$
$$f(s_0(x),\mathbf{y};\mathbf{a}) = h_0(x,\mathbf{y};\mathbf{a},f(x,\mathbf{y};\mathbf{a})),\ \textit{provided } x \neq 0$$
$$f(s_1(x),\mathbf{y};\mathbf{a}) = h_1(x,\mathbf{y};\mathbf{a},f(x,\mathbf{y};\mathbf{a})).$$

The function algebra B is defined by

$$[0, I, S_0, S_1, P, C; \text{SCOMP}, \text{SRN}].$$

Theorem 6.3.39 ([BC92]). *The polynomial time computable functions are exactly those functions of B having only normal arguments, i.e.,*

$$\mathcal{F}\text{PTIME} = \text{NORMAL} \cap B.$$

The difficult direction of the proof is the inclusion from left to right. By Theorem 6.3.6 of Cobham, PTIME functions are those in the algebra

$$[0, I, s_0, s_1, \#; \text{COMP}, \text{BRN}].$$

To see the difficulties involved, suppose that f is defined by BRN from g, h_0, h_1 and that $g(\mathbf{y}) = g'(\mathbf{y};)$, $h_0(x,\mathbf{y},z) = h_0'(x,\mathbf{y},z;)$ and $h_1(x,\mathbf{y},z) = h_1'(x,\mathbf{y},z;)$. In trying to define f' by recursion on notation, one has $f'(0,\mathbf{y};) = g'(\mathbf{y};)$ and $f'(s_i(x),\mathbf{y};) = h_i'(x,\mathbf{y},f'(x,\mathbf{y};);)$. However, this violates the requirement of SRN that the function value $f'(x,\mathbf{y};)$ be in a safe position in h_i'. For this reason a different approach is necessary.

[23] In [BC92] this scheme is called *predicative notational recursion.*

Lemma 6.3.8. *If $f \in \mathcal{F}\text{PTIME}$ then there exist $f' \in B$ and a monotone increasing polynomial p_f such that $f(\mathbf{x}) = f'(w; \mathbf{x})$ for all $|w| \geq p_f(|\mathbf{x}|)$.*

Proof. Temporarily, let's say that a function f is defined by polynomially bounded recursion on notation (PBRN) from g, h_0, h_1 if f is defined by recursion on notation from these functions, and additionally there exists a polynomial p satisfying $|f(x, \mathbf{y})| \leq p(|x|, |\mathbf{y}|)$ for all $x, \mathbf{y}$. Since *pad*, # are easily defined by PBRN [for instance, $0\#y = 1$, $s_i(x)\#y = pad(y, x\#y)$ where $|x\#y| \leq |x| \cdot |y| + 1$] it follows from Theorem 6.3.6 that $\mathcal{F}\text{PTIME} = [0, I, s_0, s_1; \text{COMP}, \text{PBRN}]$. The lemma is now proved by induction on the construction of f in the latter algebra.

If f is $0, I^n_k, s_0, s_1$ then we may take f' to be the corresponding initial function of B and p_f to be 0. Suppose that $f(\mathbf{x}) = g(h_1(\mathbf{x}), \ldots, h_n(\mathbf{x}))$ is defined by composition, where by the induction hypothesis

$$g(y_1, \ldots, y_n) = g'(w; y_1, \ldots, y_n) \text{ for } |w| \geq p_g(|y_1|, \ldots, |y_n|) \tag{6.37}$$

$$h_i(\mathbf{x}) = h'_i(w; \mathbf{x}) \text{ for } |w| \geq p_{h_i}(|\mathbf{x}|). \tag{6.38}$$

Define

$$f'(w; \mathbf{x}) = g'(w; h'_1(w; \mathbf{x}), \ldots, h'_n(w; \mathbf{x})) \tag{6.39}$$

$$p_f(|\mathbf{x}|) = p_g(p_{h_1}(|\mathbf{x}|), \ldots, p_{h_n}(|\mathbf{x}|)) + \sum_{i=1}^{n} p_{h_i}(|\mathbf{x}|). \tag{6.40}$$

It follows that $f(\mathbf{x}) = f'(w; \mathbf{x})$ for all $|w| \geq p_f(|\mathbf{x}|)$.

Suppose that f is defined from g, h_0, h_1 by PBRN as follows

$$f(0, \mathbf{y}) = g(\mathbf{y})$$
$$f(s_i(x), \mathbf{y}) = h_i(x, \mathbf{y}, f(x, \mathbf{y}))$$

where $|f(x, \mathbf{y})| \leq q(|x|, |\mathbf{y}|)$ for some polynomial q. By the induction hypothesis then there exist $g', h'_0, h'_1, p_g, p_{h_0}, p_h$ satisfying

$$g(\mathbf{y}) = g'(w; \mathbf{y}) \quad \text{for } |w| \geq p_g(|\mathbf{y}|)$$
$$h_i(x, \mathbf{y}, z) = h'_i(w; x, \mathbf{y}, z) \quad \text{for } |w| \geq p_{h_i}(|x|, |\mathbf{y}|, |z|).$$

Let $E(z, w; x)$ be the initial segment of x obtained by removing from x the $|w| \dot{-} |z|$ lowest order bits. By SRN define F by

$$F(0, w; x, \mathbf{y}) = 0$$

$$F(s_i(z), w; x, \mathbf{y}) = \begin{cases} g'(w; \mathbf{y}) & \text{if Case 1} \\ h'_0(w; E(z, w; x), \mathbf{y}, F(z, w; x, \mathbf{y})) & \text{if Case 2} \\ h'_1(w; E(z, w; x), \mathbf{y}, F(z, w; x, \mathbf{y})) & \text{if Case 3} \\ F(z, w; x, \mathbf{y}) & \text{otherwise} \end{cases} \tag{6.41}$$

where

- Case 1 holds if $|w| - |x| = |s_i(z)| \leq |x|$,

- Case 2 holds if $|w|-|x| < |s_i(z)| \leq |x|$ and the low order bit of $E(s_i(z), w; x)$ is 0,
- Case 3 holds if $|w|-|x| < |s_i(z)| \leq |x|$ and the low order bit of $E(s_i(z), w; x)$ is 1.

To see that $F \in B$, introduce the following functions. The low order bit $M(;a) = a \bmod 2$ is defined by $M(;a) = C(;a, 0, S_1(0))$. The truncation function $T(x;a) = \lfloor a/2^{|x|} \rfloor$ is defined by

$$T(0;a) = a$$
$$T(s_i(x);a) = P(;T(x;a)).$$

Let $T'(x,y;) = T(x,y;) = \lfloor y/2^{|x|} \rfloor$, and define the extraction operator

$$E(x, w; a) = T(T'(x, w;); a) = \lfloor a/2^{|w| \dot{-} |x|} \rfloor$$

so that $E(x, w; a)$ is the initial segment of a produced by removing from a the $|w| \dot{-} |x|$ lowest order bits. Define the bitwise OR function by

$$\vee(0;a) = M(;a)$$
$$\vee(s_i(x);a) = C(;\vee(x;a), M(;T(s_i(x);a)), 1).$$

Note that for $|w| - |x| \leq |y| \leq |w|$ it follows that $|w| - |x| = |y|$ if and only if $\vee(w; E(y, w; x)) = 0$ and $|w| - |x| < |y|$ iff $\vee(w; E(y, w; x)) = 1$. It is now straightforward, to give a more formal definition of F placing it in B. Now set

$$f'(w; x, \mathbf{y}) = F(w, w; x, \mathbf{y}) \tag{6.42}$$

and

$$p_f(|x|, |\mathbf{y}|) = \tag{6.43}$$
$$p_{h_0}(|x|, |\mathbf{y}|, q(|x|, |\mathbf{y}|)) + p_h(|x|, |\mathbf{y}|, q(|x|, |\mathbf{y}|)) + p_g(|\mathbf{y}|) + |x| + 1 \tag{6.44}$$

Claim 5. If u satisfies $|w| - |x| \leq |u| \leq |w|$ and $|w| \geq p_f(|x|, |\mathbf{y}|)$ then $F(u, w; x, \mathbf{y}) = f(E(u, w; x), \mathbf{y})$.

Proof. Fix w satisfying $|w| \geq p_f(|x|, |\mathbf{y}|)$. Proceed by induction on $|u|$. First suppose that $|u| = |w| - |x|$. Then $E(u, w; x) = \lfloor x/2^{|w|-|u|} \rfloor = \lfloor x/2^{|x|} \rfloor = 0$. By (6.43), $|w| \geq |x| + 1$, so $|u| \geq 1$ and Case 1 applies. It then follows that

$$F(u, w; x, \mathbf{y}) = g'(w; \mathbf{y}) = f(E(u, w; x), \mathbf{y}).$$

Now suppose that $|w| - |x| < |u| \leq |w|$, and that $u = s_0(z)$ or $u = s_1(z)$. In the definition of $F(s_i(z), w; x, \mathbf{y})$ only case 2 or case 3 can occur.

Case 1. The $(|x| + |u| - |w|)$-th bit of x from the left is 0, or equivalently the $(|w| - |z| - 1)$-st bit of x from the right is 0. Then

$$\begin{aligned}
F(s_i(z), w; x, \mathbf{y}) &= h_0'(w; E(z,w;x), \mathbf{y}, F(z,w;x,\mathbf{y}))\\
&= h_0'(w; E(z,w;x), \mathbf{y}, f(E(z,w;x), \mathbf{y}))\\
&\quad \text{by induction hypothesis}\\
&= h_0(E(z,w;x), \mathbf{y}, f(E(z,w;x), \mathbf{y}))\\
&\quad \text{by justification below}\\
&= f(s_0(E(z,w;x)), \mathbf{y})\\
&\quad \text{by definition of } f, \text{ if } E(z,w;x) \neq 0\\
&= f(E(s_i(z), w; x), \mathbf{y})
\end{aligned}$$

The last line follows, because in case 1, the low order bit of $E(s_i(z), w; x)$ is 0, so by the definition of E, $E(s_i(z), w; x) = s_0(E(z, w; x))$. The justification for the second line in the above equations is given as follows.

$$\begin{aligned}
|w| &\geq p_{h_0}(|x|, |\mathbf{y}|, q(|x|, |\mathbf{y}|))\\
&\geq p_{h_0}(|E(z,w;x)|, |\mathbf{y}|, q(|E(z,w;x)|, |\mathbf{y}|)) && \text{as } |E(z,w;x)| \leq |x| \text{ and}\\
& && q \text{ is monotonic}\\
&\geq p_{h_0}(|E(z,w;x)|, |\mathbf{y}|, |f(E(z,w;x), \mathbf{y})|) && \text{as } q \text{ bounds length of } f\\
&\geq p_{h_0}(|E(z,w;x)|, |\mathbf{y}|, |F(z,w;x,\mathbf{y})|) && \text{induction hypothesis}\\
&\text{of claim.}
\end{aligned}$$

Case 2. The $(|w| - |z| - 1)$-st bit of x from the left is 1.
This case in handled similarly to that of case 1, and so the proof of the claim is complete.

By definition of E, it is clear that for all $|u| = |v|$ we have $E(u, w; x) = E(v, w; x)$. Using this, an easy induction on notation yields that for $|u| \geq |x|$ and $|w| \geq p_f(|x|, |\mathbf{y}|)$

$$F(u, w; x, \mathbf{y}) = F(x, w; x, \mathbf{y}).$$

For $|w| \geq p_f(|x|, |\mathbf{y}|)$ then

$$\begin{aligned}
f'(w; x, \mathbf{y}) &= F(w, w; x, \mathbf{y}) && \text{by definition}\\
&= F(x, w; x, \mathbf{y}) && \text{as } |w| \geq |x| + 1\\
&= f(E(x, w; x), \mathbf{y}) && \text{by Claim 5}\\
&= f(x, \mathbf{y}) && \text{by definition of } E.
\end{aligned}$$

This completes the proof of the lemma.

To show all functions of $\mathcal{F}$PTIME are functions of B containing only normal arguments, appropriate bounding functions in B must be defined.

Theorem 6.3.40. *If $f \in \mathcal{F}$PTIME then $f(\mathbf{x};) \in B$.*

Proof. Since f is polynomially bounded, let m, c be such that

$$|f(x_1,\ldots,x_n)| \le (\sum_{i=1}^{n} |x_i|)^m + c.$$

Define

$$\begin{aligned}
&\mathrm{pad}^2(0;y) = y\\
&\mathrm{pad}^2(s_i(x);y) = S_1(;\mathrm{pad}^2(x;y))\\
&\mathrm{pad}^{k+1}(x_1,\ldots,x_k;x_{k+1}) = \mathrm{pad}^2(x_1;\mathrm{pad}^k(x_2,\ldots,x_k;x_{k+1})).
\end{aligned}$$

Define

$$\begin{aligned}
&\mathrm{smash}(0,x;) = 1\\
&\mathrm{smash}(s_i(y),x;) = \mathrm{pad}^2(x;\mathrm{smash}(y,x;))
\end{aligned}$$

Then $\mathrm{smash}(y,x;) = 2^{|x|\cdot|y|}$. Let $b_0(x;)$ be obtained by composing $\mathrm{smash}(x,x;)$ with itself so as to satisfy $|b_0(x;)| \ge |x|^m + c$ and define

$$b(x_1,\ldots,x_n;) = b_0(\mathrm{pad}^{n+1}(x_1,\ldots,x_n;1);).$$

Then $|f(x_1,\ldots,x_n)| \le |b(x_1,\ldots,x_n;)|$ and so for f' given by Lemma 6.3.8, define F by $F(\mathbf{x};) = f'(b(\mathbf{x};);\mathbf{x})$. Then $F \in B$ and $f(\mathbf{x}) = F(\mathbf{x};)$.

For the reverse inclusion, the following bounding lemma is proved by induction on the construction of f in B.

Lemma 6.3.9. *Let f belong to B. There is a monotone increasing polynomial q_f such that $|f(\mathbf{x};\mathbf{y})| \le q_f(|\mathbf{x}|) + \max_i |y_i|$ for all $\mathbf{x}, \mathbf{y}$.*

Theorem 6.3.41 ([BC92]). *If $f(\mathbf{x};\mathbf{y}) \in B$ then there is $f'(\mathbf{x},\mathbf{y}) \in \mathcal{F}\mathrm{PTIME}$ such that $f(\mathbf{x};\mathbf{y}) = f'(\mathbf{x},\mathbf{y})$ for all $\mathbf{x}, \mathbf{y}$.*

Proof. By induction on the construction of f in B. The case for initial functions and composition is straightforward. For monotonic bounding polynomial q_f as given by the preceding lemma, there is a function

$$g \in [0, I, s_0, s_1, \#; \mathrm{COMP}, \mathrm{BRN}]$$

satisfying $q_f(|\mathbf{x}|, |\mathbf{y}|) \le |g(\mathbf{x},\mathbf{y})|$. Thus SRN may be simulated using BRN.

Corollary 6.3.6. $\mathrm{PTIME} = (\mathrm{NORMAL} \cap [0, I, S_0, S_1, P, C; \mathrm{SCOMP}, \mathrm{SRN}])_*$.

This approach has led to other characterizations of familiar complexity classes using *safe* variants of unbounded recursion schemes.

Theorem 6.3.42 ([Bel92]). *Let $f(\mathbf{x})$ be a function satisfying $|f(\mathbf{x})| = O(\log |x|)$. Then $f(\mathbf{x})$ is computable by a logspace Turing machine iff*

$$f(\mathbf{x};) \in [0, I, S_1, P, C; \mathrm{SCOMP}, \mathrm{SRN}].$$

Note that the function S_0 does not belong to the above algebra, so the intuition is that for *small* functions (of logarithmic growth rate), LOGSPACE computations are arithmetized by using unary numerals on the work tape along with the same closure operators as for polynomial time. Bellantoni first proves his result for the operation of *simultaneous* safe recursion on notation, and then simulates this simultaneous scheme by SRN, a non-trivial task since the usual pairing function uses S_0. The previous theorem yields a nice characterization of LOGSPACE, to be compared with Corollary 6.3.6.

Corollary 6.3.7. $\text{LOGSPACE} = (\text{NORMAL} \cap [0, I, S_1, P, C; \text{SCOMP}, \text{SRN}])_*$.

Definition 6.3.34. *The function f is defined by* safe minimization *(SMIN) from the function g, denoted $f(\mathbf{x}; \mathbf{b}) = s_1(\mu a[g(\mathbf{x}; a, \mathbf{b}) \bmod 2 = 0)])$, if*

$$f(\mathbf{x}; \mathbf{b}) = \begin{cases} s_1(\min\{a : g(\mathbf{x}; a, \mathbf{b}) = 0\}), & \textit{if such exists,} \\ 0 & \textit{else.} \end{cases}$$

The algebra $\mu B = [0, I, S_0, S_1, P, C; \text{SCOMP}, \text{SRN}, \text{SMIN}]$. Let μB_i denote the set of functions derivable in μB using at most i applications of safe minimization.

Theorem 6.3.43 ([Bel94]). $\square_i^P = \{f(\mathbf{x};) : f \in \mu B_i\}$.

Definition 6.3.35 ([Bel92]). *The function f is defined by* safe recursion[24] *(SR) from the functions g, h if*

$$f(0, \mathbf{y}; \mathbf{a}) = g(\mathbf{y}; \mathbf{a})$$
$$f(x + 1, \mathbf{y}; \mathbf{a}) = h(x, \mathbf{y}; \mathbf{a}, f(x, \mathbf{y}; \mathbf{a})).$$

Define the following initial functions by

$$\textit{(successor)} \qquad S(; a) = a + 1 \tag{6.45}$$
$$\textit{(predecessor)} \qquad Pr(; a) = a \dot{-} 1 \tag{6.46}$$
$$\textit{(conditional)}\ K(; a, b, c) = \begin{cases} b & \textit{if } a = 0 \\ c & \textit{else.} \end{cases} \tag{6.47}$$

Recall that $\mathcal{E}^2$, the second level of the Grzegorczyk hierarchy, is the collection of linear space computable functions.

Theorem 6.3.44 ([Bel92]).
$\mathcal{E}^2 = \text{NORMAL} \cap [0, I, S, Pr, K; \text{SCOMP}, \text{SR}]$.

W. Handley (unpublished) and D. Leivant (unpublished) both independently obtained Theorem 6.3.44. Building on Bellantoni's proof, in her work on linear

[24] In [Bel92] this scheme is called *predicative primitive recursion*.

space reasoning, A.P. Nguyen [Ngu96] gave a slightly different characterization of this class. In [Clo97b] the P. Clote gave a *safe* characterization of ETIME functions of linear growth, by adapting the proof of Theorem 6.3.37.

In [Lei94] D. Leivant gave an alternative formulation of the safe characterizations of polynomial time and of linear space, by introducing a *tiering* notion to arbitrary word algebras. The idea is that one admits various copies or *tiers* W_0, W_1, ... of the word algebra W (generated from 0 by s_0, s_1),[25] and defines *ramified recurrence* by

$$f(s_i(x), \mathbf{y}) = h_i(f(x, \mathbf{y}), x, \mathbf{y})$$

where the tier of the first argument $s_i(x)$ is larger than the tier of the value $f(x, \mathbf{y})$. Comparing with the Bellantoni–Cook notation, tier 0 is safe, whereas tier 1 is normal. Leivant then shows that f is computable by a register machine over algebra A in time polynomial in the length of the inputs iff f is definable by explicit definition (corresponding essentially to safe composition) and ramified recurrence over A. This yields that f is polynomial time computable iff f is definable by explicit definition and ramified recurrence over W_0, W_1, whereas f is linear space computable (i.e., in $\mathcal{E}^2$) iff f is definable by explicit definition and ramified recurrence over $\mathbf{N}$, the unary algebra defined from 0, S.

These characterizations of complexity classes in terms of safe operations suggests the following problem.

Problem 6.3.1. Characterize the classes $\mathcal{M}^n$, for $n = 0, 1, 2$ and for each $n \geq 0$, the Grzegorczyk class $\mathcal{E}^n$ via appropriate initial functions, and safe operations. In particular, can one characterize $\mathcal{M}^2$ by

$$[0, I, S, Pr, K; \text{SCOMP}, \text{SMIN}]?$$

(Note that the conditional function $cond \in \mathcal{E}^1 - \mathcal{E}^0$.)

Turning to parallel computation, by building on Theorem 6.3.11, S. Bellantoni [Bel92] characterizes NC as those functions with normal variables in an algebra built up from 0, I, S_0, S_1, the conditional C, the bit function BIT, the length function $L(;a) = |a|$, a variant $\#'$ of the smash function, and closed under safe composition, concatenation recursion on notation and a *safe* version of weak bounded recursion on notation. Define the *half* function by $H(x) = \lfloor x/(2^{\lceil |x|/2 \rceil}) \rfloor$, and note that the least number of times which H can be iterated on x before reaching 0 is $||x||$. The function f is defined by *safe weak recursion on notation* (SWRN) [26] from the functions g, h if

$$f(0, \mathbf{y}; \mathbf{a}) = g(\mathbf{y}; \mathbf{a})$$
$$f(x, \mathbf{y}; \mathbf{a}) = h(x, \mathbf{y}; \mathbf{a}, f(H(x), \mathbf{y}; \mathbf{a})), \text{provided } x \neq 0.$$

[25] Leivant considers more general algebras defined from finitely many constructors.

[26] In [Bel92] this scheme is called *log recursion.*

Theorem 6.3.45 ([Bel92]).

$$\text{NC} = [0, I, S_0, S_1, C, L, \text{BIT}, \#'; \text{SCOMP}, \text{CRN}, \text{SWRN}].$$

Following [All91], define $\text{BH}(x) = x \bmod 2^{\lceil |x|/2 \rceil}$ and $\text{FH}(x) = msp(x, \text{BH}(x))$. The *back half* $\text{BH}(x)$ consists of the $\lceil |x|/2 \rceil$ rightmost bits of x, while the *front half* $\text{FH}(x)$ consists of the $\lfloor |x|/2 \rfloor$ leftmost bits of x. In [Blo94] S. Bloch defines two distinct safe versions of Allen's divide and conquer recursion.

Definition 6.3.36 (S. Bloch [Blo94]). *The function f is defined by* safe divide and conquer recursion *(SDCR) from the functions g, h if*

$$f(x, y, \mathbf{z}; \mathbf{a}) = \begin{cases} g(x, \mathbf{z}; \mathbf{a}) & \text{if } |x| \leq \max(|y|, 1) \\ h(x, y, \mathbf{z}; \mathbf{a}, f(\text{FH}(; x), y, \mathbf{z}; \mathbf{a}), & \\ \qquad f(\text{BH}(; x), y, \mathbf{z}; \mathbf{a})) & \text{else.} \end{cases}$$

The function f is defined by very safe divide and conquer recursion *(VSDCR) from the functions g, h if*

$$f(x, y, \mathbf{z}; \mathbf{a}) = \begin{cases} g(x, \mathbf{z}; \mathbf{a}) & \text{if } |x| \leq \max(|y|, 1) \\ h(; x, \mathbf{z}, \mathbf{a}, f(\text{FH}(; x), y, \mathbf{z}; \mathbf{a}), & \\ \qquad f(\text{BH}(; x), y, \mathbf{z}; \mathbf{a})) & \text{else.} \end{cases}$$

Note that in VSDCR the iteration function h has no normal parameters, and hence cannot itself be defined by recursion.

Theorem 6.3.46 (S. Bloch [Blo94]). *There is a collection* BASE *of initial functions, for which*

$$\text{ALOGTIME} = (\text{NORMAL} \cap [\text{BASE}; \text{SCOMP}, \text{VSDCR}])_*$$

$$\text{POLYLOGTIME} = (\text{NORMAL} \cap [\text{BASE}; \text{SCOMP}, \text{SDCR}])_*.$$

Proof. (Outline) The collection BASE of initial functions consists of NC^0 computable versions of MSP, LSP, FH, BH, a conditional function, and some string manipulating functions (see [Blo94] for details).

Only the proof of the first assertion will be sketched. Consider first the inclusion $[\text{BASE}; \text{SCOMP}, \text{VSDCR}]_* \subseteq \text{ALOGTIME}$. Show that $\text{BASE} \subseteq \text{NC}^0 \subseteq \text{ALOGTIME}$. Since the iterating function $h(; z, \mathbf{x}, \mathbf{y}, u, v)$ has only safe parameters, h must be obtained from BASE by safe composition, hence belongs to NC^0. Very safe divide and conquer recursion corresponds to the evaluation of a binary tree of logarithmic depth, whose leaves correspond to $g(x, \mathbf{z}; \mathbf{a})$ for $|x| \leq \max(|y|, 1)$, and whose internal nodes correspond to the NC^0 function h. Since the resulting circuit is of logarithmic depth, it follows that the function f defined by VSDCR belongs to ALOGTIME.

Now consider the inclusion $\text{ALOGTIME} \subseteq [\text{BASE}; \text{SCOMP}, \text{VSDCR}]_*$. Without using VSDCR, define certain string manipulating functions explicitly. Let M be a RATM. The computation tree of M corresponds to a binary branching logdepth tree, all nodes of which are encodings of the current work tape, index tapes, state and tape head positions. Without loss of generality, one may assume that a bit of the input can be queried, using the index tape, only at a leaf configuration. Depending on the current contents (say i) of the index tape, a bit (say the i-th bit) of the input is accessed. Depending on that query, evaluation of the leaves of the computation tree is defined, and evaluation of the internal nodes involves the simple evaluation of an AND-OR tree (minimax strategy). Describe the leaf nodes by a function in [BASE; SCOMP, VSDCR]. Evaluation of the AND-OR tree is very simply described, using an iterating function having only safe parameters.

It seems clear that linear time on multitape Turing machines or on random access machines can be characterized using appropriate initial functions, closure under safe composition and some form of simultaneous very safe recursion (simultaneous recursion, since a pairing function apparently cannot be defined from the initial functions using safe composition – such would be necessary for defining NEXT_M). Details have been worked out by S. Bloch [BBG96] and J. Otto [Ott94, Ott96, Ott95], the latter using category theory.

6.4 A Glimpse of Other Work

Higher type functional complexity theory is an emerging field, along with the related field of complexity theory for real-valued functions. Some aspects of higher type complexity theory are presented in the next chapter, while we mention here some references for the interested reader. In the book, [Ko91], Ker-I Ko surveyed the theory of sequential complexity theory of real valued functions, while H.J. Hoover [Hoo90] investigated parallel computable real valued functions, using the boolean circuit model. Complexity theory for type 2 functionals was initiated by R. Constable [Con73] and K. Mehlhorn [Meh76], both of whom worked with a type 2 version of Cobham's algebra, while complexity theory for all finite types was first considered by S. Buss, who in [Bus86b] introduced a polynomial time analogue of the *hereditarily recursive operations* HRO to define polynomial time functionals of all finite types decorated with runtime bounds. Using similar runtime decorated types, A. Nerode, J. Remmel and A. Scedrov [NRS89] studied a polynomially graded λ-calculus. By placing bounds on the modalities, in [GSS90], J.-Y. Girard, A. Scedrov and P. Scott introduced *bounded linear logic*, and proved a normalization theorem which yielded a characterization of a feasible class of type 2 functionals, whose type 1 section is the class of polytime computable functions. Later in [Gir98] J.-Y. Girard developed *light linear logic*, a form of linear logic without artificially bounded modalities, yet which still admitted a polynomial bound for normalization, hence a feasible class of type 2 functionals. In [CU93], S. Cook and A. Urquhart introduced an analogue of Gödel's system **T** by admitting a recursor for bounded recursion on notation for type

1 objects. Their system PV^ω provided a natural class of polynomial time higher type functionals (called the *basic feasible functionals of higher type*), whose type-2 section of PV^ω is BFF. A parallel complexity analogue of this difficult result is proved in the next chapter. In [Har92], V. Harnik extended Cook-Urquhart's functionals to levels of the polynomial time hierarchy. In [CK90] S. Cook and B. Kapron characterized the higher type functionals in PV^ω by certain kinds of programming language constructs, *typed while* programs and *bounded loop* programs. This was extended to parallel classes by P. Clote, B. Kapron and A. Ignatovic in [CIK93]. In [Set94] A. Seth extended his definition of *counter* Turing machine to all finite types, thus characterizing PV^ω by a machine model. If one additionally allows dynamic computation of indices of subprograms within this counter Turing machine model, then Seth has conjectured this class to properly contain PV^ω. Work of D. Leivant develops *tiered* lambda calculi for various complexity classes. For instance, in [LM93] D. Leivant and J.-Y. Marion gave various characterizations of PTIME by typed λ-calculi with pairing over an algebra $\mathbf{W}$ of words over $\{0, 1\}$.

6.5 Historical and Bibliographical Remarks

With the exception of Section 6.2.3, the material for this chapter, modulo small editing changes, is reprinted from "Computation Models and Function algebras", by P. Clote, *Handbook of Recursion Theory*, ed. E. Griffor, 589–681 (1999), North-Holland, with the kind permission of Elsevier Science - NIL, Sara Burgerhartstraat 25, 1055 KV Amsterdam, The Netherlands. Much of the material of Section 6.2.3, especially Algorithm 6.2.4, Algorithm 6.2.5, and their proof of correctness, derive from unpublished course notes of E. Kaltofen at Rensellaer University. The interested reader should consult [KR90] as well as the texts by S. Akl [Akl89] and F. Thomson Leighton [Lei92] for more on parallel algorithms.

Good surveys of function algebras include the monographs by H. Rose [Ros84] and K. Wagner and G. Wechsung [WW86] (chapters 2, 10). Theorem 6.3.3 was first obtained by combining the P. Clote's result [Clo90] that A_0 equals FO definable functions, and the Barrington-I-mmerman–Straubing result [BIS90] that FO = LH, an analogue of Bennett's Theorem 6.3.23.[27] The current proof is direct, influenced by A. Woods' presentation in [Woo86], and simplifies the argument of [BIS90] by using Lemma 6.3.4 and Lemma 6.3.5, both of which were generalized from Lemma 6.3.7. Lemma 6.3.7 was first proved by Nepomnjascii [Nep70] (a related result proved by Bennett [Ben62]), though R. Kannan [Kan81] later rediscovered this result. The idea of encoding a sequence of instructions rather than a sequence of configurations has been repeatedly used by a number of persons.

[27] See Exercise 6.6.18 for definition of FO.

6.6 Exercises

Exercise 6.6.1. ($\star\star$) If one replaces the smash function $x\#y = 2^{|x|\cdot|y|}$ by the slower growing, non-commutative function $x@y = 2^{|x|\cdot||y||}$, then many analogues of the function algebraic characterizations in fact go through, but with *quasi-linear* $n \cdot O(\log^k n)$ in place of polynomial n^k. Consider function algebras with both function $\#, @$, and define a measure of the *#-rank* to be the nesting depth of applications of the smash function. Define natural function algebras, using $\#, @$, which correspond to specific circuit complexity classes of certain depth *and* size bounds; e.g. can one characterize log-depth, size n^2 circuits?

Exercise 6.6.2. This exercise concerns Grzegorczyk's class $\mathcal{E}^0$.

1. Does the function $\lfloor\sqrt{x}\rfloor$ belong to $\mathcal{E}^0$?
2. Prove that the greatest common denominator function, $gcd(x, y)$, belongs to $\mathcal{E}^0$.
3. Show that $\max(x, y)$ does not belong to $\mathcal{E}^0$.

Exercise 6.6.3. Despite the fact that addition, multiplication and other fast growing functions do not belong to $\mathcal{E}^0$, suitably defined bounded versions of these functions do belong to this class.

1. Define the "modified conditional" function
$$mcond(x, y, z, w) = \begin{cases} y & \text{if } x = 0, y \leq w \\ z & \text{if } x \neq 0, z \leq w \\ w + 1 & \text{else.} \end{cases}$$
and proved that $mcond \in \mathcal{E}^0$.
2. Show that
$$add(x, y, z) = \begin{cases} x + y \text{ if } x + y \leq z \\ z + 1 \text{ else} \end{cases}$$
belongs to $\mathcal{E}^0$.
HINT. Use bounded recursion on y with the modified conditional function, *mcond*. Alternately, note that $add(x, 0, z) = s(z) \dot{-} (s(z) \dot{-} x))$ and that $add(x, y + 1, z) = s(z) \dot{-} (s(z) \dot{-} s(add(x, y, z)))$.
3. Show that
$$mult(x, y, z) = \begin{cases} x \cdot y & \text{if } x \cdot y \leq z \\ z + 1 & \text{else} \end{cases}$$
belongs to $\mathcal{E}^0$.
HINT. Use bounded recursion on y with *mcond*,*add*.
4. Show that similarly defined bounded versions of all the principal functions f_n in the Grzegorczyk hierarchy belong to $\mathcal{E}^0$. Conclude that the graphs of addition, multiplication, and of the f_n belong to $\mathcal{E}^0$.

5. ($\star$) Show that the graphs of the f_n belong in fact to the linear time hierarchy LTH.
   HINT. Use repeated squaring and sequence encoding, as in the proof of Bennett's Theorem 6.3.22 that the graph of exponentiation is in the linear time hierarchy.
6. Prove that $\text{LTH} \subseteq \mathcal{E}^0$, so this is a stronger result.

Exercise 6.6.4. Throughout our definitions of various recursion schemes, we defined the function $f(x, \mathbf{y})$ in terms of a previous value of f, the value being $f(\lfloor x/2 \rfloor, \mathbf{y})$ for recursion on notation, and $f(x \dot{-} 1, \mathbf{y})$ for recursion. This exercise is concerned with parameter-free recursion schemes, where the one-argument function $f(x)$ is defined in terms of a previous value of f. Let the bijective pairing function $\langle x, y \rangle$ be defined by $\frac{(i+j)(i+j \dot{-} 1)}{2} + i$, together with the projections $\pi_0(\langle x, y \rangle) = x$ and $\pi_1(\langle x, y \rangle) = y$.

1. Noting that
$$\langle x, y \rangle = \sum_{k<x+y} k + x$$
$$\pi_0(z) = \mu x \leq z[(\exists y \leq z)(z = \langle x, y \rangle)]$$
$$\pi_1(z) = \mu y \leq z[(\exists x \leq z)(z = \langle x, y \rangle)]$$
   prove that $\langle x, y \rangle$, π_0, π_1 belong to $\mathcal{E}^2$.
2. Prove that the collection of all primitive recursive one-variable functions is the closure of the initial functions $s, \dot{-}, \langle x, y \rangle, \pi_1, \pi_2$ by composition of one-variable functions and parameter-free iteration:
$$f(0) = 0$$
$$f(x+1) = f(f(x))$$
   HINT. See [Ros84, p. 21].
3. Define parameter-free recursion on notation. Prove or disprove that A_0 is the closure of the initial functions
$$0, I, s_0, s_1, |x|, \text{BIT}, \#, \langle x, y \rangle, \pi_0, \pi_1$$
   under composition and parameter-free recursion on notation.

Exercise 6.6.5. Show that the binomial coefficient $\binom{n}{i}$, as a function of n, i, belongs to $\mathcal{E}^3$.

Exercise 6.6.6. Consider the sequence $\langle f_n : n \in N \rangle$ of functions given by
$$g_0(x) = x + 1$$
$$g_{n+1}(x) = g_n^{(x+1)}(x)$$
and define the diagonal function G by $G(x) = g_x(x)$. Show that G is not primitive recursive.
HINT. A function f is said to *majorize* a function g if asymptotically f is

larger than g; i.e., $(\exists n_0)(\forall n \geq n_0)[f(n) \geq g(n)]$. Prove that G majorizes the Grzegorczyk principal functions f_n, for all n.

Exercise 6.6.7 (L.A.S Kirby, J.B. Paris). ($\star$) Define the functions f, g_i as follows. Let $g_1(x) = x$ and for $m > 1$, let $g_m(x)$ be the result of writing $x-1$ in base m, and then replacing all occurrences of m by $m+1$. For instance, $g_3(12) = 18$, since the base 3 representation of $11 = 12 - 1$ is $3^2 + 2 \cdot 3^0$, and 18 is $4^2 + 2 \cdot 4^0$. Define $f(0) = 0$, $f(1) = 1$ and for $x \geq 2$,

$$f(x) = \mu y \left[g_{y+1}(g_y(\cdots g_2(x))\cdots)) = g_y(g_{y-1}(\cdots g_2(x))\cdots))\right]$$

1. Show that f is a well-defined, total function.
2. Show that f is not primitive recursive, but that f is primitive recursive in the Ackermann function.
   HINT. These results are due to J.B. Paris and L.A.S. Kirby [KP81] and proved using techniques of model theory in mathematical logic. A. Cichon [Cic83] later gave a proof theoretic proof, relying on ordinal notations. See also [Ros84, pp. 178-181] for an outline of E.A. Cichon's proof.

Exercise 6.6.8. Write a program to compare values of $\Pi(x)$, the number of primes less than or equal to x, and the function $\frac{x}{\ln(x)}$ when $x = 10, 20, 50$.

Exercise 6.6.9. ($\star\star$) Is deterministic linear time closed under part-of quantifiers and under sharply bounded quantifiers?

Exercise 6.6.10. An alternate possible definition of *bounded quantifier formula* is to allow only bounded quantifiers of the form $\exists x < y$ and $\forall x < y$, where y is a variable, rather than a term. Show that a collection of sets definable by bounded quantifier formulas (alternate definition) over $0, 1, +, \times, \leq$ is exactly the linear time hierarchy LTH.
HINT. Note that $(\exists x \leq y \cdot z)\phi(x)$ is equivalent to $(\exists x_1 \leq y)(\exists x_2 \leq z)\phi(x_1 \cdot x_2)$ and argue by induction on the length of the formula.

Exercise 6.6.11 ([PW86]). Prove the following result due to Paris-Wilkie, without applying Lemma 6.3.7. If $A(n, m, \mathbf{y})$ is a predicate, then we write $A_n(\mathbf{y})$ to denote the set

$$\{m < n : A(n, m, \mathbf{y})\} \subseteq \{0, \ldots, n-1\}.$$

For ease of notation, the parameters are omitted, although they should be carried through in the proofs. Let $k \geq 1$, $0 < \alpha < 1$, and $g(n) = \log^{\frac{1-\alpha}{2}}(n)$. Suppose that A is Δ_0 definable and that $A_n \subseteq g(n)^k$ holds for all integers n. Prove that the following function is Δ_0 definable.

$$F(n) = \min(|A_n|, g(n)^k + 1)$$

HINT. Prove the assertion by induction on $k \geq 1$, for all Δ_0 definable A simultaneously. When $k = 1$, $F(n) = m$ iff

$$(m \leq g(n) \wedge (\exists f)(f \text{ is an injection from } m \text{ into } A_n))$$

or

$$(m = g(n) + 1 \wedge \text{no such injection exists.})$$

Now the injection f can be coded by

$$(1 + \max(A_n))^m + \sum_{i<m} f(i) \cdot (1 + \max(A_n))^i < 2n.$$

Using Theorem 6.3.22, one can express the encoding of the function f in a Δ_0 manner. Now proceed similarly for $k+1$: $F(A_n) = m$ if and only if $(6.48 \wedge 6.49) \vee 6.50$, where

$$m \leq g^{k+1}(n) \tag{6.48}$$

$$\begin{aligned}(\exists i_0 < \cdots < i_{g(n)}) \Big(i_0 = 0 \wedge i_{g(n)} = 2^{\log^\alpha(n)} \wedge \\ \left(\sum_{j<g(n)} |A_n \cap [i_j, i_{j+1}]| = m \right) \wedge \\ (\forall j < g(n))(|A_n \cap [i_j, i_{j+1}]| \leq g^k(n)))\end{aligned} \tag{6.49}$$

$$m = g^{k+1} + 1 \wedge \text{no such partition exists .} \tag{6.50}$$

Exercise 6.6.12. Define the function $\text{BITSUM}(x) = \sum_{i<|x|} \text{BIT}(i, x)$. It follows from Theorem 2.6.2 that $\text{BITSUM} \notin \text{AC}^0$. For fixed $k \geq 1$, define $f(x)$ to be the sum of the first $|n|^k$ low order bits of x, where $n = |x|$. Show that $f \in \text{AC}^0$.
HINT. Almost identical to the proof that $\text{SBBITSUM} \in \text{AC}^0$, where

$$\text{SBBITSUM}(x, y) = \begin{cases} \sum_{i<|y|} \text{BIT}(i, y) & \text{if } y \leq |x| \\ |x| + 1 & \text{else} \end{cases}$$

Exercise 6.6.13. Induction on notation is a method of inference of the form "if $\phi(0)$ is true and $\phi(x)$ implies that $\phi(2x)$ and $\phi(2x+1)$ hold, then $(\forall x)\phi(x)$ is true".

1. Using induction on notation, prove the correctness of the repeated squaring algorithm for exponentiation.
2. Using the conditional function *cond*, for n-ary relations P, Q whose characteristic function belongs to A_0, the characteristic functions for $\neg P$, $P \wedge Q$ and $P \vee Q$ are easily expressed. For each of the following propositional tautologies, first write out the characteristic function of the tautology and then prove by induction on notation that $c_\phi(x_0, \ldots, x_n) = 0$, where $x_0, \ldots, x_{n-1}$ are the variables occurring in the relations appearing in the tautology. For instance, $P(x) \vee \neg P(x)$ has characteristic function

$$\mathit{cond}(c_P(x), 0, \mathit{cond}(\mathit{cond}(c_P(x), 1, 0), 0, 1)),$$

and by induction on notation, one can show that $c_{P \vee \neg P}(x) = 0$ for all x.

(a) $P \to (Q \to P)$.
(b) $P \to (Q \vee P)$.
(c) $[P \to (Q \to R)] \to [(P \to Q) \to (P \to R)]$.

Exercise 6.6.14 ([Woo86]). If Γ is a class of first order formulas, then Γ^N denotes the collection of predicates definable by a formula in Γ. Let $\Sigma_{0,m}$ denote the collection of bounded quantifier formulas of the form

$$(\exists \mathbf{x}_1)(\forall \mathbf{x}_2)\ldots(Q\mathbf{x}_m)\phi$$

where ϕ is a quantifier free formula in the first order language $0, 1, +, \cdot, \leq$. Thus $\Sigma_{0,0}$ is the collection of quantifier free formulas. Prove that

$$\Sigma_{0,0}^N \subseteq \Sigma_2 - \text{TIME}(O(n)) \cap \Pi_2 - \text{TIME}(O(n)) \tag{6.51}$$

and deduce that for $m \geq 1$,

$$(\dagger)\Sigma_{0,m}^N \subseteq \Sigma_{m+1} - \text{TIME}(O(n)).$$

HINT. For (6.51), guess non-deterministically all terms and subterms appearing in the given quantifier free formula, guess the first $c \cdot \ln(n)$ many prime numbers p and the residues modulo p of the products of subterms occurring in a term, and branching universally, verify that the computations are correct. By the prime number theorem, the product of the first $c \cdot \ln(n)$ primes is larger than the values of the terms occurring in the formula, and so by the Chinese remainder theorem, the computations are correct iff they are correct modulo the primes (see [Woo86]).

Exercise 6.6.15. Show that if $0 < \epsilon < 1$ then for all k, m

$$\Sigma_m - \text{DTIMESPACE}(n^{O(1)}, n^{1-\epsilon}) \subseteq \text{LTH}.$$

HINT. See Kannan [Kan81] and Woods [Woo86] for an analogous result concerning the linear time hierarchy.

Exercise 6.6.16. It is well-known (see [AHU74]) that for multi-tape Turing machines without random access, and for $f(n) = o(\log\log(n))$,

$$\text{DSPACE}(0) = \text{DSPACE}(f) \subset \text{DSPACE}(\log\log(n)).$$

It follows from Lemma 6.3.4 that for multi-tape Turing machines with random access,

$$\text{DSPACE}(\log\log(n)) \subseteq \text{LH}.$$

However, PARITY $= \{w \in \{0,1\}^* : |w|_1 \equiv 0 \bmod 2\}$ is obviously a regular language, hence in DSPACE$(0) \in$ DSPACE$(0) \setminus$ LH. Thus DSPACE(0) on a Turing machine without random access is not contained in DSPACE$(\log\log(n))$ with random access. Why is this not a contradiction?
HINT. How many configurations are possible in the computation of a constant space bounded Turing machine without random access? How many configurations are possible in the computation of a log log space bounded Turing machine with random access?

Exercise 6.6.17 ([MR67]). Show that depth k loop programs define exactly the functions of $\mathcal{E}^n$.
HINT. For exact definitions and proof, see [MR67].

Exercise 6.6.18 ([Clo90]). In [Imm89], the concept of *first order* definable over finite models, *FO*, is defined, by adding the bit function to an earlier definition of Büchi and McNaughton. A language $L \subseteq \{0,1\}^+ = \{0,1\}^* - \{\lambda\}$ is *FO* if there is a first order sentence ϕ in the language with constant symbols $c_{\min}, c_{\max}$, unary predicate symbol Z, binary predicate symbols $\leq, BIT$, such that for any word $w \in \{0,1\}^+$

$$w \in L \Leftrightarrow \langle \{0, \ldots, |w| - 1\}, \leq, Z^w, BIT \rangle \models \phi.$$

Here $Z^w(i)$ holds when the i-th bit of w is 1, and $c_{\min}$ [resp. $c_{\max}$] is interpreted by 0 [resp. $|w| - 1$]. Relations of arity n are similarly defined. An n-ary function is defined to by *FO* if the following conditions hold.

1. The function f is *polynomially bounded*; i.e., there exists a multivariate polynomial p such that for all $x_1, \ldots, x_n \in \{0,1\}^+$, we have
$$|f(x_1, \ldots, x_n)| \leq p(|x_1|, \ldots, |x_n|).$$
2. The bitgraph B_f of f is *FO*, where
$$B_f = \{(x_1, \ldots, x_n, i) : \text{the } i\text{-th bit of } f(x_1, \ldots, x_n) \text{ is } 1\}.$$

Prove that $A_0 = FO$.
HINT. Inequality $\leq$ is explicitly defined in A_0 just after Equation (6.20), while in Lemma 6.3.1 it was shown that A_0 is closed under sharply bounded quantifiers. It follows that all *FO* relations belong to A_0. Using CRN, it follows that all *FO* functions belong to A_0. This concludes the difficult direction.

Exercise 6.6.19 (J. Krajíček). Recall the concatenation function $x * y$ defined in Equation 6.14. Let's say that the function f is defined by *strong concatenation recursion on notation* (CRN$*$) from g, h_0, h_1 if

$$\begin{aligned} f(0, \mathbf{y}) &= g(\mathbf{y}) \\ f(s_0(x), \mathbf{y}) &= f(x, \mathbf{y}) * h_0(x, \mathbf{y}) \quad \text{if } x \neq 0 \\ f(s_1(x), \mathbf{y}) &= f(x, \mathbf{y}) * h_1(x, \mathbf{y}). \end{aligned}$$

Define the algebra A_0' to be $[0, I, s_0, s_1, \text{BIT}, |x|, \#; \text{COMP}, \text{CRN}*]$. Prove that A_0 is properly contained in A_0'.
HINT. Define $f(0) = 0$, $f(s_0(x)) = f(x) * 0$, and $f(s_1(x)) = f(x) * 1$. Then the parity function $\oplus_{i=1}^n x_i \in A_0'$, since for $x = x_1 \cdots x_n$, $\oplus_{i=1}^n x_i = \text{BIT}(0, |f(x)|)$. By Theorem 6.3.3 we have $A_0 = \mathcal{F}\text{LH}$, i.e., A_0 equals logtime uniform AC^0. Finally, by Theorem 2.6.2 and Corollary 2.6.1, the parity function does not belong to AC^0.

Exercise 6.6.20. This exercise investigates alternative manners of defining the class of primitive recursive functions, defined as the smallest algebra containing $0, s, I_k^n$ for all $1 \leq k \leq n$ and closed under composition and (unbounded) recursion.

1. Prove that the collection of primitive recursive functions is the smallest algebra containing $0, I, s_0, s_1$, and closed under composition and (unbounded) recursion on notation – i.e., recursion on notation, rather than recursion.
2. Prove that the collection of primitive recursive functions is generated from the initial functions $0, I, s_0, s_1, |x|, \text{BIT}$ by composition, CRN and weak (unbounded) recursion on notation.
3. Generalize this to recursion schemes admitting definitions of functions f by $f(x, \mathbf{y}) = F(|x|^{(k)}, \mathbf{y})$, where

$$\begin{aligned} F(0, \mathbf{y}) &= g(\mathbf{y}) \\ F(2x, \mathbf{y}) &= h_0(x, \mathbf{y}, F(x, \mathbf{y})) \\ F(2x+1, \mathbf{y}) &= h_1(x, \mathbf{y}, F(x, \mathbf{y})). \end{aligned}$$

7. Higher Types

In ordinary computability theory, *computable* means computable by an algorithm, and this should be the case for higher type computability theory too. If the algorithms in question are sequential, they should yield a notion of time complexity. ... From the point of view of complexity theory, functional programming languages such as $\mathcal{L}_{DA}$ and $\mathcal{L}_{PA+\exists}$ are not completely satisfactory, because they do not provide any intrinsic notion of time complexity. *S.A. Cook [Coo92, p. 60, 61]*

Quand j'ai voulu me restreindre, je suis tombé dans l'obscurité; j'ai préféré passer pour un peu bavard. *H. Poincaré [Poi95]*

7.1 Introduction

Many programming languages allow functions to be passed as parameters to other functions or procedures. For instance, C allows function parameters (passed as pointers), C++ supports function templates, Java allows generic programs using interfaces and classes, and ML admits polymorphism, i.e., abstraction over data types (for instance, a general sorting algorithm where the data type is given at runtime). Oracle Turing machines were first introduced by A.M. Turing [Tur37] to model relative recursion, where the set A is recursive relative to set B, denoted $A \leq_T B$, if set membership in A can be decided by a Turing machine furnished with an oracle for B. Complexity analogues, such as polynomial time reducibility $A \leq_T^p B$, etc. have since been rigorously studied. The function oracle Turing machine can be used to model function parameter passing, and feasibly computable type 2 functionals have been studied using this model or a variant thereof by Constable [Con73], Mehlhorn [Meh76], Buss [Bus86b], Townsend [Tow90], Cook and Kapron [KC96], Clote, Ignjatovic and Kapron [CIK93], Seth [Set94] and others.

7.2 Type 2 Functionals

Higher type *functional* complexity theory is a relatively new area with several striking results, and some fundamental open problems.

Definition 7.2.1. *A* type 2 *functional* F *of* rank (k, ℓ) *is a total mapping* $(\mathbf{N})^k \times \mathbf{N}^\ell \to \mathbf{N}$. *If* C *is a class of type 2 functionals, then the* type 1 section *of* C *is the set of type 1 functions belonging to* C*; i.e., those type 2 functionals of rank* $(0, \ell)$*, where* $\ell \in \mathbf{N}$.

Definition 7.2.2. *A function oracle Turing machine (*OTM*) is a Turing machine* M*, which in addition to read-only input tape, distinguished output tape and finitely many work tapes, has an* oracle query tape *and* oracle answer tape, *both one-way infinite, for each function input. Additionally,* M *has a special oracle query state for each function input.*

In order to query a function input f at x, the machine M takes steps to write x in binary on the oracle query tape. When the oracle query tape head is in its leftmost square, M enters a special query state. In the next step, M erases both the oracle query[1] and answer tapes, writes the function value $f(x)$ in binary on the oracle answer tape, and leaves the oracle query and answer tape heads in their leftmost squares. Upon entering the oracle query state, there seem to be two natural measures for the time to complete the function query $f(x)$. The *unit cost,* considered by Mehlhorn [Meh76], charges unit time, while the *function length cost,* considered by Constable [Con73] and later Kapron and Cook [KC96], charges $\max\{1, |f(x)|\}$ time. The machine M computes the $rank(n, m)$ functional $F(f_1, \dots, f_n, x_1, \dots, x_m)$ if M has n oracle query states, query and answer tapes corresponding to $f_1, \dots, f_n$ and if M outputs the integer $F(f_1, \dots, f_n, x_1, \dots, x_m)$ in binary on the output tape, when started in its initial state q_0 with input tape $\underline{B}x_1 B x_2 B \cdots B x_m B$.

Definition 7.2.3. *For any* OTM M*, for any inputs* $f_1, \dots, f_n, x_1, \dots, x_m$ *and integer* t*, the query answer set* $QA_M(\mathbf{f}, \mathbf{x}, t)$ *is defined as* $\{(y, z) : M$ *on input* $\mathbf{f}, \mathbf{x}$ *queries some* $f_i(y) = z$ *within time* $S(t)$ *steps*$\}$*, where* $S(t)$ *is the least number of steps* s *for which if* M *runs* s *steps then its time complexity is at least* t.[2] *The* query set $Q_M(\mathbf{f}, \mathbf{x}, t)$ *is* $\{y : (\exists z)[(y, z) \in QA_M(\mathbf{f}, \mathbf{x}, t)]\}$*; the* answer set $A_M(\mathbf{f}, \mathbf{x}, t)$ *is* $\{z : (\exists y)[(y, z) \in QA_M(\mathbf{f}, \mathbf{x}, t)]\}$.

An OTM M *is a* polynomial time oracle Turing machine *(*POTM*) if* M *computes a total* $rank(n, m)$ *functional* F *and there is a polynomial* p *such that for all input* $f_1, \dots, f_n, x_1, \dots, x_m$ *and times* t *we have*

[1] In [DM97], Durand and More studied the possibility of dropping this requirement, thus allowing for the non-erasure of oracle query tapes. In the previous chapter, we defined LTH to be the collection of languages computable on an alternating Turing machine with a constant number of alternations. It can be shown that LTH equals $\cup_{i=0}^{\infty} \Sigma_i^L$, where Σ_0^L is deterministic linear time on a multitape Turing machine, and Σ_{i+1}^L is the collection of languages computable by a nondeterministic linear time bounded Turing machine with an oracle from Σ_i^L. In [DM97], it is proved that LTH can equivalently be defined by allowing nonerasure for the oracle tape.

[2] For the unit cost model, $S(t) = t$; for the function length cost model, $S(t)$ could be much less than t.

$$t \leq p(|\max(\{x_1, \ldots, x_m\} \cup A_M(\mathbf{f}, \mathbf{x}, t))|).$$

OPT *is the collection of type 2 functionals computable by an oracle polynomial time oracle Turing machine.*

Example 7.2.1.

(1) $F(f, x) = \max\{f(y) : y \leq |x|\}$ belongs to OPT.
(2) $G(f, x) = \max\{f(y) : |y| \leq |x|\}$ does not belong to OPT.
(3) $H(f, x) = f^{(|x|)}(x)$ belongs to OPT.

In [Meh76] K. Mehlhorn extended Cobham's function algebra to type 2 functionals. A modern presentation of Mehlhorn's definition uses the following schemes.

Definition 7.2.4 ([Tow90]). *F is defined from $H, G_1, \ldots, G_m$ by* functional composition *(*COMP*) if for all* $\mathbf{f}, \mathbf{x}$,

$$F(\mathbf{f}, \mathbf{x}) = H(\mathbf{f}, G_1(\mathbf{f}, \mathbf{x}), \ldots, G_m(\mathbf{f}, \mathbf{x}), \mathbf{x}).$$

F is defined from G by expansion *(*EXP*) if for all* $\mathbf{f}, \mathbf{g}, \mathbf{x}, \mathbf{y}$,

$$F(\mathbf{f}, \mathbf{g}, \mathbf{x}, \mathbf{y}) = G(\mathbf{f}, \mathbf{x}).$$

F is defined from $G, G_1, \ldots, G_m$ by functional substitution *(*FSUB*) if for all* $\mathbf{f}, \mathbf{x}$,

$$F(\mathbf{f}, \mathbf{x}) = H(\mathbf{f}, \lambda y.G_1(\mathbf{f}, \mathbf{x}, y), \ldots, \lambda y.G_m(\mathbf{f}, \mathbf{x}, y), \mathbf{x}).$$

F is defined from G, H, K by limited recursion on notation *(*LRN*) if for all* $\mathbf{f}, \mathbf{x}, y$,

$$F(\mathbf{f}, \mathbf{x}, 0) = G(\mathbf{f}, \mathbf{x})$$
$$F(\mathbf{f}, \mathbf{x}, y) = H(\mathbf{f}, \mathbf{x}, y, F(\mathbf{f}, \mathbf{x}, \lfloor \tfrac{y}{2} \rfloor)), \ \textit{if } y \neq 0$$

provided that $F(\mathbf{f}, \mathbf{x}, y) < K(\mathbf{f}, \mathbf{x}, y)$ holds for all $\mathbf{f}, \mathbf{x}, y$.

The natural type 2 extension of bounded recursion on notation from the previous chapter is given as follows.

Definition 7.2.5. *F is defined by* bounded recursion on notation *(*BRN*) from G, H_0, H_1, K if*

$$F(\mathbf{f}, \mathbf{x}, 0) = G(\mathbf{f}, \mathbf{x})$$
$$F(\mathbf{f}, \mathbf{x}, s_0(y)) = H_0(\mathbf{f}, \mathbf{x}, y, F(\mathbf{f}, \mathbf{x}, y)), \ \textit{if } y \neq 0$$
$$F(\mathbf{f}, \mathbf{x}, s_1(y)) = H_1(\mathbf{f}, \mathbf{x}, y, F(\mathbf{f}, \mathbf{x}, y))$$

provided $F(\mathbf{f}, \mathbf{x}, y) \leq K(\mathbf{f}, \mathbf{x}, y)$ for all $\mathbf{x}, y$.

Provided that $s_0, s_1, \lfloor x/2 \rfloor$, MOD2, *cond* are available, limited recursion on notation is equivalent with bounded recursion on notation (see Exercise 7.8.1). In this chapter, we generally use limited recursion schemes, which are notationally easier to manipulate in the proofs that follow. We'll often leave the definition of the limited or bounded version of a recursion scheme to the reader, provided one or the other has been defined, and tacitly assume the equivalence of both schemes when it suits us.

Definition 7.2.6 ([Tow90], [CK90]). *Let X be a class of type 2 functionals. The class of* basic feasible functionals *defined from X, denoted* BFF(X), *is the smallest class of functionals containing X, 0,s_0,s_1,I,$\#$ and the application functional Ap, defined by $Ap(f,x) = f(x)$, and which is closed under functional composition, functional substitution, expansion, and* LRN. *If $F \in$* BFF(X), *then F is basic feasible in X. The class* BFF *of basic feasible functionals*[3] *is* BFF$(\emptyset)$.

Definition 7.2.7. *For a function f, $x = \text{argmax}_{i \leq n} f(i)$ means that x is the smallest argument bounded by n such that $f(x) = \max_{i \leq n} f(i)$.*

Recall from Example 7.2.1 that $F(f,x) = \max\{f(y) : y \leq |x|\}$ belongs to oracle polynomial time OPT. It is not difficult to show that $F \in$ BFF, by first showing that $\text{argmax}_{i \leq |x|} f(i)$ belongs to BFF – see Exercise 7.8.2, where this functional is proved to belong to $\mathcal{A}_0$, hence *a fortiori* to BFF.

In [Meh76] Mehlhorn introduced the Turing machine model with function oracle, charging unit cost for a function oracle call, independent of the length of the function value returned. Mehlhorn's model has an oracle input tape and an oracle output tape, thus avoiding the situation where m successive iterates of a function $f(f(\ldots f(x)\ldots))$ might take m steps. Using the techniques of low-level arithmetization from the previous chapter, the following result is straightforward to prove.

Theorem 7.2.1 ([Meh76]). *For every functional F in* BFF, *there is a unit cost model* OTM *M which computes F, i.e., $M(\mathbf{f},\mathbf{x}) = F(\mathbf{f},\mathbf{x})$, and where the runtime of M on all input $\mathbf{f},\mathbf{x}$ is bounded by $|G(\mathbf{f},\mathbf{x})|$ for some G belonging to* BFF. *Conversely, if functional F is computed by* OTM *M, which on input $\mathbf{f},\mathbf{x}$ has runtime at most $|G(\mathbf{f},\mathbf{x})|$ for some G belonging to* BFF, *then $F \in$* BFF.

Definition 7.2.8 ([KC96]). *Let $\mathcal{F}$ be a class of type 2 functionals. $\mathcal{F}$ has the* Ritchie-Cobham property *if*

---

[3] Though Townsend [Tow90] calls this class **POLY**, we follow Cook and Kapron in calling this class *basic feasible*. Functional substitution is superfluous, as noted in [Tow90, Tow82]. The proof of this fact in [Tow82] is flawed, as Townsend has acknowledged in personal correspondence, and the first correct proof of this assertion to appear in the literature is due to K.-H. Niggl (see [Nig98], where Niggl proves that functional substitution can be eliminated from a particular lambda-calculus; see also Lemmas 7.3.3 and 7.3.4 from this chapter).

$$\mathcal{F} = \{F : \textit{there exist } G \in \mathcal{F} \textit{ and } \text{OTM } M \textit{ which on any input } \mathbf{f}, \mathbf{x} \textit{ computes } F(\mathbf{f}, \mathbf{x}) \textit{ within time } |G(\mathbf{f}, \mathbf{x})|\}.$$

With this definition, Theorem 7.2.1 can be rephrased by the statement that BFF has the Ritchie-Cobham property using unit cost OTM.

It is clear that OPT contains functionals which are not intuitively feasible. In particular, substituting the polynomial time computable function $\lambda y.y^2$ for f in H, where $H(f,x) = f^{(|x|)}(x)$, above yields $H(\lambda y.y^2, x) = x^{2^{|x|}}$ which is not a polynomial time computable type 1 function (example due to A. Seth [Set92]).

In [KC96], Kapron and Cook lift Cobham's characterization of polynomial time computable functions to functionals of type 2. To state their result, the notion of *length* of a function and that of second order polynomial must be introduced.

Definition 7.2.9. *The length $|f|$ of one-place function f is itself a one-place function defined by*

$$|f|(n) = \max_{|x| \leq n}\{|f(x)|\}.$$

Suppose that $f_1, \ldots, f_m$ be variables ranging over $\mathbf{N}^N$ and $x_1, \ldots, x_n$ be variables ranging over $\mathbf{N}$. The collection C of second order polynomials $P(f_1, \ldots, f_m, x_1, \ldots, x_n)$ is defined inductively as follows.

1. *Every integer c belongs to C.*
2. *For every $1 \leq i \leq n$, $x_i \in C$.*
3. *If $P, Q \in C$ then $P + Q \in C$ and $P \cdot Q \in C$.*
4. *If $P \in C$ then $f_i(P) \in C$ for $1 \leq i \leq m$.*

The depth $d(P)$ of a second order polynomial P is defined inductively by $d(c) = 0 = d(x_i)$, $d(P + Q) = d(P \cdot Q) = \max(d(P), d(Q))$, $d(f_i(P)) = 1 + d(P)$.

For example, $|f|(|f|(|x|^3 + 2 \cdot |x|)) + |f|(|f|(|f|(|x|^2)))$ is a second order polynomial (in variables $|f|, |x|$) of depth 3.

Theorem 7.2.2 ([KC96]). BFF *is the collection of functionals*

$$F(f_1, \ldots, f_n, x_1, \ldots, x_m)$$

computable in time

$$P(|f_1|, \ldots, |f_n|, |x_1|, \ldots, |x_m|)$$

for some second order polynomial P on an OTM *with function length cost.*[4]

[4] In unpublished work, by proof theoretic techniques, A. Ignjatovic has proved that this result holds as well for unit cost.

Rather than presenting the proof of Theorem 7.2.2 (see Exercise 7.8.6), we present a new result due to the first author, similarly characterizing type 2 parallel complexity classes. To that end, we must introduce the type 2 analogues of the functional classes A_0 and A, and study closure properties of these classes, in order to allow an arithmetization of parallel computations. The proof will take most of the rest of this chapter.

In [Coo92], S.A. Cook proposed that any class C of feasible type 2 functionals must satisfy the following two conditions:

1. BFF $\subseteq C \subseteq$ OPT,
2. C is closed under abstraction and application.

In [Set92] A. Seth defined a class C_2 of type 2 functionals defined by *counter* Turing machines with polynomial bounds, which satisfies the previous conditions, and proved that no recursively presentable class of functionals exists which contains C_2 and satisfies the previous conditions. From further work in [Set93], it now appears doubtful whether any natural conditions will guarantee the uniqueness of BFF or another class as *the* class of feasible (polynomial time) functionals. Nevertheless, BFF has been characterized as the type 2 section of PV^ω, a polynomial time version of Gödel's $\mathbf{T}$ (see [CU93]), as well as via function oracle Turing machines (Theorem 7.2.2), so, at least to our mind, has the strongest claim to being type 2 polynomial time.

The following example, due to S. Cook, provides a functional which belongs to OPT yet not to BFF. Let $\preceq$ quasi-order $\mathbf{N} \times \mathbf{N}$ by *length first difference*; i.e., $(a,b) \preceq (c,d)$ iff $|a| < |c|$ or ($|a| = |c|$ and $|b| \leq |d|$). Transfer this ordering to $\mathbf{N}$ by a standard polynomial time pairing function. Define the $rank(1,0)$ functional L by $L(f) = \mu i[(\exists j < i)(f(j) \preceq f(i))]$. Note that $\preceq$ defines a well quasi-ordering on $\mathbf{N} \times \mathbf{N}$, so L is well defined.

Theorem 7.2.3 ([Coo92]). *The functional L belongs to* OPT *yet not to* BFF.

S. Cook [Coo92] points out that the type-1 section of the closure of OPT with L is just the type-1 section of OPT, i.e., the class of polynomial time computable functions, and so L should be considered a feasible functional. This argument suggests that BFF should not be considered the class of all feasible type-2 functionals. Against this, in [Set92] A. Seth proves that the type-1 section of the closure of type-2 exponential time with L is not the class of exponential time computable functions, and hence L should not be considered a feasible functional.

7.3 Some Closure Properties of $\mathcal{A}_0$

The type 2 analogue of *concatenation recursion on notation* is given by the following.

Definition 7.3.1. *F is defined from G, H, K by* concatenation recursion on notation *(CRN) if for all* $\mathbf{f}, \mathbf{x}, y$,

$$F(\mathbf{f}, \mathbf{x}, 0) = G(\mathbf{f}, \mathbf{x})$$
$$F(\mathbf{f}, \mathbf{x}, s_0(y)) = s_{\text{MOD2}(H(\mathbf{f},\mathbf{x},y))}(F(\mathbf{f}, \mathbf{x}, y)), \textit{ provided that } y \neq 0$$
$$F(\mathbf{f}, \mathbf{x}, s_1(y)) = s_{\text{MOD2}(K(\mathbf{f},\mathbf{x},y))}(F(\mathbf{f}, \mathbf{x}, y)).$$

In the interests of readability, we introduce the function $x^\frown y$ defined by

$$x^\frown y = \begin{cases} s_0(x) \text{ if } y = 0 \\ x * y \text{ else} \end{cases}$$

where the concatenation function $x * y = 2^{|y|} \cdot x + y$ was defined in last chapter by

$$x * 0 = x$$
$$x * s_i(y) = s_i(x * y).$$

With this notation, the previous scheme is given as follows:

$$F(\mathbf{f}, \mathbf{x}, 0) = G(\mathbf{f}, \mathbf{x})$$
$$F(\mathbf{f}, \mathbf{x}, s_0(y)) = F(\mathbf{f}, \mathbf{x}, y)^\frown \text{MOD2}(H(\mathbf{f}, \mathbf{x}, y)), \text{ provided that } y \neq 0$$
$$F(\mathbf{f}, \mathbf{x}, s_1(y)) = F(\mathbf{f}, \mathbf{x}, y)^\frown \text{MOD2}(K(\mathbf{f}, \mathbf{x}, y)).$$

Definition 7.3.2. *The type 2 functional H is defined by* weak bounded recursion on notation *(WBRN) from G, H_0, H_1, K if*

$$F(\mathbf{f}, \mathbf{x}, 0) = G(\mathbf{f}, \mathbf{x})$$
$$F(\mathbf{f}, \mathbf{x}, s_0(y)) = H_0(\mathbf{f}, \mathbf{x}, y, F(\mathbf{f}, \mathbf{x}, y)), \textit{ if } y \neq 0$$
$$F(\mathbf{f}, \mathbf{x}, s_1(y)) = H_1(\mathbf{f}, \mathbf{x}, y, F(\mathbf{f}, \mathbf{x}, y))$$
$$H(\mathbf{f}, \mathbf{x}, y) = F(\mathbf{f}, \mathbf{x}, |y|)$$

provided that $F(\mathbf{f}, \mathbf{x}, y) \leq K(\mathbf{f}, \mathbf{x}, y)$ holds for all $\mathbf{f}, \mathbf{x}, y$.

Definition 7.3.3. *The algebra $\mathcal{A}_0$ is the smallest class of type 2 functionals containing $0, I, s_0, s_1$, BIT, $|x|$, #, Ap and closed under functional composition, expansion, functional substitution and* CRN. *The algebra $\mathcal{A}$ is the closure of $0, I, s_0, s_1$, BIT, $|x|$, #, Ap under functional composition, expansion, functional substitution,* CRN *and* WBRN, *while $\mathcal{A}_k$ is the collection of those functionals in $\mathcal{A}$ allowing at most k nested applications of* WBRN.

Remark. Note that in the definition of $\mathcal{A}_0$, etc., we did not have to include an application functional $Ap_n(f, x_1, \ldots, x_n)$ with value $f(x_1, \ldots, x_n)$, for every arity n. As explained in the previous chapter, the pairing function τ in Equation (6.21) and the projection functions π_1, π_2 in Equation (6.24) belong to the algebra $\mathcal{A}_0$, and so we can define e.g.

$$Ap_3(f, x_1, x_2, x_3) = Ap(g, \tau(x_1, \tau(x_2, x_3)))$$

where $g(\tau(x_1, \tau(x_2, x_3))) = f(x_1, x_2, x_3)$.

We now define several variants of *concatenation recursion on notation.* These are unnecessary for the main theorem of this section, but will somewhat simplify the definition of recursor C in the finite typed lambda calculus of Section 7.6.3.

Definition 7.3.4. *F is defined from H, K by* CRN1 *if for all* $\mathbf{f}, \mathbf{x}, y$,

$$\begin{aligned} F(\mathbf{f}, \mathbf{x}, 0) &= 1 \\ F(\mathbf{f}, \mathbf{x}, s_0(y)) &= F(\mathbf{f}, \mathbf{x}, y)^\frown \mathrm{MOD2}(H(\mathbf{f}, \mathbf{x}, y)), \text{ provided that } y \neq 0 \\ F(\mathbf{f}, \mathbf{x}, s_1(y)) &= F(\mathbf{f}, \mathbf{x}, y)^\frown \mathrm{MOD2}(K(\mathbf{f}, \mathbf{x}, y)). \end{aligned}$$

Recall the *truncate left* function TL defined in Equation (6.23), which truncates the leftmost bit of its argument. If F ($F1$) is defined by CRN (CRN1) from G, H, K (H, K), then we can define F from $F1$, TL, $x^\frown y$ by

$$F(\mathbf{f}, \mathbf{x}, y) = G(\mathbf{f}, \mathbf{x})^\frown \mathrm{TL}(F1(\mathbf{f}, \mathbf{x}, y)).$$

Definition 7.3.5. *F is defined from G, H by* CRN2 *if for all* $\mathbf{f}, \mathbf{x}, y$,

$$F(\mathbf{f}, \mathbf{x}, y) = \begin{cases} G(\mathbf{f}, \mathbf{x}) & \text{if } y = 0 \\ F(\mathbf{f}, \mathbf{x}, \lfloor \frac{y}{2} \rfloor)^\frown \mathrm{MOD2}(H(\mathbf{f}, \mathbf{x}, y)) & \text{else.} \end{cases}$$

Using the method outlined in Exercise 7.8.1, it is straightforward to show that a functional F defined by CRN from G, H_0, H_1 can be defined by CRN2 from G, H where H is defined from H_0, H_1 using $s_0, s_1, \lfloor x/2 \rfloor$, MOD2, *cond*, and vice-versa. It follows that in the presence of the auxiliary functions $s_0, s_1, \lfloor x/2 \rfloor$, MOD2, *cond*, TL, we can replace CRN by the following scheme CRN3. (Here, note that both $x * y$ and $x^\frown y$ are definable using CRN3.)

Definition 7.3.6. *F is defined from H by* CRN3 *if for all* $\mathbf{f}, \mathbf{x}, y$,

$$F(\mathbf{f}, \mathbf{x}, y) = \begin{cases} 1 & \text{if } y = 0 \\ F(\mathbf{f}, \mathbf{x}, \lfloor \frac{y}{2} \rfloor)^\frown \mathrm{MOD2}(H(\mathbf{f}, \mathbf{x}, y)) & \text{else.} \end{cases}$$

The pairing function τ in Equation (6.21) is defined by composition from *cond*, MSP, $|x|$, *pad*, $*$, while the projection functions π_1, π_2 in Equation (6.24) are defined by composition from TL, MSP, LSP, $|x|$. Note that LSP is defined by composition from MSP, *rev*0, $|x|$ and that *rev*0 is defined from $|x|$, BIT, hence from $|x|$, MSP, MOD2, using CRN0. With help of τ, π_1, π_2, we can encode and decode finite tuples, and hence reduce parameter lists $f_1, \ldots, f_n$ and $x_1, \ldots, x_m$ to single parameters f, x. This will be exploited in the next variant of concatenation recursion on notation, where parameters are f, x rather than $\mathbf{f}, \mathbf{x}$, the latter which abbreviate $f_1, \ldots, f_n$ and $x_1, \ldots, x_n$.

Definition 7.3.7. *F is defined from H by* CRN4 *if for all* f, x, y,

$$F(f, x, y) = \begin{cases} 1 & \text{if } y = 0 \\ F(f, x, \lfloor \frac{y}{2} \rfloor)^\frown \mathrm{MOD2}(H(f, x, y)) & \text{else.} \end{cases}$$

Note that the definition of $rev0$ in Equation (6.3) can be defined using CRN4 and MOD2, that $x * y, x^\frown y$ can be defined from using CRN4 and MOD2, and that TL can be defined from LSP, $|x|$ by composition. Summarizing this discussion, we have the following.

Theorem 7.3.1.

$$\begin{aligned}\mathcal{A}_0 &= [0, I, s_0, s_1, BIT, |x|, \#, Ap; \text{COMP}, \text{CRN}, \text{EXP}, \text{FSUB}] \\ &= [0, I, s_0, s_1, \text{MSP}, \text{MOD2}, cond, |x|, \#, Ap; \text{COMP}, \text{CRN4}, \text{EXP}, \text{FSUB}].\end{aligned}$$

Sharply bounded quantifiers are of the form $(\exists y < |T(\mathbf{f}, \mathbf{x})|)$ and $(\forall y < |T(\mathbf{f}, \mathbf{x})|)$.

Definition 7.3.8. *The functional F is defined by sharply bounded existential quantification from functionals G, H if*

$$F(\mathbf{f}, \mathbf{x}, y) = \begin{cases} 0 \text{ if } \exists i < |G(\mathbf{f}, \mathbf{x}, y)|[H(\mathbf{f}, \mathbf{x}, y, i) = 0] \\ 1 \text{ else.} \end{cases}$$

The functional F is defined by sharply bounded universal quantification from functionals G, H if

$$F(\mathbf{f}, \mathbf{x}, y) = \begin{cases} 0 \text{ if } \forall i < |G(\mathbf{f}, \mathbf{x}, y)|[H(\mathbf{f}, \mathbf{x}, y, i) = 0] \\ 1 \text{ else.} \end{cases}$$

Lemma 7.3.1. *The algebra $\mathcal{A}_0$ is closed under sharply bounded quantification.*

Proof. Suppose that $G, H \in \mathcal{A}_0$ and that F is defined by

$$F(\mathbf{f}, \mathbf{x}, y) = \begin{cases} 0 \text{ if } \exists i < |G(\mathbf{f}, \mathbf{x}, y)|[H(\mathbf{f}, \mathbf{x}, y, i) = 0] \\ 1 \text{ else.} \end{cases}$$

Define F_0 using CRN as follows:

$$F_0(\mathbf{f}, \mathbf{x}, y, 0) = 0$$

$$F_0(\mathbf{f}, \mathbf{x}, y, s_i(z)) = \begin{cases} s_0(F_0(\mathbf{f}, \mathbf{x}, y, z)) \text{ if } H(\mathbf{f}, \mathbf{x}, y, |z|) \neq 0 \\ s_1(F_0(\mathbf{f}, \mathbf{x}, y, z)) \text{ else.} \end{cases}$$

Then

$$F(\mathbf{f}, \mathbf{x}, y) = \overline{sg}(F_0(\mathbf{f}, \mathbf{x}, G(\mathbf{f}, \mathbf{x}, y), y)$$

where Kleene's opposite signum function $\overline{sg}$ is defined by $\overline{sg}(0) = 1$ and $\overline{sg}(x) = 0$ for $x > 0$. Since *cond*, *sg*, $\overline{sg}$ etc. were given explicit definitions in $\mathcal{A}_0$, it follows that F belongs to $\mathcal{A}_0$. In a similar manner, one can show closure under sharply bounded universal quantification.

Definition 7.3.9. *The functional F is defined from G, H by sharply bounded μ-operator, or sharply bounded minimization, if $F(\mathbf{f}, \mathbf{x}, y) = i_0$ provided that*

$$i_0 < |G(\mathbf{f}, \mathbf{x}, y)| \wedge H(\mathbf{f}, \mathbf{x}, y, i_0) = 0 \wedge \forall i < i_0 (H(\mathbf{f}, \mathbf{x}, y, i) \neq 0)$$

and otherwise $F(\mathbf{f}, \mathbf{x}, y) = |G(\mathbf{f}, \mathbf{x}, y)|$. This is denoted by $F(\mathbf{f}, \mathbf{x}, y) = \mu i < |G(\mathbf{f}, \mathbf{x}, y)|[H(\mathbf{f}, \mathbf{x}, y, i) = 0]$.

Lemma 7.3.2. *$\mathcal{A}_0$ is closed under sharply bounded μ-operator.*

Proof. Given $G, H \in \mathcal{A}_0$, define F_0 by CRN as follows:

$$F_0(\mathbf{f}, \mathbf{x}, y, 0) = 0$$

$$F_0(\mathbf{f}, \mathbf{x}, y, s_i(z)) = \begin{cases} s_0(F_0(\mathbf{f}, \mathbf{x}, y, z)) \text{ if } \forall i \leq |z|[H(\mathbf{f}, \mathbf{x}, y, i) \neq 0] \\ s_1(F_0(\mathbf{f}, \mathbf{x}, y, z)) \text{ else.} \end{cases}$$

Let

$$F(\mathbf{f}, \mathbf{x}, y) = |G(\mathbf{f}, \mathbf{x}, y)| \dot{-} |F_0(\mathbf{f}, \mathbf{x}, G(\mathbf{f}, \mathbf{x}, y), y)|.$$

It is straightforward to see that

$$F(\mathbf{f}, \mathbf{x}, y) = \mu i < |G(\mathbf{f}, \mathbf{x}, y)|[H(\mathbf{f}, \mathbf{x}, y, i) = 0]$$

so that $F \in \mathcal{A}_0$.

Similarly, all the derivations of the previous chapter concerning a sequence coding function β, etc. can be lifted to functionals. The proofs are essentially identical, where extra parameters occur, as in the last two lemmas.

The scheme of functional substitution is redundant, as indicated in Corollary 7.3.1, whose proof depends on the following lemmas.

Definition 7.3.10. *Let $\mathcal{A}_0^-$ be the smallest class of functionals containing $0, I, s_0, s_1$, BIT, $|x|$, $\#$, Ap and closed under functional composition, expansion and* CRN. *Similarly define $\mathcal{A}_k^-$ and $\mathcal{A}^-$ to be the corresponding classes without the functional substitution scheme.*

The following proof was generously furnished by K.-H. Niggl, using the technique from Lemma 6.1 of [Nig98].

Lemma 7.3.3 (K.-H. Niggl). *Let*

$$H(\mathbf{f}, \mathbf{g}, \mathbf{x}), G_1(\mathbf{f}, \mathbf{h}, \mathbf{x}, \mathbf{y}, z_1), \ldots, G_m(\mathbf{f}, \mathbf{h}, \mathbf{x}, \mathbf{y}, z_m)$$

belong to $\mathcal{A}_0^-$. Then there exists H^* in $\mathcal{A}_0^-$ such that for all $\mathbf{f}, \mathbf{h}, \mathbf{x}, \mathbf{y}$,

$$H^*(\mathbf{f}, \mathbf{h}, \mathbf{x}, \mathbf{y}) = H(\mathbf{f}, \overline{\lambda z_i G_i(\mathbf{f}, \mathbf{h}, \mathbf{x}, \mathbf{y}, z_i)}, \mathbf{x}).$$

Proof. By induction on the complexity of H.

Case 1. $H(\mathbf{f},\mathbf{g},\mathbf{x})$ is of the form Ap, so that $\mathbf{f}$ is empty, $\mathbf{g}$ consists of g, and $\mathbf{x}$ consists of x; thus $H(g,x) = Ap(g,x)$.

$$H(\lambda z_1 G_1(\mathbf{h},x,\mathbf{y},z_1),x) = Ap(\lambda z_1 G_1(\mathbf{h},x,\mathbf{y},z_1),x) = G_1(\mathbf{h},x,\mathbf{y},x)$$

so define $H^*(\mathbf{h},x,\mathbf{y},x) = G_1(\mathbf{h},x,\mathbf{y},x)$.

Case 2. H is one of the initial functions $0, s_0, s_1, I_k^n, |x|, \text{BIT}, \#$. In this case H has no function arguments, so take $H^* = H$.

Case 3. H is obtained from $L, K_1, \ldots, K_n$ by functional composition

$$H(\mathbf{f},\mathbf{g},\mathbf{x}) = L(\mathbf{f},\mathbf{g},K_1(\mathbf{f},\mathbf{g},\mathbf{x}),\ldots,K_n(\mathbf{f},\mathbf{g},\mathbf{x}),\mathbf{x}).$$

By the induction hypothesis, there are K_i^* in $\mathcal{A}_0^-$ satisfying

$$K_i^*(\mathbf{f},\mathbf{h},\mathbf{x},\mathbf{y}) = K_i(\mathbf{f},\overline{\lambda z_i G_i(\mathbf{f},\mathbf{h},\mathbf{x},\mathbf{y},z_i)},\mathbf{x}).$$

By expansion, define $G_i'(\mathbf{f},\mathbf{h},\mathbf{k},\mathbf{x},\mathbf{y},z_i) = G_i(\mathbf{f},\mathbf{h},\mathbf{x},\mathbf{y},z_i)$ and apply the induction hypothesis to L and the G_i' to obtain $L^* \in \mathcal{A}_0$ satisfying

$$L^*(\mathbf{f},\mathbf{h},k_1,\ldots,k_n,\mathbf{x},\mathbf{y}) = L(\mathbf{f},\overline{\lambda z_i G_i'(\mathbf{f},\mathbf{h},k_1,\ldots,k_n,\mathbf{x},\mathbf{y},z_i)},k_1,\ldots,k_n,\mathbf{x}).$$

Using functional composition, define H^* by

$$\begin{aligned} H^*(\mathbf{f},\mathbf{h},\mathbf{x},\mathbf{y}) &= L^*(\mathbf{f},\mathbf{h},K_1^*(\mathbf{f},\mathbf{h},\mathbf{x},\mathbf{y}),\ldots,K_n^*(\mathbf{f},\mathbf{h},\mathbf{x},\mathbf{y}),\mathbf{x},\mathbf{y}) \\ &= L(\mathbf{f},\overline{\lambda z_i G_i'(\mathbf{f},\mathbf{h},k_1,\ldots,k_n,\mathbf{x},\mathbf{y},z_i)}, \\ &\qquad K_1^*(\mathbf{f},\mathbf{h},\mathbf{x},\mathbf{y}),\ldots,K_n^*(\mathbf{f},\mathbf{h},\mathbf{x},\mathbf{y}),\mathbf{x}). \end{aligned}$$

Thus $H^*(\mathbf{f},\mathbf{h},\mathbf{x},\mathbf{y})$ equals

$$\begin{aligned} L(&\mathbf{f},\overline{\lambda z_i G_i(\mathbf{f},\mathbf{h},\mathbf{x},\mathbf{y},z_i)}, \\ &K_1(\mathbf{f},\overline{\lambda z_i G_i(\mathbf{f},\mathbf{h},\mathbf{x},\mathbf{y},z_i)},\mathbf{x}),\ldots,K_n(\mathbf{f},\overline{\lambda z_i G_i(\mathbf{f},\mathbf{h},\mathbf{x},\mathbf{y},z_i)},\mathbf{x}),\mathbf{x}). \end{aligned}$$

Case 4. H is obtained by expansion from L.

SUBCASE (a). $H(\mathbf{f},\mathbf{k},\mathbf{g},\mathbf{x},\mathbf{u}) = L(\mathbf{f},\mathbf{x})$.
In this case, $H^*(\mathbf{f},\mathbf{k},\mathbf{g},\mathbf{x},\mathbf{u}) = L(\mathbf{f},\mathbf{x})$, so $H^* \in \mathcal{A}_0^-$ by closure of $\mathcal{A}_0^-$ under expansion.

SUBCASE (b). $H(\mathbf{f},\mathbf{k},\mathbf{g},\mathbf{x},\mathbf{u}) = L(\mathbf{f},\mathbf{g},\mathbf{x})$.
By the induction hypothesis (grouping together $\mathbf{k},\mathbf{h}$ and $\mathbf{u},\mathbf{y}$), there is L^* in $\mathcal{A}_0^-$ satisfying

$$L^*(\mathbf{f},\mathbf{k},\mathbf{h},\mathbf{x},\mathbf{u},\mathbf{y}) = L(\mathbf{f},\overline{\lambda z_i G_i(\mathbf{f},\mathbf{k},\mathbf{h},\mathbf{x},\mathbf{u},\mathbf{y},z_i)},\mathbf{x}).$$

Define H^* by $H^*(\mathbf{f},\mathbf{k},\mathbf{h},\mathbf{x},\mathbf{u},\mathbf{y}) = L^*(\mathbf{f},\mathbf{k},\mathbf{h},\mathbf{x},\mathbf{u},\mathbf{y})$. Then

$$\begin{aligned} H^*(\mathbf{f},\mathbf{k},\mathbf{h},\mathbf{x},\mathbf{u},\mathbf{y}) &= L^*(\mathbf{f},\mathbf{k},\mathbf{h},\mathbf{x},\mathbf{u},\mathbf{y}) \\ &= L(\mathbf{f},\overline{\lambda z_i G_i(\mathbf{f},\mathbf{k},\mathbf{h},\mathbf{x},\mathbf{u},\mathbf{y},z_i)},\mathbf{x}) \\ &= H(\mathbf{f},\mathbf{k},\overline{\lambda z_i G_i(\mathbf{f},\mathbf{k},\mathbf{h},\mathbf{x},\mathbf{u},\mathbf{y},z_i)},\mathbf{x},\mathbf{u}). \end{aligned}$$

Case 5. H is obtained by CRN from L, J, K:

$$\begin{aligned} H(\mathbf{f},\mathbf{g},\mathbf{x},0) &= L(\mathbf{f},\mathbf{g},\mathbf{x}) \\ H(\mathbf{f},\mathbf{g},\mathbf{x},s_0(y)) &= H(\mathbf{f},\mathbf{g},\mathbf{x},y)^\frown \mathrm{MOD2}(J(\mathbf{f},\mathbf{g},\mathbf{x},y)), \text{ if } y \neq 0 \\ H(\mathbf{f},\mathbf{g},\mathbf{x},s_1(y)) &= H(\mathbf{f},\mathbf{g},\mathbf{x},y)^\frown \mathrm{MOD2}(K(\mathbf{f},\mathbf{g},\mathbf{x},y)). \end{aligned}$$

By expansion, let

$$\mathcal{G}_i(\mathbf{f},\mathbf{h},\mathbf{x},y,u,\mathbf{y},z_i) = G_i(\mathbf{f},\mathbf{h},\mathbf{x},u,\mathbf{y},z_i).$$

By the induction hypothesis, there are L^*, J^*, K^* in $\mathcal{A}_0^-$ satisfying

$$\begin{aligned} L^*(\mathbf{f},\mathbf{h},\mathbf{x},u,\mathbf{y}) &= L(\mathbf{f},\overline{\lambda z_i G_i(\mathbf{f},\mathbf{h},\mathbf{x},u,\mathbf{y},z_i)},\mathbf{x}) \\ J^*(\mathbf{f},\mathbf{h},\mathbf{x},y,u,\mathbf{y}) &= J(\mathbf{f},\overline{\lambda z_i \mathcal{G}_i(\mathbf{f},\mathbf{h},\mathbf{x},y,u,\mathbf{y},z_i)},\mathbf{x},y) \\ K^*(\mathbf{f},\mathbf{h},\mathbf{x},y,u,\mathbf{y}) &= K(\mathbf{f},\overline{\lambda z_i \mathcal{G}_i(\mathbf{f},\mathbf{h},\mathbf{x},y,u,\mathbf{y},z_i)},\mathbf{x},y). \end{aligned}$$

Using the projection functions I_k^n, let

$$\begin{aligned} J^{**}(\mathbf{f},\mathbf{h},\mathbf{x},u,\mathbf{y},y) &= J^*(\mathbf{f},\mathbf{h},\mathbf{x},y,u,\mathbf{y}) \\ K^{**}(\mathbf{f},\mathbf{h},\mathbf{x},u,\mathbf{y},y) &= K^*(\mathbf{f},\mathbf{h},\mathbf{x},y,u,\mathbf{y}) \end{aligned}$$

and define $F \in \mathcal{A}_0^-$ by CRN from L^*, J^{**}, K^{**}.

CLAIM. For all $\mathbf{f},\mathbf{h},\mathbf{x},u,\mathbf{y},y$, we have

$$F(\mathbf{f},\mathbf{h},\mathbf{x},u,\mathbf{y},y) = H(\mathbf{f},\overline{\lambda z_i G_i(\mathbf{f},\mathbf{h},\mathbf{x},u,\mathbf{y},z_i)},\mathbf{x},y).$$

Proof of Claim. By induction on notation on y.

$$\begin{aligned} F(\mathbf{f},\mathbf{h},\mathbf{x},u,\mathbf{y},0) &= L^*(\mathbf{f},\mathbf{h},\mathbf{x},u,\mathbf{y}) \\ &= L(\mathbf{f},\overline{\lambda z_i G_i(\mathbf{f},\mathbf{h},\mathbf{x},u,\mathbf{y},z_i)},\mathbf{x}) \\ &= H(\mathbf{f},\overline{\lambda z_i G_i(\mathbf{f},\mathbf{h},\mathbf{x},u,\mathbf{y},z_i)},\mathbf{x},0). \end{aligned}$$

Now assume that the claim holds for y, and consider $s_0(y)$. Temporarily abbreviate $\mathrm{MOD2}(J(\mathbf{f},\overline{\lambda z_i G_i(\mathbf{f},\mathbf{h},\mathbf{x},u,\mathbf{y},z_i)},\mathbf{x},y))$ by $J2u$.

$$\begin{aligned} F(\mathbf{f},\mathbf{h},\mathbf{x},u,\mathbf{y},s_0(y)) &= F(\mathbf{f},\mathbf{h},\mathbf{x},u,\mathbf{y},y)*\mathrm{MOD2}(J^{**}(\mathbf{f},\mathbf{h},\mathbf{x},u,\mathbf{y},y)) \\ &= F(\mathbf{f},\mathbf{h},\mathbf{x},u,\mathbf{y},y)*\mathrm{MOD2}(J^{*}(\mathbf{f},\mathbf{h},\mathbf{x},y,u,\mathbf{y})) \\ &= F(\mathbf{f},\mathbf{h},\mathbf{x},u,\mathbf{y},y)* \\ &\quad \mathrm{MOD2}(J(\mathbf{f},\overline{\lambda z_i \mathcal{G}_i(\mathbf{f},\mathbf{h},\mathbf{x},y,u,\mathbf{y},z_i)},\mathbf{x},y)) \\ &= F(\mathbf{f},\mathbf{h},\mathbf{x},u,\mathbf{y},y)*J2u \\ &= H(\mathbf{f},\overline{\lambda z_i G_i(\mathbf{f},\mathbf{h},\mathbf{x},u,\mathbf{y},z_i)},\mathbf{x},y)*J2u \\ &= H(\mathbf{f},\overline{\lambda z_i G_i(\mathbf{f},\mathbf{h},\mathbf{x},u,\mathbf{y},z_i)},\mathbf{x},s_0(y)). \end{aligned}$$

Finally, in an analogous manner, one establishes that

$$F(\mathbf{f},\mathbf{h},\mathbf{x},u,\mathbf{y},s_1(y)) = H(\mathbf{f},\overline{\lambda z_i G_i(\mathbf{f},\mathbf{h},\mathbf{x},u,\mathbf{y},z_i)},\mathbf{x},s_1(y)).$$

This completes the proof of the claim. From the claim, substituting u by $s_0(y)$ $(s_1(y))$, we have

$$\begin{aligned}
H^*(\mathbf{f},\mathbf{h},\mathbf{x},0,\mathbf{y}) &= H(\mathbf{f},\overline{\lambda z_i G_i(\mathbf{f},\mathbf{h},\mathbf{x},0,\mathbf{y},z_i)},\mathbf{x},0)\\
&= F(\mathbf{f},\mathbf{h},\mathbf{x},0,\mathbf{y},0)\\
H^*(\mathbf{f},\mathbf{h},\mathbf{x},s_0(y),\mathbf{y}) &= H(\mathbf{f},\overline{\lambda z_i G_i(\mathbf{f},\mathbf{h},\mathbf{x},s_0(y),\mathbf{y},z_i)},\mathbf{x},s_0(y))\\
&= F(\mathbf{f},\mathbf{h},\mathbf{x},s_0(y),\mathbf{y},s_0(y))\\
H^*(\mathbf{f},\mathbf{h},\mathbf{x},s_1(y),\mathbf{y}) &= H(\mathbf{f},\overline{\lambda z_i G_i(\mathbf{f},\mathbf{h},\mathbf{x},s_1(y),\mathbf{y},z_i)},\mathbf{x},s_1(y))\\
&= F(\mathbf{f},\mathbf{h},\mathbf{x},s_1(y),\mathbf{y},s_1(y)).
\end{aligned}$$

It follows that

$$H^*(\mathbf{f},\mathbf{h},\mathbf{x},y,\mathbf{y}) = F(\mathbf{f},\mathbf{h},\mathbf{x},y,\mathbf{y},y)$$

and since $F \in \mathcal{A}_0^-$, we have $H^* \in \mathcal{A}_0^-$.

Using Niggl's technique from the previous lemma, we have the following.

Lemma 7.3.4. *If* $H(\mathbf{f},\mathbf{g},\mathbf{x})$, $G_1(\mathbf{f},\mathbf{h},\mathbf{x},\mathbf{y},z_1),\ldots,G_m(\mathbf{f},\mathbf{h},\mathbf{x},\mathbf{y},z_m)$ *belong to* $\mathcal{A}^-$ *then there exists* H^* *in* $\mathcal{A}^-$ *such that for all* $\mathbf{f},\mathbf{h},\mathbf{x},\mathbf{y}$,

$$H^*(\mathbf{f},\mathbf{h},\mathbf{x},\mathbf{y}) = H(\mathbf{f},\overline{\lambda z_i G_i(\mathbf{f},\mathbf{h},\mathbf{x},\mathbf{y},z_i)},\mathbf{x}).$$

Proof. By induction on formation of H. Because of the previous lemma, it only remains to show closure under WBRN. Suppose that H is defined by WBRN from L, J, K, B:

$$\begin{aligned}
\widetilde{H}(\mathbf{f},\mathbf{g},\mathbf{x},0) &= L(\mathbf{f},\mathbf{g},\mathbf{x})\\
\widetilde{H}(\mathbf{f},\mathbf{g},\mathbf{x},s_0(y)) &= J(\mathbf{f},\mathbf{g},\mathbf{x},y,\widetilde{H}(\mathbf{f},\mathbf{g},\mathbf{x},y)),\ \text{if } y \neq 0,\\
\widetilde{H}(\mathbf{f},\mathbf{g},\mathbf{x},s_1(y)) &= K(\mathbf{f},\mathbf{g},\mathbf{x},y,\widetilde{H}(\mathbf{f},\mathbf{g},\mathbf{x},y))\\
H(\mathbf{f},\mathbf{g},\mathbf{x},y) &= \widetilde{H}(\mathbf{f},\mathbf{g},\mathbf{x},|y|)
\end{aligned}$$

provided that

$$\widetilde{H}(\mathbf{f},\mathbf{g},\mathbf{x},y) \leq B(\mathbf{f},\mathbf{g},\mathbf{x},y)$$

for all $\mathbf{f},\mathbf{y},\mathbf{x},y$. By expansion, let

$$\begin{aligned}
\mathcal{G}_i(\mathbf{f},\mathbf{h},\mathbf{x},h,y,u,\mathbf{y},z_i) &= G_i(\mathbf{f},\mathbf{h},\mathbf{x},u,\mathbf{y},z_i)\\
G_i'(\mathbf{f},\mathbf{h},\mathbf{x},y,u,\mathbf{y},z_i) &= G_i(\mathbf{f},\mathbf{h},\mathbf{x},u,\mathbf{y},z_i).
\end{aligned}$$

By the induction hypothesis, there are L^*, J^*, K^*, B^* in $\mathcal{A}^-$ satisfying

$$\begin{aligned}
L^*(\mathbf{f},\mathbf{h},\mathbf{x},u,\mathbf{y}) &= L(\mathbf{f},\overline{\lambda z_i G_i(\mathbf{f},\mathbf{h},\mathbf{x},u,\mathbf{y},z_i)},\mathbf{x})\\
J^*(\mathbf{f},\mathbf{h},\mathbf{x},h,y,u,\mathbf{y}) &= J(\mathbf{f},\overline{\lambda z_i \mathcal{G}_i(\mathbf{f},\mathbf{h},\mathbf{x},h,y,u,\mathbf{y},z_i)},\mathbf{x},y,h)\\
K^*(\mathbf{f},\mathbf{h},\mathbf{x},h,y,u,\mathbf{y}) &= K(\mathbf{f},\overline{\lambda z_i \mathcal{G}_i(\mathbf{f},\mathbf{h},\mathbf{x},h,y,u,\mathbf{y},z_i)},\mathbf{x},y,h)\\
B^*(\mathbf{f},\mathbf{h},\mathbf{x},y,u,\mathbf{y}) &= B(\mathbf{f},\overline{\lambda z_i G_i'(\mathbf{f},\mathbf{h},\mathbf{x},y,u,\mathbf{y},z_i)},\mathbf{x},y).
\end{aligned}$$

Using the projection functions I_k^n, let

$$\begin{aligned}
J^{**}(\mathbf{f},\mathbf{h},\mathbf{x},u,\mathbf{y},y,h) &= J^*(\mathbf{f},\mathbf{h},\mathbf{x},h,y,u,\mathbf{y})\\
K^{**}(\mathbf{f},\mathbf{h},\mathbf{x},u,\mathbf{y},y,h) &= K^*(\mathbf{f},\mathbf{h},\mathbf{x},h,y,u,\mathbf{y})\\
B^{**}(\mathbf{f},\mathbf{h},\mathbf{x},u,\mathbf{y},y) &= B^*(\mathbf{f},\mathbf{h},\mathbf{x},y,u,\mathbf{y})
\end{aligned}$$

and define $\widetilde{F}$ by BRN (not WBRN) from L^*, J^{**}, K^{**}, B^{**}.

CLAIM. For all $\mathbf{f},\mathbf{h},\mathbf{x},u,\mathbf{y},y$, we have

$$\widetilde{F}(\mathbf{f},\mathbf{h},\mathbf{x},u,\mathbf{y},y) = \widetilde{H}(\mathbf{f},\overline{\lambda z_i G_i(\mathbf{f},\mathbf{h},\mathbf{x},u,\mathbf{y},z_i)},\mathbf{x},y).$$

Proof of Claim. By induction on notation on y.

$$\begin{aligned}
\widetilde{F}(\mathbf{f},\mathbf{h},\mathbf{x},u,\mathbf{y},0) &= L^*(\mathbf{f},\mathbf{h},\mathbf{x},u,\mathbf{y})\\
&= L(\mathbf{f},\overline{\lambda z_i G_i(\mathbf{f},\mathbf{h},\mathbf{x},u,\mathbf{y},z_i)},\mathbf{x})\\
&= \widetilde{H}(\mathbf{f},\overline{\lambda z_i G_i(\mathbf{f},\mathbf{h},\mathbf{x},u,\mathbf{y},z_i)},\mathbf{x},0).
\end{aligned}$$

Now assume that the claim holds for y, and consider $s_0(y)$. Temporarily, abbreviate $\widetilde{F}(\mathbf{f},\mathbf{h},\mathbf{x},u,\mathbf{y},y)$ by $\widetilde{F}y$ and $\widetilde{H}(\mathbf{f},\overline{\lambda z_i G_i(\mathbf{f},\mathbf{h},\mathbf{x},u,\mathbf{y},z_i)},\mathbf{x},y)$ by $\widetilde{H}y$. Then

$$\begin{aligned}
\widetilde{F}(\mathbf{f},\mathbf{h},\mathbf{x},u,\mathbf{y},s_0(y)) &= J^{**}(\mathbf{f},\mathbf{h},\mathbf{x},u,\mathbf{y},y,\widetilde{F}y)\\
&= J^*(\mathbf{f},\mathbf{h},\mathbf{x},\widetilde{F}y,y,u,\mathbf{y})\\
&= J(\mathbf{f},\overline{\lambda z_i \mathcal{G}_i(\mathbf{f},\mathbf{h},\mathbf{x},\widetilde{F}y,y,u,\mathbf{y},z_i)},\mathbf{x},y,\widetilde{F}y)\\
&= J(\mathbf{f},\overline{\lambda z_i \mathcal{G}_i(\mathbf{f},\mathbf{h},\mathbf{x},\widetilde{H}y,y,u,\mathbf{y},z_i)},\mathbf{x},y,\widetilde{H}y)\\
&= J(\mathbf{f},\overline{\lambda z_i G_i(\mathbf{f},\mathbf{h},\mathbf{x},u,\mathbf{y},z_i)},\mathbf{x},y,\widetilde{H}y)\\
&= \widetilde{H}(\mathbf{f},\overline{\lambda z_i G_i(\mathbf{f},\mathbf{h},\mathbf{x},u,\mathbf{y},z_i)},\mathbf{x},s_0(y)).
\end{aligned}$$

Finally, in an analogous manner, one establishes that

$$\widetilde{F}(\mathbf{f},\mathbf{h},\mathbf{x},u,\mathbf{y},s_1(y)) = \widetilde{H}(\mathbf{f},\overline{\lambda z_i G_i(\mathbf{f},\mathbf{h},\mathbf{x},u,\mathbf{y},z_i)},\mathbf{x},s_1(y)).$$

This completes the proof of the claim.

It follows that $F \in \mathcal{A}^-$, where

$$\begin{aligned}
F(\mathbf{f},\mathbf{h},\mathbf{x},u,\mathbf{y},y) &= \widetilde{F}(\mathbf{f},\mathbf{h},\mathbf{x},u,\mathbf{y},|y|)\\
&= \widetilde{H}(\mathbf{f},\overline{\lambda z_i G_i(\mathbf{f},\mathbf{h},\mathbf{x},u,\mathbf{y},z_i)},\mathbf{x},|y|)\\
&= H(\mathbf{f},\overline{\lambda z_i G_i(\mathbf{f},\mathbf{h},\mathbf{x},u,\mathbf{y},z_i)},\mathbf{x},y).
\end{aligned}$$

Substituting y for u, we obtain

$$\begin{aligned} H^*(\mathbf{f},\mathbf{h},\mathbf{x},y,\mathbf{y}) &= H(\mathbf{f},\overline{\lambda z_i G_i(\mathbf{f},\mathbf{h},\mathbf{x},y,\mathbf{y},z_i)},\mathbf{x},y) \\ &= F(\mathbf{f},\mathbf{h},\mathbf{x},y,\mathbf{y},y). \end{aligned}$$

Thus $H^*(\mathbf{f},\mathbf{h},\mathbf{x},y,\mathbf{y})$ belongs to $\mathcal{A}^-$.

Corollary 7.3.1. *For all $k \geq 0$, $\mathcal{A}_k^- = \mathcal{A}_k$, and $\mathcal{A}^- = \mathcal{A}$.*

Proof. By induction on the complexity of $F \in \mathcal{A}_k$, show the existence of $F^* \in \mathcal{A}_k^-$ such that $\forall \mathbf{f},\mathbf{x}(F(\mathbf{f},\mathbf{x}) = F^*(\mathbf{f},\mathbf{x}))$. The only difficult case is when F is defined by functional substitution, and this is handled in lemmas 7.3.3 and 7.3.4 by taking $\mathbf{h},\mathbf{y}$ to be empty.

7.4 Square-Root and Multiple Recursion

This section consists of several technical results which show that the algebra $\mathcal{A}$ is closed under particular kinds of *simultaneous* weak bounded recursion on notation. These closure properties are crucial for proof of the main theorem of this chapter, where we show that $\mathcal{A}$ consists of exactly those type 2 functionals in the analogue of NC, i.e., polylogarithmic time with a polynomial processor bound on a parallel random access machine.

Recall from Definition 7.3.2, F is defined by weak bounded recursion on notation (WBRN) from G, H_0, H_1, K if

$$\begin{aligned} \widetilde{F}(\mathbf{f},\mathbf{x},0) &= G(\mathbf{f},\mathbf{x}) \\ \widetilde{F}(\mathbf{f},\mathbf{x},s_i(y)) &= H_i(\mathbf{f},\mathbf{x},y,\widetilde{F}(\mathbf{f},\mathbf{x},y)) \\ F(\mathbf{f},\mathbf{x},y) &= \widetilde{F}(\mathbf{f},\mathbf{x},|y|) \end{aligned}$$

provided $\widetilde{F}(\mathbf{f},\mathbf{x},y) \leq K(\mathbf{f},\mathbf{x},y)$ for all $\mathbf{x},y$. In other words, $\widetilde{F}$ is defined by bounded recursion on notation, and $F(\mathbf{f},\mathbf{x},y) = \widetilde{F}(\mathbf{f},\mathbf{x},|y|)$. In line with our remarks on *limited* versus *bounded* recursion schemes, we leave the definition of weak limited recursion on notation, and related limited schemes to the reader.

Definition 7.4.1. *F is defined by* bounded weak recursion on notation *(BWRN) from G, H_0, H_1, K if*

$$\begin{aligned} \widetilde{F}(\mathbf{f},\mathbf{x},0) &= G(\mathbf{f},\mathbf{x}) \\ \widetilde{F}(\mathbf{f},\mathbf{x},s_i(y)) &= H_i(\mathbf{f},\mathbf{x},y,\widetilde{F}(\mathbf{f},\mathbf{x},y)) \\ F(\mathbf{f},\mathbf{x},y) &= \widetilde{F}(\mathbf{f},\mathbf{x},|y|) \end{aligned}$$

provided $F(\mathbf{f},\mathbf{x},y) \leq K(\mathbf{f},\mathbf{x},y)$ for all $\mathbf{x},y$. In other words, F is defined by weak recursion on notation, and is moreover bounded.

Note that WBRN requires that $\widetilde{F}(\mathbf{f},\mathbf{x},y) \leq K(\mathbf{f},\mathbf{x},y)$, whereas BWRN requires that $F(\mathbf{f},\mathbf{x},y) = \widetilde{F}(\mathbf{f},\mathbf{x},|y|) \leq K(\mathbf{f},\mathbf{x},y)$.

Lemma 7.4.1. *Let $\mathcal{A}'$ be defined as $\mathcal{A}$ with* BWRN *in place of* WBRN. *Then* $\mathcal{A} = \mathcal{A}'$.

Proof. Clearly WBRN implies BWRN, since if $\widetilde{F}(\mathbf{f}, \mathbf{x}, y) \le K(\mathbf{f}, \mathbf{x}, y)$ then $F(\mathbf{f}, \mathbf{x}, y) \le K(\mathbf{f}, \mathbf{x}, |y|)$. Thus $\mathcal{A} \subseteq \mathcal{A}'$.

We prove the containment $\mathcal{A}' \subseteq \mathcal{A}$ by induction on formation of functionals. Suppose that $G, H_i, K \in \mathcal{A}$, and that F is defined by BWRN from $G, H_i, K \in \mathcal{A}$, i.e.,

$$\begin{aligned}\widetilde{F}(\mathbf{f}, \mathbf{x}, 0) &= G(\mathbf{f}, \mathbf{x})\\ \widetilde{F}(\mathbf{f}, \mathbf{x}, s_i(y)) &= H_i(\mathbf{f}, \mathbf{x}, y, \widetilde{F}(\mathbf{f}, \mathbf{x}, y))\\ F(\mathbf{f}, \mathbf{x}, y) &= \widetilde{F}(\mathbf{f}, \mathbf{x}, |y|)\end{aligned}$$

and

$$F(\mathbf{f}, \mathbf{x}, y) \le K(\mathbf{f}, \mathbf{x}, y)$$

for all $\mathbf{f}, \mathbf{x}, y$. Now $F \in \mathcal{A}'$, and we must show that $F \in \mathcal{A}$. Define

$$\widetilde{F}'(\mathbf{f}, \mathbf{x}, y, z) = \begin{cases} \widetilde{F}(\mathbf{f}, \mathbf{x}, y) & \text{if } 2 \cdot \lfloor y/2 \rfloor \le |z| \\ z & \text{else} \end{cases}$$

and let

$$K'(\mathbf{f}, \mathbf{x}, y, z) = \begin{cases} K(\mathbf{f}, \mathbf{x}, y) & \text{if } 2 \cdot \lfloor y/2 \rfloor \le |z| \\ z & \text{else.} \end{cases}$$

Let

$$G'(\mathbf{f}, \mathbf{x}, z) = G(\mathbf{f}, \mathbf{x})$$

and

$$H_i'(\mathbf{f}, \mathbf{x}, y, u, v) = \begin{cases} H_i(\mathbf{f}, \mathbf{x}, y, u) & \text{if } s_i(y) \le |v| \\ v & \text{else.} \end{cases}$$

Note that for $y \le |z|$, $\widetilde{F}'(\mathbf{f}, \mathbf{x}, y, z) = \widetilde{F}(\mathbf{f}, \mathbf{x}, y)$ and that $\widetilde{F}'(\mathbf{f}, \mathbf{x}, y, z) \le K'(\mathbf{f}, \mathbf{x}, y, z)$ for all $\mathbf{f}, \mathbf{x}, y, z$. Since $G, H_i, K \in \mathcal{A}$, it follows by the induction hypothesis that $G', H_i', K' \in \mathcal{A}$.

CLAIM. $\widetilde{F}'$ is defined by BRN from G', H_i', K'.

Proof of Claim. For all z, $0 \le |z|$, so

$$\begin{aligned}\widetilde{F}'(\mathbf{f}, \mathbf{x}, 0, z) &= \widetilde{F}(\mathbf{f}, \mathbf{x}, 0)\\ &= G(\mathbf{f}, \mathbf{x})\\ &= G'(\mathbf{f}, \mathbf{x}, z).\end{aligned}$$

Assume now that $2 \cdot \lfloor s_i(y)/2 \rfloor = s_0(y) \le |z|$, and so $2 \cdot \lfloor y/2 \rfloor \le |z|$.

$$
\begin{aligned}
\widetilde{F}'(\mathbf{f},\mathbf{x},s_i(y),z) &= \widetilde{F}(\mathbf{f},\mathbf{x},s_i(y)) \\
&= H_i(\mathbf{f},\mathbf{x},y,\widetilde{F}(\mathbf{f},\mathbf{x},y)) \\
&= H_i(\mathbf{f},\mathbf{x},y,\widetilde{F}'(\mathbf{f},\mathbf{x},y,z)) \\
&= H_i'(\mathbf{f},\mathbf{x},y,\widetilde{F}'(\mathbf{f},\mathbf{x},y,z),z).
\end{aligned}
$$

Assume now that $2 \cdot \lfloor s_i(y)/2 \rfloor = s_0(y) > |z|$. Then

$$
\begin{aligned}
\widetilde{F}'(\mathbf{f},\mathbf{x},s_i(y),z) &= z \\
&= H_i'(\mathbf{f},\mathbf{x},y,\widetilde{F}'(\mathbf{f},\mathbf{x},y,z),z).
\end{aligned}
$$

This establishes the claim. Define $F'(\mathbf{f},\mathbf{x},y,z) = \widetilde{F}'(\mathbf{f},\mathbf{x},|y|,z)$. Then F' is defined by WBRN from G', H_i', K', hence belongs to $\mathcal{A}$. Moreover

$$
F'(\mathbf{f},\mathbf{x},y,y) = \widetilde{F}'(\mathbf{f},\mathbf{x},|y|,y) = \widetilde{F}(\mathbf{f},\mathbf{x},|y|) = F(\mathbf{f},\mathbf{x},y)
$$

so that $F \in \mathcal{A}$.

Definition 7.4.2. *$F_1,\ldots,F_k$ are defined from* $\mathbf{G},\mathbf{H},\mathbf{K}$ *by simultaneous weak limited recursion on notation* SWLRN *if for all* $\mathbf{f},\mathbf{x},y$

$$
\begin{aligned}
&\widetilde{F}_i(\mathbf{f},\mathbf{x},0) = G_i(\mathbf{f},\mathbf{x}), \quad 1 \le i \le k \\
&\widetilde{F}_i(\mathbf{f},\mathbf{x},y) = H_i\left(\mathbf{f},\mathbf{x},y,\widetilde{F}_1\left(\mathbf{f},\mathbf{x},\left\lfloor \frac{y}{2} \right\rfloor\right),\ldots,\widetilde{F}_k\left(\mathbf{f},\mathbf{x},\left\lfloor \frac{y}{2} \right\rfloor\right)\right), \\
&\qquad y > 0, \quad 1 \le i \le k \\
&\widetilde{F}_i(\mathbf{f},\mathbf{x},y) \le K_i(\mathbf{f},\mathbf{x},y), \quad 1 \le i \le k \\
&F_i(\mathbf{f},\mathbf{x},y) = \widetilde{F}_i(\mathbf{f},\mathbf{x},|y|), \quad 1 \le i \le k.
\end{aligned}
$$

In other words, $\widetilde{F}_1,\ldots,\widetilde{F}_k$ are defined from $\mathbf{G},\mathbf{H},\mathbf{K}$ *by simultaneous limited recursion on notation* SLRN *and $F_i(\mathbf{f},\mathbf{x},y) = \widetilde{F}(\mathbf{f},\mathbf{x},|y|)$.*

Lemma 7.4.2. *If* $\mathbf{G},\mathbf{H},\mathbf{K}$ *belong to $\mathcal{A}$ and $F_1,\ldots,F_k$ are defined from* $\mathbf{G},\mathbf{H},\mathbf{K}$ *by* SWLRN *then $F_1,\ldots,F_k$ belong to $\mathcal{A}$.*

Proof. Define

$$
\widetilde{F}(\mathbf{f},\mathbf{x},0) = \langle G_1(\mathbf{f},\mathbf{x},0),\ldots,G_k(\mathbf{f},\mathbf{x},0)\rangle
$$

$$
\begin{aligned}
\widetilde{F}(\mathbf{f},\mathbf{x},y) = \Big\langle & H_1(\mathbf{f},\mathbf{x},y,\Pi_1^k(\widetilde{F}(\mathbf{f},\mathbf{x},\lfloor \tfrac{y}{2} \rfloor)),\ldots,\Pi_k^k(\widetilde{F}(\mathbf{f},\mathbf{x},\lfloor \tfrac{y}{2} \rfloor))),\ldots, \\
& H_k(\mathbf{f},\mathbf{x},y,\Pi_1^k(\widetilde{F}(\mathbf{f},\mathbf{x},\lfloor \tfrac{y}{2} \rfloor)),\ldots,\Pi_k^k(\widetilde{F}(\mathbf{f},\mathbf{x},\lfloor \tfrac{y}{2} \rfloor)))\Big\rangle,
\end{aligned}
$$

where the bottom equation above holds for $y > 0$. Then

$$
\widetilde{F}(\mathbf{f},\mathbf{x},y) \le \langle K_1(\mathbf{f},\mathbf{x},y),\ldots,K_k(\mathbf{f},\mathbf{x},y)\rangle
$$

and define

$$
F(\mathbf{f},\mathbf{x},y) = \widetilde{F}(\mathbf{f},\mathbf{x},|y|).
$$

Since the k-tupling function, here represented as $\langle Z_1, \ldots, Z_k \rangle$ and the projections $\Pi_i^k(\langle Z_1, \ldots, Z_k \rangle) = Z_i$ belong to $\mathcal{A}$, by WLRN it follows that $F \in \mathcal{A}$. Since $F_i(\mathbf{f}, \mathbf{x}, y) = \Pi_i^k(F(\mathbf{f}, \mathbf{x}, y))$, it follows that $F_1, \ldots F_k \in \mathcal{A}$.

Definition 7.4.3. *Define the function $sqrt(x)$ to be* $\text{MSP}(x, \text{TR}(|s_0(x)|))$, *where* $\text{TR}(x) = \lfloor x/2 \rfloor$.

Note that $\text{sqrt}(0) = 0$. Suppose that $|x| = n \geq 1$. Then $\text{TR}(|s_0(x)|) = \lfloor \frac{n+1}{2} \rfloor$, so $\text{MSP}(x, \text{TR}(|s_0(x)|))$ consists of the result of truncating the rightmost $\lfloor \frac{n+1}{2} \rfloor$ bits from the length n string representation of x. Thus

$$|\text{sqrt}(x)| = |x| - \left\lfloor \frac{|x|+1}{2} \right\rfloor = \left\lfloor \frac{|x|}{2} \right\rfloor.$$

This observation will be used in the proof of Theorem 7.4.1 and Lemma 7.4.4. Recall that $f^{(i)}(x)$ is the i-fold iteration of function f applied to x, so that $\text{sqrt}^{(0)}(y) = y$ and $\text{sqrt}^{(i+1)}(y) = \text{sqrt}(\text{sqrt}^{(i)}(y))$. The function $\text{sqrt}(y)$ has growth rate roughly that of $\sqrt{y}$, hence the name. By induction, it is simple to establish that the smallest value t for which $\text{sqrt}^{(t)}(y) = 0$ satisfies $t \leq 2 \cdot ||y||$.

Definition 7.4.4. *F is defined from G, H, K by* limited square-root recursion *(LSR) if*

$$F(\mathbf{f}, \mathbf{x}, 0) = G(\mathbf{f}, \mathbf{x})$$
$$F(\mathbf{f}, \mathbf{x}, y) = H(\mathbf{f}, \mathbf{x}, y, F(\mathbf{f}, \mathbf{x}, sqrt(y))), \text{ if } y \neq 0$$

provided that $F(\mathbf{f}, \mathbf{x}, y) \leq K(\mathbf{f}, \mathbf{x}, y)$ for all $\mathbf{f}, \mathbf{x}, y$.

Theorem 7.4.1. Let $\mathcal{A}'$ ($\mathcal{A}''$) be defined as $\mathcal{A}$, but with LWRN (LSR) in place of WBRN. Then $\mathcal{A}' = \mathcal{A}''$, hence both equal $\mathcal{A}$.

Proof. Consider the direction $\mathcal{A}' \subseteq \mathcal{A}''$. The proof is by induction on formation of functionals. Suppose that F is defined by LWRN from $G, H, K \in \mathcal{A}''$, so that $F(\mathbf{f}, \mathbf{x}, y) = \widetilde{F}(\mathbf{f}, \mathbf{x}, |y|)$, where

$$\widetilde{F}(\mathbf{f}, \mathbf{x}, 0) = G(\mathbf{f}, \mathbf{x})$$
$$\widetilde{F}(\mathbf{f}, \mathbf{x}, y) = H(\mathbf{f}, \mathbf{x}, y, \widetilde{F}(\mathbf{f}, \mathbf{x}, \lfloor \tfrac{y}{2} \rfloor)), \text{ if } y \neq 0$$

provided that $F(\mathbf{f}, \mathbf{x}, y) \leq K(\mathbf{f}, \mathbf{x}, y)$ for all $\mathbf{f}, \mathbf{x}, y$. Then

$$\begin{aligned} F(\mathbf{f}, \mathbf{x}, 0) &= \widetilde{F}(\mathbf{f}, \mathbf{x}, |0|) \\ &= \widetilde{F}(\mathbf{f}, \mathbf{x}, 0) \\ &= G(\mathbf{f}, \mathbf{x}) \end{aligned}$$

and for $y > 0$

$$
\begin{aligned}
F(\mathbf{f},\mathbf{x},y) &= \widetilde{F}(\mathbf{f},\mathbf{x},|y|)\\
&= H(\mathbf{f},\mathbf{x},|y|,\widetilde{F}(\mathbf{f},\mathbf{x},\lfloor\frac{|y|}{2}\rfloor))\\
&= H(\mathbf{f},\mathbf{x},|y|,\widetilde{F}(\mathbf{f},\mathbf{x},|\mathrm{sqrt}(y)|))\\
&= H(\mathbf{f},\mathbf{x},|y|,F(\mathbf{f},\mathbf{x},\mathrm{sqrt}(y))).
\end{aligned}
$$

Letting $H'(\mathbf{f},\mathbf{x},y,z) = H(\mathbf{f},\mathbf{x},|y|,z)$, it follows that F is defined by LSR from G, H', K. Now $|x| \in A_0$ and $H \in \mathcal{A}''$, so $H' \in \mathcal{A}''$, hence $F \in \mathcal{A}''$.

Consider the direction $\mathcal{A}'' \subseteq \mathcal{A}'$. Suppose that F is defined by LSR from $G, H, K \in \mathcal{A}'$, so that

$$
\begin{aligned}
F(\mathbf{f},\mathbf{x},0) &= G(\mathbf{f},\mathbf{x})\\
F(\mathbf{f},\mathbf{x},y) &= H(\mathbf{f},\mathbf{x},y,F(\mathbf{f},\mathbf{x},\mathrm{sqrt}(y)))
\end{aligned}
$$

provided that $F(\mathbf{f},\mathbf{x},y) \leq K(\mathbf{f},\mathbf{x},y)$ for all $\mathbf{f},\mathbf{x},y$. Define

$$
\begin{aligned}
\widetilde{H}(\mathbf{f},\mathbf{x},y,u,z) &= H(\mathbf{f},\mathbf{x},\mathrm{MSP}(y,|y| \dot{-} u),z)\\
\widetilde{F}(\mathbf{f},\mathbf{x},y,0) &= G(\mathbf{f},\mathbf{x})\\
\widetilde{F}(\mathbf{f},\mathbf{x},y,u) &= \widetilde{H}(\mathbf{f},\mathbf{x},y,|u|,\widetilde{F}(\mathbf{f},\mathbf{x},y,\mathrm{sqrt}(u))).
\end{aligned}
$$

CLAIM 1. $\mathrm{MSP}(y,|y| \dot{-} |\mathrm{sqrt}^{(\ell)}(y)|) = \mathrm{sqrt}^{(\ell)}(y)$, for all $\ell \geq 0$.

Proof of Claim 1. By induction on ℓ.

BASE STEP Let $\ell = 0$.

$$
\begin{aligned}
\mathrm{MSP}(y,|y| \dot{-} |\mathrm{sqrt}^{(0)}(y)|) &= \mathrm{MSP}(y,|y| \dot{-} |y|)\\
&= y\\
&= \mathrm{sqrt}^{(0)}(y).
\end{aligned}
$$

INDUCTION STEP. Let $\ell \geq 0$, and assume the claim holds for values less than or equal to ℓ.

$$
\begin{aligned}
&\mathrm{MSP}(y,|y| \dot{-} |\mathrm{sqrt}^{(\ell+1)}(y)|)\\
&= \mathrm{MSP}(y,|y| \dot{-} |\mathrm{sqrt}^{(\ell)}(y)| + |\mathrm{sqrt}^{(\ell)}(y)| \dot{-} |\mathrm{sqrt}^{(\ell+1)}(y)|)\\
&= \mathrm{MSP}(\mathrm{MSP}(y,|y| \dot{-} |\mathrm{sqrt}^{(\ell)}(y)|,|\mathrm{sqrt}^{(\ell)}(y)| \dot{-} |\mathrm{sqrt}^{(\ell+1)}(y)|)\\
&= \mathrm{MSP}(\mathrm{sqrt}^{(\ell)}(y),|\mathrm{sqrt}^{(\ell)}(y)| \dot{-} |\mathrm{sqrt}^{(\ell+1)}(y)|)\\
&= \mathrm{MSP}(\mathrm{sqrt}^{(\ell)}(y),|\mathrm{sqrt}^{(\ell)}(y)| \dot{-} |\mathrm{sqrt}(\mathrm{sqrt}^{(\ell)}(y))|)\\
&= \mathrm{MSP}(\mathrm{sqrt}^{(\ell)}(y),|\mathrm{sqrt}^{(\ell)}(y)| \dot{-} \lfloor\frac{|\mathrm{sqrt}^{(\ell)}(y)|}{2}\rfloor)\\
&= \mathrm{MSP}(\mathrm{sqrt}^{(\ell)}(y),\lceil\frac{|\mathrm{sqrt}^{(\ell)}(y)|}{2}\rceil)\\
&= \mathrm{MSP}(\mathrm{sqrt}^{(\ell)}(y),\lfloor\frac{|\mathrm{sqrt}^{(\ell)}(y)|+1}{2}\rfloor)\\
&= \mathrm{MSP}(\mathrm{sqrt}^{(\ell)}(y),\mathrm{TR}(|s_0(\mathrm{sqrt}^{(\ell)}(y))|))
\end{aligned}
$$

$$= \mathrm{sqrt}(\mathrm{sqrt}^{(\ell)}(y))$$
$$= \mathrm{sqrt}^{(\ell+1)}(y).$$

This establishes Claim 1.

CLAIM 2. $\widetilde{F}(\mathbf{f}, \mathbf{x}, y, \mathrm{sqrt}^{(j)}(y)) = F(\mathbf{f}, \mathbf{x}, \mathrm{sqrt}^{(j)}(y))$, for all $j \geq 0$.

Proof of Claim 2. Let $r(y) = \min j \quad [\mathrm{sqrt}^{(j)}(y) = 0]$. It is easily seen that $r(y) \leq 2 \cdot ||y||$.

Proof. By reverse induction on $j = r(y)$ down to 0.

BASE STEP Let $\ell = r(y)$

$$\begin{aligned}\widetilde{F}(\mathbf{f}, \mathbf{x}, y, \mathrm{sqrt}^{(\ell)}(y)) &= \widetilde{F}(\mathbf{f}, \mathbf{x}, y, 0)\\ &= G(\mathbf{f}, \mathbf{x})\\ &= F(\mathbf{f}, \mathbf{x}, 0)\\ &= F(\mathbf{f}, \mathbf{x}, \mathrm{sqrt}^{(\ell)}(y)).\end{aligned}$$

INDUCTION STEP Let $\ell < r(y)$, and assume the claim holds for values larger than ℓ.

$$\begin{aligned}\widetilde{F}(\mathbf{f}, \mathbf{x}, y, \mathrm{sqrt}^{(\ell)}(y)) &= \widetilde{H}(\mathbf{f}, \mathbf{x}, y, |\mathrm{sqrt}^{(\ell)}(y)|, \widetilde{F}(\mathbf{f}, \mathbf{x}, y, \mathrm{sqrt}^{(\ell+1)}(y))\\ &= H(\mathbf{f}, \mathbf{x}, \mathrm{MSP}(y, |y| \dot{-} |\mathrm{sqrt}^{(\ell)}(y)|),\\ &\quad \widetilde{F}(\mathbf{f}, \mathbf{x}, y, \mathrm{sqrt}^{(\ell+1)}(y))\\ &= H(\mathbf{f}, \mathbf{x}, \mathrm{sqrt}^{(\ell)}(y), \widetilde{F}(\mathbf{f}, \mathbf{x}, y, \mathrm{sqrt}^{(\ell+1)}(y))\\ &= H(\mathbf{f}, \mathbf{x}, \mathrm{sqrt}^{(\ell)}(y), F(\mathbf{f}, \mathbf{x}, \mathrm{sqrt}^{(\ell+1)}(y))\\ &= F(\mathbf{f}, \mathbf{x}, \mathrm{sqrt}^{(\ell)}(y)).\end{aligned}$$

This establishes Claim 2.

Since $F(\mathbf{f}, \mathbf{x}, y) \leq K(\mathbf{f}, \mathbf{x}, y)$ by hypothesis, it follows from Claim 2 that

$$\widetilde{F}(\mathbf{f}, \mathbf{x}, y, \mathrm{sqrt}^{(j)}(y)) \leq K(\mathbf{f}, \mathbf{x}, \mathrm{sqrt}^{(j)}(y))$$

for all $j \geq 0$. Now define

$$F^*(\mathbf{f}, \mathbf{x}, y, 0) = G(\mathbf{f}, \mathbf{x})$$

$$\begin{aligned}F^*(\mathbf{f}, \mathbf{x}, y, u) \min \{ &\widetilde{H}(\mathbf{f}, \mathbf{x}, y, u, F^*(\mathbf{f}, \mathbf{x}, y, \lfloor \tfrac{u}{2} \rfloor)),\\ &K(\mathbf{f}, \mathbf{x}, \min\{2^u - 1, s_0(y)\})\}\end{aligned}$$

and let $F'(\mathbf{f}, \mathbf{x}, y, u) = F^*(\mathbf{f}, \mathbf{x}, y, |u|)$. It follows that F' is defined by bounded limited recursion on notation from functions in $\mathcal{A}$, hence belongs to $\mathcal{A}$.

CLAIM 3. $F'(\mathbf{f}, \mathbf{x}, y, \mathrm{sqrt}^{(j)}(y)) = \widetilde{F}(\mathbf{f}, \mathbf{x}, y, \mathrm{sqrt}^{(j)}(y))$ for all $j \geq 0$.

Proof. By reverse induction on $j = r(y)$ down to 0.

BASE STEP Let $\ell = r(y)$

$$\begin{aligned} F'(\mathbf{f}, \mathbf{x}, y, \mathrm{sqrt}^{(\ell)}(y)) &= F'(\mathbf{f}, \mathbf{x}, y, 0) \\ &= F^*(\mathbf{f}, \mathbf{x}, y, |0|) \\ &= G(\mathbf{f}, \mathbf{x}) \\ &= \widetilde{F}(\mathbf{f}, \mathbf{x}, y, 0). \end{aligned}$$

INDUCTION STEP Let $\ell < r(y)$, and assume the claim holds for values larger than ℓ.

$$\begin{aligned} F'(\mathbf{f}, \mathbf{x}, y, \mathrm{sqrt}^{(\ell)}(y)) &= F^*(\mathbf{f}, \mathbf{x}, y, |\mathrm{sqrt}^{(\ell)}(y)|) \\ &= \widetilde{H}(\mathbf{f}, \mathbf{x}, y, |\mathrm{sqrt}^{(\ell)}(y)|, F^*(\mathbf{f}, \mathbf{x}, y, \lfloor \frac{|\mathrm{sqrt}^{(\ell)}(y)|}{2} \rfloor)) \\ &= \widetilde{H}(\mathbf{f}, \mathbf{x}, y, |\mathrm{sqrt}^{(\ell)}(y)|, F^*(\mathbf{f}, \mathbf{x}, y, |\mathrm{sqrt}^{(\ell+1)}(y)|)) \\ &= \widetilde{H}(\mathbf{f}, \mathbf{x}, y, |\mathrm{sqrt}^{(\ell)}(y)|, F'(\mathbf{f}, \mathbf{x}, y, \mathrm{sqrt}^{(\ell+1)}(y))) \\ &= \widetilde{H}(\mathbf{f}, \mathbf{x}, y, |\mathrm{sqrt}^{(\ell)}(y)|, \widetilde{F}(\mathbf{f}, \mathbf{x}, y, \mathrm{sqrt}^{(\ell+1)}(y))) \\ &= \widetilde{F}(\mathbf{f}, \mathbf{x}, y, \mathrm{sqrt}^{(\ell)}(y)). \end{aligned}$$

This establishes Claim 3.

Now from Claims 2 and 3, we have

$$F'(\mathbf{f}, \mathbf{x}, y, y) = \widetilde{F}(\mathbf{f}, \mathbf{x}, y, y) = F(\mathbf{f}, \mathbf{x}, y).$$

F' is defined by WBRN from functions in $\mathcal{A}$, hence belongs to $\mathcal{A}$, so $F \in \mathcal{A}$.

Corollary 7.4.1. *The class $\mathcal{A}$ can be equivalently defined by replacing* WBRN *by any of* WLRN, LWRN, BWRN, LSR.

Definition 7.4.5. *$F_1, \ldots, F_k$ are defined from $\mathbf{G}, \mathbf{H}, \mathbf{K}$ by limited simultaneous weak recursion on notation (*LSWRN*) if for all $\mathbf{f}, \mathbf{x}, y$*

$$\widetilde{F_i}(\mathbf{f}, \mathbf{x}, 0) = G_i(\mathbf{f}, \mathbf{x}), \quad 1 \le i \le k$$

$$\widetilde{F_i}(\mathbf{f}, \mathbf{x}, y) = H_i(\mathbf{f}, \mathbf{x}, y, \widetilde{F_1}(\mathbf{f}, \mathbf{x}, \lfloor \frac{y}{2} \rfloor), \ldots, \widetilde{F_k}(\mathbf{f}, \mathbf{x}, \lfloor \frac{y}{2} \rfloor))$$

$$F_i(\mathbf{f}, \mathbf{x}, y) = \widetilde{F_i}(\mathbf{f}, \mathbf{x}, |y|)$$

provided that

$$F_i(\mathbf{f}, \mathbf{x}, y) \le K(\mathbf{f}, \mathbf{x}, y)$$

*for all $\mathbf{f}, \mathbf{x}, y$. In other words, $F_1, \ldots, F_k$ are defined by simultaneous weak recursion on notation (*SWRN*) and are moreover limited.*

Definition 7.4.6. *$F_1, \ldots, F_k$ are defined from $\mathbf{G}, \mathbf{H}, \mathbf{K}$ by limited simultaneous square-root recursion (*LSSR*) if for all $\mathbf{f}, \mathbf{x}, y$*

$$F_i(\mathbf{f}, \mathbf{x}, 0) = G_i(\mathbf{f}, \mathbf{x}), \quad 1 \le i \le k$$

$$F_i(\mathbf{f}, \mathbf{x}, y) = H_i(\mathbf{f}, \mathbf{x}, y, F_1(\mathbf{f}, \mathbf{x}, sqrt(y)), \ldots, F_k(\mathbf{f}, \mathbf{x}, sqrt(y)))$$

provided that

$$F_i(\mathbf{f}, \mathbf{x}, y) \le K(\mathbf{f}, \mathbf{x}, y)$$

for all $\mathbf{f}, \mathbf{x}, y$*. In other words,* $F_1, \ldots, F_k$ *are defined by simultaneous square-root recursion (*SSR*) and are moreover limited.*

Lemma 7.4.3. *If* $F_1, \ldots, F_k$ *are defined from* $\mathbf{G}, \mathbf{H}, \mathbf{K} \in \mathcal{A}$ *by* LSWRN *or* LSSR *then* $F_1, \ldots, F_k \in \mathcal{A}$*.*

Proof. Straightforward, using the techniques of Lemmas 7.4.1, 7.4.2 and Theorem 7.4.1.

Definition 7.4.7. $F_1, \ldots, F_k$ *are defined from* $\mathbf{G}, \mathbf{H}, \mathbf{K}$ *by multiple limited square-root recursion* MLSR *if for all* $\mathbf{f}, \mathbf{x}, y$

$$\begin{aligned}
&F_i(\mathbf{f}, \mathbf{x}, 0) = G_i(\mathbf{f}, \mathbf{x}), \quad 1 \le i \le k \\
&F_i(\mathbf{f}, \mathbf{x}, y) = H_i(\mathbf{f}, \mathbf{x}, y, F_1(\mathbf{f}, \mathbf{x}, sqrt(y)), \ldots, F_k(\mathbf{f}, \mathbf{x}, sqrt(y))) \\
&\qquad\qquad\quad y > 0, \quad 1 \le i \le k, \\
&F_1(\mathbf{f}, \mathbf{x}, y) \le K_1(\mathbf{f}, \mathbf{x}, y) \\
&F_i(\mathbf{f}, \mathbf{x}, y) \le K_i(\mathbf{f}, \mathbf{x}, y, F_{i-1}(\mathbf{f}, \mathbf{x}, y)), \quad 2 \le i \le k.
\end{aligned}$$

Lemma 7.4.4 (E Lemma). *If* $F_1, \ldots, F_k$ *are defined from* $\mathbf{G}, \mathbf{H}, \mathbf{K}$ *by* MLSR *and* $\mathbf{G}, \mathbf{H}, \mathbf{K}$ *belong to* $\mathcal{A}$ *then there exist* $E_1, \ldots, E_k$ *in* $\mathcal{A}$ *such that for all* $u_1, \ldots, u_{k-1}, v_1, \ldots, v_{k-1}$ *and all* y IF *for all* j[5]

$$F_i(\mathbf{f}, \mathbf{x}, sqrt^{(j)}(y)) \le K_i(\mathbf{f}, \mathbf{x}, u_{i-1}, v_{i-1}), \quad 2 \le i \le k,$$

THEN

$$E_i(\mathbf{f}, \mathbf{x}, \mathbf{u}, \mathbf{v}, y) = F_i(\mathbf{f}, \mathbf{x}, y), \quad 1 \le i \le k.$$

Proof. Define

$$\min(x, y) = \begin{cases} x \text{ if } x \le y \\ y \text{ else.} \end{cases}$$

We've already seen that $\min \in \mathcal{A}_0$, so that $\min \in \mathcal{A}$. Define $K'_1, \ldots, K'_k$ by

$$K'_1(\mathbf{f}, \mathbf{x}, \mathbf{u}, \mathbf{v}, y) = K_1(\mathbf{f}, \mathbf{x}, y)$$

$$K'_i(\mathbf{f}, \mathbf{x}, \mathbf{u}, \mathbf{v}, y) = K_i(\mathbf{f}, \mathbf{x}, u_{i-1}, v_{i-1}), \quad 2 \le i \le k.$$

Let $H'_1, \ldots, H'_k$ be defined by

$$H'_i(\mathbf{f}, \mathbf{x}, \mathbf{u}, \mathbf{v}, y, \mathbf{t}) = \min(H_i(\mathbf{f}, \mathbf{x}, y, \mathbf{t}), K'_i(\mathbf{f}, \mathbf{x}, \mathbf{u}, \mathbf{v}, y))$$

and $G'_1, \ldots, G'_k$ be defined by

$$G'_i(\mathbf{f}, \mathbf{x}, \mathbf{u}, \mathbf{v}) = G_i(\mathbf{f}, \mathbf{x}).$$

[5] As mentioned above, it suffices to take $j \le 2 \cdot ||y||$.

By hypothesis and closure properties of $\mathcal{A}$, we have that $G'_i, H'_i, K'_i \in \mathcal{A}$ for $1 \leq i \leq k$. Define $E_1, \ldots, E_k$ by limited simultaneous square-root recursion (LSSR) from $\mathbf{G}, \mathbf{H}, \mathbf{K}$ as follows.

$$\begin{aligned} E_i(\mathbf{f}, \mathbf{x}, \mathbf{u}, \mathbf{v}, 0) &= G'_i(\mathbf{f}, \mathbf{x}, \mathbf{u}, \mathbf{v}) \\ &= G_i(\mathbf{f}, \mathbf{x}) \end{aligned}$$

and define $E_i(\mathbf{f}, \mathbf{x}, \mathbf{u}, \mathbf{v}, y)$ to be

$$H'_i(\mathbf{f}, \mathbf{x}, \mathbf{u}, \mathbf{v}, y, E_1(\mathbf{f}, \mathbf{x}, \mathbf{u}, \mathbf{v}, \mathrm{sqrt}(y)), \ldots, E_k(\mathbf{f}, \mathbf{x}, \mathbf{u}, \mathbf{v}, \mathrm{sqrt}(y)))$$

for $1 \leq i \leq k$. By definition of H'_i, it follows that

$$E_i(\mathbf{f}, \mathbf{x}, \mathbf{u}, \mathbf{v}, y) \leq K'_i(\mathbf{f}, \mathbf{x}, \mathbf{u}, \mathbf{v}, y).$$

CLAIM.. For all $\mathbf{f}, \mathbf{x}, \mathbf{u}, \mathbf{v}, y$ and $2 \leq i \leq k$, if for all $j \geq 0$, $F_i(\mathbf{f}, \mathbf{x}, \mathrm{sqrt}^{(j)}(y)) \leq K_i(\mathbf{f}, \mathbf{x}, u_{i-1}, v_{i-1})$ then $E_i(\mathbf{f}, \mathbf{x}, \mathbf{u}, \mathbf{v}, y) = F_i(\mathbf{f}, \mathbf{x}, y)$.

Proof of Claim. By induction on y. When $y = 0$,

$$\begin{aligned} E_i(\mathbf{f}, \mathbf{x}, \mathbf{u}, \mathbf{v}, 0) &= G'_i(\mathbf{f}, \mathbf{x}, \mathbf{u}, \mathbf{v}) \\ &= G_i(\mathbf{f}, \mathbf{x}) \\ &= F_i(\mathbf{f}, \mathbf{x}, 0). \end{aligned}$$

Assume the claim holds for values less than y. By hypothesis, for $i = 2, \ldots, k$, for all j, $F_i(\mathbf{f}, \mathbf{x}, \mathrm{sqrt}^{(j)}(y)) \leq K_i(\mathbf{f}, \mathbf{x}, u_{i-1}, v_{i-1}, y)$ so by the induction hypothesis

$$E_i(\mathbf{f}, \mathbf{x}, \mathbf{u}, \mathbf{v}, \mathrm{sqrt}(y)) = F_i(\mathbf{f}, \mathbf{x}, \mathbf{u}, \mathbf{v}, \mathrm{sqrt}(y))$$

for $1 \leq i \leq k$. Thus

$$\begin{aligned} F_i(\mathbf{f}, \mathbf{x}, y) &= H_i(\mathbf{f}, \mathbf{x}, y, F_1(\mathbf{f}, \mathbf{x}, \mathrm{sqrt}(y)), \ldots, F_k(\mathbf{f}, \mathbf{x}, \mathrm{sqrt}(y))) \\ &= H_i(\mathbf{f}, \mathbf{x}, y, E_1(\mathbf{f}, \mathbf{x}, \mathbf{u}, \mathbf{v}, \mathrm{sqrt}(y)), \ldots, \\ &\quad E_k(\mathbf{f}, \mathbf{x}, \mathbf{u}, \mathbf{v}, \mathrm{sqrt}(y))) \\ &= H'_i(\mathbf{f}, \mathbf{x}, \mathbf{u}, \mathbf{v}, y, E_1(\mathbf{f}, \mathbf{x}, \mathbf{u}, \mathbf{v}, \mathrm{sqrt}(y)), \ldots, \\ &\quad E_k(\mathbf{f}, \mathbf{x}, \mathbf{u}, \mathbf{v}, \mathrm{sqrt}(y))) \\ &= E_i(\mathbf{f}, \mathbf{x}, \mathbf{u}, \mathbf{v}, y). \end{aligned}$$

The penultimate equality above holds, since by assumption $F_1(\mathbf{f}, \mathbf{x}, y) \leq K_1(\mathbf{f}, \mathbf{x}, y)$ and for $2 \leq i \leq k$,

$$\begin{aligned} F_i(\mathbf{f}, \mathbf{x}, y) &\leq K_i(\mathbf{f}, \mathbf{x}, u_{i-1}, v_{i-1}) \\ &\leq K'_i(\mathbf{f}, \mathbf{x}, \mathbf{u}, \mathbf{v}, y) \end{aligned}$$

thus

$$\begin{aligned} F_i(\mathbf{f}, \mathbf{x}, y) &= H_i(\mathbf{f}, \mathbf{x}, y, \mathbf{E}(\mathbf{f}, \mathbf{x}, \mathrm{sqrt}(y))) \\ &\leq \min\{H_i(\mathbf{f}, \mathbf{x}, y, \mathbf{E}(\mathbf{f}, \mathbf{x}, \mathrm{sqrt}(y))), K'_i(\mathbf{f}, \mathbf{x}, \mathbf{u}, \mathbf{v}, y)\} \\ &= H'_i(\mathbf{f}, \mathbf{x}, \mathbf{u}, \mathbf{v}, y, \mathbf{E}(\mathbf{f}, \mathbf{x}, \mathrm{sqrt}(y))) \end{aligned}$$

hence

$$H_i'(\mathbf{f}, \mathbf{x}, \mathbf{u}, \mathbf{v}, y, \mathbf{E}(\mathbf{f}, \mathbf{x}, \mathbf{u}, \mathbf{v}, \mathrm{sqrt}(y))) = H_i(\mathbf{f}, \mathbf{x}, y, \mathbf{E}(\mathbf{f}, \mathbf{x}, \mathbf{u}, \mathbf{v}, \mathrm{sqrt}(y))).$$

This concludes the proof of the claim. Since $E_1, \ldots, E_k$ are defined by LSSR, they belong to $\mathcal{A}$.

Lemma 7.4.5. *If F_1, F_2 are defined from $\mathbf{G}, \mathbf{H}, \mathbf{K}$ by multiple limited square-root recursion (MLSR) and $\mathbf{G}, \mathbf{H}, \mathbf{K} \in \mathcal{A}$ then F_1, F_2 belong to $\mathcal{A}$.*

Proof. Consider the parameters $\mathbf{f}, \mathbf{x}$ as fixed throughout the argument. This will somewhat simplify notation, where we will later write $t(w), z(w)$ instead of $t(\mathbf{f}, \mathbf{x}, w), z(\mathbf{f}, \mathbf{x}, w)$. Given $G_1, G_2, H_1, H_2, K_1, K_2$, recall the definitions of $G_1', G_2', H_1', H_2', K_1', K_2'$ from Lemma 7.4.4. Since $G_1, G_2, H_1, H_2, K_1, K_2 \in \mathcal{A}$, we have $G_1', G_2', H_1', H_2', K_1', K_2' \in \mathcal{A}$. Let $\langle\ ,\ \rangle$ denote the pairing function. Using limited simultaneous square-root recursion (LSSR), define E_1, E_2, P from $G_1, G_2, H_1', H_2', K_1, K_2$ as follows:

$$\begin{aligned} E_1(\mathbf{f}, \mathbf{x}, u, v, 0) &= G_1'(\mathbf{f}, \mathbf{x}, u, v) = G_1(\mathbf{f}, \mathbf{x}) \\ E_2(\mathbf{f}, \mathbf{x}, u, v, 0) &= G_2'(\mathbf{f}, \mathbf{x}, u, v) = G_2(\mathbf{f}, \mathbf{x}) \\ P(\mathbf{f}, \mathbf{x}, 0) &= \langle 0, G_1(\mathbf{f}, \mathbf{x})\rangle\rangle. \end{aligned}$$

If $y \neq 0$, then let $E_1(\mathbf{f}, \mathbf{x}, u, v, y)$ equal

$$H_1'(\mathbf{f}, \mathbf{x}, u, v, y, E_1(\mathbf{f}, \mathbf{x}, u, v, \mathrm{sqrt}(y)), E_2(\mathbf{f}, \mathbf{x}, u, v, \mathrm{sqrt}(y))),$$

let $E_2(\mathbf{f}, \mathbf{x}, u, v, y)$ equal

$$H_2'(\mathbf{f}, \mathbf{x}, u, v, y, E_1(\mathbf{f}, \mathbf{x}, u, v, \mathrm{sqrt}(y)), E_2(\mathbf{f}, \mathbf{x}, u, v, \mathrm{sqrt}(y))),$$

and let $P(\mathbf{f}, \mathbf{x}, y)$ equal

$$\begin{cases} \langle y, z(y)\rangle & \text{if } K_2(\mathbf{f}, \mathbf{x}, y, z(y)) > K_2(\mathbf{f}, \mathbf{x}, \Pi_1^2(t(y)), \Pi_2^2(t(y))) \\ P(\mathbf{f}, \mathbf{x}, \mathrm{sqrt}(y)) & \text{else.} \end{cases}$$

Here, for fixed $\mathbf{f}, \mathbf{x}$, we use the abbreviations when $w \neq 0$:

$$\begin{aligned} t(w) &= P_1(\mathbf{f}, \mathbf{x}, \mathrm{sqrt}(w)) \\ z(w) &= H_1(\mathbf{f}, \mathbf{x}, w, E_1(\mathbf{f}, \mathbf{x}, \Pi_1^2(t(w)), \Pi_2^2(t(w)), \mathrm{sqrt}(w)), \\ &\qquad E_2(\mathbf{f}, \mathbf{x}, \Pi_1^2(t(w)), \Pi_2^2(t(w)), \mathrm{sqrt}(w))). \end{aligned}$$

The intuition is that $P(\mathbf{f}, \mathbf{x}, y)$ *picks out* that argument $\mathrm{sqrt}^{(j_0)}(y)$ and functional value $F_1(\mathbf{f}, \mathbf{x}, \mathrm{sqrt}^{(j_0)}(y))$ for which the bounding function K_2 for F_2 achieves a maximum.

For given y (and fixed $\mathbf{f}, \mathbf{x}$), let $\ell = \ell(y) \leq 2 \cdot ||y||$ be the *largest* index[6] such that for all j

[6] Here and elsewhere, we intend that index $\ell \leq r(y) \leq 2 \cdot ||y||$, where $r(y)$ is the smallest index ℓ satisfying $\mathrm{sqrt}^{(\ell)}(y) = 0$.

$$K_2(\mathbf{f},\mathbf{x},\mathrm{sqrt}^{(\ell)}(y),F_1(\mathbf{f},\mathbf{x},\mathrm{sqrt}^{(\ell)}(y))) \geq$$
$$K_2(\mathbf{f},\mathbf{x},\mathrm{sqrt}^{(j)}(y),F_1(\mathbf{f},\mathbf{x},\mathrm{sqrt}^{(j)}(y))).$$

CLAIM 1. $P(\mathbf{f},\mathbf{x},y) = \langle \mathrm{sqrt}^{(\ell)}(y), F_1(\mathbf{f},\mathbf{x},\mathrm{sqrt}^{(\ell)}(y))\rangle$

CLAIM 2.. For all j,

$$F_2(\mathbf{f},\mathbf{x},\mathrm{sqrt}^{(j)}(y)) \leq K_2(\mathbf{f},\mathbf{x},\Pi^2_1(P(\mathbf{f},\mathbf{x},y)),\Pi^2_2(P(\mathbf{f},\mathbf{x},y))).$$

We prove both claims simultaneously by induction on y.

BASE STEP

$$\begin{aligned} P(\mathbf{f},\mathbf{x},0) &= \langle 0, G_1(\mathbf{f},\mathbf{x})\rangle \\ &= \langle 0, F_1(\mathbf{f},\mathbf{x},0)\rangle. \end{aligned}$$

Clearly, $\mathrm{sqrt}^{(j)}(0) = 0$ for all j. This establishes the base step of Claim 1.

By hypothesis of MLSR, $F_2(\mathbf{f},\mathbf{x},0) \leq K_2(\mathbf{f},\mathbf{x},0,F_1(\mathbf{f},\mathbf{x},0))$. Thus

$$\begin{aligned} F_2(\mathbf{f},\mathbf{x},0) &\leq K_2(\mathbf{f},\mathbf{x},0,F_1(\mathbf{f},\mathbf{x},0)) \\ &= K_2(\mathbf{f},\mathbf{x},0,G_1(\mathbf{f},\mathbf{x})) \\ &= K_2(\mathbf{f},\mathbf{x},\Pi^2_1(P(\mathbf{f},\mathbf{x},0)),\Pi^2_2(P(\mathbf{f},\mathbf{x},0))). \end{aligned}$$

This establishes the base step for Claim 2.

INDUCTION STEP Assume $y > 0$, and that both claims hold for smaller values than y.

Now by definition $z(y)$ equals

$$\begin{aligned} &H_1(\mathbf{f},\mathbf{x},y,E_1(\mathbf{f},\mathbf{x},\Pi^2_1(t(y)),\Pi^2_2(t(y)),\mathrm{sqrt}(y)), \\ &E_2(\mathbf{f},\mathbf{x},\Pi^2_1(t(y)),\Pi^2_2(t(y)),\mathrm{sqrt}(y))) \end{aligned} \tag{7.1}$$

By the induction hypothesis for Claim 2, it is the case that for all $j \geq 0$,

$$\begin{aligned} F_2(\mathbf{f},\mathbf{x},\mathrm{sqrt}^{(j)}(\mathrm{sqrt}(y))) &\leq K_2(\mathbf{f},\mathbf{x},\Pi^2_1(P(\mathbf{f},\mathbf{x},\mathrm{sqrt}(y))), \\ &\qquad \Pi^2_2(P(\mathbf{f},\mathbf{x},\mathrm{sqrt}(y)))) \\ &= K_2(\mathbf{f},\mathbf{x},\Pi^2_1(t(y)),\Pi^2_2(t(y))). \end{aligned}$$

Thus by Lemma 7.4.4,

$$\begin{aligned} F_1(\mathbf{f},\mathbf{x},\mathrm{sqrt}(y)) &= E_1(\mathbf{f},\mathbf{x},\Pi^2_1(t(y)),\Pi^2_2(t(y)),\mathrm{sqrt}(y)) \\ F_2(\mathbf{f},\mathbf{x},\mathrm{sqrt}(y)) &= E_2(\mathbf{f},\mathbf{x},\Pi^2_1(t(y)),\Pi^2_2(t(y)),\mathrm{sqrt}(y)). \end{aligned}$$

It follows from (7.1) that

$$\begin{aligned} z(y) &= H_1(\mathbf{f},\mathbf{x},y,F_1(\mathbf{f},\mathbf{x},\mathrm{sqrt}(y)),F_2(\mathbf{f},\mathbf{x},\mathrm{sqrt}(y))) \\ &= F_1(\mathbf{f},\mathbf{x},y) \end{aligned} \tag{7.2}$$

Case 1. $K_2(\mathbf{f},\mathbf{x},y,z(y)) > K_2(\mathbf{f},\mathbf{x},\Pi^2_1(t(y)),\Pi^2_2(t(y)))$.

In this case, by definition,

$$\begin{aligned} P(\mathbf{f},\mathbf{x},y) &= \langle y, z(y)\rangle \\ &= \langle y, F_1(\mathbf{f},\mathbf{x},y)\rangle. \end{aligned}$$

Hence Claim 1 is satisfied with $\ell(y) = 0$. To verify Claim 2 in the case under consideration, proceed as follows.

$$\begin{aligned} F_2(\mathbf{f},\mathbf{x},y) &\le K_2(\mathbf{f},\mathbf{x},y,F_1(\mathbf{f},\mathbf{x},y)) \\ &= K_2(\mathbf{f},\mathbf{x},\Pi_1^2(P(\mathbf{f},\mathbf{x},y)),\Pi_2^2(P(\mathbf{f},\mathbf{x},y)). \end{aligned} \tag{7.3}$$

By the induction hypothesis of Claim 2, for $j \ge 0$,

$$\begin{aligned} F_2(\mathbf{f},\mathbf{x},\mathrm{sqrt}^{(j)}(\mathrm{sqrt}(y))) \le K_2(\mathbf{f},\mathbf{x},&\,\Pi_1^2(P(\mathbf{f},\mathbf{x},\mathrm{sqrt}(y))), \\ &\Pi_2^2(P(\mathbf{f},\mathbf{x},\mathrm{sqrt}(y)))). \end{aligned}$$

Under the case assumption, we have

$$K_2(\mathbf{f},\mathbf{x},\Pi_1^2(P(\mathbf{f},\mathbf{x},y)),\Pi_2^2(P(\mathbf{f},\mathbf{x},y))) = K_2(\mathbf{f},\mathbf{x},y,z(y))$$

and

$$\begin{aligned} &K_2(\mathbf{f},\mathbf{x},y,z(y)) \\ &> K_2(\mathbf{f},\mathbf{x},\Pi_1^2(t(y)),\Pi_2^2(t(y))) \\ &= K_2(\mathbf{f},\mathbf{x},\Pi_1^2(P(\mathbf{f},\mathbf{x},\mathrm{sqrt}(y)),\Pi_2^2(P(\mathbf{f},\mathbf{x},\mathrm{sqrt}(y)))). \end{aligned}$$

By the induction hypothesis of Claim 2,

$$\begin{aligned} F_2(\mathbf{f},\mathbf{x},\mathrm{sqrt}^{(j)}(\mathrm{sqrt}(y))) \le K_2(\mathbf{f},\mathbf{x},&\,\Pi_1^2(P(\mathbf{f},\mathbf{x},\mathrm{sqrt}(y)), \\ &\Pi_2^2(P(\mathbf{f},\mathbf{x},\mathrm{sqrt}(y)))) \end{aligned} \tag{7.4}$$

for all $j \ge 0$. Thus by (7.3,7.4,7.4), we have

$$F_2(\mathbf{f},\mathbf{x},\mathrm{sqrt}^{(j)}(y)) \le K_2(\mathbf{f},\mathbf{x},\Pi_1^2(P(\mathbf{f},\mathbf{x},y),\Pi_2^2(P(\mathbf{f},\mathbf{x},y)))$$

for all $j \ge 0$. This completes the induction step for Case 1.

Case 2. $K_2(\mathbf{f},\mathbf{x},y,z(y)) \le K_2(\mathbf{f},\mathbf{x},\Pi_1^2(t(y),\Pi_2^2(t(y)))$.

Then by definition of P,

$$P(\mathbf{f},\mathbf{x},y) = P(\mathbf{f},\mathbf{x},\mathrm{sqrt}(y)).$$

Temporarily write ℓ_0 for $\ell(\mathrm{sqrt}(y))$. By the induction hypothesis for Claim 1,

$$P(\mathbf{f},\mathbf{x},\mathrm{sqrt}(y)) = \langle \mathrm{sqrt}^{(\ell_0)}(\mathrm{sqrt}(y)), F_1(\mathbf{f},\mathbf{x},\mathrm{sqrt}^{(\ell_0)}(\mathrm{sqrt}(y)))\rangle.$$

Taking $\ell = \ell(y) = \ell_0 + 1$, we have

$$P(\mathbf{f},\mathbf{x},y) = \langle \mathrm{sqrt}^{(\ell)}(y), F_1(\mathbf{f},\mathbf{x},\mathrm{sqrt}^{(\ell)}(y))\rangle.$$

This establishes the induction step for Claim 1. Now let's consider Claim 2. By hypothesis of MLSR,

$$F_2(\mathbf{f},\mathbf{x},y) \leq K_2(\mathbf{f},\mathbf{x},y,F_1(\mathbf{f},\mathbf{x},y)).$$

In (7.2) we determined that

$$z(y) = F_1(\mathbf{f},\mathbf{x},y)$$

so that

$$\begin{aligned} F_2(\mathbf{f},\mathbf{x},y) &\leq K_2(\mathbf{f},\mathbf{x},y,z(y)) \\ &\leq K_2(\mathbf{f},\mathbf{x},\Pi^2_1(t(y)),\Pi^2_2(t(y))) \\ &= K_2(\mathbf{f},\mathbf{x},\Pi^2_1(P(\mathbf{f},\mathbf{x},\mathrm{sqrt}(y)),\Pi^2_2(P(\mathbf{f},\mathbf{x},\mathrm{sqrt}(y)))) \\ &= K_2(\mathbf{f},\mathbf{x},\Pi^2_1(P(\mathbf{f},\mathbf{x},y),\Pi^2_2(P(\mathbf{f},\mathbf{x},y))). \end{aligned}$$

This establishes the induction step for Claim 2 under the current case hypothesis, and hence concludes the proof of both claims.

From Claim 1, it follows that

$$\begin{aligned} P(\mathbf{f},\mathbf{x},y) &\leq \max_{j\leq 2\cdot\|y\|} \left\{\langle \mathrm{sqrt}^{(j)}(y), F_1(\mathbf{f},\mathbf{x},\mathrm{sqrt}^{(j)}(y))\rangle\right\} \\ &\leq \max_{j\leq 2\cdot\|y\|} \left\{\langle \mathrm{sqrt}^{(j)}(y), K_1(\mathbf{f},\mathbf{x},\mathrm{sqrt}^{(j)}(y))\rangle\right\}. \end{aligned}$$

Using limited square-root recursion (LSR), it is easy to define a bounding function $B \in \mathcal{A}$ for which $P(\mathbf{f},\mathbf{x},y) \leq B(\mathbf{f},\mathbf{x},y)$, for all $\mathbf{f},\mathbf{x},y$. At the beginning of the proof of the current lemma, we defined E_1, E_2, P by simultaneous square-root recursion. Bounding functions for E_1, E_2 were shown in Lemma 7.4.4 to belong to $\mathcal{A}$. It follows that E_1, E_2, P are defined by limited simultaneous square-root recursion (LSSR), and so by Lemma 7.4.3 belong to $\mathcal{A}$. By Lemma 7.4.4,

$$\begin{aligned} F_1(\mathbf{f},\mathbf{x},y) &= E_1(\mathbf{f},\mathbf{x},\Pi^2_1(P(\mathbf{f},\mathbf{x},y)),\Pi^2_2(P(\mathbf{f},\mathbf{x},y)),y) \\ F_2(\mathbf{f},\mathbf{x},y) &= E_2(\mathbf{f},\mathbf{x},\Pi^2_1(P(\mathbf{f},\mathbf{x},y)),\Pi^2_2(P(\mathbf{f},\mathbf{x},y)),y) \end{aligned}$$

and so $F_1, F_2 \in \mathcal{A}$. This concludes the proof of the Lemma 7.4.5.

Lemma 7.4.6. *If $F_1,\ldots,F_k$ are defined from $\mathbf{G},\mathbf{H},\mathbf{K} \in \mathcal{A}$ by multiple limited square-root recursion (*MLSR*) then $F_1,\ldots,F_k \in \mathcal{A}$.*

Proof. The proof is by induction on k, where the base case $k = 2$ was treated in Lemma 7.4.5. Assume the lemma holds for $k-1$. Let $G'_1, G'_2, H'_1, H'_2, K'_1, K'_2$ be as in the proof of Lemma 7.4.4. By LSSR we will define $E_1,\ldots,E_k$ and $P_1,\ldots,P_{k-1}$ from functionals in $\mathcal{A}$ such that for all $\mathbf{f},\mathbf{x},y,\ 2 \leq i \leq k$ and all $j \geq 0$

$$F_i(\mathbf{f},\mathbf{x},\mathrm{sqrt}^{(j)}(y)) \leq K_i(\mathbf{f},\mathbf{x},\Pi^2_1(P_{i-1}(\mathbf{f},\mathbf{x},y)),\Pi^2_2(P_{i-1}(\mathbf{f},\mathbf{x},y))).$$

By Lemma 7.4.4, we will then have

$$F_i(\mathbf{f},\mathbf{x},y) = E_i(\mathbf{f},\mathbf{x},\Pi_1^2(P_1(\mathbf{f},\mathbf{x},y)),\ldots,\Pi_1^2(P_{i-1}(\mathbf{f},\mathbf{x},y)),\\ \Pi_2^2(P_1(\mathbf{f},\mathbf{x},y)),\ldots,\Pi_2^2(P_{i-1}(\mathbf{f},\mathbf{x},y)),y)$$

for $2 \le i \le k$, so that $F_1,\ldots,F_k$ will belong to $\mathcal{A}$.

Define $P_1,\ldots,P_{k-1}$ and $E_1,\ldots,E_k$ by LSSR as follows. For $1 \le i \le k-1$,

$$P_i(\mathbf{f},\mathbf{x},0) = \langle 0, G_i(\mathbf{f},\mathbf{x})\rangle$$

and for $1 \le i \le k$,

$$E_i(\mathbf{f},\mathbf{x},u,v,0) = G_i'(\mathbf{f},\mathbf{x},u,v) = G_i(\mathbf{f},\mathbf{x}).$$

For $y > 0$ and $1 \le i \le k-1$, define $P_i(\mathbf{f},\mathbf{x},y)$ to be

$$\begin{cases} \langle y, z_i(y)\rangle & \text{if } K_{i+1}(\mathbf{f},\mathbf{x},y,z_i(y)) > K_{i+1}(\mathbf{f},\mathbf{x},\Pi_1^2(t_i(y)),\Pi_2^2(t_i(y))) \\ P_i(\mathbf{f},\mathbf{x},\mathrm{sqrt}(y)) & \text{else} \end{cases}$$

and for $1 \le i \le k$ define $E_i(\mathbf{f},\mathbf{x},u,v,y)$ to be

$$H_i'(\mathbf{f},\mathbf{x},u,v,y,E_1(\mathbf{f},\mathbf{x},u,v,\mathrm{sqrt}(y)),\ldots,E_k(\mathbf{f},\mathbf{x},u,v,\mathrm{sqrt}(y)))$$

where

$$\begin{aligned} t_i(y) &= P_i(\mathbf{f},\mathbf{x},\mathrm{sqrt}(y)) \\ z_i(y) &= H_i(\mathbf{f},\mathbf{x},y,\mathcal{E}_1,\ldots,\mathcal{E}_k) \\ \mathcal{E}_i &= E_i(\mathbf{f},\mathbf{x},\Pi_1^2(t_i(y)),\ldots,\Pi_1^2(t_{k-1}),\Pi_2^2(t_i(y)),\ldots, \\ &\quad \Pi_2^2(t_{k-1}),\mathrm{sqrt}(y)). \end{aligned}$$

Now the E_i are bounded by K_i', as in Lemma 7.4.4. Hence the P_i are bounded by functionals in $\mathcal{A}$ and so the $P_i \in \mathcal{A}$.

For given y and $1 \le i \le k-1$ (and fixed $\mathbf{f},\mathbf{x}$), let $\ell_i = \ell_i(y) \le 2 \cdot ||y||$ be the *largest*[7] index such that for all j

$$K_{i+1}(\mathbf{f},\mathbf{x},\mathrm{sqrt}^{(\ell_i)}(y),F_i(\mathbf{f},\mathbf{x},\mathrm{sqrt}^{(\ell_i)}(y))) \ge K_{i+1}(\mathbf{f},\mathbf{x},\mathrm{sqrt}^{(j)}(y),\\ F_i(\mathbf{f},\mathbf{x},\mathrm{sqrt}^{(j)}(y))).$$

CLAIM 1. For each $1 \le i \le k-1$,

$$P_i(\mathbf{f},\mathbf{x},y) = \langle \mathrm{sqrt}^{(\ell_i)}(y), F_i(\mathbf{f},\mathbf{x},\mathrm{sqrt}^{(\ell_i)}(y))\rangle.$$

CLAIM 2. For each $1 \le i \le k-1$ and all $j \ge 0$,

$$F_{i+1}(\mathbf{f},\mathbf{x},\mathrm{sqrt}^{(j)}(y)) \le K_{i+1}(\mathbf{f},\mathbf{x},\Pi_1^2(P_i(\mathbf{f},\mathbf{x},y)),\Pi_2^2(P_i(\mathbf{f},\mathbf{x},y))).$$

[7] Again, the largest index less than or equal to $r(y)$, where $r(y)$ is the smallest ℓ for which $\mathrm{sqrt}^{(\ell)}(y) = 0$.

Proof of the claims. We prove both claims simultaneously by induction on y. When y is 0, we have $P_i(\mathbf{f},\mathbf{x},0) = \langle 0, G_i(\mathbf{f},\mathbf{x})\rangle = \langle 0, F_i(\mathbf{f},\mathbf{x},0)\rangle$. Moreover $F_{i+1}(\mathbf{f},\mathbf{x},0) = G_{i+1}(\mathbf{f},\mathbf{x}) \leq K_{i+1}(\mathbf{f},\mathbf{x},0,F_i(\mathbf{f},\mathbf{x},0))$. Thus both claims hold when y is 0. Assume that $y > 0$ and that Claim 1 and Claim 2 hold for values smaller than y. For $1 \leq i \leq k-1$,

$$P_i(\mathbf{f},\mathbf{x},y) = \begin{cases} \langle y, z_i(y)\rangle & \text{if } K_{i+1}(\mathbf{f},\mathbf{x},y,z_i(y)) > \\ & \quad K_{i+1}(\mathbf{f},\mathbf{x},\Pi_1^2(t_i(y)),\Pi_1^2(t_i(y))) \\ P_i(\mathbf{f},\mathbf{x},\mathrm{sqrt}(y)) & \text{else.} \end{cases}$$

By the induction hypothesis for Claim 2, for $1 \leq i \leq k-1$ and $j \geq 0$,

$$F_{i+1}(\mathbf{f},\mathbf{x},\mathrm{sqrt}^{(j)}(\mathrm{sqrt}(y))) \leq K_{i+1}(\mathbf{f},\mathbf{x},\Pi_1^2(P_i(\mathbf{f},\mathbf{x},\mathrm{sqrt}(y))), \\ \Pi_2^2(P_i(\mathbf{f},\mathbf{x},\mathrm{sqrt}(y)))).$$

hence by applying Lemma 7.4.4 with

$$u_i = \Pi_1^2(P_i(\mathbf{f},\mathbf{x},\mathrm{sqrt}(y))),$$
$$v_i = \Pi_2^2(P_i(\mathbf{f},\mathbf{x},\mathrm{sqrt}(y)))$$

we have

$$E_{i+1}(\mathbf{f},\mathbf{x},u_1,\ldots,u_{k-1},v_1,\ldots,v_{k-1},\mathrm{sqrt}(y)) = F_{i+1}(\mathbf{f},\mathbf{x},\mathrm{sqrt}(y)).$$

Thus for $1 \leq i \leq k-1$,

$$\begin{aligned} z_i(y) &= H_i(\mathbf{f},\mathbf{x},y,E_1(\mathbf{f},\mathbf{x},\mathbf{u},\mathbf{v},\mathrm{sqrt}(y)),\ldots,E_k(\mathbf{f},\mathbf{x},\mathbf{u},\mathbf{v},\mathrm{sqrt}(y))) \\ &= H_i(\mathbf{f},\mathbf{x},y,F_1(\mathbf{f},\mathbf{x},\mathrm{sqrt}(y)),\ldots,F_k(\mathbf{f},\mathbf{x},\mathrm{sqrt}(y))) \\ &= F_i(\mathbf{f},\mathbf{x},y). \end{aligned}$$

It follows that for $1 \leq i \leq k-1$

$$P_i(\mathbf{f},\mathbf{x},y) = \langle y, F_i(\mathbf{f},\mathbf{x},y)\rangle$$

provided that

$$K_{i+1}(\mathbf{f},\mathbf{x},y,F_i(\mathbf{f},\mathbf{x},y)) > K_{i+1}(\mathbf{f},\mathbf{x},\Pi_1^2(P_i(\mathbf{f},\mathbf{x},\mathrm{sqrt}(y))), \\ \Pi_1^2(P_i(\mathbf{f},\mathbf{x},\mathrm{sqrt}(y))))$$

and otherwise

$$P_i(\mathbf{f},\mathbf{x},y) = P_i(\mathbf{f},\mathbf{x},\mathrm{sqrt}(y)).$$

From this and the inductive hypothesis of Claim 1, we obtain the induction step for Claim 1, namely that

$$P_i(\mathbf{f},\mathbf{x},y) = \langle \mathrm{sqrt}^{(\ell_i)}(y), F_i(\mathbf{f},\mathbf{x},\mathrm{sqrt}^{(\ell_i)}(y))\rangle.$$

Similarly it follows that for $1 \leq i \leq k-1$ and all $j \geq 0$,

$$F_{i+1}(\mathbf{f},\mathbf{x},\mathrm{sqrt}^{(j)}(y)) \leq K_{i+1}(\mathbf{f},\mathbf{x},\Pi_1^2(P_i(\mathbf{f},\mathbf{x},y)),\Pi_1^2(P_i(\mathbf{f},\mathbf{x},y))).$$

This establishes the induction step of Claim 2, and concludes the proof of both claims.

Define $L_1, \ldots, L_k$ by

$$L_1(\mathbf{f}, \mathbf{x}, y) = K_1(\mathbf{f}, \mathbf{x}, y)$$

and for $1 \leq i \leq k-1$

$$L_{i+1}(\mathbf{f}, \mathbf{x}, y) = K_{i+1}(\mathbf{f}, \mathbf{x}, \Pi_1^2(P_i(\mathbf{f}, \mathbf{x}, y)), \Pi_2^2(P_i(\mathbf{f}, \mathbf{x}, y))).$$

It now follows that $F_1, \ldots, F_k$ can be defined by LSSR from $\mathbf{G}, \mathbf{H}, \mathbf{L}$, and so $F_1, \ldots, F_k \in \mathcal{A}$. This completes the proof of the lemma.

Definition 7.4.8. *$F_1, \ldots, F_k$ are defined from $\mathbf{G}, \mathbf{H}, \mathbf{K}$ by strong multiple limited square-root recursion* (SMLSR) *if for all $\mathbf{f}, \mathbf{x}, y$ and $1 \leq i \leq k$,*

$$F_i(\mathbf{f}, \mathbf{x}, 0) = G_i(\mathbf{f}, \mathbf{x})$$
$$F_i(\mathbf{f}, \mathbf{x}, y) = H_i(\mathbf{f}, \mathbf{x}, y, F_1(\mathbf{f}, \mathbf{x}, sqrt(y)), \ldots, F_k(\mathbf{f}, \mathbf{x}, sqrt(y))), \quad \text{for } y > 0$$

provided that for all $\mathbf{f}, \mathbf{x}, y$ and $1 \leq i \leq k-1$

$$F_1(\mathbf{f}, \mathbf{x}, y) \leq K_1(\mathbf{f}, \mathbf{x}, y)$$
$$F_{i+1}(\mathbf{f}, \mathbf{x}, y) \leq K_{i+1}(\mathbf{f}, \mathbf{x}, y, F_1(\mathbf{f}, \mathbf{x}, y), \ldots, F_i(\mathbf{f}, \mathbf{x}, y)).$$

Theorem 7.4.1. *If $F_1, \ldots, F_k$ are defined by* SMLSR *from $\mathbf{G}, \mathbf{H}, \mathbf{K} \in \mathcal{A}$ then $F_1, \ldots, F_k$ belong to $\mathcal{A}$.*

Proof. Define $L_1, \ldots, L_k$ by limited simultaneous square-root recursion (LSSR) as follows. For $1 \leq i \leq k$ set

$$L_i(\mathbf{f}, \mathbf{x}, 0) = \langle G_1(\mathbf{f}, \mathbf{x}), \ldots, G_i(\mathbf{f}, \mathbf{x}) \rangle$$

and for $y > 0$

$$L_i(\mathbf{f}, \mathbf{x}, y) = \langle \mathcal{H}_1(y), \ldots, \mathcal{H}_i(y) \rangle,$$

where $\mathcal{H}_i(y)$ equals

$$H_i(\mathbf{f}, \mathbf{x}, y, \Pi_1^k(L_1(\mathbf{f}, \mathbf{x}, \mathrm{sqrt}(y))), \ldots, \Pi_k^k(L_k(\mathbf{f}, \mathbf{x}, \mathrm{sqrt}(y)))).$$

Then $L_i(\mathbf{f}, \mathbf{x}, y) \leq \langle K_1(\mathbf{f}, \mathbf{x}, y) \rangle$, and for $2 \leq i \leq k$, $L_i(\mathbf{f}, \mathbf{x}, y)$ is less than or equal to

$$\Big\langle K_1(\mathbf{f}, \mathbf{x}, y), K_2(\mathbf{f}, \mathbf{x}, y, \Pi_1^1(L_1(\mathbf{f}, \mathbf{x}, y)))$$
$$, \ldots,$$
$$K_i(\mathbf{f}, \mathbf{x}, y, \Pi_1^{i-1}(L_{i-1}(\mathbf{f}, \mathbf{x}, y)), \ldots, \Pi_{i-1}^{i-1}(L_{i-1}(\mathbf{f}, \mathbf{x}, y)))\Big\rangle.$$

Thus the $\mathbf{L}$ are defined from functionals in $\mathcal{A}$ by LSSR and so belong to $\mathcal{A}$. It is easy to see that

$$L_i(\mathbf{f}, \mathbf{x}, y) = \langle F_1(\mathbf{f}, \mathbf{x}, y), \ldots, F_i(\mathbf{f}, \mathbf{x}, y) \rangle$$

for $1 \leq i \leq k$. It follows that $F_1, \ldots, F_k$ belong to $\mathcal{A}$.

7.5 Parallel Machine Model

In this section, we study parallel complexity classes of higher type functionals. To define type 2 parallel computable functionals, we introduce the oracle concurrent random access machine OCRAM, which allows simultaneous oracle calls to type 1 functions by different active processors. Several related fundamental questions for our model are:

1. What cost should the model charge for a function oracle call $f(y)$?
2. Should active processors be allowed to execute oracle calls $f(x)$, where x is the value in a local memory register, or should the model require active processors to access a global memory device when executing oracle calls?
3. How does the model simulate the application functional $Ap(f, x) = f(x)$, where the integer input x is (by convention) given in binary with each bit in a different global register?

We take *unit cost* for oracle calls $f(x)$, rather than a measure dependent on the length $|f(x)|$ of the returned value. This is because in one step of computation, many processors can simultaneously execute an oracle function call (on possibly different arguments), as explained later. I/O specification requires that input and output be given bitwise in the global memory. This seems reasonable, since one would like to allow different processors to work on different bits of the problem, in order to allow non-trivial computations in constant parallel time. If processors perform an oracle call $f(x)$, where x is a local memory variable, then in order to simulate $Ap(f, x)$, one would first have to collect the bits $x_1, \ldots, x_n$ of the input into an integer x to be stored in local memory (requiring $\log n$ operations). Thus $Ap(f, x)$ could not be executed in constant parallel time. For these reasons, the approach we adopt is to allow any processor in one step to retrieve the function value

$$f(x_i \cdots x_j) = f(\sum_{k=i}^{j} x_k \cdot 2^{j-k})$$

where i, j, ℓ are current values of local registers, and $i \leq j$. The oracle is called from special oracle registers, and the formal details of the our model of computation ensure that the size of the function value on any oracle call will be bounded by the product of the number of active processors and the total computation time (this product is sometimes called the *work* performed by the parallel computer). At the start of the computation, all oracle registers are empty, and one can always modify a computation, so that when the machine halts, all oracle registers are empty. This is formalized as follows.

An OCRAM has a sequence $R_0, R_1, \ldots$ of random access machines RAM's which operate in a synchronous fashion in parallel. Each R_i knows its *processor identity number* or PIN, and has its own local memory, an infinite collection of registers, each of which can hold an arbitrary non-negative integer. There is an infinite collection of global or common registers accessible to all RAM's, which are used for reading the input, processor intercommunication, and output. Global registers are denoted $M_0^g, M_1^g, M_2^g, \ldots$, and local registers by $M_0, M_1, M_2, \ldots$. At times, we may write $M_{i,j}$ to indicate the local register M_i of processor P_j. In the case that more than one RAM attempts to write in the same global memory location, the lowest numbered processor succeeds (*priority resolution* of write conflict). An input $x = x_1 \cdots x_n$ is initially given in the global registers, the register M_i^g holding the i-th bit x_i of x. All other registers initially contain the special symbol \$, which is different from 0, 1 and designates that the register is empty. Similarly at termination, the output y is specified in the global memory, the register M_i^g holding the i-th bit of y. At termination of a computation all other global registers contain the symbol \$.

For each k-ary function argument f, there are k infinite collections of *oracle registers*, the i-th collection labeled $M_0^{o,i}, M_1^{o,i}, M_2^{o,i}, \ldots$, for $1 \leq i \leq k$. As with global memory, in the event of a write conflict the lowest numbered processor succeeds in writing to an oracle register. Let *res* (result), *op*0 (operand 0) and *op*1 (operand 1) be non-negative integers, as well as *op*2,*op*3,...,*op*(2*k*). If any register occurring on the right side of an instruction contains '\$', then the register on the left side of the instruction will be assigned the value '\$' (undefined). For instance, if a unary oracle function f is called in the instruction

$$M_{res} := f([M_{op1} \cdots M_{op2}])$$

and if some register M_i contains '\$', where $op1 \leq i \leq op2$, then M_{res} is assigned the value '\$'.

Instructions are as follows.

$$\begin{aligned}
M_{res} &:= \text{constant} \\
M_{res} &:= \text{PIN (processor identity number)} \\
M_{res} &:= M_{op1} \\
M_{res} &:= M_{op1} + M_{op2} \\
M_{res} &:= M_{op1} \dot{-} M_{op2} \\
M_{res} &:= MSP(M_{op1}, M_{op2}) \\
M_{res} &:= LSP(M_{op1}, M_{op2}) \\
M_{res} &:= *M_{op1} \\
M_{res} &:= *M_{op1}^g \\
*M_{res} &:= M_{op1} \\
*M_{res}^g &:= M_{op1} \\
*M_{res}^o &:= 0 \\
*M_{res}^o &:= 1 \\
*M_{res}^o &:= \$ \\
M_{res}^o &:= 0
\end{aligned}$$

$$
\begin{aligned}
M^o_{res} &:= 1 \\
M^o_{res} &:= \$ \\
M_{res} &:= *M^o_{op1} \\
M_{res} &:= f([M_{op1} \cdots M_{op2}]_1, [M_{op3} \cdots M_{op4}]_2, \ldots, \\
&\qquad [M_{op(2k-1)} \cdots M_{op(2k)}]_k)
\end{aligned}
$$

GOTO label

GOTO label IF$M_{op1} = M_{op2}$

GOTO label IF$M_{op1} \leq M_{op2}$

HALT

Cutoff subtraction is defined by $x \dot{-} y = x - y$, provided that $x \geq y$, else 0. The shift operators MSP and LSP are defined as follows.

- $MSP(x, y) = \lfloor x/2^y \rfloor$, *provided* that $y < |x|$, otherwise '\$'.
- $LSP(x, y) = x - 2^y \cdot (\lfloor x/2^y \rfloor)$, *provided* that $y \leq |x|$, otherwise '\$'.

In [Imm89] N. Immerman defined the CRAM of Chapter 6 to be essentially the PRAM of [SV84], but additionally allowed the SHIFT operator

$$M_{res} := \text{SHIFT}(M_{op_1}, M_{op_2})$$

where "SHIFT(x, y) causes the word x to be shifted y bits to the right". For non-negative x, y, SHIFT(x, y) is thus the function $MSP(x, y)$. If one assumes the CRAM of [Imm89] operates on positive and negative integers (this is explicitly stated for the PRAM in [SV84] though not for the CRAM in [Imm89]) and if one assumes that for $y \geq 0$,

$$\text{SHIFT}(x, -y) = x \cdot 2^y$$

i.e., SHIFT by a negative value means *shift left*, *provided* that $y \leq |x|$, then Immerman's Theorem 1.1 of [Imm89] will hold. Without this interpretation, it is likely that Theorem 1.1 of [Imm89] fails.[8]

Instructions with $*$ concern indirect addressing. The instruction $M_{res} := *M_{op1}$ reads the current contents of local memory whose address is the value of M_{op1} into local memory M_{res}. This can be interpreted as $M_{res} := M_{M_{op1}}$, i.e., read the contents of local memory, whose register index is M_{op1}, into M_{res}. Similarly, $M_{res} := *M^g_{op1}$ performs an indirect read from global memory into local memory. This operation can be interpreted as $M_{res} := M^g_{M_{op1}}$; in other words, $M_{res} := *M^g_{op1}$ reads the value from global memory whose address is given by the current contents of local memory M_{op1}. The instruction $*M_{res} := M_{op1}$, interpreted as $M_{M_{res}} := M_{op1}$, writes the value of

[8] In personal correspondence N. Immerman [Imm93] indicated that he indeed intended to allow both right and left shifts, up to $\log(x)$ bits. Since this was not explicitly stated in [Imm89], and may be a source of confusion, we have outlined this point in detail.

local memory M_{op1} into local memory whose address is given by the current contents of local memory M_{res}. Similarly, $*M^g_{res} := M_{op1}$, interpreted as $M^g_{M_{res}} := M_{op1}$, performs an indirect write into global memory. Namely, $*M_{res} := M_{op1}$ writes the value of local memory M_{op1} into global memory whose address is given by the current contents of local memory M_{res}. The notation $[M_{op(2i-1)} \cdots M_{op(2i)}]_i$ denotes the integer whose binary notation is given in oracle registers $M^{o,i}_{M_{op(2i-1)}}$ through $M^{o,i}_{M_{op(2i)}}$. In other words,

$$[M_{op(2i-1)} \cdots M_{op(2i)}]_i = \sum_{m=M_{op(2i-1)}}^{M_{op(2i)}} M^{o,i}_m \cdot 2^{op(2i)-m}$$

so that a binary representation of an integer is allowed to have leading 0's. The instruction $*M^o_{res} := 0$ sets the contents of the oracle register whose address is given by the current contents of local memory M_{res} to 0. Similarly for the instruction $*M^o_{res} := 1$. The instruction $M_{res} := *M^o_{op1}$ sets the contents of local memory M_{res} to be the current contents of the oracle register whose address is given by the current contents of local memory M_{op1}. With these instructions, it will be the case that oracle registers hold a 0 or 1 but no larger integer.

In summary, the OCRAM has instructions for *(i)* local operations — addition, cutoff subtraction, shift, *(ii)* oracle calls, *(iii)* global and local indirect reading and writing, *(iv)* control instructions — GOTO, conditional GOTO and HALT.

A program is a finite sequence of instructions. Each individual RAM of an OCRAM has the same program — this machine model is uniform, and can be made non-uniform by having a different program for each input size n. Each instruction has unit-cost. Note well that we have not yet specified which processors of M are *active*, i.e., should execute a program instruction. We now discuss this issue.

In characterizing AC^k in the non-oracle case, Stockmeyer and Vishkin [SV84] require a polynomial bound $p(n)$ on the processor identity number (PIN) on inputs of length n. With the above definition of OCRAM one might hope to characterize the class of type 2 functionals computable in constant parallel time with a second order polynomial[9] number of processors as exactly the type 2 functionals in the algebra $\mathcal{A}_0$. Unfortunately, this is false. To illustrate the definitions given so far, we give an OCRAM which computes a functional H running in constant time with $|f|(|x|)$ as processor bound, where H satisfies $|H(f,x)| \geq |f|(|x|)$ for all f, x. Theorem 5.5 from [KC96] (see Exercise 7.8.2) states that no such H is basic feasible (in BFF), so certainly $H \notin \mathcal{A}_0$.

```
1  M1 = PID
2  M2 = *M1^g
```

[9] Second order polynomials were defined in Definition 7.2.9.

```
3  if (M_1 > 0 and M_2 = $) then *M_1^g = 1
4  HALT
```

For any input f, x, provided that $|f|(|x|) \geq |x|$, the $|f|(|x|) - |x|$ many \$-symbols at the end of input x will be written over by 1's, so that $|H(f,x) = |f|(|x|)$. If $|f|(|x|) < |x|$, then nevertheless $|H(f,x)| = |x|$. Thus in constant time with $|f|(|x|)$ many processors, we have computed a function $H(f,x)$ such that $|H(f,x)| \geq |f|(|x|)$. By Exercise 7.8.2, no such function H can be basic feasible, let alone belong to $\mathcal{A}_0$. To rectify this situation, we proceed as follows.

Definition 7.5.1. *For every* OCRAM M, *functions* f, g *and integers* x, t *the query set* $Q(M, f, x, t, g)$ *is defined as*

$$\{y : M \text{ with inputs } f, x \text{ queries } f \text{ at } y \text{ in } < t \text{ steps, where for each } i < t \text{ the active processors are those with } PIN\ 0, \ldots, g(i)\}.$$

Definition 7.5.2. *Let* M *be an* OCRAM, P *a functional of rank (1,1),* f *a function and* x, t *integers. For any* f *and* $Q \subseteq \mathbf{N}$, *let* f_Q *be defined by*

$$f_Q(x) = \begin{cases} f(x) & \text{if } x \in Q \\ 0 & \text{else.} \end{cases}$$

$\mathcal{M} = \langle M, P \rangle$ *is called a* fully specified OCRAM *if for all* f, x, t *the* OCRAM M *on input* f, x *either is halted at step* t *or executes at step* t *with active processors* $0, \ldots, P(|f_{Q_t}|, |x|)$ *where*

$$Q_t = Q(M, f, x, t, P(|f_{Q_{t-1}}|, |x|))$$

is the collection of queries made by $\mathcal{M}$ *before step* t.

If $\mathcal{M} = \langle M, P \rangle$ is a fully specified OCRAM with input f, x define

$$Q_{\mathcal{M}}(f, x, t) = \{y : \mathcal{M} \text{ queries } y \text{ at some time } i < t \text{ on input } f, x\}.$$

Often, in place of stating that $\mathcal{M} = \langle M, P \rangle$ is fully specified, we may simply say that M runs with processor bound P. From definitions, the following lemma is clear.

Lemma 7.5.1. *The fully specified* OCRAM $\mathcal{M} = \langle M, P \rangle$ *makes identical moves in the first* t *steps on inputs* f_{Q_t}, x *as on* f, x.

Remark 7.5.1. The following operations can be simulated by a constant number of basic OCRAM instructions, and hence will be allowed in our OCRAM programs:

```
*M_res^g = constant
*M_res = constant
M_res^g = M_op
M_res^g = constant
```

```
Mres = Mg_op
goto label if Mop1 ≠ Mop2
goto label if Mop1 ≰ Mop2
if <boolean condition> then <instruction>
if <boolean condition> then
   <instruction1>
else
   <instruction2>
```

Here `<boolean condition>` is a simple combination of $=, \leq$ using connectives AND, OR, NOT, and `<instruction>` is a basic OCRAM instruction (similarly for `<instruction1>`, `<instruction2>`).

The input to an OCRAM is stipulated by $\$X_1X_2\cdots X_n\$\$\cdots$ where $X_i = x_{n-i}$ for $i \leq n$. This corresponds to the usual convention that the leftmost bit of the binary representation of $x = \sum_{i<n} x_i 2^i$ is the most significant bit. Multiple inputs are separated by a single symbol '\$' as in $\$x_{n-1}\cdots x_0\$y_{m-1}\cdots y_0\$\$$.

If $F(\mathbf{f},\mathbf{x})$ abbreviates $F(f_1,...,f_m,x_1,...,x_n)$ and P is a second order polynomial, then $P(|\mathbf{f}|,|\mathbf{x}|)$ abbreviates $P(|f_1|,...,|f_m|,|x_1|,...,|x_n|)$.

Theorem 7.5.1. *If $F(\mathbf{f},\mathbf{x})$ is a functional belonging to $\mathcal{A}_0$ then there exists a second order polynomial $P(|\mathbf{f}|,|\mathbf{x}|)$ such that $F(\mathbf{f},\mathbf{x})$ is computable by an* OCRAM *in constant time using $P(|\mathbf{f}|,|\mathbf{x}|)$ many processors.*

Proof. By definition, $\mathcal{A}_0$ is the closure of $0, s_0, s_1, I_k^n, \mathrm{BIT}, |x|, \#, Ap$ under functional composition, functional substitution, expansion and concatenation recursion on notation. Moreover, by Lemma 7.3.3, functional substitution is redundant. First, for each initial function, we give OCRAM program running in constant time with a second order polynomial bound on the number of processors.

- OCRAM program for the zero-place constant function 0.

```
1  M0 = $
2  M1 = 0
3  M2 = 1
4  M3 = PID
5  *Mg_3 = M0          % erase global memory
6  *Mg_2 = M1          % Mg_1 = 0
7  HALT
```

The previous program is pedantic, using only the originally given OCRAM instructions. From now on, we will use instructions from remark 7.5.1 without further remark. Using this convention, we give the following program for computing the function 0.

```
1  M1 = PID
2  *M1 = $
3  Mg_1 = 0
4  HALT
```

Processor bound: $P(|x|) = |x|$.

• OCRAM program for $s_0(x) = 2 \cdot x$.

```
1 if (M_1^g = 0 and M_2^g = $) then goto 12
2 M_1 = $
3 M_2 = PID
4 M_3 = *M_2^g                     % in P_i, M_3 = X_i
5 if (M_3 ≠ M_1) then goto 12      % if X_i ≠ $, then P_i halts
6 M_4 = M_1^g                      % M_4 = X_0
7 M_0^g = M_2                      % M_0^g = i_0 = μi [X_i = $]
8 M_0 = M_0^g                      % M_0 = i_0
9 M_0^g = $                        % put back in standard I/O format
10 if (M_0 ≠ M_2) then goto 11     % for i ≠ i_0, P_i halts
11 *M_2^g = 0                      % M_{i_0}^g = 0
12 HALT
```

Processor bound: $P(|x|) = |x| + 1$.

The program for $s_1(x) = 2 \cdot x + 1$ is similar and left to the reader.

• OCRAM program for $I_k^n(x_1, \ldots, x_n) = x_k$.

```
1 M_1 = PID
2 M_2 = k-1
3 if (M_2 = 0) then goto 12
4 M_3 = *M_1^g
5 if (M_1 > 0 and M_3 = $) then M_0^g = M_1
6 M_4 = M_0^g              % M_4 holds position of next $
7 M_5 = M_1 + M_4
8 M_6 = *M_5^g
9 *M_1^g = M_6
  % shift input $x_i$···$x_n$ to $x_{i+1}$···$x_n$
10 M_2 = M_2 ∸ 1
11 goto 3
  % input is now of form $x_k$···$x_n$
12 M_3 = *M_1^g
13 if (M_1 > 0 and M_3 = $) then M_0^g = M_1
14 M_4 = M_0^g
15 if (M_1 ≥ M_4) then *M_1^g = $
16 M_0^g = $
17 HALT
```

Processor bound: $P(|x|) = \sum_{i=1}^{n} |x_i| + n = O(|x|)$.

• OCRAM program for $|x| = \lceil \log_2(x+1) \rceil$. As a preliminary step, we present a program for the instruction

$$M_{res} = \text{BIT}(M_{op1}, M_{op2})$$

which, for $i = M_{op1}$ computes the coefficient of 2^i in the binary representation of the integer stored in M_{op2}, *provided* that $i < |M_{op2}|$ and returns the value

'\$' otherwise. Note that this is <u>not</u> the initial function $\text{BIT}(x, y)$ of $\mathcal{A}_0$ which shall be treated later. Specific differences between BIT and BIT are that in computing $\text{BIT}(x, y)$ by convention x, y are stored bit-wise in different registers of global memory, whereas in the function BIT the arguments are current values of local memory. As well, for $x \geq |y|$, $\text{BIT}(x, y)$ is defined to be 0, whereas if $x = M_{op1}$, $y = M_{op2}$ and $x > |y|$ then $\text{BIT}(x, y) = \$$.

Assume that $res, op1, op2$ are distinct from $0, 1, 2, 3$.

- OCRAM program for $M_{res} = \text{BIT}(M_{op1}, M_{op2})$.

```
1  M_0 = MSP(M_op2, M_op1)
2  M_1 = MSP(M_op1, 1)
3  M_res = M_0 ∸ (M_1 + M_2)
4  HALT
```

Processor bound: 5.

We can define RBIT as the "reverse bit" function, where for x (y) the contents of local memory M_{op1} [resp M_{op2}], $\text{RBIT}(x, y)$ is taken to be $\text{BIT}(|y| \dot{-} (x + 1), y)$ *provided* that $x < |y|$, and taken to be '\$' otherwise.

- OCRAM program for $|x|$.

```
1  M_1 = PID
2  M_2 = *M_1^g      % in P_i, M_i = X_i
3  if (M_1 > 0 and M_2 = $) then M_0^g = M_1
4  M_3 = M_0^g ∸ 1      % in P_i, M_3 = |x|
5  *M_1^g = $           % erase global memory
6  if (M_1 > 0) then M^4 = BIT(M_3 ∸ M_1, M_3)
7  if (M_1 > 0) then *M_1^g = M_4
8  HALT
```

Processor bound: $P(|x|) = |x| + 1$.

- OCRAM program for $\text{BIT}(x, y) = \lfloor x/2^y \rfloor - 2 \cdot (\lfloor x/2^{y+1} \rfloor)$.

As a preliminary step, we leave to the reader the design of a program to output 0 if $|x| > 2||y||$, and 0 otherwise. If $|x| > 2||y||$ then

$$|x| = \lceil \log(x + 1) \rceil \geq \log(x)$$

and

$$2 \log x \geq 2 \lceil \log(|y| + 1) \rceil \geq 2 \log(|y|)$$

hence $x \geq |y|$. In this case, $\text{BIT}(x, y) = 0$, so outputting 0 is correct.

Now assume that $|x| \leq 2||y||$. The following program computes $\text{BIT}(x, y)$ under this assumption with $O(|y|^2)$ processors. The intuition is given as follows. For fixed input x, y and i a processor number, think of $\text{LSP}(i, |x|)$ and $\text{MSP}(i, |x|)$ as the right and left projection functions applied to decode the integer i into $\langle left(i), right(i) \rangle$. Processor P_i attempts to verify that $\text{BIT}(left(i), right(i)) = \text{BIT}(left(i), x)$. If for all values j,

$$\text{BIT}(j, right(i)) = \text{BIT}(j, x)$$

then $right(i) = x$. Using the priority resolution of write conflicts on the PRAM model, we can find those processors P_i for which $right(i) = x$, and then can obtain $\text{BIT}(x, y)$.

```
1  M_1 = PID
2  M_2 = *M_1^g
3  if (M_1 > 0 and M_2 = $) then *M_0^g = M_1
4  M_3 = *M_0^g ∸ 1  %  M_3 = |x|
5  if (M_1 > M_3 + 1 and M_2 = $) then *M_0^g = M_1
6  M_4 = *M_1^g ∸ (M_3 + 1)  %  M_4 = |y|
7  *M_1^g = M_2  % restore input
8  M_5 = LSP(M_1, M_3)  % in P_i, M_5 = LSP(i, |x|)
9  M_6 = MSP(M_1, M_3)  % in P_i, M_6 = MSP(i, |x|)
10 M_7 = M_3 ∸ M_6
11 M_8 = *M_7^g  % in P_i, M_7 = BIT(MSP(i, |x|), x)
12 *M_1^g = $  % erase global input
13 *M_5^g = 1  % P_i writes M^g_{LSP(i,|x|)} = 1
14 if (1 ≤ M_5 ≤ M_3) and (M_8 ≠ BIT(M_6, M_5)) then *M_5^g = 0
          % M^g_{LSP(i,|x|)} = 1 iff LSP(i, |x|) = x
15 M_9 = *M_5^g  % in P_i, M_9 = 1 iff LSP(i, |x|) = x
16 *M_1^g = M_2  % restore input
17 M_10 = M_3 + 1 + M_4 ∸ M_5
18 if M_9 = 1 then M_11 = *M_10^g
          % in those P_i where M_9 = 1, M_11 = BIT(x, y)
19 *M_1^g = $  % erase input
20 if M_9 = 1 then M_1^g = M_11
21 HALT
```

If we replace instructions 17–20 by instruction (17′) below

```
17' if M_0 = 1 then M_0^g = M_5
```

then we have placed the argument x into global memory register M_0^g, where x was originally specified bitwise in the global registers. This approach works only when x is small, i.e., bounded by $|y|$ for a second argument y.

- OCRAM program for $x\#y = 2^{|x|\cdot|y|}$.

Rather than giving a direct program, we proceed indirectly, as in our proof that $\# \in \mathcal{F}\text{LH}$ in Theorem 6.3.3. To compute $x\#y = 2^{|x|\cdot|y|}$ on a CRAM, when given x and y bitwise in global memory, we first compute $|x|$ and $|y|$ as previously explained, then compute $|x| \cdot |y|$ and output in parallel 1 followed by $|x| \cdot |y|$ many 0's. Since $|x| \cdot |y| \in \text{DSPACE}(\log\log(n))$ on a RATM, and $\text{DSPACE}(\log\log(n)) \subseteq \text{DTIMESPACE}(\log^k(n), \log^{1-1/m}(n))$ on a RATM, it suffices to show that $\text{DTIMESPACE}(\log^k(n), \log^{1-1/m}(n)) \subseteq \text{CRAM}(O(1), n^{O(1)})$, i.e., computable on a CRAM in constant time with a number of processors polynomial in n. Let M be a Turing machine with random access which runs in time $O(\log^k(n))$ and space $O(\log^{1-1/m}(n))$. As in Lemma 6.3.3 it is easy to see that the functions $initial_M$ and NEXT_M are computed in

CRAM$(O(1), n^{O(1)})$, where $initial_M(x) = \alpha$, if α codes the initial configuration of M, and $\text{NEXT}_M(x, \alpha) = \beta$ if β codes that configuration which immediately follows the configuration coded by α in the computation of M on input x. Since M allows random access to its input, it is essential here that MSP, LSP be atomic operations of the processors of the CRAM. By the method in the proof of Lemma 6.3.4, it is not difficult to see that the functions $\text{NEXT}_{M,k+1}(x, \alpha) = \beta$ belong to CRAM$(O(1), n^{O(1)})$, for all $k \geq 0$. Indeed, the idea is to encode a sequence $\alpha_1, \ldots, \alpha_r$ of configurations such that $\text{NEXT}_{M,k}(\alpha_i) = \alpha_{i+1}$ for $1 \leq i < r$ and the total space for the sequence is bounded by $O(\log n)$, i.e., $r = \log^{1/m}(n)$ and each $|\alpha_i| \leq \log^{1-1/m}(n)$ so that $r \cdot \max\{|\alpha_1|, \ldots, |\alpha_m|\} = O(\log n)$. This can be done by checking, in parallel, all possible integers i, where $|i| = O(\log n)$, to determine whether i encodes a valid sequence as just described – for the least i which encodes such a sequence $\alpha_1, \ldots, \alpha_r$, output in parallel the bits of α_m in global memory. There are $2^{O(\log n)} = n^{O(1)}$ many integers i to check, so in parallel have processors $P_{i,j}$, $1 \leq j \leq p(n)$ for some polynomial p, check whether i encodes a valid sequence. Clearly only a polynomial number of processors are required.

- OCRAM program for $Ap(f, x) = f(x)$.

```
1  M0 = 1
2  M1 = PID
3  M2 = *M1^g
4  if (M1 > 0 and M2 = $) then M0^g = M1
5  M3 = M0^g ∸ 1  %  M3 = |x|
6  *M1^g = $  % erase global memory
7 *M1^o = M2  %  in Pi, Mi^o = Xi
8  M4 = f([M0...M3])
9  M5 = BIT(M1, M4)
10  if (M5 = $) then M0^g = M1
11 M5 = M0^g ∸ 1  %  M5 = |f(x)|
12 M6 = M5 ∸ (M1 + 1)
13 M7 = BIT(M6, M4)
14 *M1^o = $
15 *M1^g = $   % erase oracle and global memory
16 M8 = M1 + 1
17 *M8^g = M7
18 if (M1 = 0 or M1 > M5) then M1^g = $
19 HALT
```

- functional composition

$$F(\mathbf{f}, \mathbf{x}) = H(\mathbf{f}, \mathbf{x}, G, (\mathbf{f}, \mathbf{x}), \ldots, G_n(\mathbf{f}, \mathbf{x}))$$

Assume OCRAM programs $\mathcal{P}_F, \mathcal{P}_H, \mathcal{P}_{G_i}$ computing the functionals F, H, G_i, for $i = 1, \ldots, n$. Run $\mathcal{P}_{G_1}, \ldots \mathcal{P}_{G_n}$ and then $\mathcal{P}_H$ on the outputs. If T_F, T_H, T_{G_i} and P_F, P_H, P_{G_i} denote respectively the time and processor bounds for programs $\mathcal{P}_F, \mathcal{P}_H, \mathcal{P}_{G_i}$ then

$$
\begin{aligned}
T_F(|\mathbf{f}|,|\mathbf{x}|)O\ (\ &\sum_{i=1}^{n} T_{G_i}(|\mathbf{f}|,|\mathbf{x}|) \\
&+ T_H(|\mathbf{f}|,|\mathbf{x}|,|G_1(\mathbf{f},\mathbf{x})|,\ldots,|G_n(\mathbf{f},\mathbf{x})|)) \\
= O\ (\ &\sum_{i=1}^{n} T_{G_i}(|\mathbf{f}|,|\mathbf{x}|) \\
&+ T_H(|\mathbf{f}|,|\mathbf{x}|,P_{G_1}(|\mathbf{f}|,|\mathbf{x}|),\ldots,P_{G_n}(|\mathbf{f}|,|\mathbf{x}|)))
\end{aligned}
$$

and a similar expression holds for $\mathcal{P}_F(|\mathbf{f}|,|\mathbf{x}|)$.

- expansion

$$F(\mathbf{f},\mathbf{g},\mathbf{x},\mathbf{y}) = G(\mathbf{f},\mathbf{x})$$

Trivial.

- concatenation recursion on notation CRN

$$
\begin{aligned}
F(\mathbf{f},\mathbf{x},0) &= G(\mathbf{f},\mathbf{x}) \\
F(\mathbf{f},\mathbf{x},s_i(y)) &= F(\mathbf{f},\mathbf{x},y)^\frown \mathrm{BIT}(0,H_i(\mathbf{f},\mathbf{x},y))
\end{aligned}
$$

We sketch the idea of an OCRAM program.

STEP 1. Determine $|G(\mathbf{f},\mathbf{x})|$
STEP 2. Find

$$i_0 = \max i < |y|[\mathrm{BIT}(0,H_{\mathrm{BIT}(i,y)}(\mathbf{f},\mathbf{x},\mathrm{MSP}(y,i+1)) = 1]$$

STEP 3.

```
if (|G(f,x)|=0) then
  begin
  P_i writes
     BIT(0, H_BIT(i_0 ∸ i,y)(f, x, MSP(y, i_0 + 1 − i)))
  if (i > i_0) then P_i writes $ in M^g_{i+1}
  end
else
  begin
  L = |G(f,x)|
  if (i < L) then
    P_i writes BIT(L ∸ (i + 1), (G(f, x)) in M^g_i
  if (i ≥ L) then
    P_i writes
       BIT(0, H_BIT(|y|−(i+1−L),y)(f, x, MSP(y, |y| − (i − L))))
    in M^g_{i+1}
  if (i ≥ L + |y|) then
    P_i writes $ in M^g_{i+1}
  end
```

A small example might render the above code in the case of CRN a bit easier to understand.

Illustrative Example

Suppose that F has been defined by CRN from G, H_0, H_1, where for notational ease we drop the parameters $\mathbf{f}, \mathbf{x}$ and also assume that the value of H_0, H_1 is either 0 or 1, so we need not apply the function BIT. Thus

$$\begin{aligned} F(0) &= G \\ F(s_i(y)) &= F(y)^\frown H_i(y). \end{aligned}$$

Suppose that $y = 2 = s_0(s_1(0))$. Then

$$\begin{aligned} F(s_0(s_1(0))) &= F(s_1(0))^\frown h_0(s_1(0)) \\ &= F(0)^\frown H_1(0)^\frown H_0(s_1(0)) \\ &= G^\frown H_{\text{BIT}(1,y)}(\text{MSP}(y,2)^\frown H_{\text{BIT}(0,y)}(\text{MSP}(y,1)) \end{aligned}$$

Working through this example should convince the reader of the correctness of the OCRAM program for CRN given above. Clearly the program runs in constant time with a second order polynomial bound on the number of processors. This completes the proof of Theorem 7.5.1.

To obtain the converse, we must arithmetize the computation of an OCRAM within $\mathcal{A}_0$. We begin with the following lemma.

Lemma 7.5.2. *Suppose that* OCRAM M *computes in time* $T(|f|, |x|)$ *using at most* $P(|f|, |x|)$ *processors. For each* f, x*, designating* $T(|f|, |x|)$ *by* T *and* $P(|f|, |x|)$ *by* P*, the size of every oracle function value returned during the computation of* M *is at most* $|f|(T \cdot P)$*, and for all* i*,*

$$\begin{aligned} M_i &= \$ \textit{ or } |M_i| \leq 2^T \cdot \max\{|f|((P+1) \cdot T), 1\} \\ M_i^g &= \$ \textit{ or } |M_i^g| \leq 2^T \cdot \max\{|f|((P+1) \cdot T), 1\}. \end{aligned}$$

Proof. Recall that M_i^0 is either $\$, 0, 1$. The length ℓ of the longest contiguous sequence $M_\alpha^0 \ldots M_{\alpha+(\ell-1)}^0$ of oracle registers, none of which contains \$ is at most $(P+1) \cdot T$. To see this bound, reason as follows. One can imagine every processor writing one bit at each step to a different oracle register (producing an bitwise representation of length $(P+1) \cdot T$), then calling the function on subwords of the oracle contents. Apart from oracle calls, one can at most double the contents of a register, since the atomic operations are $+, \dot{-}$, MSP, LSP. Thus after T steps of doubling a number of size at most $|f|((P+1) \cdot T)$, one obtains a number of size at most $2^T \cdot |f|((P+1) \cdot T)$.

Recall that depth $d(P)$ of a second order polynomial P was defined in Definition 7.2.9.

Definition 7.5.3. *The class of multivariate non-negative polynomials is the smallest class* $[0, 1, I, +, \times\ ; \text{COMP}]$ *of functions containing the constants* $0, 1$*, the projection functions* I_k^n*, addition* $+$*, multiplication* $\times$*, and closed under composition.*

Lemma 7.5.3. *For any multivariate non-negative polynomial* $p(n_1, \ldots, n_m)$*, there exists a function* $f(x_1, \ldots, x_n) \in A_0$ *such that* $p(|x_1|, \ldots, |x_n|) = |f(x_1, \ldots, x_n)|$.

Proof. By induction on formation of polynomials. Constants and projections belong to A_0. Suppose now that p, q are polynomials, and $f, g \in A_0$ and

$$
\begin{aligned}
p(|x_{i_1}|, \ldots, |x_{i_n}|) &= |f(x_{i_1}, \ldots, x_{i_n})|, \\
q(|x_{j_1}|, \ldots, |x_{j_m}|) &= |g(x_{j_1}, \ldots, x_{j_m})|
\end{aligned}
$$

Then

$$
\begin{aligned}
p+q &= |f| + |g| \\
&= |\lfloor \frac{f \cdot 2^{|g|+1} + 2^{|g|}}{2} \rfloor| \\
&= |\lfloor \frac{(2 \cdot f + 1) \cdot 2^{|g|}}{2} \rfloor| \\
&= |\mathrm{TR}(pad(s_1(f), g))| \\
p \cdot q &= |f| \cdot |g| \\
&= |\lfloor \frac{2^{|f| \cdot |g|}}{2} \rfloor| \\
&= |\mathrm{TR}(f \# g)|.
\end{aligned}
$$

Remark 7.5.2. Let $P(|f|, |x|)$ be a depth d second order polynomial. Let $P_1^c, \ldots, P_{k_c}^c$ be an enumeration of all depth c subpolynomials of P of the form $L(Q)$, where L is a second order variable, and the depth of Q is $c-1$. Since P_i^c is of the form $L(Q)$, denote the corresponding Q by Q_i^c. For $1 \leq c \leq d$, and $1 \leq i \leq k_c$, let $q_{c,i}^t$ denote the *maximizing query argument* (called in short *maxquery*) for Q_i^c at time t or before; i.e., that smallest value satisfying the following conditions:

$$
\begin{aligned}
|q_{c,i}^t| &\leq Q_i^c(|f_{Q_t}|, |x|) && (7.5) \\
|f(q_{c,i}^t)| &= P_i^c(|f_{Q_t}|, |x|). && (7.6)
\end{aligned}
$$

The following lemma is immediate.

Lemma 7.5.4. *Let* P *be a depth* d *second order polynomial with subpolynomials* Q_i^c*, for* $1 \leq c \leq d$*, and* $1 \leq i \leq k_c$*. Then there are first order polynomials* $\widehat{Q}_i^c$ *such that* $Q_i^c(|f_{Q_t}|, |x|)$ *equals*

$$\widehat{Q}_i^c(|f(q_{1,1}^t)|, \ldots, |f(q_{1,k_1}^t)|, \cdots, |f(q_{c-1,1}^t)|, \ldots, |f(q_{c-1,k_{c-1}}^t)|, |x|). \quad (7.7)$$

and $P(|f_{Q_t}|, |x|)$ *equals*

$$\widehat{P}(|f(q_{1,1}^t)|, \ldots, |f(q_{1,k_1}^t)|, \cdots, |f(q_{d,1}^t)|, \ldots, |f(q_{d,k_d}^t)|, |x|). \quad (7.8)$$

Lemma 7.5.5. *For depth* d *second order polynomial* P *and subpolynomials* Q_i^c*, there are functionals* $\overline{Q}_i^c, \overline{P} \in A_0$ *for which* $Q_i^c(|f_{Q_t}|, |x|)$ *equals*

$$|\overline{Q}_i^c(f, q_{1,1}^t, \ldots, q_{1,k_1}^t, \cdots, q_{c-1,1}^t, \ldots, q_{c-1,k_{c-1}}^t, x)| \tag{7.9}$$

for each $1 \leq i \leq k_c$ *and* $P(|f_{Q_t}|, |x|)$ *equals*

$$|\overline{P}(f, q_{1,1}^t, \ldots, q_{1,k_1}^t, \cdots, q_{d,1}^t, \ldots, q_{d,k_d}^t, x)|. \tag{7.10}$$

Proof. This follows from Lemmas 7.5.4,7.5.3.

Let's illustrate the last lemma with a small example. Suppose that $P(L, n) = L(L(n^2) + L(n))$, so that

$$P(|f|, |x|) = |f|\,(|f|(|x| \cdot |x|) + |f|(|x|))\,.$$

Then $k_1 = 2$, $k_2 = 1$, and

$$\begin{aligned}
Q_1^1(|f_{Q_t}|, |x|) &= |x| \cdot |x| \\
\widehat{Q}_1^1(|x|) &= |x| \cdot |x| \\
\overline{Q}_1^1(f, x) &= \text{TR}(x\#x) \\
Q_2^1(|f_{Q_t}|, |x|) &= |x| \\
\widehat{Q}_2^1(|x|) &= |x| \\
\overline{Q}_2^1(f, x) &= x \\
|q_{1,1}^t| &\leq |x| \cdot |x| \\
|f(q_{1,1}^t)| &= P_1^1(|f_{Q_t}|, |x|) \\
&= |f_{Q_t}|(|x| \cdot |x|) \\
|q_{1,2}^t| &\leq |x| \\
|f(q_{1,2}^t)| &= P_2^1(|f_{Q_t}|, |x|) \\
&= |f_{Q_t}|(|x|) \\
Q_1^2(|f_{Q_t}|, |x|) &= |f|(|x| \cdot |x|) + |f|(|x|) \\
\widehat{Q}_1^2(|f(q_{1,1}^t)|, |f(q_{1,2}^t)|, |x|) &= |f(q_{1,1}^t)| + |f(q_{1,2}^t)| \\
\overline{Q}_1^2(f, q_{1,1}^t, q_{1,2}^t, x) &= \text{TR}(pad(s_1(f(q_{1,1}^t)), f(q_{1,2}^t))).
\end{aligned}$$

Then

$$\begin{aligned}
|\overline{Q}_1^2(f, q_{1,1}^t, q_{1,2}^t, x)| &= |2^{|f(q_{1,1}^t)| + |f(q_{1,2}^t)|}| \dot{-} 1 \\
&= |f(q_{1,1}^t)| + |f(q_{1,2}^t)|
\end{aligned}$$

as required. Now

$$\begin{aligned}
|q_{2,1}^t| &\leq Q_1^2(|f_{Q_t}|, |x|) \\
|f(q_{2,1}^t)| &= P(|f_{Q_t}|, |x|).
\end{aligned}$$

Finally,

$$\begin{aligned}
\widehat{P}(|f(q_{1,1}^t)|, |f(q_{1,2}^t)|, |f(q_{2,1}^t)|, |x|) &= |f(q_{2,1}^t)| \\
\overline{P}(f, q_{1,1}^t, q_{1,2}^t, q_{2,1}^t, x) &= f(q_{2,1}^t).
\end{aligned}$$

Theorem 7.5.2. *Let P be a second order polynomial. If a functional $F(\mathbf{f}, \mathbf{x})$ is computable by a fully specified* OCRAM $\mathcal{M} = \langle M, P \rangle$ *in constant time, then $F \in \mathcal{A}_0$.*

Proof. Let's begin by a proof sketch that (type 1) functions computable on a CRAM in constant time with a polynomial bound on the indices of active processors belong to A_0. The general approach is to arithmetize the computation of a CRAM M. A first thought is to encode the global and local memories as sequences of integers, but the following consideration indicates problems with this approach. It is possible that processor p of M doubles its contents of a local register M_0 in each step, so that after t steps, M_0 contains the integer 2^t. Processor p may then execute a global write instruction, such as $*M_0^g = 5$, thus writing 5 in $M_{M_0}^g$, i.e., in $M_{2^t}^g$. With this arithmetization, one would have to encode the sequence

$$\langle M_0^g, \ldots, M_{2^{|p(n)|^k}}^g \rangle$$

in order to arithmetize a polylogarithmic computation running in $|p(n)|^k$ steps. This, however, requires integers of length $2^{|p(n)|^k}$, which is not polynomial in n. In proving the equivalence between boolean circuit families and parallel random access machines, Stockmeyer and Vishkin [SV84] used the following trick, which we apply to the function algebra case.

Introduce arrays $A^g, V^g, A^\ell, V^\ell, I$ such that

- $A^g(p,t)$ ($A^\ell(p,t)$) is the address in global (local) memory to which processor p writes at time t, if p writes at this time, and ∞ otherwise,
- $V^g(p,t)$ ($V^\ell(p,t)$) is the value that processor p writes at time t, else ∞,
- $I(p,t)$ is the instruction (between 1 and N, for a program of N lines) which processor p carries out at time t.

It turns out that $A^g, V^g, A^\ell, V^\ell, I$ are definable by a suitable form of simultaneous recursion, and that from these values, one can compute M_i for each processor p, as well as global memory register contents M_i^g for each time t.

Suppose that on input of size n, the active processors of CRAM M are those with index at most $P(n)$, for some polynomial P. Note first that the instruction $I(p,t_0)$, which processor p should carry out at time t_0, is easily determined by definition by cases, given previous values $I(p,t_0-1)$, and global/local memory register contents at time t_0-1. Now the values M_i for every active processor $p \le P(n)$ at time t_0-1, and the values of global memory registers M_i^g at time t_0-1 can be determined as follows. We describe the case for M_i^g.

Let t_1 be the last time some processor wrote to M_i^g, so that

$$t_1 = \begin{cases} \max t < t_0\, [\exists p \le P(n)(A^g(p,t) = i)] & \text{if such exists} \\ \infty & \text{else.} \end{cases}$$

The processor $p(t_1)$ having smallest identity number, which wrote to global memory register M_i^g must satisfy

$$p(t_1) = \begin{cases} \min p \leq P(n)\,[A^g(p,t_1) = i] & \text{if such exists} \\ \infty & \text{else.} \end{cases}$$

It now follows that the value of M_i^g at time $t_0 - 1$ must be

$$M_i^g = \begin{cases} A^g(p(t_1),t_1) \text{ if } t_1 \neq \infty \\ \text{BIT}(n-i,x) \text{ if } t_1 = \infty \text{ and } 1 \leq i \leq n \\ \$ \qquad\qquad \text{else.} \end{cases}$$

In a similar manner, the value of every M_i of every active processor $p \leq P(n)$ at time $t_0 - 1$ can be determined.

The arithmetic functions $+, \dot{-}, \text{MSP}, \text{LSP}$ which occur on the right side of assignment statements are all definable in the algebra A_0. The current contents of global and local memory registers at time $t_0 - 1$ are determined as above, and the indices of global and local memory registers for indirect reads and writes can be evaluated. From this, one can define the new values $A^g(p,t_0)$, $V^g(p,t_0)$, $A^\ell(p,t_0)$, $V^\ell(p,t_0)$.

In this manner, by t steps of simultaneous recursion, the values of $A^g, V^g, A^\ell, V^\ell, I$ can be determined for all active processors $p \leq P(n)$ at time t. For t bounded by an absolute constant T_0, in the case of constant time computation, it follows that the above functions can be defined using composition.

In the type 2 case of constant parallel time bounded by absolute constant T_0, with a second order polynomial bound $P(|f|,|x|)$ on the indices of active processors, we additionally note that there exist maxqueries

$$q_{1,1}^{T_0}, \ldots, q_{1,k_1}^{T_0}, \ldots, q_{d,1}^{T_0}, \ldots, q_{d,k_d}^{T_0}$$

so that

$$|\overline{P}(f, q_{1,1}^{T_0}, \ldots, q_{1,k_1}^{T_0}, \ldots, q_{d,1}^{T_0}, \ldots, q_{d,k_d}^{T_0}, x)| = P(|f|,|x|).$$

Since $\mathcal{A}_0$ is closed under sharply bounded minimization and sharply bounded quantification, the particular applications of max and min required to compute M_i^g from A_M^g, V_M^g, A_M^ℓ, V_M^ℓ are available in $\mathcal{A}_0$. This concludes the proof of Theorem 7.5.2.

Summarizing, we now have a characterization of constant parallel time.

Theorem 7.5.3. *A functional $F(\mathbf{f},\mathbf{x}) \in \mathcal{A}_0$ if and only if it is computable on an* OCRAM *in constant time with at most $P(|\mathbf{f}|,|\mathbf{x}|)$ many processors, for some second-order polynomial P.*

We turn now to the characterization of type 2 NC functionals via algebra $\mathcal{A}$. Suppose that $\mathcal{M} = \langle M, P \rangle$ is a fully specified OCRAM running in time $|P(|f|,|x|)|^k$ on input f, x. Let $\overline{P}, \overline{Q}_i^c$, for $1 \leq c \leq d$ and $1 \leq i \leq k_c$, be functionals in $\mathcal{A}_0$ satisfying the conclusion of Lemma 7.5.5. Let $\overline{q_c^{||s||}}$ abbreviate

$$q_{c,1}^{||s||}, \ldots, q_{c,k_c}^{||s||}$$

for $1 \leq c \leq d$, and let $\overline{P}_{||s||}$ abbreviate

$$P(f, \overline{q_1^{||s||}}, \ldots, \overline{q_d^{||s||}}, x).$$

Then $|\overline{P}_{||s||}|$ is a bound on the indices of active processors at time $||s||$ or before.

Using strong multiple limited square-root recursion, define the following functions in $\mathcal{A}$ simultaneously: $Instr_{\mathcal{M}}(f,x,p,s)$, $A^g_{\mathcal{M}}(f,x,p,s)$, $V^g_{\mathcal{M}}(f,x,p,s)$, $A^\ell_{\mathcal{M}}(f,x,p,s)$, $V^\ell_{\mathcal{M}}(f,x,p,s)$, $A^o_{\mathcal{M}}(f,x,p,s)$, $V^o_{\mathcal{M}}(f,x,p,s)$, $\overline{\overline{MQ}}^c_{\mathcal{M},i}$ for $1 \leq c \leq d$ and $1 \leq i \leq k_c$, and $\overline{\overline{MQ}}^{d+1}_{\mathcal{M},1}$.

The intent is that $Instr_{\mathcal{M}}(f,x,p,s)$ is the instruction processor p executes at time $||s||$, $A^g_{\mathcal{M}}(f,x,p,s)$ is the address in global memory to which processor p writes at step $||s||$ (else ∞), $V^g_{\mathcal{M}}(f,x,p,s)$ is the value in global memory which processor p writes at step $||s||$ (else ∞), and similarly $A^\ell_{\mathcal{M}}(f,x,p,s)$, $V^\ell_{\mathcal{M}}(f,x,p,s)$ for local registers. In the type 2 case, we additionally have oracle registers, so that $A^o_{\mathcal{M}}(f,x,p,s)$ ($V^o_{\mathcal{M}}(f,x,p,s)$) represent the address (value) of oracle register to which (which) processor p writes at step $||s||$ (else ∞). The maxquery $q_{c,i}^{||s||}$ at step $||s||$ for the i-th depth c subpolynomial of P will be given by $\overline{\overline{MQ}}^c_{\mathcal{M},i}(f,x,s)$, for $1 \leq c \leq d$ and $1 \leq i \leq k_c$.

Now $A^g_{\mathcal{M}}(f,x,p,s)$ is obtained by appending to $A^g_{\mathcal{M}}(f,x,p,\mathrm{sqrt}(s))$ a new column of length $|\overline{P}_{||s||}| + 1$, where for processor p, with $0 \leq p \leq |\overline{P}_{||s||}|$, $A^g_{\mathcal{M}}(f,x,p,s)$ is the global address (if any) to which processor p writes, and ∞ else. Consult Figure 7.5 for a picture of the address array.

The instruction counter $I_{\mathcal{M}}(f,x,p,s)$ is bounded by an absolute constant (the number of lines of the program for M), and since P is monotonic, $|A^g_{\mathcal{M}}(f,x,p,s')| \leq |\overline{P}_{||s||}|$ for all $s' \leq s$, and similarly for $A^\ell_{\mathcal{M}}$ and $A^o_{\mathcal{M}}$. If the processor identity numbers of active processors are bounded by m, then by Lemma 7.5.2, in t steps of computation, the size of any integer in any register is bounded by $2^t \cdot |f_{Q_t}|(m \cdot t)$. It follows that $|V^g_{\mathcal{M}}(f,x,p,s')| \leq (|s|+1) \cdot |f_{Q_{||s||}}|(|\overline{P}_{||s||}|)$, and similarly for $V^\ell_{\mathcal{M}}$, and $V^o_{\mathcal{M}}$. Thus $|\overline{P}_{||s||}| + (|s|+1) \cdot |f_{Q_{||s||}}|(|\overline{P}_{||s||}|)$ is an upper bound for the length of $A^g_{\mathcal{M}}$, $V^g_{\mathcal{M}}$, $A^\ell_{\mathcal{M}}$, $V^\ell_{\mathcal{M}}$, $A^o_{\mathcal{M}}$, $V^o_{\mathcal{M}}$, $I_{\mathcal{M}}$.

Let $R(|f|,|x|,|s|) = P(|f|,|x|) + (|s|+1) \cdot |f|(P(|f|,|x|))$. Then R is a second order polynomial of degree $d+1$, which bounds $|A^g_{\mathcal{M}}(f,x,p,s')|$, $|V^g_{\mathcal{M}}(f,x,p,s')|$ for $s' \leq s$, and similarly for $A^\ell_{\mathcal{M}}$, $V^\ell_{\mathcal{M}}$, $A^o_{\mathcal{M}}$, $V^o_{\mathcal{M}}$, and $I_{\mathcal{M}}$. By Lemma 7.5.5, there is a functional $\overline{R} \in \mathcal{A}_0$ and a maxquery $q_{d+1,1}^{||s||}$, such that

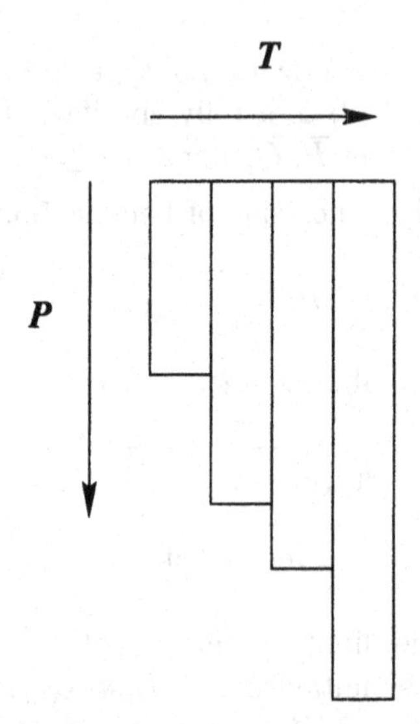

Fig. 7.1. Address array $A^g_{\mathcal{M}}(f,x,p,t)$

$$R(|f_{Q_{||s||}}|,|x|,|s|) \leq \left|\overline{R}(f,\overline{q_1^{||s||}},\ldots,\overline{q_d^{||s||}},q_{d+1,1}^{||s||},x)\right|.$$

Since $\mathcal{A}_0$ and $\mathcal{A}$ are closed under sharply bounded min, sharply bounded max, and sharply bounded quantification, $\overline{\overline{MQ}}^c_{\mathcal{M},i}(f,x,s)$ can inductively be defined so as to pick the maxquery $q_{c,i}^{||s||}$. Hence

$$\begin{aligned}
|\overline{\overline{MQ}}^1_{\mathcal{M},i}(f,x,s)| &\leq |\overline{Q}^1_i(f,x)| \\
&\quad \text{for } 1 \leq i \leq k_1 \\
|\overline{\overline{MQ}}^2_{\mathcal{M},i}(f,x,s)| &\leq |\overline{Q}^2_i(f,\overline{\overline{MQ}}^1_{\mathcal{M},1}(f,x,\text{sqrt}(s)),\ldots, \\
&\quad \overline{\overline{MQ}}^1_{\mathcal{M},k_1}(f,x,\text{sqrt}(s)),x)|, \text{ for } 1 \leq i \leq k_2 \\
&\vdots \\
|\overline{\overline{MQ}}^c_{\mathcal{M},i}(f,x,s)| &\leq |\overline{Q}^c_i(f,\overline{\overline{MQ}}^1_{\mathcal{M},1}(f,x,\text{sqrt}(s)),\ldots, \\
&\quad \overline{\overline{MQ}}^c_{\mathcal{M},k_c}(f,x,\text{sqrt}(s)),x)|, \text{ for } 1 \leq i \leq k_c \\
&\vdots \\
|\overline{\overline{MQ}}^d_{\mathcal{M},i}(f,x,s)| &\leq |\overline{Q}^d_i(f,\overline{\overline{MQ}}^1_{\mathcal{M},1}(f,x,\text{sqrt}(s)),\ldots, \\
&\quad \overline{\overline{MQ}}^d_{\mathcal{M},k_d}(f,x,\text{sqrt}(s)),x)|, \text{ for } 1 \leq i \leq k_d \\
|\overline{\overline{MQ}}^{d+1}_{\mathcal{M},1}(f,x,s)| &\leq |\overline{Q}^{d+1}_1(f,\overline{\overline{MQ}}^1_{\mathcal{M},1}(f,x,\text{sqrt}(s)),\ldots, \\
&\quad \overline{\overline{MQ}}^{d+1}_{\mathcal{M},1}(f,x,\text{sqrt}(s)),x)|.
\end{aligned}$$

It now follows that all these functions can be defined by strong multiple limited square-root recursion from functions in $\mathcal{A}$, hence by Theorem 7.4.1 the functions belong to $\mathcal{A}$.

Definition 7.5.4. *A configuration z at time t of the fully specified* OCRAM $\mathcal{M} = \langle M, P\rangle$ *is a sequence number, encoding the values* $Instr_{\mathcal{M}}(f,x,p,s)$, $A^{g}_{\mathcal{M}}(f,x,p,s)$, $V^{g}_{\mathcal{M}}(f,x,p,s)$, $A^{\ell}_{\mathcal{M}}(f,x,p,s)$, $V^{\ell}_{\mathcal{M}}(f,x,p,s)$, $A^{o}_{\mathcal{M}}(f,x,p,s)$, $V^{o}_{\mathcal{M}}(f,x,p,s)$, $\overline{\overline{MQ}}^{c}_{\mathcal{M},i}$ *for* $1 \leq c \leq d$ *and* $1 \leq i \leq k_c$, *and* $\overline{\overline{MQ}}^{d+1}_{\mathcal{M},1}$ *for all* $p \leq P(|f_{Q_{||s||}}|, |x|)$ *and for* $s = 0, 1, 2^{2^0}, 2^{2^1}, \ldots, 2^{2^{t-1}}$.

Recalling that $||2^{2^{t-1}}|| = t$, it follows that the configuration z at time t encodes the entire computation of $\mathcal{M} = \langle M, P\rangle$ at steps $0, 1, \ldots, t$. From the previous discussion, we have proved the following.

Lemma 7.5.6 (Run Lemma). *Suppose that $\mathcal{M} = \langle M, P\rangle$ is a fully specified* OCRAM *computing a rank $(1,1)$ functional. Then there exists a function $Run_{\mathcal{M}} \in \mathcal{A}$ such that for all f, x, s, $Run_{\mathcal{M}}(f,x,s)$ encodes the configuration of $\mathcal{M}$ at time $||s||$.*

We state as follows without proof the generalization of Lemma 7.5.6.

Lemma 7.5.7 (Generalized Run Lemma). *Suppose that P is a second order polynomial, and that $\mathcal{M} = \langle M, P\rangle$ is fully specified* OCRAM *computing a rank $(1,1)$ functional. Then there is a functional $GenRun_{\mathcal{M}} \in \mathcal{A}$ such that for all f, x, s and configurations z, $GenRun_{\mathcal{M}}(f,x,z,s)$ encodes the configuration of $\mathcal{M}$ after $||s||$ computation steps, when starting from configuration z.*

Lemma 7.5.8 (k-Generalized Run Lemma). *Suppose that P is a second order polynomial, and that $\mathcal{M} = \langle M, P\rangle$ is a fully specified* OCRAM *computing a rank $(1,1)$ functional. Then there exists a function $GenRun^{k}_{\mathcal{M}} \in \mathcal{A}$ such that for all f, x, s and configuration z, $GenRun^{k}_{\mathcal{M}}(f,x,z,s)$ encodes the configuration of $\mathcal{M}$ after $||s||^{k}$ computation steps, when starting from configuration z.*

Proof. The proof is by induction on k. The base case $k = 1$ is treated in Lemma 7.5.7. Suppose that $GenRun^{k}_{\mathcal{M}} \in \mathcal{A}$ satisfies the present lemma for k. Define

$$F(f,x,z,s,0) = z$$
$$F(f,x,z,s,t) = GenRun^{k}_{\mathcal{M}}(f,x,F(f,x,z,\mathrm{sqrt}(t)),s).$$

F is defined by limited square-root recursion from functionals in $\mathcal{A}$, hence belongs to $\mathcal{A}$. Taking

$$GenRun^{k+1}_{\mathcal{M}}(f,x,z,s) = F(f,x,z,s,s)$$

encodes the configuration of $\mathcal{M}$ after $||s||^{k+1}$ steps.

Corollary 7.5.1. *Suppose that $\mathcal{M} = \langle M, P \rangle$ is a fully specified* OCRAM *computing a rank* $(1,1)$ *functional. Then there exists a function* $Run^k_{\mathcal{M}} \in \mathcal{A}$ *such that for all* f, x, s, $Run^k_{\mathcal{M}}(f, x, s)$ *encodes the configuration of* $\mathcal{M}$ *at time* $||s||^k$.

Definition 7.5.5. *A* query step *in the execution of fully specified* OCRAM $\mathcal{M} = \langle M, P \rangle$ *is a computation step where at least one processor queries the oracle.*[10] *The total number of computation steps in the computation of* $\mathcal{M}$ *on arguments* f, x *will be denoted by* $S_{\mathcal{M}}(f, x)$. *Let* $t(\mathcal{M}, f, x, r)$ *be the least number of steps in the computation of* $\mathcal{M}$ *on inputs* f, x *such that exactly* r *query steps have been made, or the computation of* $\mathcal{M}$ *is terminated. When parameters are clear from context,* $t(\mathcal{M}, f, x, r)$ *will be abbreviated by* $t(r)$.

Note that though the number $|Q_{\mathcal{M}}(f, x, t)|$ of oracle queries may be quite large, the number of query steps is $< t$.

Fact 7.5.1. Suppose that P, P' are second order polynomials, $\mathcal{M} = \langle M, P' \rangle$ is a fully specified OCRAM and the run time of $\mathcal{M}$ on inputs f, x is bounded by $|P(|f|, |x|)|^k$.

1. Suppose that $Q_t = Q_{\mathcal{M}}(f, x, t)$ and that $t > |P(|f_{Q_t}|, |x|)|^k$. Then $S_{\mathcal{M}}(f_{Q_t}, x) < t$; i.e., $\mathcal{M}$ on inputs f_{Q_t}, x halts in less than t steps.
2. Suppose that $Q_t = Q_{\mathcal{M}}(f, x, t)$, $Q_{t'} = Q_{\mathcal{M}}(f, x, t')$ and that $t' > |P(|f_{Q_t}|, |x|)|^k$. Then either $S_{\mathcal{M}}(f_{Q_{t'}}, x) < t'$ or $|Q_t| < |Q_{t'}|$. In other words, either $\mathcal{M}$ halts in less than $|P(|f_{Q'_t}|, |x|)|^k$ steps or Q_t is properly contained in Q'_t.
3. Suppose that $t = t(\mathcal{M}, f, x, r)$ and that $Q_t = Q_{\mathcal{M}}(f, x, t)$. If $t < S_{\mathcal{M}}(f, x)$ then $t \leq S_{\mathcal{M}}(f_{Q_t}, x)$.

Proof. The OCRAM $\mathcal{M}$ on inputs f, x makes the same moves as $\mathcal{M}$ on f_{Q_t}, x at steps $0, 1, \ldots, t-1$. Since $t > |P(|f_{Q_t}|^k, |x|)|$ and the runtime of $\mathcal{M}$ is bounded by $|P|^k$, it follows that $\mathcal{M}$ halts in less than t steps on inputs f_{Q_t}, x. This proves (1).

If $|Q_t| = |Q_{t'}|$, then $Q_t = Q_{t'}$, so that $P(|f_{Q_t}|, |x|) = P(|f_{Q_{t'}}|, |x|)$. Since $t' > |P(|f_{Q_t}|, |x|)|^k$, it follows that $t' > |P(|f_{Q_{t'}}|, |x|)|^k$. But $|P|^k$ bounds the runtime of $\mathcal{M}$, so $S_{\mathcal{M}}(f_{Q_{t'}}, x) < t'$. This proves (2).

Since $t < S_{\mathcal{M}}(f, x)$, the OCRAM $\mathcal{M}$ does not halt in t steps on input f, x, so it follows that $\mathcal{M}$ must execute at least r query steps before halting. By Lemma 7.5.1, $\mathcal{M}$ makes identical moves in the first t steps on inputs f_{Q_t}, x as on f, x. By definition, $t = t(\mathcal{M}, f, x, r)$ is the smallest number of steps for which either $\mathcal{M}$ halts, or executes r query steps. By minimality of t, it follows that $t \leq S_{\mathcal{M}}(f_{Q_t}, x)$. This proves (3).

[10] Note that many processors may in parallel execute a different query during *one* query step.

Lemma 7.5.9 (Maxquery at time $t(||r||)$). *Let $\mathcal{M} = \langle M, P \rangle$ be a fully specified* OCRAM *whose runtime on arguments f, x is bounded by $|P(|f|, |x|)|^k$, where $P(|f|, |x|)$ is a depth d second order polynomial. For $1 \leq c \leq d$, and $1 \leq i \leq k_c$, let P_i^c, Q_i^c, and $q_{c,i}^t$ be as given in Remark 7.5.2. Let $t(r) = t(\mathcal{M}, f, x, r)$, the least number of steps in which $\mathcal{M}$ executes r query steps. Then there exist functionals $\overline{\overline{MQS}}^c_{\mathcal{M},i} \in \mathcal{A}$ such that for $1 \leq c \leq d$ and $1 \leq i \leq k_c$*

$$\overline{\overline{MQS}}^c_{\mathcal{M},i}(f, x, r) = q_{c,i}^{t(||r||)}.$$

Proof. Before we begin the proof, note the distinction between the auxiliary functions $\overline{\overline{MQ}}^c_{\mathcal{M},i}(f, x, s)$, which yield the maxquery in $||s||$ steps of computation, and $\overline{\overline{MQS}}^c_{\mathcal{M},i}(f, x, r)$, which yield the maxquery in $t(\mathcal{M}, f, x, ||r||)$ steps of computation.

Using strong multiple limited square-root recursion, simultaneously define $\overline{\overline{MQS}}^c_{\mathcal{M},i}$ for $1 \leq c \leq d$, and $1 \leq i \leq k_c$, in $\mathcal{A}$ as follows.

BASE CASE. $\overline{\overline{MQS}}^c_{\mathcal{M},i}(f, x, 0) = 0$.

INDUCTIVE CASE: Suppose that $\overline{\overline{MQS}}^c_{\mathcal{M},i}(f, x, \text{sqrt}(r))$ has been defined for all $1 \leq c \leq d$, and $1 \leq i \leq k_c$. Temporarily, let

$$\begin{aligned} t &= t(\mathcal{M}, f, x, ||\text{sqrt}(r)||) \\ &= t(\mathcal{M}, f, x, ||\lfloor \frac{|r|}{2} \rfloor||) \\ &= t(\mathcal{M}, f, x, ||r|| - 1). \end{aligned}$$

Abbreviate

$$\overline{P}(f, \overline{\overline{MQS}}^1_{\mathcal{M},1}(f, x, \text{sqrt}(r)), \ldots, \overline{\overline{MQS}}^d_{\mathcal{M},k_d}(f, x, \text{sqrt}(r)), x)$$

by $\overline{P}$ and note that $|\overline{P}| = P(|f_{Q_t}|, |x|)$ and that up to time $t-1$, there have been $||r|| - 1$ query steps. Define

$$P^*(f, \overline{\overline{MQS}}^1_{\mathcal{M},1}(f, x, \text{sqrt}(r)), \ldots, \overline{\overline{MQS}}^d_{\mathcal{M},k_d}(f, x, \text{sqrt}(r)), x)$$

(abbreviated as P^*) by $2\#\overline{P}$. Then $|P^*| = 2 \cdot |\overline{P}| + 1$, and $||P^*|| > ||\overline{P}|| = |P(|f_{Q_t}|, |x|)|$.

Case 1. $t \geq S_{\mathcal{M}}(f, x)$.

This case can be effectively decided by testing, using $Run_{\mathcal{M}}(f, x, \overline{P})$, whether all processors have halted. In this case, $\mathcal{M}$ altogether makes $||\text{sqrt}(r)||$ many query steps before halting, so define

$$\overline{\overline{MQS}}^c_{\mathcal{M},i}(f, x, r) = \overline{\overline{MQS}}^c_{\mathcal{M},i}(f, x, \text{sqrt}(r)).$$

Case 2. $t < S_{\mathcal{M}}(f, x)$.

Then by (3) of Fact 7.5.1, $t \leq S_{\mathcal{M}}(f_{Q_t}, x)$. Define $t' = ||P^*||^k$, so we have

$$t' = ||P^*||^k > ||\overline{P}||^k = |P(|f_{Q_t}|, |x|)|^k.$$

By (2) of Fact 7.5.1 either $S_{\mathcal{M}}(f_{Q_{t'}}, x) < t'$ or $|Q_t| < |Q_{t'}|$.

SUBCASE A: $S_{\mathcal{M}}(f_{Q_{t'}}, x) < t'$

Thus $\mathcal{M}$ halts within t' steps on input f, x. From $Run^k_{\mathcal{M}}(f, x, P^*)$ one can determine whether there is a $||r||$-th query step before $\mathcal{M}$ halts, and so accordingly define the values of $\overline{\overline{MQS}}^c_{\mathcal{M},i}(f, x, r)$ for $1 \leq c \leq d$ and $1 \leq i \leq k_c$.

SUBCASE B: $|Q_t| < |Q_{t'}|$

Thus $||r|| - 1 = |Q_t| < |Q_{t'}|$, so that $|Q_{t'}| \geq ||r||$. The values of

$$\overline{\overline{MQS}}^c_{\mathcal{M},i}(f, x, r)$$

can then be obtained from $Run^k_{\mathcal{M}}(f, x, P^*)$. This completes the inductive case.

To complete the proof of the lemma, note that

$$|\overline{\overline{MQS}}^1_{\mathcal{M},i}(f, x, r)| \leq |\overline{Q}^1_i(f, x)| \quad \text{for } 1 \leq i \leq k_1$$

$$|\overline{\overline{MQS}}^2_{\mathcal{M},i}(f, x, s)| \leq |\overline{Q}^2_i(f, \overline{\overline{MQS}}^1_{\mathcal{M},1}(f, x, s), \ldots, \overline{\overline{MQS}}^1_{\mathcal{M},k_1}(f, x, s), x)|, \text{ for } 1 \leq i \leq k_2$$

$$\vdots$$

$$|\overline{\overline{MQS}}^d_{\mathcal{M},i}(f, x, s)| \leq |\overline{Q}^d_i(f, \overline{\overline{MQS}}^1_{\mathcal{M},1}(f, x, s), \ldots, \overline{\overline{MQS}}^{d-1}_{\mathcal{M},k_{d-1}}(f, x, s), x)|, \text{ for } 1 \leq i \leq k_d.$$

Thus the $\overline{\overline{MQS}}^c_{\mathcal{M},i}$ can be defined using simultaneous limited recursion on notation, and so belong to $\mathcal{A}$.

Lemma 7.5.10 (Maxquery at time $t(|r|)$). *Under the same hypotheses as Lemma 7.5.9, there exist functionals $\overline{MQS}^c_{\mathcal{M},i} \in \mathcal{A}$ such that*

$$\overline{MQS}^c_{\mathcal{M},i}(f, x, r) = q^{t(|r|)}_{c,i}.$$

Proof. Using strong multiple limited square-root recursion, simultaneously define $\overline{MQS}^c_{\mathcal{M},i}$, for $1 \leq c \leq d$, and $1 \leq i \leq k_c$, in $\mathcal{A}$ as follows.

BASE CASE. $\overline{MQS}^c_{\mathcal{M},i}(f, x, 0) = 0$.

INDUCTIVE CASE: Suppose that $\overline{MQS}^c_{\mathcal{M},i}(f, x, \mathrm{sqrt}(r))$ has been defined for all $1 \leq c \leq d$, and $1 \leq i \leq k_c$. Temporarily, let

$$t = t(\mathcal{M}, f, x, |\text{sqrt}(r)|)$$
$$= t(\mathcal{M}, f, x, \lfloor \frac{|r|}{2} \rfloor).$$

Abbreviate

$$\overline{P}(f, \overline{MQS}^1_{\mathcal{M},1}(f, x, \text{sqrt}(r)), \ldots, \overline{MQS}^d_{\mathcal{M},k_d}(f, x, \text{sqrt}(r)), x)$$

by $\overline{P}_0$ and note that $|\overline{P}_0| = P(|f_{Q_t}|, |x|)$ and up to time $t-1$ there have been $\lfloor \frac{|r|}{2} \rfloor$ query steps. From $Run^k_{\mathcal{M}}(f, x, \overline{P}_0)$ extract maxqueries $q_{c,i}^{||\overline{P}||^k}$, and define

$$\widetilde{P} = \overline{P}(f, q_{1,1}^{||\overline{P}_0||^k}, \ldots, q_{d,k_d}^{||\overline{P}_0||^k}, x).$$

Define $P^* = 2\#\widetilde{P}$. Then $|P^*| = 2 \cdot |\widetilde{P}| + 1$, and $||P^*|| > ||\widetilde{P}|| = |P(|f_{Q_{t^k}}|, |x|)|$.

Case 1. $t^k \geq S_{\mathcal{M}}(f, x)$.
This case can be effectively decided by testing, using $Run^k_{\mathcal{M}}(f, x, \overline{P}_0)$, whether all processors have halted. From $Run^k_{\mathcal{M}}$, one can determine whether $|r|$ query steps were made before $\mathcal{M}$ halts, and correspondingly define $\overline{MQS}^c_{\mathcal{M},i}(f, x, r)$.

Case 2. $t^k < S_{\mathcal{M}}(f, x)$.

Then by (3) of Fact 7.5.1, $t^k \leq S_{\mathcal{M}}(f_{Q_{t^k}}, x)$, and by (1) of Fact 7.5.1, $t^k \leq |P(|f_{Q_{t^k}}|, |x|)|^k$. Now $||P^*|| > |P(|f_{Q_{t^k}}|, |x|)|$, so $||P^*||^k > t^k$. Since $t = t(\mathcal{M}, f, x, \lfloor \frac{|r|}{2} \rfloor)$, it follows that $t \geq \lfloor \frac{|r|}{2} \rfloor$, so $||P^*||^k > t^k \geq \lfloor \frac{|r|}{2} \rfloor^k$ and hence $||P^*|| \geq \lfloor \frac{|r|}{2} \rfloor$. For all $x \geq 1$,

$$\log_2(x) + 1 \geq |x| = \lceil \log_2(x+1) \rceil \geq \log_2(x+1) \geq \log_2(x)$$

and so

$$2 \cdot x \geq 2^{|x|} \geq x + 1 \geq x.$$

Letting $P^{**} = s_1(P^*\#P^*\#2^7)$, we have that $P^{**} \in \mathcal{A}$. Now $2 \cdot ||P^*|| + 1 \geq |r|$, so $2^{2||P^*||+1} \geq r$, and

$$\begin{aligned} |P^{**}| &> 8 \cdot |P^*|^2 \\ &= 2(2 \cdot |P^*|)^2 \\ &\geq 2(2^{||P^*||})^2 \\ &= 2^{2||P^*||+1} \\ &\geq 2^{|r|} \\ &\geq r. \end{aligned}$$

Hence $P^{**} \geq 2^{r-1}$. From the techniques of the previous chapter (see Exercises), the function $Exp2(a, b) = \min(2^a, b) \in A_0$, so it follows for $r > 0$ that

$$r = |Exp2(r-1, P^{**})|.$$

Now define

$$\overline{MQS}^c_{\mathcal{M},i}(f,x,r) = \overline{\overline{MQS}}^c_{\mathcal{M},i}(f,x,Exp2(r-1,P^{**})).$$

It follows that

$$\begin{aligned}\overline{MQS}^c_{\mathcal{M},i}(f,x,r) &= q_{c,i}^{t(||Exp2(r-1,P^{**})||)} \\ &= q_{c,i}^{t(|r|)}.\end{aligned}$$

This completes the inductive case.

Appropriate bounds for the $|\overline{MQS}^c_{M,i}(f,x,r)|$ can be given, so that the functions $\overline{MQS}^c_{\mathcal{M},i}$ can be defined using simultaneous limited recursion on notation, and hence belong to $\mathcal{A}$. Namely,

$$\begin{aligned}
|\overline{MQS}^1_{\mathcal{M},i}(f,x,r)| &\le |\overline{Q}^1_i(f,x)| \\
&\qquad \text{for } 1 \le i \le k_1 \\
|\overline{MQS}^2_{\mathcal{M},i}(f,x,s)| &\le |\overline{Q}^2_i(f, \overline{MQS}^1_{\mathcal{M},1}(f,x,s),\ldots, \\
&\qquad \overline{MQS}^1_{\mathcal{M},k_1}(f,x,s),x)|, \text{ for } 1 \le i \le k_2 \\
&\ \ \vdots \\
|\overline{MQS}^d_{\mathcal{M},i}(f,x,s)| &\le |\overline{Q}^d_i(f, \overline{MQS}^1_{\mathcal{M},1}(f,x,s),\ldots, \\
&\qquad \overline{MQS}^{d-1}_{\mathcal{M},k_{d-1}}(f,x,s),x)|, \text{ for } 1 \le i \le k_d.
\end{aligned}$$

This completes the proof of the lemma.

Lemma 7.5.11 (Maxquery at time $t(r)$). *Under the same hypotheses as Lemma 7.5.9, there exist functionals $MQS^c_{\mathcal{M},i} \in \mathcal{A}$ such that*

$$MQS^c_{\mathcal{M},i}(f,x,r) = q_{c,i}^{t(r)}.$$

Proof. Using strong multiple limited square-root recursion, simultaneously define $MQS^c_{\mathcal{M},i}$, for $1 \le c \le d$, and $1 \le i \le k_c$, in $\mathcal{A}$ as follows.

BASE CASE. $MQS^c_{\mathcal{M},i}(f,x,0) = 0$.

INDUCTIVE CASE: Suppose that $MQS^c_{\mathcal{M},i}(f,x,\text{sqrt}(r))$ has been defined for all $1 \le c \le d$, and $1 \le i \le k_c$. Temporarily, let

$$t = t(\mathcal{M}, f, x, \text{sqrt}(r)).$$

Abbreviate

$$\overline{P}(f, MQS^1_{\mathcal{M},1}(f,x,\text{sqrt}(r)),\ldots, MQS^d_{\mathcal{M},k_d}(f,x,\text{sqrt}(r)),x)$$

by $\overline{P}_1$ and note that $|\overline{P}_1| = P(|f_{Q_t}|,|x|)$ and that up to time $t-1$ there have been $\text{sqrt}(r)$ query steps. From $Run^k_{\mathcal{M}}(f,x,\overline{P}_1)$ extract maxqueries $q_{c,i}^{||\overline{P}_1||^k}$, and define

$$\widetilde{P} = \overline{P}(f, q_{1,1}^{||\overline{P}_1||^k}, \ldots, q_{d,k_d}^{||\overline{P}_1||^k}, x).$$

Define $P^* = 2\#\widetilde{P}$. Then $|P^*| = 2 \cdot |\widetilde{P}| + 1$, and $||P^*|| > ||\widetilde{P}|| = |P(|f_{Q_{t^k}}|, |x|)|$.

Case 1. $t^k \geq S_{\mathcal{M}}(f, x)$.
This case can be effectively decided by testing, using $Run^k_{\mathcal{M}}(f, x, \overline{P}_1)$, whether all processors have halted. From $Run^k_{\mathcal{M}}$, one can determine whether r query steps were made before $\mathcal{M}$ halts, and correspondingly define $MQS^c_{\mathcal{M},i}(f, x, r)$.

Case 2. $t^k < S_{\mathcal{M}}(f, x)$.

Then by (3) of Fact 7.5.1, $t^k \leq S_{\mathcal{M}}(f_{Q_{t^k}}, x)$, and by (1) of Fact 7.5.1, $t^k \leq |P(|f_{Q_{t^k}}|, |x|)|^k$. Now $||P^*|| > |P(|f_{Q_{t^k}}|, |x|)|$, so $||P^*||^k > t^k$. Since $t = t(\mathcal{M}, f, x, \text{sqrt}(r))$, it follows that $t \geq \text{sqrt}(r)$, so $||P^*||^k > t^k \geq \text{sqrt}(r)^k$ and hence $||P^*|| \geq \text{sqrt}(r)$. Let $K_0 = 2^{2^{10}}$. Clearly, for $n \geq K_0$, $|n| \geq ||n||^2$. It follows that for $P^* \geq K_0$,

$$|P^*| \geq r.$$

Setting $P^{**} = P^* + K_0$, we have $|P^{**}| \geq r$ and so

$$r = |Exp2(r-1, P^{**})|.$$

Now define

$$MQS^c_{\mathcal{M},i}(f, x, r) = \overline{MQS}^c_{\mathcal{M},i}(f, x, Exp2(r-1, P^{**})).$$

It follows that

$$\begin{aligned} MQS^c_{\mathcal{M},i}(f, x, r) &= q_{c,i}^{t(|Exp2(r-1,P^{**})|)} \\ &= q_{c,i}^{t(r)}. \end{aligned}$$

This completes the inductive case. As before, appropriate bounds for the $|MQS^c_{\mathcal{M},i}(f, x, r)|$ can be given, so that the functions $MQS^c_{\mathcal{M},i}$ can be defined using simultaneous limited recursion on notation, and hence belong to $\mathcal{A}$. This completes the proof of the lemma.

From the preceding, we finally have the desired function algebraic characterization of type 2 NC.

Theorem 7.5.4 (P. Clote). *Suppose that $\mathcal{M} = \langle M, P \rangle$ is a fully specified* OCRAM *and the runtime of $\mathcal{M}$ on inputs f, x is bounded by $|P(|f|, |x|)|^k$. Then the rank $(1,1)$ functional F computed by $\mathcal{M}$ belongs to $\mathcal{A}$.*

Proof. Suppose that the second order polynomial P has degree d. Recall from Definition 7.5.5 that $t(\mathcal{M}, f, x, r)$ is the least number of steps in the computation of $\mathcal{M}$ on inputs f, x. Since $\mathcal{M}, f, x$ are fixed in context, we'll write $t(r)$ in place of $t(\mathcal{M}, f, x, r)$.

Let K_0 be an integer such that $|N + K_0| \geq ||N||^k$ for all $N \geq 0$. Define

$$\begin{aligned}
r_1 &= \max\left\{\overline{Q}_i^1(f, x) : 1 \leq i \leq k_1\right\} \\
T_1 &= \overline{P}(f, MQS^1_{\mathcal{M},1}(f, x, r_1), \ldots, MQS^d_{\mathcal{M},k_d}(f, x, r_1), x) + K_0 \\
q^\omega_{1,i} &= \operatorname{argmax}\left\{|f(y)| : y \leq \overline{Q}_i^1(f, x) \leq |T_1|\right\}, \\
&\quad \text{for } 1 \leq i \leq k_1 \\
r_c &= \max\left\{\overline{Q}_i^c(f, q^\omega_{1,1}, \ldots, q^\omega_{c-1,k_{c-1}}, x)\right\} \\
&\quad \text{for } 2 \leq c \leq d \text{ and } 1 \leq i \leq k_c \\
T_c &= \overline{P}(f, q^\omega_{1,1}, \ldots, q^\omega_{c-1,k_{c-1}}, MQS^c_{\mathcal{M},1}(f, x, r_c), \ldots, \\
&\quad MQS^d_{\mathcal{M},k_d}(f, x, r_c), x) + K_0, \text{ for } 2 \leq c \leq d \\
T_{d+1} &= \overline{P}(f, q^\omega_{1,1}, \ldots, q^\omega_{d,k_d}, x) \\
q^\omega_{c,i} &= \operatorname{argmax}\left\{|f(y)| : y \leq \overline{Q}_i^c(f, q^\omega_{1,1}, \ldots, q^\omega_{c-1,k_{c-1}}, x)\right\}, \\
&\quad \text{for } 2 \leq c \leq d.
\end{aligned}$$

Let

$$G(f, x) = \max\{T_1, \ldots, T_{d+1}\}.$$

By standard techniques of the last chapter, $A(f, x) = \operatorname{argmax}_{y \leq |x|} |f(y)|$ belongs to $\mathcal{A}_0$ (see Exercise 7.8.2), so by a modification of that argument the $q^\omega_{c,i}$ belong to $\mathcal{A}$. It follows that $G \in \mathcal{A}$.

CLAIM. $\mathcal{M}$ halts on input f, x within $||G(f, x)||^k$ steps.

Proof of Claim. If $\mathcal{M}$ halts within

$$p_1 = ||\overline{P}(f, MQS^1_{\mathcal{M},1}(f, x, r_1), \ldots, MQS^d_{\mathcal{M},k_d}(f, x, r_1), x)||^k$$

steps, then certainly $\mathcal{M}$ halts within $||G(f, k)||^k \geq ||T_1||^k \geq p_1$ steps. Otherwise,

$$\begin{aligned}
|T_1| &= |p_1 + K_0| \\
&\geq ||p_1||^k \\
&\geq t(r_1) \\
&\geq r_1 \\
&= \max\{\overline{Q}_i^1(f, x) : 1 \leq i \leq k_1\}
\end{aligned}$$

hence $q^\omega_{1,i} = q^\infty_{1,i}$ for $1 \leq i \leq k_1$. Now

$$r_2 = \max\{\overline{Q}_i^2(f, q_{1,1}^{\omega}, \ldots, q_{1,k_1}^{\omega}, x)\}$$
$$= \max\{\overline{Q}_i^1(f, q_{1,1}^{\infty}, \ldots, q_{1,k_1}^{\infty}, x)\}.$$

If $\mathcal{M}$ halts within

$$p_2 = ||\overline{P}(f, q_{1,1}^{\omega}, \ldots, q_{1,k_1}^{\omega}, MQS_{\mathcal{M},1}^2(f, x, r_2), \ldots, MQS_{\mathcal{M},k_d}^d(f, x, r_2), x)||^k$$

steps, then certainly $\mathcal{M}$ halts within $||G(f,k)||^k \geq ||T_2||^k \geq p_2$ steps. Otherwise,

$$|T_2| = |p_2 + K_0|$$
$$\geq ||p_2||^k$$
$$\geq t(r_2)$$
$$\geq r_2$$
$$= \max\{\overline{Q}_i^2(f, q_{1,1}^{\omega}, \ldots, q_{1,k_1}^{\omega}, x) : 1 \leq i \leq k_2\}$$

hence $q_{2,i}^{\omega} = q_{2,i}^{\infty}$ for $1 \leq i \leq k_2$. Continuing, we argue that either $\mathcal{M}$ halts within p_c steps, where p_c equals

$$||\overline{P}(f, q_{1,1}^{\omega}, \ldots, q_{c-1,k_{c-1}}^{\omega}, MQS_{\mathcal{M},1}^c(f, x, r_c), \ldots, MQS_{\mathcal{M},k_d}^d(f, x, r_c), x)||^k,$$

hence within $||G(f,x)||^k$ steps, or that

$$q_{c,i}^{\omega} = q_{c,i}^{\infty}$$

holds for $1 \leq i \leq k_c$. If $\mathcal{M}$ has not halted within $||T_c||^k$ steps, for some $1 \leq c \leq d$, then it follows that

$$q_{d,i}^{\omega} = q_{d,i}^{\infty}$$

for $1 \leq i \leq k_d$. But then $\mathcal{M}$ must terminate in

$$||T_{d+1}||^k = ||\overline{P}(f, q_{1,1}^{\omega}, \ldots, q_{d,k_d}^{\omega}, x)||^k$$
$$= ||\overline{P}(f, q_{1,1}^{\infty}, \ldots, q_{d,k_d}^{\infty}, x)||^k$$

steps. This concludes the proof of the claim.

It now follows that $F(f,x)$ can be extracted from $Run_{\mathcal{M}}^k(f, x, G(f,x))$ using functions in $\mathcal{A}$, so $F \in \mathcal{A}$.

The converse of the theorem is straightforward (see Exercise 7.8.4), since, in view of Theorem 7.5.1, all that remains to be shown is how to program the scheme of limited square-root recursion on an OCRAM in (second order) polylogarithmic time with a second order polynomial number of processors.[11] Hence we have the following principal result.

[11] Recall that functional substitution has been shown to be superfluous.

Theorem 7.5.5 (P. Clote). *Suppose that $F(f,x)$ is a rank $(1,1)$ functional. Then $F \in \mathcal{A}$ if and only if F is computable by a fully specified* OCRAM $\mathcal{M} = \langle M, P \rangle$ *with runtime bound* $|P(|f|,|x|)|^k$ *and processor bound* $P(|f|,|x|)$.

Clearly, the class of functionals characterized in this theorem equals the class of functionals computable on a fully specified OCRAM $\mathcal{M} = \langle M, R \rangle$ with runtime $|P(|f|,|x|)|^k$, for second order polynomials P, R.

7.6 λ-Calculi for Parallel Computable Higher Type Functionals

We now consider parallel computable functionals of all finite types. In our notation and definitions, we follow the presentation [CU89, CU93, CK90] as well as the classic text [Bar84]. The collection TP of all *finite types* is defined inductively as follows: 0 is a finite type, if σ and τ are finite types then $(\sigma \to \tau)$ is a finite type. The *level* of a type is defined as follows:

- $level(0) = 0$
- $level(\rho_1 \to \cdots \to \rho_k \to 0) = 1 + \max_{1 \le i \le k}\{level(\rho_i)\}$

By abuse of notation, often we say type instead of the more correct type level. For instance, natural numbers are of type (level) 0, number theoretic functions are of type (level) 1, etc. By induction on τ, it is simple to prove that every finite type $\sigma \to \tau$ can be put in the unique form

$$(\sigma_1 \to (\sigma_2 \to (\cdots \to (\sigma_n \to 0)\cdots))).$$

In the following, we will omit parentheses when possible, with association understood to the right.

By induction on type formation, define the collection HT of hereditarily total functionals of finite type, as follows: $HT_0 = \mathbf{N}$, $HT_{\sigma \to \tau} = HT_\tau^{HT_\sigma}$, the collection of all total functions with domain HT_σ and range HT_τ. Finally, let $HT = \cup_\sigma HT_\sigma$, where the union is taken over all finite types σ. A higher type functional F is said to be of type σ if $F \in HT_\sigma$. A functional of type

$$0 \to 0 \to \cdots \to 0$$

is called a functional of type (level) 1, and corresponds to a number theoretic function having n arguments if there are $n+1$ occurrences of 0 in the above. If F is of type ρ, where $\rho = \rho_1 \to \cdots \to \rho_k \to 0$, then often $F(X_1, \ldots, X_k)$ is written in place of $F(X_1)(X_2)\cdots(X_k)$. The type 2 (type 1) *section* of a class $\mathcal{C}$ of higher type functionals is the set of type 2 (type 1) functionals belonging to $\mathcal{C}$.

7.6.1 Introduction to Higher Types

In his attempted proof of the continuum hypothesis, D. Hilbert [Hil25] studied classes of higher type functionals defined by the operations of composition and primitive recursion. Hilbert's general scheme ([Hil25], p. 186) was of the form

$$\begin{aligned} \mathcal{F}(G,H,0) &= H \\ \mathcal{F}(G,H,n+1) &= G(\mathcal{F}(G,H,n),n) \end{aligned} \tag{7.11}$$

where $\mathcal{F}, G, H$ are higher type functionals of appropriate types possibly having other parameters not indicated. Illustrating the power of primitive recursion over higher type objects, Hilbert gave a simple higher type primitive recursive definition of (essentially) the Ackermann function, known not to be primitive recursive. For example, define the iterator IT_1 of type

$$(0 \to 0) \to 0 \to (0 \to 0)$$

by

$$\begin{aligned} IT_1(g,0) &= g \\ IT_1(g,n+1) &= g(IT_1(g,n)) \end{aligned}$$

so that $IT_1(g,n) = g^{(n+1)}$ is a primitive recursive functional, which iterates type $0 \to 0$ objects. Recall that s designates the successor function $s(x) = x+1$, and that $g^{(0)}(x) = x$ and $g^{(n+1)}(x) = g(g^n(x))$. Define the diagonal function D of type

$$(0 \to 0) \to 0 \to 0$$

by $\lambda g, n.\, D(g)(n) = \lambda g, n.\, IT_1(g,n)(n) = \lambda g, n.\, g^{(n+1)}(n)$ and f of type $0 \to (0 \to 0)$ by

$$\begin{aligned} f(0) &= s \\ f(n+1) &= D(f(n))(n). \end{aligned} \tag{7.12}$$

If we designate $g_0(x) = s(x)$ and $g_{n+1}(x) = g_n^{(x+1)}(x)$ to be (essentially) the principal functions in the Grzegorczyk hierarchy, then

$$\begin{aligned} f(n+1)(x) &= D(f(n))(n)(x) \\ &= IT_1(f(n),x)(n)(x) \\ &= (f(n))^{(n+1)}(x) \\ &= g_n^{(x+1)}(x) \\ &= g_{n+1}(x) \end{aligned}$$

so f cannot be primitive recursive.

The previous definition (7.12) of f does not quite fit into the syntactic form required by Hilbert's higher type primitive recursion scheme (7.11), so let's reformulate this as follows. Define the iterator IT_2 of type

$$((0 \to 0) \to 0 \to 0) \to (0 \to 0) \to (0 \to (0 \to 0))$$

by

$$IT_2(G, f, 0) = f$$
$$IT_2(G, f, n+1) = G(IT_2(G, f, n)).$$

Then $IT_2(G, f, n) = G^{(n)}(f)$ is a type level 3 primitive recursive functional which iterates type level 2 functionals. Now

$$\lambda n, x.\, f(n, x) = \lambda n, x.\, IT_2(D, s, n)(x),$$

so that f indeed belongs to the type 1 section of the primitive recursive functionals, yet is not a primitive recursive function.

In [Göd58], K. Gödel developed a formal system **T** of primitive recursive functionals of higher type, by including, for all finite types σ, a constant R_σ of type

$$\sigma \to \tau \to 0 \to \sigma$$

where $\tau = (\sigma \to 0 \to \sigma)$, satisfying

$$R_\sigma(u, v, 0) = u \tag{7.13}$$
$$R_\sigma(u, v, n+1) = v(R_\sigma(u, v, n))(n). \tag{7.14}$$

This scheme clearly generalizes primitive recursion of the form

$$f(0) = m$$
$$f(n+1) = h(f(n), n)$$

by taking in this case $\sigma = 0$, $u = m$, $v = h$, so that $v(R(u, v, n))(n) = h(f(n), n)$.[12] The following systems AV^ω, NCV^ω and PV^ω develop analogues of system **T** for well-studied sequential and parallel complexity classes. However, before giving the formal development, we present an alternative approach.

7.6.2 *p*-Types

A very natural complexity theoretic approach to developing a theory of *feasible* higher type functionals was given by S. Buss [Bus86b], described as follows. From a programming standpoint, it seems reasonable to require that for every computable, type 2 functional F with arguments f (function input) and x (integer input), there is a program M, for which $M(f, x) = F(f, x)$; moreover, the program M should be allowed to call a module for f on intermediate results (i.e., an oracle querying access for $f(y)$, where y is an intermediate result for which the module is called). Such a functional F might be

12 In [Göd58], K. Gödel proved that the type 1 section of **T** is the set of functions provably recursive in Peano arithmetic.

considered feasible if its runtime is bounded by a polynomial in the length $|x|$ of integer input x and the length $|f|$ of the *description* (or source code) of the function f. This intuition can be formalized as follows.

Let *suitable* polynomial mean a one-variable polynomial having nonnegative integer coefficients.

Definition 7.6.1 ([Bus86b]). *The collection of p-types is defined by:*[13]

- *0 is a p-type,*
- *if σ, τ are p-types and r is a suitable polynomial, then $\sigma \to_r \tau$ is a p-type.*

Let M_e denote the Turing machine whose Gödel number is e.[14] *For each p-type σ, define the set $HPOC_\sigma$ of codes of polynomial time operators of type σ as follows.*

- *For all $e \in \mathbf{N}$, $\langle 0, e\rangle$ belongs to $HPOC_0$.*
- *Given the p-type $\rho = \sigma \to_r \tau$ and $e \in \mathbf{N}$, if for all $x \in HPOC_\sigma$ there exists $y \in HPOC_\tau$, for which $M_e(x)$ outputs y in at most $r(|x|)$ steps, then $\langle \rho, e\rangle$ belongs to $HPOC_\rho$.*

A unary function f is defined to be a hereditarily polynomial time operator of p-type σ, denoted $f \in HPO_\sigma$, if there is a code $\langle \sigma, e\rangle \in HPOC_\sigma$ for which $M_e(x) = f(x)$ holds for all $x \in \mathbf{N}$. The collection HPO of all hereditarily polynomial time computable operators[15] *is the set of (partial) functionals,*[16] *whose code belongs to $\bigcup_\sigma HPOC_\sigma$.*

Note that, as defined, hereditarily polynomial time operators may be partial, since we have not specified the value of a functional of p-type $\sigma \to_r \tau$ on inputs not belonging to $HPOC_\sigma$. Using fully constructible runtime polynomials r, etc. one could define the related class of total functionals. In any case, it is clear that parallel versions based on AC^k and NC can be developed.

A major drawback of this approach, despite its naturalness from a programming standpoint, is that the application functional $Ap(f, x) = f(x)$ is not (equivalent to) a functional in HPO. Indeed, letting $\uparrow$ denote "undefined", the function

$$A(m, x) = \begin{cases} M_e(x) & \text{if } m = \langle 0 \to_p 0, e\rangle \in HPOC \\ \uparrow & \text{else.} \end{cases}$$

[13] Our definition is equivalent, but notationally different from that of [Bus86b]. In particular, functionals in HPO_σ are there called $\square_1^P$ functionals of p-type σ, and Buss includes an additional tupling for p-types.

[14] Here, we assume that Gödel numbering is done in a polynomial time computable manner.

[15] We have renamed the Buss p-type $\square_1^P$ operators as HPO, in analogy with Kreisel's HRO, the class of hereditarily recursive operators.

[16] The functionals may be partial, since we have not specified the value of a functional of p-type $\sigma \to_r \tau$ on inputs not belonging to $HPOC_\sigma$.

cannot belong to HPO, since its runtime is not bounded by any single polynomial.

A second drawback of this approach is that types are necessarily decorated by a polynomial runtime. In contrast, Gödel's system **T** contains the Ap functional, has the usual finite types (undecorated by polynomials), and can be extended into complexity theory in a straightforward manner, provided one has a function algebraic characterization of a given complexity class – namely, in place of general recursors R_σ from (7.13), one requires $\sigma = 0$, and allows appropriate forms of bounded recursion on notation.[17]

7.6.3 Finite Typed Lambda Calculus

In our notation and definitions, we rather closely follow Chapter 5 of [CU89, CU93], where we modify their definition of PV^ω to our smaller systems AV^ω and NCV^ω, based on the function algebraic characterizations of AC^0 and NC from the last chapter.

Definition 7.6.2. *Let $\mathcal{F}$ be a set of function symbols of arbitrary type. Terms of the finite typed $\Lambda(\mathcal{F})$-calculus are built up from variables*

$$X_0^\sigma, X_1^\sigma, X_2^\sigma, \ldots$$

for all finite types σ, symbols from $\mathcal{F}$, together with left '(' and right ')' parentheses and the abstraction operator λ. The collection $\Lambda(\mathcal{F})$ of λ-terms over $\mathcal{F}$ is the smallest set of terms satisfying the following.

1. *For each $i \in \mathbf{N}$ and finite type σ, X_i^σ belongs to $\Lambda(\mathcal{F})$ and is of type σ.*
2. *For each symbol $F \in \mathcal{F}$ of type σ, F belongs to $\Lambda(\mathcal{F})$ and is of type σ.*
3. *If $S \in \Lambda(\mathcal{F})$ is of type τ, and variable X is of type σ, then $(\lambda X.S) \in \Lambda(\mathcal{F})$ is of type $(\sigma \to \tau)$.*
4. *If $S, T \in \Lambda(\mathcal{F})$ are of types $\sigma \to \tau$ and σ, respectively, then $(ST) \in \Lambda(\mathcal{F})$ is of type τ.*

Let $\mathcal{F} = \{0, s_0, s_1, s, |x|, \text{TL}, \text{TR}, \text{MOD2}, msp, pad, *, ^\frown, \#, cond, C\}$ The terms of AV^ω are those of $\Lambda(\mathcal{F})$. For clarity, we define the system in detail.

Definition 7.6.3. *By induction, we define the function symbols and typed terms of the system AV^ω.*

1. *The constant 0 is a term of type 0.*
2. *For each finite type σ, there are infinitely many variables $X_0^\sigma, X_1^\sigma, X_3^\sigma, \ldots$ and each such variable is a term of type σ.*
3. *s_0, s_1, s, $|x|$, TL, TR, MOD2 are terms of type $0 \to 0$.*
4. *msp, pad, $*$, $^\frown$, $\#$ are terms of type $0 \to 0 \to 0$.*

[17] See Exercise 7.8.9 for an interesting research direction concerned with forms of bounded higher type recursion on notation.

5. *cond is a term of type* $0 \to 0 \to 0 \to 0$.
6. *If* F *is a term of type* $\sigma \to \tau$ *and* G *is a term of type* σ, *then* (FG) *is a term of type* τ *(application).*
7. *If* T *is a term of type* τ *and* X *is a variable of type* σ, *then* $(\lambda X.T)$ *is a term of type* $(\sigma \to \tau)$ *(abstraction).*
8. C *is a term of type* $(0 \to 0 \to 0) \to 0 \to 0 \to 0$ *and of type level* 2.

Remark 7.6.1. The intended interpretation of the function symbols is given by: $s_0(x) = 2 \cdot x$, $s_1(x) = 2 \cdot x + 1$, $s(x) = x + 1$, $|x| = \lceil log_2(x+1) \rceil$, $\text{TL}(x) = x - 2^{|x| \dot{-} 1}$ if $x \neq 0$, else 0, $\text{TR}(x) = \lfloor x/2 \rfloor$, $\text{MOD2}(x) = x \bmod 2 = x - 2 \cdot \text{TR}(x)$, $msp(x,y) = \lfloor x/2^{|y|} \rfloor = \text{MSP}(x,|y|)$, $pad(x,y) = 2^{|y|} \cdot x$, $x * y = 2^{|y|} \cdot x + y$, $x \frown y = x * y$, if $y \neq 0$, else $s_0(x)$, and $x \# y = 2^{|x| \cdot |y|}$. The *conditional* function satisfies

$$cond(x,y,z) = \begin{cases} y & \text{if } x = 0 \\ z & \text{else} \end{cases}$$

and the recursor C, formalizing CRN4 from Definition 7.3.7, satisfies

$$C(H,u,x) = \begin{cases} 1 & \text{if } x = 0 \\ C(H,u,\lfloor x/2 \rfloor) \frown \text{MOD2}(H(u,x)) & \text{else} \end{cases}$$

and so implements concatenation recursion on notation.

Note that the projection functions I^n_k can be defined by $[\lambda x_1 \ldots x_n . x_k]$ and that the unary constant function $z(x) = 0$ can be defined by $[\lambda x.0]$. We define the notion of a *free occurrence* of a variable in a term by induction on term formation. The variable X is *free* in the term X. If the term S (T) has type $\sigma \to \tau$ resp. σ, and if X is free in S,T then X is free in (ST). If X is free in the term T, then X is not free in the term $\lambda X.T$. A term of the above λ-calculus is *closed* if it has no free variables.

We generally use lowercase letters $x, y, z, \ldots$ and $t, u, v, \ldots$ to denote type 0 variables and terms, respectively. Uppercase letters $X, Y, Z, \ldots$ and $S, T, U, V, \ldots$ are generally used to denote variables and terms, respectively, of arbitrary type.

If S and T are terms of types σ and τ, respectively, and X is a variable of type τ then $S[T/X]$ is the term resulting from substituting T for all free occurrences of X in S, where bound occurrences of X in S have first been renamed to a new variable not occurring in S or T (this is to avoid free variables of T becoming bound in $S[T/X]$ after substitution of X by T). We say that term T is *closed* if T contains no free variables.

Definition 7.6.4. *A term of the form* $((\lambda X.S)T)$ *is a* β-redex *and* $S[T/X]$ *is its contractum. Replacement of a* β*-redex by its contractum is called* β*-reduction. A term of the form* $(\lambda X.SX)$ *is an* η*-redex, provided that there are no free occurrences of* X *in* T, *and* S *is its contractum. Replacement of an* η*-redex by its contractum is called* η*-reduction. Replacement of the term*

$(\lambda X.S)$ by the term $(\lambda Y.S[Y/X])$, where Y is a new variable not occurring in S, is called α-reduction (i.e., renaming bound variables).

We define conversion rules $A \Rightarrow B$, for terms A, B for the λ-calculus AV^ω. Here 1 abbreviates $s_1(0)$.

1. $(\lambda X.S)T \Rightarrow S[T/X]$ (β-reduction), provided X is not free in T.
2. $(\lambda X.SX) \Rightarrow S$ (η-reduction).
3. $(\lambda X.S) \Rightarrow (\lambda Y.S[Y/X])$ (α-reduction), provided Y is not free in S.
4. $s_0(0) \Rightarrow 0$.
5. $\text{MOD2}(0) \Rightarrow 0$.
6. $\text{MOD2}(s_0(x)) \Rightarrow 0$.
7. $\text{MOD2}(s_1(x)) \Rightarrow 1$.
8. $\text{TR}(0) \Rightarrow 0$.
9. $\text{TR}(s_0(x)) \Rightarrow x$.
10. $\text{TR}(s_1(x)) \Rightarrow x$.
11. $\text{TL}(0) \Rightarrow 0$
12. $\text{TL}(s_0(x)) \Rightarrow s_0(\text{TL}(x))$
13. $\text{TL}(s_1(x)) \Rightarrow cond(x, 0, s_1(\text{TL}(x)))$
14. $cond(0, y, z) \Rightarrow y$.
15. $cond(s_0(x), y, z) \Rightarrow cond(x, y, z)$.
16. $cond(s_1(x), y, z) \Rightarrow z$.
17. $pad(x, 0) \Rightarrow x$.
18. $pad(x, s_0(y)) \Rightarrow cond(y, x, s_0(pad(x, y))$.
19. $pad(x, s_1(y)) \Rightarrow s_0(pad(x, y))$.
20. $msp(x, 0) \Rightarrow x$.
21. $msp(x, s_0(y)) \Rightarrow cond(y, x, \text{TR}(msp(x, y)))$.
22. $msp(x, s_1(y)) \Rightarrow \text{TR}(msp(x, y))$.
23. $x\#0 \Rightarrow 1$.
24. $x\#s_0(y) \Rightarrow cond(y, 1, pad(x\#y, x))$.
25. $x\#s_1(y) \Rightarrow pad(x\#y, x)$.
26. $x * 0 \Rightarrow x$.
27. $x * s_0(y) \Rightarrow cond(y, x, s_0(x * y))$.
28. $x * s_1(y) \Rightarrow s_1(x * y)$.
29. $x^\frown y \Rightarrow cond(y, s_0(x), x * y)$
30. $s(0) \Rightarrow s_1(0)$.
31. $s(s_0(x)) \Rightarrow s_1(x)$.
32. $s(s_1(x)) \Rightarrow s_0(s(x))$.
33. $|0| \Rightarrow 0$.
34. $|s_0(x)| \Rightarrow cond(x, 0, s(|x|))$.
35. $|s_1(x)| \Rightarrow s(|x|)$.
36. $C(H, u, 0) \Rightarrow 1$.
37. $C(H, u, s_0(x)) \Rightarrow$
    $cond(x, 1, cond(H(u, s_0(x)), s_0(C(H, u, x)), s_1(C(H, u, x))))$.
38. $C(H, u, s_1(x)) \Rightarrow cond(H(u, s_1(x)), s_0(C(H, u, x)), s_1(C(H, u, x)))$.

Definition 7.6.5. *A binary relation $\mathcal{R}$ on the set $\Lambda(\mathcal{F})$ of λ-terms over $\mathcal{F}$ is compatible if the following hold.*

1. *If S, T are of type $\sigma \to \tau$, and U of type σ, and $S\mathcal{R}T$ holds, then $(SU)\mathcal{R}(TU)$ holds.*
2. *If S, T are of type σ, and U of type $\sigma \to \tau$, and $S\mathcal{R}T$ holds, then $(US)\mathcal{R}(UT)$ holds.*
3. *If S, T are of type σ and $S\mathcal{R}T$ holds, then $(\lambda X.S)\mathcal{R}(\lambda X.T)$ holds.*

Definition 7.6.6. *A binary relation on $\Lambda(\mathcal{F})$ is a reduction if it is reflexive, transitive and compatible. The relation $\Rightarrow^*$ is the reflexive, transitive, compatible closure of $\Rightarrow$. A term of $\Lambda(\mathcal{F})$ having no subterm which is a β- or η-redex is said to be in normal form.*

The next two theorems follow from the well-known Church-Rosser and strong normalization theorems of λ-calculus (see, the classic text [Bar84]), or, for a newer, simpler proof, see [JM99]).

Theorem 7.6.1 (Church-Rosser). *If S, U, V are terms of AV^ω and $S \Rightarrow^* U$ and $S \Rightarrow^* V$, then there is a term T of AV^ω for which $U \Rightarrow^* T$ and $V \Rightarrow^* T$.*

Theorem 7.6.2 (Strong normalization). *Given any term S in AV^ω, any sequence of β- and η-contractions of S leads to a normal form T, which is unique modulo renaming bound variables.*

Let us now consider the following example. $C\,msp\,0\,s_1(0)$ is a closed term of type 0, $(C\lambda X_1^0.\lambda X_2^0.X_3^0)X_4^0X_5^0$ is a term of type 0 having type 0 free variables X_3^0, X_4^0, X_5^0, and $CX^{0\to0\to0}Y^0Z^0$ is a term of type 0 having type 1 free variable $X^{0\to0\to0}$ and type 0 free variables Y^0, Z^0.

Definition 7.6.7 (Semantics). *An assignment ϕ is a type-preserving mapping from the set $\{X_i^\sigma : \sigma \in \mathrm{TP}, i \in \mathbf{N}\}$ of variables of all finite types into the class $\{HT_\sigma : \sigma \in \mathrm{TP}\}$ of all hereditarily total functionals of finite type. Relative to an assignment ϕ, the value $\nu_\phi(T)$ of an AV^ω term T is defined by induction on term formation. If T is a variable X_i^σ, then*

$$\nu_\phi(T) = \nu_\phi(X_i^\sigma) = \phi(X_i^\sigma).$$

If T is a function symbol of AV^ω, then $\nu_\phi(T)$ is the corresponding function, as given in Remark 7.6.1. If $T = (UV)$, then $\nu_\phi(T) = \nu_\phi(U)\,(\nu_\phi(V))$. If $T = \lambda X^\sigma.S$ and $F \in HT_\sigma$, then $(\nu_\phi(T)F) = \nu_{\phi'}(S)$, where the assignment ϕ' maps X^σ to F, and otherwise is identical to ϕ.

If ϕ is an assignment, then it is straightforward to show that if $T \Rightarrow^* S$, then $\nu_\phi(T) = \nu_\phi(S)$. Moreover, if T is a closed term, then its value $\nu_\phi(T)$ is independent of every assignment ϕ. In such a case, we simply write $\nu(T)$.

Definition 7.6.8 **(AV^ω-definability).** *A functional $F \in V$ is defined by an AV^ω term T relative to assignment ϕ, if $\nu_\phi(T) = F$. A functional $F \in V$ is AV^ω-definable if it is definable by an AV^ω term T, which is closed.*

Theorem 7.6.3. *The type 2 section of AV^ω is $\mathcal{A}_0$. In particular, the following holds.*

1. *If $F \in \mathcal{A}_0$ then F is AV^ω-definable.*
2. *If T is a closed type 2 term of AV^ω, then $\nu(T) \in \mathcal{A}_0$.*

Proof. (1) By Theorem 7.3.1, we can define $\mathcal{A}_0$ using CRN4 in place of CRN. By induction, it is straightforward to show that every functionals in $\mathcal{A}_0$ is AV^ω-definable.

(2) Let T be an AV^ω term of type σ, where σ is of type level 2 with the form

$$\sigma_1 \to \cdots \to \sigma_m \to 0.$$

Let $X_1^{\sigma_1}, \ldots, X_m^{\sigma_m}$ be fresh variables not appearing in T. By the Strong Normalization Theorem7.6.2, let U be the normal form of the type 0 term $T(X_1^{\sigma_1}, \ldots, X_m^{\sigma_m})$. It can be shown that U has no bound variables of type greater than 0, apart from the given free variables $X_1^{\sigma_1}, \ldots, X_m^{\sigma_m}$. Fix an assignment ϕ, and for $1 \le i \le m$ let the type 0 or 1 functional $F_i \in HT_{\sigma_i}$ be defined by $F_i = \phi(X_i^{\sigma_i})$. By induction on term formation, it is straightforward to show that for every subterm S of U, the value $\nu_\phi(S)$ belongs to $\mathcal{A}_0(F_1, \ldots, F_m)$ (the scheme of functional substitution can be used in the case of lambda abstraction over type 0 variables). It follows that $\nu(T) \in \mathcal{A}_0$.

Corollary 7.6.1. *The type 1 section of AV^ω equals A_0.*

To characterize NC computable functionals, we add the two new function symbols sqrt,MSP to those of AV^ω, where sqrt has type $0 \to 0$ and MSP has type $0 \to 0 \to 0$. In addition to the recursor C of AV^ω, we have the recursor T of type

$$(0 \to 0 \to 0 \to 0) \to (0 \to 0 \to 0) \to 0 \to 0 \to 0 \to 0$$

whose intended interpretation is

$$T(H, B, u, y, x) = \begin{cases} y & \text{if } x = 0 \\ H(u, T(H, B, u, y, \text{sqrt}(x)), B(u, x)) & \text{else if } (\dagger) \\ B(u, x) & \text{else} \end{cases}$$

where ($\dagger$) is the condition

$$|H(u, T(u, H, B, y, \text{sqrt}(x)), B(u, x))| < |B(u, x)|.$$

This scheme clearly corresponds to limited square-root recursion. To formally define NCV^ω, we need to add appropriate conversion rules to those of AV^ω, in order to define MSP, sqrt and the action of the recursor T. Recalling that $\text{sqrt}(x) = \text{MSP}(x, \text{TR}(|s_0(x)|))$, where the interpretation of $\text{TR}(x)$ is $\lfloor x/2 \rfloor$. we add the following conversions.

1. $\text{MSP}(x, 0) \Rightarrow x$.
2. $\text{MSP}(x, s_0(y)) \Rightarrow cond(y, x, \text{MSP}(\text{MSP}(x, y), y))$.
3. $\text{MSP}(x, s_1(y)) \Rightarrow \text{TR}(\text{MSP}(\text{MSP}(x, y), y))$.
4. $\text{sqrt}(x) \Rightarrow cond(x, 0, \text{MSP}(x, \text{TR}(|s_0(x)|)))$.
5. $T(H, B, u, y, 0) \Rightarrow y$.
6. $T(H, B, u, y, s_0(x)) \Rightarrow$
   $cond(x, y, cond(msp(t_0, B(u, s_0(x))), t_0, B(u, s_0(x))))$.
7. $T(H, B, u, y, s_1(x)) \Rightarrow cond(msp(t_1, B(u, s_1(x))), t_1, B(u, s_1(x)))$.

In the last two conversions, we made the abbreviations

$$t_0 \text{ is } H(u, T(H, B, u, y, \text{sqrt}(s_0(x))), s_0(x))$$
$$t_1 \text{ is } H(u, T(H, B, u, y, \text{sqrt}(s_1(x))), s_1(x)).$$

As with the case of AV^ω, we have the following result.

Theorem 7.6.4. *The type 2 (type 1) section of NCV^ω is $\mathcal{A}$ (A, hence* NC*).*

In [CU93], S.A. Cook and A. Urquhart developed a polynomial time version of Gödel's system **T**. The function symbols of PV^ω are those of AV^ω, with the exception of $s, |x|, *, ^\frown, \text{TL}, C$.[18] In place of the recursor C of AV^ω, the recursor R of [CU93] has type

$$0 \to (0 \to 0 \to 0) \to (0 \to 0) \to 0 \to 0$$

and satisfies

$$R(y, H, B, x) = \begin{cases} y & \text{if } x = 0 \\ t & \text{else if } |t| \geq |B(x)| \\ B(x) & \text{else} \end{cases}$$

where $t = H(y, R(y, H, B, \lfloor x/2 \rfloor))$.

Theorem 7.6.5 ([CU93]). *Type* 2 *functionals of PV^ω are exactly those in* BFF*, hence computable by a function oracle Turing machine in second order polynomial time.*

From the preceding results, we now have the following separation.

Theorem 7.6.6 ([CIK93]). *$AV^\omega \subset NCV^\omega \subset PV^\omega$, where all inclusions are proper.*

[18] There are some notational differences with Definition 3.2 of [CU93]: our MOD2 is their *Parity*, our TR is their $\lfloor \frac{1}{2} \rfloor$, our *cond* is their *Cond*, our $pad(x, y)$ is their $x \boxplus y$, our $msp(x, y)$ is their $x \div y$. We have attempted to develop a uniform notation, consistent with other authors. For instance, $\text{MSP}(x, y) = \lfloor \frac{x}{2^y} \rfloor$ has been in use since [Bus86a], and $msp(x, y) = \text{MSP}(x, |y|)$, a unary form of the latter function.

Proof. By Theorems 7.6.3, 7.6.4 and 7.6.5, the type 2 section of AV^ω (NCV^ω resp. PV^ω) equals $\mathcal{A}_0$ [resp. $\mathcal{A}$ resp. BFF]. We separate the corresponding type 2 classes.

C. Wilson [Wil87] constructed an oracle $A \subseteq \mathbf{N}$ for which NC^A is a proper subset of P^A. Thus, let $B = M_e^A \in P^A - NC^A$, where M_e is a polynomial time bounded oracle Turing machine. Let $F(f,x) = M_e^{sg(f)}(x)$, so $F \in$ BFF $\subseteq PV^\omega$ by one direction of Theorem 7.2.2. Letting c_A [resp. c_B] denote the characteristic function of A [resp. B], if $F \in NCV^\omega$, then by composition, the characteristic function $\lambda x. F(c_A, x) = 1$ [resp. 0] iff $x \in B$ [resp. $x \notin B$] belongs to the type 1 section of NCV^ω. This contradicts Wilson's theorem, so that $NCV^\omega \subset PV^\omega$. From the lower bounds in Chapter 2, we know that $AC^0 \subset NC$, so a similar argument with oracle $\emptyset$ shows that $AV^\omega \subset NCV^\omega$.

Finally, it should be mentioned that using the techniques of [CK90], Clote, Ignjatovic and Kapron [CIK93] introduced a class of loop programs known as *bounded typed loop programs* (BTLP's), which compute exactly the functionals in AV^ω and NCV^ω.

7.7 Historical and Bibliographical Remarks

The OCRAM machine model was defined in [CIK93], where type 2 constant parallel time was characterized in terms of the algebra $\mathcal{A}_0$. There, a type of Cobham–Ritchie theorem was proved, which characterized the functionals in $\mathcal{A}$ as those computable on an OCRAM, with runtime bound $||G(f,x)||^k$ and processor bound $|G(f,x)|$, for some functional $G \in \mathcal{A}$. Theorem 7.5.5, due to P. Clote, is a far-reaching extension of the main result in [CIK93], and is the exact type 2 analogue of our characterization of type 1 NC functions via the function algebra A, given in the previous chapter. The main results of Sections 7.3 through 7.5 are all new, due to P. Clote (unless otherwise specified), and do not appear elsewhere in the literature.

Despite the fact that these results are new, the proof technique borrows heavily from the elegant and ground-breaking characterization of type 2 polynomial time via the type 2 Cobham algebra, given by B.M. Kapron and S.A. Cook in [KC96]. Theorem 7.5.5 can clearly be extended to characterize all rank (k,ℓ) type 2 functionals. The results of Section 7.6.3 are modifications (or extensions in the case of NCV^ω) of results of P. Clote, B.M. Kapron, and A. Ignjatovic [CIK93]. Thanks to K.-H. Niggl for a critical reading and correction of our initially flawed proof that functional substitution is superfluous.

Recently, S. Bellantoni, K.-H. Niggl and H. Schwichtenberg [BNS00] have developed a new lambda calculus whose type 1 section is the set of polynomial time computable functions. Unlike PV^ω, they admit a higher type recursor at all type levels, which, however, is restricted by a *linearity* condition fashioned after the work of M. Hofmann [Hof99]. Using the techniques of this chapter and the previous chapter, there is current work in progress to extend this linearly constrained lambda calculus in order to characterize the parallel complexity classes AC^0, NC, etc.

7.8 Exercises

Exercise 7.8.1 (Folklore). Assuming that the functions

$$s_0, s_1, \lfloor x/2 \rfloor, \text{MOD2}, \textit{cond}$$

are available, prove that bounded recursion on notation and limited recursion on notation are equivalent.
HINT. Suppose that for $x > 0$

$$\begin{aligned} f(s_i(x), \mathbf{y}) &= h_i(x, \mathbf{y}, f(x, \mathbf{y})) \\ &= \overline{h}(s_i(x), \mathbf{y}, f(\lfloor s_i(x)/2 \rfloor, \mathbf{y})) \end{aligned}$$

where

$$\overline{h}(x, \mathbf{y}, z) = \begin{cases} h_0(\lfloor x/2 \rfloor, \mathbf{y}, z) & \text{if MOD2}(x) = 0 \\ h_1(\lfloor x/2 \rfloor, \mathbf{y}, z) & \text{if MOD2}(x) = 1. \end{cases}$$

Thus, provided *cond*, MOD2, $\lfloor x/2 \rfloor$ are available, LRN simulates BRN. Now suppose that for $x > 0$

$$\begin{aligned} f(x, \mathbf{y}) &= h(x, \mathbf{y}, f(\lfloor x/2 \rfloor, \mathbf{y})) \\ &= h(2 \cdot \lfloor x/2 \rfloor + \text{MOD2}(x), y, f(\lfloor x/2 \rfloor, \mathbf{y})). \end{aligned}$$

Let

$$\begin{aligned} h_0(x, \mathbf{y}, z) &= h(s_0(x), \mathbf{y}, z) \\ h_1(x, \mathbf{y}, z) &= h(s_1(x), \mathbf{y}, z). \end{aligned}$$

Then

$$\begin{aligned} f(s_0(x), \mathbf{y}) &= h_0(x, \mathbf{y}, f(x, \mathbf{y})) \\ f(s_1(x), \mathbf{y}) &= h_1(x, \mathbf{y}, f(x, \mathbf{y})). \end{aligned}$$

Thus, provided s_0, s_1 are available, BRN simulates LRN. The extension to type 2 is routine.

Exercise 7.8.2 (P. Clote). Define $A(f, x) = \text{argmax}_{y \le |x|} |f(y)|$, so that $f(A(f, x)) = \max_{y \le |x|} |f(y)|$. Prove that the functional $A \in \mathcal{A}_0$.
HINT. Note that $A(f, x) = z$ if and only if

$$z \le |x| \wedge \forall y \le |x| [f(y) \le f(x)] \wedge \forall y < z [f(y) < f(z)].$$

Sharply bounded quantifiers can be handled by CRN, and from the previous chapter it follows that inequality $\le$ and $<$ belong to $\mathcal{A}_0$.

Exercise 7.8.3 (P. Clote). Prove that function $Exp2(a,b) = \min(2^a, b) \in \mathcal{A}_0$.

HINT. $2^{\min(a,|b|)} = \text{TR}(rev(exp0(x,y,s_1(y))))$ where the auxiliary function $exp0$ is defined by

$$exp0(x,y,0) = 1$$

$$exp0(x,y,s_i(z)) = \begin{cases} s_1(exp0(x,y,z)) \text{ if } |z| = x \leq |y| \vee |z| = |y| \leq x \\ s_0(exp0(x,y,z)) \text{ else.} \end{cases}$$

Exercise 7.8.4. Suppose that $F(f,x)$ is a type 2 functional in $\mathcal{A}$. Prove that there is a fully specified OCRAM $\mathcal{M} = \langle M, P \rangle$ which computes $F(f,x)$ with runtime bound $|P(|f|,|x|)|^k$ and processor bound $P(|f|,|x|)$.

Exercise 7.8.5 ([KC96]). Prove Theorem 7.2.2, which asserts that type 2 functionals computable in (second order) polynomial time on a function oracle Turing machine are exactly those in BFF.

HINT. See [KC96].

Exercise 7.8.6 ([KC96]). Prove that no BFF functional exists for which $|H(f,x)| \geq |f|(|x|)$ for all f, x.

HINT. See Theorem 5.5 of [KC96]. Proof sketch is as follows. Suppose that f is a 0-1 valued function. If $H \in$ BFF, then there is a (first order) polynomial p and an OTM M such that $M(f,x)$ is computed by M in time $p(|x|)$, for all f, x. There must then be an input x_0 with y, such that $|y| \leq |x_0|$ and M does not query y on input f, x_0. If one defines

$$f'(x) = \begin{cases} 2 \cdot H(f,x) + 1 \text{ if } x = y \\ f(x) \qquad\qquad\quad \text{else} \end{cases}$$

then $M(f,x_0) = M(f',x_0)$, so, since M computes H,

$$|H(f,x_0)| = |H(f',x_0)| \geq |f'|(|x_0|) = |H(f,x_0)| + 1$$

a contradiction.

Exercise 7.8.7 (K.-H. Niggl). Prove Townsend's assertion that BFF is the same class of functionals, whether or not the scheme of functional substitution is included.

HINT. Use the technique from Lemmas 7.3.3 and 7.3.4.

Exercise 7.8.8. Using Hilbert's scheme of higher type primitive recursion, we defined a function with the same growth rate as that of the Ackermann function. Prove that the Ackermann function

$$A(n,m) = \begin{cases} n+1 & \text{if } m = 0 \\ A(m-1,1) & \text{if } n = 0 \\ A(m-1, A(m,n-1)) & \text{else} \end{cases}$$

belongs to Gödel's $\mathbf{T}$.

Exercise 7.8.9 (S.A. Cook). ($\star$) Consider σ of type level 1, and define the bounded recursion on notation analogue of Gödel's scheme (7.13). Prove that the characteristic function of every set in the polynomial time hierarchy PH can be defined in PV^ω, augmented by this recursor.

Exercise 7.8.10 (P. Clote). Define system ALV^ω from AV^ω, by adding, for each k, new terms S_k of type $0 \to (0 \to 0 \to 0 \to 0) \to 0 \to 0 \to 0$. The intended interpretation of the recursor S_k is

$$S_k(u, H, y, x) = \begin{cases} y & \text{if } x = 0 \\ \min\{k, H(u, S_k(y, H, \lfloor x/2 \rfloor), y, x)\} & \text{otherwise.} \end{cases}$$

Prove that the type 1 section of ALV^ω consists exactly of the ALOGTIME computable functions.

Exercise 7.8.11. ($\star\star$) Define $\widetilde{ALV^\omega}$ as in ALV^ω, but where we admit the recursor S_4, in place of S_k, for $k \in \mathbf{N}$. It follows from Theorem 6.3.10 that the type 1 section of $\widetilde{ALV^\omega}$ equals that of ALV^ω. Does ALV^ω equal $\widetilde{ALV^\omega}$?

References

[AAHK88] K. Abrahamson, A. Adler, L. Higham, and D. Kirkpatrick. Randomized evaluation on a ring. In J. van Leeuwen, editor, *Proceedings of 2nd International Workshop on Distributed Algorithms*, volume 312 of *Springer Lecture Notes in Computer Science*, pages 324–331, 1988.

[AB87] N. Alon and R. B. Boppana. The monotone circuit complexity of boolean functions. *Combinatorica*, 7: 1–22, 1987.

[ABO84] M. Ajtai and M. Ben-Or. A theorem on probabilistic constant depth circuits. In *Proceedings of 16th Annual ACM Symposium on Theory of Computing*, pages 471–474, 1984.

[Ach97] D. Achlioptas. A simple upper bound for satisfiability. Personal Communication, 1997.

[Ach00] D. Achlioptas. Setting two variables at a time yields a new lower bound for random 3-SAT. In *Proceedings of 32nd Annual ACM Symposium on Theory of Computing*, pages 28–37, 2000.

[Ack28] W. Ackermann. Zum Hilbertschen Aufbau der reellen Zahlen. *Mathematische Annalen*, 99: 118–133, 1928.

[AF00] D. Achlioptas and J. Franco. Lower bounds for random 3-SAT via differential equations. Unpublished preprint, 2000.

[AGG00] A. Atserias, N. Galesi, and R. Gavaldà. Monotone proofs of the pigeon hole principle. In *Proceedings of the 27th ICALP*, volume 1853 of *Springer Lecture Notes in Computer Science*, pages 151–162, 2000.

[AHU74] A. Aho, J. Hopcroft, and J. Ullman. *The Design and Analysis of Computer Algorithms*. Addison–Wesley, Reading, MA, 1974.

[Ajt83] M. Ajtai. Σ_1^1-formulae on finite structures. *Annals of Pure and Applied Logic*, 24: 1–48, 1983.

[Ajt89] M. Ajtai. First-order definability on finite structures. *Annals of Pure and Applied Logic*, 45: 211–225, 1989.

[Ajt94a] M. Ajtai. The complexity of the pigeon hole principle. *Combinatorica*, 14: 4, 1994.

[Ajt94b] M. Ajtai. On the existence of modulo p cardinality functions. In [CR94], pages 1–14.

[AK96] F. Ablayev and M. Karpinski. On the power of randomized branching programs. In *Proceedings of ICALP'96*, volume 1099 of *Springer Lecture Notes in Computer Science*, pages 348–356, 1996.

[AKK+97] D. Achlioptas, L. M. Kirousis, E. Kranakis, D. Krizanc, M. Molloy, and Y. C. Stamatiou. Random constraint satisfaction: A more accurate picture. In *Proceedings of the Third International Conference on Principles and Practice of Constraint Programming (CP'97)*, volume 1330 of *Springer Lecture Notes in Computer Science*, pages 107–120, 1997.

[AKKK97] D. Achlioptas, L. M. Kirousis, E. Kranakis, and D. Krizanc. Rigorous results for random $(2+p)$-SAT. Manuscript. Preliminary version presented in RALCOM 97, Santorini, Greece, Oct. 6–11, 1997.

[Akl89] S. Akl. *The Design and Analysis of Parallel Algorithms.* Prentice-Hall, 1989.

[AKS83] M. Ajtai, E. Komlós, and E. Szemerédi. An $O(n \log n)$ sorting network. In *Proceedings of 15th Annual ACM Symposium on Theory of Computing*, pages 1–9, 1983.

[Ald89] D. J. Aldous. The harmonic mean formula for probabilities of unions: applications to sparse random graphs. *Discrete Mathematics*, 76: 167–176, 1989.

[All91] B. Allen. Arithmetizing uniform NC. *Annals of Pure and Applied Logic*, 53(1): 1–50, 1991.

[AM97a] D. Achlioptas and M. Molloy. Almost all graphs with $2.53n$ edges are 4-chromatic. Manuscript, 1997.

[AM97b] D. Achlioptas and M. Molloy. The analysis of a list-coloring algorithm on a random graph. In *Proceedings of 38th Annual IEEE Symposium on Foundations of Computer Science*, pages 202–212, 1997.

[AM99] D. Achlioptas and M. Molloy. Almost all graphs with $2.522n$ edges are not 3-colorable. *The Electronic Journal of Combinatorics*, 6 (Research paper R29), 1999. http://www.combinatorics.org/

[And85] A. E. Andreev. On a method for obtaining lower bounds for the complexity of individual monotone functions. *Doklady Akademii Nauk SSSR*, 285(5): 1033–1037, 1985. English translation in *Soviet Math. Doklady*, 31: 530–534.

[Ang80] D. Angluin. Local and global properties in networks of processors. In *Proceedings of 12th Annual ACM Symposium on Theory of Computing*, pages 82–93, 1980.

[Apo76] T.M. Apostol. *Introduction to Analytic Number Theory.* Undergraduate Texts in Mathematics, Springer, 1976, 2nd edition, corr. 5th printing 1998.

[APT79] B. Aspvall, M. P. Plass, and R. E. Tarjan. A linear-time algorithm for testing the truth of certain quantified boolean formulas. *Information Processing Letters*, 8: 121–123, 1979.

[Ara96] N. Arai. Tractability of cut-free Gentzen type propositional calculus with permutation inference I. *Theoretical Computer Science*, 170: 129–144, 1996.

[Ara00] N. Arai. Relative efficiency of propositional proof systems: Resolution and cut-free LK. *Annals of Pure and Applied Logic*, 104: 3–16, 2000.

[Arb68] M.A. Arbib. *The Algebraic Theory of Machines, Languages, and Semigroups.* Academic Press, 1968.

[ASE92] N. Alon, J.H. Spencer, and P. Erdős. *The Probabilistic Method.* John Wiley and Sons, New York, 1992.

[Ass87] G. Asser. Primitive recursive word-functions of one variable. In E. Börger, editor, *Computation Theory and Logic*, volume 270 of *Springer Lecture Notes in Computer Science*, pages 14–19, 1987.

[ASW88] H. Attiya, M. Snir, and M. K. Warmuth. Computing on an anonymous ring. *Journal of the ACM*, 35(4): 845–875, Oct. 1988.

[Avi61] A. Avizienis. Signed-digit number representations for fast parallel arithmetic. *IRE Transactions on Electronic Computers*, 10: 389–400, 1961. Reprinted in E. E. Swartzlander, *Computer Arithmetic*, Vol. 2, IEEE Computer Society Press Tutorial, Los Alamitos, CA, 1990.

[BA80] M. Ben-Ari. A simplified proof that regular resolution is exponential. *Information Processing Letters*, 10: 96–98, 1980.

[Bab81] L. Babai. On the order of uniprimitive permutation groups. *Annals of Mathematics*, 113: 553–568, 1981.

[Bar80] T. C. Bartee. *Digital Computer Fundamentals.* McGraw-Hill, 5th edition, 1980.

[Bar84] H.P. Barendregt. *The Lambda Calculus: Its Syntax and Semantics.* North-Holland, 1984. Revised edition.

[Bar89] D.A. Barrington. Bounded-width polynomial-size branching programs recognize exactly those languages in NC^1. *Journal of Computer and System Sciences*, 38: 150–164, 1989.

[Bar92] D. A. Mix Barrington. Some problems involving Razborov-Smolensky polynomials. In M.S. Paterson, editor, *Boolean Function Complexity*, volume 169 of *London Mathematical Society Lecture Notes Series.* Cambridge University Press, 1992.

[BB89] P. W. Beame and H. L. Bodlaender. Distributed computing on transitive networks: The torus. In B. Monien and R. Cori, editors, *6th Annual Symposium on Theoretical Aspects of Computer Science, STACS*, volume 349 of *Springer Lecture Notes in Computer Science*, pages 294–303, 1989.

[BB94] M.-L. Bonet and S.R. Buss. Size-depth tradeoffs for boolean formulae. *Information Processing Letters*, 11: 151–155, 1994.

[BB96] D.G. Bobrow and M. Brady. Phase transitions and complexity. *Artificial Intelligence*, 81, 1996. Special Volume on Frontiers in Problem Solving, T. Hogg, B.A. Hubermann, and C.P. Williams, Guest Editors.

[BB98] P. W. Beame and S. R. Buss, editors. *Proof Complexity and Feasible Arithmetics*, volume 39 of *DIMACS Series in Discrete Mathematics and Theoretical Computer Science.* American Mathematical Society, 1998. Proceedings of DIMACS Workshop held at Rutgers University, April 21–23, 1996.

[BBG96] S. Bloch, J. Buss, and J. Goldsmith. Sharply bounded alternation within P. In *ECCCTR: Electronic Colloquium on Computational Complexity, technical reports*, 1996.

[BBR92] D. A. Mix Barrington, R. Beigel, and S. Rudich. Representing boolean functions as polynomials modulo composite numbers (extended abstract). In *Proceedings of 24th Annual ACM Symposium on Theory of Computing*, pages 455–461, 1992.

[BBTN92] L. Babai, R. Beals, and P. Takacsi-Nagy. Symmetry and complexity. In *Proceedings of 24th Annual ACM Symposium on Theory of Computing*, pages 438–449, 1992.

[BC92] S. Bellantoni and S. Cook. A new recursion-theoretic characterization of the polytime functions. *Computational Complexity*, 2: 97–110, 1992.

[BC96] S. R. Buss and P. Clote. Cutting planes, connectivity, and threshold logic. *Archive for Mathematical Logic*, 35(1): 33–62, 1996.

[BCE91] N.H. Bshouty, R. Cleve, and W. Eberly. Size-depth tradeoffs for algebraic formulae. In *Proceedings of 32th Annual IEEE Symposium on Foundations of Computer Science*, pages 334–341, 1991.

[BCE+95] P. Beame, S. Cook, J. Edmonds, R. Impagliazzo, and T. Pitassi. The relative complexity of NP search problems. In *Proceedings of 27th Annual ACM Symposium on Theory of Computing*, pages 303–314, 1995. Journal version: *Journal of Computer and System Sciences*, 57(1): 3–19, 1998.

[BCH86] P.W. Beame, S.A. Cook, and H.J. Hoover. Log depth circuits for division and related problems. *SIAM Journal on Computing*, 15: 994–1003, 1986.

[BDG+99] M. Luisa Bonet, C. Domingo, R. Gavalda, A. Maciel, and T. Pitassi. Non-automatizability of bounded-depth Frege proofs. In *Proceedings of the 14th Annual IEEE Conference on Computational Complexity*, 1999.

[Bea] P. Beame. Exponential size lower bounds for bounded-depth Frege proofs of the pigeonhole principle, Part I and Part II. Typeset manuscript for overhead transparencies.

[Bea94] P.W. Beame. A switching lemma primer. Technical report, University of Washington, November 1994.

[BEGJ98] M. Luisa Bonet, J.L. Estaban, N. Galesi, and J. Johannsen. Exponential separations between restricted resolution and cutting planes proof systems. In *Proceedings of 39th Annual IEEE Symposium on Foundations of Computer Science*, 1998.

[Bei92] R. Beigel. When do extra majority gates help? Polylog(n) majority gates are equivalent to one. In *Proceedings of 24th Annual ACM Symposium on Theory of Computing*, pages 450–454, 1992.

[Bei93] R. Beigel. The polynomial method in circuit complexity. In *Proceedings of 8th Annual IEEE Conference on Structure in Complexity Theory*, pages 82–95, 1993.

[Bel82] A. Bel'tyukov. A computer description and a hierarchy of initial Grzegorczyk classes. *Journal of Soviet Mathematics*, 20: 2280–2289, 1982. Translation from Zap. Nauk. Sem. Lening. Otd. Mat. Inst., *V. A. Steklova AN SSSR*, 88: 30–46, 1979.

[Bel92] S. Bellantoni. Predicative recursion and computational complexity. Technical Report 264/92, University of Toronto, Computer Science Department, September 1992.

[Bel94] S. Bellantoni. Predicative recursion and the polytime hierarchy. In [CR94], pages 15–29.

[Ben62] J.H. Bennett. *On Spectra*. PhD thesis, Princeton University, Department of Mathematics, 1962.

[Ber71] C. Berge. *Principles of Combinatorics*. Academic Press, 1971.

[BF91] L. Babai and L. Fortnow. Arithmetization: a new method in structural complexity theory. *Computational Complexity*, 1: 41–66, 1991.

[BGIP99] S.R. Buss, D. Grigoriev, R. Impagliazzo, and T. Pitassi. Linear gaps between degrees for the polynomial calculus modulo distinct primes. In *Proceedings of 31st Annual ACM Symposium on Theory of Computing*, 1999.

[BIK+96] P. Beame, R. Impagliazzo, J. Krajíček, T. Pitassi, and P. Pudlák. Lower bounds on Hilbert's Nullstellensatz and propositional proofs. *Proc. London Math. Soc.*, 73(3): 1–26, 1996.

[BIK+97] S. Buss, R. Impagliazzo, J. Krajíček, P. Pudlák, A.A. Razborov, and J. Sgall. Proof complexity in algebraic systems and constant depth Frege systems with modular counting. *Computational Complexity*, 6: 256–298, 1997.

[BIP98] P. Beame, R. Impagliazzo, and T. Pitassi. Improved depth lower bounds for small distance connectivity. *Computational Complexity*, 7: 325–345, 1998.

[BIS90] D. Mix Barrington, N. Immerman, and H. Straubing. On uniformity in NC^1. *Journal of Computer and System Sciences*, 41(3): 274–306, 1990.

[BKPS] P. Beame, R. Karp, T. Pitassi, and M. Saks. On the complexity of unsatisfiability proofs for random kCNF formulas. Submitted.

[Bla37] A. Blake. *Canonical Expressions in Boolean Algebra*. PhD thesis, University of Chicago, 1937.

[Blo94] S. Bloch. Function-algebraic characterizations of log and polylog parallel time. *Computational Complexity*, 4(2): 175–205, 1994.

[BLS87] L. Babai, E. Luks, and A. Seress. Permutation groups in NC. In *Proceedings of 19th Annual ACM Symposium on Theory of Computing*, 1987.

[Blu84] N. Blum. A boolean function requiring $3n$ network size. *Theoretical Computer Science*, 28: 337–345, 1984.

[BM76] J. A. Bondy and U. S. R. Murty. *Graph Theory with Applications*. North-Holland, 1976.

[BNS00] S. Bellantoni, K.-H. Niggl, and H. Schwichtenberg. Higher type recursion, ramification and polynomial time. *Annals of Pure and Applied Logic*, 104: 17–30, 2000.

[Bol85] B. Bollobás. *Random Graphs*. Academic Press, London, 1985.

[Boo58] G. Boole. *The Laws of Thought*. Dover, 1958. Original edition 1853.

[Bop84] R. Boppana. Threshold functions and bounded depth monotone circuits. In *Proceedings of 16th Annual ACM Symposium on Theory of Computing*, pages 475–479, 1984.

[Bor73] A. Borodin. On relating time and space to size and depth. *SIAM Journal on Computing*, 6: 733–744, 1973.

[BP55] R. A. Beaumont and R. P. Peterson. Set-transitive permutation groups. *Can. J. Math.*, 7: 35–42, 1955.

[BP96] P. Beame and T. Pitassi. Simplified and improved resolution lower bounds. In *Proceedings of 37th Annual IEEE Symposium on Foundations of Computer Science*, pages 274–282, 1996.

[BP98] P. Beame and T. Pitassi. Propositional proof complexity: Past, present and future. *EATCS Bulletin*, 65: 66–89, June 1998. The Computational Complexity Column (ed. E. Allender).

[BPU92] S. Bellantoni, T. Pitassi, and A. Urquhart. Approximation and small depth Frege proofs. *SIAM Journal on Computing*, 21(6): 1161–1179, 1992.

[BR98] P.W. Beame and S. Riis. More on the relative strength of counting principles. In [BB98].

[Bre74] R.P. Brent. The parallel evaluation of general arithmetic expressions. *Journal of the ACM*, 21: 201–208, 1974.

[Bry86] R.E. Bryant. Graph-based algorithms for boolean function manipulation. *IEEE Transactions on Computers*, C-35(8), 1986.

[BS90] R. Boppana and M. Sipser. The complexity of finite functions. In J. van Leeuwen, editor, *Handbook of Theoretical Computer Science*, volume A, pages 759–804. Elsevier and MIT Press, 1990.

[BS90a] S.R. Buss and P.J. Scott, editors. *Feasible Mathematics*, volume 9 of *Progress in Computer Science*. Birkhäuser, Boston, 1990.

[BSI99] E. Ben-Sasson and R. Impagliazzo. Random CNF's are hard for the polynomial calculus. In *Proceedings of 40th Annual IEEE Symposium on Foundations of Computer Science*, 1999.

[BSIW99] E. Ben-Sasson, R. Impagliazzo, and A. Wigderson. Optimal separation of treelike and general resolution. Typescript dated 24 Sep. 1999.

[BSS89] L. Blum, M. Shub, and S. Smale. On a theory of computation and complexity over the real numbers: NP-completeness, recursive functions and universal machines. *Bulletin of the American Mathematical Society*, 21: 1–46, 1989.

[BSW99] E. Ben-Sasson and A. Wigderson. Short proofs are narrow – Resolution made simple. In *Proceedings of the 14th Annual IEEE Conference on Computational Complexity (CCC-99)*, page 2, Los Alamitos, 1999. IEEE Computer Society.

[BSW01] E. Ben-Sasson and A. Wigderson. Short proofs are narrow – Resolution made simple. *Journal of the ACM*, 48(2), 2001.

[BT88a] D. A. Mix Barrington and D. Thérien. Finite monoids and the fine structure of NC^1. *Journal of the ACM*, 35(4): 941–952, 1988.

[BT88b] S. Buss and G. Turán. Resolution proofs of generalized pigeonhole principles. *Theoretical Computer Science*, 62: 311–317, 1988.

[Bus86a] S. Buss. *Bounded Arithmetic*, volume 3 of *Studies in Proof Theory*. Bibliopolis, 1986.

[Bus86b] S. Buss. The polynomial hierarchy and intuitionistic bounded arithmetic. In A.L. Selman, editor, *Structure in Complexity Theory*, volume 223 of *Springer Lecture Notes in Computer Science*, pages 77–103, 1986.

[Bus87a] S. Buss. The boolean formula value problem is in ALOGTIME. In *Proceedings of 19th Annual ACM Symposium on Theory of Computing*, pages 123–131, 1987.

[Bus87b] S. Buss. The propositional pigeonhole principle has polynomial size Frege proofs. *Journal of Symbolic Logic*, 52: 916–927, 1987.

[Bus92] S.R. Buss. The graph of multiplication is equivalent to counting. *Information Processing Letters*, 41: 199–201, 1992.

[Bus93] S. R. Buss. Algorithms for boolean formula evaluation and for tree contraction. In [CK93], pages 96–115.

[Bus98a] S.R. Buss. An introduction to proof theory. In S.R. Buss, editor, *Handbook of Proof Theory*, pages 1–78. Elsevier, 1998.

[Bus98b] S.R. Buss. Lower bounds on Nullstellensatz proofs via designs. In [BB98].

[BW93] T. Becker and V. Weispfenning. *Gröbner Bases.* Volume 141 of *Graduate Texts in Mathematics.* Springer, 1993.

[CCT87] W. Cook, C.R. Coullard, and G. Turan. On the complexity of cutting plane proofs. *Discrete Applied Mathematics*, 18: 25–38, 1987.

[CDL99] A. Chiu, G. Davida, and B. Litow. NC^1 division. Typescript, 22 Nov. 1999.

[CEI96] M. Clegg, J. Edmonds, and R. Impagliazzo. Using the Groebner basis algorithm to find proofs of unsatisfiability. In *Proceedings of 28th Annual ACM Symposium on Theory of Computing*, pages 174–183, 1996.

[CF86] M.-T. Chao and J. Franco. Probabilistic analysis of two heuristics for the 3-satisfiability problem. *SIAM Journal on Computing*, 15: 1106–1118, 1986.

[CF88] J.-Y. Cai and M.L. Furst. *PSPACE* survives three-bit bottlenecks. In *Proceedings of 3th Annual IEEE Conference on Structure in Complexity Theory*, pages 94–102, 1988.

[CF90] M.-T. Chao and J. Franco. Probabilistic analysis of a generalization of the unit-clause literal selection heuristic for the k-satisfiability problem. *Information Science*, 51: 289–314, 1990.

[Chu36] A. Church. An unsolvable problem in elementary number theory. *American Journal of Mathematics*, 58: 345–363, 1936.

[Chu56] A. Church. *Introduction to Mathematical Logic.* Princeton University Press, Princeton, New Jersey, 1956.

[Cic83] E.A. Cichon. A short proof of two recently discovered independence results using recursion theoretic methods. *Proceedings of the American Mathematical Society*, 87: 704–706, 1983.

[CIK93] P. Clote, A. Ignjatovic, and B. Kapron. Parallel computable higher type functionals. In *Proceedings of 34th Annual IEEE Symposium on Foundations of Computer Science*, Nov. 3–5, 1993, pages 72–83.

[CJ98] P. Clote and J. Johannsen. On threshold logic and cutting planes proofs. Typeset manuscript, dated 14 December 1998.

[CK73] C. C. Chang and H. J. Keisler. *Model Theory*, volume 73 of *Studies in Logic and the Foundations of Mathematics.* North-Holland, 1973.

[CK90] S. A. Cook and B.M. Kapron. Characterizations of the feasible functionals of finite type. In [BS90a], pages 71–98.

[CK91] P. Clote and E. Kranakis. Boolean functions invariance groups and parallel complexity. *SIAM Journal on Computing*, 20(3): 553–590, 1991.

[CK93] P. Clote and J. Krajíček, editors. *Arithmetic, Proof Theory and Computational Complexity*. Oxford University Press, 1993.

[CKP+96] D. Culler, R. Karp, D Patterson, A. Sahay, E. Santos, K. Schauser, R. Subramonian, and Th. von Eicken. LogP, a practical model of parallel computation. *Communications of the ACM*, 39(11): 78–85, 1996.

[CKS81] A. Chandra, D. Kozen, and L. J. Stockmeyer. Alternation. *Journal of the ACM*, 28: 114–133, 1981.

[CL88] S.A. Cook and M. Luby. A simple parallel algorithm for finding a satisfying truth assignment to a 2-CNF formula. *Information Processing Letters*, 27: 141–145, 1988.

[Cla91] M. Clausen. Almost all boolean functions have no linear symmetries. *Information Processing Letters*, 41(6): 291–292, 1991.

[Clo88] P. Clote. A sequential characterization of the parallel complexity class NC. Technical Report BCCS-88-07, Boston College, Computer Science Department, 1988.

[Clo89] P. Clote. A first order theory for the parallel complexity class NC. Technical Report BCCS-89-01, Boston College, Computer Science Department, January 1989.

[Clo90] P. Clote. Sequential, machine-independent characterizations of the parallel complexity classes $ALOGTIME, AC^k, NC^k$ and NC. In [BS90a], pages 49–70.

[Clo92a] P. Clote. $ALOGTIME$ and a conjecture of S.A. Cook. *Annals of Mathematics and Artificial Intelligence*, 6: 57–106, 1992.

[Clo92b] P. Clote. A time-space hierarchy between P and PSPACE. *Math. Systems Theory*, 25: 77–92, 1992.

[Clo93] P. Clote. Polynomial size Frege proofs of certain combinatorial principles. In [CK93], pages 162–184.

[Clo95] P. Clote. Cutting plane and Frege proofs. *Information and Computation*, 121(1): 103–122, 1995.

[Clo96] P. Clote. A note on the relation between polynomial time functionals and Constable's class $\mathcal{K}$. In H. Kleine Büning, editor, *Computer Science Logic*, volume 1092 of *Springer Lecture Notes in Computer Science*, pages 145–160, 1996.

[Clo97a] P. Clote. Nondeterministic stack register machines. *Theoretical Computer Science*, 178(1–2): 37–76, 1997.

[Clo97b] P. Clote. A safe recursion scheme for exponential time. In L. Beklemishev, editor, *Logical Foundations of Computer Science, LFCS'97*, volume 1234 of *Springer Lecture Notes in Computer Science*, pages 44–52, 1997.

[Clo98] P. Clote. Computation models and function algebras. In E. Griffor, editor, *Handbook of Computability Theory*, pages 589–681. Elsevier, 1998.

[CM97] S. A. Cook and D. G. Mitchell. Finding hard instances of the satisfiability problem: A survey. In D. Du, J. Gu, and P. M. Pardalos, editors, *Satisfiability Problem: Theory and Applications*, volume 35 of *DIMACS Series in Discrete Mathematics and Theoretical Computer Science*, pages 1–17. American Mathematical Society, 1997.

[Cob65] A. Cobham. The intrinsic computational difficulty of functions. In Y. Bar-Hillel, editor, *Logic, Methodology and Philosophy of Science II*, pages 24–30. North-Holland, 1965.

[Com70] L. Comtet. *Analyse Combinatoire*. Presses Universitaires de France, Collection SUP, 1970.

[Con73] R. Constable. Type 2 computational complexity. In *Proceedings of 5th Annual ACM Symposium on Theory of Computing*, pages 108–121, 1973.

[Coo25] J. L. Coolidge. *An Introduction to Mathematical Probability*. Oxford, Clarendon Press, London, 1925.

[Coo71] S. A. Cook. The complexity of theorem proving procedures. In *Proceedings of 3rd Annual ACM Symposium on Theory of Computing*, pages 151–158, 1971.

[Coo76] S. A. Cook. A short proof of the pigeonhole principle using extended resolution. *SIGACT News*, 8(4): 28–32, 1976.

[Coo92] S. A. Cook. Computability and complexity of higher type functions. In Y. N. Moschovakis, editor, *Logic from computer science*, volume 21 of *Mathematical Sciences Research Institute Publications*, pages 51–72. Springer, 1992. Proceedings of the workshop held in Berkeley, California, November 13–17, 1989.

[CR74] S. A. Cook and R. Reckhow. On the lengths of proofs in propositional calculus. In *Proceedings of 6th Annual ACM Symposium on Theory of Computing*, pages 135–148, 1974.

[CR77] S. A. Cook and R. Reckhow. On the relative efficiency of propositional proof systems. *Journal of Symbolic Logic*, 44: 36–50, 1977.

[CR92] V. Chvátal and B. Reed. Mick gets some (the odds are on his side). In *Proceedings of 33th Annual IEEE Symposium on Foundations of Computer Science*, pages 620–627, 1992.

[CR94] P. Clote and J. Remmel, editors. *Feasible Mathematics II*, volume 13 of *Progress in Computer Science*. Birkhäuser, Boston, 1994.

[CS88] V. Chvátal and E. Szemerédi. Many hard examples for resolution. *Journal of the ACM*, 35(4): 759–768, 1988.

[CS98] P. Clote and A. Setzer. On PHP, *st*-connectivity and odd charged graphs. In [BB98], pages 93–118.

[Csi47] P. Csillag. Eine Bemerkung zur Auflösung der eingeschachtelten Rekursion. *Acta Sci. Math. Szeged.*, 11: 169–173, 1947.

[CSV84] A. Chandra, L. J. Stockmeyer, and U. Vishkin. Constant depth reducibility. *SIAM Journal on Computing*, 13: 423–439, 1984.

[CT86] P. Clote and G. Takeuti. Exponential time and bounded arithmetic. In A. L. Selman, editor, *Structure in Complexity Theory*, volume 223 of *Springer Lecture Notes in Computer Science*, pages 125–143, 1986.

[CT92] P. Clote and G. Takeuti. Bounded arithmetics for NC, $ALOGTIME$, L and NL. *Annals of Pure and Applied Logic*, 56: 73–117, 1992.

[CT94] P. Clote and G. Takeuti. First order bounded arithmetic and small boolean circuit complexity classes. In [CR94], pages 154–218.

[CU89] S.A. Cook and A. Urquhart. Functional interpretations of feasibly constructive arithmetic. In *Proceedings of 21st Annual ACM Symposium on Theory of Computing*, pages 107–112, 1989.

[CU93] S.A. Cook and A. Urquhart. Functional interpretations of feasibly constructive arithmetic. *Annals of Pure and Applied Logic*, 63(2): 103–200, 1993.

[Dav58] M. Davis. *Computability and Unsolvability*. McGraw Hill, 1958.

[DB97] O. Dubois and Y. Boufkhad. A general upper bound for the satisfiability threshold of random r-SAT formulae. *Journal of Algorithms*, 24(3): 395–420, 1997.

[DBM00] O. Dubois, Y. Boufkhad, and J. Mandler. Typical random 3-SAT formulae and the satisfiability threshold. In *Proceedings of 11th Annual ACM-SIAM Symposium on Discrete Algorithms*, San Francisco, CA, 2000.

[Deg95] J.W. Degen. Pigeonhole principles and choice principles. Universität Erlangen-Nürnberg, 1995.

[Deu85] D. Deutsch. Quantum theory, the Church-Turing principle and the universal quantum computer. *Proc. R. Soc. Lond.*, A 400: 73–90, 1985.

[DGS84] P. Duris, Z. Galil, and G. Schnitger. Lower bounds on communication complexity. In *Proceedings of 16th Annual ACM Symposium on Theory of Computing*, pages 81–91, 1984.

[DM97] A. Durand and M. More. Non-erasing, counting and majority over the linear hierarchy. Typescript, 1997.

[Dow85] M. Dowd. Model theoretic aspects of P $\neq$ NP. Unpublished preprint, 1985.

[DP60] M. Davis and H. Putnam. A computing procedure for quantification theory. *Journal of the ACM*, 1: 201–215, 1960.

[Ede92] E. Eder. *Relative Complexities of First Order Calculi.* Verlag Vieweg, 1992.

[EIS76] S. Even, A. Itai, and A. Shamir. On the complexity of timetable and multicommodity flow problems. *SIAM Journal on Computing*, 5: 1691–703, 1976.

[ER60] P. Erdős and A. Rényi. On the evolution of random graphs. *Publications of the Mathematical Institute of the Hungarian Academy of Sciences*, 56: 17–61, 1960.

[Fag74] R. Fagin. Generalized first-order spectra and polynomial-time recognizable sets. In R. M. Karp, editor, *Complexity of Computation*, volume 7 of *SIAM-AMS Proceedings*, pages 43–73, 1974.

[Fag90] R. Fagin. Finite-model theory – a personal perspective. In S. Abiteboul and P. C. Kanellakis, editors, *Proc. 1990 International Conference on Database Theory*, volume 470 of *Springer Lecture Notes in Computer Science*, pages 3–24, 1990. Journal version: *Theoretical Computer Science*, 116(1): 3–31, 1993.

[Fel68] W. Feller. *An Introduction to Probability Theory and Its Applications*, Volume 1. J. Wiley and Sons, 3rd edition, 1968.

[FHL80] M. Furst, J. Hopcroft, and E. Luks. Polynomial-time algorithms for permutation groups. In *Proceedings of 21st Annual IEEE Symposium on Foundations of Computer Science*, 1980.

[FK98] H. Fournier and P. Koiran. Are lower bounds easier over the reals? In *Proceedings of 30th Annual ACM Symposium on Theory of Computing*, pages 507–513, 1998.

[FKPS85] R. Fagin, M. Klawe, N. Pippenger, and L. Stockmeyer. Bounded-depth polynomial-size circuits for symmetric functions. *Theoretical Computer Science*, 36: 239–250, 1985.

[FP83] J. Franco and M. Paull. Probabilistic analysis of the Davis–Putnam procedure for solving the satisfiability problem. *Discrete Applied Mathematics*, 5: 77–87, 1983.

[Fre93] G. Frege. *Grundgesetze der Arithmetik*, volume 1. Jena, 1893. Reprinted by Georg Olms Verlagsbuchhandlung, Hildesheim, 1968.

[Fre67] G. Frege. Begriffsschrift, eine der arithmetischen nachgebildete Formelsprache des reinen Denkens, Halle, 1879. English translation in J. van Heijenoort, editor, *From Frege to Gödel, a Sourcebook in Mathematical Logic*, pages 1–82. Harvard University Press, Cambridge, MA, 1967.

[Fri97] E. Friedgut. Necessary and sufficient conditions for sharp thresholds. Unpublished preprint, 1997.

[FS96] A. Frieze and S. Suen. Analysis of two simple heuristics on a random instance of k-SAT. *Journal of Algorithms*, 20: 312–355, 1996.

[FSS84] M. Furst, J. B. Saxe, and M. Sipser. Parity circuits and the polynomial time hierarchy. *Mathematical Systems Theory*, 17: 13–27, 1984.

[Fu94] LiMin Fu. *Neural Networks in Computer Intelligence.* McGraw Hill, 1994.

[FW78] S. Fortune and J. Wyllie. Parallelism in random access machines. In *Proceedings of 10th Annual ACM Symposium on Theory of Computing*, pages 114–118, 1978.

[Gal77a] Z. Galil. On resolution with clauses of bounded size. *SIAM Journal of Computing*, 6: 444–459, 1977.

[Gal77b] Z. Galil. On the complexity of regular resolution and the Davis–Putnam procedure. *Theoretical Computer Science*, 4: 23–46, 1977.

[Gen34] G. Gentzen. Untersuchungen über das logische Schliessen. *Mathematische Zeitschrift*, 39: 176–210, 405–431, 1934.

[Gir87] J.-Y. Girard. *Proof Theory and Logical Complexity*, volume 1 of *Studies in Proof Theory*. Bibliopolis, 1987.

[Gir98] J.-Y. Girard. Light linear logic. *Information and Computation*, 143(2): 175–204, 15 June 1998.

[Göd31] K. Gödel. Über formal unentscheidbare Sätze der Principia Mathematica und verwandter Systeme. *Monatshefte Math. Phys.*, 38: 173–198, 1931.

[Göd58] K. Gödel. Über eine bisher noch nicht benutzte Erweiterung des finiten Standpunktes. *Dialectica*, 12: 280–287, 1958.

[Goe92] A. Goerdt. Cutting plane versus Frege proof systems. In E. Börger, editor, *Computer Science Logic 1990*, volume 552 of *Springer Lecture Notes in Computer Science*, pages 174–194, 1992.

[Gol77] L. Goldschlager. Synchronous parallel computation. Technical Report 114, University of Toronto, December 1977.

[Gol82] L. Goldschlager. A unified approach to models of synchronous parallel machines. *Journal of the ACM*, 29(4): 1073–1086, October 1982.

[GR90] G. Gasper and M. Rahman. *Basic Hypergeometric Series, Encyclopedia of Mathematics and its Applications*, Vol. 25. Cambridge University Press, Cambridge, 1990.

[Gri98] D. Grigoriev. Nullstellensatz lower bounds for Tseitin tautologies. In *Proceedings of 39th Annual IEEE Symposium on Foundations of Computer Science*, pages 648–652, 1998.

[Grz53] A. Grzegorczyk. Some clases of recursive functions. *Rozprawy Matematyczne*, 4, 1953.

[GS86] Y. Gurevich and S. Shelah. Fixed-point extensions of first-order logic. *Annals of Pure and Applied Logic*, 32: 265–280, 1986.

[GS89] Y. Gurevich and S. Shelah. Nearly linear time. *Symposium on Logical Foundations of Computer Science*, volume 363 of *Springer Lecture Notes in Computer Science*, pages 108–118, 1989.

[GSS90] J.-Y. Girard, A. Scedrov, and P. Scott. Bounded linear logic. In [BS90a], pages 195–210.

[HA50] D. Hilbert and W. Ackermann. *Principles of Mathematical Logic.* Chelsea, 1950. Translated from the 1937 German edition.

[Hak85] A. Haken. The intractability of resolution. *Theoretical Computer Science*, 39: 297–305, 1985.

[Hak95] A. Haken. Counting bottlenecks to show monotone $P \neq NP$. In *Proceedings of 36th Annual IEEE Symposium on Foundations of Computer Science*, pages 117–132, 1995.

[Hal74] P. Halmos. *Lectures on Boolean Algebras.* Springer, 1974. Reprinted from the edition published by Van Nostrand.

[Han94a] W.G. Handley. LTH plus nondeterministic summation mod M_3 yields ALINTIME. Personal Communication, 22 December 1994.

[Han94b] W.G. Handley. Deterministic summation modulo $\mathcal{B}_n$, the semi-group of binary relations on $\{0, 1, \ldots, n-1\}$. Personal Communication, May 1994.

[Har04] G.H. Hardy. A theorem concerning the infinite cardinal numbers. *Quarterly Journal Math.*, pages 87–94, 1904.

[Har75] K. Harrow. Small Grzegorczyk classes and limited minimum. *Zeitschr. Math. Logik*, 21: 417–426, 1975.

[Har79] K. Harrow. Equivalence of some hierarchies of primitive recursive functions. *Zeitschr. Math. Logik*, 25: 411–418, 1979.

[Har92] V. Harnik. Provably total functions of intuitionistic bounded arithmetic. *Journal of Symbolic Logic*, 57(2): 466–477, 1992.

[Hås87] J. T. Håstad. *Computational Limitations for Small Depth Circuits.* ACM Doctoral Dissertation Award (1986). MIT Press, 1987.

[HC99] A. Haken and S.A. Cook. An exponential lower bound for the size of monotone real circuits. *Journal of Computer and System Sciences*, 58, 1999.

[HH87] B. A. Haberman and T. Hogg. Phase transitions in artificial intelligence systems. *Artificial Intelligence*, 33: 155–171, 1987.

[Hil25] D. Hilbert. Über das Unendliche. *Mathematische Annalen*, 95: 161–190, 1925.

[HKP84] H.J. Hoover, M.M. Klawe, and N.J. Pippenger. Bounding fan-out in logical networks. *Journal of the ACM*, 31: 13–18, 1984.

[HMT88] A. Hajnal, W. Maass, and G. Turán. On the communication complexity of graph properties. In *Proceedings of 20th Annual ACM Symposium on Theory of Computing*, pages 186–191, 1988.

[Hof99] M. Hofmann. Linear types and non-size-increasing polynomial time computation. In *Proceedings of 14th Annual IEEE Symposium on Logic in Computer Science*, 1999.

[Hoo88] J.N. Hooker. Generalized resolution and cutting planes. *Annals of Operations Research*, pages 217–239, 1988.

[Hoo90] H.J. Hoover. Computational models for feasible real analysis. In [BS90a], pages 221–238.

[Hop82] J.J. Hopfield. Neural networks and physical systems with emergent collective computational abilities. *Proc. Natl. Acad. Sci. USA*, 79: 2554–2558, April 1982.

[HP93] P. Hájek and P. Pudlák. *Metamathematics of First-Order Arithmetic.* Perspectives in Mathematical Logic. Springer-Verlag, Berlin, 1993.

[HPW84] W. Handley, J. B. Paris, and A. J. Wilkie. Characterizing some low arithmetic classes. In *Theory of Algorithms*, pages 353–364. Akadémiai Kiadó, Budapest, 1984. Colloquia Societatis János Bolyai.

[Hro97] J. Hromkovic. *Communication Complexity and Parallel Computing.* Texts in Theoretical Computer Science, An EATCS Series. Springer, 1997.

[HS86] W.D. Hillis and G.L. Steele, Jr. Data parallel algorithms. *Communications of the ACM*, 29(12): 1170–1183, December 1986.

[HT86] J.J. Hopfield and D.W. Tank. Computing with neural circuits: A model. *Science*, 233(4764): 625–633, 1986.

[HU79] J.E. Hopcroft and J.D. Ullman. *Introduction to Automata Theory.* Addison–Wesley, 1979.

[HW79] G.H. Hardy and E.M. Wright. *An Introduction to the Theory of Numbers.* Oxford, Clarendon Press, 5th edition, 1979.

[Imm87] N. Immerman. Languages that capture complexity classes. *SIAM Journal of Computing*, 16: 760–778, 1987.

[Imm89] N. Immerman. Expressibility and parallel complexity. *SIAM Journal of Computing*, 18(3): 625–638, 1989.

[Imm93] N. Immerman. Email dated 26 March 1993. Personal communication.

[Imp93] R. Impagliazzo. On satisfiable sets. Personal correspondence, 1993.

[IPS] R. Impagliazzo, P. Pudlák, and J. Sgall. Lower bounds for the polynomial calculus and the Groebner basis algorithm. Personal Communication.

[Jan98] S. Janson. New versions of Suen's correlation inequality. *Random Structures and Algorithms*, 13: 467–483, 1998.

[JK70] N. Johnson and S. Kotz. *Discrete Distributions.* John Wiley and Sons, 1970.

[JM82] J.P. Jones and Y. Matijasivič. A new representation for the symmetric binomial coefficient and its applications. *Ann. Sc. Math., Quebec*, 6(1): 81–97, 1982.

[JM99] F. Joachimski and R. Matthes. Short proofs of normalization for the simply-typed λ-calculus, permutative conversions and Gödel's T. November 1999, to appear in *Arch. Math. Log.*

[Joh96] J. Johannsen. *Schwache Fragmente der Arithmetik und Schwellwertschaltkreise beschränkter Tiefe.* PhD thesis, Universität Erlangen-Nürnberg, 13 May 1996.

[Joh98] J. Johannsen. Lower bounds for monotone real circuit depth and formula size and tree-like cutting planes. *Information Processing Letters*, 67(1): 37–41, 1998.

[JS74] N.D. Jones and A.L. Selman. Turing machines and the spectra of first-order formulas. *Journal of Symbolic Logic*, 39: 139–150, 1974.

[Juk97] S. Jukna. Finite limits and monotone computations: The lower bounds criterion. In *Proceedings of 12th Annual IEEE Conference on Computational Complexity*, 14–17 June 1997, pages 302–313.

[Juk99] S. Jukna. Combinatorics of monotone computations. *Combinatorica*, 19(1): 65–85, 1999.

[Kal43] L. Kalmár. Egyszerü példa eldönthetetlen aritmetikai problémára. *Mate és Fizikai Lapok*, 50: 1–23, 1943. In Hungarian with German abstract.

[Kan81] R. Kannan. Towards separating nondeterministic time from deterministic time. In *Proceedings of 22nd Annual IEEE Symposium on Foundations of Computer Science*, pages 235–243, 1981.

[Kar93] M. Karchmer. On proving lower bounds for circuit size. In *Proceedings of 8th Annual IEEE Conference on Structure in Complexity Theory*, pages 112–118, 1993.

[Kau70] S.A. Kauffman. Behavior of randomly constructed genetic nets: Binary element nets. In C.H. Waddington, editor, *Towards a Theoretical Biology*, pages 18–37. Aldine, Chicago, 1970.

[KC96] B. Kapron and S. Cook. A new characterization of type-2 feasibility. *SIAM Journal on Computing*, 25(1): 117–132, 1996.

[Kir74] D. G. Kirkpatrick. *Topics in the Complexity of Combinatorial Algorithms.* PhD thesis, University of Toronto, December 1974. Tech. Rep. No. 74.

[Kis82] G. S. Kissin. Modeling energy consumption in VLSI circuits: A foundation. In *Proceedings of 14th Annual ACM Symposium on Theory of Computing*, pages 99–104, 1982.

[Kis87] G. S. Kissin. *Modeling Energy Consumption in VLSI Circuits.* PhD thesis, University of Toronto, 1987.

[Kis90] G. S. Kissin. Models of multiswitch energy. *CWI-Quarterly*, 3(1): 45–66, 1990.

[Kis99] A. Kisielewicz. Symmetry groups of boolean functions and constructions of permutation groups. *Journal of Algebra*, 199(2), 1999.

[KK92] E. Kranakis and D. Krizanc. Computing boolean functions on Cayley networks. In *Proceedings of the 4th IEEE Symposium on Parallel and Distributed Processing*, pages 222–229, 1992. Arlington, Texas, 1–4 Dec. 1992.

[KK97] E. Kranakis and D. Krizanc. Distributed computing on anonymous hypercube networks. *Journal of Algorithms*, 23: 32–50, 1997. Also in *Proceedings of the 3rd IEEE Symposium on Parallel and Distributed Processing*, Dallas, 2–5 Dec. 1991, pages 722–729.

[KKK96] L. M. Kirousis, E. Kranakis, and D. Krizanc. Approximating the unsatisfiability threshold of random formulas. In Josep Diaz and Maria Serna, editors, *European Symposium on Algorithms*, volume 1136 of *Springer Lecture Notes in Computer Science*, pages 27–38, 1996.

[KKK97] L. M. Kirousis, E. Kranakis, and D. Krizanc. A better upper bound for the unsatisfiability threshold. In D. Du, J. Gu, and P. M. Pardalos, editors, *Satisfiability Problem: Theory and Applications*, volume 35 of *DIMACS Series in Discrete Mathematics and Theoretical Computer Science*, pages 643–648. American Mathematical Society, 1997.

[KKKS98] L. M. Kirousis, E. Kranakis, D. Krizanc, and Y. Stamatiou. Approximating the unsatisfiability threshold of random formulas. *Random Structures and Algorithms*, 1998.

[KKL88] J. Kahn, G. Kalai, and N. Linial. The influence of variables on boolean functions. In *Proceedings of 29th Annual IEEE Symposium on Foundations of Computer Science*, pages 68–80, 1988.

[KL88] P. B. Kleidman and M. W. Liebeck. A survey of the maximal subgroups of the finite simple groups. In M. Aschbacher, A. M. Cohen, and W. M. Kantor, editors, *Geometries and Groups*. Reidel, Dordrecht, 1988. Reprinted from *Geometriae Dedicata*, 25(1–3): 375–389.

[Kle36a] S.C. Kleene. General recursive functions of natural numbers. *Math. Ann.*, 112: 727–742, 1936.

[Kle36b] S.C. Kleene. λ-definability and recursiveness. *Duke Mathematical Journal*, 2: 340–353, 1936.

[KLP89] Z. M. Kedem, G. M. Landau, and K. V. Palem. Optimal parallel suffix-prefix matching algorithm and applications. In *Proceedings of the 1st Annual ACM Symposium on Parallel Algorithms and Architectures*, pages 388–398, 1989.

[KMPS95] A. Kamath, R. Motwani, K. Palem, and P. Spirakis. Tail bounds for occupancy and the satisfiability threshold conjecture. *Random Structures and Algorithms*, 7: 59–80, 1995.

[Ko91] K. Ko. *Complexity Theory of Real Functions*. Progress in Theoretical Computer Science, Birkhäuser, Boston, 1991.

[Koi94] P. Koiran. Computing over the reals with addition and order. *Theoretical Computer Science*, 133: 35–48, 1994.

[Koi97] P. Koiran. A weak version of the Blum, Shub and Smale model. *Journal of Computer and System Sciences*, 54: 177–189, 1997.

[KP81] L. Kirby and J. Paris. Accessible independence results for Peano arithmetic. *Bulletin of the London Mathematical Society*, 14: 285–293, 1981.

[KP83] R. M. Karp and J. Pearl. Searching for an optimal path in a tree with random costs. *Artificial Intelligence*, 21: 99–116, 1983.

[KP89] J. Krajíček and P. Pudlák. Propositional proof systems, the consistency of first order theories and the complexity of computations. *Journal of Symbolic Logic*, 54(3):1063–1079, 1989.

[KP90] J. Krajíček and P. Pudlák. Quantified propositional calculi and fragments of bounded arithmetic. *Zeitschrift für Mathematische Logik und Grundlagen der Mathematik*, 36: 29–46, 1990.

[KP95] J. Krajíček and P. Pudlák. Some consequences of cryptographic conjectures for S_2^1 and EF. In D. Leivant, editor, *Logic and Computational Complexity*, volume 960 of *Springer Lecture Notes in Computer Science*, pages 210–220, 1995.

[KPW95] J. Krajíček, P. Pudlák, and A. Woods. An exponential lower bound to the size of bounded depth Frege proofs of the pigeonhole principle. *Random Structures and Algorithms*, 7(1): 15–39, 1995.

[KR87] R. M. Karp and M. O. Rabin. Efficient randomized pattern matching algorithms. *IBM J. Res. Develop.*, 31(2): 249–260, 1987.

[KR90] R.M. Karp and V. Ramachandran. Parallel algorithms for shared-memory machines. In J. van Leeuwen, editor, *Handbook of Theoretical Computer Science*, volume A, pages 871–942. Elsevier and MIT Press, 1990.

[Kra] J. Krajíček. On the degree of ideal membership proofs from uniform families of polynomials over a finite field. To appear.

[Kra94a] J. Krajíček. Lower bounds to the size of constant-depth propositional proofs. *Journal of Symbolic Logic*, 59(1): 73–86, 1994.

[Kra94b] J. Krajíček. On Frege and extended Frege systems. In [CR94], pages 284–319.

[Kra95] J. Krajíček. *Bounded Arithmetic, Propositional Logic, and Complexity Theory*. Cambridge University Press, 1995.

[Kra97a] J. Krajíček. Interpolation theorems, lower bounds for proof systems, and independence results for bounded arithmetic. *Journal of Symbolic Logic*, 62(2): 457–486, 1997.

[Kra97b] E. Kranakis. Symmetry and computability in anonymous networks. In N. Santoro and P. Spirakis, editors, *SIROCCO 96 (3rd Annual International Conference on Structure Information and Communication Complexity)*, pages 1–16, Ottawa, 1997. Carleton University Press.

[Kra98] J. Krajíček. Interpolation by a game. *Mathematical Logic Quarterly*, 44(4), 1998. Available as ECCC TR97-015.

[Kri85] B. Krishnamurthy. Short proofs for tricky formulas. *Acta Informatica*, 22: 253–275, 1985.

[KS81] R. M. Karp and M. Sipser. Maximum matchings in sparse random graphs. In *Proceedings of 22nd Annual IEEE Symposium on Foundations of Computer Science*, pages 364–375, 1981.

[KS94] S. Kirkpatrick and B. Selman. Critical behavior in the satisfiability of random boolean expressions. *Science*, 264: 1297–1301, 1994.

[KS96] L. M. Kirousis and Y.C. Stamatiou. An inequality for reducible, increasing properties of randomly generated words. Technical Report TR-96.10.34, Computer Technology Institute, University of Patras, Greece, 1996.

[Kuc82] L. Kucera. Parallel computation and conflicts in memory access. *Information Processing Letters*, 14(2): 93–96, 1982.

[Kut88] M. Kutyłowski. Finite automata, real time processes and counting problems in bounded arithmetics. *Journal of Symbolic Logic*, 53(1): 243–258, 1988.

[KW90] M. Karchmer and A. Wigderson. Monotone circuits for connectivity require super-logarithmic depth. *J. Discrete Mathematics*, 3: 255–265, 1990.

[LC91] T. Luczak and J.E. Cohen. Stability of vertices in random boolean cellular automata. *Random Structures and Algorithms*, 2: 237–334, 1991.

[Lee59] C.Y. Lee. Representation of switching functions by binary decision programs. *Bell Systems Technical Journal*, 38: 985–999, 1959.

[Lei89] D. Leivant. Stratified polymorphism. In *Proceedings of 4th Annual IEEE Symposium on Logic in Computer Science*, pages 39–47, 1989. Journal version: Finitely stratified polymorphism. *Information and Computation* 93: 93–113, 1991.

[Lei91] D. Leivant. A foundational delineation of computational feasibility. In *Proceedings of 6th Annual IEEE Symposium on Logic in Computer Science*, 1991.

[Lei92] F. Thomson Leighton. *Introduction to Parallel Algorithms and Architectures: Arrays, Trees, Hypercubes.* Morgan Kaufmann Publishers, 1992.

[Lei93] D. Leivant. Stratified functional programs and computational complexity. In *Conference Record of the 20th Annual ACM Symposium on Principles of Programming Languages*, 1993.

[Lei94] D. Leivant. Ramified recurrence and computational complexity I: Word recurrence and poly-time. In [CR94], pages 320–343.

[Len90] Th. Lengauer. *Combinatorial Algorithms for Integrated Circuit Layout.* Wiley-Teubner, 1990.

[LF80] R. E. Ladner and M. J. Fischer. Parallel prefix computation. *Journal of the ACM*, 27(4): 831–838, 1980.

[Lin74] J.C. Lind. Computing in logarithmic space. Technical Report, Project MAC Technical Memorandum 52, Massachusetts Institute of Technology, September 1974.

[LM93] D. Leivant and J.-Y. Marion. Lambda-calculus characterizations of poly-time. *Fundamenta Informaticae*, 19: 167–184, 1993.

[Lov70] D. Loveland. A linear format for resolution. In Proc. of the IRIA 1968 Symposium on Automatic Demonstration, volume 125, New York, 1970.

[Lov79] L. Lovász. *Combinatorial Problems and Exercises.* North-Holland, Amsterdam, 1979.

[LP81] H.R. Lewis and C.H. Papadimitriou. *Elements of the Theory of Computation.* Prentice-Hall, Englewood Cliffs, 1981.

[LPS88] M. W. Liebeck, C. E. Praeger, and J. Saxl. On the O'Nan-Scott theorem for finite primitive permutation groups. *Journal of the Australian Mathematical Society (Series A)*, 44: 389–396, 1988.

[LS95] L. Lovasz and A. Schrijver. Cones of matrices and set-functions and 0-1 optimization. *SIAM J. Optimization*, 1(2): 166–190, 1995.

[LSH65] P. Lewis, R.E. Stearns, and J. Hartmanis. Memory bounds for recognition of context-free and context-sensitive languages. In *Proceedings of 6th Annual IEEE Symposium on Switching Circuit Theory and Logical Design*, pages 191–202, 1965.

[Lup58] O. Lupanov. A method of circuit synthesis. *Izv. V.U.Z. Radiofiz.*, 1(1): 120–140, 1958.

[Lup61a] O. Lupanov. Implementing the algebra of logic in terms of constant depth circuits in the basis $\wedge, \vee, \neg$. *Dokl. Akad. Nauk SSSR*, 136: 1041–1042, 1961. English translation in *Sov. Math. Dokl.*, 6: 107–108, 1961.

[Lup61b] O. Lupanov. On the reduction of functions of logical algebra by formulae of finite classes (formulae of limited depth). *Proble. Kiber.*, 6: 5–14, 1961. English translation in *Probl. Cybern.*, 6: 1–14, 1965.

[Lyn93] J.F. Lynch. A criterion for stability in random boolean cellular automata. *Ulam Quarterly*, 2: 32–44, 1993.

[Lyn95] J.F. Lynch. On the threshold of chaos in random boolean cellular automata. *Random Structures and Algorithms*, 6: 239–260, 1995.

[Mar73] G. A. Margulis. Explicit construction of concentrators. *Problems of Information Transmission*, 9: 325–332, 1973.

[Mar80] S.S. Marčenkov. A superposition basis in the class of Kalmár elementary functions. *Matematicheskie Zametki*, 27(3): 321–332, 1980. Translation in *Mathematical Notes of the Academy of Sciences of the USSR*, Plenum.

[Mas76] W. Masek. A fast algorithm for the string editing problem and decision graph complexity. Master's thesis, Massachusetts Institute of Technology, Department of EECS, May 1976.

[MC80] C. Mead and L. Conway. *Introduction to VLSI Systems*. Addison–Wesley, 1980.

[MC85] P. McKenzie and S. A. Cook. The parallel complexity of abelian permutation group problems. Technical Report 181/85, Department of Computer Science, University of Toronto, 1985.

[McD92] C. McDiarmid. On a correlation inequality of Farr. *Combinatorics, Probability and Computing*, 1: 157–160, 1992.

[Mee87] L. Meertens. Personal communication, 1987.

[Meh76] K. Mehlhorn. Polynomial and abstract subrecursive classes. *Journal of Computer and System Sciences*, 12: 147–178, 1976.

[Mon77] B. Monien. A recursive and grammatical characterization of exponential time languages. *Theoretical Computer Science*, 3: 61–74, 1977.

[MP68] M. Minsky and S. Papert. *Perceptrons*. MIT Press, 1968. 2nd revised edition, 1988.

[MPV87] M. Mézard, G. Parisi, and M. Virasoro. *Spin Glass Theory and Beyond*. World Scientific, Singapore, 1987.

[MR67] A.R. Meyer and D. Ritchie. The complexity of loop programs. *Proc. ACM Nat. Conf.*, pages 465–469, 1967.

[MR95] M. Molloy and B. Reed. The dominating number of a random cubic graph. *Random Structures and Algorithms*, 7: 209–221, 1995.

[Muc76] S.S. Muchnick. The vectorized Grzegorczyk hierarchy. *Zeitschr. Math. Logik*, 22: 441–80, 1976.

[Mul56] D.E. Muller. Complexity in electronic switching circuits. *IRE Transactions on Electronic Computers*, 5: 15–19, 1956.

[Mül74] H. Müller. *Klassifizierungen der primitiv-rekursiven Funktionen*. PhD thesis, Universität Münster, 1974.

[Mul86] K. Mulmuley. A fast parallel algorithm to compute the rank of a matrix over an arbitrary field. In *Proceedings of 18th Annual ACM Symposium on Theory of Computing*, pages 338–339, 1986.

[Mun82] D. Mundici. Complexity of Craig's interpolation. *Annales Societatis Mathematicae Poloniae, Series IV: Fundamenta Informaticae*, 3/4: 261–278, 1982.

[Mun83] D. Mundici. A lower bound for the complexity of Craig's interpolants in sentential logic. *Archiv für Mathematische Logik*, 23: 27–36, 1983.

[MW86] S. Moran and M. Warmuth. Gap theorems for distributed computation. In *5th Annual ACM Symposium on Principles of Distributed Computation*, pages 131–140, 1986.

[MZK+] R. Monasson, R. Zecchina, S. Kirkpatrick, B. Selman, and L. Troyansky. Phase transition and search cost in the $2+p$-SAT problem. Boston University, 22–24 November 1996. (PhysComp96)

[Nec66] E. I. Nechiporuk. A boolean function. *Dokl. Akad. Nauk SSSR*, 169: 765–766, 1966. English translation in *Soviet Math. Dokl.* 7: 999–1000, 1966.

[Nep70] V.A. Nepomnjascii. Rudimentary predicates and Turing calculations. *Dokl. Akad. Nauk SSSR*, 195: 29–35, 1970. Translated in *Soviet Math. Dokl.* 11: 1462–1465, 1970.

[Ngu96] A.P. Nguyen. A formal system for linear space reasoning. Technical Report 300/96, University of Toronto, 1996.

[Nig98] K.-H. Niggl. A restricted computation model on Scott domains and its partial primitive recursive functionals. *Archive for Mathematical Logic*, 1998.

[NRS89] A. Nerode, J. Remmel, and A. Scedrov. Polynomially graded logic I-a graded version of system T. In *Proceedings of 4th Annual IEEE Symposium on Logic in Computer Science*, 1989.

[NS92] N. Nisan and M. Szegedy. On the degree of boolean functions as polynomials. In *Proceedings of 24th Annual ACM Symposium on Theory of Computing*, pages 462–467, 1992.

[O'D85] M.J. O'Donnell. *Equational Logic as a Programming Language.* MIT Press, 1985.

[Ott94] J. Otto. Tiers, tensors, and Δ_0^0. Talk at meeting LCC, Indianapolis, organizer D. Leivant, October 13–16, 1994.

[Ott95] J. Otto. *Complexity Doctrines.* PhD thesis, Department of Mathematics and Statistics, McGill University, June 13, 1995.

[Ott96] J. Otto. Half tiers and linear space (and time). Talk at DIMACS Workshop on Computational Complexity and Programming Languages, organized by B.M. Kapron and J. Royer, July 25–26, 1996.

[Pak79] S.V. Pakhomov. Machine independent description of some machine complexity classes (in Russian). *Issledovanija po konstrukt. matemat. i mat. logike*, VIII:176–185, LOMI 1979.

[Pat92] R. Paturi. On the degree of polynomials that approximate symmetric boolean functions. In *Proceedings of 24th Annual ACM Symposium on Theory of Computing*, pages 468–474, 1992.

[Pau77] W. J. Paul. A $2.5n$ lower bound on the combinational complexity of boolean functions. *SIAM Journal on Computing*, 6(3): 427–443, 1977.

[PBI93] T. Pitassi, P. Beame, and R. Impagliazzo. Exponential lower bounds for the pigeonhole principle. *Computational Complexity*, 3(2): 97–140, 1993.

[Pen89] R. Penrose. *The Emperor's New Mind.* Oxford University Press, 1989.

[Pét36] R. Péter. Über die mehrfache Rekursion. *Mathematische Annalen*, 113: 489–526, 1936.

[Pet61] W. W. Peterson. *Error Correcting Codes.* MIT Press, 1961. (Appendix 1)

[Pit97] T. Pitassi. Algebraic propositional proof systems. In N. Immerman and Ph. Kolaitis, editors, *Descriptive Complexity and Finite Models*, volume 31 of *DIMACS Series in Discrete Mathematics and Theoretical Computer Science*, pages 214–244. American Mathematical Society, 1997.

[Pit98] F. Pitt. A quantifier-free theory based on a string algebra for NC^1. In [BB98], pages 229–252.

[Poi95] H. Poincaré. Analysis situs. *Journal de l'Ecole Polytechnique*, 1895.

[Poi52] H. Poincaré. *Science et Méthode.* Dover, 1952. Translated by F. Maitland from the original 1906 edition.

[Poi95] B. Poizat. *Les Petits Cailloux.* Alias Editeur, 1995.

[Pos21] E. Post. Introduction to a general theory of elementary propositions. *American Journal of Mathematics*, 43: 163–185, 1921.

[Pos41] E. Post. Two-valued iterative systems of mathematical logic. *Annals of Mathematical Studies*, 5: 163–185, 1941.

[PR87] G. Pólya and R. C. Read. *Combinatorial Enumeration of Groups, Graphs and Chemical Compounds.* Springer, 1987.

[PS82a] C. H. Papadimitriou and M. Sipser. Communication complexity. In *Proceedings of 14th Annual ACM Symposium on Theory of Computing*, pages 196–200, 1982.

[PS82b] C. H. Papadimitriou and K. Steiglitz. *Combinatorial Optimization.* Prentice-Hall, 1982.

[PSW96] B. Pittel, J. Spencer, and N. Wormald. Sudden emergence of a giant k-core in a random graph. *Journal of Combinatorial Theory, Series B*, 67: 111–151, 1996.

[Pud87] P. Pudlák. The hierarchy of boolean circuits. *Computers and Artificial Intelligence*, 6(3): 449–468, 1987.

[Pud91] P. Pudlák. Ramsey's theorem in bounded arithmetic. In E. Börger et al., editor, *Computer Science Logic 1990*, volume 533 of *Springer Lecture Notes in Computer Science*, pages 308–317, 1991.

[Pud94] P. Pudlák. Complexity theory and genetics. In *Proceedings of 9th Annual IEEE Conference on Structure in Complexity Theory*, 1994.

[Pud97] P. Pudlák. Lower bounds for resolution and cutting planes proofs and monotone computations. *Journal of Symbolic Logic*, pages 981–998, 1997.

[Pud98] P. Pudlák. The length of proofs. In S.R. Buss, editor, *Handbook of Proof Theory*, pages 547–637. Elsevier, 1998.

[PW85] J. B. Paris and A. J. Wilkie. Counting problems in bounded arithmetic. In C. A. di Prisco, editor, *Methods in Mathematical Logic*, volume 1130 of *Springer Lecture Notes in Mathematics*, pages 317–340, 1985. Proceedings of the 6th Latin American Symposium on Mathematical Logic, Caracas, Venezuela, 1983.

[PW86] J.B. Paris and A.J. Wilkie. Counting δ_0 sets. *Fundamenta Mathematicae*, 127: 67–76, 1986.

[PWW88] J. B. Paris, A. J. Wilkie, and A. R. Woods. Provability of the pigeonhole prnciple and the existence of infinitely many primes. *Journal of Symbolic Logic*, 53(4): 1235–1244, 1988.

[Rab60] M. Rabin. Degree of difficulty of computing a function and a partial ordering of recursive sets. Tech. Rep. No. 1, O.N.R., Jerusalem, 1960.

[Raz87a] A. A. Razborov. Lower bounds for the size of circuits of bounded depth in basis $\{\wedge, \oplus\}$. *Mathematicheskie Zametki*, 41(4): 598–607, 1987. English translation in *Math. Notes Acad. Sci. USSR*, 41(4): 333–338.

[Raz87b] A. A. Razborov. Lower bounds on the monotone complexity. *Doklady Akademii Nauk SSSR*, 281(4): 798–801, 1987. English translation in *Soviet Math. Doklady*, 31: 354–357.

[Raz89] A. A. Razborov. On the method of approximations. In *Proceedings of 21st Annual ACM Symposium on Theory of Computing*, pages 112–118, 1989.

[Raz91] A. A. Razborov. Lower bounds for deterministic and nondeterministic branching programs. In L. Budach, editor, *Proceedings of FCT'91*, volume 529 of *Springer Lecture Notes in Computer Science*, pages 47–60, 1991.

[Raz93] A.A. Razborov. An equivalence between second order bounded domain bounded arithmetic and first order bounded arithmetic. In [CK93], pages 247–277.

[Raz94] A. A. Razborov. Bounded arithmetic and lower bounds in boolean complexity. In [CR94], pages 344–386.

[Raz98] A.A. Razborov. Lower bounds for the polynomial calculus. *Computational Complexity*, 7: 291–324, 1998.

[Rec75] R. Reckhow. *On the Lengths of Proofs in the Propositional Calculus.* PhD thesis, University of Toronto, 1975.

[Red94] D. Redfern. *The Maple Handbook: Maple V Release 3, Programming*, volume 1. Springer, 2nd edition, 1994.

[Rit63] R.W. Ritchie. Classes of predictably computable functions. *Trans. Am. Math. Soc.*, 106: 139–173, 1963.

[Rob47] R.M. Robinson. Primitive recursive functions. *Bulletin of the Amer. Math. Society*, 53: 923–943, 1947.

[Rob49] J. A. Robinson. Definability and decision problems in arithmetic. *Journal of Symbolic Logic*, 14: 98–114, 1949.

[Rob65a] J. A. Robinson. A machine oriented logic based on the resolution principle. *Journal of the ACM*, 12: 23–41, 1965.

[Rob65b] J.A. Robinson. Automatic deduction with hyper-resolution. *International Journal of Computer Mathematics*, 1: 227–234, 1965.

[Ros84] H. E. Rose. *Subrecursion: Function and Hierarchies*, volume 9 of *Oxford Logic Guides*. Clarendon Press, Oxford, 1984.

[Ros97] A. Rosenbloom. Monotone real circuits are more powerful than monotone boolean circuits. *Information Processing Letters*, 61: 161–164, 1997.

[Ruz80] W.L. Ruzzo. Tree-size bounded alternation. *Journal of Computer and System Sciences*, 21: 218–235, 1980.

[Ruz81] W.L. Ruzzo. On uniform circuit complexity. *Journal of Computer and System Sciences*, 22: 365–383, 1981.

[Sav70] W. J. Savitch. Relationship between nondeterministic and deterministic tape complexities. *Journal of Computer and System Sciences*, 4: 177–192, 1970.

[SCDM86] J.-C. Simon, J. Carlier, O. Dubois, and O. Moulines. Étude statistique de l'existence de solutions de problèmes SAT, application aux systèmes-experts. *C.R. Acad. Sci. Paris, Sér. I Math.*, 302: 283–286, 1986.

[Sch52] H. Scholz. Ein ungelöstes Problem in der symbolischen Logik. *Journal of Symbolic Logic*, 17: 160, 1952.

[Sch69] H. Schwichtenberg. Rekursionszahlen und die Grzegorczyk-Hierarchie. *Arch. Math. Logik*, 12: 85–97, 1969.

[Sch78] C. P. Schnorr. Satisfiability is quasilinear complete in NQL. *Journal of the ACM*, 25(1): 136–145, 1978.

[Sch97] Uwe Schöning. Resolution proofs, exponential bounds, and Kolmogorov complexity. In I. Prívara and P. Ruzicka, editors, *Mathematical Foundations of Computer Science 1997*, Bratislava, Slovakia, 25–29 August 1997, volume 1295 of *Springer Lecture Notes in Computer Science*, pages 110–116.

[Sed83] R. Sedgewick. *Algorithms*. Addison–Wesley, Reading, MA, 1983.

[Set92] A. Seth. There is no recursive axiomatization for feasible functionals of type 2. In *Proceedings of 7th Annual IEEE Symposium on Logic in Computer Science*, pages 286–295, 1992.

[Set93] A. Seth. Some desirable conditions for feasible functionals of type 2. In *Proceedings of 8th Annual IEEE Symposium on Logic in Computer Science*, 1993.

[Set94] A. Seth. Turing machine characterizations of feasible functionals of all finite types. In [CR94], pages 407–428.

[SH86] G. Steele, Jr. and W.D. Hillis. Connection machine Lisp: fine-grained parallel symbolic processing. *Proceedings of the 1986 ACM Conference on Lisp and Functional Programming*, pages 279–297, August 4–6, 1986.

[Sha38] C. E. Shannon. A symbolic analysis of relay and switching circuits. *Transactions of AIEE*, 57: 713–723, 1938.

[Sha49] C. E. Shannon. The synthesis of two-terminal switching circuits. *Bell Systems Technical Journal*, 28: 59–98, 1949.

[Sho97] P. Shor. Polynomial-time algorithms for prime factorization and discrete logarithms on a quantum computer. *SIAM Journal on Computing*, 26: 1484–1509, 1997.

[Sik69] R. Sikorski. *Boolean Algebras*. Springer, 3rd edition, 1969.

[Sim88] H. Simmons. The realm of primitive recursion. *Archive for Mathematical Logic*, 27: 177–188, 1988.

[Sip83] M. Sipser. Borel sets and circuit complexity. In *Proceedings of 15th Annual ACM Symposium on Theory of Computing*, pages 61–69, 1983.

[Sip85a] M. Sipser. Lecture notes on complexity theory. Notes of advanced course on computational complexity theory given at MIT, 1985.

[Sip85b] M. Sipser. A topological view of some problems in complexity theory. *Colloquia Mathematica Societatis János Bolyai*, 44: 387–391, 1985.

[Sip92] M. Sipser. The history and status of the P versus NP question. In *Proceedings of 24th Annual ACM Symposium on Theory of Computing*, pages 603–618, 1992.

[Sko23] T. Skolem. Begründung der elementaren Arithmetik durch die rekurrierende Denkweise ohne Anwendung scheinbarer Veränderlichen mit unendlichem Ausdehnungsbereich. *Skrifter utgit av Videnskapsselskapet, I. Mate. Klasse*, 6, 1923. Oslo.

[Smo87] R. Smolensky. Algebraic methods in the theory of lower bounds for boolean circuit complexity. In *Proceedings of 19th Annual ACM Symposium on Theory of Computing*, pages 77–82, 1987.

[Smu61] R. Smullyan. *Theory of Formal Systems.* Annals of Mathematical Studies, Vol. 47. Princeton University Press, 1961.

[Spe94] J.H. Spencer. *Ten Lectures on the Probabilistic Method.* SIAM, Philadelphia, 2nd edition, 1994.

[Spi71] P.M. Spira. On time hardware complexity tradeoffs for Boolean functions. In *Proceedings of the 4th Hawaii International Symposium on System Sciences*, pages 525–527. IEEE, 1971.

[SR88] N. Shankar and V. Ramachandran. Efficient parallel circuits and algorithms for division. *Information Processing Letters*, 29(6): 307–313, 1988.

[Sta78] R. Statman. Bounds for proof search and speed-up in the predicate calculus. *Annals of Mathematical Logic*, 15: 225–287, 1978.

[Ste89] I. Stewart. *Does God Play Dice?* Penguin, 1989.

[Sue90] W. C. Suen. A correlation inequality and a poisson limit theorem for nonoverlapping balanced subgraphs of a random graph. *Random Structures and Algorithms*, 1: 231–242, 1990.

[SV81] S. Skyum and L. G. Valiant. A complexity theory based on boolean algebra. In *Proceedings of 22nd Annual IEEE Symposium on Foundations of Computer Science*, pages 244–253, 1981.

[SV82] Y. Shiloach and U. Vishkin. Finding the maximum, merging and sorting in a parallel computation model. *Journal of Algorithms*, 3: 57–67, 1982.

[SV84] L. Stockmeyer and U. Vishkin. Simulation of parallel random access machines by circuits. *SIAM Journal on Computing*, 13: 409–422, 1984.

[Tak75] G. Takeuti. *Proof Theory.* Studies in Logic and the Foundations of Mathematics. North-Holland, 1975, 2nd edition 1987.

[Tho72] D.B. Thompson. Subrecursiveness: machine independent notions of computability in restricted time and storage. *Math. Systems Theory*, 6: 3–15, 1972.

[Tow82] M. Townsend. *The Polynomial Jump Operator and Complexity for Type Two Relations.* PhD thesis, University of Michigan, 1982.

[Tow90] M. Townsend. Complexity for type-2 relations. *Notre Dame Journal of Formal Logic*, 31: 241–262, 1990.

[Tsa93] S.-C. Tsai. Lower bounds on representing boolean functions as polynomials in Z_m. In *Proceedings of 8th Annual IEEE Conference on Structure in Complexity Theory*, pages 96–101, 1993.

[Tse68] G. S. Tseitin. On the complexity of derivation in the propositional calculus. In A. O. Slisenko, editor, *Structures in Constructive Mathematics and Mathematical Logic*, volume II, pages 115–125. Consultants Bureau, New York and London, 1968. Translated from the Russian.

[Tse83] G. S. Tseitin. On the complexity of derivation in propositional calculus. In J. Siekmann and G. Wrightson, editors, *Automation of Reasoning*, volume II, pages 466–483. Symbolic Computation, Springer, 1983.

[Tur37] A.M. Turing. On computable numbers, with an application to the Entscheidungsproblem. *Proc. Lond. Math. Soc., Series 2*, 42: 230–265, 1936/37.

[UF96] A. Urquhart and X. Fu. Simplified lower bounds for propositional proofs. *Notre Dame Journal of Formal Logic*, 37(4): 523–544, 1996.

[Ull84] J. D. Ullman. *Computational Aspects of VLSI.* Computer Science Press, 1984.

[Urq87] A. Urquhart. Hard examples for resolution. *Journal of the ACM*, 34(1): 209–219, 1987.

[Urq95] A. Urquhart. The complexity of propositional proofs. *Bulletin of Symbolic Logic*, 1: 425–467, 1995.

[Urq99] A. Urquhart. The symmetry rule in propositional logic. *Discrete Applied Mathematics*, 96/97: 177–193, 1999.

[Val79] L. Valiant. The complexity of computing the permanent. *Theoretical Computer Science*, 8: 189–201, 1979.

[vEB90] P. van Emde Boas. Machine models and simulations. In J. van Leeuwen, editor, *Handbook of Theoretical Computer Science*, volume A, pages 1–66. Elsevier and MIT Press, 1990.

[vN58] J. von Neumann. *The Computer and the Brain.* Yale University Press, New Haven, 1958.

[VW96] H. Vollmer and K. Wagner. Recursion theoretic characterizations of complexity classes of counting functions. *Theoretical Computer Science*, 163: 245–258, 1996.

[Wag79] K. Wagner. Bounded recursion and complexity classes. In J. Bečvář, editor, *Mathematical Foundations of Computer Science 1979*, Olomouc, Czechoslovakia, 3–7 September 1979, volume 74 of *Springer Lecture Notes in Computer Science*, pages 492–498.

[Wal64] C.S. Wallace. A suggestion for a fast multiplier. *IEEE Transactions on Computers*, C-13: 14–17, 1964.

[WCR64] L. Wos, D. Carson, and G. A. Robinson. The unit preference strategy in theorem proving. In *Fall Joint Computer Conference, AFIPS, Washington D.C.*, volume 26, pages 615–621. Spartan Books, 1964.

[Weg87] I. Wegener. *Complexity of Boolean Functions.* Wiley-Teubner, 1987.

[Wel88] D. Welsh. *Codes and Cryptography.* Oxford University Press, 1988.

[Whi25] A. N. Whitehead. *Science and the Modern World.* The Macmillan Company, 1925.

[Wie64] H. Wielandt. *Finite Permutation Groups.* Academic Press, 1964.

[Wil83] A.J. Wilkie. Modèles non standard de l'arithmétique et complexité algorithmique. In A.J. Wilkie and J.-P. Ressayre, editors, *Modèles non standard en Arithmétique et en Théorie des Ensembles*, pages 5–45. Publications Mathématiques de l'Université Paris VII, 1983.

[Wil87] C. Wilson. Relativized NC. *Math. Systems Theory*, 20: 13–29, 1987.

[Woo86] A. Woods. Bounded arithmetic formulas and Turing machines of constant alternation. In J.B. Paris, A.J. Wilkie, and G.M. Wilmers, editors, *Logic Colloquium 1984*. North-Holland, 1986.

[Wor95] N. C. Wormald. Differential equations for random processes and random graphs. *Annals of Applied Probability*, 5: 1217–1235, 1995.

[Wra76] C. Wrathall. Complete sets and the polynomial time hierarchy. *Theoretical Computer Science*, 3: 23–33, 1976.

[WRC65] L. Wos, G. A. Robinson, and D. F. Carson. Efficiency and completeness of the set of support strategy in theorem proving. *Journal of the ACM*, 12(4): 536–541, October 1965.

[WW86] K. Wagner and G. Wechsung. *Computational Complexity*. Reidel, 1986.

[Xia00] W. Xiao. Representation problems of boolean functions. Preprint, Department of Mathematics, Xiamen University, P. R. China, 2000.

[Yao85] A. Yao. Separating the polynomial-time hierarchy by oracles. In *Proceedings of 26th Annual IEEE Symposium on Foundations of Computer Science*, pages 1–10, 1985.

[Zam97] D. Zambella. Forcing in finite structures. *Math. Logic Quart.*, 43: 401–412, 1997.

Index

Back Forward Reload Home Search Guide Images Print Security

Location: http://www.springer.de/comp/

You are just one **click** *away*
from a **world of computer science**

Come and visit **Springer's**

Computer Science Online Librar

Books

- Search the Springer website catalogue
- Subscribe to our free alerting service for new books
- Look through the book series profiles

You want to order? Email to: orders@springer.de

Journals

- Get abstracts, ToC´s free of charge to everyone
- Use our powerful search engine LINK Search
- Subscribe to our free alerting service LINK *Alert*
- Read full-text articles (available only to subscribers of the paper version of a journal)

You want to subscribe? Email to: subscriptions@springer.de

Electronic Media

- Get more information on our software and CD-ROMs

You have a question on an electronic product? Email to: helpdesk-em@springer.de

Bookmark now:

http://www.springer.de/comp/

Springer · Customer Service
Haberstr. 7 · D-69126 Heidelberg, Germany
Tel: +49 6221 345 217/218 · Fax: +49 6221 345229
d&p. · 006700_001x_1c